Molecular Electrochemistry of Inorganic, Bioinorganic and Organometallic Compounds

NATO ASI Series

Advanced Science Institutes Series

A Series presenting the results of activities sponsored by the NATO Science Committee, which aims at the dissemination of advanced scientific and technological knowledge, with a view to strengthening links between scientific communities.

The Series is published by an international board of publishers in conjunction with the NATO Scientific Affairs Division

A Life Sciences **B Physics**	Plenum Publishing Corporation London and New York
C Mathematical and Physical Sciences **D Behavioural and Social Sciences** **E Applied Sciences**	Kluwer Academic Publishers Dordrecht, Boston and London
F Computer and Systems Sciences **G Ecological Sciences** **H Cell Biology** **I Global Environmental Change**	Springer-Verlag Berlin, Heidelberg, New York, London, Paris and Tokyo

NATO-PCO-DATA BASE

The electronic index to the NATO ASI Series provides full bibliographical references (with keywords and/or abstracts) to more than 30000 contributions from international scientists published in all sections of the NATO ASI Series.
Access to the NATO-PCO-DATA BASE is possible in two ways:

– via online FILE 128 (NATO-PCO-DATA BASE) hosted by ESRIN, Via Galileo Galilei, I-00044 Frascati, Italy.

– via CD-ROM "NATO-PCO-DATA BASE" with user-friendly retrieval software in English, French and German (© WTV GmbH and DATAWARE Technologies Inc. 1989).

The CD-ROM can be ordered through any member of the Board of Publishers or through NATO-PCO, Overijse, Belgium.

Series C: Mathematical and Physical Sciences - Vol. 385

Molecular Electrochemistry of Inorganic, Bioinorganic and Organometallic Compounds

edited by

A. J. L. Pombeiro

Instituto Superior Técnico,
Lisboa, Portugal

and

J. A. McCleverty

University of Bristol,
School of Chemistry,
Bristol, U.K.

Kluwer Academic Publishers

Dordrecht / Boston / London

Published in cooperation with NATO Scientific Affairs Division

Proceedings of the NATO Advanced Research Workshop on
Molecular Electrochemistry of Inorganic, Bioinorganic and
Organometallic Compounds
Sintra, Portugal
March 25–29, 1992

ISBN 0-7923-2077-8

Published by Kluwer Academic Publishers,
P.O. Box 17, 3300 AA Dordrecht, The Netherlands.

Kluwer Academic Publishers incorporates the publishing programmes of
D. Reidel, Martinus Nijhoff, Dr W. Junk and MTP Press.

Sold and distributed in the U.S.A. and Canada
by Kluwer Academic Publishers,
101 Philip Drive, Norwell, MA 02061, U.S.A.

In all other countries, sold and distributed
by Kluwer Academic Publishers Group,
P.O. Box 322, 3300 AH Dordrecht, The Netherlands.

Printed on acid-free paper

Printed in the Netherlands

TABLE OF CONTENTS

PREFACE xiii

ORGANIZING COMMISSION xvii

LOCAL EXECUTIVE COMMISSION xvii

LIST OF PARTICIPANTS xix

PART 1. **OVERVIEW** 1

Molecular Electrochemistry of Coordination Compounds. Present State and New Trends.
A.A. Vlcek 3

PART 2. **ELECTROSYNTHESIS** 13

Direct Electrochemical Synthesis of Inorganic and Organometallic Compounds
D.G. Tuck 15

Electrochemical Generation of Substrate-Binding Sites in Organometallic Complexes with the $\{Mo_2(\mu\text{-}SR)_n\}$ Core. Electrosynthesis of Cyanide and Isocyanide Derivatives.
F. Gloaguen, C. Le Floc'h, F.Y. Pétillon, P. Schollhammer, J. Talarmin, M. El Khalifa and J.Y. Saillard 33

PART 3. **REDOX PROPERTIES and RELATIONSHIPS with STRUCTURE and REACTIVITY** 39

3.1 **INORGANIC and ORGANOMETALLIC COMPLEXES** 39

The Parameterisation of Metal Centred Redox Couples
A.B.P. Lever 41

Electrochemical Ligand Parameters for Phosphonium-Functionalized Isocyanides and Derived Carbenes and Indoles of Group VI Transition Metal Carbonyls
M.T.A. Ribeiro, A.J.L. Pombeiro, G. Facchin, M. Mozzon and R.A. Michelin 57

Redox Properties and Ligand Effects for the Hydridotechnetium-Dinitrogen, -Carbonyl and -Isocyanide Complexes *trans*-$[TcH(L)(Ph_2PCH_2CH_2PPh_2)_2]$ ($L=N_2$, CO or CNR)
Yu Wang, A.J.L. Pombeiro, L. Kaden and M. Wahren 63

Electrochemical Studies of Coordination Complexes of Osmium
K.J. Taylor and L.J. Yellowlees 69

Molybdenum Macrocycles and Molecular Electronics
J.A. McCleverty 77

An Electrochemical Evaluation of Long Range Metal-Metal Interactions Through Hydrocarbon Bridges
C. J. Jones, S. L. W. McWhinnie, F. S. McQuillan and J. A. McCleverty 89

Electrochemical Properties of Molybdenum Complexes Containing a Bulky Tris(Pyrazolyl)Borate Ligand
M. Cano, J.V. Heras, C.J. Jones and J.A. McCleverty 95

Redox Properties of Symmetrical and Unsymmetrical Sandwich Complexes of Lutetium(III) with Phthalocyaninato and Naphthalocyaninato Ligands
M. L'Her, A. Pondaven, Y. Cozien and F. Guyon 101

New Polydentate Ligands and Complexes: Protic and Topographic Effects on Redox Properties
M. Gross, A. Nurhadi and E. Graf 107

Redox-Structural Correlations in Metal Thioether Macrocyclic Complexes: The Stabilisation of Mononuclear Silver(II) and Gold(II)
A.J. Blake, R.O. Gould, A.J. Holder, T.I. Hyde, G. Reid, A. Taylor, M. Schröder and D. Collison 121

3.2 **BIOINORGANIC COMPOUNDS** 131

Bioelectrochemistry: Making Use of the Electrochemical Behaviour of Proteins
H.A.O. Hill 133

Electrochemical Studies of Synthetic Models of Copper Proteins
J.O. Cabral 151

Cyclic Voltammetric Behaviour of some Copper Complexes with a Bis(Pyrazole)Bis(Thioether) Ligand
M.F. Cabral, J.O. Cabral, W.G. Haanstra, W.L. Driessen and J. Reedijk 171

PART 4. **DYNAMICS OF ELECTRON TRANSFER** 177

Correlation Between Structure and Heterogeneous Electron Transfer Rates of Coordination Compounds
F.A. Schultz 179

Some Electronic and Dynamical Factors in Electron-Transfer Chemistry: Metal Complexes, Clusters and Surfaces
M.J. Weaver 193

Theoretical Study of Electron Transfer Reactions in Porphyrins and Cytochromes
L.G. Arnaut 207

PART 5. **POLYNUCLEAR and SUPRAMOLECULAR COORDINATION COMPOUNDS** 215

Supramolecular Electrochemistry: Polynuclear Metal Compunds.
S. Roffia, C. Paradisi and C.A. Bignozzi 217

The Redox Propensity of High Nuclearity Metal Clusters
P. Zanello 229

Redox-Active Tetrapalladia Units $[Pd_4]^{n+}$ (n=0,1,2,3,4) from Double Oxidative Carbon-Carbon Coupling
P. Espinet and G. García-Herbosa 247

PART 6. **MOLECULAR ELECTROACTIVATION and ELECTROCATALYSIS** 253

Using Electrochemical Information for Organometallic Coordination Chemistry: Stoichiometric and Catalytic Reactions
W. Bruns, W. Kaim, M. Ladwig, B. Olbrich-Deussner, T. Roth and B. Schwederski 255

Oxidatively Induced, Electrocatalytic Ligand Substitution in 17-Electron Manganese Complexes: an Investigation of the Kinetics by Derivative Cyclic Voltammetry
M. Tilset, V. Skagestad and V.D. Parker 269

Electron-Transfer-Chain Catalysis: Role of Transition-Metal Radicals, Side Reactions and Coupling with Organometallic Catalysis
D. Astruc, M.-H. Delville and J. Ruiz 277

Molecular Catalysis of Electrochemical Reactions by Low Valent Transition Metal Complexes: Metalloporphyrins.
D. Lexa and J.-M. Saveant 295

One Two-Electron *vs.* Two One-Electron Mechanisms in the Catalysis of Electrocarboxylation of Aryl Halides by Nickel and Palladium Complexes
C. Amatore, A. Jutand and M.F. Nielsen 307

Electrochemical Reduction of CO_2 Catalysed by Transition Metal Complexes
A. Gennaro, A.A. Isse and E. Vianello 311

Organometallic Electrochemistry of Metal Alkyne and Related Complexes
N.G. Connelly 317

Electrochemical Behaviour of Complexes Derived from the Activation of Alkynes, Isocyanides and Nitriles
A.J.L. Pombeiro 331

Electrochemical Behaviour of Aminooxy-, Dioxy- and Diaminocarbene Complexes of Palladium(II) and Platinum(II)
T. J. Castilho, M. F. C. Guedes da Silva, A. J. L. Pombeiro, R. Bertani, M. Mozzon and R.A. Michelin 345

Electrochemistry of Platinum Phosphine Complexes: C-H and C-X Activation by Highly Reactive Intermediates
J.A. Davies, L. Chen, C.T. Eagle and R.J. Staples 351

Azametallic Electrochemistry
C.J. Pickett 357

Electrocatalysis of the Reduction of O_2 within Polyelectrolyte Coatings on Electrodes
C. Shi and F.C. Anson 381

Electrochemical Studies on Activation of Dioxygen at Cobalt Macrocycles Derived from Cobaloxime
G. Costa and C. Tavagnacco 391

Behaviour of the Ionomer / Metal Interface Towards the Electroreduction of Oxygen
C. Paliteiro and M.L. Pereira 405

Redox Properties of the Amavadine Models $[V(HIDA)_2]^{2-}$ and $[V(HIDPA)_2]^{2-}$ and their Electroinduced Reactivity Towards Activated-Thiols and -Phenols
J.J.R. Fraústo da Silva, M.F.C. Guedes da Silva, J.A.L. da Silva and A.J.L. Pombeiro 411

Electrochemical Study of the Oxidation Chemistry of Chromium Carbonyl Thiolates $RSCr(CO)_5^-$ and $RSCr_2(CO)_{10}^-$
J. Spring, C.P. Janzen, M. Darensbourg, J.C. Calabrese, P.J. Krusic, J.N. Verpeaux and C. Amatore 417

New Developments in Voltammetric Behaviour and Electrocatalytic Activity of Ni(II) Tetraazamacrocyclic Complexes
J. Taraszewska and G. Roslonek 423

Electrooxidation as a Route to Dihydrogen Complexes
B.E. Hauger, M. Tilset and K.G. Caulton 431

PART 7. **MECHANISTIC STUDIES (INCLUDING NUMERICAL and COMPUTING METHODS)** 435

Mechanistic Studies in Organo-Metallic Electrochemistry. The Role of Computer Based Techniques & Simulation
N. Taylor, T. Boddington, P. Halford-Maw, R. Luo and D. Mills 437

Numerical Methods in Synthesis and Analysis of Electrochemical Responses
R. Seeber, M.I. Pilo and G. Sanna 453

Molecules with Identical Electroactive Centres as a Probe for Dielectric Properties of Electrolytes
N. Taylor, T. Boddington, I. Dobson, B. Gill, R. Luo and B. Shaw 467

A Comparative Study of Numerical Methods for Cyclic Voltammetry Digital Simulation of an Electrochemical Process with a Coupled Chemical Reaction
M.A.N.D.A. Lemos and A.J.L. Pombeiro 477

Mechanism of the Electroactivation of the Metal- Hydride Bond in $[ReClH(NCR)(Ph_2PCH_2CH_2PPh_2)_2][BF_4]$
M.F.C. Guedes da Silva, J.J.R. Fraústo da Silva, A.J.L. Pombeiro, C. Amatore and J.N. Verpeaux 483

Electrochemistry and *cis/trans*-Isomerization of $Cr^{III/II}$-Aminocarboxylate Complexes
R. Meier, M. Boddin, M. Hecht and F.A. Schultz 489

Three Dimensional Electrochemistry: an Alternative Experimental Way for the Study of Electrochemical Processes
N. Papadopoulos, K. Hasiotis and S. Sotiropoulos 495

PART 8. **SPECTROELECTROCHEMISTRY** 501

Spectroelectrochemical (UV/Vis/near IR, EPR) Studies of Coordination Complexes
S.A. Macgregor, E. McInnes, R.J. Sorbie and L.J. Yellowlees 503

One- and Two-Electron Oxidations of Bimetallic Fulvalene Complexes Studied by Voltammetry and IR Spectroelectrochemistry
W.E. Geiger, C.G. Atwood and T.T. Chin 519

Spectro-Elctrochemistry and Electrochemical Spectroscopy
G.A. Heath 533

Application of "in situ" Reflectance Spectroscopy to Molecular Electrocatalysis
C. Lamy 549

Low Temperature Electrochemistry and Spectroelectrochemistry of Catalytically Important Tungsten(0) Complexes
M. Wilgocki, T. Szymanska-Buzar, M. Jaroszewski and J.J. Ziolkowski 573

Electrochemistry and Spectroelectrochemistry of Dinuclear Ruthenium and Osmium Complexes
R. Hage, J.G. Haasnoot, J. Reedijk and J.G. Vos 583

The Optical Spectra and Voltammetry of Mixed Halide/Nitrile/Carbonyl Osmium(III) Complexes. Additive and Non-Additive Effects
G.A. Heath and D.G. Humphrey 589

Buckminsterfullerene and its Electrogenerated Anions $C_{60}{}^{1-}$, $C_{60}{}^{2-}$ and $C_{60}{}^{3-}$
G.A. Heath, J.E. McGrady and R.L. Martin 595

PART 9. **EXPERIMENTAL ASPECTS** 601

Pratical Aspects of Electrochemistry and Applications Towards the Solving of Chemical Problems
K.M. Kadish and D. Dubois 603

Charge Transport Properties of Electrodes Modified with Osmium Containing Polymer Films
R.J. Forster and J.G. Vos 615

PART 10. **ELECTROCHEMISTRY in NON-CONVENTIONAL CONDITIONS. USE of ULTRAMICROELECTRODES** 623

Unconventional Electrochemistry at Ultramicroelectrodes: New Approaches for the Investigation of Chemical Reactivity
C. Amatore 625

A Microelectrode Study of Metal Hexacarbonyls in Aprotic Solvents
M.I. Montenegro and M.A. Queirós 645

The Use of Microelectrodes in Kinetic Studies of Homogeneous Chemical Reactions of Molybdenum Complexes
A.M. Fonseca, M.A. Queirós and A.R. Dias 649

PART 11. **GENERAL DISCUSSION, CONCLUSIONS and PROSPECTS** 653

Report 1 - Round Table Discussion
D. Tuck 655

Report 2 - General Discussions and Prospects
A.A. Vlcek 657

Report 3 - General Comments
C. Amatore 659

INDEX 661

PREFACE

The use of electrochemical techniques by chemists, particularly those who regard themselves as "inorganic" coordination chemists, has undergone a very rapid growth in the last 15-20 years. The techniques, as classically applied to inorganic species, had their origins in analytical chemistry, and the methodology had assumed, until the mid 60s, more importance than the chemistry. However, the growth of interest in coordination compounds (including organometallic complexes) having unusually rich redox properties, and in the understanding of electron-transfer in bio-inorganic species, has propelled electro-chemistry into the foreground of potentially readily available techniques for application to a very wide range of problems of interest to those chemists. This growth has been fuelled additionally by the availability of relatively cheap equipment of growing sophistication and by an increase in the "inorganic" chemists' general knowledge of physical electrochemistry. In particular, with increasing availability and sophistication of equipment, kinetic problems are now being addressed, and the range of electrode types and configuration and solvents has been greatly expanded. Furthermore, the rapid expansion of interest in biological problems has opened new avenues in functionalisation of electrodes, in the development of sensory devices and, in a sense, a return to the analytical base of the science, using novel and multi-disciplinary techniques drawing on synthesis chemistry and electronic micro-engeneering. The drive towards increasing use of microcomputer-controlled data analysis and the development of microelectrodes has opened exciting new avenues for the exploration of chemical reactions involving electron-transfer processes. For these reasons, electrochemistry as applied to "inorganic" coordination systems is now at a cross-roads, and the time is ripe for a review of the state-of-the-art.

"Inorganic" coordination chemists use electrochemistry in a variety of ways which include: (a) as a method of screening compounds for overall electrochemical activity (partly in the way that NMR and IR spectroscopies are used for characterisation purposes); (b) determining efficient ways of synthesising an oxidised or reduced species by "wet chemical" processes, or preparing a species (either of limited life or which is inaccessible by conventional routes) by controlled potential electrolysis; (c) for studying the electronic properties of compounds, *e.g.* how neighbouring redox sites interact, sensing properties, modification of reactivity by the presence of single or multiple redox sites; (d) for studying electrode reactions and the fate of oxidised or reduced species following initial electron-transfer, *i.e.*, kinetics studies; (e) initiation of novel chemical reactions involving ligand species bound to metals; (f) electrosynthesis of inorganic or organic products, and electro-polymerisation.

As a result of the explosion of interest in the use of electrochemical techniques in the almost routine study of coordination and organometallic compounds (some liken the technique to IR spectroscopic characterisation), catalysis, bio-inorganic chemistry and in the development of inorganic materials, a workshop surveying and permitting high-level discussion of electrochemistry as applied to molecular "inorganic" systems (coordination compounds, organometallics, transition and non-transition elements), to problems in bio-inorganic chemistry, to the study of reaction mechanisms and towards an appreciation of the newest techniques was thought to be highly appropriate.

The contents of this book survey the field of "molecular" electrochemistry as presented at the **NATO Advanced Research Workshop (ARW)** on **"Molecular Electrochemistry of Inorganic, Bioinorganic and Organometallic Compounds"** held in Sintra, Portugal, in March, 1992.

The purpose of the ARW was to assess, in an interdisciplinary way, the state-of-the-art in the application of electrochemistry to the investigation of the properties of inorganic, bioinorganic and organometallic complexes, by providing a forum for discussion involving not only electrochemists who have turned their attention to chemical problems, but also chemists who have already applied electrochemistry to the investigation of properties of such compounds. The contibutions are ordered according to the following **main topics which have been covered in the Workshop**:

(1) Overview, (2) electrosynthesis, (3) redox properties (and relationships with structure and reactivity), (4) dynamics of electron-transfer, (5) polynuclear and supramolecular complexes, (6) molecular electroactivation and electrocatalysis, (7) mechanistic studies, (8) spectroelectrochemistry, (9) experimental aspects, (10) electrochemistry in non-conventional conditions, and (11) conclusions and prospects.

The ARW began by an overall review (by A.A.Vlcek) of the **current state** of molecular electrochemistry of coordination compounds with emphasis on the elucidation of electrode processes, their significance in terms of chemical species, and structural interpretation at the molecular or submolecular level.

After a presentation of various methods of direct **electrochemical systhesis** (D.G. Tuck) the **redox properties** of those types of complexes and their **relationships with structure and reactivity** have been discussed. In particular, the parameterisation of metal centred redox couples has been evaluated (A.B.P. Lever), synthetic, structural and electronic consequences of electron-transfer reactions have been presented (M. Gross and K. Wieghardt) and redox-active complexes discussed (J.A. McCleverty) as components of secondary harmonic generators in non-linear optical materials.

Moreover, within the recent developments in **Bioelectrochemistry**, relevant attention has been devoted to the electrochemistry of redox proteins and redox enzymes (H.A.O. Hill), as well as of synthetic models of copper proteins (J.O. Cabral).

The **dynamics of electron-transfer** have also been treated (M.J. Weaver and F.A. Schultz), namely in the framework of the control of the electron-transfer rates by the interplay of donor-acceptor coupling and nuclear dynamics, the comparison of the kinetic properties of molecular redox reagents and electrode surfaces, and the relationships between structure and electrochemical reactivity.

The electrochemistry of **polynuclear** complexes has also been considered (S. Roffia and P. Zanello) and that of **supramolecular** systems has been outlined.

In the fields of **molecular electroactivation** and **electrocatalysis**, electroinduced ligand exchange, rearrangement and dimerization processes were presented (W. Kaim and D. Astruc), whereas low valent transition metalloporphyrins were applied in the catalytic electrochemical reduction of electrophilic molecules such as organohalides and carbon dioxide, and the mechanisms discussed (D. Lexa and J.-M. Savéant). The electrochemical behaviour of complexes (*e.g.*, with multiple metal-carbon bonds), derived from the activation of **small molecules** (such as alkynes and isocyanides), was presented (N.G. Connelly and A.J.L. Pombeiro), whereas in the field of azametallic electrochemistry the electroactivation of a number of N-bonded species was considered, namely N_2 and derived diazenides, diazoalkanes, nitride or nitric oxide (C.J. Pickett), and nitrile-derived methyleneamido species (A.J.L. Pombeiro). A variety of electroinduced reactions (involving, *e.g.*, bond cleavage or formation) was recognized in these electrode processes. Electrochemical studies on the activation of dioxygen by transition metal centres have allowed, *e.g.*, the estimate of thermodynamic and kinetic parameters for its coordination, and the electrocatalysis of its reduction to H_2O_2 or H_2O has also been discussed at various types of cobalt macrocyclic centres (F.C. Anson and G. Costa).

Mechanistic studies have constituted a matter of concern and the application of a variety of electrochemical methods has been discussed, including the valuable combination of electrochemistry with spectroscopy (uv, vis, near IR, EPR) in **spectroelectrochemical** studies (L. J. Yellowlees, W.E. Geiger, G.A. Heath and C. Lamy). Moreover, the role of **computer based** and simulation techniques has been analysed (R. Seeber and N. Taylor).

Practical aspects of electrochemistry applied towards the solving of chemical problems have been considered (K.M. Kadish), as well as several **unconventional electrochemical** approaches made possible through the development of ultramicroelectrodes (C. Amatore) allowing, in particular, to work within a submicrosecond time scale, in poorly conducting solvents or even in the absence of added supporting electrolyte.

The ARW ended with a general discussion focused on various aspects of molecular electrochemistry which raised particular interest and on promising **prospects** of the theme of the conference and these subjects are summarized by scientists with complementary expertises (A.A. Vlcek, D. Tuck and C. Amatore) in the final section of this book.

A word of thanks is due to the 57 authors who have contributed to this book (both the key speakers at the NATO ARW, who have been mentioned above, and the poster presenters, who have provided the full texts of their contributions). Moreover, we also gratefully acknowledge NATO and its Scientific and Environmental Affairs Division, as well as the following co-sponsors, for their generous financial support to this Workshop: Fundação Luso-Americana para o Desenvolvimento (FLAD, Luso-American Foundation for Development), Banco Nacional Ultramarino (BNU, Ultramarine National Bank) and Instituto Nacional de Investigação Científica (INIC, National Institute for Scientific Research). In addition, we are also indebted to the following institutions or companies for their additional support: The British Council, Junta Nacional de Investigação Científica e Tecnológica (JNICT, National Board for Scientific and Technological Research), Spanish Embassy, Calouste Gulbenkian Foundation, Sintra County Hall, Instituto Superior Técnico, Academy of Sciences of Lisbon, Dias de Sousa Lda., Sotel, Aliança Seguradora, Elnor, Refrige (Soc. Ind. Refrigerantes), Centralcer (Central de Cervejas) and Ana. Appreciation must also been expressed to the following members of the Local Executive Commission for their generous effort and full dedication: Luísa M.D. Ribeiro, Maria Amélia N.D.A. Lemos, M. Fátima C. Guedes da Silva, Elsa M.P.R.P. Branco, Yu Wang, M. Teresa A. Ribeiro, Luciana M.G. Costa, and José Armando L. Silva.

Armando J. L. Pombeiro
Editor

Jon A. McCleverty
Co-editor

ORGANIZING COMMISSION

Armando J.L. POMBEIRO	(DIRECTOR)
Jon A. Mc.CLEVERTY	(CO-DIRECTOR)
Jean-Michel SAVÉANT	
Karl M. KADISH	

LOCAL EXECUTIVE COMMISSION

Armando J.L. POMBEIRO
Maria Amélia N.D.A. LEMOS
Maria de Fátima C. GUEDES da SILVA
Yu WANG
Maria Teresa A. RIBEIRO
Luísa M. D. RIBEIRO
Elsa M.P.R.P. BRANCO
Luciana M.G. COSTA
José Armando L. SILVA

LIST OF PARTICIPANTS

Christian AMATORE*
École Normale Superieure, Laboratoire de Chimie, URA CNRS 1110, 24 rue Lhomond, F-75231 Paris cedex 05, France

Fred ANSON*
California Institute of Technology, Division of Chemistry & Chemical Engeneering, Pasadena, California 91125, U.S.A

Luís G. ARNAUT
Universidade de Coimbra, Departamento de Química, 3000 Coimbra, Portugal

Didier ASTRUC *
Université de Bordeaux I, Laboratoire de Chimie Organique, URA CNRS, 33405 Talance cedex , France

Elsa M. BRANCO
Instituto Superior Técnico, Complexo I, Av. Rovisco Pais, 1096 Lisboa codex, Portugal

João CABRAL*
Faculdade de Ciências do Porto, Departamento de Química, Universidade do Porto, 4000 Porto, Portugal

M. Fernanda CABRAL
Faculdade de Ciências do Porto, Departamento de Quimica, Universidade do Porto, 4000 Porto, Portugal

Mercedes CANO
Facultad Quimica, Departamento de Quimica Inorgânica, Universidad Complutense de Madrid, 28040 Madrid , Spain

Neil G. CONNELLY *
University of Bristol, School of Chemistry, Cantock's Close, Bristol BS8 1TS, U. K.

Giacomo COSTA*
Universitá degli Studi di Trieste, Dipartamiento Scienze Chimiche, Piazzale Europa 1, 34127 Trieste, Italy

Luciana M.G. COSTA
Instituto Superior Técnico, Complexo I, Av. Rovisco Pais, 1096 Lisboa codex, Portugal

Julian A. DAVIES
University of Toledo, Department of Chemistry, Toledo, Ohio 43606, U.S.A.

Gabriel GARCIA-HERBOSA
Colegio Universitario de Burgos, Departamento de Quimica Inorgânica, Apartado 231, 9080 Burgos, Spain

William GEIGER*
University of Vermont, Department of Chemistry, Burlington, VT 05405-0125, U.S.A.

Maurice GROSS*
Université Louis Pasteur, Department de Chimie, 4 Rue Blaise Pascal, 6700 Strasbourg, France

M.Fátima C.GUEDES da SILVA
Instituto Superior Técnico, Complexo I, Av. Rovisco Pais, 1096 Lisboa codex, Portugal

Bryan HAUGER
Indiana University, Department of Chemistry, Bloomington, Indiana 47405, U.S.A.

Graham A. HEATH*
Research School of Chemistry, Inorganic Electrochemistry Laboratory, G.P.O. Box 4, Camberra, ACT 2601, Australia

José V. HERAS
Facultad Ciências Quimicas, Departamento de Quimica Inorgânica, Universidad Complutense de Madrid, 28040 Madrid, Spain

H. Allen O. HILL*
University of Oxford, Inorganic Chemistry Laboratory, South Parks Road, Oxford 0X1 3QR, U.K.

Christopher J. JONES
School of Chemistry, University of Birmingham, Edgbaston, Birmingham B15 2TT, U.K.

Karl KADISH*
University of Houston, Department of Chemistry, Houston, Texas 77204-5641, U.S.A.

Wolfgang KAIM*
Universitat Stuttgart, Institut fur Anorganische Chemie, Pfaffenwaldring 55, 7000 Sttugart 80, Germany

Maurice L' HER
Université de Bretagne Occidentale, URA CNRS 322, Faculté des Sciences, 6 Av. Victor Le Gorgeu, 29287 Brest cedex, France

Claude LAMY*
Université de Poitiers, CNRS Laboratoire de Chimie I, 40 Av.du Recteur Pineau, 86022 Poitiers, France

Christine LE FLOC'H
University of Sussex, AFRC Nitrogen Fixation Laboratory, Falmer, Brighton BN1 9RQ, U.K.

M. Amélia N.D.A. LEMOS
Instituto Superior Técnico, Complexo I, Av. Rovisco Pais, 1096 Lisboa codex, Portugal

Alfred B. P. LEVER*
York University, Department of Chemistry, 4700 Keele Street, North York, Ontario, Canada M3J 1P3

Doris LEXA*
Université de Paris, Laboratoire d'Electrochimie Moleculaire, 2 Place Jussieu, 75251 Paris cedex 05, France

Maria Helena LOPES
Departamento de Química, Faculdade de Ciências do Porto, Universidade do Porto, 4000 Porto, Portugal

Jon A. McCLEVERTY*
University of Bristol, School of Chemistry, Cantock's Close,Bristol BS8 1TS, U. K.

Roland MEIER
Technical University of Leipzig, Department of Natural Sciences, P.O. Box 66, 7030 Leipzig, Germany

Rino A. MICHELIN
Facoltá di Ingegneria,Università degli studi di Padova, Istituto di Chimica Industriale, Via F. Marzolo 9, 35131 Padova, Italy

Maria Irene MONTENEGRO
Universidade do Minho, Departamento de Química, Largo do Paço, 4719 Braga codex, Portugal

Giovanni MORI
Istituto di Chimica Generale, Universitá di Parma, Viale della Scienze, I - 43100 Parma, Italy

Carlos PALITEIRO
Faculdade de Ciências e Tecnologia, Departamento de Química, Universidade de Coimbra, 3000 Coimbra, Portugal

Nikos PAPADOPOULOS
University of Thessaloniki, Laboratory of Physical Chemistry, Thessaloniki, 54006, Greece

Christopher J. PICKETT*
University of Sussex, AFRC Nitrogen Fixation Laboratory, Falmer, Brighton BN1 9RQ, U.K.

Lucinda PINHEIRO
University of Oxford, Inorganic Chemistry Laboratory, South Parks Road, Oxford 0X1 3QR, U.K.

Armando J. L. POMBEIRO*
Instituto Superior Técnico, Complexo I, Av. Rovisco Pais, 1096 Lisboa codex, Portugal

M. Arlete QUEIRÓS
Universidade do Minho, Departamento de Química, Largo do Paço, 4719 Braga codex, Portugal

Luísa M.D. RIBEIRO
Instituto Superior Técnico, Complexo I, Av. Rovisco Pais, 1096 Lisboa codex, Portugal

M. Teresa A. RIBEIRO
Instituto Superior Técnico, Complexo I, Av. Rovisco Pais, 1096 Lisboa codex, Portugal

Anne RIEGER
Department of Chemistry, Brown University, Providence, R I 02912, U.S.A.

Philip RIEGER
Department of Chemistry, Brown University, Providence, R I 02912, U.S.A.

Sergio ROFFIA*
Università di Bologna, Dipartimento di Chimica "G. Ciamician", Via Selmi 2, 40126 Bologna, Italy

Jean-Michel SAVEANT
Université de Paris 7, Laboratoire d'Electrochimie Moleculaire, 2 Place Jussieu, 75251 Paris cedex 05, France

Martin SCHRODER
University of Edinburgh, Department of Chemistry, West Mains Road, Edinburgh EH9 3JJ Scotland, U.K.

Franklin A. SCHULTZ*
Department of Chemistry, Indiana University-Purdue University at Indianapolis, 1125 East 38th Street, Indianapolis, Indiana 46205-2810, U.S.A

Renato SEEBER*
Università di Sassari, Dipartimento di Chimica, Via Vienna 2, 07100 Sassari, Italy

J. Armando L. SILVA
Instituto Superior Técnico, Complexo I, Av. Rovisco Pais, 1096 Lisboa codex, Portugal

Jean TALARMIN
Université de Bretagne Occidentale, URA CNRS 322, Faculté des Sciences, 6 Av. Victor Le Gorgeu, 29287 Brest cedex, France

Joanna TARASZEWSKA
Polish Academy of Sciences, Institut of Physical Chemistry, ul. Kasprzaka 44-52, 01-224 Warsaw, Poland

Norman TAYLOR*
University of Leeds, School of Chemistry, Leeds, U.K.

Mats TILSET
University of Oslo, Department of Chemistry, P.O.Box 1033 Blindern, N-0315 Oslo 3, Norway

Dennis TUCK*
University of Windsor, Department of Chemistry and Biochemistry, University of Windsor, 401 Sunset Ontario, Canada N9B 3P4

Jean-Nöel VERPEAUX
École Normale Superieure, Laboratoire de Chimie, URA CNRS 1110, 24 rue Lhomond, 75231 Paris cedex 05, France

Elio VIANELLO
Università di Padova, Dipartimento di Chimica Fisica, Via Loredan 2, 35100 Padova, Italy

Antonin A. VLCÉK*
Czechoslovak Academy of Sciences, J. Heyrovsky Institute of Physical Chemistry and Electrochemistry, Dolejskova 3, 18223 Prague 8, Czechoslovakia

Johannes G. VOS
Dublin City University, School of Chemical Sciences, Dublin 9, Ireland

Yu WANG
Instituto Superior Técnico, Complexo I, Av. Rovisco Pais, 1096 Lisboa codex, Portugal

Michel J. WEAVER*
Purdue University, Department of Chemistry, West Lafayette, Indiana 47907, U.S.A.

Karl WIEGHARDT *
Fakultat fur Chemie, Lehrstuhl fur Anorganische Chemie I, Ruhr-Universitat Bochum, Postfach 102148, 4630 Bochum 1, Germany

Michal WILGOCKI
Institute of Chemistry, University of Wroclaw, F. Joliot-Curie 14 Street, 50383 Wroclaw, Poland

Lesley YELLOWLEES*
University of Edinburgh, Chemistry Department, West Mains Road, Edinburgh EH9 3JJ, U.K.

Piero ZANELLO*
Universitá di Siena, Dipartimento di Chimica, Pian dei Mantellini, 53100 Siena, Italy

* Key Lecturer

PART 1. **OVERVIEW**

MOLECULAR ELECTROCHEMISTRY OF COORDINATION COMPOUNDS -PRESENT STATE AND NEW TRENDS

A.A.VLCEK
Czechoslovak Academy of Sciences,
Prague, Czechoslovakia

The term molecular electrochemistry appears in early seventies unifying gradually the studies of electrochemical behavior of inorganic and organic compounds. This newly coined term, and the newly developing approach connected with it became of importance when it was fully realized that to get a deeper insight into the mechanism of electrode process it was necessary to study whole series of compounds and to get new information on the mechanism and chemistry of reactions investigated. This trend brought the attention to the improvement of experimental techniques and introduced new ones to extract as much information as possible from the electrochemical experiment.

There has not been, to my knowledge, any precise definition of molecular electrochemistry given. Summing up the papers dealing with the subject and extracting from the most widely used approaches the general pattern, it might be possible to specify the essential steps in molecular electrochemical studies:

i) Elucidation of the complete mechanism of the overall process in electrochemical terms.

This elucidation is based essentially on the wide use of various electrochemical methods with the goal to find out the electrochemical parameters of the compounds studied, including all the coupled chemical reactions. This has to be carried in a very broad range of experimental conditions, such as concentration, supporting electrolyte, solvent, temperature etc. The onset of molecular electrochemistry is marked, among others, by a wide use of nonaqueous solvents for the study of electrode processes and by using methods excluding not only oxygen but mainly traces of moisture.[1] On the ground of these investigations the mechanism is described generally and the electrochemical and coupled chemical steps are specified, however without much knowledge of the actual chemistry taking place in the individual steps.

ii) Electrochemical terms are to be then, in the second step, translated into electrochemical species.

Without the knowledge of the composition of the intermediates, most of the description of the electrode reaction mechanism is merely a guess, even if qualified, as regards the chemical identity of individual products and intermediates. This drawback was early realized and led to an increased effort to couple the electrochemical techniques with other methods which would make it possible to identify independently various intermediates and products formed in the course of electrode process. These attempts seem to underlay the development of molecular electrochemistry into its present shape and represent one of the major driving forces of its further development.
Even if the tendency to identify the products and intermediates is much older and goes through many decades of history of electrochemistry, the onset of molecular electrochemistry was closely connected with the concentrated efforts to solve this problem.

A. J. L. Pombeiro and J. A. McCleverty (eds.),
Molecular Electrochemistry of Inorganic, Bioinorganic and Organomettallic Compounds, 3–11.

Mainly various spectroscopic methods were applied[2] to the study of intermediates and products of electrode reactions in situ, however, they are predominantly based on the electrolysis of the solution as a prerequisite for any other investigation. Even with special experimental set up the time of electrolysis cannot be shortened beyond a certain limit due to the detection limits of the spectroscopic techniques used. It is hardly possible to investigate intermediates with life times shorter than about few seconds.

iii)Explanation of the observed mechanism on the basis of molecular and submolecular structure.

The first attempts to get relationships between structure and electrochemical behavior go back to early fifties together with wider use of quantum chemical description of molecules and especially complexes[3]. The same holds for the theoretical interpretation of the rates of electron transfer in the solution as well as at the electrode[4]. Difficulties met with detailed description of complexes and changes in structure induced by the change in the overall number of electrons led to the search for empirical correlations and definition of parameters reflecting the change in structure along a given series of complexes[5]. The main goal, however, still the Holy Grail, is to get such an understanding of structure - electrode behavior relation as to make it feasible to predict, on the basis of known structure , the electrochemical behavior or, on the other hand, to deduce from the known mechanism of electrode process information on the structure. This knowledge would help greatly to tailor molecules for specific purposes and could be also of great value as a feedback to correct the interpretation of electrochemical descriptions and predict new useful experiments.

To sum up, the main goal of molecular electrochemistry is to find out, describe and understand the chemistry of an electrode process and interpret it on the basis of structure. The knowledge of the changes in the structure incurred by the electron transfer should be very detailed, i.e. the charge change localization should be precisely described as to make it possible to deduce not only the pathway of the electrode process but also the eventual specific reactivity of the species formed. General conclusions, valid for series of compounds of the given structural type, are to be sought for to make the prediction and feedback possible.

Coordination compounds and organometallics represent a vast group of compounds with a unique property: There are two types of completely different redox centers - the metal based ones and the ligand based ones. These different redox centers have, of course, different properties, especially in the effect the charge change localized on them during electron transfer exerts upon the structure of the species. There is, however, a wide range of combination of these redox centers, from fully localized ones to fully delocalized, which provides not only intellectual interest for the study of complexes but also possibilities, how to use these properties for a broad variety of purposes.

Having specified the main aims of molecular electrochemistry with special emphasis to coordination compounds, we may compare them with the present situation.

The electrochemical fundamentals proper are reasonably developed[6,7] and in most cases it is possible to find a solution of the electrochemical problem. Complicated mechanisms may be solved unambiguously by comparison of dependencies of experimental measurables with those predicted from the theoretical solution of the corresponding rate law of a model mechanism. The most widely used method for these comparisons is a plot of a measurable, or a quantity derived from it by simple algebraic transformation, against a variablc. Such plots usually give adequate answers; however, in many cases predicted differences

in plots for various mechanisms are smaller than experimental errors. Broadly used method to overcome this difficulty is the direct simulation[8,9] of current potential curves and comparison of these with experimental results. However , it is still unadvisable to limit the comparison to one simulated curve or dependence and it is preferable to investigate the correctness of the fit by comparing a set of experimental and simulated curves dependent on certain different experimental variables.
The problem which still lays open is the electrochemical (as distinct from chemical!)[10] irreversibility. For some cases it might be interpreted using the theory of electron transfer[4,11] (especially for one atom or one group localized processes), in other cases, such as bond rupture or bond formation accompanying the electrode reaction, it might be explained on the basis of energy necessary to bring the species into the state in which the components are "preprepared" for the given process in the activated state[12]. However, there are many cases in which the internal mechanism of irreversible electron transfer is not known (sometimes even not predictable at the present state of knowledge). One of the possibilities is the concerted electron transfer and hydrogen abstraction from solvation shell, however, no proof of such a concerted process has been given. A much closer investigation will be needed to build up a theory, possibly a quantum chemical one, covering these situations and to make such an electrochemical process predictable.

The present state-of-the-art of detection of intermediates indicates that this problem is by far more complicated, at present, than that of the electrochemical description outlined above. Even so, mostly reasonable results are obtained for intermediates with larger life times. Most of the usual spectroscopic methods, such as ESR,UV-VIS,IR,Raman and resonance Raman are being frequently used to get an insight into the chemistry of electrode processes[13]. The newest type of OTTLE cells[14] make it possible to carry the experiment using various spectroscopic techniques in the same solution and at the same set up of electrodes. The greatest advantage of these cells is the possibility to keep the cell and fill it with solution out of contact with the atmosphere and prevent thus any contamination with oxygen or moisture. This, in connection with the use of aprotic solvents , broadens the possibilities of spectroelectrochemistry to rather very sensitive intermediates and products. Compounds with redox potentials as negative as about -2.5 V are nowadays easily accessible, so far they are intrinsically stable. However, experimental data show that great care in analysing the measurables is needed to get proper answers. Any analysis has to be carried out, if possible, in a quantitative way and the study of positions as well as intensities of the responses is necessary. It becomes evident that in many cases the use of one single spectroscopic method does not give the proper answer, especially if detailed information is required on the nature of the species formed. However not in all situations are various spectroscopic methods usable or sensitive enough to chemical changes accompanying the electron transfer process. It is to be required that full consistency is reached between electrochemical and spectrochemical as well as among various spectrochemical interpretations of the chemistry of the process. This statement might seem trivial, however some recent studies substantiate the importance of stressing it.

This problem might be demonstrated on the group of compounds carrying several redox centers in weak interaction. These complexes, forming multi electron redox series[15], were studied for some decades with ever increasing accuracy and ever increasing precision of description of the redox process. The studies started with investigations of cobalt- tris bipyridine[16] and chromium- tris bipyridine [17] complexes and became of immense importance when ruthenium polypyridine complexes attracted the attention of electrochemistry[18]. The main problem which is still subject of discussion, is connected with the localization of the redox change on the individual redox centers. With homoleptic species, like $[Ru(BPY)_3]^{2+}$, electrochemistry by itself is not capable to give the answer whether the electron is accepted by one ligand or whether it gets delocalized over the whole coordination sphere. Some of the

ESR studies[19] indicate one ligand localization with subsequent hopping of the electron from one ligand to the other. To solve the problem we attempted[20] to investigate heteroleptic complexes in genetic series, i.e. from A_3 through A_2B, AB_2 to B_3, comparing the changes in redox potentials along the redox series of individual compounds and correlating them throughout the genetic series. Using a rather simplified quantum chemical model it was possible to calculate the expected differences in redox potentials ΔE^o for various localizations of electrons on individual redox centers. If the values of site (or orbital) energies of individual ligands and the values of spin-pairing (intraligand) and interligand electronic interaction energies are reasonably different it should be in principle possible to see the expected changes in ΔE^o for various electron configurations. Thus e.g. for complex AB_2 ($e_A < e_B$)there are three possible sets of configurations, the calculation for two of them giving the following sequences of ΔE^o values :

	ΔE^o		ΔE^o
	ABB		
A^-BB	J_A	A^-BB	$\Delta e + v_{AB}$
$A^{2-}BB$	$\Delta e + 2v_{AB} - J_A$	A^-B^-B	v_{BB}
$A^{2-}B^-B$	v_{BB}	$A^-B^-B^-$	$-\Delta e + J_A + v_{AB} - V_{BB}$
$A^{2-}B^-B^-$	J_B	$A^{2-}B^-B^-$	$\Delta e + \Delta J_i + v_{BB}$
$A^{2-}B^{2-}B^-$	v_{BB}	$A^{2-}B^{2-}B^-$	v_{BB}
	$A^{2-}B^{2-}B^{2-}$		

ΔE^o is the difference between two successive redox potentials, i.e. a double difference of energies of configurations(e.g. $e_1 - e_2 = E^o{}_1$, $e_2 - e_3 = E^o{}_2$; $\Delta E_1 = E^o{}_2 - E^o{}_1$), The values are given at the middle configuration of the three from which it has been calculated. Δe is the difference in site energies of A and B, J_A and J_B the spin-pairing energies in ligands A and B, respectively, v_{AA}, v_{BB}, v_{AB} the interaction energies between ligands A and B. These parameters can be extracted from independent experimental data and all of them contain contributions from solvation energy change as well as from the corresponding entropy changes. From above given expressions for ΔE^o it is seen immediately that the pattern of redox series differs for the two sequences of configurations and comparison with experimental data might give the answer which sequence of configurations is being followed in the individual redox steps.

Using Δe =0.62V, J_A = 0.45V, J_B =0.68V, v_{AA} =0.12V, v_{BB} = 0.1V for $[Ru(5\text{-}DCEB)(BPY)_2]^{2+}$ (DCEB standing for di-carboxyethylbipyridine), we find that the first sequence of configurations fits very closely the experimental data. Comparison with UV-VIS spectra taken for the first five states along the redox series of this particular complex indicates a complete consistency of electrochemical and spectroelectrochemical data so that it can be unambiguously concluded that the 5-DCEB ligand gets completely reduced to the dianion before the first reduction of BPY takes place. The same holds fully also for the bis-5-DCEB complex.

However, the reduction pattern of the analogous 4-DCEB complexes differs and the pathway of the electron transfer along the redox series is more complicated. For bis-4-DCEB complex the energies calculated (using Δe=0.4V, J_B=0.43V and v_{BB}=0.13V) for configurations $A^-B^-B^-$ and $AB^{2-}B^-$ and for configurations $A^{-}B^{2-}B^{-}$ and $AB^{2-}B^{2-}$, respectively, are, within experimental error, almost identical (-0.47V vs -0.51V and -0.21V vs -0.22V, respectively). This result indicates that it is impossible to distinguish solely on ground of electrochemical results the sequence of configurations along the corresponding redox series. Analysis of spectra, based predominantly on the intensity of intramolecular transitions in unreduced coordinated ligands (around 280-290 nm) and those of monoanions (around 350 nm) would be expected to yield the extinction of $5.3*10^4$ in the 350 nm region for configuration $A^{-}B^{-}B^{-}$ whereas extinction of $5.0*10^4$ in 290 nm region and $2.85*10^4$ in the 350 nm region for isoelectronic $AB^{2-}B^{-}$ configuration would be anticipated. Experimental data show extinction of about $4.7*10^4$ in the latter region and $3.2*10^4$ in the 290 nm region, in complete disagreement with any of the simulated values. This deviation of experimental and simulated curves is fully consistent with the calculation of the energies which coincide for configurations $A^{-}B^{-}B^{-}$ and $AB^{2-}B^{-}$. The only conclusion which can be drawn out of these results is that after the third electron is accepted (and, analogously , the fourth) not any single electronic configuration can describe the electronic state of the reduced species, the actual situation being best expressed by their linear combination. The spectra as well as the ΔE^{o} values fit best for the combination $0.65(A^{-}B^{-}B^{-}) + 0.35(AB^{2-}B^{-})$ for the complex with overall charge -1. A very similar situation, much closer to 1:1 mixing, exists for the next reduction product with charge -2. It can thus be concluded that reduction of 4-DCEB mixed complexes does not follow a single sequence of configurations and that states with 3 and 4 electrons are best represented as mixed configurations. Similar results were obtained for the mono-4-DCEB redox series as well as for other genetic series, e.g. that derived from BPYM and BPY. Most importantly, this combined approach to specify the structure of reduction products has helped to solve the puzzling problem of some heteroleptic complexes in which it appears that some of the reduced states cannot be described by one specific localization of electrons on individual ligands.

These findings open new ways to describe the electron transfer. Up to now, the concept of redox orbital has been predominantly used.[15,21-23] This concept assumes that the electron is entering (or leaving) a more or less well defined orbital localized in a specified part of the molecule and takes also into ac count, when necessary, the possible deviation from Koopmans theorem, i.e. the rearrangement energy and the changes in structure related to this rearrangement. However, already the first considerations on detailed mechanism along redox series led to the prediction of multiplets of redox orbitals[15]. This prediction seems now to be proven by the above described experiments and further development in the interpretation of redox processes should follow this path of dealing more with configurations (and their combinations) than with rigid redox orbitals. Of course, the redox orbital concept is much simpler to use, more easily accessible to quantum chemical calculations and, especially, very easy to picture the electronic changes.

In the above discussion on present state of molecular electrochemistry of coordination compounds some pending problems were already outlined needing further deeper study. Structure based theory of irreversible processes was mentioned as one of the open questions of molecular electrochemistry. New or improved methods for detection of intermediates represent also a very challenging field. Either it will be made possible to reduce the time of electrolysis by better design of cells or by increasing the detection limits of spectroscopic methods used or some new approaches will to be looked for. It might be possible to use, in special cases, induced electrochemiluminescence with time of polarization pulses as short as micro seconds for detection of intermediates. The other, not yet much explored, possibility, is the trapping of very short lived intermediates by chemical reactions and the determination of the products formed. This latter approach would need, of course, not only chemical intuition to design the proper reagent but also the necessary equations would have to be solved describing simultaneous transport of two or three components in comparable concentration to (or from) the electrode. Computer simulation seems to be the only possible way how to deal with this sort of problem. The knowledge of the nature of intermediates, especially their specific description in terms of charge change localization is of immense importance for further use of electrochemistry in several fields. This particular knowledge , together with the understanding of other factors determining the electrode behavior, should lead to the tailoring of new processes in which the reactants would be designed not only with the respect of their thermodynamics but also with respect to their selective reactivity (as mediators of electron transfer) or specific coordination ability (as homogeneous catalysts). The lack of general understanding, and in many cases just purely empirical approach, hinders the broader exploitation of electrochemical methods in homogeneous catalysis, a field for which electrochemistry is essential. However, it should be mentioned that up to now not a single electrochemically induced homogeneous catalytic reaction has been industrially applied which renders further research in this important and very challenging area.

Other, also extremely challenging purpose-oriented application of the knowledge of localization of electron changes and of transfer of electrons is the use of compounds with multiple redox centers as parts of molecular electronic devices. This field was up to now mostly speculated on from the point of view of photochemically induced electron transfer properties [24] and various devices have been proposed able to perform certain functions using molecular assemblies. Starting with these considerations, photochemistry got one of their most important pushes and continues to solve gradually and successfully individual problems. However it should be remembered that mostly electron transfer processes thermodynamically allowed are rather fast and it seems almost impossible to keep an information in the form of a localized charge for any longer time. The photochemical charge separation lasts very shortly even in very complicated molecular assemblies and even at rather long distances[4,11]. There seems, however, to be a possibility in combining the photochemical and electrochemical processes to make the desired functions in supramolecular systems possible. The systems under consideration have to allow only vectorial electron transfer along the assembly of properly chosen molecules and there has to be at a certain point an energy barrier in the system, big enough as to prevent any further electron transfer in the given direction. This barrier can be crossed by a light pulse(or possibly even electric pulse) and the electron thus transferred should travel very rapidly further along the molecular assembly to the electronic "well". The supramolecule permitting such a process has thus to carry several distinct redox centers, or groups of such centers, in different parts of the space with properly chosen redox processes and should be also well matched for any intramolecular electron transfer . Some of these conditions seem to be fulfilled in complexes carrying multiple distinct redox centers, however, constructed in an array with other molecules so that the whole assembly meets both the requirement of the vectorial transfer and that of the energy barrier.

Let us, for simplicity, assume a very simple arrangement

L_1--------M-------L_2----------A

in which L_1 and L_2 are redox centers (ligands) with E^o for L_2 at least 0,6 - 0.8V more negative than that of L_1. A is the electron acceptor ("well"), the electron affinity of which is higher than that of L_1. M is the metal center not directly involved in the redox process. Furthermore, let us assume L_1 to be bonded to the electrode, the whole array to be linear with A attached to an other electrode. In this assembly L_2 forms the barrier without crossing of which the electron cannot be transferred from left to right. If the electrode to which L_1 is attached is brought to the potential corresponding to the reduction of L_1 the electron should be transferred from the electrode to L_1 and, at the same time, an electron transferred from A to the second electrode to keep with the principle of electroneutrality. The electron at L_1 cannot pass along the system as L_2 forms the energy barrier and so far the electrode at L_1 position is held at the appropriate potential the electron stays localized on L_1 . To induce an intramolecular transport of the electron a light pulse of appropriate energy is to be applied to the system. L_1 -L_2 transition (LLCT process, essentially) should take place transferring the electron from L_1 to L_2. A is supposed to be an electron "well" deep in energy so that the probability of the intramolecular transfer L_2 - A is expected to be much faster than the back reaction L_2 - L_1. Both L_1 and A being held at the given potentials, the situation before the light pulse is immediately restored and a current pulse is registered to flow through the system. This picture, given here on a molecular level, is, however, supported by several observations and findings[25,26,27] which, even if not related directly to the above picture, point to the possibility to set up such an arrangement of molecules.

Above conditions must be met to make a process fully one ligand localized (i.e. difference in site energies are to be bigger than corresponding pairing energies). Preliminary estimations of the rate of intramolecular electron transfer suggest that the two centers L_1 and L_2 should operate better when localized on two different metal centers connected by a nonreducible bridging ligand which, however, does not prevent an interaction between both metal centers and thus also between the two ligands. The efficiency of the barrier would be increased when the crossing of it would be accompanied by an additional structural change.
There are, however, several questions to be raised for consideration when dealing with this problem. Whereas the efficiency of an electrochemical transfer is very high, the LLCT transitions, which have not been studied in enough detail up to now, are known to be of medium or low intensity so that the probability of passing the barrier after the light pulse is considerably smaller than that of the electron transfer from the electrode. This problem might be solved by making use of a stepwise transition , all the steps of which would be of substantially higher efficiency. Furthermore, not much being known on ligand-intervalence transitions (as opposed to metal-intervalence ones) , it might be expected that the investigation of complexes in question might bring new results which would help to tailor the necessary combination of L_1-L_2. There is, however, a possibility to overcome this difficulty by using an increased very short electrical pulse which would bring the electron over to L_2 and then immediately to the electron "well" at A.

The other questions related to the use of redox multicentered complexes as part of molecular electronic devices seem to be common with those of analogous devices working solely on photochemical principles. They include e.g. the problem to which extent it might be possible to address one specific molecule with a photon (or electron), i.e. to let the whole system work on a really molecular level. Recent estimations expect at least 5000 molecules to be necessary for an effective process to occur[28]. Furthermore, overhearing from one molecule (or set of molecules) is to be prevented as well as the quenching of the signal by neighboring molecules. The whole assembly should have be packed into a very small space, much smaller than are todays ultramicroelectrodes. At the present state of knowledge of vectorial transfer along supramolecular aggregates it is an open question to which extent the separation of charges, induced electrochemically and kept by holding the appropriate centers at given potentials, would meet the requirements of electroneutrality - formally the supramolecule would be neutral, however, highly polarised and the charges being rather far apart they might require independent neutralization by counterions. For the charge transfer being very fast along the molecule and the original situation being quickly restored not much rearrangement of counterions is to be expected.

The described process, even if hypothetical, outlines the possible fundamentals of the use of redox multicentered complexes as parts of molecular electronic devices. The complexes under consideration seem to be extremely well suited to meet the requirements of molecular electronics in other aspects too, however, in all cases only under the assumption of the vectorial electron transfer . As an other example let us consider a complex with two identical metal centers connected by a bridging ligand, the bridge being non reducible, however, well transmitting the electronic effects. Each of the metal centers is attached to one of the electrodes . When a controlled bias is applied across the "sandwiched" assembly, on one side the metal gets oxidized, on the other the complex gets reduced. If the energy difference between the two states thus formed is big enough the centers differing in the oxidation states by two units would immediately interact under the formation of an excited state of the original oxidation state. This excited state, with properly chosen metal and ligands, would convert the electrical energy into the energy of a photon and radiation is expected to be emitted. This "energy convertor" seems to be much easier to build than the above mentioned system with the energy barrier.

The use of these principles, and some other derived from them, might "build" in the redox multicentered complexes as components of molecular switches, memories, diodes or energy convertors. However, much more understanding of the processes taking place in these systems is needed and much deeper knowledge of the vectorial electron transfer is required. The present knowledge of electrochemistry of spatially ordered depolarizers is very small and the transfer of experience from disordered systems in solution to ordered ones sandwiched between two electrodes might not be completely reliable. Of course the challenges are big and it might be expected that all will be met in future development of molecular electrochemistry of coordination compounds.

References.

1) E.Geiger and M.D.Hawley in "Physical Methods of Chemistry" (Sec. Ed.), R.W.Rossiter and J.F.Hamilton editors, J.Wiley., New York, Vol. II,p.1
2) R.L.McCreeny, ibid., p.591
3) A.A.Vlcek,Progr. Inorg. Chem, 1963,5, 211
4) R.A.Marcus and N.Sutin, Bioch. et Biophys. Acta,1985,811, 265
5) A.B.P.Lever, Inorg. Chem., 1990,29,1271
6) A.J.Bard and L.R.Faulkner, Electrochemical Methods, J.Wiley ,New York, 1980.
7) a)Z.Galus in "Physical Methods of Chemistry" (Sec. Ed.), R.W.Rossiter and J.F.Hamilton editors,J.Wiley , New York, Vol. II, p.191, (1988)
b) E.R.Brown and J.C.Sandifer, ibid., p.273
c) R.W.Murray, ibid. p. 525
8) S.W.Feldberg, Electroanal. Chem. 1969,3,199.
9) S.W.Feldberg in "Computers in Chemistry and Instrumentation", Vol. II, "Electrochemistry",J.S.Mattson,H.B.Mark Jr. and H.C.MacDonald Jr., editors, Marcel Dekker, New York, 1972.
10) A.A.Vlcek,R.Kalvoda, L.Pospisil and J.Volke in "Physical Methods of Chemistry" (Sec. Ed.), R.W.Rossiter and J.F.Hamilton editors, J.Wiley, New York, Vol. II,p.797,(1988).
11) T.J.Meyer, Prog. Inorg. Chem. 1983,30,389.
12) R.G.Pearson, Symmetry Rules for Chemical Reactions,J.Wiley,New York,1976.
13) R.L.McCreeny in "Physical Methods of Chemistry" (Sec. Ed.), R.W.Rossiter and J.F.Hamilton editors, J.Wiley, New York, Vol. II,p. 591 (1988).
14) M.Krejcik, M.Danek and F.Hartl, J .Electroanal. Chem.,1991,317,179.

15) A.A.Vlcek, Coord .Chem .Rev.,1982,43,39.
16) A.A.Vlcek, Z .Elektrochem,.1958,61,1014.
17) a)A.A.Vlcek, Nature,1961,189,393.
b) T.Saji and S.Aoyagui, J .Electroanal .Chem.,1975,63,405.
18) T.Saji and S.Aoyagui, J .Electroanal .Chem.,1975,60,1.
19) D.Morris,K.W.Hanck and M.K.DeArmond, J. Electroanal .Chem., 1983,147,115.
20) S.Zalis,M.Krejcik,V.Drchal and A.A.Vlcek, J .Am .Chem .Soc., in press
21) A.A.Vlcek, Electrochim. Acta,1968,13,1063.
22) M.K.DeArmond,K.W.Hanck and D.W.Wertz, Coord .Chem .Rev. 1985,64,65.
23) A.A.Vlcek, Rev .de Chim. minerale,1983,20,612.
24) V.Balzani,L.Moggi and F.Scandola in "Supramolecular Photochemistry", V.Balzani editor, D.Riedl Publish. Co.,1987.
25) Xung Zhang and A.J.Bard, J .Am .Chem .Soc.,1989,111,8098.
26) J.C.Jernigan and R.W.Murray, J. Am. Chem. Soc.,1987,109,1738.
27) P.G.Pickup,W.Kutner,C.R.Leidner and R.W.Murray, J .Am .Chem .Soc.,1984,106,1991.
28) J.J.Hopfield,J.N.Onuchie and D.N.Beratan, J .Phys .Chem. 1989,93,6350.

PART 2. **ELECTROSYNTHESIS**

DIRECT ELECTROCHEMICAL SYNTHESIS OF INORGANIC AND ORGANOMETALLIC COMPOUNDS

DENNIS G. TUCK
Department of Chemistry and Biochemistry
University of Windsor
Windsor, Ontario N9B 3P4
Canada

ABSTRACT. The method of direct electrochemical synthesis consists of oxidizing a metal anode in a non-aqueous solution containing a ligand (or ligand precursor) to produce the appropriate inorganic or organometallic compound. In many cases, the product precipitates directly in the cell, making for easy isolation, so that the technique is both direct and simple, and in addition the product yields are very high. One advantage of the technique is that the products are often derivatives of a low oxidation state of the metal; examples of this include chromium(III) bromide, tin(II) and lead(II) diolates and thiolates, hexahalogenodigallate(II) anions, thorium diiodide, copper(I) thiolate complexes, and indium(I) derivatives of thiols, dithiols, and diols. In some systems, the low oxidation state compound undergoes subsequent reaction; for example, in the synthesis of $RInX_2$ the reaction sequence involves the oxidation of indium metal to give InX, which then reacts with RX to give $RInX_2$. Another possible post-electrolysis process is disproportionation. Examples of these various preparative routes will be discussed.

Details of two recent investigations are also reported. One of these depends on the oxidation of indium in solutions containing CH_2X_2 (X = Br,I), to give X_2InCH_2X or $X_2InCX_2InX_2$ species as the final products. A different system involves the oxidation of indium in liquid ammonia solutions of NH_4X or aromatic 1,2-diols, and the reactions in liquid ammonia are discussed.

1. Introduction

A basic tenet of many of the papers presented at this meeting is that the electrochemical technique represents the optimum method of carrying out oxidation or reduction reactions, in large measure because the removal or addition of electrons to a given solute species can be achieved without the complications attendant upon the addition of redox reagents to the reaction mixture. In this paper, we are particularly concerned with preparative electrochemistry, which is a subject which has developed considerably in recent years in inorganic, organometallic and biochemistry. The work which we have done in Windsor has concentrated on the use of non-aqueous solvents, and we have been able to prepare a wide range of compounds by the use of some simple apparatus, and some equally simple ideas. In particular, we are concerned with electrochemical systems in which the anode serves not only as a sink for electrons but also as a reagent towards species which are present in solution, or which are generated in solution as the electrolysis proceeds. The use of a sacrificial electrode is particularly important, because a high-purity metal serves as the starting point of the synthesis. In addition, we have established that

A. J. L. Pombeiro and J. A. McCleverty (eds.),
Molecular Electrochemistry of Inorganic, Bioinorganic and Organomettallic Compounds, 15–31.

the absence of detailed electrochemical parameters such as E_o in preparative non-aqueous electrochemistry is no more a hindrance to the use of the method than is the absence of thermodynamic data to the chemist who simply wishes to heat substances together in a flask.

Inorganic chemists were rather slow to recognize the advantages inherent in the use of metals as reagents, which is surprising given the fact that organozinc halides were first synthesized 130 years ago from Zn + RX, and that the use of magnesium to yield Grignard reagents is fundamental in organic and organometallic chemistry. The use of high vacuum techniques, which allow the evaporation of gram quantities of metals into a high temperature vapour, and the development of the associated equipment for collecting the products, led to a new interest in the use of metals as synthetic reagents. Our use of metals in direct electrochemical synthesis was entirely coincidental, but it is worth noting that some of the compounds which we have prepared have also been obtained by the more exacting methods of vapour phase synthesis. In principle, one can use the metal as either cathode or anode in direct electrochemical synthesis, but most of our work has been concerned with anodic oxidation, and the discussion in this paper will be confined to such experiments. This is not to ignore the many interesting reports on the use of sacrificial cathodes, some of which have been reviewed elsewhere [1]. I shall also outline the general directions which our research has taken, and identify the particular ligand systems for which we believe direct electrochemical synthesis provides advantages of yield and purity.

The final point to be emphasized is that because one starts with a metal which is by definition in the zero oxidation state, the experimental technique will necessarily give preferential access to the lower oxidation states if these can be stabilized in the solvent system in question. In particular, for a number of metals in the Main Group section of the Periodic Table, direct electrochemical synthesis is a simple and attractive way of getting to compounds which otherwise may not easily prepared, and hence provides an entrée to the study of their chemistry.

2. Experimental Outline

As with most electrochemical systems, our work has been conducted at or near room temperature and with solvent systems which are readily accessible in most laboratories. The cells which we have used are unsophisticated in the extreme, and the electrical power can be derived either from a relatively cheap AC/DC rectifier power pack, or in the extreme case from chemical storage batteries [2]. The typical simple cell shown in Fig. 1 is based on a 100 mL tall-form beaker; a stream of dry nitrogen passing through the solution allows one to work in oxygen- and moisture-free conditions. The apparatus can be modified if the products or reagents are very air-sensitive, and one such system has been described by Casey *et al.* [3]. The favoured solvent in our laboratory is acetonitrile, but other basic solvents, or mixtures thereof, have also been used. The main criteria are solubility for ligand and background electrolyte, little or no reactivity towards the product, ease of purification, and stability to an applied voltage of ca. 20 V cm^{-1}. The background electrolyte has generally been Et_4NClO_4 (ca. 50 mg per 50 mL of solution), but tetraalkylammonium salts of PF_6^- or BF_4^- have also been used, especially when oxidation by ClO_4^- is a real or suspected problem.

We normally run experiments at a current of 20-30 mA, and the applied voltage is then that required to achieve such a current; typical values would be in the range 10-30 V, depending on the solutes in the electrolyte phase. The applied voltage required obviously depends on the electrode potential for the reaction, but more importantly on the EMF needed to drive the current carriers through a medium of low dielectric constant. Given the use of electrodes with surface areas of a few cm^2, the current density at the anode is in the order 10-20 mA cm^{-2}. These

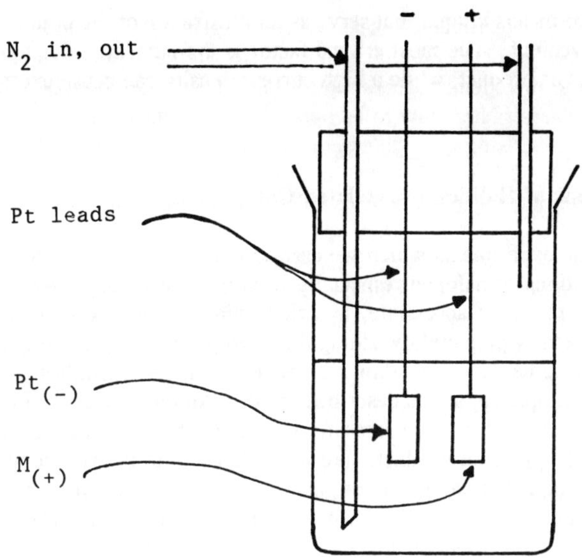

Figure 1. Diagrammatic representation of electrochemical cell

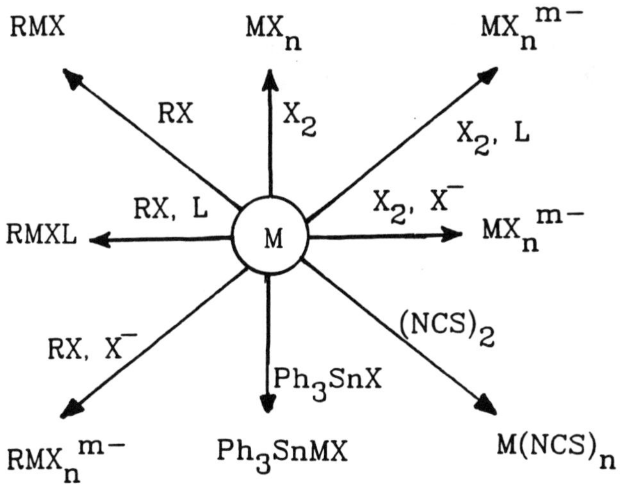

Scheme 1

parameters are by no means unique, but serve as an illustration of the practical conditions which we have found convenient. The most critical factor is the current, since too low a value gives only small quantities of product, while a high current density can cause excessive heating of the cell.

3. Direct Synthesis of Halides and Related Compounds

The first systematic experiments which we carried out involved the oxidation of metal anodes in non-aqueous solutions of halogen, especially bromine and iodine. We found that the metal halides themselves are easily accessible, and that anionic complexes or adducts with neutral ligands can also be readily prepared by adding the appropriate species to the solution. This work is summarized in Scheme 1, which shows a range of different syntheses, all of which were carried out with the simple apparatus described above. An obvious extension was to investigate the reactions of organic halides, and successful syntheses of RMX, RMX_m^{n-}, etc., were found to be possible, with good yields of product. We were also able to prepare compounds by inserting a metal into an existing metal-halogen bond, and derivatives of the type $Ph_3SnZnCl$.tmed (tmed = N,N,N',N'-tetramethylethanediamine) were obtained from Ph_3SnCl [4].

Another extension is that pseudohalogens can replace halogen, so that neutral and anionic isothiocyanates are also accessible by this method [5], as are heterometallic carbonyls [6]. A simple preparation of $(Ph_3PH)_2CoCl_4$ makes a useful undergraduate laboratory experiment [2].

All of these processes can be represented as the electrochemically driven oxidative addition of a metal atom to an X-X, R-X, M'-X or M-M bond. The metals for which such reactions have been observed in our laboratory in one or all of these systems include Mg, Ti, Zr, Hf, V, Cr, Mo, Mn, Fe, Co, Ni, Pd, Cu, Ag, Au, Zn, Cd, Hg, Ga, In, Sn, Th and U [7]-[31].

We have not followed this work recently as much as we would have wished, but the range of experiments, which are discussed in more detail in an earlier paper [1], already shows that both transition metal and Main Group elements can be used as the anode in such systems, and that a wide range of useful products can be prepared in a very simple and straightforward way. I do not propose to discuss this area further, other than to emphasize the particular advantage of such syntheses, which are that one can prepare the anhydrous halides without resorting to the high temperature methods which are otherwise required: that removal of water from hydrated materials is not necessary: that the derivatives (*e.g.*, MX_n^{m-}, RMX_n^{m-}, RMX_nL_m) are as easily obtained in the one-step synthesis as are the parent halides themselves; and that the products are obtained in high yield and purity.

4. Thiolates and Related Compounds

It has been known for many years that thiols or disulfides can be reduced electrochemically to the corresponding RS^- anions. This is the first step in the direct electrochemical synthesis of metal thiolates and their derivatives, since these anions, or more probably the radicals produced when the anions discharge at the anode, react with a variety of metals. We have carried out successful syntheses with the elements Co, Ni, Cu, Ag, Au, Zn, Cd, Hg, In, Tl, Sn and Pb to give $M(SR)_n$, with R = Et, t-Bu, n-Bu, C_5H_{11}, Ph, o-, m-, p-tolyl, 2-naphthyl, etc. (not all combinations) [32]-[37]. As with the halide systems, one can equally well produce the compounds themselves, or their derivatives, by appropriate adjustment of the solution phase conditions. The synthesis of these substances is experimentally simple and straightforward, especially since the products are often insoluble in the solvent systems used, and so can be conveniently collected beneath the anode.

This is a convenient point at which to introduce the method of chemical accounting which gives some insight into the mechanism of these and related electrochemical syntheses. For a thiol, the sequence is

$$\text{cathode: } \underline{n}RSH + \underline{n}e \rightarrow \underline{n}RS^- + {}^{n}/_{2}H_2 \quad (1)$$

$$\text{anode: } \underline{n}RS^- + M \rightarrow M(SR)_n + \underline{n}e \quad (2)$$

giving as the overall stoichiometry

$$nRSH + M \rightarrow M(SR)_n + {}^{n}/_{2}H_2 \quad (3)$$

If a disulfide is used, eq. (1) is replaced by

$$\text{cathode: } {}^{n}/_{2}R_2S_2 + \underline{n}e \rightarrow \underline{n}RS^- \quad (1a)$$

and the overall process is

$$ {}^{n}/_{2}R_2S_2 + M \rightarrow (SR)_n \quad (3a)$$

There is one noteworthy difference between the two systems, in that none of the metals listed above react directly with thiols, whereas a number of Main Group metals will undergo a thermal reaction with Ph_2S_2 to give reactions which are stoichiometrically equivalent to eq. (3a) [38].

The sum of eqs. (1) + (2), or (1a) + (2), corresponds to the obvious fact that electrons flow through the cell, and leads to an important experimental parameter which we term the electrochemical efficiency (E_F), defined as moles of metal dissolved from the anode per Faraday of electricity flowing through the cell. The weight loss at the anode is readily determined; the total quantity of electricity can be measured either by placing a silver coulometer in series with the cell and power supply, or by maintaining a constant current by manual control for a given period of time. Under the typical conditions which we have used, a current of 20 mA over 2 h leads to the dissolution of approx. 50-200 mg of metal, depending on the atomic mass of the latter, so that E_F is easily determined to $\pm$ 0.02 mol F^{-1}. For the elements listed above, the E_F values are invariably 0.5 mol F^{-1} for Zn, Cd, Sn and Pb, and 1.0 mol F^{-1} for Cu, Ag, In and Tl. We shall return to a discussion of some of the values below, but for the case of Zn or Cd, this result is in accord with the formation of $M(SR)_2$ compounds or their derivatives.

Following the syntheses of thiolato compounds, we also successfully prepared some analogous $M(SePh)_n$ compounds (M = Cu, Ag, Zn, Cd, Tl, Sn) and adducts such as $Cd(SePh)_2$.phen and $CuSePh.1.5Ph_3P$ by electrolysis with solutions of Ph_2Se_2 in toluene/CH_3CN mixtures [39]. Similarly, solutions of Ph_2PH in CH_3CN gave $M(PPh_2)_n$ (M = Co, Cu, Ag, Au, Zn, Cd), and in an extension of this work, solutions of Ph_2PH and S_8 in toluene/CH_3CN yielded derivatives of $M(S_2PPh_2)_2$ for Co, Ni, Zn and Cd [40]. These straightforward syntheses lead to studies of the oxidative and structural investigation of some of these compounds. Attempts to extend the diphenylphosphido work by using ($\underline{c}$-$C_6H_{11})_2PH$ were not successful, since the products were extremely unstable.

These syntheses are all characterized by simplicity of procedure, by high yield, and by the formation of pure products. Since the one-step method requires the use of essentially only metal plus ligand precursor, the chances of contaminating the product are minimized, and we believe that this approach is an improvement on those in the literature for thiolates and selenolates. The generality of the methods for phosphido complexes is open to doubt in view of the problems encountered with $(\underline{c}\text{-}C_6H_{11})_2PH$, and further work is obviously required here. The potential significance of these syntheses lies not only in their simplicity, or in the interesting structures identified, but also because some of these products may prove to be precursors of the III-V and IV-VI compounds which are so important in the microelectronics industry.

One interesting and unexpected aspect of these syntheses is that the products of the electrochemical oxidation of copper included some molecules with unusual cage structure based on Cu_4S_4, Cu_2S_2, Cu_4P_4, etc., rings [35], [41]-[44]. This is not the place to attempt a discussion of such structures, nor those of the $[M_4(SPh)_{10}]^{2-}$ anions (M = Zn, Cd), since good reviews have already been published [45,46], but their easy accessibility by direct electrochemical synthesis is yet another feature of this versatile method.

Some special mention should also be made of the electrochemical synthesis of amido complexes. In some early unpublished work, we used solutions of $\underline{i}\text{-}Pr_2NH$, 2,2,6,6-tetramethylpiperidine or $(Me_3Si)_2NH$ attempts to prepare $M(NR_2)_2$ compounds (M = Zn, Cd, Hg), but although solid products were obtained, their properties were not those of the expected properties, and this work was not pursued in the light of competing interests. With the amine py_2NH, on the other hand, the experiment proceeded smoothly to give $M(Npy_2)_n$ (M = Cu, Ag, Zn, Cd, Tl) [47]. It seems likely that the anions of the amines used earlier, or the radicals derived from them, react with CH_3CN to give species of the type $R_2NC(CH_3)N^-$, and that the products are in fact derivatives of such ligands.

Finally, we may note that if one views the parent ligand precursor (*e.g.*, thiol, phosphine, etc.) as a weak acid which is reduced cathodically to yield the corresponding anion, then R_3CH compounds can also be included in this group of syntheses. The successful production of PhCCCu, which is itself a useful synthetic reagent in organic chemistry, from PhCCH (pK ~25) is one example of this [48], and in related studies of wider implication, Casey [3] and Lehmkuhl [49,50] have obtained cyclopentadienyl and related compounds. The range of syntheses achieved with such weak acids, which do not react with metals under non-electrochemical conditions, is illustrated in Scheme 2.

5. Bidentate Ligands

A different group of weak acids which lend themselves readily to the direct electrochemical synthesis of metallic derivatives is illustrated in Scheme 3. These differ from those in Scheme 2 essentially in that the acidic group (OH, SH, etc.) is part of a molecule which contains a second donor atom, so that cathodic reduction gives an anionic ligand which is a potential chelating agent. In early studies [51,52], it was found that $acac^-$ (2,4-pentandionate) derivatives were readily accessible by this route, and as in the work described above, the products may be $M(acac)_n$ or $M(acac)_nL_m$ depending on the solutes present during the electrolysis [53]-[55]. A number of related bidentate oxygen donors have been studied, including catecholates [53] and other aromatic 1,2-diolates [57,58] and carboxylic acids [59]. Bidentate sulphur donor ligands represent a simple extension of this aspect of direct electrochemical synthesis, and dithiolato [60], [44], dialkylthiocarbamato [61] and diethyldithiophosphato [61] derivatives have been prepared. The metals which have been successfully used in this aspect of the work include Cr, Mn, Fe, Co, Ni, Cu, Ag, Zn, Cd, Hg, Ga, In, Tl, Sn, Th and U.

Once again, it is appropriate to emphasize that the simplicity and directness of the method offer

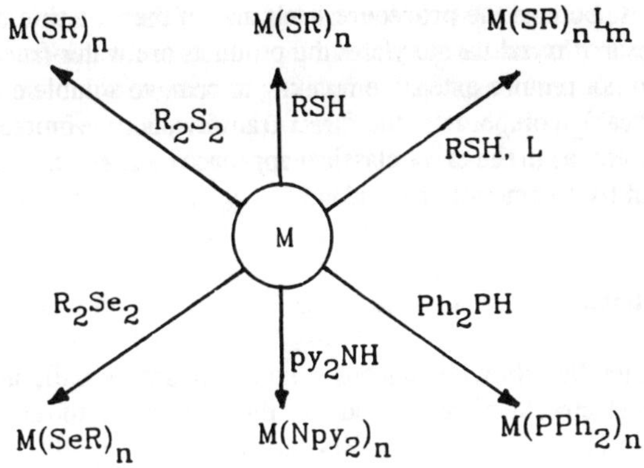

Scheme 2

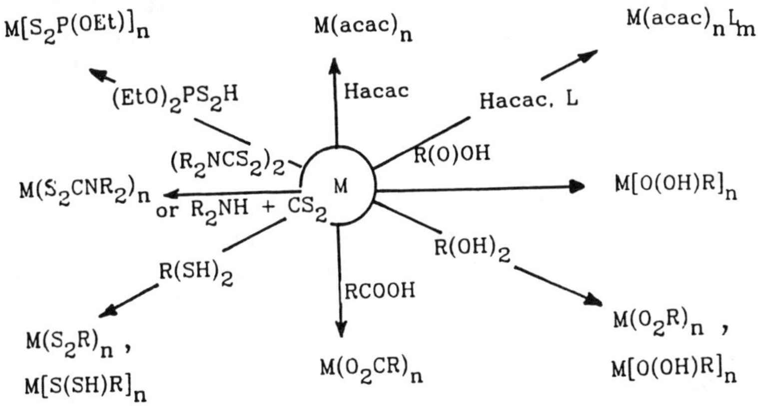

Scheme 3

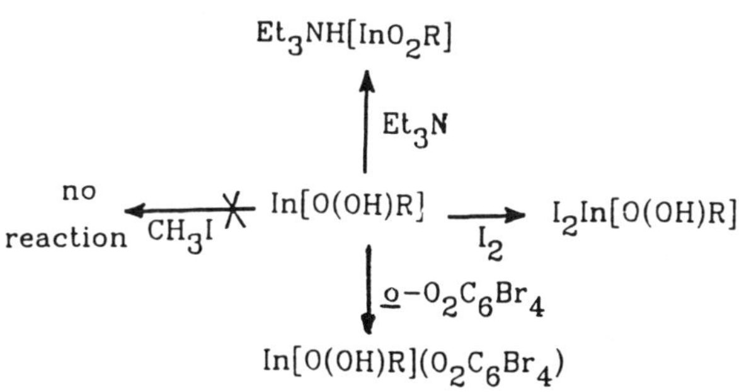

Scheme 4

considerable advantages, both in the procedure itself and in the isolation of pure products. For example, in the synthesis of metal carboxylates the products are water-free, since aqueous media are not used, and so do not require extensive washing to remove soluble contaminants. Equally, in the synthesis of $M(acac)_n$ compounds, the direct transformation of metal into complex avoids dissolution, extraction, etc. as in the more classical approach. Large scale preparations have been successfully carried out by Lehmkuhl *et al.* [51].

6. Strong Acid Systems

Since many metals react directly with aqueous solutions of strong mineral acids, there has been little incentive to apply electrochemical methods for the preparation of derivatives of these acids. A brief investigation showed that $[Cr(dmso)_6]Br_3$ can be obtained by this one-step route [15], and later efforts showed that other metals can serve as the starting point for dmso or CH_3CN complexes of the type $[ML_6](BF_4)_n$ for M = V, Cr, Mn, Fe, Co, Ni, Zn, Cd and In [62].

Some work which we were not able to develop fully as would have been wished, using the heavy elements thorium and uranium, showed that the application of the method is not restricted to lighter metals, since $[Th(dmso)_8](NO_3)_4$ could be prepared from solutions of nitric acid in tri-n-butyl phosphate; other media used in this work included N_2O_4/EtOAc/CH_3CN mixtures [63]. An efficient and compact method of treating spent fuel rods from a nuclear reactor might be developed around the electrochemical oxidation and dissolution of the metal fuel element.

7. Low Oxidation State Products

One of the most interesting aspects of the work, as noted earlier, is that low oxidation state compounds are often produced by the dissolution of a metal anode. A list of some such syntheses is given in Table 1, and the appropriate papers should be consulted for details. The particular interest in Main Group metals in our laboratory has again lead us to concentrate on the elements Ga, In, Tl, Sn and Pb, although the early syntheses on copper(I) species [25], of $CrBr_3$ [15], and of ThI_2 [28], show that these are examples of interest in transition and heavy element chemistry.

The contrast between the electrochemical method and the method in the literature is very striking in one particular case, namely the formation of the tin(II) compounds by the electrochemical oxidation of the metal in solutions of 1,2-aromatic diols in acetonitrile ($R(OH)_2$ = catechol, 2,3-dihydroxynaphthalene, $Br_4C_6(OH)_2$, 2,2'-dihydroxybiphenyl). The room temperature, high yield, electrochemical synthesis of $Sn(O_2R)$ compounds is a great improvement over the high temperature methods used in the earlier syntheses of such compounds [64]. The ready accessibility of the $Sn(O_2R)$ materials lead to a study of their redox reactions, and their coordination chemistry. Not surprisingly, the direct synthesis of $Pb(O_2R)$, and its redox chemistry, follow from the tin(II) system [65]. In each case the E_F value of 0.5 mol F^{-1} can be understood by the sequence

$$\text{cathode: } R(OH)_2 + 2e \rightarrow RO_2^{2-} + H_2 \quad (4)$$

$$\text{anode: } RO_2^{2-} + M \rightarrow M(O_2R) + 2e \quad (5)$$

giving the overall stoichiometry

TABLE 1. Low oxidation state products

Product	Ref.	Product	Ref.
$CrBr_3$	15	$Ga_2X_6^{2-}$	27,31
$Ti(acac)_3$	52	InX, InI_2^-	7,67
CuX	23	$In(SR)_n$ (n = 1,2)	37
CuSR	35,41-45	In[O(OH)R]	68
ThI_2	28	In[S(SH)R]	69
$Sn(SR)_2$	33	TlSR	37
$Sn(O_2R)$	64	Tl_2S_2R	69
$Pb(O_2R)$	65		

TABLE 2. Electrochemical efficiencies (in mol F^{-1}) for the oxidation of elemental indium.

Solute	E_F	Product	Ref.
RX	~1	$RInX_2$, etc.	20
$(R_2NCS_2)_2$	1.0 ± 0.1	InL_3	61
RSH	1.00 ± 0.03	$In(SR)_n$ n = 1,2,3	37
Ph_2Se_2	1.00	$In(SePh)_3$(?)	39
py_2NH	1.10	no stable product	47
$R(OH)_2$	1.01 ± 0.01	In[O(OH)R]	68
$R(SH)_2$	1.00 ± 0.03	In[S(SH)R]	69
R(O)OH	1.02	$In(O_2R)_3$	57,58
Ph_2PH	1.05	no stable product	40
CH_2X_2	1.02 ± 0.01	X_2InCH_2X, etc.	67

$$R(OH)_2 + M \rightarrow M(O_2R) + H_2 \tag{6}$$

As with the cases discussed above, this reaction does not occur under normal conditions. We also found that with the aromatic thiols (RSH), the primary products were $Sn(SR)_2$ or $Pb(SR)_2$, although subsequent (unidentified) oxidation processes in the presence of bidentate donors gave $Sn(SR)_4$.bpy etc. for tin, despite the fact that the E_F of 0.5 mol F^{-1} showed that the electrochemical processes are essentially unchanged by these donors from eqs. (1) + (2) [33].

The results with the elements gallium, indium and thallium are particularly interesting in this context. With thallium + RSH or $R(SH)_2$, as with other ligands, the products are invariably those of thallium(I), and E_F = 1.0 mol F^{-1} [37], which is in keeping with the known stability of this oxidation state; in the case of solutions of $R(OH)_2$, the product is contaminated with thallium metal formed by cathodic reduction of the slightly soluble product [66]. With gallium, one can obtain either the $Ga_2X_6^{2-}$ anionic complexes of gallium(II) from solutions containing HX (X = Cl, Br, I), with $E_F \sim 0.6$, or GaX_4^- from solutions of X^- + X_2, although the E_F values show that a gallium(II) species is again formed at the anode, with subsequent oxidation to gallium(III) in the electrolyte phase [27].

Indium yields indium(III) compounds as the products in a number of cases, but the E_F value of 1.0 indicates clearly that the product of anodic oxidation is an indium(I) species. (See Table 2.) Some special cases are worth discussing. In the oxidation of this element in solutions of RX (R = Me, Et, Ph, Bz, C_6F_5; X = Cl, Br, I; not all combinations) the isolated products are $RInX_2$, $RInX_2$.bpy or $R_4N[RInX_3]$ depending on the composition of the electrolyte, but in every instance one finds E_F = 1.0 mol F^{-1}, implying the sequence

$$\text{cathode:} \quad RX + e \rightarrow R^{\bullet} + X^- \tag{7}$$

$$\text{anode:} \quad X^- + In \rightarrow InX + e \tag{8}$$

followed by oxidative insertion and complexation

$$InX + RX \rightarrow RInX_2 \tag{9}$$

$$RInX_2 + X^- \rightarrow RInX_3^- \tag{10}$$

Similarly, solutions of CH_2X_2/CH_3CN (X = Cl, Br, I) give E_F = 1.0 mol F^{-1}, and in the case of X = Br or I, the products are derivatives of X_2InCH_2X; InCl disproportionates in these systems to give $In^\circ + InX_3$. The oxidative reactions of InX have been sufficiently well studied in non-electrochemical work for the chemical processes (9) and (10), and the disproportionation, to be well understood [67].

A similar situation has been found in attempts to synthesize indium complexes of a number of the ligands identified in Schemes 3 and 4. With solutions of bidentate oxygen donors [57,58], or R_2NCSH [61], E_F = 1.0 mol F^{-1}, but in each case the product recovered from the reaction mixture is InL_3, while for Ph_2PH or py_2NH but the presumed InL species is too reactive to allow any identifiable product to be isolated. These results suggest that there are again post-electrolytic oxidative processes, possibly of the type

$$InL + HL \rightarrow HInL_2 \tag{11}$$

$$HInL_2 \rightarrow \dot{H} + \dot{In}L_2 \tag{12}$$

$$\mathrm{In\dot{n}L_2 + HL \rightarrow \dot{H} + InL_3} \tag{13}$$

$$\mathrm{2I\dot{n}L_2 \rightarrow In_2L_4} \tag{14}$$

$$\mathrm{2\dot{H} + 2InL \rightarrow 2HL + 2In} \tag{15}$$

Research into such reactions, in which one-electron transfer is the essence of the presumed mechanism, is planned. The only conclusion possible at the moment is that indium(I) species are the favoured product of the anodic oxidation of this element. One system in which InL, In_2L_4 and InL_3 species were identified was for L = SR, where the nature of R appears to have a strong influence on the sequence of reactions such as eqs. (11)-(15) [37].

Finally, we note the unusual compounds prepared from indium and solutions of aromatic 1,2-diols [68] or aliphatic dithiols [69], where the interest is on both the low oxidation state of the metal and the structure of the ligand. In each system, the product is an indium(I) complexes of the type In[O(OH)R] or In[S(SH)R] in which only one of the two acidic hydrogens of the parent has been lost. The electrochemical efficiency is unity, which can be explained in at least two ways. In the first of these, the sequence eq. (4) + (5) (see above) is followed by

$$\mathrm{In_2(O_2R) + R(OH)_2 \rightarrow 2In[O(OH)R]} \tag{16}$$

with an overall stoichiometry analogous to eq. (6), namely

$$\mathrm{R(OH)_2 + In \rightarrow In[O(OH)R] + \tfrac{1}{2}H_2} \tag{17}$$

and as before it should be noted that this reaction only occurs in the electrochemical cell.

An alternative to eqs. (16)-(18) is to replace eq. (4) by

$$\text{cathode:}\quad \mathrm{R(OH)_2 + e \rightarrow R(OH)O^- + \tfrac{1}{2}H_2} \tag{18}$$

and the anode process is then obviously

$$\mathrm{R(OH)O^- + In \rightarrow In[O(OH)R]} \tag{19}$$

This would entail a revision of eq. (5) to allow for a sequence such as

$$\text{solution:}\quad \mathrm{R(OH)O^- + R(OH)_2 \rightarrow [R(OH)O(OH)_2R]^-} \tag{20}$$

$$\text{anode:}\quad \mathrm{2[R(OH)O(OH)_2R]^- + M \rightarrow M(O_2R) + 2e + 3R(OH)_2} \tag{21}$$

and a number of other variants can also be invented. There is no experimental evidence as to the species in solution, but since the existence of the products is unquestionable, further work on this matter is indicated.

The chemistry of these species involves two types of reaction, namely removal of the hydrogen to give (*e.g.*) $Et_3NH^+[InO_2R]^-$ and the oxidation to the corresponding indium(III) species by iodine or o-quinone. These are summarized in Scheme 4. We have not been able to obtain crystals of either In[O(OH)R] or In[S(SH)R] compounds, but fortunately a related thallium(I) compound, $Tl_2[O(OH)C_{12}H_8]_2$, was prepared many years ago [70], and this has a dimeric structure based on a Tl_2O_2 ring with a lone pair of electrons on thallium(I) [66]. Generally, the chemistry of In[O(OH)R] and In[S(SH)R] compounds can be satisfactorily explained on the basis of an analogous structure, especially since compounds with In_2O_2 rings have been identified in other studies. There is still much to be done to develop the chemistry of indium(I), and direct electrochemical synthesis has provided a most useful entry in a new aspect of this work.

The structure of the singly protonated ligand, as well as the nature of the species generated the electrolyte phase, is another challenging aspect of this work. The indium(I) compounds are not unique in this respect, since studies with copper anodes and solutions of aromatic 1,2-diols gave a copper(I) derivative of $R(OH)O^-$ as the adduct $Cu_2[OC_6Cl_4(OH)]_2$.diphos [71]. Further work involving other metals and such ligands is planned.

8. Solutions in Liquid Ammonia

We have not attempted to carry out any direct electrochemical syntheses in aqueous media, but there are reports in the literature of work carried out some 40 years ago on the electrochemical oxidation of a number of metals in liquid ammonia [72,73], and since two of the metals in question were gallium and indium, this seemed a natural area for further investigation. The immediate conclusion, which we established by measuring E_F, is that indium is oxidized in liquid ammonia solutions of ammonium halides at -35°C to the +II state, but unfortunately we were not able to isolate any compounds of this oxidation state from the resultant solution, although in one particular case we were able to show by Raman spectroscopy that species with the characteristic ν(In-In) stretching mode were present in the solution [74]. When we attempted to work up these solutions both indium (I) and (III) halide derivatives of ammonia were obtained, and a mass balance, taking into account the quantity of material isolated and the quantity of indium dissolved, showed that the typical disproportionation reactions of indium +II were indeed being reproduced under these conditions. We concluded that the overall stoichiometry is

$$In + 2NH_4X\ (sol) \rightarrow 2NH_3(l) + H_2(g) + \tfrac{1}{2}In_2X_4\ (sol) \qquad (24)$$

but that unfortunately the inherent instability of In_2X_4 in these solutions, even in the presence of ligands known to stabilize these species in other circumstances, means that this is not a useful preparative route to indium(II) complexes.

We also investigated solutions of quinones and substituted catechols, in the latter case using mixtures of ammonia and an organic solvent to achieve sufficient solubility, and here the electrochemical efficiency shows that indium goes to the +III oxidation state. In the presence of an o-quinone, oxidation of the lower oxidation state halide leads to InX(catecholate), and hence by substitution to InX_3. When a 1,2-diol is used, there are again questions as to the solute species which are generated at the cathode, but the overall reaction can be written as

$$In + 4NH_4^+ + 2\ cat^{2-} \rightarrow NH_4[In(cat)_2] + 3NH_3 + {}^3/_2H_2 \qquad (25)$$

with $E_F = 0.33$ mol F^{-1}. The indium(III) derivative of 3,5-di-tert-butyl-catechol is in fact the dimeric anion $[In_2(dbcatechol)_4(NH_3)_4]^{2-}$, whose crystal structure confirms the presence of the substituted catecholate ligand, and shows that the dimer is dependent upon an In_2O_2 ring.

It seems likely that the use of liquid ammonia media may offer some advantages in the direct electrochemical synthesis of low oxidation state complexes. We hope to investigate this matter in the future.

9. Conclusions

I have tried to show that the use of a metal as the sacrificial anode in electrochemical oxidation can form the basis of a very wide range of syntheses of inorganic and organometallic compounds. We believe that the method is indeed a choice one for a series of ligands derived from weak mono- and dibasic organic acids, and that while we have not investigated all metals ourselves, there seems every reason to assume that the range of syntheses could be extended to most areas of the Periodic Table.

The emphasis in our work on the metallic elements Zn, Cd, Hg, Ga, In, Tl, Sn and Pb is a logical consequence of the synergistic effect of other work going on in our laboratory. These electropositive metals do not generally react with weak organic acids, but the driving force provided by the applied potential brings about reactions which proceed with an understandable stoichiometry, with high yield, and under mild conditions. I believe the range of syntheses discussed in this paper has served to establish direct electrochemical synthesis as a useful and readily accessible experimental technique for those who are not afraid to use electricity in chemistry.

10. Acknowledgement

Much of the work reported in this and other papers has been supported by Operating Grants from the Natural Sciences and Engineering Research Council of Canada. It is also a pleasure to acknowledge my gratitude to the large number of co-workers whose names are recorded in the papers which I have quoted here and elsewhere.

References

1. Tuck, D.G. (1979) 'Direct electrochemical synthesis of inorganic and organometallic compounds', *Pure and Appl. Chem.*, 51, 2005-2018.
2. Oldham, C. and Tuck, D.G. (1982) 'The direct electrochemical synthesis of $[(C_6H_5)_3PH]_2[CoCl_4]$', *J. Chem. Ed.*, 59, 420-421.
3. Casey, A.T. and Vecchio, A.M. (1990) 'The electrochemical synthesis of metallocenes, polymethylmetallocenes, open metallocenes and metal phenylacetylides', *Appl. Organometal. Chem.*, 4, 513-522.
4. Habeeb, J.J., Osman, A., and Tuck, D.G. (1979) 'The electrochemical synthesis of some Ph_3SnMCl adducts (M = Zn, Cd, Hg)', *Inorg. Chim. Acta.*, 35, 105-108.
5. Oldham, C., Taylor, M.J., and Tuck, D.G. (1985) 'Direct electrochemical synthesis of metal thiocyanate complexes', *Inorg. Chem. Acta*, 100, L9-10.
6. Habeeb, J.J., Tuck, D.G. and Zhandhire, S. (1979) 'The electrochemical synthesis of some heteronuclear metal carbonyls', *Can. J. Chem.*, 57, 2196-2199.
7. Habeeb, J.J., and Tuck, D.G. (1975) 'Direct electrochemical synthesis of indium compounds', *Chem. Comm.*, 808-809.
8. Habeeb, J.J., Neilson, L. and Tuck, D.G. (1976) 'Direct electrochemical synthesis of anionic complexes of metallic elements', *Synth. React. Inorg. Metal-org. Chem.*, 6, 105-113.
9. Habeeb, J.J., Osman, A. and Tuck, D.G. (1976) 'The direct electrochemical synthesis of neutral and anionic organocadmium halides', *Chem. Comm.*, 379-380.

10. Habeeb, J.J. and Tuck, D.G. (1977) 'Electrochemical preparation of some organotin compounds', *J. Organometal. Chem.*, 134, 363-379.
11. Habeeb, J.J., Said, F.F. and Tuck, D.G. (1977) 'Direct electrochemical synthesis of neutral and anionic chloro- and bromo-complexes of titanium, zirconium and hafnium', *Can. J. Chem..*, 55, 3882-3886.
12. Habeeb, J.J. and Tuck, D.G. (1977) 'The electrochemical preparation of organo-nickel and palladium halides', *J. Organometal. Chem.*, 139, C17-20.
13. Habeeb, J.J., Neilson, Lynn and Tuck, D.G. (1978) 'Electrochemical preparation of anhydrous halides of transition metals (Mn-Zn)', *Inorg. Chem.*, 17, 306-310.
14. Habeeb, J.J. and Tuck, D.G. (1978) 'Electrochemical preparation of neutral organocadmium halides', *J. Organometal. Chem.*, 146, 213-219.
15. Habeeb, J.J. and Tuck, D.G. (1979) 'The direct electrochemical synthesis of chromium(III) bromide; a facile route to chromium(III) complexes', *Inorg. Synth.*, 19, 123-127.
16. Habeeb, J.J. and Tuck, D.G. (1979) 'The direct electrochemical synthesis of indium(III) complexes', *Inorg. Synth.*, 19, 257-261 (1979).
17. Osman, A. and Tuck, D.G. (1979) 'The direct electrochemical synthesis of anionic organodihalogenocadmate(II) complexes', *J. Organometal. Chem.*, 169, 255-258 (1979).
18. Habeeb, J.J., Osman, A. and Tuck, D.G. (1980) 'The electrochemical synthesis of neutral and anionic organozinc halides', *J. Organometal. Chem.*, 185, 117-127.
19. Habeeb, J.J., Said, F.F. and Tuck, D.G. (1980) 'Coordination compounds of indium. Part XXXV. The direct electrochemical synthesis of adducts of indium(III) halides', *J. Chem. Soc., Dalton Trans.*, 1161-1163.
20. Habeeb, J.J., Said, F.F. and Tuck, D.G. (1980) 'Coordination compounds of indium. Part 36. The direct electrochemical synthesis of neutral and anionic organoindium halides', *J. Organometal. Chem.*, 190, 325-335.
21. Said, F.F. and Tuck, D.G. (1980) 'The electrochemical synthesis of organometallic halides of titanium, zirconium and hafnium' *Can. J. Chem.*, 58, 1673-1677.
22. Khan, M., Oldham, C., Taylor, M.J. and Tuck, D.G. (1980) 'Preparative and structural studies of triphenylphosphonium salts' *Inorg. Nucl. Chem. Lett.*, 16, 469-474.
23. Said, F.F. and Tuck, D.G. (1981) 'The fortuitous direct electrochemical synthesis of some copper(I) complexes', *Can. J. Chem.*, 59, 62-64.
24. Habeeb, J.J., Oldham, C. and Tuck, D.G. (1981) 'The electrochemical oxidation of molybdenum into non-aqueous solutions of halogen', *J. Inorg. Nucl. Chem.*, 43, 2087-2088.
25. Khan, M., Oldham, C. and Tuck, D.G. (1981) 'The direct electrochemical synthesis of triphenylphosphine adducts of Group IB monohalides', *Can. J. Chem.*, 59, 2714-2718.
26. Said, F.F. and Tuck, D.G. (1982) 'The direct electrochemical synthesis of organometallic halides of zinc and cadmium with substituted phenyl halides', *J. Organometal. Chem.*, 224, 121-124.
27. Taylor, M.J. and Tuck, D.G. (1983) 'Electrochemical synthesis of hexahalogenodigallate(II) and tetrahalogenogallate(III) anions', *Inorg. Synth.*, 22, 135-142.
28. Kumar, N. and Tuck, D.G. (1983) 'The direct electrochemical synthesis of thorium diiodide, and some derivatives', *Inorg. Chem.*, 22, 1951-1952.
29. Kumar, N. and Tuck, D.G. (1984) 'The direct electrochemical synthesis of neutral and anionic halogeno complexes of uranium(IV) and uranium(VI)', *Inorg. Chem. Acta*, 95, 211-215.
30. Hayes, P.C., Osman, A., Seudeal, N. and Tuck, D.G. (1985) 'The direct electrochemical synthesis of 2,2'-bipyridine adducts of organomagnesium halides, and of salts of organodihalogenomagnesium(II) anions', *J. Organometal. Chem.*, 291.

31. Khan, M.A., Taylor, M.J., Tuck, D.G. and Rogers, D.A. (1986) 'Crystal structures and vibrational spectra of triphenylphosphonium hexahalogenodigallates(II), $(Ph_3PH)_2Ga_2X_6$ (X = Cl, Br, I)', *J. Crystal. Spectro. Res.*, 16, 895-905.
32. Said, F.F. and Tuck, D.G. (1982) 'The direct electrochemical synthesis of some thiolates of zinc, cadmium and mercury' *Inorg. Chim. Acta*, 59, 1-4.
33. Hencher, J.L., Khan, M.A., Said, F.F., Sieler, R. and Tuck, D.G. (1982) 'The direct electrochemical synthesis of phenylthiolato complexes of tin(II), tin(IV) and lead(II); The molecular structure of the 2,2'-bipyridine adduct of $Sn(SC_6H_5)_4$', *Inorg. Chem.*, 21, 2787-2791.
34. Hencher, J.L., Khan, M.A., Said, F.F. and Tuck, D.G. (1985) 'The direct electrochemical synthesis, and crystal structure, of salts of $[M_4(SC_6H_5)_{10}]^{2-}$ anions (M = Zn, Cd)', *Polyhedron*, 4, 1263-1267.
35. Chahda, R.K., Kumar, R. and Tuck, D.G. (1987) 'The direct electrochemical synthesis of thiolato complexes of copper, silver, and gold; molecular structure of $[Cu(SC_6H_4CH_3\text{-}\underline{o})$ (1,10-phenanthroline)$]_2.CH_3CN$', *Can. J. Chem.*, 65, 1336-1342.
36. Chadha, R.K., Kumar, R., Lopez-Grado, J.R. and Tuck, D.G. (1988) 'The direct electrochemical synthesis of thiolato complexes of cobalt and nickel, and the crystal structure of bis(phenylthiolato)bis(1,10-phenanthroline)cobalt(III) perchlorate', *Can. J. Chem.*, 66, 2151-2156.
37. Green, J.H., Kumar, R., Seudeal, N. and Tuck, D.G. (1989) 'The direct electrochemical synthesis of alkyl and aryl thiolato complexes of indium and thallium', *Inorg. Chem.*, 28, 123-127.
38. Mabrouk, H.E., Kumar, R. and Tuck, D.G. (1988) 'Studies of the reactions of some Main Group metals with diphenylsulphide and diphenylselenide', *J. Chem. Soc., Dalton Trans.*, 1045-1047.
39. Kumar, R. and Tuck, D.G. (1989) 'The direct electrochemical synthesis of phenylselenato complexes of some Main Group and late transition elements', *Can. J. Chem.*, 67, 127-129.
40. Annan, T.A., Kumar, R. and Tuck, D.G. (1991) 'Direct electrochemical synthesis and crystallographic characterization of metal-diphenylphosphido and diphenylthiophosphinato compounds, and some derivatives', *J. Chem. Soc., Dalton Trans.*, 11-18.
41. Chadha, R.K., Kumar, R. and Tuck, D.G. (1986) 'The preparation and crystal structure of the unusual copper-sulphur cage complex $Cu_8(SC_5H_{11})_4(S_2CSC_5H_{11})_4$', *J. Chem. Soc., Chem. Commun.*, 188-189.
42. Khan, M.A., Kumar, R. and Tuck, D.G. (1988) 'The direct electrochemical synthesis of adducts of bis(diphenylphosphino)methane(dppm) with copper(I) thiolates, and the molecular structure of $Cu_4(\mu\text{-}SC_5H_{11})_4(dppm)_2$', *Polyhedron*, 7, 49-55.
43. Chadha, R.K., Kumar, R. and Tuck, D.G. (1988) 'The synthesis, properties, and crystal structure of the copper(I)-thiolato-thioxanthato complex $Cu_8(SC_5H_{11})_4(S_2CSC_5H_{11})_4$', *Polyhedron*, 7, 1121-1128.
44. Annan, T.A., Kumar, R. and Tuck, D.G. (1990) 'The direct electrochemical synthesis of copper and silver derivatives of alkanedithiols, and the crystal structure of $[Cu_2S_2C_3H_6\text{-}1,2\cdot(C_6H_5)_2PCH_2P(C_6H_5)_2]_4$', *Inorg. Chem.*, 29, 2475-2479.
45. Dance, I.G. (1986) 'The structural chemistry of metal thiolate complexes', *Polyhedron*, 5, 1037-1104.
46. Blower, P.J., and Dilworth, J.R. (1986) 'Thiolato complexes of the transition metals', *Coord. Chem. Revs.*, 76, 121-185?
47. Kumar, R. and Tuck, D.G. (1989) 'The direct electrochemical synthesis of metal complexes of 2,2'-dipyridylamine', *Inorg. Chim. Acta*, 157, 51-56.

48. Kumar, R. and Tuck, D.G. (1985) 'An efficient electrochemical synthesis of phenylethynyl copper(I), $C_6H_5C{\equiv}CCu$', *J. Organometal. Chem.*, 281, C47-48.
49. Eisenbach, W., and Lehmkuhl, H. (1982) 'Electrosynthesis of ferrocene', *Chem.-Ing.-Tech.*, 54, 690-691.
50. Eisenbach, W. and Lehmkuhl, H. (1985) 'Electrosynthesis of ferrocene; development from laboratory to industry', *DECHEMA-Monogr.*, 98, 269-289.
51. Lehmkuhl, H. and Eisenbach, W. (1975) 'Electrosynthesis of alkoxides and acetylacetonates of iron, cobalt and nickel', *Justus Liebig Ann. Chem.*, 672-691.
52. Habeeb, J.J., Tuck, D.G. and Walters, F.H. (1978) 'Direct electrochemical synthesis of some metal chelate complexes', *J. Coord. Chem.*, 8, 27-33.
53. Kumar, N. and Tuck, D.G. (1982) 'The direct electrochemical synthesis of neutral and anionic complexes of thorium(IV)', *Can. J. Chem.*, 60, 2579-2582.
54. Bustos, L., Green, J.H., Khan, M.A. and Tuck, D.G. (1983) 'The electrochemical synthesis of β-diketonato complexes of cadmium(II), and the crystal and molecular structure of $Cd(acac)_2$phen', *Can. J. Chem.*, 61, 2141-2146.
55. Matassa, L., Kumar, N. and Tuck, D.G. (1985) 'Direct electrochemical synthesis of chelate complexes of uranium(IV) and uranium(VI)', *Inorg. Chim. Acta*, 109, 19-21.
56. Mabrouk, H.E., Tuck, D.G. and Khan, M.A. (1987) 'The direct electrochemical synthesis of zinc and cadmium catecholates and related compounds', *Inorg. Chim. Acta*, 129, 75-80.
57. Annan, T.A., Peppe, C. and Tuck, D.G. (1990) 'The direct electrochemical synthesis of d^{10} metal ion complexes of some anionic bidentate oxygen donors', *Can. J. Chem.*, 68, 423-430.
58. Annan, T.A., Peppe, C. and Tuck, D.G. (1990) 'The direct electrochemical synthesis of some metal derivatives of 3-hydroxy-2-methyl-4-pyrone', *Can. J. Chem.*, 68, 1598-1605.
59. Kumar, N., Tuck, D.G. and Watson, K.D. (1987) 'The direct electrochemical synthesos of some transition metal carboxylates', *Can. J. Chem.*, 65, 740-743.
60. Mabrouk, H.E. and Tuck, D.G. (1988) 'The direct electrochemical synthesis of zinc and cadmium derivatives of α,ω-dithiols, and their reaction with carbon disulphide', *Inorg. Chim. Acta*, 145, 237-241.
61. Geloso, C., Kumar, R., Lopez-Grado, J.R. and Tuck, D.G. (1987) The direct electrochemical synthesis of dialkyldithiocarbamates and diethyldithiophosphate complexes of Main Group and transition metals', *Can. J. Chem.*, 65, 928-932.
62. Habeeb, J.J., Said, F.F. and Tuck, D.G. (1981) 'The direct electrochemical synthesis of cationic complexes', *J. Chem. Soc., Dalton Trans.*, 118-120.
63. Kumar, N. and Tuck, D.G. 'The direct electrochemical synthesis of some thorium(IV) nitrate complexes', *Can. J. Chem.*, 62, 1701-1704.
64. Mabrouk, H.E. and Tuck, D.G. (1988) 'The direct electrochemical synthesis of tin(II) derivtaives of aromatic 1,2-diols, and a study of their oxidative addition reactions', *J. Chem. Soc., Dalton Trans.*, 2539-2543.
65. Barnard, G., Mabrouk, H.E., and Tuck, D.G. Unpublished results.
66. Kickham, J.E., Taricani, L. and Tuck, D.G. Unpublished results.
67. Annan, T.A., Tuck, D.G., Khan, M.A. and Peppe, C. (1991) 'Direct electrochemical synthesis of X_2InCH_2X compounds (X = Br,I), and a study of their coordination chemistry, *Organometallics*, 10, 2159-2166.
68. Mabrouk, H.E. and Tuck, D.G. (1989) 'Coordination compounds of indium. Part 45. Indium(I) derivatives of aromatic diols' *Can. J. Chem.*, 67, 746-750.
69. Geloso, C., Mabrouk, H.E. and Tuck, D.G. (1989) 'Coordination compounds of indium. Part 47. Indium(I) and thallium(I) derivatives of alkanedithiols' *J. Chem. Soc., Dalton*

Trans., 1759-1763.
70. Brady, O.L., and Hughes, E.D. (1933) 'Coordination compounds of 2,2'-diphenol', *J. Chem. Soc.*, 1227-1230.
71. Annan, T.A., Kickham, J.E. and Tuck, D.G. (1991) The direct electrochemical synthesis of the novel copper(I) complex $Cu_2[OC_6Cl_4(OH)]_2[(C_6H_5)_2PCH_2P(C_6H_5)_2]_2$' *Can. J. Chem.*, 69, 251-256.
72. McElroy, A.D., Kleinberg, J., and Davidson, A.W. (1952) 'The anodic oxidation of higher members of the aluminum family in liquid ammonia', *J. Am. Chem. Soc.*, 74, 735-739.
73. Davidson, A.W., and Kleinberg, J. (1953) 'Unfamiliar oxidation states in liquid ammonia', *J. Phys. Chem.*, 57, 571-576.
74. Annan, T.A., Gu, J., Tian, Z. and Tuck, D.G. Unpublished results.

ELECTROCHEMICAL GENERATION OF SUBSTRATE-BINDING SITES IN ORGANOMETALLIC COMPLEXES WITH THE $\{Mo_2(\mu\text{-}SR)_n$ CORE. ELECTROSYNTHESIS OF CYANIDE AND ISOCYANIDE DERIVATIVES.

F. GLOAGUEN, C. Le FLOC'H, F.Y.PÉTILLON, P. SCHOLLHAMMER and J. TALARMIN
URA CNRS 322, "Chimie, Electrochimie Moleculaires et Chimie Analytique",
Université de Bretagne Occidentale,
6 avenue V. Le Gorgeu, 29287 Brest-cedex, France
M. EL KHALIFA and J.Y. SAILLARD
URA CNRS 254, "Chimie du Solid et Inorganic Moleculaire",
Université de Rennes I, 35042 Rennes-cedex, France

Abstract. The electrochemical activation of dinuclear molybdenum thiolate complexes and the reactivity of electrogenerated intermediates toward CO, RNC and RCN are described.

Introduction

The reduction of various substrates (N_2, RCN, RNC, CN^-, NO_3^-) by metalloenzymes with molybdenum-sulfur sites (nitrogenase, nitrate reductase) has prompted active studies of the chemistry and electrochemistry of complexes with Mo-S assemblies which could mediate or catalyse the reduction of these substrates.

The first step of a transition-metal catalysed reduction process, the substrate-binding step, can be achieved electrochemically (Scheme 1), either by the reductive cleavage of M-X bonds, or *via* the substitution of a ligand initially present in the coordination sphere of the metal centre (this can be induced or catalysed by an electron-transfer step).

1

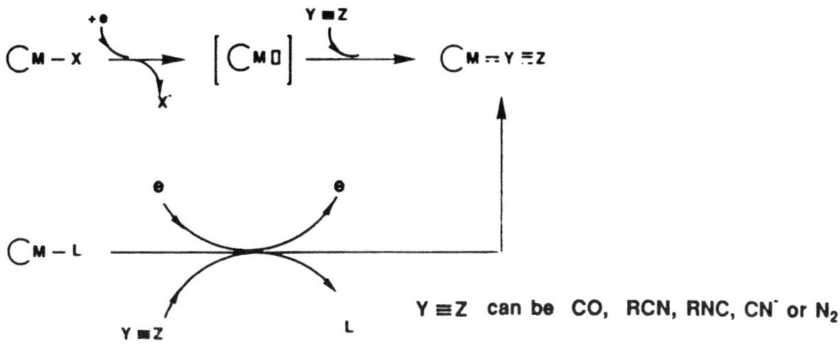

The work described herein is concerned with the initial step of a ligand activation process mediated by molybdenum complexes with the $\{Mo_2(\mu\text{-}SR)_n\}$ core.

A. J. L. Pombeiro and J. A. McCleverty (eds.),
Molecular Electrochemistry of Inorganic, Bioinorganic and Organometallic Compounds, 33–38.

Oxidatively-Induced and Electron-Transfer Catalysed Substitution of Isocyanide or Cyanide ligands for CO.

The electrochemical oxidation of $[Mo_2Cp_2(CO)_4(\mu\text{-}SR)_2]$ or the electrochemical reduction of $[Mo_2Cp_2(CO)_4(\mu\text{-}SR)_2]^{2+}$ (R= Me, Ph) at 40°C in the presence of the substrate MeCN or t-BuNC, lead to the formation of the substituted derivative $[Mo_2Cp_2(CO)_3(L)(\mu\text{-}SR)_2]^{2+}$ (L= MeCN or t-BuNC);[1,2] the substitution process is either oxidatively-induced (transfer of two electrons) when carried out from the neutral tetracarbonyl species, or catalysed by an Electron-Transfer Chain process when performed from a solution of the parent dication (Scheme 2). Under these conditions, the substitution process is triggered by the passage of a small amount of charge (0.1F/mol of starting material) and proceeds after the electrolysis has been interrupted .

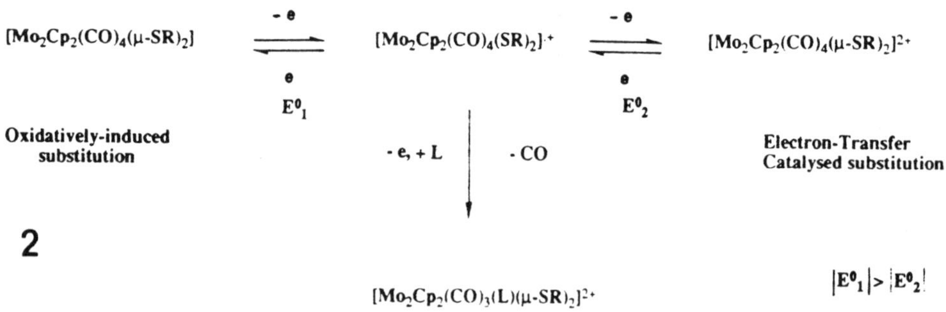

Neither the neutral complex nor the dication react with the substrate on the time-scale of the electrochemical experiment under the experimental conditions used; therefore, the species responsible for the catalytic process is an unstable radical cation intermediate in the single step, reversible two-electron transfer (Scheme 2). EHMO calculations suggest that the substrate-binding site may result from the opening of one of the thiolate bridges at the radical cation stage.[3]

The comparison of the electrochemical behaviour of the complexes $[Mo_2Cp_2(CO)_3(L)(\mu\text{-}SR)_2]^{2+}$ with L= CO, t-BuNC and MeCN indicates that the two-electron reduction results in an electron flux from molybdenum filled d orbitals towards the substrate vacant π^* orbitals,[4] and consequently to the activation of the bound substrate.

Reductive Cleavage of Mo-X bonds (X= SMe, Cl, Br, I) or of M-L bond (L= MeCN)

The electrochemical reduction of complexes derived from $[Mo_2Cp_2(CO)_4(\mu\text{-}SMe)_2]^{2+}$ by substitution of one CO group and of two CO groups by electron-releasing or by poorly accepting ligands has been investigated in the absence of substrate, in the presence of CO, and in the presence of isocyanides RNC (R= t-Bu, xyl, Bz).

Reduction in the absence of substrate. The two types of complexes, *e.g.* singly substituted $[Mo_2Cp_2(CO)_3(X)(\mu\text{-}SMe)_2]^+$ (X= Br, I) and $[Mo_2Cp_2(CO)_3(MeCN)(\mu\text{-}SMe)_2]^{2+}$, and disubstituted $[Mo_2Cp_2(CO)_2(\mu\text{-}SMe)_3]^+$ and $[Mo_2Cp_2(CO)_2(X)_2(\mu\text{-}SMe)_2]$, reduce according to two different mechanisms involving two different intermediates. However, in the absence of substrate (N_2 or Ar atmosphere), the same stable product is formed from the exhaustive electrolyses (Schemes 3 and 4).[5]

Intermediate 1 arising from the reductive cleavage of two Mo-S or of two Mo-X bonds, namely *cis*-[$Mo_2Cp_2(CO)_2(\mu\text{-}SMe)_2$], is detected by cyclic voltammetry and converts to its stable *trans* isomer on the electrolysis time-scale (Scheme 3).[6]

3

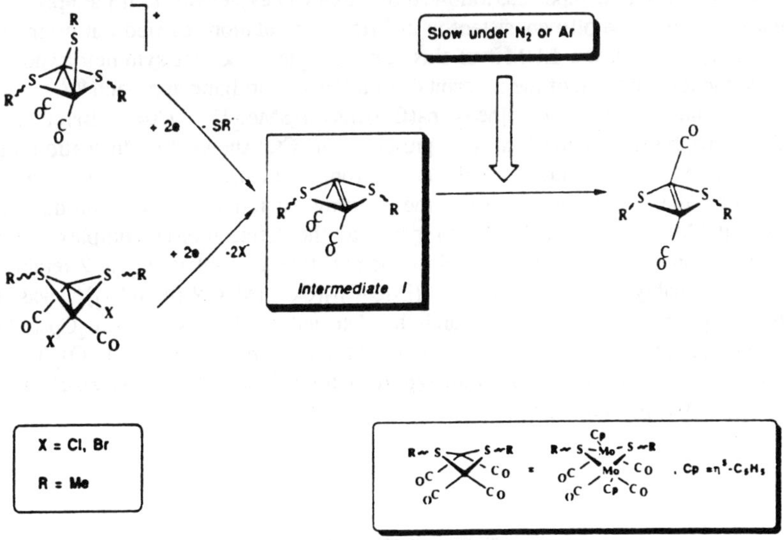

The formation of *trans*-[$Mo_2Cp_2(CO)_2(\mu\text{-}SMe)_2$] from the reduction of the singly substituted complexes [$Mo_2Cp_2(CO)_3(X)(\mu\text{-}SMe)_2$]$^+$ and [$Mo_2Cp_2(CO)_3(MeCN)(\mu\text{-}SMe)_2$]$^{2+}$ is much faster since it is observed on the c.v. time-scale. The transient existence of intermediate 2 (Scheme 4), which is not detected by cyclic voltammetry under the experimental conditions used ($v < 2\ Vs^{-1}$), is unambiguously established by trapping experiments, see below.

4

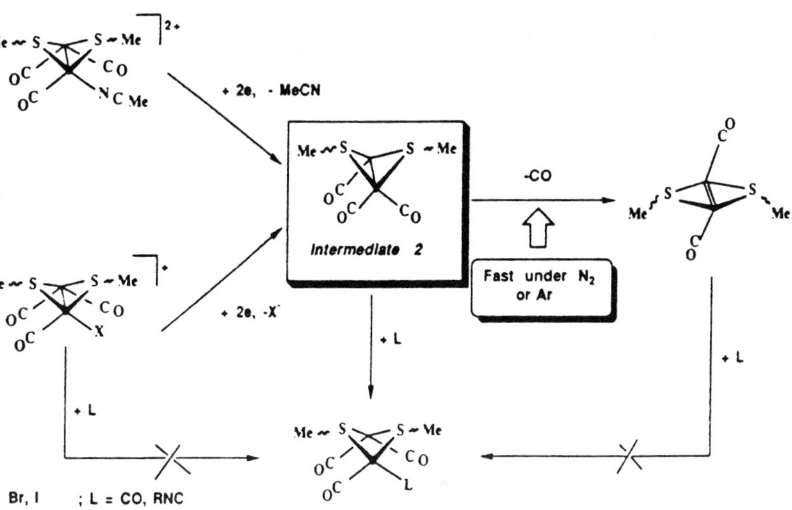

Reduction in the presence of carbon monoxide. The formation of *trans*-$[Mo_2Cp_2(CO)_2(\mu\text{-}SMe)_2]$ from the reduction of the disubstituted complexes ***is catalysed by CO***: the redox processes of the *cis* intermediate are replaced by those of the stable *trans* isomer when the cyclic voltammetry of $[Mo_2Cp_2(CO)_2(\mu\text{-}SMe)_3]^+$ and $[Mo_2Cp_2(CO)_2(X)_2(\mu\text{-}SMe)_2]$ is run under CO. We believe that the catalytic effect results from CO insertion into the Mo=Mo double bond, leading to a CO-bridged species, followed by rapid CO expulsion from the opposite face of the complex.[5,6] This is wholly consistent with EHMO calculations carried out on the model *cis*-$[Mo_2Cp_2(CO)_2(\mu\text{-}SH)_2]$: the LUMO of this complex shows perfect symmetry and energy for a strong interaction with one of the π^* orbitals of a CO in a bridging approach.[6,7]
The cyclic voltammetry of $[Mo_2Cp_2(CO)_3(X)(\mu\text{-}SMe)_2]^+$ (X= Br, I) and $[Mo_2Cp_2(CO)_3(MeCN)(\mu\text{-}SMe)_2]^{2+}$ in the presence of CO shows the formation of *cis*-$[Mo_2Cp_2(CO)_4(\mu\text{-}SMe)_2]$, Scheme 4. This is a strong evidence for the existence of the undetected intermediate 2, since neither the starting material nor the product *trans*-$[Mo_2Cp_2(CO)_2(\mu\text{-}SMe)_2]$ react with CO to generate the tetracarbonyl complex. Trapping experiments in the presence of RNC (see below) demonstrate that intermediate 2 retains three carbonyl groups, probably in a *cis* arrangement ($[Mo_2Cp_2(CO)_3(MeCN)(\mu\text{-}SPh)_2]^{2+}$ was shown to adopt the *cis* geometry in the solid,[2] and the detected product is *cis*-$[Mo_2Cp_2(CO)_4(\mu\text{-}SMe)_2]$[8,9]). The conversion of $\{[Mo_2Cp_2(CO)_3(\mu\text{-}SMe)_2]\}$ to *trans*-$[Mo_2Cp_2(CO)_2(\mu\text{-}SMe)_2]$ can be regarded ***as the intramolecular counterpart of the CO-catalysed isomerisation of cis-$[Mo_2Cp_2(CO)_2(\mu\text{-}SMe)_2]$*** (Scheme 5).

5

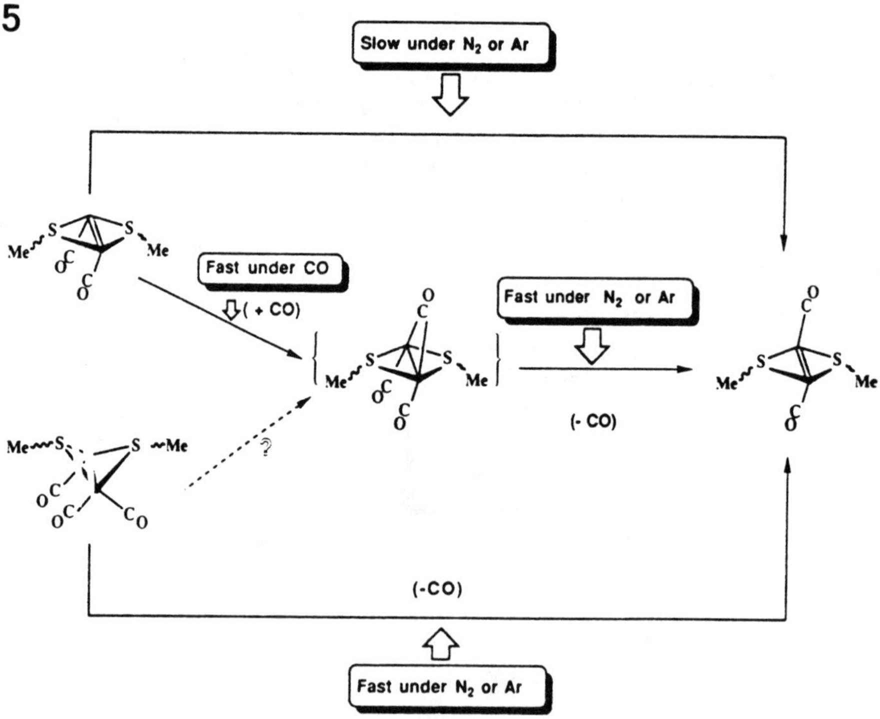

Reduction in the presence of isocyanide ligands. Trapping of intermediate 2 with RNC (Scheme 4) generates $[Mo_2Cp_2(CO)_3(CNR)(\mu\text{-}SMe)_2]$; this demonstrates that CO loss is not concomitant with the cleavage of the Mo-X or Mo-L bond: intermediate 2 must be a tricarbonyl species.

Cyclic voltammetry of $[Mo_2Cp_2(CO)_2(\mu\text{-}SMe)_3]^+$ and $[Mo_2Cp_2(CO)_2(X)_2(\mu\text{-}SMe)_2]$ in the presence of RNC demonstrates that the electrogenerated *cis*-$[Mo_2Cp_2(CO)_2(\mu\text{-}SMe)_2]$ intermediate reacts rapidly with the isocyanide ligand to yield a new complex that we assign as *trans*-$[Mo_2Cp_2(CO)(CNR)(\mu\text{-}SMe)_2]$.[6] A second isomer of this complex is formed on the longer time-scale of electrolyses, from the reaction of *trans*-$[Mo_2Cp_2(CO)_2(\mu\text{-}SMe)_2]$ with RNC; the *trans* dicarbonyl complex arises from the reaction of CO, released upon RNC binding to *cis*-$[Mo_2Cp_2(CO)_2(\mu\text{-}SMe)_2]$, with either *cis*-$[Mo_2Cp_2(CO)_2(\mu\text{-}SMe)_2]$ or *trans*-$[Mo_2Cp_2(CO)(CNR)(\mu\text{-}SMe)_2]$ (Scheme 6).

6

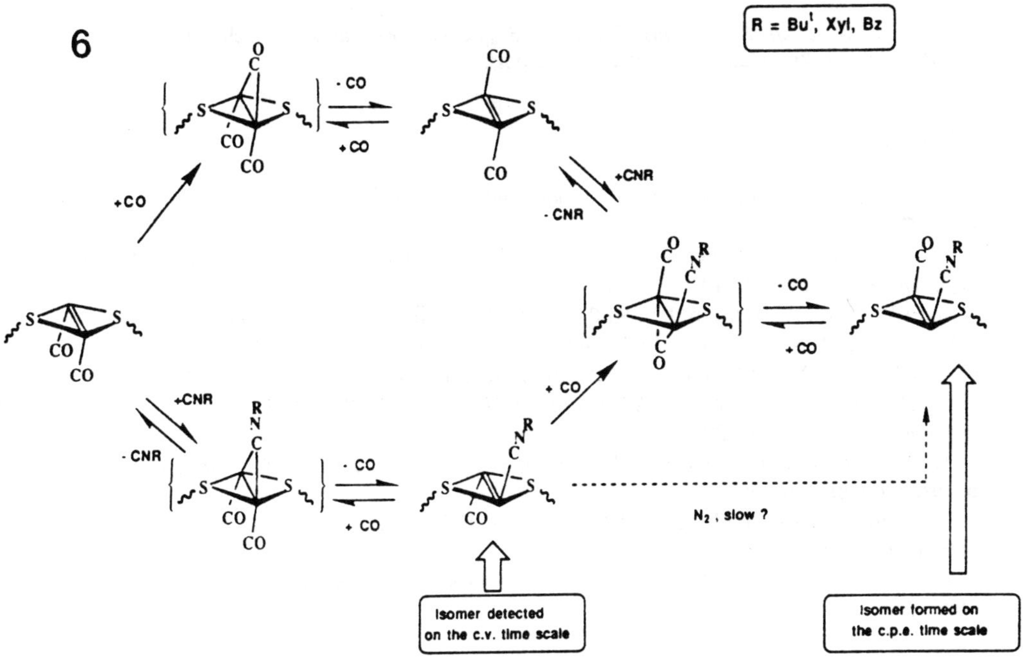

Both isomers of $[Mo_2Cp_2(CO)_2(\mu\text{-}SMe)_2]$ react with RNC to produce two isomers of $[Mo_2Cp_2(CO)(CNR)(\mu\text{-}SMe)_2]$, which, in turn, both react with CO to generate **only** *trans*-$[Mo_2Cp_2(CO)_2(\mu\text{-}SMe)_2]$ (Scheme 6). Consequently, the isomerisation of *trans*- to *cis*-$[Mo_2Cp_2(CO)(CNR)(\mu\text{-}SMe)_2]$ is ***induced by carbon monoxide***.[6]

Conclusion

The work reported in this communication illustrates two different electrochemical routes leading to the generation of RCN and RNC binding sites, at formally Mo^{III} and Mo^{II} metal centres, in complexes possessing a $\{Mo_2(\mu\text{-}SR)_n\}$ core.

The electrochemical cleavage of Mo-X bonds in complexes derived from $[Mo_2Cp_2(CO)_4(\mu\text{-}SMe)_2]^{2+}$ by substitution of one or two donor ligands for CO produces two different intermediates. These convert to a single product, although at quite different rates, in the absence of substrate; in the presence of RNC, these intermediates produce two different types of isocyanide complexes. Interestingly, CO also reacts at these sites and inhibits the binding of isocyanide, as it is observed in the case of nitrogenase. We have also demonstrated that CO can *catalyse or induce an isomerisation* about the Mo=Mo double bond of $[Mo_2Cp_2(CO)(L)(\mu\text{-}SMe)_2]$ (L= CO and RNC, respectively).

Acknowledgements. The C.N.R.S. (Centre National de la Recherche Scientifique) is acknowledged for financial support.

References

1- M. Guéguen, J.E. Guerchais, F.Y. Pétillon, and J. Talarmin, *J.Chem.Soc., Chem.Commun.*, **1987**,557.

2- M. El Khalifa, M. Guéguen, R. Mercier, F.Y. Pétillon, J.Y. Saillard and J. Talarmin, *Organometallics*, **1989**,*8*,140.

3- M. El Khalifa, F.Y. Pétillon, J.Y. Saillard and J. Talarmin, *Inorg.Chem.*, **1989**,*28*,3849.

4- M. Guéguen, F.Y. Pétillon, and J. Talarmin, *Organometallics*, **1989**,*8*,148.

5- C. Le Floc'h, F.Y. Pétillon, C.J. Pickett, and J. Talarmin, *J.Organomet.Chem.*, **1990**,*390*,C39.

6- F. Gloaguen, C. Le Floc'h, F.Y. Pétillon, J. Talarmin, M. El Khalifa, and J.Y. Saillard, *Organometallics*, **1991**,*10*,2004.

7- M. El Khalifa, F. Gloaguen, C. Le Floc'h, F.Y. Pétillon, J.Y. Saillard, and J. Talarmin, manuscript in preparation.

8- J. Courtot-Coupez, M. Guéguen, J.E. Guerchais, F.Y. Pétillon, J. Talarmin, and R. Mercier, *J.Organomet.Chem.*, **1986**,*312*,81.

9- M. Guéguen, F.Y. Pétillon, J. Talarmin, P. Jaitner, and W. Winder, *Inorg.Chim.Acta*, **1988**,*144*,5.

PART 3. REDOX PROPERTIES and RELATIONSHIPS with STRUCTURE and REACTIVITY

3.1 INORGANIC and ORGANOMETALLIC COMPLEXES

THE PARAMETERISATION OF METAL CENTRED REDOX COUPLES

A. B. P. Lever
Dept. of Chemistry, York University,
North York (Toronto), Ontario, Canada, M3J IP3

1. Introduction

Crystal field theory played a dominant role, from the 1950s, in the renaissance of inorganic chemistry. Its ability to reproduce the d-d (crystal field) spectra of almost any metal complex, through combination of geometry and the use of simple equations involving the crystal field parameter, Dq, made a major impact on the field [1].

The success of this theory is owed to the Spectrochemical Series of ligands in which:-

1) The relative position of any ligand in the Series is independent of the metal ion to which it is attached, and

2) The contribution of a set of n ligands is additive, i.e. is obtained by summing the Dq contribution of each ligand.

The question then arises - could one create an Electrochemical Series of ligands, with parameter, say, $E_L(L)$, which could be used to reproduce any metal-centred redox potential, E(M(m+1/m)) in a general complex $MX_xY_yZ_z$, from a weighted sum of $E_L(L)$ parameters,

i.e. in this case, $[\Sigma E_L(L))] = x E_L(X) + y E_L(Y) + z E_L(Z)$.

Such a Series would clearly be invaluable for reasons which will be detailed later. On the face of it, this sounds like a difficult proposition, because the Electrochemical Series would have to meet both the criteria expounded for the Spectrochemical Series, otherwise it would not work, or would have many exceptions.

One may expect, for example, that the sequence of $E_L(L)$ values for a soft metal ion, such as Cr(0) would surely be different than for a hard metal ion such as Ta(V), i.e. the old idea that soft ligands prefer to bind to soft metal ions, and hard ligands to hard metal ions, would surely mitigate against criterion number 1 above, being valid for electrochemical potentials. That this does not appear to be the case, at least within the framework of this model, is one of the more important and revolutionary ideas to come out of this analysis.

Historically there is a precedent for the additive nature of electrochemical potentials stemming from the work of Pickett, Pletcher, Fenske, Chatt and Treichel between 1972 and 1980 [2-8] and summarised by Bursten [9] and Chatt [10], with more recent contributions from Pombeiro [11].

Basically, they demonstrated that if one successively substituted a carbonyl species, say $Cr(CO)_5$, by ligands L, to form $Cr(CO)_4L$, $Cr(CO)_3L_2$ etc., one may write an equation for the oxidation potential, in this case E(Cr(I/0)):-

$$E(ox) = A + n(dE^o/dn)_L + Cy \qquad (1)$$

where dE^o/dn is the change in potential upon replacement of n CO groups by n ligands, and A and C are constants. A ligand parameter, P_L was defined [10,12,13] where, for example,

A. J. L. Pombeiro and J. A. McCleverty (eds.),
Molecular Electrochemistry of Inorganic, Bioinorganic and Organomettallic Compounds, 41–55.

$$P_L = E_{1/2}[Cr(CO)_6] - E_{1/2}[Cr(CO)_5L] \quad (2)$$

with some additional refinements to which the reader is referred [11]. We also note an early recognition that Ru(III/II) potentials correlate with P_L values [14]. Thus, ligand additivity does have apparent validity, at least within a fairly narrow group of organometallic complexes. Yet the model, represented by equations (1,2) has not been developed in any detail, and to date, there are only about 20 ligands for which P_L parameters have been derived.

2. Standardization

We therefore seek a more broadly based procedure, standardisation, which would permit this theory to be used widely both in organometallic and coordination chemistry.

A standard must have certain characteristics:-

i) be available for a very wide range of ligands, ii) be electrochemically reversible, or at least quasi-reversible, and be largely solvent independent (except for the special case of water), iii) have potentials which are largely independent of stereochemistry (e.g. cis, trans, mer, fac etc) and independent of overall charge i.e. it should not matter whether we deal with $[ML_6]^{2+}$ or $[ML_6]^{3-}$.

The "obvious" choice to fit these criteria, is the E(Ru(III/II)) couple which is known for a very large number of complexes and usually meets the other criteria listed above.

Thus we define a parameter for the infamous $[Ru(bpy)_3]^{2+}$ ion (bpy = 2,2'-bipyridine) such that:-

$$E_L(bpy) = 1/6\, E(Ru(III/II))[Ru(bpy)_3]^{2+} \quad (3)$$

and where all potentials refer to NHE. The factor 1/6 is employed because there are 6 identical Ru-N bonds in the $[Ru(bpy)_3]^{2+}$ ion. It is evident that the sum, 6 E_L(bpy), reproduces the Ru(III/II) potential for the $[Ru(bpy)_3]^{2+}$ ion.

If we now consider species such as $Ru(bpy)_2L_2$, $Ru(bpy)L_4$ or indeed RuL_6, their Ru(III/II) potentials are reproduced by:

$$\begin{aligned} E(obs) &= 4\, E_L(bpy) + 2\, E_L(L) \\ E(obs) &= 2\, E_L(bpy) + 4\, E_L(L) \\ E(obs) &= 6\, E_L(L) \quad (4) \end{aligned}$$

respectively. Thus, the observed electrochemical potential for the general complex $RuX_xY_yZ_z$ is given by:

$$E(obs) = x\, E_L(X) + y\, E_L(Y) + z\, E_L(Z) = [\Sigma E_L(L)] \quad (5)$$

and it is clear that from an extensive database of ruthenium complexes, with known Ru(III/II) potentials, one may derive E_L(L) values for a large number of different ligands. Where a given L appears in many different complexes, values of E_L(L) are averaged to give the best fit to the data. In this fashion, E_L(L) values have now been reported for over 200 different ligands [15].

3. Procedure and Experimental Results

We may now take any general metal complex, $MX_xY_yZ_z$ (not necessarily ruthenium) and write, for any observed potential (e.g. M(III/II), or M(II/I) etc.):

$$E(obs) = S_M[\Sigma E_L(L)] + I_M \quad (6)$$

where S_M, the slope, and I_M, the intercept would be 1 and zero for the specific case of the Ru(III/II) couple in a ruthenium complex (in an organic solvent), but differ from 1 and zero in the general case. Equation (6) represents a plot of any observed metal-centred redox potential for the general species $MX_xY_yZ_z$ against the calculated Ru(III/II) potential for the corresponding $RuX_xY_yZ_z$ species (which does not in fact have to exist).

If this procedure is to be successful, then equation (6) must be valid, linear, for a wide range of metal complexes, in many oxidation states. This is seen to be the case. Data for Os(III/II), Fe(III/II), Cr(III/II), Mn(II/I), Cr(I/0), Mo(I,0), Nb(V/IV), Nb(IV/III), Ta(V/IV), Re(IV/III) are shown in Figures 3 - 10 of the original publication [15], with slopes and intercepts listed in Table 1. Some additional data for Ru(IV/III) and Os(IV/III) are shown here in Figures 1,2.

Our study shows that for most complexes, good linear correlations are indeed observed (with equation (6)), with no significant dependence upon the organic solvent, unless special solvatochromic phenomena are present, and no significant dependence upon the overall net charge (e.g. between $[MX_6]^m$ and $[MY_6]^{m-}$) in organic solvents.

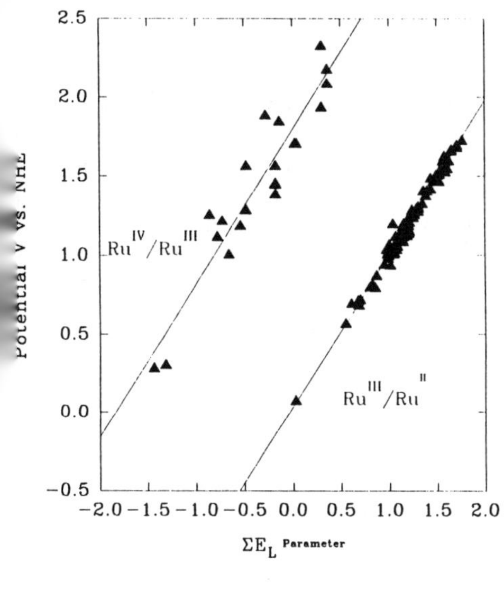

Figure 1 A plot of Ru(IV/III) and Ru(III/II) potentials in organic phase, vs $[\Sigma E_L(L)]$.

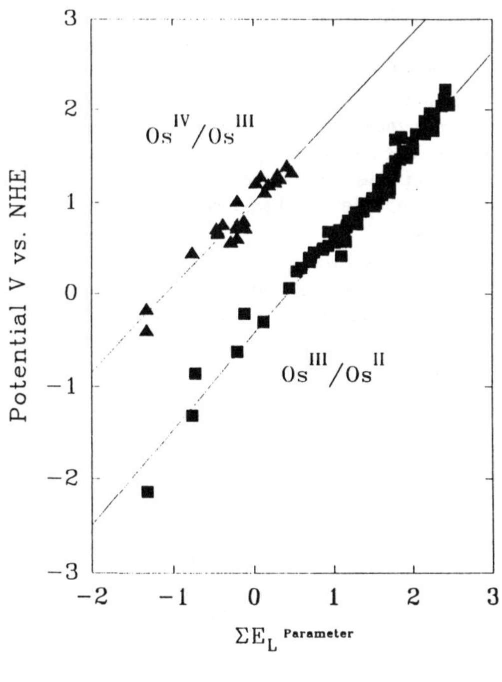

Figure 2 A plot of Os(IV/III) and Os(III/II) potentials in organic phase, vs $[\Sigma E_L(L)]$.

Water, however, is a special case. Thus S_M and I_M values, listed in Table 1, are different in water from their values in a general organic solvent, and are strongly dependent upon net charge. Thus if the net charge is kept constant, (e.g. all $[MX_xY_yZ_z]^{2+}$) good linear correlations are also observed for aqueous phase couples.

Since the electrode potential is determined, at least in part (see section D, (a) below) by the relative binding energies of the ligands L to the metal ion in the lower and upper oxidation states, the S_M and I_M values in equation (6) should be dependent upon:
i) the coordination number, and the stereochemistry of the two species ML_n^{m+1} and ML_n^{m}, ii) the spin states of ML_n^{m+1} and ML_n^{m}, and iii) the pair of oxidation states involved. i.e. the value of m.

Data have been collected almost exclusively for six coordinate (n = 6) octahedral species. Thus the major variable is spin state and indeed (see Table 1) the magnitudes of the S_M and I_M parameters do depend upon spin state, where, for example, the low spin Fe(III)/high spin Fe(II) potentials have different S_M and I_M values from the low spin Fe(III)/low spin Fe(II) potentials.

They also, obviously, must depend upon the specific pair of oxidation states involved, but this raises some intriguing questions, to which we return later.

A very large body of data was presented in ref. [15] clearly legitimising the approach and showing indeed that an Electrochemical Series of ligands can be derived.

3.1 RANGE OF LEGITIMACY

Most of the complexes studied to date involve redox processes occurring at the t_{2g} sub-shell, i.e. involving $d\pi$ electrons. High spin Fe(II), Fe(III) and Cr(II) are the only species to involve $d\sigma^*$ electrons for which S_M and I_M values are so far reported. It was also a primary requirement in collecting the data for this analysis that the redox couples were all electrochemically reversible; the absence of many $d\sigma^*$ redox processes from the correlation reflects the fact that these processes are often irreversible and that extended sets of electrochemical $d\sigma^*$ data are absent from the literature (because they are irreversible ?!).

Using the S_M and I_M values for a specific redox couple with the appropriate $E_L(L)$ values (vide infra) then allows one to calculate the redox potential, vs NHE, for any metal complex and any redox couple where the necessary parameters have been reported.

However there are circumstances where the calculated result may differ from that experimentally observed. These might be sub-divided into two sets, intrinsic or extrinsic. Intrinsic situations are those where the actual magnitude of the potential does differ from the calculated value because for example:

a) there are strong synergistic interactions between between pairs of ligands via the metal such that their $E_L(L)$ values are modified, b) there are redox active ("non-innocent") ligands present, e.g. quinones, NO, c) a bulky ligand causes a steric interaction which interferes with the E_L contribution of another ligand (e.g. 2,2':6',2"-terpyridine is a problem in this respect), d) the case of macrocycles where the hole size is an important factor in determining the metal ligand binding energy, and hence $E_L(L)$ is not transferable.

Extrinsic factors are those where one is, in fact, calculating the true redox potential, but the measured experimental value reflects complicating external physical or chemical factors, the most obvious one being a kinetic contribution, from whatever cause, to the measured redox potential. Such problems might arise because: a) the redox potentials are particularly solvent dependent, b) equilibria are present in either oxidation state M^{m+1}, or M^{m}, say where MX_xY_y depends upon the bulk concentration of X or Y, e.g. where X and/or Y are solvolysed rapidly, or where, for example, a supporting electrolyte ion, or solvent molecule, might bind to the metal ion in one of its oxidation states, or in general, any coupled chemical reactions which change the inner shell of the metal ion in either oxidation state. Clearly one should also take the positive view that such deviations between observed and calculated redox potentials provide a clue that one or more of the above effects is present and may lead to an estimate of its energy contribution.

Table 1 Slope and Intercept Data[a], Volts vs NHE

	Slope	Intercept	R.[b]	#[c]	Solvent
Chromium Cr(III)/Cr(II)(LS)	1.18	-1.72	0.98	14	Organic
Chromium Cr(III)/Cr(II)(LS)	0.575	-1.12	0.98	8	Water
Chromium Cr(III)/Cr(II)(HS)	0.84	-1.18	0.98	14	Organic
Chromium Cr(I)/Cr(0)	0.52	-1.75	0.97	39	Organic
Iron Fe(III)/Fe(II)(LS)	0.68	0.24	0.99	24	Water
Iron Fe(III)/Fe(II)(LS)	1.10	-0.43	0.99	14	Organic
Iron Fe(III)/Fe(II)(HS)	0.89	-0.25	0.99	8	Organic
Manganese Mn(II)/Mn(I)	0.81	-1.76	0.99	23	Organic
Molybdenum Mo(I)/Mo(0)	0.74	-2.25	0.99	24	Organic
Niobium Nb(V)/Nb(IV)	0.76	1.24	0.999	3	Organic
Niobium Nb(IV)/Nb(III)	0.75	-0.12	0.999	3	Organic
Osmium Os(IV)/Os(III)	0.93	1.0	0.97	25	Organic
Osmium Os(III)/Os(II)	1.01	-0.40	0.98	80	Organic
Osmium Os(III)/Os(II)	1.61	-1.30	0.99	18	Water
Rhenium Re(IV)/Re(III)	0.85	0.50	0.98	18	Organic
Rhenium Re(III)/Re(II)	1.17	-0.88	0.97	22	Organic
Rhenium Re(II)/Re(I) upper	0.76	-0.95	0.96	59	Organic
Rhenium Re(II)/Re(I) lower	0.27	-1.43	0.76	12	Organic
Ruthenium Ru(IV)/Ru(III)	0.98	1.80	0.94	22	Organic
Ruthenium Ru(III)/Ru(II)	1	0	1	--	Organic[d]
Ruthenium Ru(III)/Ru(II)	1.14	-0.35	0.97	44	Water
Tantalum Ta(V)/Ta(IV)	0.78	0.64	0.98	5	Organic

a) All six coordinate. LS = low spin, HS = high spin. b) Correlation coefficient. c) Number of data points. e) by definition.

3.2 MAGNITUDE OF THE $E_L(L)$ PARAMETERS

The $E_L(L)$ parameters range in magnitude from about -0.6 for very soft anionic species, to 0 to 0.3 for most amines, to 0.99 for carbon monoxide and greater than 2 for the nitrosonium cation. In particular, one may establish relatively narrow ranges for specific groups of ligands:

$E_L(L)$ -0.63 ----> 0 V
OH^-, most X^{n-} ions, including S anions, strong π-bases.
$E_L(L)$ 0 -----> 0.1 V
Saturated amines fall into a fairly narrow range, weakly π-acid unsaturated amines.
$E_L(L)$ 0.1 -----> 0.40 V
Unsaturated amines of stronger π-acid character, pyridines, bipyridines etc.
$E_L(L)$ 0.30 -----> 0.40 V
Hard thioethers, nitriles, softer phosphines.
$E_L(L)$ 0.35 -----> 0.50 V
Isonitriles, harder phosphines, arsine, stibines, softer phosphites.
$E_L(L)$ 0.50 -----> 0.65 V
harder phosphites.
$E_L(L)$ 0.65 -----> 0.75 V
Dinitrogen, nitrites.
$E_L(L)$ 0.70 -----> 0.95 V
Positively charged ligands, π-acid olefins.
$E_L(L)$ > 0.9 V
CO, NO^+

Therefore, even if a particular ligand does not have a tabulated $E_L(L)$ value (in [15]), a rough estimate can generally be made.

3.3 PROBLEMS WITH STEREOCHEMISTRY

In the case of most coordination complexes (as distinct from organometallic complexes), cis/trans, mer/fac isomers etc usually have similar potentials. Thus equation (6) can be used without concern as to the mutual arrangements of the ligands. This is often not true for organometallic species where differences may be 0.1 - 0.5 V. Fenske [4] developed a correction for the HOMO energy of a carbonyl complex in terms of the number of carbonyl π^* orbitals interacting with the HOMO. In the case of a d^6 $M(CO)_6$ species, all three t_{2g} orbitals (degenerate HOMO) interact with 4 CO π^* orbitals, while for trans-$ML_4(CO)_2$ species, the t_{2g} will split commonly to place d(xy) (HOMO) above d(xz,yz). This d(xy) will not interact with CO π^*. Similarly, for cis-$ML_4(CO)_2$ the d(xz,yz)(HOMO) will have 1 CO π^* interaction per orbital. Since the stabilisation of the HOMO is related both to the number of CO groups bonded thereto and the relevant potential [4,9], one introduces, into equation (6), a variable x such that for the carbonyl species under discussion, equation (6) is replaced by [15b]:-

$$E_{obs} = S_M[\Sigma E_L(L) + qx] + I_M \qquad (7)$$

where q = 4, 0 and 1 respectively for the hexacarbonyl, trans and cis dicarbonyl.

In practice, only CO and the isonitrile ligand appear to require this type of stereochemical correction. For complexes containing both these ligands, a double correction must be imposed, viz:

$$E_{obs} = S_M[\Sigma E_L(L) + qx + q'y] + I_M \qquad (8)$$

A Table of corrections (q and q' values) for all common stereochemistries is shown in ref. [15, Table III]. Commonly, for low valent carbonyl species, x is of the order of 0 - 0.1 V per HOMO-interacting CO, while y can be larger, as much as 0.2 - 0.3 V per HOMO-interacting isonitrile group. The isonitrile is a special case since they may be linear or bent and the electron distribution on the species does depend upon this geometry, i.e. one may suppose that the y correction allows for some variation of E_L(isonitrile) value with the geometry of the metal-isonitrile bonding.

4. Significance and Application

4.1 SLOPE AND INTERCEPT

4.1.1. Preamble

Electrode potentials, for say ML_6^{m+1}/ML_6^{m}, may be thought of as being determined by the relative stability constants for formation, in the gas phase, of the ML_6 species in each oxidation state, and by the differential solvation free energies of each species in solution. The lack of dramatic variations in redox potential from one organic solvent to another, and the lack of any major dependence, in organic solvent, of the redox potential on the net charge of an ion argues for this differential solvation energy to be rather small, or relatively constant from one organic solvent to another, or possibly with magnitude scaled by the $[\Sigma E_L(L)]$ value. Thus arguments based upon changes in relative binding energy are probably valid.

This issue has been considered in some detail [16-18], where, when dealing with aqueous phase electrode potentials, on the other hand, it is emphasized [18] that differential solvation free energy terms can be very large, and certainly very dependent upon net charge.

4.1.2. Slope data in organic solvents.

A slope, S_M, of unity, for dependence of the given $[ML_6]^{m+1}/[ML_6]^{m}$ couple upon $[\Sigma E_L(L)]$ requires that the ratio of the stability constants (binding energies) for formation of $[ML_6]^{m+1}$ and $[ML_n]^{m}$ is the same as the ratio for binding of the same ligands to Ru(III) and Ru(II), for all ligands.

Slopes greater than unity require that, relative to the Ru(III/II) ratio, there is preferential binding to the lower oxidation state, $[ML_6]^{m}$, and conversely for slopes less than unity.

4.1.3. Slope data in aqueous phase.

It is difficult to draw conclusions because of the dominance of the solvation free energy term. However, it is probable that the argument used above for the organic solvents, can be used if one, say, compares aqueous phase data for the general species $[ML_6]^{2+}$ with data for dipositively charged ruthenium species in aqueous medium. Thus S_M for the Os(III/II) couple, (derived from dipositively charged Os(II) species) in water is 1.61, compared with 1.14 for the Ru(III/II) couple (derived from dipositively charged Ru(II) species).

Thus, in water, $[Os(II)L_6]^{2+}$ species are more stable than $[Ru(II)L_6]^{2+}$ species, relative to their M(III) partners. The reason is not known but one may speculate that hydrogen bonding is more important to the lower oxidation state, because of overall charge considerations, and that the larger size of the Os(II) core allows water to penetrate the coordination sphere to a greater degree than in the case of Ru(II).

4.1.4. Intercept.

We may choose to write the potential:-

$$E_{obs} = a + nb + c + S_M[\Sigma E_L(L)] \tag{9}$$

where a is the M^{m+1}/M^{m} ionisation energy in the gas phase. The variable b (always negative) could be defined similarly to the spherical electrostatic term in crystal field theory raising all the energy levels to the same degree determined by n, the number of ligands and a consequence of the electrostatic repulsion between ligand lone pair and metal d electrons. Note, however, that in this development, nb will be defined as the difference between the spherical terms for the oxidised and reduced species. Variable c (either sign) contains contributions from the reference electrode and the differential solvation free energies of the two redox states, thus $I_M = a + nb + c$.

For a given redox couple, a is a constant. The lack of solvent dependence (in organic solvents) suggests that c is approximately constant (for a given system) and therefore the existence of linear correlations (Equation (7)) suggests that the differential spherical term nb is also reasonably constant, for a given redox couple. Alternatively, the invariance of I_M may reflect that the term nb is itself linearly dependent upon $[\Sigma E_L(L)]$. Further information concerning this term can be derived from redox potentials of other coordination numbers (varying n) once sufficient data are available.

The value of b will vary with different metals and with different oxidation states of the same metal. The variation in a, with oxidation couple, makes a major contribution to the magnitude of the intercept, and is responsible, for example, for the larger negative values derived for low oxidation state couples.

4.1.5. Slope as a function of oxidation state

Clearly the slope, S_M, (of eqn. (6)) can be expected to change from one metal centre to another, but what about when changing oxidation state but keeping the metal centre constant ? In the initial publication [15], we reported that Nb(V/IV) and Nb(IV/III) do have the same slope, but only very limited data were available. Two data sets were reported for Cr(III/II) for low spin and high spin Cr(II). The S_M value for low spin d^4 Cr(III/II) (0.57) is experimentally almost indistinguishable from that for Cr(I/0) (0.52).

Is this a general result ? We report here data for the Ru(IV/III) and Os(IV/III) couples, (Figures 1,2). The lines are indeed almost parallel with the corresponding M(III/II) lines. (Figure 1).

Meanwhile an analysis of the electrochemistry of rhenium couples [21] (discussed further below) (Figure 3), shows that the Re(IV/III) (S_M = 0.86) and (upper) Re(II/I) (S_M = 0.76) correlations are approximately parallel but with greater deviation for Re(III/II) (S_M = 1.17).

Clarke reported data for Tc(IV/III), Tc(III/II) and Tc(II/I) [20], with S_M values of 1.0. 1.3 amd 1.4, the last pair, in particular, being almost parallel given the experimental scatter (see Figure 2 in [21]).

Thus there are sufficient data to tentatively conclude that pairs of redox couples of the same metal ion, will often have similar S_M values, i.e. be approximately parallel, though this will not exclusively be true.

This is a significant observation which indicates that ratios of the binding constants to the upper and lower oxidation states of pairs of couples, e.g. Ru(V/IV), Ru(IV/III) and Ru(III/II), are closely similar, for all ligands (except those for which there are special problems (identified above)).

4.2 THE RHENIUM STORY

A considerable body of data exists for rhenium redox couples [21]. There are additional complexities in that seven coordinate Re(II) (d^4) species are known. In a few cases [21], six coordinate Re(II) complexes may bind solvent or electrolyte to generate a seven coordinate species whose formation will lead, obviously, to an anomalous potential. Nevertheless, the Re(IV/III) line is well-behaved and the Re(III/II), subject to the above proviso, is also quite well behaved (Figure 3); however they are not parallel.

Strikingly, the Re(II/I) correlation breaks down into two separated lines, with a break near $[\Sigma E_L(L)]$ = ca. 1.5 (Figure 3). Had the upper line been extended to values of $[\Sigma E_L(L)]$ ca 0, then the two lines would have crossed and two-electron couples (Re(III/I)) would have been observed. This is avoided by complexes with $[\Sigma E_L(L)] < 1.5$ falling on the lower Re(II/I) line.

Complexes on the lower line are substantially more stable in the Re(II) oxidation state (or less stable in the Re(I) oxidation state) than they would have been, had they fallen on the upper line. The reason for this is by no means clear. Arguments should be based upon spin state, coordination number or stereochemistry. The most likely possibility is a change in stereochemistry, perhaps from octahedral to trigonal prismatic, from the upper to the lower line. However this explanation is only favoured in the sense that arguments based upon spin state or coordination number variations seem less tenable.

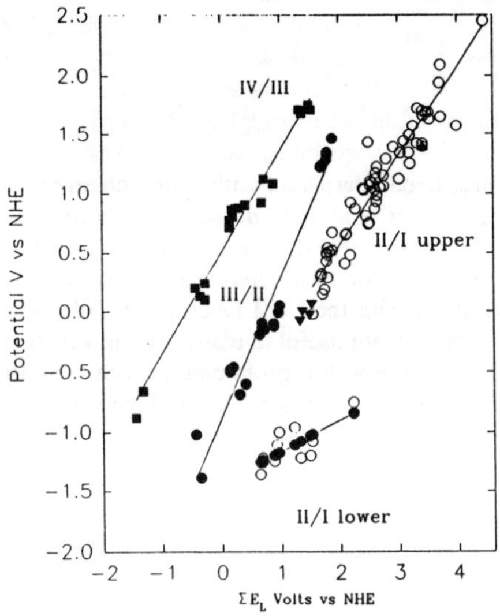

Figure 3. A plot of Re(IV/III), Re(III/II) and Re(II/I) potentials in organic phase, vs $[\Sigma E_L(L)]$.

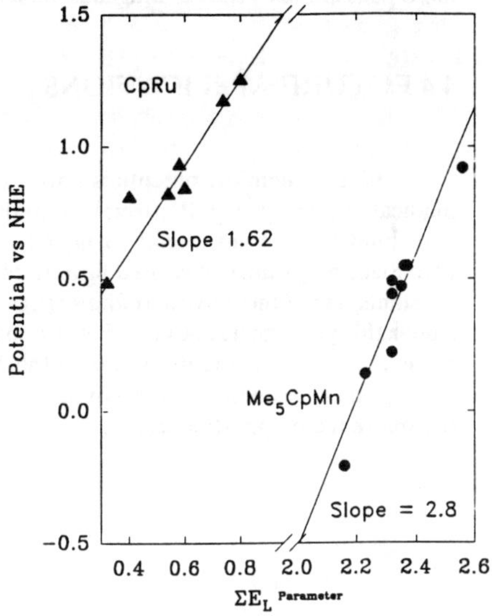

Figure 4. A plot of cylopentadienylruthenium (Ru(III/II)) and pentamethylcyclopentadienyl manganese (Mn(II/I)) redox potentials versus $[\Sigma E_L(L)]$ in organic phase (data from [23-26]).

4.3 CYCLOPENTADIENYL SYSTEMS

Redox potentials of cyclopentadienyl (and other arene ligand) complexes have been studied in some depth by Strelets [22] who documented well behaved variations with properties such as the ionisation potential of the metal. No intensive analysis of such electrochemical data has yet been attempted with $E_L(L)$. Clearly the Cp unit will modify the metal ion dramatically and one does not expect CpMLX complexes to fall on the same correlation lines as non-Cp containing organometallic or coordination complexes of the same metal.

As examples we show in Figure 4 [data from 23-26], a plot of data for ruthenium and manganese Cp species, with variation of the other ligands bound to these RuCp and MnCp cores. Clearly they are well behaved, but the slopes are dramatically greater than for any other correlations yet demonstrated. Further studies are necessary, but it appears that the Cp unit renders the central metal atom very much more polarisable by the other ligands, than in its absence.

4.4 FUTURE APPLICATIONS

Electrochemical potentials have, in the literature, been correlated with many chemical and physical properties, e.g. IR stretching frequencies in nitrosyls or dinitrogen derivatives, photoelectron core binding energies, metal to ligand charge transfer transition (MLCT) energies, ligand redox potentials, pK_a values of related ligands, Hammett σ-parameters for substituted ligands and some rate constants, etc. One may therefore expect similar correlations with the $E_L(L)$ parameter. We do not labour this point here, but note that one may be able to derive more useful fundamental information if the correlation is made directly with the $E_L(L)$ parameter than with a gross metal redox potential. Some possible examples are shown here, in a very preliminary form, in order to encourage researchers to explore these opportunities.

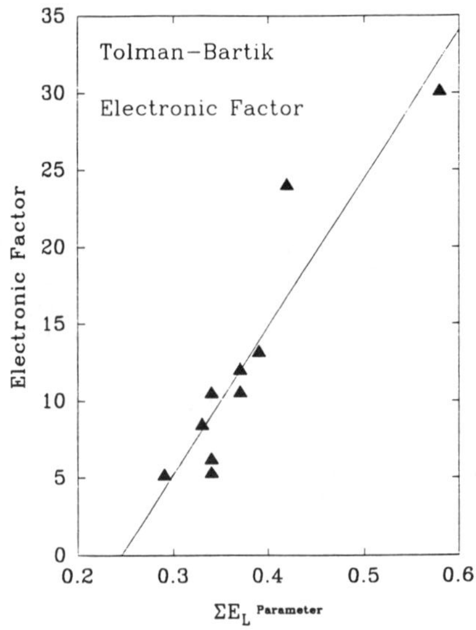

Figure 5. A plot of Tolman-Bartik electronic factors vs. $[\Sigma E_L(L)]$.

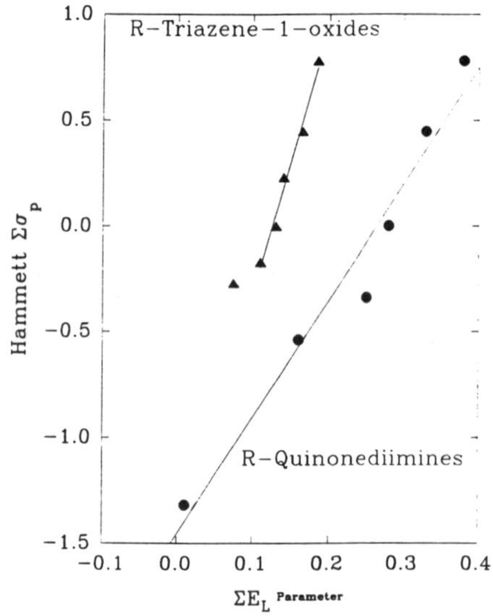

Figure 6. A plot of Hammett σ_p values versus $[\Sigma E_L(L)]$ for ruthenium complexes (Ru(III/II)) of triazene-1-oxides and quinonediimine species (data from [14,30]).

4.4.1. Tolman-Bartik electronic factors

Tolman introduced [27] electronic factors, χ_L, for phosphine and phosphite ligands, based upon the totally symmetric CO stretching vibration in a phosphine nickel tricarbonyl derivative. More recently Bartik and co-workers have established this correlation more accurately [28]. These parameters, together with Tolman's cone angle data, have been of great utility in understanding the chemistry and dynamics of phosphine complexes.

One can expect a correlation between $E_L(L)$ and χ_L and this is shown in Figure 5. This is a preliminary analysis to show the validity of this correlation which will be explored in more detail [29]. Once the connection between the two parameters has been well established, it can be used to derive electronic factors for a range of other related ligands, such as chelating diphosphines for which the Tolman analysis based on a monophosphine nickel tricarbonyl is unusable.

4.4.2. Hammett σ-factors.

Clearly one can expect a good correlation with $E_L(L)$. Two examples are shown in Figure 6, for some substituted triazene-1-oxides [14] and quinonediimines [30]. The value of such plots lies in the analysis of their slope providing a two-dimensional picture of the changes of electronic structure which occur upon changing the substituent.

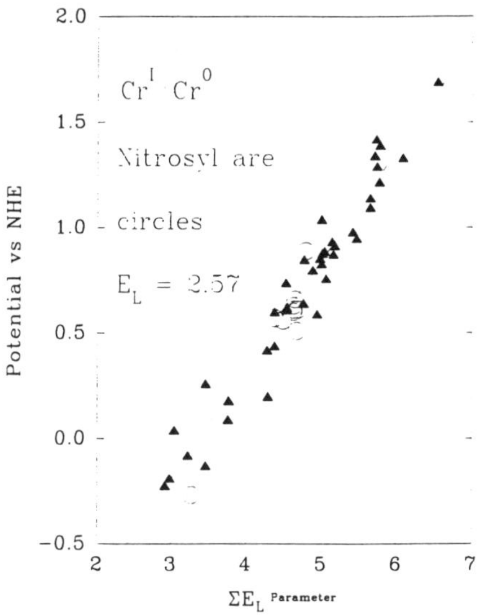

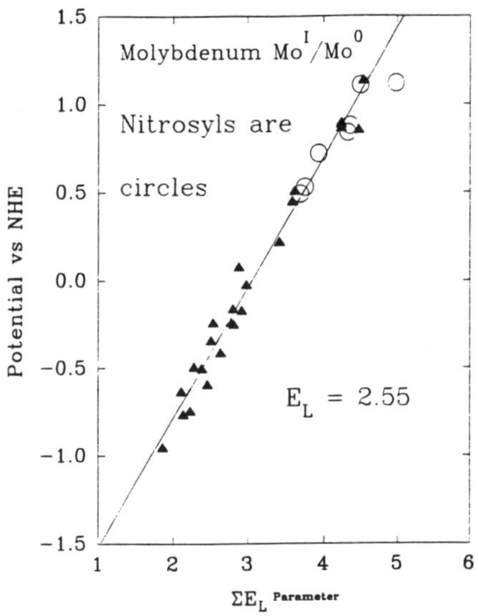

Figures 7,8. Plots of Cr(I/0) and Mo(I/0) redox data for for some organometallic species, showing nitrosyl-containing species as larger open circles, for clarity (data from [10,31,32]).

4.4.3. Nitrosyl complexes

Nitrosyls present a special challenge since they are non-innocent (redox active) and the ligand orbitals generally mix strongly with metal orbitals such that redox processes are often not clearly centered on metal or NO. Given that NO can bind, formally, as NO^+, NO or NO^-, a wide variation in $E_L(NO)$ can be expected. Thus the $E_L(L)$ value could be a useful new tool to understand nitrosyl chemistry, though its application may be difficult to handle.

Figures 7,8 shows plots of data for chromium and molybdenum nitrosyl species; the slope with $E_L(L)$ is essentially the same as for non-NO species. Thus, in Figure 7,8 [data from 10,31,32] , a value for $E_L(NO)$ is used which will cause the nitrosyl complexes to fit the correlation line. This yields $E_L(NO^+)$ = 2.55 - 2.57 for both metal ions. Values in the range 1.2 - 1.8, for $E_L(NO^+)$, have been described by Clarke [33] for some technetium nitrosyl derivatives. This area would benefit from a detailed analysis.

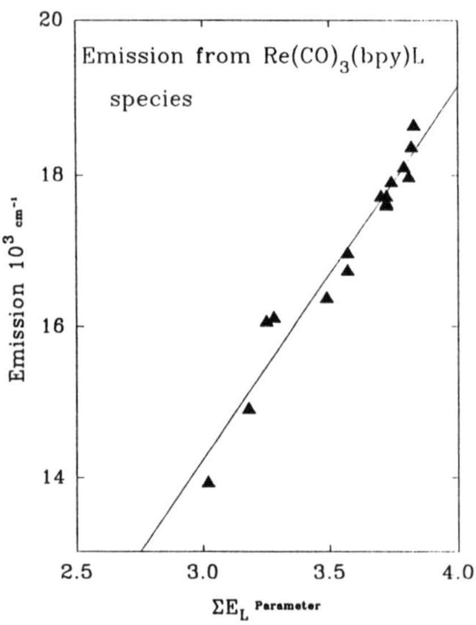

Figure 9. A plot of the emission energies (corrected) for $Re(CO)_3(bpy)L$ with varying L vs $[\Sigma E_L(L)]$ (data from [35-37]).

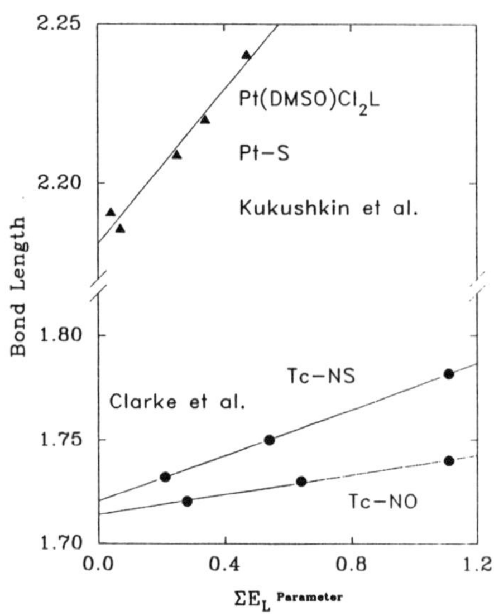

Figure 10. A plot of Pt-S bond lengths in $Pt(DMSO)Cl_2L$ vs. $E_L(L)$ [38] and of Tc-N bond lengths in some technetium NO and NS complexes versus $[\Sigma E_L(L)]$ [39].

4.4.4 Emission

The many correlations of electrochemical potentials with MLCT transition energies [e.g. 34] could be transcribed into $E_L(L)$ plots to yield rather obvious relationships which would allow the prediction of excited state redox potentials based upon $E_L(L)$ parameters and ligand reduction potentials. Of more potential value is the analysis of emission energies, necessarily for a given ligand, when bound to a metal atom to which is attached a series of other (spectator) ligands. An example is shown in Figure 9 for emission from $Re(CO)_3(bpy)L$ species as a function of widely varying ligand L [35-37]. Unlike the case with the MLCT excitation energies, which are Franck-Condon in nature, this plot correlates the equilibrated CT excited state energy versus the $[\Sigma E_L(L)]$ parameter.

The slope is dependent both on the relative binding of ligand L to the lower and upper oxidation states of the metal, and also upon the relative binding of ligand L, and its reduced species, L^-, to the metal upper oxidation state.

4.4.5. Metal-ligand bond distances

Two groups have reported noting a correlation between certain metal-ligand bond distances, and $E_L(L)$. Thus the Pt-S bond in $Pt(DMSO)Cl_2L$ varies linearly with $E_L(L)$ [38] (Figure 9) and the Tc-NX (X = O,S) varies linearly with $[\Sigma E_L(L)]$ (for complex, all ligands excluding $E_L(NO)$) in technetium nitrosyl derivatives [39] (Figure 10). These observations are noted with great interest and one expects further examples to be found. Such structural relationships with $E_L(L)$ are of especial importance in the further understanding of these useful parameters.

4.4.6. Hydrogen and Hydride Complexes.

The hydride ion is an interesting ligand but its analysis by this $E_L(L)$ method is made difficult by its non-innocent nature and by the observation, generally, of irreversible redox processes. Jia and Morris [40] have studied the chemistry of coordinated dihydrogen complexes and have shown a relationship between the $pK_a(Ru(H_2)^+)$ and the anodic potential for oxidation of the complex. They used the $E_L(L)$ parameters to show that available (irreversible) redox data for CpRuH(X)(Y) species could be fitted on the basis of the sum of non-hydrogen ligands (assuming a value for $E_L(Cp)$). With wide variation of ligands X,Y, the H_2 complexes may exhibit characteristics of either coordinated dihydrogen or di-hydride. There is a narrow range of possible energies for the t_{2g} electrons, as probed by electrochemistry, independent of the metal ion, where stable bonding of the η^2-H_2 ligand is possible, at 25° C. Using the $E_L(L)$ model, Morris [40,41] has reationalised this chemistry in terms of the stability of the di-hydrogen complex and its chemical reactivity.

5. Summary

$E_L(L)$ parameters permit one to:-

a) predict the redox potential of given metal couples when structural and spin state information are available. In a complex voltammogram this will aid in redox couple assignment;

b) predict structure and spin state for a metal complex through fitting its observed redox potential to a previous correlation;

c) calculate the thermodynamic value for a redox couple when kinetic effects or coupled chemical reactions, etc, prevent it from being experimentally derived;

d) design a metal complex to have a specific redox potential;

e) obtain bonding or structural information where predicted values disagree with the experimental value, (synergism, non-innocence etc);

f) through detailed analysis of the slopes and intercepts, obtain additional insight into the nature of the metal-ligand bond.

Extension to the types of example shown above in section 4.4 suggests that it may be possible to derive new information which cannot readily be obtained with the previous type of electrochemical analysis. In particular, further studies of the dependence of $[\Sigma E_L(L)]$ upon excited state energies, bond distances and non-innocent ligand behaviour is clearly desirable.

6. Acknowledgements.

The author is grateful to the Natural Sciences and Engineering Research Council (Ottawa) and the Office of Naval Research (Washington) for continued support of his work.

7. Bibliography

1. A. B. P. Lever, "Inorganic Electronic Spectroscopy", Elsevier, Amsterdam, 1st Edn 1968, 2nd Edn. 1984.
2. P. M. Treichel, G. E. Dureen, and H. J. Mueh, J.Organomet.Chem., 1972, 44, 339.
3. P. M. Treichel, H. J. Mueh, and B. E. Bursten, Isr.J.Chem., 1976, 15, 253-257.
4. A. Sarapu, and R. F. Fenske, Inorg.Chem., 1975, 14, 247-253.
5. C. J. Pickett, and D. Pletcher, J. Organomet. Chem., 1975, 102, 327-333.
6. J. Chatt, C. T. Kan, G. J. Leigh, C. J. Pickett, and D. R. Stanley, J. Chem. Soc. Dalton, 1980, 2032.
7. P. M. Treichel, and G. J. Essenmacher, Inorg. Chem., 1976, 15, 146-150.
8. P. M. Treichel, H. J. Mueh, and B. E. Bursten, J. Organomet. Chem., 1976, 110, C49.
9. B. E. Bursten, and M. R. Green, Progress Inorg.Chem., 1988, 36, 474.
10. J. Chatt, Coord. Chem. Rev., 1982, 43, 337-348.
11. A. J. L. Pombeiro, Portug. Electrochim. Acta, 1985, 3, 41-66; A. J. L. Pombeiro, Inorg. Chim. Acta, 1985, 103, 95-103; A. J. L. Pombeiro; C. J. Pickett, and R. L. Richards, R.L. J. Organomet. Chem., 1982, 224, 285-294; M. Fernanda, N. N. Carvalho, and A. J. L. Pombeiro, J. Chem. Soc. Dalton, 1989, 1209.
12. J. Chatt, W. Hussain, G. J. Leigh, H. M. Ali, C. J. Pickett, and D. A. Rankin, J.Chem.Soc. Dalton, 1985, 1131-1136.
13. G. Hussain, G. J. Leigh, H. M. Ali, C. J. Pickett, and D. A. Rankin, J.Chem.Soc. Dalton, 1984, 1703.
14. S. Goswami, R. Mukherjee and A. Chakravorty, Inorg. Chem, 1983, 22, 2825.
15 a) A. B. P. Lever, Inorg. Chem., 1990, 29, 1271;
15 b) Note that Eqs.(10) and (11) in ref.[15a] were incorrectly set in print; they should appear as Eqs. (7) and (8) in this version.
16. D. D. Perin, Rev. Pure Appl. Chem. 1959, 9, 257.
17. A. A. Vlcek, Progr. Inorg. Chem., 1963, 5, 211.
18. D. E. Richardson, Inorg. Chem., 1990, 30, 3213.
19. C. M. Duff and G. A. Heath, J. Chem. Soc., Dalton Trans., 1990, 2401.
20. J. Lu, A. Yaman and M. Clarke, Inorg. Chem. 1990, 29, 3483.
21. A. B. P. Lever, Inorg. Chem. 1991, 30, 1980.
22. V. V. Strelets and S. V. Kukharenko, Dokl. Akad. Nauk. SSSR, 1984, 275, 894; V. V. Strelets, Coord. Chem. Rev., 1992, 000, 0000.
23. S. K. Mandal and A. R. Chakravarty, Ind. J. Chem., 1990, 29A, 18.
24. L. F. Szczepura and K. J. Takeuchi, Inorg. Chem., 1990, 29, 1772.
25. N. G. Connelly and M. D. Kitchen, J. Chem. Soc., Dalton, 1977, 931.
26. J. W. Hershberger and J. K. Kochi, Polyhedron, 1983, 2, 929.
27. C. A. Tolman, Chem. Rev., 1977, 77, 313.
28. T. Bartik, T. Himmler, H-G. Schulte and K. Seevogel, J. Electroanal. Chem., 1984, 272, 29.
29. A. B. P. Lever, work in progress.
30. H. Masui and A. B. P. Lever, Inorg. Chem. to be submitted shortly.
31. D. E. Wigley and R. A. Walton, Inorg. Chem., 1983, 22, 3138.

32. W. E. Robinson, D. E. Wigley and R. A. Walton, Inorg. Chem., 1985, 24, 918.
33. M. J. Clarke, personal communication 1991.
34. E. S. Dodsworth and A. B. P. Lever, Chem. Phys. Lett., 1986, 124, 152.
35. A. J. Lees, Chem. Revs., 1987, 87, 711.
36. J. W. Caspar, B. P. Sullivan and T. J. Meyer, Inorg. Chem., 1984, 23, 2104.
37. L. A. Stacksteder, A. P. Zipp, E. A. Brown, J. Streich, J. N. Demas and B. A. De Graff, Inorg. Chem., 1990, 29, 4335.
38. V. Y. Kukushkin, V. E. Konovalov, N. S. Panina and V. N. Yakovlev, personal communication, 1991.
39. J. Lu and M. J. Clarke, J. Chem. Soc. Dalton. 1992, 000, and personal communication.
40. G. Jia and R. H. Morris, J. Am. Chem. Soc., 1991, 113, 875.
41. R. H. Morris, Inorg. Chem. 1992, 31, 0000.

ELECTROCHEMICAL LIGAND PARAMETERS FOR PHOSPHONIUM-FUNCTIONALIZED ISOCYANIDES AND DERIVED CARBENES AND INDOLES OF GROUP VI TRANSITION METAL CARBONYLS

M.T.A. RIBEIRO [a], A.J.L. POMBEIRO [a], G. FACCHIN [b], M. MOZZON [b], and R.A. MICHELIN [b]

[a] *Centro de Quimica Estrutural, Complexo I, Instituto Superior Tecnico, Av. Rovisco Pais, 1096 Lisboa Codex, Portugal, and* [b] *Centro di Chimica e Tecnologia dei Composti Metallorganici degli Elementi di Transizione del C.N.R. and Istituto di Chimica Industriale, Facoltà di Ingegneria, Università di Padova, Via F. Marzolo 9, 35131 Padova, Italy.*

ABSTRACT. A series of $(CO)_5ML$ (M = Cr, Mo, W) complexes, where L is a phosphonium-substituted phenylisocyanide ligand of the type *o*-$R_3PCH_2C_6H_4NC$ [PR_3 = PMe_3 , PPh_2Bz , PPh_3], a carbene-type $(CO)_5M$=[*o*-$\overline{CN(H)C_6H_4C}(PR_3)$], indole-type $(CO)_5M$-[o-$\overline{NC(H)C(PR_3)(C}_6H_4)$] or protonated indole-type $(CO)_5M$-[*o*-$\overline{N(H)C(H)C(PR_3)(C}_6H_4)$] species, have been studied by cyclic voltammetry and controlled potential electrolysis in aprotic media. They undergo a single-electron (reversible or irreversible) anodic wave and their oxidation potential has been applied to estimate the P_L and E_L ligand parameters for the isocyanide and derived ligands which, according to their net electron donor ability, can then be ordered as follows: indoles > carbenes > protonated carbenes or protonated indoles > isocyanides. For the metal-centres $\{M(CO)_5\}$ (M = Mo or W), the electrochemical parameter β (polarisability) has also been estimated.

1. Introduction

Following our interest on the redox properties of transition metal isocyanide and carbene complexes [1], we report the investigation of the electrochemical behaviour of new phosphonium-functionalized isocyanide (**A**), and derived carbene (**B**), indole (**C**) and protonated indole (**D**) complexes of Cr, Mo and W pentacarbonyls. These studies appear to have been undertaken for the first time for complexes with such types of ligands. It was also our object to correlate the redox properties of these compounds with the electron donor/acceptor ability of these ligands. Moreover, this study would also extend to novel carbene complexes the rather limited electrochemical investigation reported [2] for compounds with multiple metal-carbon bonds.

2. Electrochemical Studies

2.1. GENERAL

The redox properties of most of the above mentioned complexes have been studied by cyclic voltammetry (C.V.) or controlled potential electrolysis, in 0.2 mol·dm^{-3} $[Bu_4N]$/NCMe, at a

A. J. L. Pombeiro and J. A. McCleverty (eds.),
Molecular Electrochemistry of Inorganic, Bioinorganic and Organomettallic Compounds, 57–62.

A 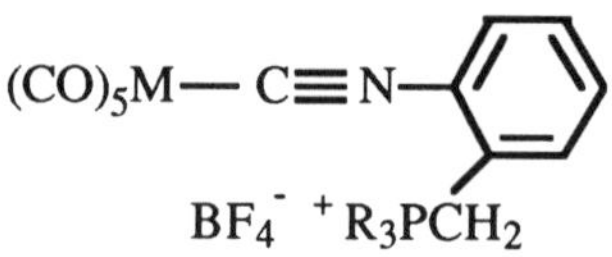

B 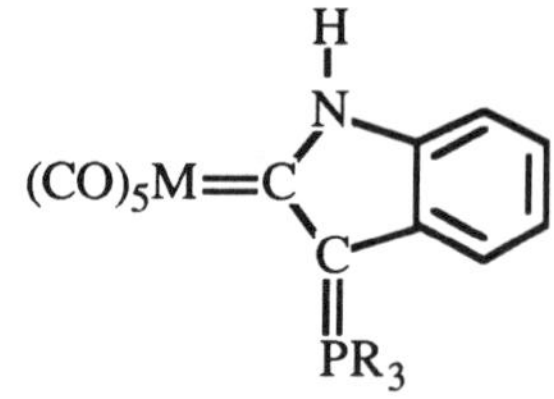

C 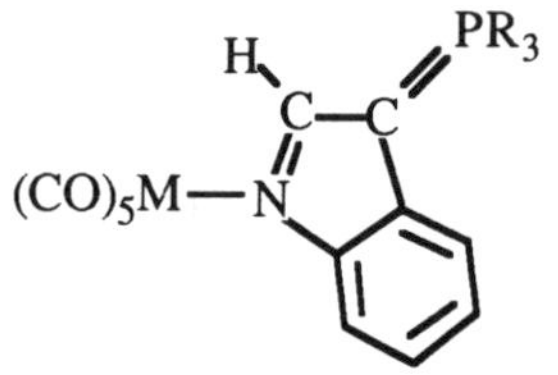

D 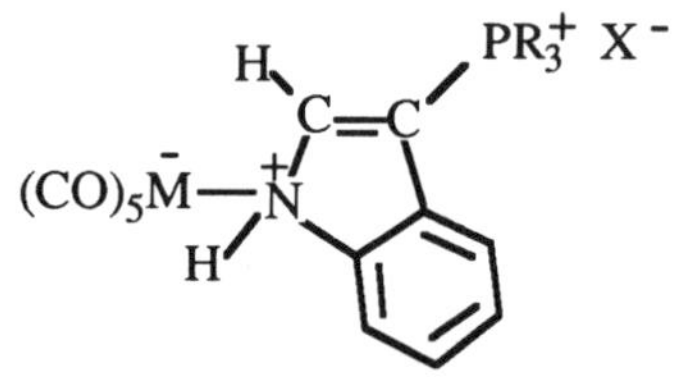

Pt electrode. They undergo a single-electron, reversible or irreversible, anodic process (Fig.) which, as known [3-7] to occur for many other 18-electron complexes with the $\{M(CO)_5\}$ (M = Cr, Mo or W) metal d^6 sites, should correspond to the expected M(0) —> M(I) oxidation; this is commonly followed, at a higher potential, by a second (irreversible) anodic process which has not been investigated further. The values of the oxidation potential (Table) for the first wave ($E_{1/2}{}^{ox}$ or $E_{p/2}{}^{ox}$/V *vs.* S.C.E., the latter for the irreversible processes) lie in the following ranges: 1.3-1.1 (isocyanide complexes) > 1.1-0.9 (protonated indole and protonated carbene complexes) > 0.8-0.6 (carbene complexes) > 0.6-0.5 V (indole complexes), therefore suggesting the following order for the net electron donor character of the corresponding ligands: indoles > carbenes > protonated indoles, protonated carbenes > isocyanides.

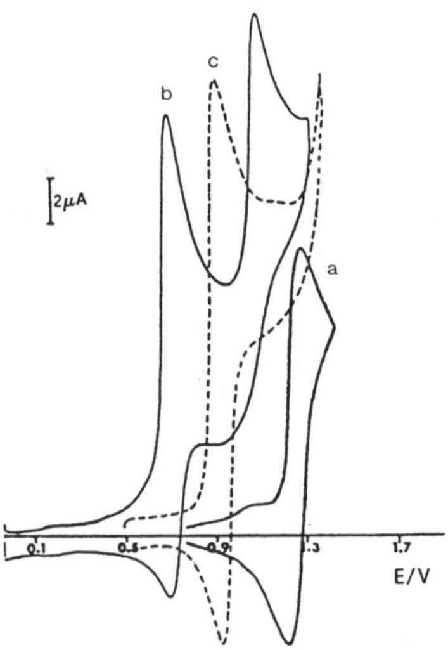

Fig. Cyclic voltammograms for $(CO)_5Cr(o\text{-}Me_3PCH_2C_6H_4NC)$ (**a**), $(CO)_5Cr[o\text{-}\overline{CN(H)C_6H_4C}(PMe_3)]$ (**b**) and the derived protonated species (**c**), at a Pt electrode, in 0.2 M $[n\text{-}Bu_4N][BF_4]$/NCMe (values of potential in V *vs.* S.C.E.).

Table. Summary of electrochemical data ($E_{1/2}^{ox}$ or $E_{p/2}^{ox}$)[a] for complexes [$(CO)_5ML$] (M = Cr, Mo or W) and estimated P_L and E_L parameters for their L ligands.

L	$E_{1/2}^{ox}$ or ($E_{p/2}^{ox}$) (V *vs.* S.C.E.)	P_L (V)	E_L (V *vs.* N.H.E.)
Isocyanides			
o-$R_3PCH_2C_6H_4NC$	1.25 - 1.13	-0.28 to -0.36	0.50 - 0.43
Carbenes			
o-$\overline{CN(H)C_6H_4C}(PR_3)$	0.79 - 0.63	-0.79 to -0.87	0.06 to -0.01
Indoles			
o-$\overline{N=C(H)C(PR_3)(C}_6H_4)$	0.65 - 0.55	-0.92 to -1.0	-0.05 to -0.14
Protonated carbenes	1.07 - 0.90	-0.51 to -0.60	0.30 - 0.22
Protonated indoles	1.06 - 0.88	-0.44 to -0.62	0.36 - 0.20

[a] $E_{1/2}^{ox}$ or $E_{p/2}^{ox}$ for the reversible or irreversible processes, respectively. Values measured at 100 mV s^{-1}, in 0.2 M $[Bu_4N][BF_4]$/NCMe.

For the functionalized isocyanide compounds $[(CO)_5M(o\text{-}R_3PCH_2C_6H_4NC)]BF_4$, the oxidation potentials are only marginally higher than those commonly found in related non-functionalized aromatic isocyanide complexes, thus indicating that the phosphonium group has not a substantial effect on the metal-centred anodic process, possibly as a result of its distance to the metal.

Deprotonation of the isocyanide complexes generates the neutral cycloaminocarbene or indole compounds $(CO)_5M[o\text{-}\overline{CN(H)C_6H_4C}(PR_3)]$ or $(CO)_5M[o\text{-}\overline{N=C(H)C(PR_3)(C}_6H_4)]$, respectively, which present much lower oxidation potentials (by *ca.* 0.5 or 0.6~0.7 V, respectively). Therefore, both carbenes and indoles present a much stronger net electron releasing ability than their parent isocyanides.

The Cr carbene complexes of this study present oxidation potentials which are somewhat lower than those reported [4] for other aminocarbenes of the type $(CO)_5Cr[C(NRR')Y]$ (R,R' or Y = H, alkyl, aryl or related group), thus suggesting a considerable weight of the phosphonium form with localization of a negative charge at the metal.

Both the carbene and the indole complexes are susceptible to protonation, the former possibly at the β-carbon of the carbene ring and the latter at the indole-nitrogen atom (the latter giving products with the type **D** structure).

The products were generated *in situ* (by addition of $[Et_2OH]BF_4$, 1 equiv., to the electrolytic solutions of the corresponding parent complexes in NCMe) and detected by their C.V. anodic waves which replaced those of the parent complexes, which were regenerated upon subsequent addition of a base (such as $[n\text{-}Bu_4N]OH$). Thus, in the case of the protonated-indole/indole system, a reversible protonation/deprotonation process was observed, which, however, does not occur in the case of the protonated carbene complexes. The oxidation potentials of the protonated carbene or protonated indole complexes are considerably more anodic (1.1-0.9 V) than those of the parent non-protonated compounds (0.9-0.5 V), but they still lie below those exhibited by the starting isocyanide complexes (1.3-1.1 V). Therefore the phosphonium-functionalized isocyanide ligands behave as stronger net electron acceptors than the protonated carbene or the protonated indole ligands.

2.2. ESTIMATE OF THE ELECTROCHEMICAL P_L LIGAND AND METAL-CENTRE β PARAMETERS

The P_L ligand parameter, defined [8] by equation (1), has been proposed as a measure of the net electron σ-donor minus π-acceptor ability of a ligand: the greater is this character of a ligand, the lower is its P_L value.

$$P_L = E_{1/2}{}^{ox}[Cr(CO)_5L] - E_{1/2}{}^{ox}[Cr(CO)_6] \qquad (1)$$

From the application of this equation to $[Cr(CO)_5L]$ for which we have measured $E_{1/2}{}^{ox}$ and considering the known [6] value ($E_{1/2}{}^{ox}$ = 1.50 V) for $[Cr(CO)_6]$, we have estimated the P_L parameter for a number of our ligands (Table).
When the complexes with the $\{Cr(CO)_5\}$ site were not available, the P_L values of the ligands have been estimated as follows. We have considered the analogous complexes with the $\{Mo(CO)_5\}$ or the $\{W(CO)_5\}$ centres and the expected linear dependences of the oxidation potential on P_L - eq. (2), where E_S and β are the electron-richness and the polarisability of the corresponding binding metal site M_S, the former defined by eq. (3) [8].

$$E_{1/2}{}^{ox}[M_SL] = E_S + \beta \cdot P_L \qquad (2)$$

$$E_S = E_{1/2}{}^{ox}[M_S(CO)] \qquad (3)$$

E_S is known for these Mo and W sites since $E_{1/2}{}^{ox}[M(CO)_6]$ = 1.50 V (M = Cr, Mo or W) [6], but the polarisability (β) has not been reported. Therefore we have estimated this parameter by plotting $E_{1/2}{}^{ox}$ (or $E_{p/2}{}^{ox}$) of the members of each series of the complexes $[M(CO)_5L]$ (M = Mo or W) *versus* the corresponding P_L values for the cases they were known from this study or from literature data [6,7].
The linear plots are expressed by equations (4) and (5) (with a correlation coefficient of 0.99) which give the following β values for the metal sites: 0.86 or 0.90 for $\{Mo(CO)_5\}$ or $\{W(CO)_5\}$, respectively.
Moreover, from these expressions, the P_L values have been estimated for other ligands, by taking also into account the oxidation potentials, measured in this study, of their complexes

with $\{M(CO)_5\}$ (M = Mo or W).

$$E_{1/2}^{ox}[Mo(CO)_5L] = 1.44 + 0.86\ P_L \qquad (4)$$

$$E_{1/2}^{ox}[W(CO)_5L] = 1.52 + 0.90\ P_L \qquad (5)$$

The phosphonium-functionalized isocyanides present P_L values (-0.36 to -0.28 V) which are only slightly higher than those of other non-functionalized aromatic isocyanides, in agreement with the long distance of the phosphonium group to the metal. In addition, the carbene ligands behave as much stronger net electron-donors (P_L in the range -0.87 to -0.79 V), even slightly stronger than other related aminocarbenes (see above); they present net electron-donor abilities close to those of anionic ligands such as NCS^- (P_L = -0.88 V [8]). Indoles are even more effective net electron-donors with P_L values in the range -1.0 to -0.92 V, as known [8] for the CN^- ligand (P_L = -1.00 V).
Moreover, either the protonated indoles or the protonated carbenes are much weaker net electron-donors (P_L values in the range -0.44 to *ca.* -0.6 V).

2.3. ESTIMATE OF THE ELECTROCHEMICAL E_L LIGAND PARAMETER

A distinct electrochemical ligand parameter, E_L, has been proposed recently [9] and used to predict the redox potential of complexes by assuming an additive contribution of all their ligands. The empirically derived relationship can be given by eq. (6), expressed in volt *vs.* N.H.E., where S_M and I_M depend upon the metal and redox couple, the spin state and stereochemistry.

$$E^{ox} = S_M[\Sigma E_L] + I_M \qquad (6)$$

Moreover, the linear relationship (7) has been experimentally observed [9] (for a considerable number of ligands, but not for CO) between the E_L and the P_L parameters, thus allowing us to estimate one of them from the knowledge of the other one.

$$P_L = 1.17\ E_L - 0.86 \qquad (7)$$

Comparisons with other ligands with known [9] E_L values can then be made, similarly to those based on the P_L parameter, but the conclusions are not entirely identical in view of possible deviations from eq. (7).
However, for ligands with an extensive π-influence on the HOMO energy (CO or isocyanides), it has been recognized [9] the need for the introduction of corrections to E_L, by replacing eq. (7) by the corrected form (8), where *m* and *m'* are the numbers of ligands that interact directly with the HOMO (CO and CNR), whereas *x* and *y* are the corresponding corrections. Their values have been quoted [9] for CO and for some alkyl or aryl isocyanides ligating the $\{Cr(CO)_5\}$ centre, but are still unknown for the analogous Mo or W complexes.

$$E^{ox} = S_M[\Sigma E_L + m\,x + m'\,y] + I_M \qquad (8)$$

By applying eq. (8) to our $\{Cr(CO)_5\}$ isocyanide complexes, taking into account the known [9] values of S_M (0.52), I_M (-1.75 V *vs.* N.H.E.) and x (0.15 for CO), and assuming that three CO ligands (m = 3) and one isocyanide ligand (m' = 1) interact directly with the HOMO, we have estimated the correction term (y) for some isocyanides and the values (0.19 to 0.27 V) are comparable with that reported [9] for CNPh (0.19 V) at the same Cr site and appear to increase with the π-acceptance of the phosphonium group in accord with the expected relevant role of π-effects on such a correction.

In this study are quoted, for the first time, the P_L and E_L electrochemical ligand parameters for phosphonium-functionalized isocyanides and derived carbene, indole and corresponding protonated ligands. Therefore their net electron donor properties have been investigated and compared with those of other related ligands. The polarisability β has also been estimated for the metal centres $\{M(CO)_5\}$ (M = Mo or W), but we did not obtain a sufficiently wide ΣE_L range of values to allow us to estimate the unknown values of S_M and I_M for $\{W(CO)_5\}$.

Moreover, the interconversion of those types of species has been conveniently monitored by C.V., by following their reactions with base or acid.

3. Acknowledgements

This work has been partially supported by CNR (Italy)/INIC (Portugal) co-operative programme to R.A.M and A.J.L.P. and also by MURST (R.A.M.) and JNICT (A.J.L.P.).

4. References

[1] For an overall review, see: Pombeiro, A.J.L. (1992) 'Electrochemical Behaviour of Complexes Derived from the Activation of Alkynes, Isocyanides and Nitriles', in A.J.L. Pombeiro and J.A. McCleverty (eds.), Molecular Electrochemistry of Inorganic, Bioinorganic and Organometallic Compounds, NATO ASI Series, Kluwer Academic Publishers, Dordrecht.

[2] Lemos, M.A.N.D.A.; and Pombeiro, A.J.L. (1988), J. Organometal. Chem. 356, C79, and references cited therein.

[3] Connor, J.A.; Jones, E.M.; McEwen, G.K.; Lloyd, M.K.; and McCleverty, J.A. (1972), J.Chem. Soc., Dalton Trans., 1246.

[4] Lloyd, M.K.; McCleverty, J.A.; Orchard, D.G.; Connor, J.A.; Hall, M.B.; Hillier, I.H.; Jones, E.M.; and McEwen, G.K. (1973), J. Chem. Soc., Dalton Trans., 1743.

[5] El-Shihi, T.; Sijlmuller, F.; Herrmann, R.; Carvalho, M.F.N.N.; and Pombeiro, A.J.L. (1987), J. Organometal. Chem. 335, 239.

[6] Pickett, C.J.; and Pletcher, D. (1975), J. Chem. Soc., Dalton Trans., 879.

[7] Bond, A.M.; Bowden, J.A.; and Cotton, R. (1974), Inorg. Chem. 13, 602.

[8] Chatt, J.; Kan, C.T.; Meigh, G.J.; Pickett, C.J.; and Stanley, D.R. (1980), J. Chem. Soc., Dalton Trans., 2032.

[9] Lever, A.B.P. (1990), Inorg. Chem. 29, 1271.

REDOX PROPERTIES AND LIGAND EFFECTS FOR THE HYDRIDO-TECHNETIUM-DINITROGEN, -CARBONYL AND -ISOCYANIDE COMPLEXES *trans*-[TcH(L)($Ph_2PCH_2CH_2PPh_2$)$_2$] (L = N_2, CO or CNR)

YU WANG and ARMANDO J. L. POMBEIRO
Centro de Química Estrutural, Complexo I,
Instituto Superior Técnico, 1096 Lisbon Codex, Portugal
L. KADEN and M. WAHREN
KAI e. v., WIP Isotope Chemistry, Permoserstr. 15, 0-7050
Leipzig, Germany

ABSTRACT. The redox properties of trans-[TcH(L)(dppe)$_2$] (L = CO, N_2, $CNBu^t$ or CNC_6H_{11}; dppe = $Ph_2PCH_2CH_2PPh_2$) have been studied by cyclic voltammetry and controlled potential electrolysis at a Pt electrode in 0.2 M [Bu_4N][BF_4] / THF. They undergo a single-electron quasi-reversible or irreversible anodic process at an oxidation potential which relates linearly with that exhibited by the corresponding rhenium complexes trans-[ReCl(L)(dppe)$_2$]. The electron-richness (E_s) and polarisability (β) parameters have been obtained for the {TcH(dppe)$_2$} centre (0.34 V vs. SCE and 4.0, respectively), whereas the isocyanide ligands appear to present P_L values ca. 0.3 V above those exhibited at {Cr(CO)$_5$}. A correction for the " normal " E_L values is also suggested for these ligands at the Tc site.

1. INTRODUCTION

The electrochemical investigation of technetium coordination compounds has attracted considerable attention[1, 2] in light of the application of complexes with the isotope Tc-99m in diagnostic nuclear medicine and in view of the expected important role of the redox properties in their biological activity.

The electrochemical studies have focused on the Tc (IV / III) Tc (III / II), or Tc (II / I) redox pairs, the reduction potentials of which have been discussed [1, 2] in terms of ligand and metal site effects. In the present study we extend this investigation to the oxidation of the complexes trans-[TcH(L)(dppe)$_2$] (1; L = CO,

A. J. L. Pombeiro and J. A. McCleverty (eds.),
Molecular Electrochemistry of Inorganic, Bioinorganic and Organomettallic Compounds, 63–67.

N_2, $CNBu^t$ or CNC_6H_{11}; dppe = $Ph_2PCH_2CH_2PPh_2$), which present a low (I) metal oxdation state and are expected to exhibit a high electron-rich character. This would provide an insight into the possible effects of such electronic properties of the metal site on the coordination of their ligands and on the redox behaviour of their complexes.

2. RESULTS AND DISCUSSION

The complexes trans-[TcH(L)(dppe)$_2$] (1; L = CO, N_2, $CNBu^t$ or CNC_6H_{11}), which have been prepared according to literature methods [3], in 0.2 M [Bu_4N][BF_4] / THF undergo, by cyclic voltammetry at a Pt electrode, a single-electron quasi-reversible (1; L = $CNBu^t$ or CNC_6H_{11}) or irreversible (1; L = CO or N_2) anodic process at an oxidation potential considerably low ($E_{p/2}^{ox}$ in the range 0.34 to -0.03 V vs. SCE) (Table) in agrement with the high electron-rich character of the metal site.

TABLE. Cyclic voltammetric data for trans-[TcH(L)(dppe)$_2$] (1) and corresponding P_L and E_L ligand parameters for the ligand L

L	$E_{p/2}^{ox}$ (a)		P_L /V		E_L / V vs NHE	
	V vs (SCE)	V vs (NHE)	Corrected	Uncorrected [4 ,5]	Corrected	Uncorrected [8]
CO	0.34	0.59	0	0	0.99	0.99
N_2	0.06	0.31	-0.07	-0.07	0.68	0.68
$CNBu^t$	- 0.02 (0.00)	0.23 (0.25)	-0.09	-0.44	0.59	0.36
CNC_6H_{11}	-0.03 (0.01)	0.22 (0.26)	-0.09	-0.42	0.58	0.32

(a) values measured in 0.2 M [Bu_4N][BF_4] / THF, at a Pt electrode, at 100 mv s^{-1}, by using the redox couple [Fe(η5-C_5H_5)$_2$]$^{0/+}$ ($E_{1/2}^{ox}$=0.55 V vs. SCE) as internal reference. For the reversible process, $E_{1/2}^{ox}$ values are given in parentheses

In fact, the " electron-richness " (E_s) of the {TcH(dppe)$_2$} metal centre, defined [4] as the oxidation potential of its carbonyl complex (ca. 0.34 V vs. SCE , as measured in this study), is substantially higher than that known [5] for the related site

$\{ReCl(dppe)_2\}$ (E_s = 0.68 V) and well above those reported [4], e.g., for $\{FeH(dppe)_2\}^+$ (1.04 V) or $\{ Cr(CO)_5 \}$ (1.50 V). However, the technetium centre is less electron-rich than some related Mo(0) sites, such as $\{ MoL(dppe)_2\}$ (E_s = -0.11 or -0.13 V for L = CO or N_2, respectively [4]).

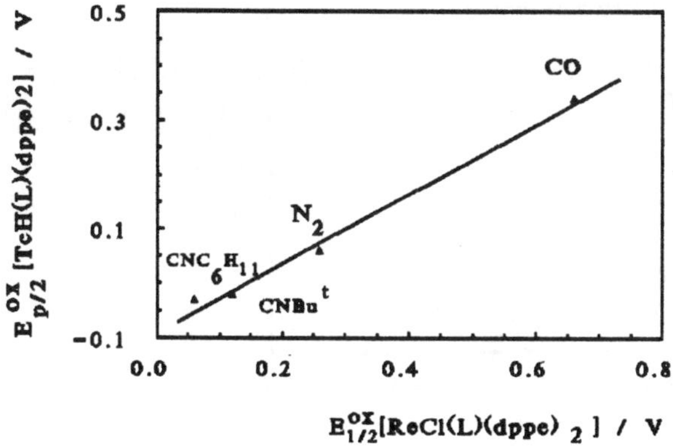

Figure 1. Plot of $E_{p/2}^{ox}$ for trans-[TcH(L)(dppe)$_2$] (1) versus $E_{1/2}^{ox}$ for trans-[ReCl(L)(dppe)$_2$] (values in volt vs. SCE)

In accord with the higher electron-rich character of the technetium centre relative to $\{ ReCl(dppe)_2 \}$, the oxidation potentials of complexes (1) are lower (by ca. 0.3 - 0.1 V) than those for the corresponding compounds trans-[ReCl(L)(dppe)$_2$] [5]. Moreover, since the technetium(I) complexes are expected to present substantially higher oxidation potentials than their analogous rhenium(I) compounds -- since Tc species, in the oxidized form, are reduced at higher potentials than their Re analogues, e.g., by ca. 0.16 V for a number of M(II/I) couples [6] -- one can anticipate that the still unknown complexes trans-[TcCl(L)(dppe)$_2$] should exhibit considerably higher oxidation potentials (by ca. 0.5-0.2 V) than the corresponding hydride complounds (1).

For complexes (1), the oxidation potential increases, as expected, with the net electron acceptor ability of the ligand L, i.e., in the order CNR < N_2 < CO, following the trend known [5] for trans-[ReCl(L)(dppe)$_2$] (a linear correlation is observed between their corresponding oxidation potentials --- Fig. 1). However, from the plot of $E_{p/2}^{ox}$ vs the P_L ligand parameter (a mesure of the net electron π-acceptor minus σ-donor character of a ligand [4]) (Fig. 2), one observes a clear deviation from the linearity, and, as previously discussed [5] for those Re complexes, this suggests that the isocyanides at the high electron-rich centre $\{ TcH(dppe)_2 \}$ behave as stronger net electron acceptors than expected on the basis of the comparison with other ligands. Therefore, their " normal " or " low " P_L values (those observed at much less electron-rich sites, e.g., { $Cr(CO)_5$ } which has been used to define P_L) should be corrected, in the current case by adding ca. 0.34 V. This correction is even slightly higher than that reported (ca. 0.26 V) [5, 7] for the slightly less electron-rich { $ReCl(dppe)_2$ } centre.

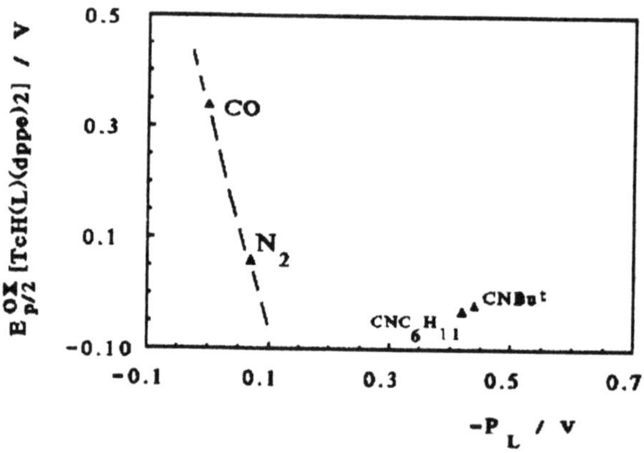

Figure 2. Plot of $E_{p/2}^{ox}$ (V vs. SCE) for trans-[$TcH(L)(dppe)_2$](1) verus - PL (V)

In addition, from the plot of Fig. 2, and under the abovementioned assumptions, the polarisability (β) of the centre { $TcH(dppe)_2$ } can be estimated as ca. 4.0 (slope of the line), a value which is extremely high [even higher than that known, 3.4 [5], for { $ReCl(dppe)_2$ }] , indicating that the redox potential (and the HOMO) in the Tc series is highly sensitive to a ligand change. This is consistent with the high reported S_M value for the Tc(II/I) couple [1] (see below), and also suggests that the hydride ligand presents a low ability to buffer electronic changes at the metal upon a variation of the ligand in trans position.

Linear correlations between the Tc reduction potentials and the summation of the electrochemical ligand additive parameter (ΣE_L) [8] have been recognized [1] for the Tc(II,I) couple --- apart from the higher oxidation state Tc(III,II) and Tc(IV,III) redox pairs - with observed least-squares slope and intercept values [1, 9] of S_M = 1.40 and I_M = -2.08 V . Our experimental results [$E_{p/2}^{ox}$ vs. NHE (Table)], although considerably away from this correlation (e.g., with a difference of +0.34 or -0.24 V between the expected and the observed redox potential for the CO or CNC_6H_{11} complex, respectively), are still within the scatter of the data used (over a wider ΣE_L range) to determine [1] such a relationship.

A closer inspection of our data (plot of $E_{p/2}^{ox}$ V vs. NHE versus ΣE_L) would show that the isocyanide E_L parameter should probably be corrected, by adding ca. 0.24 V to the " normal " values, in order to get an expected linear plot. This is consistent with the above discussion on P_L for the isocyanide ligands and conceivably results from the substantial participation of an isocyanide π^* orbital in HOMO. Moreover, the need for a related correction has also been recognized [8] for other metal centres.

ACKNOWLEDGEMENTS

This work has been partially supported by the Orient Foundation, JNICT and INIC (Portugal)

3. REFERENCES

[1] Lu, J.; Yamano, A.; and Clark, M. J. (1990), Inog. Chem. 29, 3483; and refs. cited therein.
[2] Kirchhoff, J. R.;Heneman, W. R.; and Deutsch, E. (1988), Inorg. Chem. 27, 3608; and refs. cited therein.
[3] (a) Kaden, L.; Lorenz, B.; Schmidt, K.; Sprinz, H.; and Wahren, M. (1979), Z. Chem. 19, 305.
(b) Kaden, L.; Findeisen, M.; Lorenz, B.; Schmidt, K.; and Wahren, M. (1992), Inorg. Chim. Acta, 193, 213.
[4] Chatt, J.; Kan, C. T.; Leigh, G. J.; Pickett, C. J.; and Stanley, D. R. (1980), J. Chem. Soc., Dalton Trans., 2032
[5] Pombeiro, A. J. L.; Pickett, C. J.; and Richards, R. L. (1982), J. Organometal. Chem. 224, 285.
[6] Kirchoff, J. R.; Heineman, W. R.; and Deutsch, E. (1987), Inorg. Chem. 26, 3108.
[7] Pombeiro, A. J. L. (1985), Inorg. Chim. Acta 103, 95.
[8] Lever, A. B. P. (1990), Inorg. Chem. 29, 1271.
[9] Lever, A. B. P. (1991), Inorg. Chem. 30, 1980.

ELECTROCHEMICAL STUDIES OF COORDINATION COMPLEXES OF OSMIUM

K.J. TAYLOR AND L.J. YELLOWLEES
Department of Chemistry
University of Edinburgh
West Main Road
Edinburgh EH9 3JJ, Scotland, U.K.

ABSTRACT. Several electrochemical techniques have been employed in the study of the transition metal complexes $[OsCl_{6-n}py_n]^{z+}$, n = 0-6., py = pyridine. Cyclic voltammetry has confirmed that $[OsCl_6]^{2-}$ undergoes an EC process at room temperature in CH_2Cl_2/py to form $[OsCl_5py]^{2-}$. The rate determining step is chloride loss from $[OsCl_6]^{3-}$ and further investigations of other hexahalide osmate(III) species shows the rate of halide loss increases $[OsCl_6]^{3-}<[OsBr_6]^{3-}<[OsI_6]^{3-}$. Bulk electrogeneration experiments at precise potentials and temperatures result in formation of all members of the series $[OsCl_{6-n}py_n]^{z+}$, n = 0-6, except for the *fac*-isomer of the trisubstituted species. Each member of the series exhibits a rich redox chemistry and the voltammetric data for 22 closely related couples is presented. The experimental $E_{1/2}$ values are compared with predicted half-wave potentials from two models with satisfactory agreement for the Os(IV)/Os(III) and Os(V)/Os(IV) couples. We conclude that the measured $E_{1/2}$ values for the Os(III)/Os(II) couple is dependent on n, the isomeric form and the geometric arrangement of the pyridine ligands.

1. Introduction

Electrochemical measurements on transition metal complexes are performed for a wide variety of reasons. In this paper we discuss the synthesis of new species by electrogeneration, the measurement of rates of chemical reaction by cyclic voltammetry and the parameterisation of redox potentials within the related series of complexes $[OsX_{6-n}py_n]^{z+}$, where X = Cl, Br, I, py = pyridine and n = 0-6. A further goal realised by this work was to increase the limited number of $E_{1/2}$ values for mononuclear osmium complexes reported in the literature[1].

2. Experimental

$K_2[OsCl_6]$ (Johnson Matthey) was metathesised to the tetrabutylammonium salt using $[^nBu_4N]Cl$. The py substituted complexes were prepared by redox-induced subsitution reactions of $[^nBu_4N]_2[OsCl_6]$ in CH_2H_2/py as indicated in Table 3. CH_2Cl_2 was purified by storing over KOH for two weeks followed by reflux and distillation over P_2O_5. All other solvents were used as supplied. The inert electrolyte $[^nBu_4N]BF_4$ was recrystallised from H_2O/CH_3OH and was used in

A. J. L. Pombeiro and J. A. McCleverty (eds.),
Molecular Electrochemistry of Inorganic, Bioinorganic and Organomettallic Compounds, 69–75.

all electrochemical experiments at a concentration of 0.5M.

Redox studies and EC reactions were performed in a conventional 3-electrode cell, a Pt micro-working electrode, a Pt counter electrode and a Ag/AgCl reference electrode against which the ferrocene/ferricinium couple was measured at + 0.56V. Electrosynthesis experiments utilised an H-type cell with a Pt basket w.e., Pt gauze c.e. (doubly fritted from the w.e. compartment) and r.e. as above. All solutions were purged with Ar for 20minutes prior to study. Cells were thermostated using a Haake Model Q thermostated bath circulator. Electrochemical measurements were carried out using a Hi-Tek Instruments potentiostat DT2101 coupled to a Hi-Tek waveform generator PPR1 recorded on a Hewlett Packard 7045A X-Y recorder. Positive feedback ohmic drop compensation was applied for all cyclic voltammograms recorded. Kinetic convolution mechanistic studies were performed using a PAR 273 potentiostat/galvanostat coupled to the Condecon^{TM} software package. The scan rate used for all reported results was $100 mVs^{-1}$.

3. Results and Discussion

3.1 MECHANISTIC MEASUREMENTS

The redox behaviour of $[OsCl_6]^{2-}$ in pyridine as measured by cyclic voltammetry is temperature dependent as shown in Figue 1. Thus at 240K the Os(IV)/Os(III) electron transfer step is both chemically and electrochemically reversible whereas at 290K the electron tranfer step is followed by an irreversible chemical reaction, an EC process.

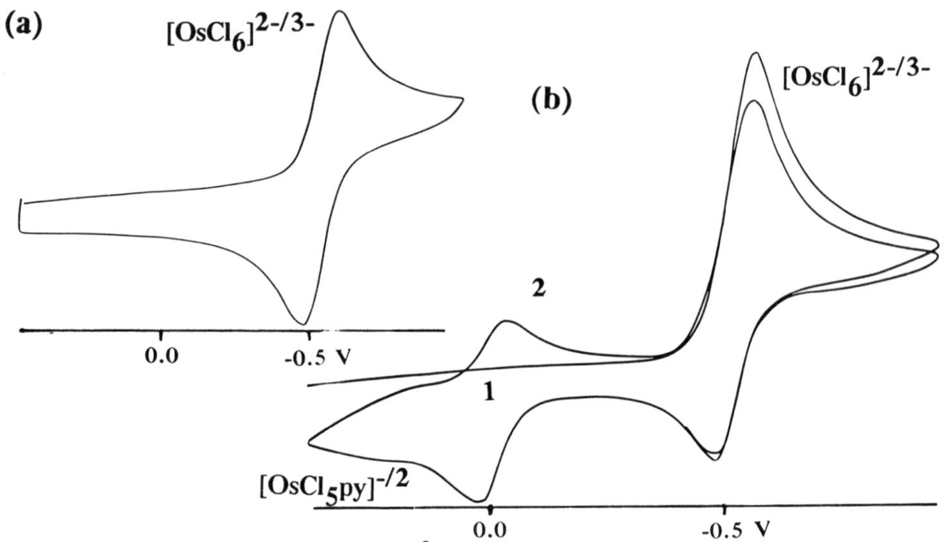

Figure 1. Cyclic voltammograms of $[OsCl_6]^{2-}$ in py/0.1M$[^nBu_4N]BF_4$ at (a) 240K and (b) 290K, 1 = first scan, 2 = second scan

The chemical reaction following the reduction step results in the formation of the daughter product $[OsCl_5py]^{2-}$ which exhibits it's own redox behaviour as shown in Figure 1. The $E_{1/2}$ values for a variety of $[OsCl_5L]^{2-/-}$ complexes are given in Table 1.

TABLE 1. $E_{1/2}$ values for $[OsCl_5L]^{2-/-}$ complexes in $CH_2Cl_2/0.5M[^nBu_4N]BF_4$ at 290K

L	$E_{1/2}{}^a$(V)
Cl^b	-0.56
py	-0.02
PhCN	+0.16
DMF	+0.20
CO	+1.06

(a) $E_{1/2} = (E_{pa}-E_{pc})/2$
(b) $E_{1/2}$ value for $[OsCl_6]^{3-/2-}$

The half-wave potential for the Os(IV)/Os(III) couple in the monosubstituted species shifts to more positive potentials as the π-donating Cl^- ligand is replaced by π-acceptor ligands.

The EC mechanism proposed in **(1)** was investigated by cyclic voltammetry following the

$$[OsCl_6]^{2-} + e^- \rightleftharpoons [OsCl_6]^{3-} \xrightarrow[k_f]{L} [OsCl_5L]^{2-} + Cl^- \qquad \textbf{(1)}$$

method of Nicholson and Shain[2] (NS) and by kinetic convolution methods (KC). The value of the pseudo-first order rate constant (k_f) is found to be independent of L and hence the rate determining step for the substitution reaction appears to be chloride loss from $[OsCl_6]^{3-}$. Both methods gave similar values for k_f - see Table 2. The rate constants for the loss of halide from the corresponding hexahalide osmate(III) parent complexes are also given in Table 2.

TABLE 2. Pseudo-first order rate constants, $k_f(s^{-1})$ for the reductive substitution of $[OsX_6]^{2-}$ in CH_2Cl_2 with a ten molar excess of pyridine

Temp(K)		k_f(NS)		k_f(KC)
X =	Cl	Br	I	Cl
273	0.09	0.20		0.08
283	0.17	0.40	18.1	0.16
293	0.31	0.71		0.34
303	0.54			

The rate constant, k_f, for halide loss increases $[OsCl_6]^{3-}<[OsBr_6]^{3-}<[OsI_6]^{3-}$ which is the expected order since iodide is a better leaving group than chloride, and iodide is more easily solvated in CH_2Cl_2 than bromide or chloride.

3.2 ELECTROSYNTHESIS

The monosubstituted species $[OsCl_5py]^{2-}$ can be efficiently prepared from $[OsCl_6]^{2-}$ by bulk electrolysis at -0.7V in pyridine or CH_2Cl_2 doped with a stoichiometric amount of pyridine. Note that the rich redox chemistry of both the starting and daughter complexes limits the choice of chemical reducing agents. $[OsCl_5py]^-$ can be prepared by heating $[OsCl_6]^{2-}$ in refluxing pyridine but the reaction is not quantitative and the desired product is difficult to isolate. The rate constant for production of the monosubstituted complex from the +4 oxidation state is 10^{10} times slower than from the +3 oxidation state. Thus the electosynthetic preparative route is the preferred method for production of the monosubstituted derivative.

If a solution of $[OsCl_5py]^{2-}$ in CH_2Cl_2/py(7%) is stirred under N_2 for 96 hours then $[OsCl_4py_2]^-$ is formed in quantitative amounts. The disubstituted species is produced in both the *cis*- and *trans*-isomeric forms. The ratio of *trans:cis* isomers in the final solution is governed by the temperature, thus at 300K the kinetically favoured *trans*-isomer is present in a nine-fold excess over the *cis*-isomer whereas at 330K there is a 50:50 mixture of the two forms. To date it has been impossible to convert one isomeric form to the other. The *trans*- and *cis*-$[OsCl_4py_2]^-$ can be seperated by bulk electrooxidation at +0.80V, when the non-polar *trans*-$[OsCl_4py_2]$ precipitates as an orange solid leaving the yellow *cis*-isomer in solution. *Trans*-$[OsCl_4py_2]^-$ was recrystallised from a CH_2Cl_2/CH_3OH mixture and *cis*-$[OsCl_4py_2]$ from slow diffusion of diethylether into a CH_2Cl_2 solution. Independent single crystal X-ray diffraction studies of both isomers has been undertaken[2] and the structure of both forms is shown in Figure 2.

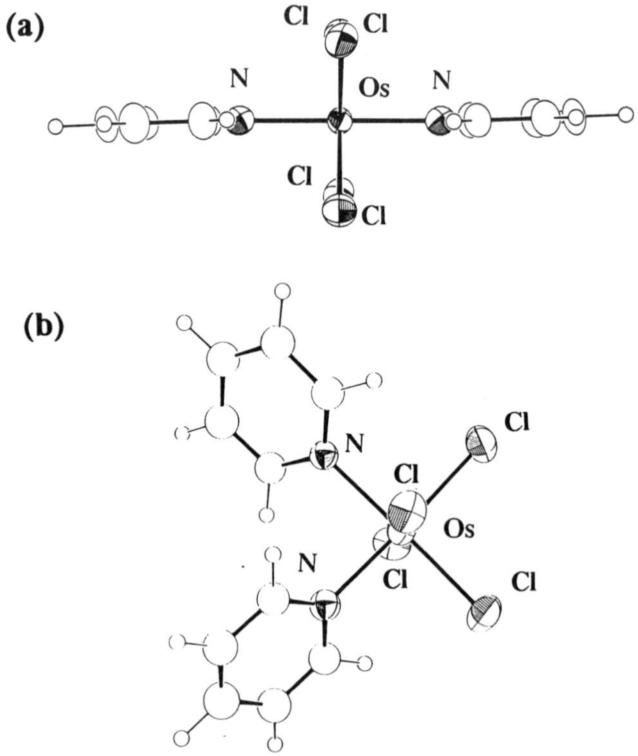

Figure 2. Structure of (a) *trans*-$[OsCl_4py_2]^-$ and (b) *cis*-$[OsCl_4py_2]$

Note that in the *trans*-isomer the py rings are coplanar but in the *cis*-isomer the planes of the two pyridine rings are at an angle of 59.62^o, presumably due to the interaction of the α-hyrogens on the two *cis*-pyridine rings.

All members of the series $[OsCl_{6-n}py_n]^{z+}$, n = 0-6 can be prepared by redox-induced substitution reactions of $[OsCl_6]^{2-}$ by alteration of temperature and electrosynthesis potential as described in

Table 3. Electrogeneration potentials were generally 150mV more negative than the $E_{1/2}$ values quoted in the table. Individual members of the series, n = 0-6, have been isolated and fully characterised by C,H,N analysis, FAB mass spectroscopy, ir spectroscopy and single crystal X-ray determinations of *trans*-$[OsCl_4py_2]^-$, *cis*-$[OsCl_4py_2]$, *mer*-$[OsCl_3py_3]$, *trans*-$[OsCl_2py_4]$ and $[Ospy_6]^{2+}$ have also been successfully undertaken[3,4].

TABLE 3. Synthetic details and electrochemical data for the series $[OsCl_{6-n}py_n]^{z+}$

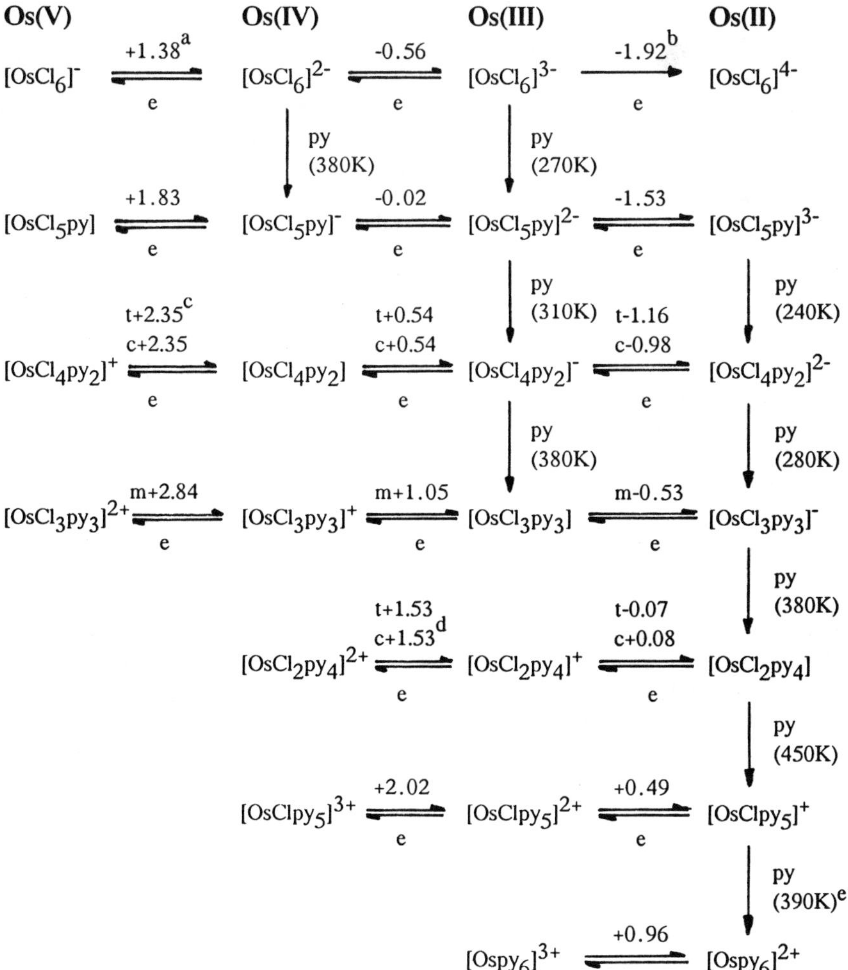

(a) $E_{1/2}$ data in $CH_2Cl_2/0.5M[^nBu_4N]BF_4$

(b) irreversible process, cathodic peak quoted

(c) t = *trans*, c = *cis*, m = *mer*

(d) *cis*-$[OsCl_2py_4]$ is prepared by the redox-induced substitution reaction of $[Ospy_6]^{3+}$ in $CH_2Cl_2/0.5M[^nBu_4N]Cl$

(e) substitution reaction done in the presence of a chloride abstractor such as Na^+

The voltammetric data presented for the $[OsCl_{6-n}py_n]^{Z+}$ series is one of the most complete sets of data available for a closely related family of complexes. There are 31 different species identified all of which have been studied by spectroelectrochemical techniques[3] with half-wave potentials ranging from +2.84V to -1.92V. Note the general trend of the redox potentials, as π-donor chloride ligands are replaced by π-accepting pyridine ligands so corresponding metal-based Os redox couples are shifted to more positive potentials. Duff and Heath have recently published the voltammetric data for the seven membered Ru series $[RuCl_{6-n}PhCn_n]^{Z+}$ but they detail only the *trans*-di and tetra substituted complexes[5].

3.3 LIGAND PARAMETERISATION

Prediction of the precise values of redox couples in transition metal complexes with an octahedral ligand set is quantified by two current models[6,7]. Our extended set of redox couples for a related series of complexes is an excellent test of the applicability of the two models.

Lever's **Ligand Parametrisation** model[6] will estimate the value of a particular redox couple using equation (2) where E_{calc} is the calculated $E_{1/2}$ value, S_M and I_M are constants for a specific

$$E_{calc} = S_M[\Sigma E_L] + I_M + C_{ref} \tag{2}$$

redox couple of the complexes of a given metal with the same stereochemistry and spin state, ΣE_L is the sum of the electronic contributions of each ligand around the metal and C_{ref} is a reference electrode correction term (C_{ref} for our experimental setup is -0.105). Inserting the values calculated by Lever for Os(III)/Os(II) couples (S_M = 1.01, I_M = -0.40) and the ligand parameter values for Cl^- and py (E_{Cl^-} = -0.24V, E_{py} = 0.25V) into equation 2 gives a good correlation between the calculated half-wave potentials and the experimental values. Thus, E_{calc} is always observed to fall within 200mV of E_{expt}. There are no values currently available for S_M and I_M for Os(IV)/Os(III) and Os(V)/Os(IV) couples. The results presented in Table 3 yield the following values: S_M = 1.05, I_M = 1.095 for Os(IV)/Os(III) couples and S_M = 1.00, I_M = 2.905 for Os(V)/Os(IV) couples. We find that Lever's model gives a good first approximation for a half-wave potential, however it does not take into account the isomeric form of the complex and as we observe in Table 3 different isomers show different redox potentials for the Os(III)/Os(II) couple.

Bursten's **Ligand Additivity** model[7] addresses the problem of the dependence of the redox potential on the isomeric form. Ligand additivity arguments predict that a plot of $E_{1/2}$ (for the Os d^5/d^6 couple) against n would be linear for n = 0,1,*trans*-2 and for n = *trans*-2,*mer*-3,4,5,6 but that the linear relationships would have different slopes. Furthermore, the *cis*- and *trans*-disubstituted complexes should exhibit very different electronic behaviour but the *cis*- and *trans*-tetrasubstituted species should have indistinguishable redox potentials. Our results confirm the first linear relationship but not the second and agree with the predicted $E_{1/2}$ behaviour for the *cis*- and *trans*-2 isomers but not for the *cis*- and *trans*-4 isomers. We suggest that the breakdown of the argument results from steric considerations. Bursten's model assumes that the π^* orbitals on the π-acceptor ligands will be aligned in the most favourable electronic positions. However in our case this is not always possible for large planar pyridine ligands (see figure 2) and hence as the number of coordinated py ligands increases so the deviation from the predicted behaviour also increases. We therefore propose that $E_{1/2}$ values for Os(III)/Os(II) couples within the series $[OsCl_{6-n}py_n]^{Z+}$ are dependent on n, the isomeric form and the geometric arrangement of the pyridine ligands.

4. Acknowledgements

We thank SERC (KJT) and the University of Edinburgh for support.

5. References

1. (a) W. Preetz and W. Fichtner, Z. Anorg. Allg. Chem., (1978), 445, 5 (b) R.A. Cipriano, W.Levason, R.A.S. Mould, D. Pletcher, M. Webster, J. Chem. Soc., Dalton Trans., (1990), 339 (c) R.A. Cipriano, W. Levason, R.A.S. Mould, D. Pletcher, M. Webster, J. Chem. Soc., Dalton Trans., (1990), 2609.
2. R.S. Nicholson and I. Shain, Anal. Chem., (1964), 36, 706.
3. K.J. Taylor, PhD Thesis, University of Edinburgh (1990).
4. A.J. Blake, G.A. Heath, G. Smith, L.J. Yellowlees, D.W.A. Sharp, Acta Crystallogr., Sect. C, (1988), 44, 1836.
5. C.M. Duff, G.A. Heath, Inorg. Chem., (1991), 30, 2528.
6. A.B.P. Lever, Inorg. Chem., (1990), 29, 1271.
7. B.E. Bursten, M.R. Green, Prog. Inorg. Chem., (1988), 36, 474.

MOLYBDENUM MACROCYCLES AND MOLECULAR ELECTRONICS

J.A.McCLEVERTY
School of Chemistry
University of Bristol
Cantocks Close
Bristol BS8 1TS
UK

ABSTRACT. The 16-electron six-coordinate nitrosyls[M(NO)L*XY], where M = Mo or W, L* = tris(3,5-dimethylpyrazolyl)borate; X, Y = halide, alkoxide, aryloxide, alkyl or arylamide, thiolate) undergo one-electron reduction whose potential is influenced by M, X and Y, and by the substituents attached to X and Y. These species have been incorporated into cyclic polyethers, their reduction potential being sensitive to added alkali metals. The electronegative {M(NO)L*Cl} may be incorporated into dipolar polarisable molecules capable of exhibiting, as solids, strong non-linear optical effects, and they may also be attached to the periphery of porphyrins, causing efficient charge-separation.

1. Introduction

Halogenation of [Mo(CO)$_2$(NO)Cp] affords the halide-bridged [{Mo(η-C$_5$H$_5$)X(NO)}$_2$(μ-X)$_2$] in which each metal atom attains the normal 18-electron configuration. The chemistry of these species is dominated by bridge-splitting reactions and some of the complexes obtained from such reactions are redox-active. In a search for ligands equivalent to the cyclopentadienide ion but having substantially greater bulk, we chose the tris(3,5-dimethylpyrazolyl)borate, [HB(3,5-Me$_2$pyz)$_3$]$^-$, L*$^-$, which had earlier been discovered by Trofimenko. He had pointed out that this and closely related monoanionic tripodal ligands could be regarded as analogues of $C_5H_5^-$, and he was able to indicate that significant "half-sandwich" chemistry parallel to that of the cyclopentadienyl system could be developed [1]. By halogenation of [M(CO)$_2$(NO)L*], we were able to prepared ***monomeric*** species, [M(NO)L*X$_2$], (M = Mo or W; X = Cl, Br or I) in which the metal is formally coordinatively unsaturated (16-electrons). We assumed that these species did not dimerise because of the steric hindrance presented by the methyl group on the pyrazolyl rings which would inhibit approach of potentially bridging halogen atoms [2].

Treatment of these monomeric dihalides with alcohols, primary amines or thiols led to the formation of monomeric mono- and di-substituted species of the type [M(NO)L*X(Q)] (M = Mo or W; X = halide' Q = OR, NHR or SR) and [M(NO)L*(Q)$_2$]. The closely related aryl species can also be obtained from phenols, anilines and arene thiolates. Furthermore, it is also possible to synthesise mixed ligand species, *e.g.* [Mo(NO)L*(OR)(OR')], [Mo(NO)L*(OR)(NHR)], [Mo(NO)L*(OR)(SR)] and

A. J. L. Pombeiro and J. A. McCleverty (eds.),
Molecular Electrochemistry of Inorganic, Bioinorganic and Organometallic Compounds, 77–88.

[Mo(NO)L*(SR)(NHR)]. These species are also monomeric, characteristically coloured, diamagnetic, sufficiently volatile for EI mass spectral characterisation, and exhibit ν_{NO} at frequencies characteristic of the donor atom set X, Y and Q. X-ray crystallographic studies of representative examples of these compound confirmed the six-coordination of these species, and the general structure is illustrated in **1** [3].

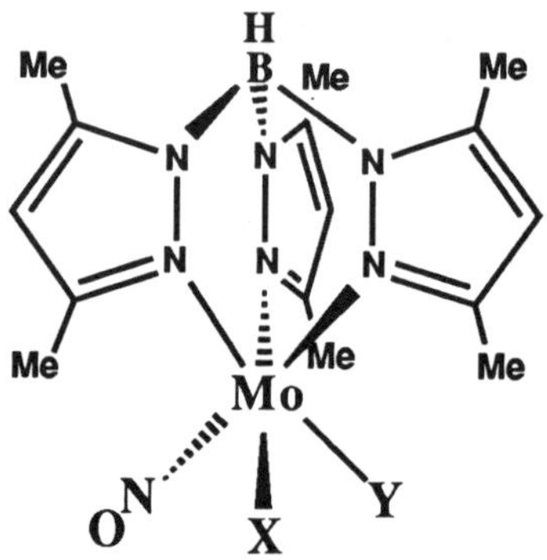

1

The monomeric nature of these compounds can be understood in terms of the particular steric and electronic properties of the L* ligand when compared to cyclopentadienyl groups. The cone angle of $C_5H_5^-$ is *ca.* 100° whereas that of the {HB(3,5-Me_2pyz)$_3$} ligand is nearly 225° and the latter can obviously afford much greater steric protection than the former. Electronically, L* and its analogues function primarily as σ-donor ligands although they have weak π-acceptor capabilities, whereas the C_5H_5 group can act as a π-donor and π-acceptor. Furthermore, the tripodal ligands strongly favour six- over seven-coordination in transition metal complexes which is due, in part, to the steric effects of the bulky ligand mentioned above but is also related to the strong directional effect of the N σ orbitals. Extended Hückel molecular orbital calculations have confirmed that {HB(pyz)$_3$} is much more effective in hybridising the metal fragment orbitals into an octahedral array than the cyclopentadienyl group [4].

The Mo-O, Mo-N and Mo-S bonds are all unusually short (Mo-O *ca.* 1.86Å; Mo-N ca. 1.9Å) and these observations can be rationalised in terms of the electron "deficiency" of the metal in these compounds. The metal atom has a 16-electron configuration, and to it is attached an extremely powerful π-acceptor (NO) and electronegative ligands (halides). The inability of the metal to expand its coordination sphere from six to seven is a consequence of the steric and electronic requirements of L*, and the only source from which the metal may obtain additional electron density to satisfy its apparent "need" is from the hetero-atom, O, N or S. This can be achieved by p_π-d_π donation, resulting in bond contraction, and the consequences of this are also evident in the Mo-E-R bond angles which in the case of alkoxide species opens up to 132°. The colours of these species are due, in large part, to the charge-transfer from the heteroatoms to Mo.

Nitrosyl compounds containing the group {$Mo(NO)^{3+}$} are expected to undergo one-electron reduction, affording species containing the {$Mo(NO)^{2+}$} core. These tris(pyrazolyl)borato species are no exception, and the reduction potential is dependent on the co-ligands. Thus, [Mo(NO)L*Cl_2] undergoes an electrochemically and chemically reversible one-electron transfer process, affording the stable paramagnetic monoanion, [Mo(NO)L*Cl_2]$^-$, which can be isolated as a tetra-alkylammonium salt. However, its iodo analogue, while undergoing reversible one-electron addition in scrupulously dry solvents, affording a paramagnetic monoanion of limited stability, readily dissociates I^- when even minute traces of water were present. The species formed, [Mo(NO)L*I(solvent)] is also electro-active in its own right, undergoing an irreversible one-electron oxidation. These reduced species have been identified spectroscopically, affording characteristic epr spectra, and the paramagnetic μ-iodo compound [$^-$Mo(NO)L*(μ-I)$_2$Li$^+$(OEt)$_2$] has been isolated[5].

The substituted species, [Mo(NO)LX(ER)] X = halide, E = O, S, NH or NR) also undergo one-electron reduction, many of the electron-transfer processes being reversible when X = Cl. The reduction potentials are strongly influenced by the nature of the donor atoms X and E [6]. Thus, the $E_{1/2}$ values for the reduction of the species where X = Cl and ER = OPh, SPh and NHPh are -0.42, -0.37 and -0.85 V (in MeCN, vs SCE), and for the species [Mo(NO)L*(ER)$_2$] where ER = OPh, SPh and NHPh, they are -0.74, -0.64 and -1.32 V, respectively. Furthermore, the reduction potential of these species can be fine-tuned by appropriate substitution of anilido or phenoxy species at the *meta* or *para* positions [7]. Thus, the $E_{1/2}$-value for the reduction of [Mo(NO)L*Cl(NHC$_6$H$_4$Z-*p*)] becomes more cathodic in the order NO_2 < CN < CO_2Me < I < Br < Cl < H < Me < OMe < NH_2, over the range -0.47 to -1.05 V. An excellent linear correlation can be made between the value of $E_{1/2}$ and the Hammett σ_{para} constant, which reveals that as the inductive and/or mesomeric electron releasing capability of the *para* group, Z, in species such as [Mo(NO)L*Cl(NHC$_6$H$_4$Z-*p*) increases, so the reduction potential becomes more cathodic. These data also show that the {Mo(NO)L*} group, and its tungsten analogue, is an extremely powerful electron-withdrawing group, whose electronegativity can be tuned by varying the associated ligands X and Y. It is also feasible, by appropriate choice of the metal atom (Mo or W), donor atom (halide and/or O, NH or S) and donor atom substituent, to *design* a molybdenum or tungsten nitrosyl complex having a redox potential between *ca.* +0.30 and -1.80V. For those compounds have reduction potentials between +0.30 and -0.50V, *chemical* reduction, using, for example, cobaltocene as the reducing agent, can afford isolable cobalticinium salts of the reduced 17-electron species. Epr data reveal that the unpaired electron in these complexes is located primarily in a metal orbital (g = ca. 1.97; a_{Mo} = *ca.* 4.7 mT).

2. Macrocyclic Molybdenum Complexes as Sensors for Alkali Metal Ions

Interest in molecules which contain both a redox-active transition metal centre and a host site at which a guest ion or molecule might bind encouraged us to seek methods of attaching the redox-active $\{Mo(NO)\}^{3+}$ centre to crown ethers. We have prepared two types of compounds, one group containing the $\{Mo(NO)L^*X\}^+$ attached to the periphery of the cyclic polyether, and the other group containing the $\{Mo(NO)L^*\}^{2+}$ within the cyclic polyether framework. The first group was prepared by reaction of [Mo(NO)L*Cl$_2$] with appropriate amine-substituted benzo-crown ethers, the second by reaction of [Mo(NO)L*I$_2$] with tetra-, penta- or hexa-ethylene glycols in the presence of triethylamine [8]. Representative compounds are **2**, **3**, **4** and **5** ({Mo} = {Mo(NO)L*Cl}).

Cyclic voltammograms of these species were recorded in acetonitrile solutions with [Bu$^n_4$N][BF$_4$] as supporting electrolyte, the reduction potentials being consistent with the formulation of the complexes as chloro-monoarylamides or bis-alkoxides. However, on addition of NaBPh$_4$ in 1:1 molar rations, anodic shifts were detected, in the range

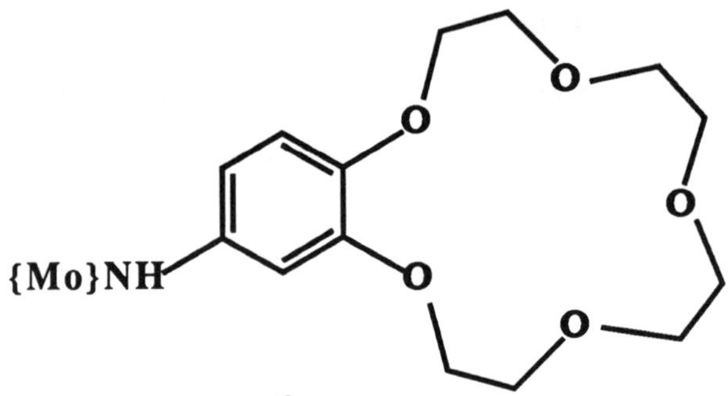

2

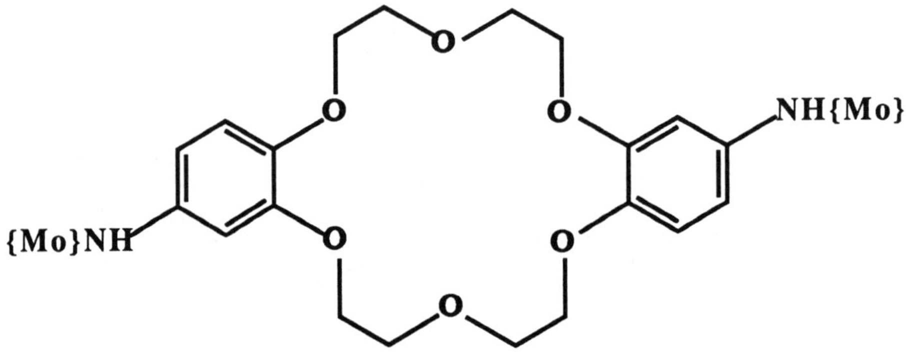

3

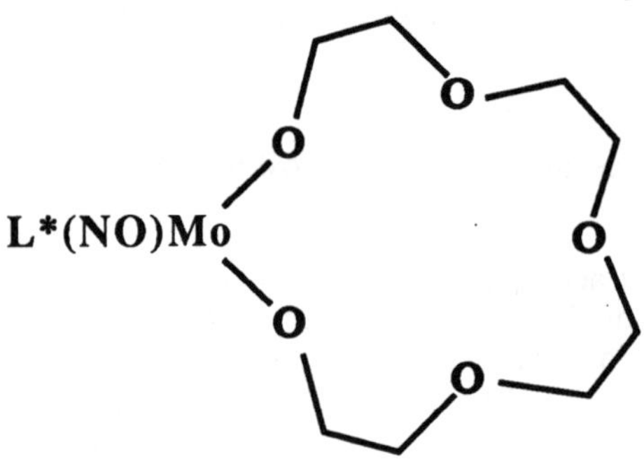

4

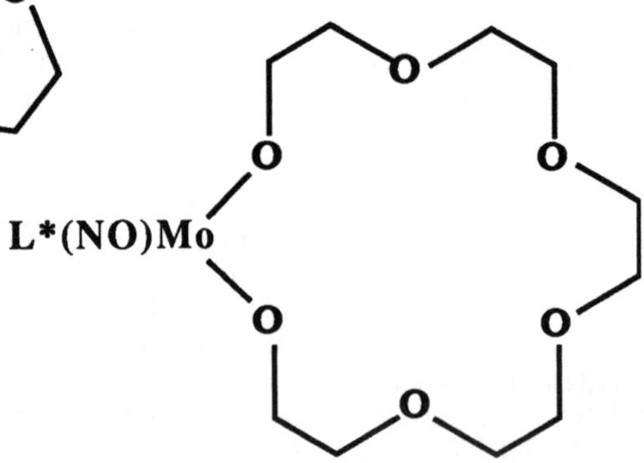

5

60-70 mV for the benzo crown ether complexes of types **2** and **3**. More substantial effects were observed with the cyclic polyether complexes of types **4** and **5**, the smallest polyether ring giving rise to the largest anodic shift of 320 mV in $E_{1/2}$. The addition of $NaBPh_4$ in 1:1 molar ratio to solutions containing [Mo(NO)L*$(OEt)_2$] produced a shift of only 40 mV while no detectable shift was found under similar conditions using [Mo(NO)L*Cl-{$NHC_6H_4(OMe)_2$-3,4}]. The magnitude of the potential shift caused by addition of Na^+ to the cyclic polyethers **4**, **5**, and closely rated compounds was relatively insensitive to the size of the cyclic polyether rings, but is substantially reduced when K^+ is used in place of Na^+. The larger potassium ion, having the smaller charge:radius ratio, produces about half the shift of the Na^+ ion, although there is clear evidence that both alkali metal cations form 1:1 adducts with the metallated polyethers. The results obtained suggest that the polarising power of the cation is of great importance in determining the magnitude of the anodic shift of $E_{1/2}$ produced by this cation binding.

3. Redox-active Molybdenum and Tungsten Complexes as Components of Secondary Harmonic Generators in Non-Linear Optic Materials

The emergence of optical data transmission and processing technologies has generated considerable interest in materials which exhibit non-linear optic (NLO) properties. Among the necessary criteria for a material to exhibit large second order optical non-linearity are that it should consist of polarisable dipolar molecules which crystallise in a non-centrosymmetric polar space group. Using such guidelines, a number of organic materials have been synthesised which exhibit substantial NLO properties. The incorporation of *d*-block metals into molecular NLO materials is of interest because it offers the potential for combining magnetic and electron transfer properties with optical properties. Further, transition metals offer a greater variety of "architectural" design possibilities when compared to carbon which is constrained to a maximum of 4-coordination.

The occurrence of organometallic and metal-organic compounds which, as solids, have significant NLO properties is relatively small, the first such material being *cis*-[CpFe(η^5-$C_5H_4CH{=}CHC_6H_4NO_2$-*p*)] which has a relative harmonic intensity for Second Harmonic Generation (SHG) of 62 times that of the urea standard used at 1906 nm [9]. Since that first discovery, improved SHG efficiencies have been detected in similar organo-ferrocenyl compounds.

Our earlier electrochemical work with the tris(pyrazolyl)borates had revealed that the 16-electron {M(NO)L*ClZ} (Z = O or NH) moiety was a very strong electron-withdrawing substituent while the reduced, 17-electron $\{M(NO)L^*ClZ\}^-$ can act as an electron-donating substituent. Furthermore, the metal centres in these molecules constitute chiral centres so that, if resolved, they must form materials which crystallise in non-centrosymmetric space groups.

These observations suggested that compounds containing the {M(NO)L*Cl} moiety, as a substituent in a polarisable molecule which also contained an electron-releasing group, might meet the molecular requirements for NLO properties. The isolation of resolved materials containing chiral metal centres might also meet the solid state structural requirements for NLO activity. In order to test these ideas, we have investigated the properties of several complexes containing chiral {M(NO)L*X} (M = Mo, X = Cl or I; M = W, X = Cl) linked to chiral ligands. We have also synthesised a variety of complexes in which this electronegative group is linked to polarisable molecules containing a ferrocenyl electron donor substituent, although these compounds were not deliberately resolved [10].

The NLO properties of the molybdenum and tungsten complexes were assessed qualitatively using the Kurtz powder test to determine the relative harmonic intensity for SHG relative to a urea standard at 1906 nm. The resolved complexes [Mo(NO)LCl(3-*O*-1,2;5,6-di-isopropylidene-α-D-glucofuranoate)], [Mo(NO)L*Cl{(1*S*, 2*R*, 5*S*)-(+)-mentholate)] and [W(NO)LCl{(*R*)-(+)-methylbenzylamide)] containing chiral ligand and chiral metal centres [11], were tested in the form of single pure diastereoisomers. These materials exhibit only weak or non-existent SHG. This can be rationalised in terms of the saturated nature of the attached ligands, the molecules not being polarisable in the region of the acceptor substituent. However, there may also be structural reasons for the low SHG intensities observed. Single crystal studies of the three complexes mentioned above revealed that they crystallise in monoclinic space groups, $P2_1$, with two independent molecules in the asymmetric unit. It is entirely possible that in such circumstances, the molecular hyperpolarisability, β, of one molecule in the unit cell may be so orientated as to oppose that of the other molecule. This produces a pseudo-centrosymmetric situation with respect to the hyperpolarisability, resulting in low SHG intensities despite the presence of a chiral molecule.

Much more substantial SHG intensities have been obtained with the some of the compounds containing polarisable ligands attached to the electron-withdrawing {M(NO)L*ClZ} group. Thus, [Mo(NO)L*Cl{NHC_6H_4-4-N=N-C_6H_4-4'-Fc}], [W(NO)L*Cl{NHC_6H_3-(3-Me)-4-N=NC_6H_4Fc}] and [Mo(NO)L*Cl{NHC_6H_3-(3-Me)-4-N=NC_6H_4Fc}] {Fc = CpFe(η^5-C_5H_4-)}, in which the ferrocenyl substituent is present as an electron-releasing group on the polarisable NHC_6H_3-(3-R)-4-N=NC_6H_4 (R = H or Me) bridge, give SHG intensities of 59, 53 and 123 times that of urea, respectively. The reduction potentials of the Group VI metal centres in these complexes are -0.57, -1.01 and -0.58 V vs SCE in MeCN, respectively, and the oxidation potentials of the ferrocene moiety are +0.57, +0.62 and +0.62 v, respectively. Since these compounds were not resolved these findings depend on the compounds fortuitously crystallising in non-centrosymmetric space groups. It should be noted that neither $O_2NC_6H_4$-4-N=N-C_6H_4-4'-Fc nor $O_2NC_6H_4$-(3-Me)-4-N=NC_6H_4-4'-Fc, in which the nitro group has replaced the {M(NO)L*Cl} group, exhibit measureable SHG intensities. The compound [Mo(NO)L*Cl{NHC_6H_4-4-N=NC_6H_4-4'-NMe_2] does not exhibit SHG either. It appears, therefore, that both the {M(NO)L*Cl} and {$FcC_6H_4N_2C_6H_4$-} moieties are necessary for large SHG intensities to be obtained in this type of compound.

More detailed studies of the SHG efficiencies of these compounds indicate strongly that the electronic properties of the {M(NO)L*X} group contribute significantly to the NLO properties of the material. The most substantial effect can be detected when Mo is replaced by W, the efficiencies in comparable compounds decreasing by factors of 2-3, and since there are no major structural differences between these compounds (confirmed crystallographically), it is tempting to correlate this effect with the observation that the {W(NO)L*Cl} group is a significantly poorer electron-acceptor than {Mo(NO)L*Cl}, the reduction potentials of the tungsten complexes [W(NO)L*XY] being *ca.* 450 mV more cathodic than their Mo-containing counterparts. Smaller effects are observed in the series [Mo(NO)L*X{NHC_6H_3-(3-Me)-4-N=NC_6H_4Fc}], (X = F, Cl, Br or I), the SHG efficiency of the bromide being ca. 50% that of the chloride, whereas the fluoride and iodide exhibit negligible frequency doubling at 1907 nm. An interesting effect is observed when the diazo link N=N is replaced by the ethenyl group C=C, when SHG efficiencies in comparable compounds increase significantly. Similar observations have been made in NLO-active organic materials containing these groups.

As indicated above, the crystal structures of certain key compounds have been determined [12]. The complexes [M(NO)L*Cl{NHC_6H_3-(3-Me)-4-N=NC_6H_4Fc}] (M = Mo or W) crystallise in the monoclinic space group $P2_1$ and are isostructural (for Mo, a = 10.0954(2), b = 20.6345(5), c = 10.3965(5)Å and β = 91.9109(3)°; for W, a = 10.147(7), b = 20.455(7), c = 10.366(3)Å, β = 91.79(4)°). The lattices in both complexes contain a molecule of included solvent (CH_2Cl_2), and although the crystals were obtained from solutions containing a racemic mixture, they contained only one enantiomer of the chiral Group VI metal centre. The angles between the vector defined by the Mo atom and the most remote C atom of the cyclopentadienyl ring coplanar with the two phenylene rings in [Mo(NO)L*Cl{NHC_6H_3-(3-Me)-4-N=NC_6H_4Fc}], which is taken as an approximation to the major axis of the molecular hyperpolarisability tensor of the molecule, and the crystal axes were calculated to be 133.95 (a), 58.39(b) and 118.03° (c), so that the angle with the b axis is close to the ideal value for SHG of 54.74°. The orientations of the possible major axes of the hyperpolarisability tensor of the isostructural tungsten-containing molecules are close to those for their Mo-containing analogues, and so a direct comparison of the SHG results is possible. Thus, the fact that the tungsten species exhibit less efficient frequency doubling as NLO materials than their molybdenum analogues can indeed be correlated with the electronic properties of the Group VI metal atom. The species [$O_2NC_6H_3$-(3-Me)-4-N=NC_6H_4Fc] does not exhibit SHG, even though one might reasonably expect the molecular hyperpolarisibility factor β to be high, and this inactivity may be attributed to structural features in that the compound crystallises in the centrosymmetric space group $P2_1/c$. Thus the crystal packing in the unit cell (4 molecules in two pairs) indicates clearly that the molecular hyperpolarisability, β, of each molecule in a pair in the unit cell is orientated to oppose that of the each molecule in the other pair. Hence the resultant molecular hyperpolarisibility is virtually zero.

It is also possible to design polarisable molecules which may exhibit significant second order molecular hyperpolarisabilities, γ, which may lead to materials capable of displaying third order linear electric susceptibilities, χ_3. Towards this end, we have synthesised the bimetallic bipyridyl complexes [{Mo(NO)L*Cl}$_2$-{$NC_5H_4(CH{=}CH)_nC_5H_4N$}] (n = 0 – 5) [13]. These complexes undergo two one-electron reductions, the precise potentials being determined by n. They are paramagnetic (1 unpaired electron per metal atom), the unpaired electrons having large exchange interactions at 300K. These species are described in more detail in the contribution presented by C.J.Jones.

4. Charge Separation in Extra-ring Metallated Porphyrin Complexes

The control of energy and/or charge distribution within molecules or molecular arrays is of fundamental importance in the development of molecular electronic systems. In the previous section we addressed an aspect of this general problem in our description of polarisable molecules containing donor and acceptor functions whose excited states could be readily accessed by ir irradiation. The resulting charge separation, equivalent to transfer of electron density from the ferrocenyl group to the Group VI metal acceptor centre (*i.e.* reduction of the Mo or W centre by {CpFe(η^5-C_5H_4-)}, is obviously facilitated by the polarisable *trans*-C_6H_4-N=N-C_6H_4 and $C_6H_4CH{=}CHC_6H_4$ bridges.

There have been many attempts to devise simple models of the early steps in photosynthesis, which involves substantial charge separation. There are many example of intramolecular electron transfer from porphyrinyl photo-donors to organic acceptors, such as quinones, methyl viologen or pyromellitimide. However, examples of electron transfer to transition metal-based redox centres are very rare. The ferricinium centres in the oxidised form of 5,10, 15, 20-tetra-(4'-ferrocenylphenyl)-21,23-H_2-porphine), [(*p*-$FcC_6H_4)_4PorphH_2$], proved to be ineffective acceptors for an electron from the excited singlet state of the porphyrinyl moiety [14]. One explanation offered was that metal-centred electron acceptors may exhibit electron transfer rates too slow to compete with fluorescence decay or intersystem crossing. However, electron transfer to a d^{10} metal centre has been demonstrated in the case of Ag^+ bound to a cyclic polyether attached to a porphyrin group [15]. During a study of the interaction between an organic redox centre (tetraphenylporphyrin) and the redox-active {Mo(NO)L*Cl} group, we have discovered that the transition metal centre could apparently act as an electron acceptor towards singlet porphinyl.

We had prepared the mono-metallated species [Mo(NO)L*X-{$EC_6H_4(Ph)_3PorphH_2$}], **6**, (E = O or NH; X = Cl, I, OH, NHCHMePh) in order to study the interaction of the organic redox centre with the well-characterised electro-active Group VI metal nitrosyl system. These compounds were obtained by reaction of [Mo(NO)L*Cl_2] with *ortho*, *meta*, and *para* $HEC_6H_4(Ph)_3PorphH_2$, and single crystal X-ray structural studies have been carried out on the species where E = *para*-O, *ortho*-O and *ortho*-NH, which have confirmed our formulations of these molecules [12]. The Mo-porphine centroid distances are respectively 9.203, 6.689 and 6.524Å in these three complexes. The internally-metallated species [Mo(NO)L*Cl{$EC_6H_4(Ph)_3PorphM$}] (M = Zn, Ni, Pt and FeCl) have also been prepared.

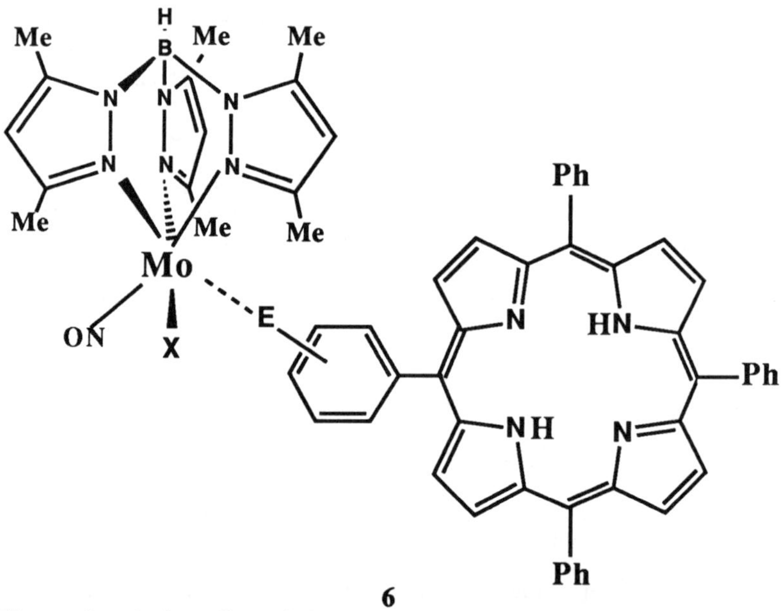

6

Electrochemical studies of these complexes reveal that there is no significant interaction between the two redox centres in the ground state [16]. Thus, by comparison with the unmetallated species, $MeOC_6H_4(Ph)_3PorphH_2$ which exhibits two one-electron oxidation and two one-electron reduction processes, the presence of the extra-ring metal groups hardly influence the $E_{1/2}$-values of the porphyrinyl group. Concomitantly, the reduction potentials of the molybdenum centre are not significantly different from those of the unsubstituted phenoxide or anilide [Mo(NO)L*Cl(EC_6H_5)].

However, the observation that the normal fluorescence of the porphyrin species is quenched in the peripherally metallated species suggested that the {Mo(NO)L*Cl} group could function as an electron acceptor with respect to the excited porphyrin. The efficiency of fluorescence quenching in the species where E = *para*-O was apparently in the order Cl > OH > NHCHMePh, which parallels the reduction potential of the molybdenum centre. Although the reduction potential of the singlet excited porphyrin would certainly be sufficient to reduce the species containing X = Cl and OH, these results were eventually recognised to be spurious. Nevertheless, they encouraged us to think that photo-electron transfer from the porphyrin to {Mo(NO)L*Cl}, *i.e.* charge-separation, was a potential explanation for the fluorescence-quenching.

Preliminary picosecond spectroscopic studies have been carried out on **6** (E =*para*-, *meta*-, and *ortho*-O and *para*-NH) to determine whether this charge-separation does indeed occur. Time-correlated single photon counting studies of DMF solutions indicate that the fluorescence lifetimes of the molybdenated porphyrins are <15 ps, compared to *ca.* 12 ns measured for the non-molybdenated models (*e.g.* $MeOC_6H_4(Pg)_3PorphH_2$), although there is evidence of a very minor long-lived (*i.e.* 10 ns) component which is ascribed to a trace (non-molybdenated) impurity. Picosecond time-resolved absorption spectra have been obtained, and the transient species is non-luminescent so that it cannot be attributed to a short-lived singlet state. The characteristic absorption at 740 nm in the spectrum provides evidence for the fast (τ < 30 ps) formation of the porphyrinyl radical cation which decays with a lifetime in the range 120 – 290 ps following first-order kinetics. The lifetimes of the charge-separated species obtained from **6** (E =*para*-, *meta*-, and *ortho*-O and *para*-NH) were 220+/-7, 220+/–44, 120+/–9 and 290+/–110 ps, respectively, and the reduction potentials of the Mo fragment, as determined by cyclic voltammetry in MeCN were -0.31, -0.32, -0.36 and -0.75 V, respectively. The estimated ΔG^o for the conversion of [Mo(NO)L*Cl{$EC_6H_4(Ph)_3PorphH_2$*] into [$^-$Mo(NO)L*Cl{$EC_6H_4(Ph)_3PorphH_2^{+\cdot}$] was -50, -50, -50 and -8 kJ mole^{-1}, respectively.

It therefore appears that very fast charge-separation occurs over a distance of *ca.* 1 nm in these peripherally metallated porphyrins. However, charge recombination is relatively slow suggesting that structural changes associated with the oxidation of the peripheral redox centre may be rate limiting, and more detailed studies are needed to clarify this.

In an extension of this work, primarily intended to avoid tedious synthesis of mono-functionalised tetra-aryl porphyrins, we have investigated the possibilities of attaching the acceptor {Mo(NO)L*Cl} group to internally-metallated porphyrins where the metal is spectroscopically "innocent" and/or "transparent". Thus, we have prepared zinc and germanium tetraphenylporphyrin complexes, and attached the molybdenum acceptor to the former via pyridyl linkages, **7**, and to the latter via phenolic links, **8** [17]. These complexes, like their predecessors, **6**, exhibit ground-state electron transfer behaviour characteristic of the constituent redox-centres which are virtually unperturbed by the presence of the other group. However, both types of molecule exhibit fluorescence quenching similar to that of **6**, and although detailed spectroscopic characterisation remains to be completed, it is quite possible that photo-electron transfer from the excited porphyrin to the {Mo(NO)L*Cl} group can also occur in these complexes.

5. Acknowledgments

I am grateful to a number of talented colleagues and associates who have helped not only with the conception of the work described above, but also to translate our ideas into realities. Foremost among these is Chris Jones, friend of many years, and close colleague for 10 in Birmingham. I am also grateful to Paul Beer who introduced me to crown ethers, and particularly to Mike Ward, who has begun to develop new N-heterocyclic systems to elaborate electron transfer behaviour in Mo, W and other metal complexes. Our much valued post-doctoral and post-graduate associates are mentioned in the references, and to them I owe a profound debt of gratitude. Their skills, patience and invention permitted us to go further even than our imaginations had led us. I wish also to thank the SERC for financial support, via the Chemistry Committee and the Initiative for Synthesis of Materials for the 21st Century, the British Council, the French CNRS, Italian CNR and Spanish DGCyT for assistance to colleagues from abroad who have contributed to our investigations. I am also grateful to ICI Fine Colours and Chemicals for a CASE award.

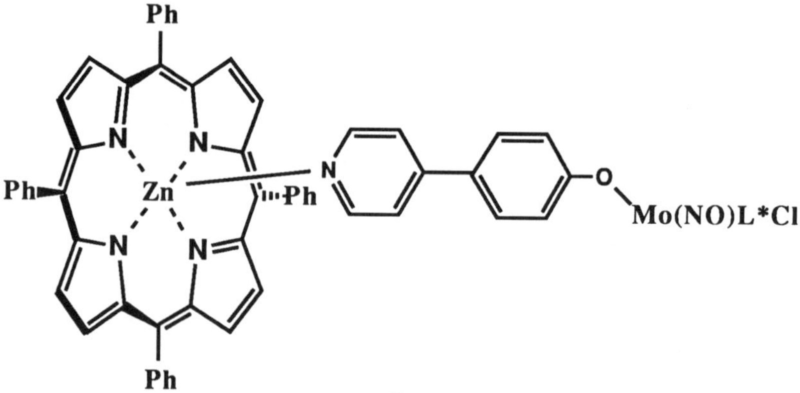

7

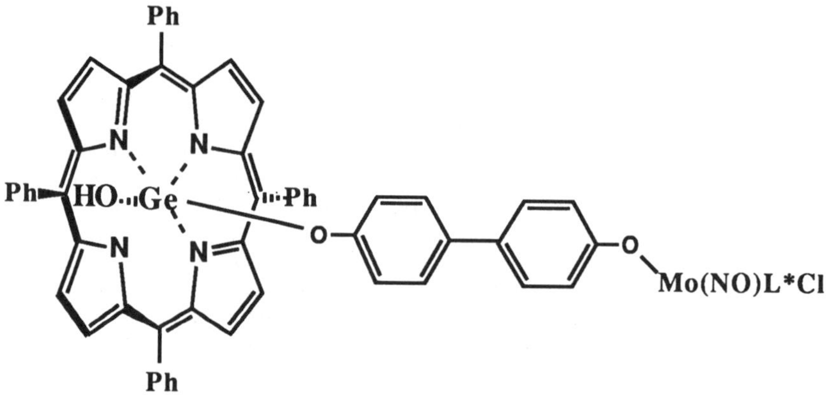

8

6. References

1. Trofimenko, S., (1971) 'Polypyrazolylborate, a new class of ligands', *Acc.Chem.Res.*, **4**, 17-22.

2. McCleverty, J.A., Seddon, D., Bailey, N.A., and Walker, N.W. (1976) 'The chemistry of cyclopentadienyl and related nitrosyl complexes of molybdenum, part 5. Dihalogenonitrosyl{tris(pyazolyl)borato]molybdenum complexes, their alcoholysis, and the crystal and molecular structure of chloronitrosylisopropoxo[tris(4-chloro-3,5-dimethylpyrazoyl)borato]molybdenum', *J.Chem.Sco., Dalton Trans*, 898-908.

3. McCleverty, J.A., (1983) 'Alkoxy, amido, hydrazido and related complexes of molybdenum and tungsten', *Chem.Soc.Rev.*, **12**, 331-60.

4. Curtis, M.D., Shiu, K.-B., and Butler, W.M. (1986) 'Synthesis, structural and theoretical studies of η^2-acyl complexes of molybdenum', *J.Am.Chem.Soc.*, **108**, 1550-61.

5. Briggs, T.N., Jones, C.J., Colquhoun, H., El Murr, N., McCleverty, J.A., Neaves, B.D., Adams, H., and Bailey, N.A. (1985) 'Electrochemical and chemical properties of di-iodonitrosyltris(3,5-dimethypyrazolyl)borato molybdenum, $[Mo(HB(Me_2pyz)_3\}(NO)I_2]$ and related complexes', *J.Chem.Soc., Dalton Trans.*, 1249-54.

6. Al Obaidi, N., Chaudhury, M., Clague, D., Jones, C.J., Pearson, J.C., McCleverty, J.A., and Salam S.S. (1987) 'Monometallic, homo- and hetero-bimetallic complexes based on redox-active tris(3,5-dimethylpyrazolyl)borato molybdenum and tungsten nitrosyls, part 4. The effects of ligating atom type on reduction potentials of monometallic complexes', *J.Chem.Soc., Dalton Trans.*, 1733-6.

7. Al Obaidi, N., Charsley, S.M., Jones, C.J., McCleverty, J.A., Neaves, B.D., and Reynolds, S.J. (1987) 'Monometallic, homo- and hetero-bimetallic complexes based on redox-active tris(3,5-dimethylpyrazolyl)borato molybdenum and tungsten nitrosyls, part 5. *Para*-substituent effects on the reduction potentials of arylamide complexes', *Trans.Met.Chem.*, **12**, 143-8.

8. Al Obaidi, N., Beer, P.D., Bright, J.P., Charsley, S.M., Jones, C.J., McCleverty, J.A., and Salam S.S. (1986) 'The synthesis and electrochemistry of novel redox-responsive molybdenum complexes containing cyclic polyether cation binding sites', *J.Chem.Soc.Chem.Comm.*, 239-41; Beer, P.D., Jones, C.J., McCleverty, J.A. and Salam, S.S. (1987) 'Redox-responsive metal complexes containing cation binding sites', *J.Incl.Phenom.*, **5**, 521-4.

9. Green, M.L.H., Marder, S.R., Thompson, M.E., Bandy, J.A., Bloor, D., Kolinsky, P.V., and Jones, R.J. (1987). 'Synthesis and structure of (*cis*)-[1-ferrocenyl-2-(4-nitrophenyl)ethylene], an organotransition metal compound with a large second-order optical non-linearity ', *Nature*, **330**, 360-2.

10. Coe, B.J., Jones, C.J., McCleverty, J.A., Bloor, D., Kolinsky, P.V., and Jones, R.J. (1989) 'Second harmonic generation by bimetallic complexes containing

ferrocenyl and molybdenum or tungsten mononitrosyl redox centres', *J.Chem.Soc., Chem.Comm.*, 1485-7

11. Wlodarczyk, A., Kurek, S.S., Foulon, J.-D.F., Hamor, T.A. and McCleverty, J.A. (1992) 'Optically resolved 1-methylbenzylamido complexes of molybdenum and tungsten, and the crystal and molecular structure of *S*-iodo[tris(3,5-dimethylpyrazolyl)borato}{(+)-1-phenylethylamido}nitrosylmolybdenum, *S*-[Mo(NO)-{HB(3,5-$Me_2C_3HN_2)_3$}I{(+)-NHCHMePh}]', *Polyhedron*, **11**, 217-26.

12. Coe, B.J., Kurek, S.S., Rowley, N.M., Foulon, J.-D.F., Hamor, T.A., Harman, M.E., Hursthouse, M.B., Jones, C.J., McCleverty, J.A and Bloor, D., (1991) 'Structural studies of some metal-organic complexes which exhibit non-linear optical properties', *Chemtronics*, **5**, 23-8.

13. McWhinnie, S.L.M., Jones, C.J., McCleverty, J.A., Collison, D., and Mabbs, F.E., (1990) 'EPR and electrochemical studies of bimetallic seventeen-electron molybdenum species exhibiting unusually large interactions across bipyridyl bridging ligands', *J.Chem.Soc., Chem.Comm.*, 940-2.

14. Schmidt, E.S., Calderwood, T.S., and Bruice, T.C., (1986) 'Synthesis and characterisation of a *meso*-tetrakis(4-ferrocenylphenyl)porphyrin and examination of its ability to undergo intramolecular photocatalysed electron transfer', *Inorg.Chem.*, **25**, 3718-20.

15. Gubelmann, M., Harriman, A., Lehn, J.-M., and Sessler, J.L., (1990) 'Quenching of porphyrin excited states by silver(I) ions and charge separation in bimolecular systems and in macrocyclic coreceptors', *J.Phys.Chem.*, **94**, 308-15.

16. Rowley, N.M., Kurek, S.S., George, M.W., Beer, P.D., Jones, C.J., Hubig, S., Harriman, A., Kelly, J.M., and McCleverty, J.A. (1992) 'Intramolecular interactions between ground and excited state porphinyl moieties and molybdenum mononitrosyl redox centres in peripherally metallated tetraphenylporphyrin derivatives', *J.Chem.Soc., Chem.Comm.*, 497-9.

17. Schiller, H., Culpin, J., Das, A., Ward, M.D., Jones, C.J., and McCleverty, J.A., work to be published.

AN ELECTROCHEMICAL EVALUATION OF LONG RANGE METAL-METAL INTERACTIONS THROUGH HYDROCARBON BRIDGES

Christopher J. JONES[a], Sean L.W. McWHINNIE*, Ferida S. McQUILLAN[a] and Jon A. McCLEVERTY[b]
[a]*School of Chemistry, The University of Birmingham, Edgbaston, Birmingham B15 2TT, U.K.*
[b]*School of Chemistry, The University of Bristol, Bristol BS8 1TS, U.K.*

ABSTRACT A number of new bimetallic complexes have been synthesised which contain two molybdenum mononitrosyl redox centres. Electrochemical studies of these systems indicate that the interaction between 16-electron $\{Mo(NO)\}^{3+}$ centres linked by two bifunctional ligands of the type O-hydrocarbyl-O differed little from the interaction when only one bridging ligand is present. The saturation of aryl linking groups has a large effect on the metal-metal interaction only when mesomeric contact between the donor atoms is removed as in converting OC_6H_4O-1,4 to $OC_6H_{10}O$-1,4.

Introduction

The now famous Creutz-Taube ion, and complexes related to it [1], have played a major role in developing our understanding [2,3] of mixed oxidation state complexes and of metal-metal interactions through hydrocarbyl bridges. The electrochemical properties of such bimetallic species provide one means of assessing the extent of metal-metal interactions and two formally $Ru^{2+}(d^6)/Ru^{3+}(d^5)$ redox couples separated by 390 mV are observed for $[\{(NH_3)_5Ru\}_2(NC_4H_4N\text{-}1,4)]^{z+}$ (z = 4, 5, 6) [4]. In the case of the related complex $[\{(NH_3)_5Ru\}_2(NC_5H_4\text{-}C_5H_4N\text{-}4,4')]^{z+}$ this separation falls to 76 mV which may be compared with the theoretical value of 36 mV expected for non-interacting equivalent metal centres [5].

The $\{Mo(NO)\}^{z+}$ (z = 1, 2, 3) moieties contain formally d^6, d^5, and d^4 Mo centres and an extensive chemistry has been developed in which each of these species is represented [6]. In particular we have previously shown that the 16-electron $\{Mo(NO)\}^{3+}$ moieties in complexes of the general formula $[\{LCl(NO)M\}_2BL]$ (L^- = hydrotris(3,5-dimethylpyrazol-1-yl)borate; M = Mo or W; BL is a bridging ligand of the form E-hydrocarbyl-E and E = O, NH) exhibit metal-metal interactions which may be detected electrochemically [7-13]. Thus the complexes exhibit two one electron reduction waves separated by a potential, ΔE_f, which may be used as a measure of the strength of these interactions. We have found that interactions also exist between the 17-electron $\{Mo(NO)\}^{2+}$ centres in $[\{LCl(NO)Mo\}_2(PP)]$ (PP = $NC_5H_4\text{-}X\text{-}C_5H_4N$-4,4'; X = nothing, CH=CH, CH_2CH_2: and $NC_5H_4\text{-}C_5H_4N$-3,3') [14].

* Present address, Department of Chemistry, Brunel University, Uxbridge UB8 3PH, UK

A. J. L. Pombeiro and J. A. McCleverty (eds.), Molecular Electrochemistry of Inorganic, Bioinorganic and Organomettallic Compounds, 89–93.

In $[\{LCl(NO)M\}_2(E_2C_6H_4)]$ the value of ΔE_f varies with both the nature of E, interactions through NH being larger than through O (Figure 1), and with the nature of M, interactions between tungsten centres being smaller than between molybdenum centres.

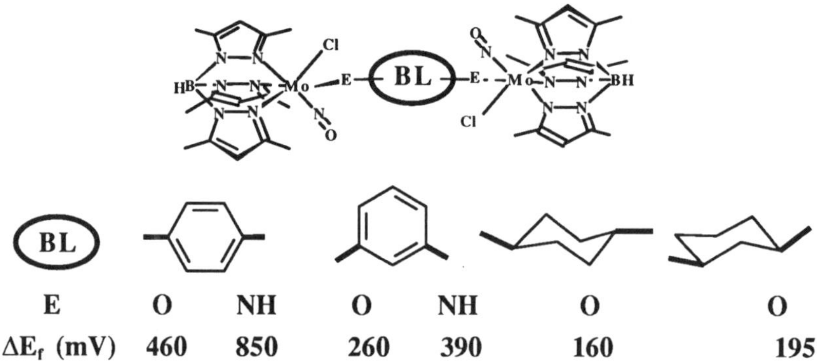

Figure 1. Variations in the separations of the reduction potentials of some bimolybdenum complexes with varying bridging groups.

The replacement of unsaturated aromatic hydrocarbyl bridges by their saturated counterparts has only a modest effect on the metal-metal interactions if the metal centres are not in mesomeric contact through the aromatic bridge. This amounts to a difference in ΔE_f of about 65 mV in the 1.3-dihydroxybenzene and 1,3-dihydroxycyclohexane derivatives as summarised in Figure 1. However, a much larger effect is found if saturation results in the removal of a mesomeric interaction. Thus in the case of 1,4-dihydroxybenzene and 1,4-dihydroxycyclohexane derivatives a difference in ΔE_f of 300 mV is found. Bimetallic complexes containing 1,3-dioxycyclopentanyl bridges give similar values of ΔE_f to their cyclohexyl containing counterparts.

We have also found evidence that steric factors reduce the Mo-Mo interaction observed in $[\{LCl(NO)Mo\}_2\{1,2\text{-}(NH)_2C_6H_4\}]$ to a value less than that observed in its 1,4-disubstituted aryl counterpart despite the proximity of the metal centres in the 1,2-disubstituted compound. Thus the respective ΔE_f values for $[\{LCl(NO)Mo\}_2\{1,X\text{-}(NH)_2\text{-}C_6H_4\}]$ are 850 (X = 4), 390 (X = 3) and 770 mV (X = 2). The crystal structure of $[\{LI(NO)Mo\}_2\{1,2\text{-}(NH)_2\text{-}C_6H_4\}]$ [15] shows that steric interactions impose a non-planar geometry on the Mo-N-C-C-N-Mo bridge thus reducing opportunities for mesomeric interactions between the aryl nitrogen substituents. We believe that this effect results in the lower ΔE_f value for the 1,2-disubstituted aryl derivative than in its 1,4-substituted counterpart.

We have extended our work on 16-electron complexes to examine the effect of incorporating two bridging ligands between the metal centres to produce $[\{L(NO)Mo\}_2(BL)_2]$. The reactions of $[Mo(NO)LI_2]$ with the bifunctional ligands $HOCH_2C_6H_4CH_2OH$-4, $HOC_6H_4CH_2C_6H_4OH$-4,4' and $C_{10}H_8(OH)_2$-2,7 (H_2BL), which cannot form chelate complexes, lead to the doubly bridged macrocyclic complexes **1** to **3**

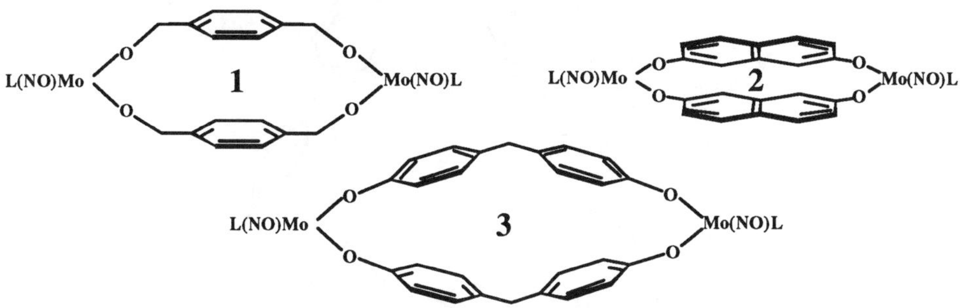

which have been characterised by elemental analyses, ^1H NMR and IR spectroscopy. Mass spectrometry confirms the presence of the binuclear complexes. The corresponding bimetallic singly bridged complexes $[\{LCl(NO)Mo\}_2(BL)]$ (BL = $OCH_2C_6H_4CH_2O$-4, $OC_{10}H_6O$-2,7 and $OC_6H_4CH_2C_6H_4O$-4,4') have also been synthesised and characterised. Electrochemical studies of these compounds reveal that there are weak interactions between the molybdenum centres and two reduction processes are present. The separations between these processes, ΔE_f, have been measured using differential pulse voltammetry [16]. Values of ΔE_f = 85 (**1**),185 (**2**) and 90 mV (**3**). were obtained for the doubly bridged complexes, $[\{L(NO)Mo\}_2(BL)_2]$, compared with respective values of 85 (BL = $OCH_2C_6H_4CH_2O$-4), 170 (BL = $OC_{10}H_6O$-2,7) and 65 mV (BL = $OC_6H_4CH_2C_6H_4O$-4,4') in the corresponding singly bridged species $[\{LCl(NO)Mo\}_2(BL)]$. These results indicate that the presence of the second bridging ligand in **1** to **3** does not substantially increase the extent of metal-metal interaction. However, these findings are not open to simple interpretation since the properties of the metal centres are also affected by the substitution of Cl by O-hydrocarbyl in going from the singly to the doubly bridged species. This leads to a cathodic shift in the first reduction potential of 55 mV in the case of **1** and 360 mV for **2** and **3** which reflects an increase in the electron density at the metal centre.

The reaction between $[Mo(NO)LI_2]$ and the more rigid bifunctional ligand 4,4'-dihydroxybiphenyl does not afford doubly bridged bimetallic species. Mass spectral and ^1H NMR data indicate that, instead, the trimeric and tetrameric metallocyclic complexes

4 and **5** are produced. These compounds contain respectively three and four redox centres, each of which should interact with its two neighbours. Thus the electrochemical properties of these novel species are an important topic for future study.

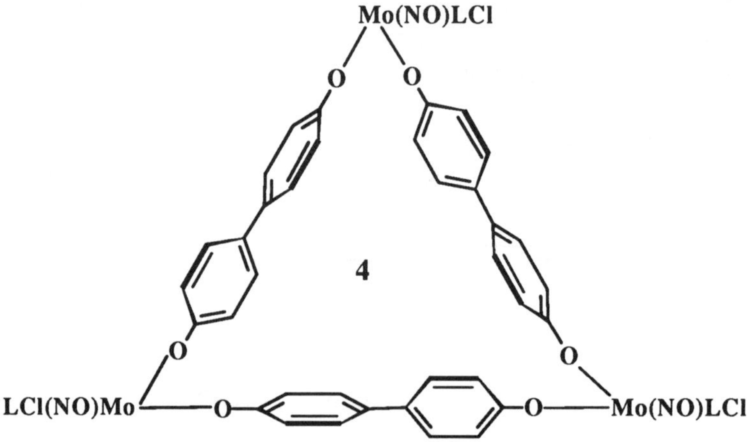

4

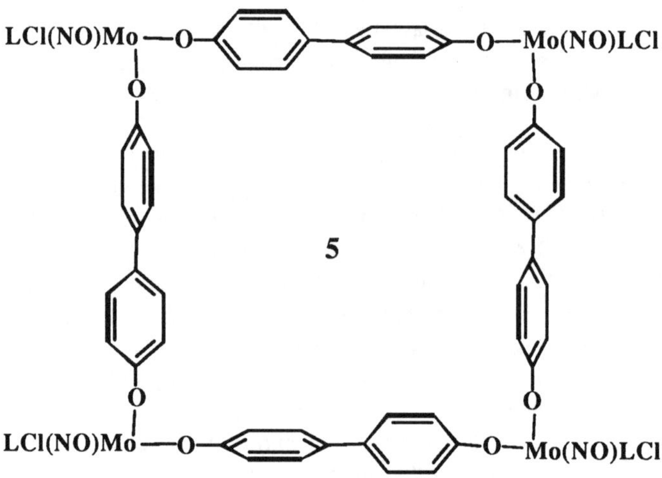

5

Acknowledgement

We are grateful to SERC (UK) for supporting this work (F.S.McQ.).

References

1 Creutz, C., (1983), *Prog. Inorg. Chem.,* **30**, 1-73.
2 Riemers, J.R., and Hush, N.S., (1990), *Inorg. Chem.,* **29**, 3686 -3697 and 4510-4513.
3 Joachim, C., Launay, J.P., and Woitellier, S., (1990), *Chem. Phys.,* **147**, 131-141.
4 McCleverty, J.A., (1983), *Chem. Soc. Rev.,* **12**, 331-360.
5 Sutton, J.E., and Taube, H., (1981), *Inorg. Chem.*, **20**, 3125-3134.
6 Richardson, D.E., and Taube, H., (1984), *Coord. Chem. Rev.*, **60**, 107-129.
7 Denti, G., Jones, C.J., McCleverty, J.A., Neaves, B.D., and Reynolds, S.J., (1983), *J. Chem. Soc., Chem. Commun.,* 474-476.
8 Charsley, S.M., Jones, C.J., McCleverty, J.A., Neaves, B.D., and Reynolds, S.J., (1988), *J. Chem. Soc., Dalton Trans.,* 301-307.
9 Charsley, S.M., Jones, C.J., McCleverty, J.A., Neaves, B.D., and Reynolds, S.J., (1986), *Trans. Met. Chem.*, **11**, 329-334.
10 Charsley, S.M., Jones, C.J., McCleverty, J.A., Neaves, B.D., Reynolds, S.J., and Denti, G., (1988), *J. Chem. Soc., Dalton Trans.,* 293-299.
11 AlObaidi, N., Hamor, T.A., Jones, C.J., McCleverty, J.A., and Paxton, K., (1987), *J. Chem. Soc., Dalton Trans.,* 2653-2660.
12 AlObaidi, N.F., Jones, C.J., and McCleverty, J.A., (1990), *Polyhedron,* **9,** 693-696.
13 McWhinnie, S.L.W., Ph.D. thesis, The University of Birmingham, 1989.
14 McWhinnie, S.L.W., Jones, C.J., McCleverty, J.A., Mabbs, F.E., and Collison, D., (1990), *J. Chem. Soc., Chem. Commun.,* 940-942.
15 Hamor, T.A., personal communication.
16 Richardson, D.E., and Taube, H., (1981), *Inorg. Chem.*, **20**, 1278-1285.

ELECTROCHEMICAL PROPERTIES OF MOLYBDENUM COMPLEXES CONTAINING A BULKY TRIS(PYRAZOLYL)BORATE LIGAND

M.CANO[a], J.V.HERAS[a], C.J.JONES[b] and J.A.McCLEVERTY[c]
[a] Dpto. de Química Inorgánica
Facultad de Ciencias Químicas
Universidad Complutense
Madrid, Spain
[b] School of Chemistry
University of Birmingham
PO Box 363
Birmingham B15 2TT, UK
[c] School of Chemistry
University of Bristol
Cantocks Close
Bristol BS8 1TS
UK

ABSTRACT. Two series of tris(pyrazolyl)borato complexes of the type [Mo(NO)LXY] where (L = L' = $HB(3,5\text{-}Pr^iMeC_3N_2H)_3$; X = Y = CO, Cl, I; X = Cl, Y = OC_6H_5, OC_6H_4Me-o and NHC_6H_4Me-o and L = L" = $HB\{(3\text{-}p\text{-}MeOC_6H_4)C_3N_2H_2\}_3$; X = Y = CO, Cl; X = Cl, Y = $OC_6H_4H_5$, OC_6H_4Me-o and NHC_6H_4Me-o) have been synthesised. All the new complexes with exception of the dicarbonyls undergo a reversible or quasi-reversible one-electron reduction. No significant differences between the $E_{1/2}$ values for the one electron reduction of complexes containing L' to those containing $HB(3,5\text{-}Me_2C_3N_2H)$ are observed. However the complexes containing L' exhibit more anodic $E_{1/2}$ potential values than their corresponding complexes containing L'. An unexpected cathodic shift of the reduction potential ($E_{1/2}$ = 0.60V vs SCE) occurs with [Mo(NO)L"Cl(OC_6H_5)]. This may be due to the enhancement of *pπ-dπ* interaction between the oxygen and the metal arising from the unusual position of phenoxide ligand as detected by X-ray structural studies.

1. Introduction

The six-coordinate, coordinatively unsaturated nitrosyl tris(3,5-dimethylpyrazolyl)borato complexes [Mo(NO){$HB(3,5\text{-}Me_2pyz)_3$}XY], where X = Y = Cl or I; X = Cl or I, Y = OR, OAr, NHR, NHAr, *etc*) are redox active, undergoing a one-electron reduction to form the paramagnetic monoanionic species $[Mo(NO)\{HB(3,5\text{-}Me_2pyz)_3XY]^-$. The neutral precursor formally contains a metal atom having a 16-electron configuration, whereas the metal in the reduced species has a 17-electron configuration.

The reduction potentials for the reduction of [Mo(NO){$HB(3,5\text{-}Me_2pyz)_3XY$] are dependent on the nature of the donor atoms in Y, but are also significantly influenced by the substituents attached to these donor atoms. Thus, the $E_{1/2}$-values are shifted to increasingly cathodic potentials in the order Y = Cl < OPh < NHPh [1].

A. J. L. Pombeiro and J. A. McCleverty (eds.),
Molecular Electrochemistry of Inorganic, Bioinorganic and Organomettallic Compounds, 95–99.

A detailed analysis of these electrochemical and related spectroscopic data lead to the conclusion that the $\{Mo(NO)L^2\}^+$ group is a powerful electron acceptor. Taking into account the electronic and steric preferences of the tris(pyrazolyl)borato ligand for six-coordination we have begun to explore the consequences of increasing the spatial requirements of substituents on the pyrazolyl rings themselves. Our interest is to establish whether pyrazolyl ring substituents can act as a further fine-tuning agent for the reduction potentials of this class of compound, and also to explore whether, by increasing the bulk of the tripodal ligand, exceptional properties can be conferred on the metal complex, *e.g.* lower coordination numbers, substantial alteration of redox properties.

2. Characterisation and Structural Results

Two series of complexes have been prepared, derived from $[HB(3,5\text{-}Pr^iMepyz)_3]^-$, L', and $[HB\{(3\text{-}p\text{-}MeOC_6H_4)pyz)\}_3]^-$, L", of the general type [Mo(NO)L'XY] (where X = Y = CO. Cl, I; X = Cl, Y = OPh, OC_6H_4Me-*o* and NHC_6H_4Me-*o* , and [Mo(NO)L"XY] (where X = Y = CO, Cl; and X = Cl, Y = OPh, OC_6H_4Me-o)[2]. The complexes have been characterised spectroscopically (^1H nmr, mass, ir) and their formulations are satisafactorily established. The structures of representative complexes have been determined crystallographically and are shown in Figures 1, 2 and 3.

3. Electrochemical Results

All the new complexes, with the exception of the dicarbonyls, formally contain molybdenum having a 16-electron configuration, and most of those undergo a reversible or quasi-reversible one-electron reduction, affording the 17-electron monoanion.

The electrochemical data are summarised in the Table.

X	Y	EPa (V)	$E_{1/2}$ (V)	X	Y	EPa (V)	$E_{1/2}$ (V)
CO	CO	1.03		CO	CO	1.04	
Cl	Cl		0.14	Cl	Cl		0.082
I	I		0.24				
I	OPh		-0.34	Cl	OPh		-0.60
Cl	OC_6H_4Me-*o*		-0.41	Cl	OC_6H_4Me-*o*		-0.43
Cl	NHC_6H_4Me-*o*		-0.85	Cl	NHC_6H_4Me-*o*		-0.89

Values obtained in CH_2Cl_2 solution

The major influence on the reduction potentials is the nature of the donor atoms associated with X and Y, as has been reported elsewhere. The effect of placing *ortho* substituents in the phenyl rings of the phenoxides and anilides does not appear, initially, to influence the values of the reduction potential

There are no major quantitative differences between the $E_{1/2}$ values for the one-electron reduction of the complexes containing $[HB(3,5\text{-}Pr^iMepyz)_3]^-$ (L') and those containing $HB(3,5\text{-}Me_2pyz)_3$, which is perhaps not too surprising, since there are no major structural differences between complexes of the two ligands. Furthermore, the

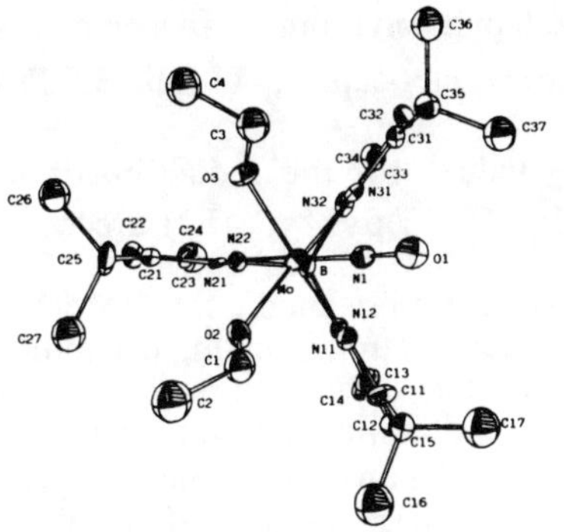

Figure 1:structure of [Mo(NO)L'(OEt)2]

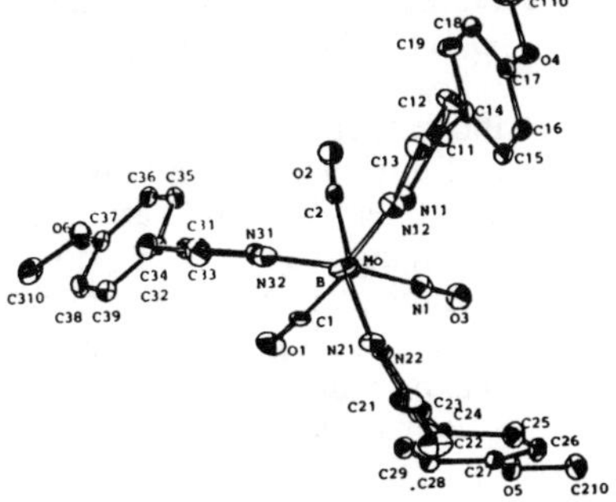

Figure 2:structure of [Mo(NO)L"(CO2)2]

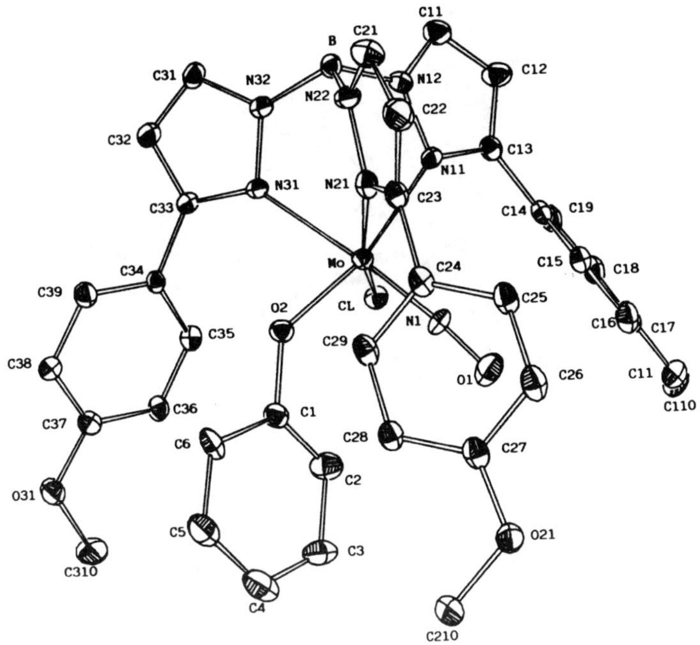

Figure 3: structure of[Mo(NO)L"Cl(OPh)](Cano, Heras *et al*, 1992).

replacement of one methyl group in each pyrazolyl ring by an isopropyl group is not going to make a large difference to the electron donor capacity of $[HB(3,5\text{-}Pr^iMepyz)_3]^-$ versus $[HB(2,3\text{-}Me_2pyz)_3]^-$.

However, when the the $E_{1/2}$ values for the reduction of the corresponding complexes containing $[HB\{(3\text{-}p\text{-}MeOC_6H_4)pyz\}_3]^-$ (L") are compared with those containing L', some significant differences can be detected. Thus, reduction of $[Mo(NO)L"Cl_2]$ occurs at a potential 60 mV more anodic than $[Mo(NO)L'Cl_2]$, which is surprising in view of the expected stronger electron donating capabilities of L'versus L". However, structural studies indicate that this difference may reflect steric interactions between the tripodal ligand and the co-ligands which reduce the strength of binding of the tripodal ligand[3]. The effect is relatively small, however. A much larger divergence occurs with $[Mo(NO)L'Cl(OC_6H_5)]$, where the $E_{1/2}$-value is -0.60 V vs. SCE. Chlorophenoxides containing L' or $\{HB(3,5\text{-}Mepyz)_3\}$ normally reduce at *ca.* -0.4 V, and this substantial cathodic shift is quite unexpected. The related *o*-methylphenoxy species $[Mo(NO)L"Cl(OC_6H_4Me\text{-}o)]$ and $[Mo(NO)L'Cl\text{-}(OC_6H_4Me\text{-}o)]$ are reduced at *ca.* -0.42 V

The unexpected cathodic shift of the reduction potential in [Mo(NO)L"Cl(OPh)] prompted our examination of the structure of this species. In most respects, the structure exhibits expected characteristics, there being a linear Mo–N–O arrangement and the Mo–O distance is relatively short (1.90 Å) as expected because of p_π–d_π donation from O to Mo. What is unusual however, is the orientation of the plane of the atoms defined by the Mo–O and the phenyl ring of the OPh group, in that it is almost coplanar with one of the three pyrazolyl rings and parallel to the H–B–Mo axis. This arrangement may be induced by the relative orientations of the *p*-$MeOC_6H_4$ groups attached to the pyrazolyl rings (this is illustrated in Figure 4).

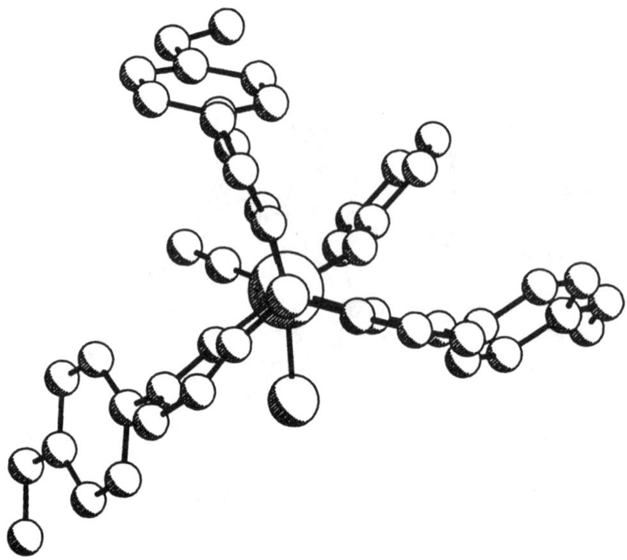

Figure 4. A view of the structure down the B-Mo axis showing the orientation of the anisoyl substituents.

The relative orientations of these groups in the dicarbonyl $[Mo(CO)_2(NO)L'']$ are similar but have no apparent steric consequences. It is not possible to make a direct structural comparison with phenolate complexes of other tris(pyrazolyl)borato complexes, but this strong orientational effect is absent in the formally isoelectronic anilido species $[Mo(NO)\{HB(3,5\text{-}Me_2pyz)_3\}I(NHC_6H_4R\text{-}p)]$ (R = Me or OMe)[4]. Certainly other work has indicated that orientational effects can have a pronounced effect on reduction potentials in these systems[5].

We are drawn to the conclusion that, because of the steric arrangements of the *p*-$MeOC_6H_4$ groups, the phenoxide ligand is forced into a position which effectively enhances the opportunity for p_π–d_π interaction between the oxygen atom and the metal. This, in effect, transfers more charge than normally encountered in phenoxide species into the metal, and consequently the reduction potential is more cathodic than expected. Further correlation between structure and electrochemical properties is clearly necessary.

4. Acknowledgments

We are grateful to MEC (Spain D.G.I.C.Y.T. PB/890127-project, and to the British Council/Acciones Integradas Programme for support of this, and related work.

5. References

1. N.Al Obaidi, M. Chaudhury, D. Clague, C.J. Jones, J.C. Pearson, J.A. McCleverty, and S.S. Salam, *J.Chem.Soc., Dalton Trans.*, 1987, 1733-1736; N.Al Obaidi, M. Charsley, W.Hussain, C.J. Jones, J.A. McCleverty, B.D. Neaves, and S. Reynolds,*Trans. Met.Chem.*, 1987, **12**, 143-148.

2. M. Cano, J.V. Heras, C.J. Jones, and J.A. McCleverty, *Polyhedron*, 1990, **9**, 4, 619-621; M. Cano, J.V. Heras, S. Trofimenko, A. Monge, E. Gutierrez, C.J. Jones, and J.A. McCleverty, *J.Chem.Soc., Dalton Trans.*, 1990, 3577-3582.

3. M. Cano, J.V. Heras, A. Monge, E. Gutierrez, C.J. Jones, and J.A. McCleverty, *Polyhedron*, to be published.

4. J.A.McCleverty, G.Denti, S.J.Reynolds, A.S.Drane, N.El Murr, A.E.Rae, N.A.Bailey, H.Adams and J.M.A.Smith, *J.Chem.Soc. Dalton Trans.*, 1983, 81-89.

5. N.Al Obaidi, C.J. Jones, and J.A. McCleverty, *Polyhedron*, 1989, **8**, 1033-1037.

REDOX PROPERTIES OF SYMMETRICAL AND UNSYMMETRICAL SANDWICH COMPLEXES OF LUTETIUM(III) WITH PHTHALOCYANINATO AND NAPHTHALOCYANINATO LIGANDS

M. L'HER, A. PONDAVEN, Y. COZIEN and F. GUYON
URA CNRS 322, Faculté des Sciences et Techniques,
Université de Bretagne Occidentale,
6 avenue Le Gorgeu, 29287 BRESI-CEDEX, France

ABSTRACT. Phthalocyanines, naphthalocyanines and their sandwich complexes with the lanthanoids are of great interest as precursors of materials. The macroscopic properties of these materials are of course related to the electronic states of the molecular units. Many sandwich complexes have been synthetized, symmetrically or unsymmetrically substituted with electron donating and withdrawing groups. The effect of the substituants have been studied by voltammetry and spectroelectrochemistry. The unpaired electron of these neutral radical species is delocalized over the sandwich complex, even for the most unsymmetrical compounds.

Since a long time, phthalocyanines are considered as being good candidates for molecular materials; they are thermally, chemically and photochemically stable. Solid phases can be obtained from the molecular units by vacuum deposition, spin-coating, electrocrystallization and by the Langmuir-Blodgett technique.

These macrocyclic ligands and their metallic complexes are potentially interesting for many purposes : as photocatalysts for light conversion to electricity [1], photoconductive materials, semi-conductors, non-linear optical materials.. [2-5].

The conjugated system of the phthalocyanine macrocycle is at the origin of their properties. The electronic structure can be tailored for special purposes by metallation, substitution, extension of the conjugation. The sandwich complexes with lanthanoid cations are also good models to study the interactions between two adjacent macrocycles of a highly conjugated system [6,7]. Their electrochromic properties was at the origin of our interest for this class of compounds [8,9].

Adding substituents on the phthalocyanines macrocycles of the sandwich complexes is not only the way to modify the electronic properties, but is also useful for increasing the solubility of these complexes. The synthesis of unsymmetrical molecules is also important for orientation of the molecules for building films by the Langmuir-Blodgett technique. But the aim of the present work is to study the modification of the electronic structure induced by the asymmetry of the molecule.

$LuPc_2$, the model compound, is a neutral radical where the Lu^{3+} cation is associated to the phthalocyaninato ligand (Pc^{2-}) and to its oxidized form ($Pc^{\bullet-}$). The unpaired electron is delocalized over the two moieties of the complex.

A. J. L. Pombeiro and J. A. McCleverty (eds.),
Molecular Electrochemistry of Inorganic, Bioinorganic and Organometallic Compounds, 101–106.

1. Synthesis

Lanthanoids bis-phthalocyaninato complexes are usually synthetized by condensation of dicyanobenzenes in the presence of Lu(III) salts. This synthetic way is not very convenient for the preparation of unsymmetrical sandwiches as it leads to many substituted derivatives which separation from th: reaction medium is very difficult. The original synthetic route followed is :

4 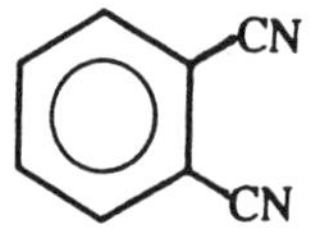 $+ 2\ MeOLi \xrightarrow[ROH]{\Delta} PcLi_2$

$$PcLi_2 + Pc'Li_2 + Lu(OAc)_3 \xrightarrow[C_6H_5Cl]{\Delta} Lu[Pc][Pc']$$

Many symmetrical and unsymmetrical derivatives were synthetized :

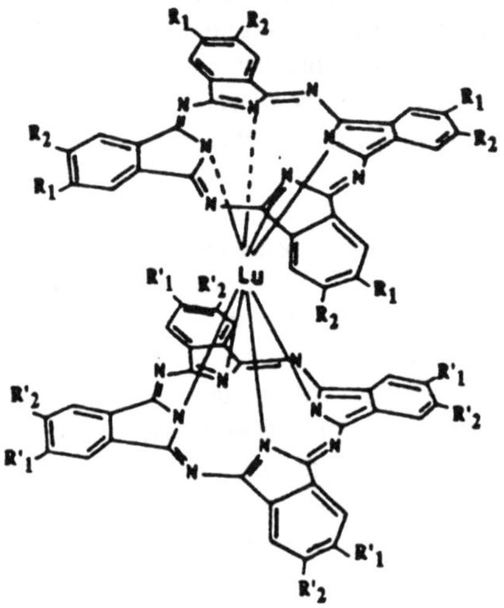

	R_1		R_2	R'_1		R'_2
1. $Lu[Pc]_2$	H		H	H		H
2	CH_3		CH_3	CH_3		CH_3
2'	CH_3		CH_3	H		H
3	OCH_3		OCH_3	OCH_3		OCH_3
3'	OCH_3		OCH_3	H		H
4		tBu		tBu		
4'		tBu	H	tBu	H	
5	Cl		Cl	H		H
6		tBu		Cl		Cl
7 Lu[1,2-Nc]		c-but.*			c-but.*	
8. Lu[2,3-Nc]		c-but.*			c-but.*	
9. Lu[Pc][1,2-Nc]		c-but.*	H	H		

*cis-butadiène

Figure 1 : Lutetium sandwich complexes

2. Electrochemistry and spectroelectrochemistry of $LuPc_2$

The redox properties of the sandwich complexes have been studied by cyclic-, rotating disk electrode voltammetry and electrolysis, in different solvents (CH_2Cl_2,DMF,DMSO,benzonitrile). The reference for the potentiel scale is the ferri/ferrocenium couple.

The cyclic voltammogram for $LuPc_2$ in CH_2Cl_2 is presented on figure 2 and the potentials in table 1. All the system are monoelectronic and reversible.

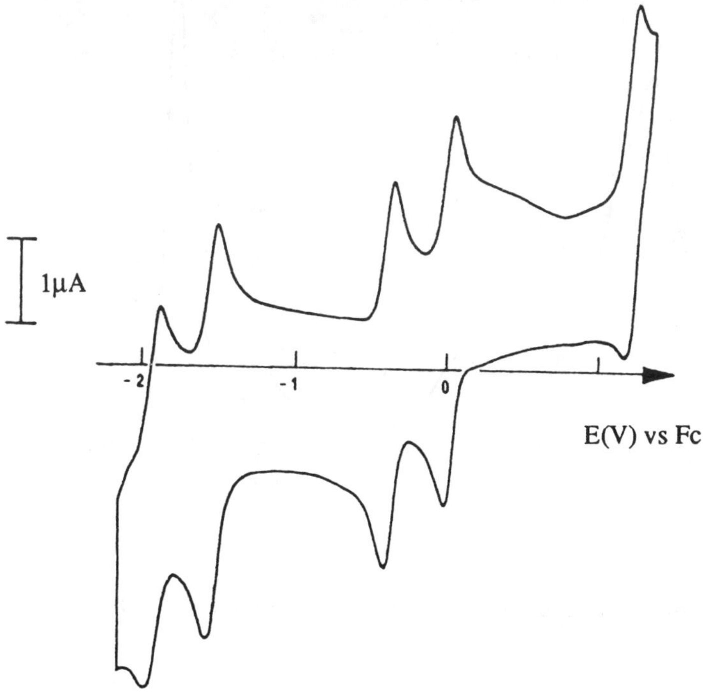

Figure 2 : Cyclic voltammogram of $LuPc_2(CH_2Cl_2)$

Table 1 : potentials of the $LuPc_2$ redox system

-2.37V -1.92V -1.54V -0.46V -0.05V 1.05V 1.43V

$LuPc_2^{4-}$	$LuPc_2^{3-}$	$LuPc_2^{2-}$	$LuPc_2^{-}$	$LuPc_2$	$LuPc_2^{+}$	$LuPc_2^{2+}$	$LuPc_2^{3+}$
	purple	dark blue	turquoise	green	yellow	red	red

The electronic spectra of $LuPc_2$ or its mono-oxidized and-reduced forms are shown on figure 3.

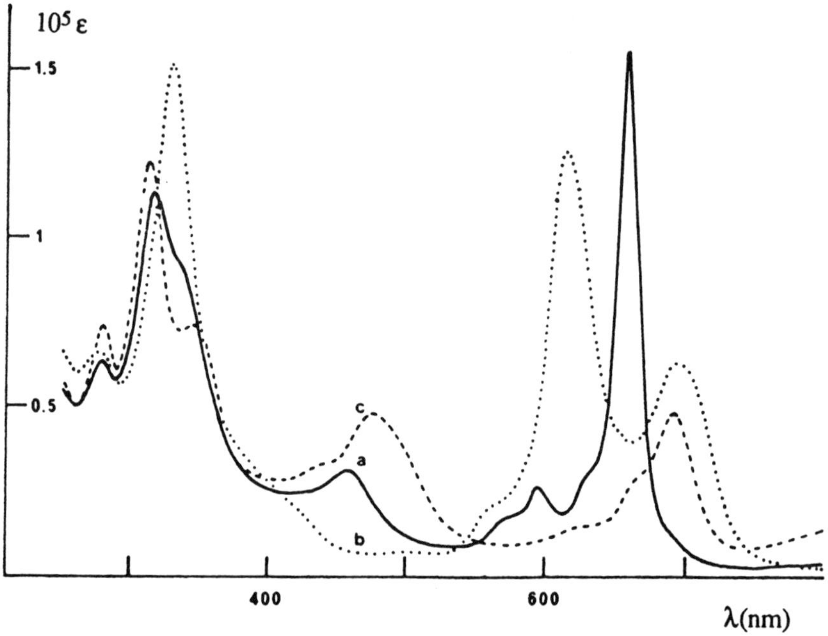

Figure 3 : Spectra of $LuPc_2$ (a), $LuPc_2^-$ (b), $LuPc_2^+$ (c).

3. Effect of the substituents

The first oxidation (E^o_1) and reduction (E^o_2) potentials of the substituted bis-phthalocyanines appear on table 2, with their difference (ΔE) and the sum of the Hammett s constants (Σσ). Figure 4 illustrates the effect of the electron donating or withdrawing power of the substituents.

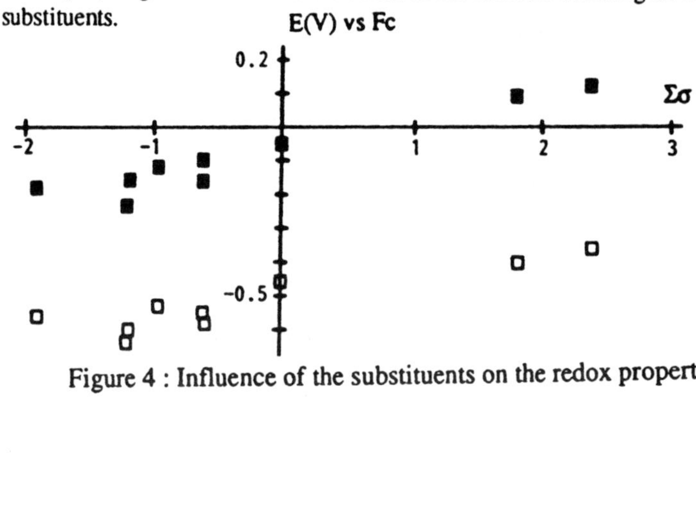

Figure 4 : Influence of the substituents on the redox properties.

Table 2 : Redox potentials (V vs Fc) of the sandwich complexes

	$\Sigma\sigma$	E^o_2	E^o_1	ΔE
1. $Lu[Pc]_2$	0	-0.05	-0.46	-0.41
2. $Lu[(CH_3)_8Pc]_2$	-1.90	-0.18	-0.56	-0.38
2'. $Lu[(CH_3)_8Pc][Pc]$	-0.95	-0.12	-0.53	0.41
3. $Lu[(CH_3O)_8Pc]_2$	-1.22	-0.23	-0.64	0.41
3'. $Lu[(CH_3O)_8Pc][Pc]$	-0.61	-0.16	-0.55	0.39
4. $Lu[(tBu)_4Pc]_2$	-1.18	-0.16	-0.60	-0.44
4'. $Lu[tBu)_4Pc][Pc]$	-0.59	-0.10	-0.58	0.48
5. $Lu[Cl)_8Pc][Pc]$	2.40	0.12	-0.36	0.48
6. $Lu[(tBu)_4Pc][(Cl)_8Pc]$	1.81	0.10	-0.40	0.50
7. $Lu[1,2\text{-}Nc]_2$		0.07	-0.26	0.33
8. $Lu[2,3\text{-}Nc]_2$		-0.34	-0.62	0.28
9. $Lu[1,2\text{-}Nc][Pc]$		0.05	-0.35	0.40

4. Discussion

As predicted by VEH calculation [6,7], interaction between the two rings induces splitting of the molecular orbitals. The HOMO of $LuPc_2$ and related compounds is half occupied, as one oxidized phthalocyanine ligand (Pc^-) provides one unpaired electron. The band observed at about 460nm (figure 3) is attributed to electronic transitions from lower levels to the HOMO. One major band is observed on the spectrum of the sandwich complexes in the Q-band region (650-700nm, figure 3). The Soret, or B-band (320nm), is much more complex. In $LuPc_2$, the difference between the HOMO and LUMO is close to 0.4V. Dimerization of monomeric phthalocyanines like H_2Pc induces a decrease of the gap between the oxidation and reduction steps, from 1.9V to 1.5V. In the sandwich complexes the redox state equivalent to a dimeric species is $LuPc_2^{-}$; the difference between the oxidation and reduction of this redox form is even lower : 1.1V.

The effect of the substituents is as expected from the influence of their electron donating or withdrawing properties. The same trend is observed when one considers the Q band of the visible spectra. From the spectral data, the delocalization of the hole on the two rings is evident, even for the more asymmetric derivatives : for example, the electronic spectra of $Lu[Pc][Pc(Cl)_8]$ and $Lu[Pc(tBu)_4][Pc(Cl)_8]$ are different from the ones expected for compounds including two non interacting different phthalocyanine moieties

5. References

[1] Wöhrle D. and Meissner D. (1991) "Organic solar cells" Advanced material,3,129-138.

[2] Simon J., André J.J. and Skoulios A. (1986) "Molecular Materials I. Generalities". New J.Chem.,10,295-311.

[3] Simon J., Tournilhac F. and André J.J. (1987) "Molecular Materials II. Towards Electronics Finalities". New J.Chem.,11,383-399.

[4] Simon J., Bassoul P. and Norvez Z. (1989) "Molecular Materials III. Towards Opto-Electronics Finalities". New J.Chem.,13,13-31.

[5] Marks T.J. (1990) "Interfaces between Molecular and Polymeric "Metals" : Electrically conductive, Structure-Enforced Assemblies of Metallomacrocycles". Angew.Chem.Int.Ed.Engl.,29,857-879.

[6] Orti E., Piqueras M.C., Crespo R. and Bredas J.L. (1990) "Influence of Annelation on the Electronic Properties of Phthalocyanine Macrocycles".Chem.Materials,2,110-116.

[7] Orti E., Brédas J.L. and Clarisse C. (1990) "Electronic Structure of Phthalocyanines : Theoretical Investigation of the Optical Properties of Phthalocyanine Monomers, Dimers and Crystals".

[8] L'Her M., Cozien Y. and Courtot-Coupez J. (1983) "Electrochemical Behaviour of Lutetium diphthalocyanine in Methylene Chloride". J.Electroanal.Chem.,157,183-187.

[9] L'Her M., Cozien Y. and Courtot-Coupez J. (1986) "Etude électrochimique de la réduction de la diphtalocyanine de lutétium en solution". C.R.Acad.Sc.Paris,302-II,9-14.

NEW POLYDENTATE LIGANDS AND COMLEXES: PROTIC AND TOPOGRAPHIC EFFECTS ON REDOX PROPERTIES

M. GROSS, A. NURHADI and E. GRAF
Laboratoire d'Electrochimie et de Chimie Physique du Corp Solid
U.R.A. au C.N.R.S. n° 405
Université Louis Pasteur, 4 Rue Blaise Pascal,
67000 Strasbourg, France

ABSTRACT

New ligands and complexes have been designed and synthesized with the purpose of improving the understanding of the relationships between their chemical composition or topography and their redox reactivity . Specifically , the Fe(II) , Co(II) , Ni(II) complexes with these new ligands exhibit the interesting characteristics of having , in each molecule , two distinct redox active sites involving separately the ligand and the coordinated metal and, simultaneously, these ligands and their complexes exhibit redox potentials adjustable through the protic acidity of the solution . Additionally, the Iron(II) complexes , upon reduction in protic conditions, were able to achieve a fast, catalytic reduction of molecular oxygen with reliable indications that this homogeneous reaction is a four electron conversion of O_2 to H_2O .

INTRODUCTION

A constant trend towards a better understanding of the relationship between redox - and more generally physicochemical - properties of the metallo-organic molecules and their composition and topography , involved in the recent years the building of ligands with specific architecture .

This approach led us to design , and further to synthesize , a new family of tetradentate open chain ligands L_{1-4} **(Fig. 1)** . These ligands **(** 1 **)** undergo a one electron or a two electron reduction, depending on the concentration of

A. J. L. Pombeiro and J. A. McCleverty (eds.),
Molecular Electrochemistry of Inorganic, Bioinorganic and Organomettallic Compounds, 107–119.

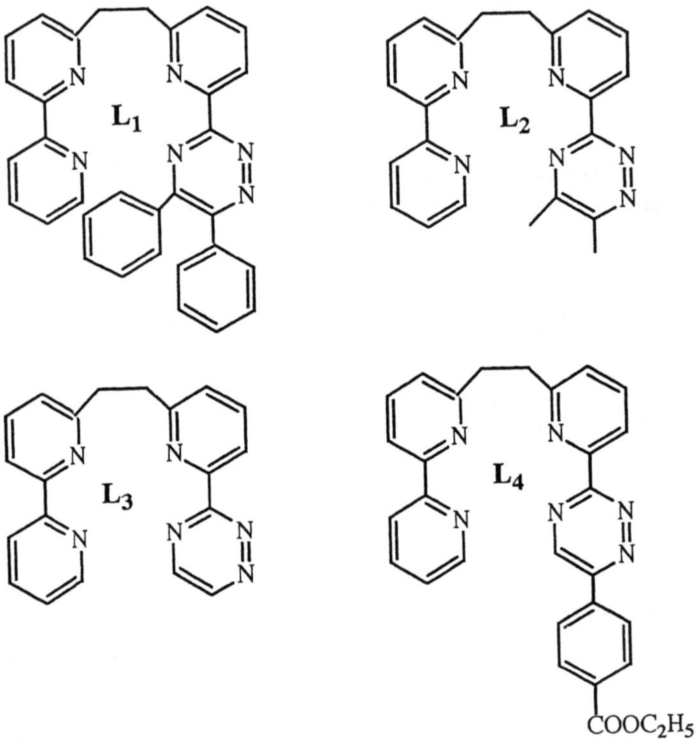

Figure 1 . The ligands studied

the protons in the solution (2) . Due to the presence in the molecule of the two moieties bipyridyl and pyridyl-1,2,4-triazine , the ligands L_{1-4} are expected to behave as non cyclic chelating ligands towards transition metals . The ligands L_{1-4} were designed so as to exhibit a good coordination flexibility , due to the ethyl bridge linking the bipyridyl to the pyridyl-1,2,4-triazine subunit . Thus, the nitrogen heteroatoms can coordinate from either a planar or a non planar arrangement.

MATERIALS AND METHODS

The synthesis and the main chemical characteristics of the ligands have been published (1) (2) . The electrochemical methods and the experimental procedures were the same than indicated for the study of synthons of the present molecules (3) .

RESULTS AND DISCUSSION

LIGANDS

In DMF + 0.1 M TBAP , the ligands L_{1-4} are electroreducible via several distinct steps as indicated in the following table **(Table 1)** . Noteworthy is the quasi reversible one electron transfer in the first reduction step of all four ligands . It is striking that the ligand L_1 merely capitalizes the contributions of the two subunits bipyridyl and pyridyl-1,2,4-triazine , whereas the ligands L_3 and L_2 are reducible via only three distinct steps , although the total amount of Faradays transferred per mole of ligand remains identical and equal to 13 . This effect is clearly due to the substituents in the 5 and 6 positions on the triazine , the separation of the reduction steps being favoured by the electron-attracting substituents . This significant effect of the substituents on the 5 and 6 positions of the pyridyl-1,2,4-triazine is illustrated by the linear correlation between the first reduction potential ($E_{1/2}$) and the frequency of the $n \rightarrow \pi^*$ transition in the diazo group of the triazine . Further , this correlation provides a clue to the involvment of the diazo-N=N-group in this first electron transfer . These first indications will receive full confirmation from the results reported hereafter on the redox behaviour of the ligands L_{1-4} in the presence of protic or electrophilic reagents.

Upon addition of protic acids to the study solution of L_{1-4} in DMF (+ 0.1 M TBAP) , the first reduction step is facilitated and it changes from a 1 e^- transfer to a 2 e^- , 2 H^+ reaction.

This is illustrated by the following figure corresponding to L_1 in the presence of $HClO_4$ **(Fig. 2)** .

Noteworthy is also that , beyond 2 H^+ / 1 L , the reduction of the bipyridyl moiety at - 2.20 V/SCE changes also from a 1 e^- process to a $2e^-$ reaction much facilitated at - 0.91 V/SCE .

Cross examination of the effects of the addition of H^+ to the solution and of the effects of pyrazine-substituents on the energy required (i.e. $E_{1/2}$) for the first reduction step of the ligands L_{1-4} indicated that :

Table 1 : Reduction of the ligands studied and of synthons .

Bipy.-2,2'				Ligand L_1 ($R_1 = R_2 = Ph$)				PDT (3)			
Tast Pol.			CV	Tast Pol.			C V	Tast Pol.			C V
$E_{1/2}$	p	n / n#		$E_{1/2}$	p	n / n#		$E_{1/2}$	p	n / n#	
				-1,44	70	1 / 2	q.rév.	-1,45	70	1 / 2	q.rev.
				-2,06	105	1 / 1	irrév.	-2,17	100	1 / *	irrev.
-2,10	60	1 / 1	rév.	-2,20	60	1 / 1	rev.				
				-2,56	90	4 / 4	*	-2,40	50	3 / *	*
				-2,64	110	3 / 3	*	-2,63	100	4 / *	*
-2,67	95	2 / *	irrev.	-2,76	105	3 / 3	*				

Ligand L_2 ($R_1 = R_2 = CH_3$)				Ligand L_3 ($R_1 = R_2 = H$)				Ligand L_4 ($R_1 = H$, $R_2 = PhCO_2C_2H_5$)			
Tast Pol.			C V	Tast Pol			C V	Tast Pol			C V
$E_{1/2}$	p	n/n#		$E_{1/2}$	p	n/n#		$E_{1/2}$	p	n/n#	
-1,73	95	1/2	q.rev.	-1,54	58	1/2	q.rév.	-1,19	68	1/2	q.rev.
								-1,66	75	1/1	irrev.
-2,18	57	2/2	irrév.	-2,17	65	4/4	irrev	-2,17	63	2/2	q.rév.
								-2,33	37	2/*	*
-2,70	95	10/*	*	-2,70	105	8/*	*	-2,72	67	7/*	*

Solvent :
DMF
+ 0.1 M TBAP

Electrode :
Hg

$E_{1/2}$: V/SCE

p : mV / log

I_{lim} :
μA / mmole

n# :
from coulometry

Figure 2

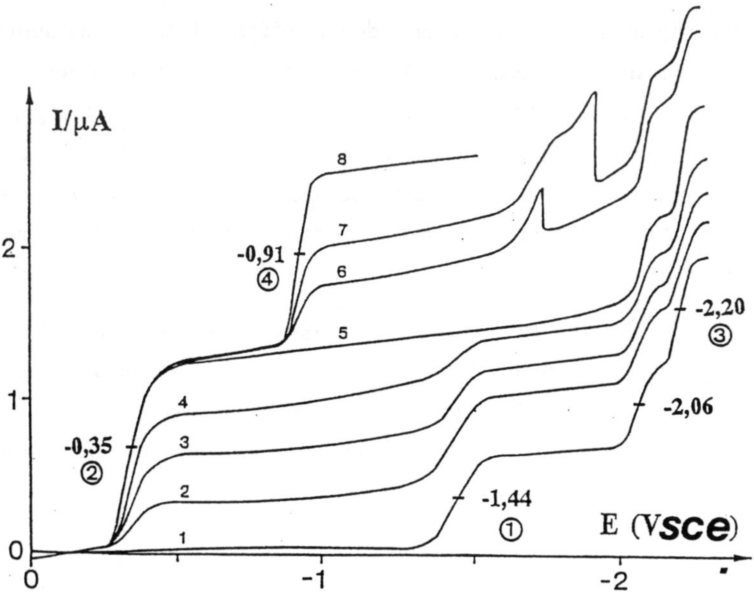

Tast polarography of L_1 ($5 . 10^{-4}$ M) in DMF (+ 0.1 M TBAP) in the presence of variable amounts of $HClO_4$.
($HClO_4$)/ (L_1) = 0 (curve 1) ; 0.5 (2) ; 1.0 (3) ; 1.5 (4) ; 2.0 (5) ; 2.5 (6) ; 3.0 (7) ;4.0 (8) .

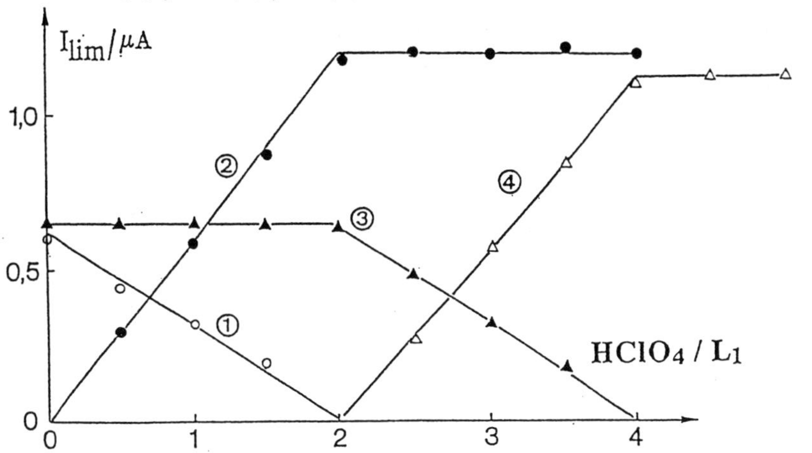

- Strong acids significantly facilitate the first reduction of L_{1-4} , whilst weak acids only slightly do it .

- Electron inductive attractor / donor effects from substituents on the pyrazine (positions 5 and 6) allow discrimination between respective basicities of the L_1 to L_4 ligands , in the presence of strong protic acidity . As this discrimination (" titration ") is not effective in the presence of a weak acid , it may reasonably be speculated that the basicities of all four ligands is weak , no matter the substituents are electron donor or electron attractors .

- Consistent with this result is the observation that , although Hammett's plots are linear in the absence and in the presence of a protic acid added to the solution , the slope of the linear plots is much smaller in the presence of a strong acid than in the absence of added H^+ or in the presence of a weak protic acid . This levelling of the effects of substituents on the energy required to achieve the first electron transfer to the ligands L_1 to L_4 , is consistent with the identification of the diazo group - N = N - as the protonating electroactive site involved in the first reduction .

Thus , the conversion of the 1 e^- reduction to the 2 e^- / 2 H^+ process is schematically described by the following reduction scheme **(Fig. 3)** .

IRON , COBALT , NICKEL COMPLEXES

The ligands L_{1-4} spontaneously generate stable coloured 1/1 complexes with the above M(II) cations . The 1/1 stoichiometry resulted unambiguously from optical density measurements at the peak wavelengths characterizing respectively the uncoordinated ligands and the M(II) complexes . Identical conclusions were obtained upon addition of M(II) salt to an initial solution of uncoordinated ligand L_1 and also upon addition of L_1 to an initial solution of M(II) salt . The stoichiometry 1/1 was also demonstrated from monitoring the limiting currents corresponding to the first reduction of the complexes (M(II) L_{1-4} $)^{2+}$, when increasing amounts of an M(II) salt were added to a solution of ligand L .

Figure 3 : Reduction scheme of the ligands L_{1-4} .

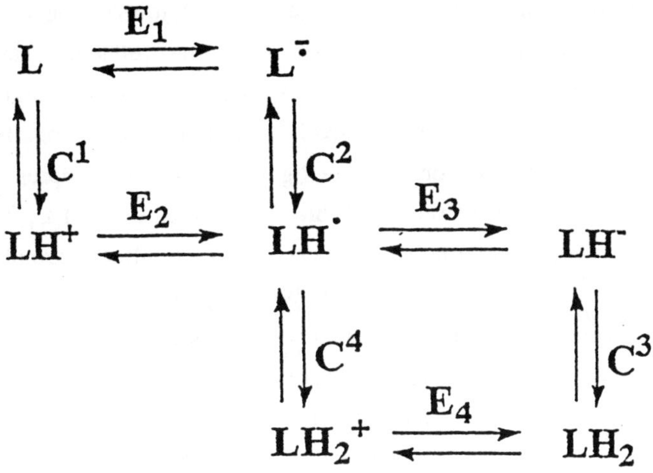

WEAK ACID :

$E_1C^2E_3C^3$ $\quad |E_3| \leq |E_1|$

OR

$E_1C^2C^4E_4$ $\quad |E_4| \leq |E_1|$

global : $L + 2\,HA + 2\,e^- \longrightarrow LH_2 + 2\,A^-$

STRONG ACID :

$C^1E_2E_3C^3$ $\quad |E_3| \leq |E_2|$

OR

$C^1E_2C^4E_4$ $\quad |E_4| \leq |E_2|$

{with L_1, L_3, L_4}

$C^1E_2E_3C^3$ $\quad |E_3| > |E_2|$ {with L_2}

global : $L + 2\,H^+ + 2\,e^- \longrightarrow LH_2$

It shall be noticed here that the M(II) coordinating sites are clearly not affected by the protonation of the ligand . This indicates that the diazo group is not involved in the coordination process .

The Iron(II) complexes exhibit specific and interesting characteristics , reported hereafter . All four iron(II) complexes are reducible via four distinct steps , of which only the first is reversible . Each of the three first reductions is a one electron process , which involves actually the complex , whereas the fourth reduction step occurs on the ligand . As an example the corresponding characteristics are reported in **Table 2** for the (Fe(II) L_1 $)^{2+}$ reduction. Electrochemical measurements were carried out on these complexes , taking advantage of the wide time scale available (cyclic voltammetry , stationary voltammetry , exhaustive potentiostatic coulometry) and of the coupling of results from electronic absorption spectra with the electrochemical results . The results obtained indicate that the first , one electron reduction generate a stable complex in which the electron is accepted in a molecular orbital largely based on the coordinated iron . In all cases also the second electron introduced into the complex generates iron(0) and free , uncoordinated ligand. Thus , the third , one electron reduction results logically in the generation of the reduced , uncoordinated ligand .

Table 2 : Reduction of [Fe^{II} L_1 $]^{2+}$ in DMF + 0.1 M TBAP

Reduction	$E_{1/2}$ (V/SCE)	# (mV/unit log)	I_{lim} (μA/mmol)	n (polaro)	n (coulo)
[FeL_1 $]^{2+}$					
1st	- 0.82	60	1.30	1	1
2d	- 1.41	70	1.02	1	1
3d	- 1.58	100	0.84	1	1
4th	- 2.20	80	2.80	3	*
L_1					
1st	- 1.44	70	1.32	1	2
$Fe^{2+} \rightarrow Fe^0$	- 1.37	40	1.20 x 2	2	2

(#) : $E = f (\log I / I_{lim} - I)$.

In the presence of strong protic acids , the first reduction is facilitated (for instance in $[Fe^{II} L_1]^{2+}$ $\Delta E_{1/2}$ = + 0.12 Volt) and , simultaneously , it changes from a " pure " one electron transfer to a two-electron + two protons reaction . When compared with the free , uncoordinated ligands L_{1-4} , the first reduction potential of the complexes $[Fe^{II} L_{1-4}]^{2+}$ exhibits the following specific features :

- The slope of the Hammett's $E_{1/2}$ vs . σ_m linear plot remains unaffected by the presence of protic acids (whilst strong changes were observed for the free ligands in the presence of strong protic acid)
- The shift of the first reduction potentials of $[Fe^{II} L_{1-4}]^{2+}$ induced by the added protons is moderate and almost constant in the series ($\Delta E_{1/2}$ = + 0.115 $\pm$ 0.015 Volt) . This is another striking difference with the results obtained with the free ligands L_{1-4} in the presence of strong protic acid .

This behavior of the complexes $Fe^{II} [L_{1-4}]^{2+}$ in the presence of strong protic acids is quite similar to the behavior observed for the free ligands L_{1-4} in the presence of weak protic acids .

These results thus indicate that , whereas $HClO_4$ was able to discriminate between the various basicities of the ligands L_1 to L_4 , this strong protic acid was not able to discriminate between the Iron(II) complexes in the series . Taking into account that the diazo group is the site of this $2 e^- / 2 H^+$ reaction (3) , the above results may reasonably be explained in terms of weakened basicity of the diazo group in the complexes, due to the coordination of Iron(II) Lewis acid in the ligands L_1 to L_4 .

A common characteristics for the Iron(II) complexes $[Fe^{II} L_{1-4}]^{2+}$ and for the corresponding free ligands L_{1-4} shall be noticed : the $2 e^- / 2 H^+$ reduction observed in the presence of protons added to the solution is chemically reversible , thus allowing the reduced species $[Fe^{II} L H_2]^{2+}$ to quantitatively revert to the original complex $[Fe^{II} L]^{2+}$. The chemical stability of the reduced species $[Fe^{II} L H_2]^{2+}$ is especially good when the ligand L_1 is involved .

The reactivity of the reduced species $[Fe^{II} L_1 H_2]^{2+}$ is remarkably high with molecular oxygen and this fact desserves specific comments .

This behavior is illustrated by the followings . First , the complex $[Fe^{II} L_1]^{2+}$ was quantitatively reduced (potentiostatic coulometry) to $[Fe^{II} L_1 H_2]^{2+}$, in the presence of $HClO_4$ (stoichiometry 2 $HClO_4$ / 1 $[Fe^{II} L_1]^{2+}$) . The obtained $[Fe^{II} L_1 H_2]^{2+}$ was identified by its oxidation signal at $E_{1/2} = -0.20$ V/SCE (**Fig. 4** curve (3)) .
This curve (3) was unambiguously distinct from the voltammetric signal linked to the initial complex $[Fe L_1]^{2+}$ in the absence (curve (1)) and in the presence (curve (2)) of added protons . Bubbling molecular oxygen in the solution containing the reduced complex $[Fe L_1 H_2]^{2+}$ regenerates the original non-reduced complex $[Fe L_1]^{2+}$ (**Fig. 4** curve (4)) . This reaction occurs almost instantly on the time scale of stationary voltammetry .

On an other hand , it was observed that the limiting current for the first reduction of the complex $[Fe L_1]^{2+}$ + 2 $HClO_4$ increases with the O_2 content in the solution , and also that , in a given solution of $[Fe L_1]^{2+}$ saturated with O_2 , the limiting current linearly increases with $[H^+]$ on the range $2 < (HClO_4) / [Fe L_1]^{2+} < 4$. Thus the chemical reaction with O_2 is catalytic and it involves the hydrogens from $[Fe^{II} L_1 H_2]^{2+}$. Many chemical and electrochemical analyses were carried out in the solution after this reaction with O_2 had occured . None was able to detect any trace of H_2O_2 . Also , no chemical reaction occurred when H_2O_2 was added to the solution containing $[Fe^{II} L_1 H_2]^{2+}$. Thus , the observed reaction of O_2 with $[Fe^{II} L_1 H_2]^{2+}$ seems to be an homogeneous, catalytic four electrons reduction of molecular oxygen by $[Fe^{II} L_1 H_2]^{2+}$, according the scheme :

$$2\,[Fe^{II} L_1 H_2]^{2+} + O_2 \longrightarrow [Fe^{II} L_1]^{2+} + 2\,H_2O$$

This reaction may be considered as an abiotic mimicking of the classical biological oxidation reaction involving the oxidized and the reduced form of a substrate . However , at variance with the biological pattern , observed for instance in the redox cross reaction between NAD^+ / NADH and O_2 / H_2O in the respiratory mitochondrial chain , no additional redox intermediate is required in the above reaction between $[Fe^{II} L_1 H_2]^{2+}$ and molecular O_2 .

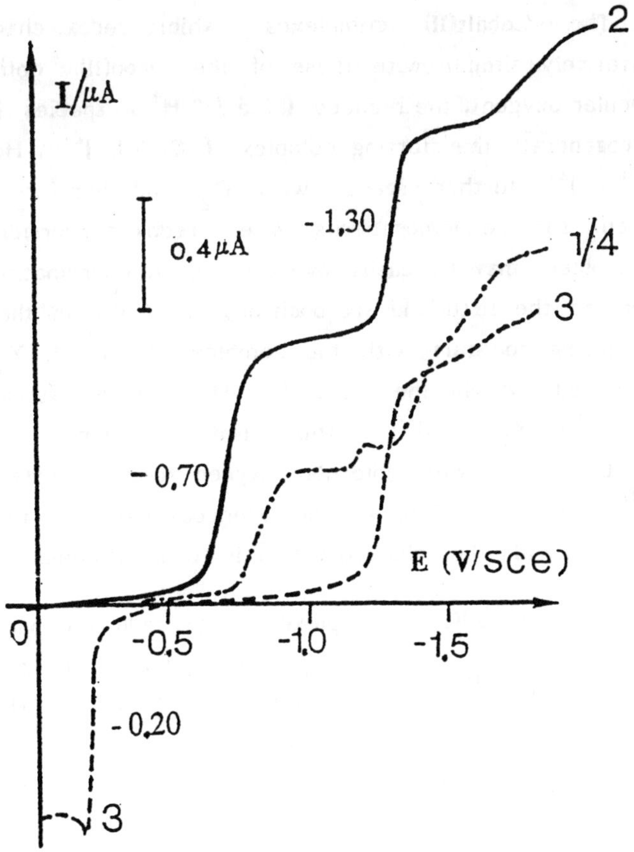

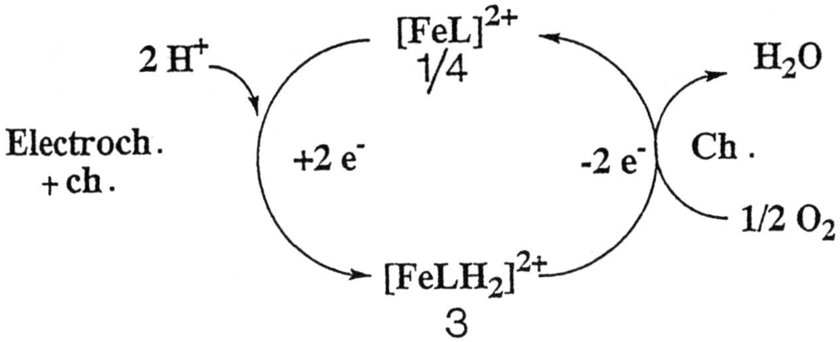

Figure 4 : Reduction of [Fe L_1]$^{2+}$ in the presence of strong protic acid , and reaction with molecular oxygen .

The Cobalt(II) complexes exhibit redox characteristics which are qualitatively similar with those of the Iron(II) complexes . Also , with molecular oxygen , the reduced (2 e / 2 H^+) species [Co^{II} L H_2 $]^{2+}$ react and regenerate the starting complex [Co^{II} L $]^{2+}$. However , this complex [Co^{II} L $]^{2+}$ further reacts with O_2 and therefore it cannot be used , as such , for the homogeneous 4 e^- redox conversion of O_2 to H_2O . This problem may be easily overcome upon coordinating proper monodentate ligands to the fifth and six positions of Co^{II} in the complex Co^{II} L_1 , thus leading to start with the complex [Co^{II} L X_2 $]^{2+}$. This complex may be reduced via the 2 e / 2 H^+ above described scheme , leading to [Co^{II} L X_2 H_2 $]^{2+}$. This reduced complex further reacts , like [Fe^{II} L H_2 $]^{2+}$, with molecular oxygen to regenerate the starting complex [Co^{II} L X_2 $]^{2+}$. Thus the homogeneous catalytic four electrons reduction of O_2 may also be achieved with this cobalt(II) complex .

The Nickel(II) complexes are reducible via three distinct steps (For instance $E_{1/2}$ = - 0.69 V/SCE ; - 0.90 V/SCE and - 1.63 V/SCE for [Ni(II) L_1 $]^{2+}$ in DMF + 0.1 M TBAP) . In all [Ni(II) L_{1-4} $]^{2+}$ complexes each of the two first reductions involves one electron. These two first one electron reductions correspond to the formal conversions Ni(II) / (Ni(I) / Ni(0) in the complexes . After two electrons have been transferred - thus generating [Ni^O L] species - this species decoordinates slowly Ni^O from L . This decoordination was observable on the time scale of the potentiostatic coulometry whereas it was not detected by voltammetry : in this method Ni^O L was further reduced and significant decoordination occured only after this third reduction step .

Addition of protic acids to the solution of Ni(II) complexes triggered significant changes in their electrochemical patterns. For instance, upon addition of a strong acid ($HClO_4$) the first reduction of [Ni^{II} L_1 $]^{2+}$ is facilitated from - 0.69 to - 0.53 V/SCE and it changes from a 1 e^- to a 2 e^- reaction . This 2 e reduction at - 0.53 V/SCE is further followed by two distinct one electron reductions at - 0.71 and at - 1.24 V/SCE respectively . The two first reductions at - 0.53 and at - 0.71 are diffusion controlled . At variance with the above results , the addition of a weak acid (CH_3COOH) to solutions of $[Ni^{II} L_1]^{2+}$ does not affect the two first one electron reductions of $[Ni L_1]^{2+}$, unless the stoechiometry (CH_3COOH) / $[Ni^{II} L_1]^{2+}$ exceeds 2 .

The complexes $[Ni^{II} L_1]^{2+}$ and $[Ni^{II} L_4]^{2+}$ exhibited the interesting characteristics that their reduction in the presence of H^+ (from a strong acid) occurs with a simultaneous consumption of ne and n H^+ (when (H^+) / $[Ni^{II} L]^{2+} < 2$) the initial complex $[Ni^{II} L]^{2+}$ remaining unaltered at the end of the reduction reaction. This catalytic reduction of H^+ to hydrogen is still observed beyond (H^+) / $[Ni^{II} L]^{2+} = 2$, with however parallel competing reactions .

REFERENCES

(1) Graf E. , (1992) , Synthesis n° 6 .

(2) Nurhadi A. , Graf E. , Gross M. , (1991) , Electrochimica Acta 36, 1997 - 2001 .

(3) El Jammal A. , Graf E. , Gross M. , (1986) , Electrochimica Acta 31, 1457 - 1465 .

REDOX-STRUCTURAL CORRELATIONS IN METAL THIOETHER MACROCYCLIC COMPLEXES: THE STABILISATION OF MONONUCLEAR SILVER(II) and GOLD(II)

A.J. BLAKE, R.O. GOULD, A.J. HOLDER, T.I. HYDE, G. REID, A. TAYLOR and M. SCHRÖDER*
Department of Chemistry, West Mains Road, University of Edinburgh, Edinburgh, EH9 3JJ, Scotland
D. COLLISON
Department of Chemistry, Oxford Road, University of Manchester, Manchester, M13 9PL, England.

ABSTRACT: The complexation of Ag(I) and Au(I) by tri-, tetra-, penta- and hexathia crown thioethers is described. The mismatch between metal ion and ligand co-ordination leads to complexes which often incorporate unusual stereochemical features and therefore show novel redox properties in solution. The stabilisation of mononuclear Ag(II) and Au(II) radical complexes has been investigated spectroelectrochemically using electronic and esr spectroscopy. In particular, the single crystal X-ray structures and redox interconversion of $[Au([9]aneS_3)_2]^{+/2+/3+}$ are described.

1. Introduction

Macrocyclic ligands are well known to form complexes that show remarkable thermodynamic stability and kinetic inertness.[1] The macrocyclic ligand can, therefore, be regarded as a protecting group for the metal centre controlling its stereochemistry, electronic and redox properties. We have investigated previously the co-ordination chemistry of tri-, tetra- penta- and hexadentate thioether crowns with the d^8 metal ions Ni(II), Pd(II) and Pt(II).[2] The introduction of inherent mismatches between the metal and donating ligands leads to the formation of strained complexes of unusual stereochemistry, with the resultant co-ordination geometries being a compromise between the preferences of the metal ion and the encapsulating ligand. This has led to the stabilisation of mononuclear d^7 Ni(III),[3] Pd(III)[4] and Pt(III)[5] species such as $[M([9]aneS_3)_2]^{3+}$ and $[M([18]aneS_6)]^{3+}$. Related d^7 Rh(II)[4,6] and Ir(II)[7] have also been reported. We were interested in extending this work to the stabilisation of other unusual oxidation state metal radical species and report herein the synthesis, structures and spectroelectrochemistry of Au(I)/(II)/(III) and Ag(I)/(II) complexes of thioether macrocycles.

2. Silver

Reaction of Ag(I) salts with L affords the complexes of stoichiometry $[Ag(L)]^+$ (L = $[12]aneS_4$, $[14]aneS_4$, $[16]aneS_4$, $[15]aneS_5$, $[18]aneS_6$,[8] $[18]aneN_2S_4$, $Me_2[18]aneN_2S_4$[9]) and $[Ag(L)_2]^+$ (L = $[9]aneS_3$).[10] The structures of $[Ag([9]aneS_3)_2]^+$[10] and $[Ag([18]aneS_6)]^+$[8] have been reported previously and show homoleptic thioether co-ordination at octahedral Ag(I) centres. *Rac*-$[Ag([18]aneN_2S_4)]^+$ and *meso*-$[Ag(Me_2[18]aneN_2S_4)]^+$ show highly distorted six co-ordinate geometries, one form of *meso*-$[Ag(Me_2[18]aneN_2S_4)]^+$ being five co-ordinate in the solid state.[9] We were interested in studying the interaction of Ag(I) with $[15]aneS_5$ since this crown does not have sufficient S-donors

A. J. L. Pombeiro and J. A. McCleverty (eds.),
Molecular Electrochemistry of Inorganic, Bioinorganic and Organomettallic Compounds, 121–129.

to enable simple octahedral geometry at Ag(I). The structure of the $[Ag([15]aneS_5)]^+$ cation is indeed unusual and reflects the mismatch between Ag(I) and $[15]aneS_5$. The solid state structure of the cation is dependent upon the counter-anion. Thus, the structure of $[Ag([15]aneS_5)]PF_6$ consists of parallel, polymeric chains of $[Ag([15]aneS_5)]^+$ units. The Ag(I) centre is bound to all five S-donors of the crown, Ag-S(1) = 3.219(5), Ag-S = 2.564(6), 2.659(5), 2.651(6), 3.075(7)Å (Figure 1a). S(1) interacts with another Ag(I) centres, Ag'-S(1) = 2.742(5)Å, thus linking the $[Ag([15]aneS_5)]^+$ units to form the polymer chain (Figure 1b). In contrast, the structure of $[Ag([15]aneS_5)]BPh_4$ shows (Figure 2) a dimeric structure in which two Ag(I) centres are bridged by two $[15]aneS_5$ ligands.[11] One Ag(I) is co-ordinated *via* one long and three short Ag-S bonds, while the other Ag(I) has one long and four short Ag-S contacts. However, $[Ag([15]aneS_5)]B(C_6F_5)_4$, incorporating the redox-inactive fluorinated derivative of the BPh_4^- anion, shows (Figure 4) a genuine monomeric structure with five co-ordinate Ag(I), Ag-S = 2.4712(19), 2.5621(19), 2.7262(20), 2.6847(21), 2.8813(19)Å.

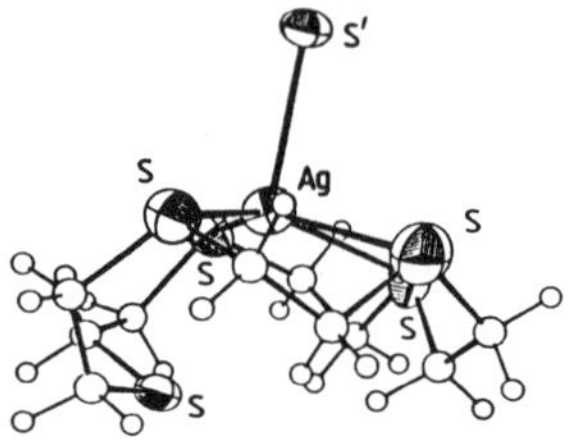

Figure 1a

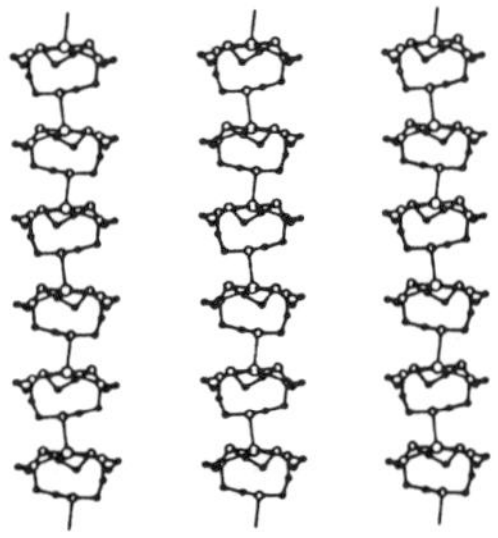

Figure 1b

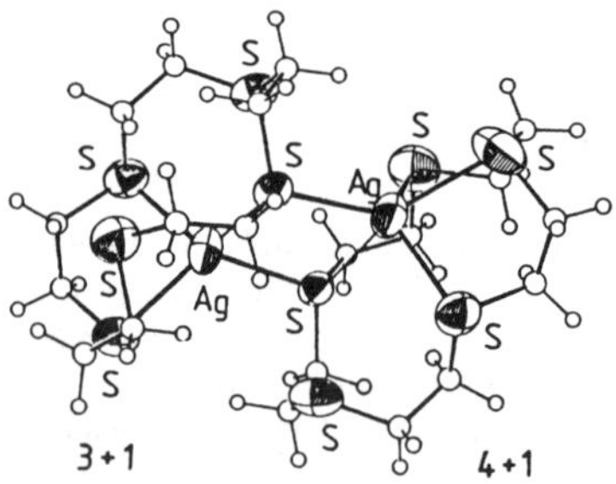

Figure 2

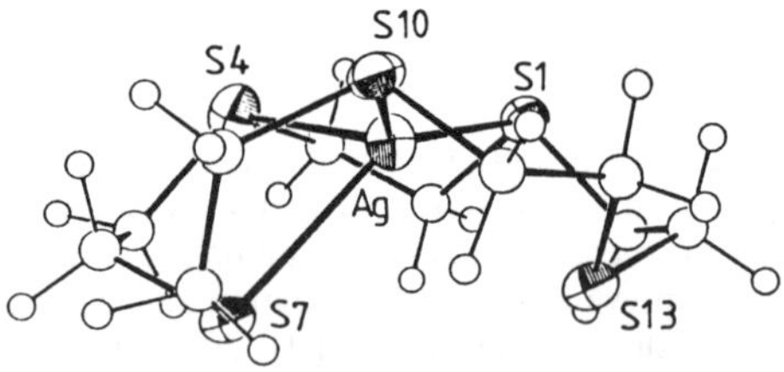

Figure 3

The above complexes probably exist as fluxional monomers in polar, co-ordinating solvents such as MeCN. Cyclic voltammetric data for the Ag(I) complexes are summarised in Table 1. The Ag(I)/(II) redox couples for $[Ag([9]aneS_3)_2]^{+/2+}$, $[Ag([18]aneS_6)]^{+/2+}$, $[Ag([15]aneS_5)]^{+/2+}$ and $[Ag([18]aneN_2S_4)]^{+/2+}$ are reversible on the cyclic voltammetric time-scale, although electrogeneration of the Ag(II) species even at low temperature is hampered by competing decomposition in organic solvents. The Ag(II) products can, however, be stabilised in acid (H_2SO_4, $HClO_4$) suggesting that one possible route for decomposition of these species is *via* deprotonation at an α-carbon centre followed by S-C bond cleavage to give vinyl thioether species.[12] Importantly, the electronic and esr spectroscopic data for chemically ($HClO_4$, H_2SO_4) and electrochemically (MeCN) generated solutions of Ag(II) are identical. Although we have not obtained a crystal structure of these Ag(II) products, it seems likely that such species would show Jahn-Teller distorted six co-ordination as found for the related Au(II) complex $[Au([9]aneS_3)_2]^{2+}$ (see below). This would explain the high instability of the Ag(II) products of $[12]aneS_4$ and $[14]aneS_4$ since the additional co-ordination number required on going from Ag(I) to Ag(II) would have to be taken up by labile solvent molecules, the resultant Ag(II) species would therefore be highly reactive.

TABLE 1. Cyclic Voltammetric Data for $[Ag(L)]^{0/+/2+}$ [a] and Electronic and *X*-Band Esr Spectal Data for Ag(II) Products.

Complex	$E_{½}$ [b] (AgI/0)	(AgI/II)	λ_{max} [d]	g-values[c] 77K	298K	A_{iso}
$[Ag([12]aneS_4)]^+$	-	+0.94[i]	648	2.062	2.034	32G
				2.036		
				2.014		
$[Ag([14]aneS_4)]^+$	-	+0.83[i]	514	2.057	2.031	36G
				2.035		
				2.007		
$[Ag[15]aneS_5)]^+$	-0.37[q]	+0.81	568	2.061	2.030	29G
				2.038		
				2.011		
$[Ag([18]aneS_6)]^+$	-0.42[q]	+1.00	566	2.067	2.036	31G
				2.033		
				2.010		
$[Ag([18]aneN_2S_4)]^+$	-0.74[q]	+0.65	530	2.050		
$[Ag(Me_2[18]aneN_2S_4)]^+$	-0.78[i]	+0.78[i]	-	-	-	-
$[Ag([9]aneS_3)_2]^+$	-0.57[i]	+0.75	570	2.063	2.034	-
				2.032		
				2.009		

i: irreversible, E_{pa} q: quasi-reversible a: measured in MeCN (0.1M $^nBu_4NPF_6$) at 243K at Pt electrodes b: potentials quoted in V vs Fc/Fc^+ c: esr spectra measured at 77K as a frozen glass and at 298K as a solution in organic or acidic media d: electronic absorption spectra quoted in nm.

The *X*-band and *Q*-band esr spectra of the Ag(II) product of $[15]aneS_5$ are shown in Figure 4 with hyperfine coupling to ^{107}Ag (I = ½, 51.4%), ^{109}Ag (I = ½, 48.6%) being clearly discernible in the *Q*-band spectrum, although the difference in hyperfine coupling between ^{107}Ag and ^{109}Ag is too small to be observed. Similar features are observed in the *X*-band and *Q*-band esr spectra of all the other Ag(II) species listed in Table 1.

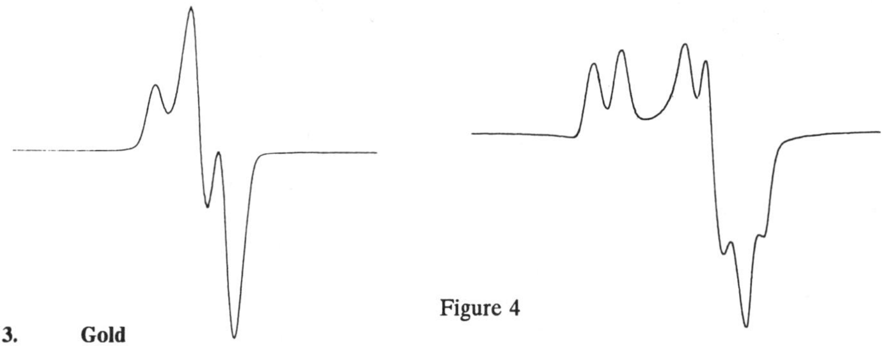

Figure 4

3. Gold

Au(I) is well known to prefer linear co-ordination. However, treatment of $[Au(tht)_2]^+$ (tht = tetrahydrothiophene) with $[9]aneS_3$, $[15]aneS_5$, $[18]aneS_6$ and $[18]aneN_2S_4$ leads to the formation of complexes incorporating highly unusual geometries. The complex $[Au([9]aneS_3)_2]^+$ illustrates this well. The trithia crown $[9]aneS_3$ may be regarded as pre-organised for facial co-ordination to a metal centre.[13] The single crystal X-ray structure of $[Au([9]aneS_3)_2]^+$ shows (Figure 5) one $[9]aneS_3$ co-ordinated as a monodentate ligand and is hence relatively labile, Au-S(1') = 2.3026(6)Å, while the other $[9]aneS_3$ is bound asymmetrically with one short and two long Au-S distances, Au-S(1) = 2.350(7), Au-S(4) = 2.733(8), Au-S(7) = 2.825(8)Å, <S(1')-Au-S(1) = 153.98(23)°.[14] Thus, the Au(I) centre in $[Au([9]aneS_3)_2]^+$ shows 2+2 S-co-ordination and can therefore be regarded to be in a tetrahedrally-distorted linear environment.[14]

The related complexes $[Au([18]aneS_6)]^+$ and $[Au([18]aneN_2S_4)]^+$ show similar geometries at Au(I) with 2 short Au-S distances, and additional long-range interactions to other donors within the ring. The structure of $[Au([18]aneS_6)]^+$ shows (Figure 6) two conformers for the cation in the solid state (Figure 7) with Au-S(1) = 2.321(3), Au-S(10) = 2.320(4), Au-S(7) = 2.856(4), Au-S(4) = 2.870(4)Å; for $[Au(Me_2[18]aneN_2S_4)]^+$ (Figure 8), Au-S(1) = 2.304(2), Au-S(10) = 2.314(1), Au-S(4) = 2.888(2), Au···S(13) = 3.624(1), Au···N(7) = 3.795(2), Au···N(16) = 3.660(1)Å, <S(1)-Au-S(10) = 177.30(5)°. These complexes, therefore, incorporate large chelate rings that span 150 to almost 180°.

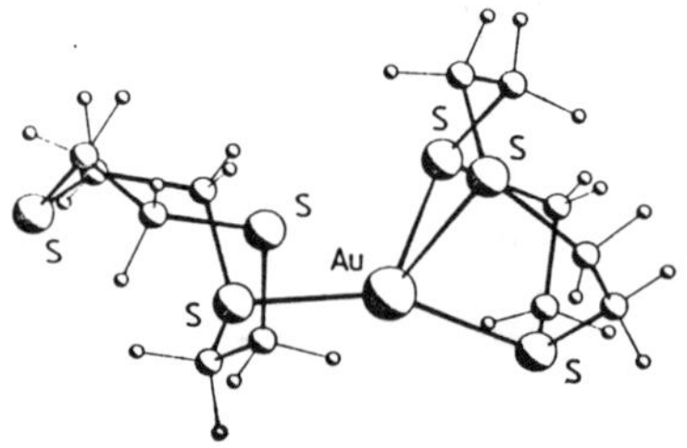

Figure 5

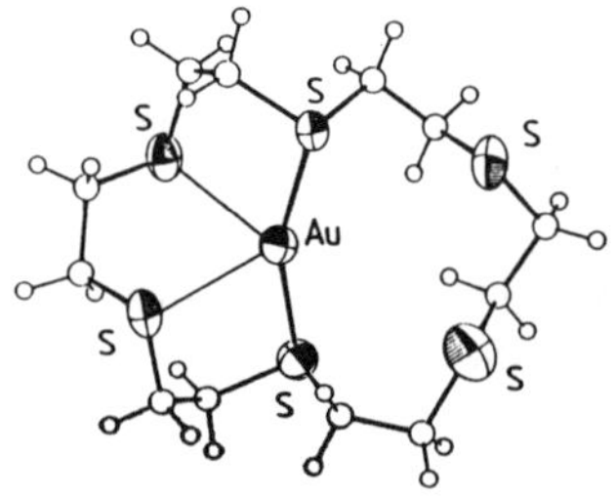

Figure 6

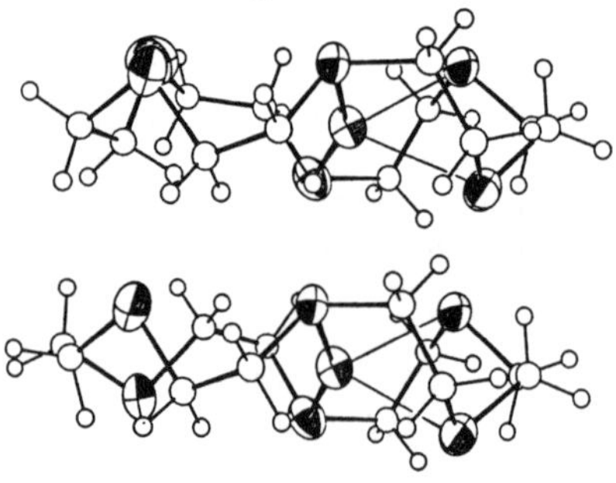

Figure 7

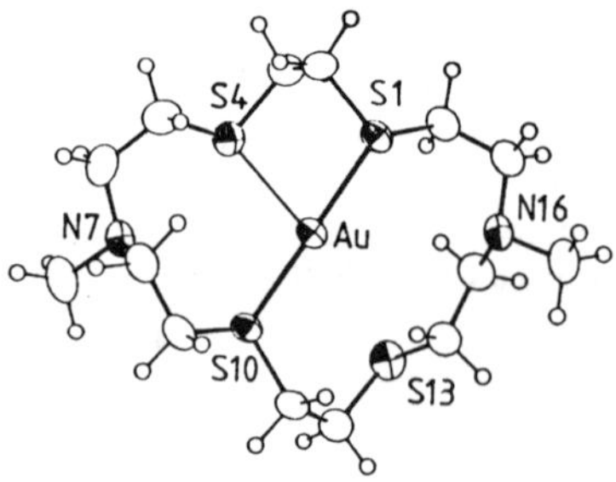

Figure 8

As for the Ag(I) complexes of $[15]aneS_5$, the Au(I) species $[Au([15]aneS_5)]^+$ shows unusual structural features. Encapsulation of Au(I) by $[15]aneS_5$ to give the same structure as that observed in $[Au[18]aneS_6]^+$ would be unlikely given the inherent strain that would be imposed on the 15-membered ring crown to form a 155° chelate as in Figure 6. The single crystal X-ray structures of $[Au([15]aneS_5)]PF_6$ and $[Au([15]aneS_5)]B(C_6F_5)_4$ both show the cation to be dimeric with Au(I) centres bridging $[15]aneS_5$ crowns in a macrobicyclic structure (Figure 9); for $[Au([15]aneS_5)]PF_6$, Au-S(1) = 2.345(7), Au-S(10') = 2.345(7), Au-S(4) = 2.887(6), Au-S(13) = 2.992(6)Å, S(1)-Au-S(10') = 158.82(22)°.

Reaction of two molar equivalents of $[Au(tht)_2]^+$ (tht = tetrahydrothiophene) with one molar equivalent of $[24]aneS_8$ or $[28]aneS_8$ (L) affords the corresponding binuclear complexes $[Au_2(L)]^{2+}$. The analogous binuclear Ag(I) complexes can also be prepared. The single crystal X-ray structure of $[Au_2([28]aneS_8)]^{2+}$ shows (Figure 10) two Au(I) centres complexed tetrahedrally within the octathia macrocycle in a 2+2 manner, Au-S(1) = 2.3301(19), Au-S(11) = 2.3378(18), Au-S(4) = 2.7891(20), Au-S(8) = 2.7629(20), Au⋯Au = 5.6977(6)Å. The <S(1)-Au-S(11) angle of 155.58(6)° contrasts to the value of 133.39(20)° as observed previously for $[Cu_2([28]aneS_8)]^{2+}$ [15] and reflects the preferences of Au(I) for linear and Cu(I) for tetrahedral co-ordination.

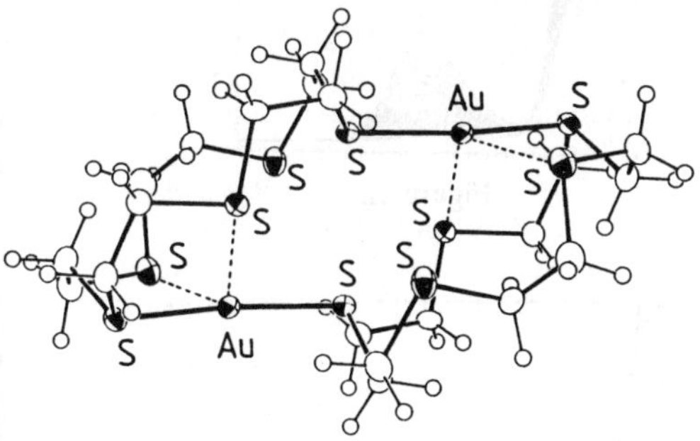

Figure 9

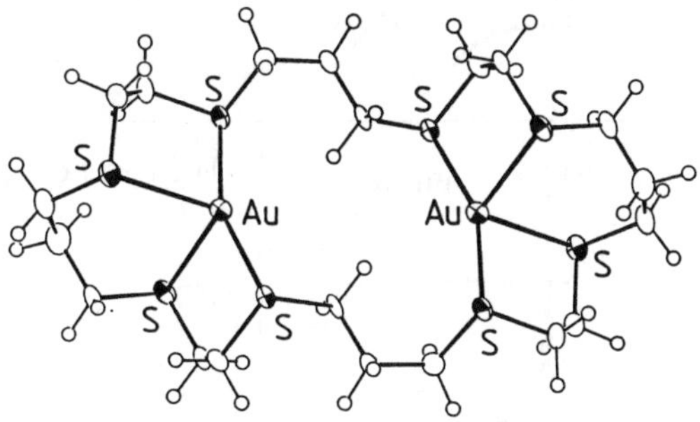

Figure 10

Significantly therefore, the structures of $[Au([9]aneS_3)_2]^+$, $[Au_2([15]aneS_5)_2]^{2+}$, $[Au([18]aneS_6)]^+$ and $[Au_2([28]aneS_8)]^{2+}$ ***all*** show 2+2 co-ordination at the Au(I) centre with two long and two short Au-S distances. $[Au(Me_2[18]aneN_2S_4)]^+$ shows 2+1 co-ordination with the fourth S-donor and the two N-donors at much longer distance from the Au(I) centre. These structural similarities are reflected in the similar redox characteristics observed for these cations in solution, with $[Au_2([15]aneS_5)_2]^{2+}$ existing as a monomer in polar, co-ordinating solvents. Table 2 summarises cyclic voltammetric and coulometric data for these Au(I) thioether complexes.

TABLE 2: Cyclic Voltammetric[a] and Coulometric Data for Au(I) Thioether Complexes

Complex	$^1E_{1/2}$	n^b	$^2E_{1/2}$	n^b
$[Au([9]aneS_3)_2]^+$	+0.12[i,c]	1	+0.46	1
$[Au([15]aneS_5)]^+$	+0.36	1	+0.54	1
$[Au([18]aneS_6)]^+$	+0.37	1	+0.56	1
$[Au(Me_2[18]aneS_2S_4)]^+$	+0.14	1	+0.43	1
$[Au([18]aneS_2S_4)]^+$	+0.85[i]	2		
$[Au_2([28]aneS_8)]^{2+}$	+0.55[br]	4		
$[Au([16]aneS_4)]^+$	+0.14	2		

i: irreversible a: measured in MeCN (0.1M $^nBu_4NPF_6$) at 293K at platinum electrodes. Potentials quoted in V vs Fc/Fc$^+$ b: number of electrons per cation obtained coulometrically at a Pt basket at 243K br: broad wave with peak-to-peak separation of 140mV.sec^{-1} c: irreversible wave (E_{pc}) due to slow Au(I) → Au(II) electron-transfer.

The oxidation of $[Au([9]aneS_3)_2]^+$, $[Au([15]aneS_5)]^+$, $[Au([18]aneS_6)]^+$ and $[Au(Me_2[18]aneN_2S_4)]^+$ to form mononuclear Au(II) and Au(III) species has been monitored spectroelectrochemically using *in situ* electronic spectroscopy. These redox processes have been found to be fully chemically reversible. Figure 11 shows the cyclic voltammogram of $[Au([18]aneS_6)]^+$ in MeCN, while Figures 12, 13, 14 and 15 show the interconversions of $[Au([9]aneS_3)_2]^{2+/+}$, $[Au([9]aneS_3)_2]^{2+/3+}$ and $[Au([18]aneS_6)]^{+/2+}$ and $[Au([18]aneS_6)]^{2+/3+}$ using an optically transparent electrode (O.T.E) system.

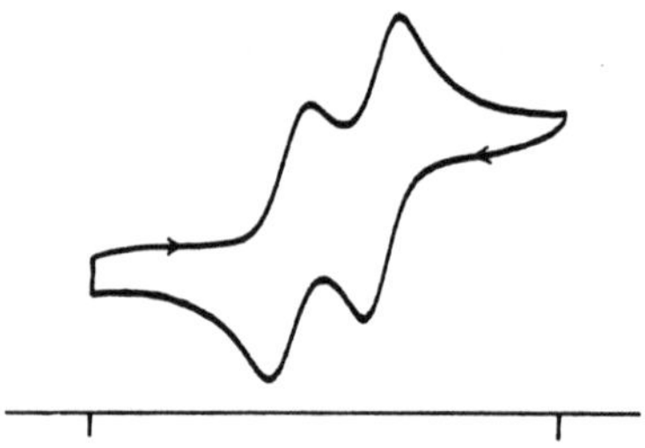

Figure 11

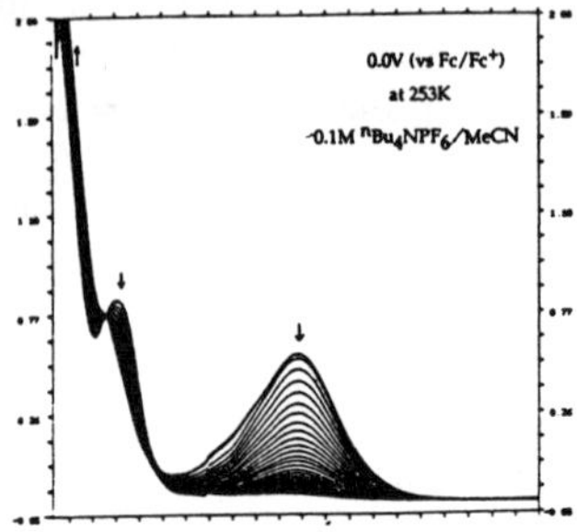

Figure 12

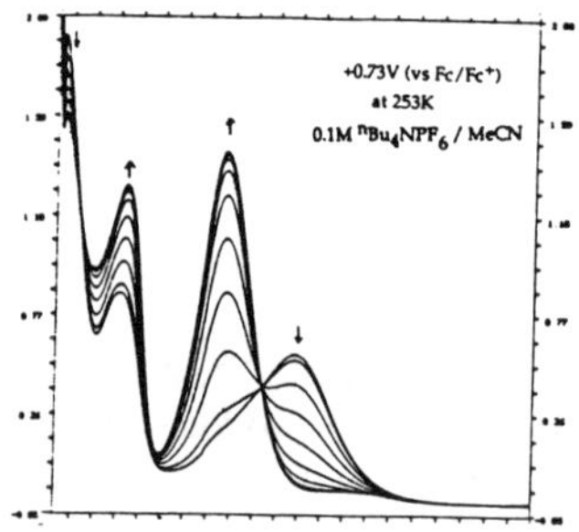

Figure 13

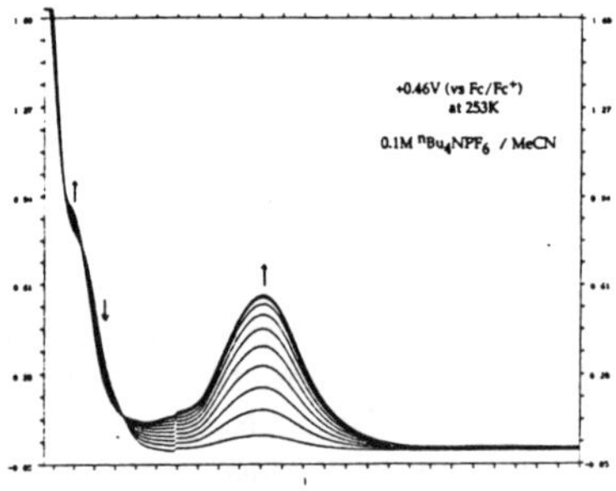

Figure 14

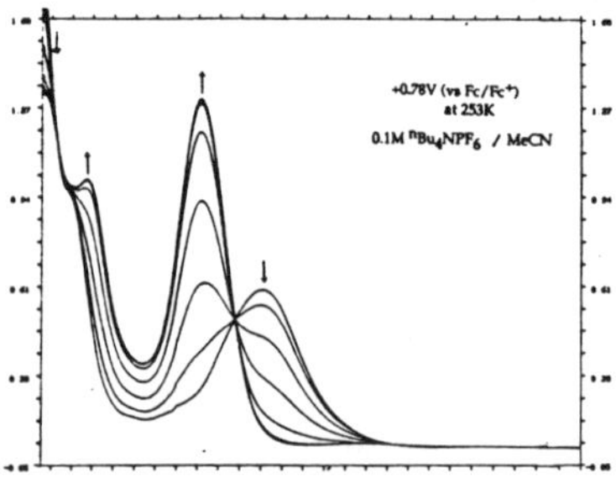

Figure 15

The one-electron oxidation products are esr-active and show four-line spectra due to hyperfine coupling to ^{197}Au (I = 3/2, 100%). Interestingly, the complexes of homoleptic thioether crowns that incorporate non-interacting S-donors in the precursor Au(I) species afford the most stable Au(II) species. This reflects the requirement of inceasing co-ordination number on going from Au(I) (two co-ordinate) to Au(II) (six co-ordinate, see below). Thus, these precursor Au(I) complexes may be regarded as stereochemically strained in the sense that the extra S-donors destabilise the d^{10} Au(I) configuration and provide a driving force for the formation of mononuclear Au(II). Esr spectroscopic data for the Au(II) complexes in fluid solution are summarised in Table 3. The esr spectra of the Au(II) species in frozen glass are much more complicated and current work is aimed at simulating and interpreting these spectra to obtain accurate g-values. Figures 16 and 17 show representative esr spectra of $[Au([9]aneS_3)_2]^{2+}$ (doped in diamagnetic $[Pd([9]aneS_3)_2]^{2+}$ at 290K) and of $[Au([18]aneS_6)]^{2+}$

(as a solution in $MeNO_2$ at 293K).

TABLE 3: *X*-Band Esr Spectral Data (293K, $MeNO_2$ solution) for Mononuclear Au(II) Thioether Complexes

Complex	g_{iso}	A_{iso}
$[Au([9]aneS_3)_2]2^+$	2.016	44.3G
$[Au([15]aneS_5)]^{2+}$	2.014	43.3G
$[Au([18]aneS_6)]^{2+}$	2.026	45.5G

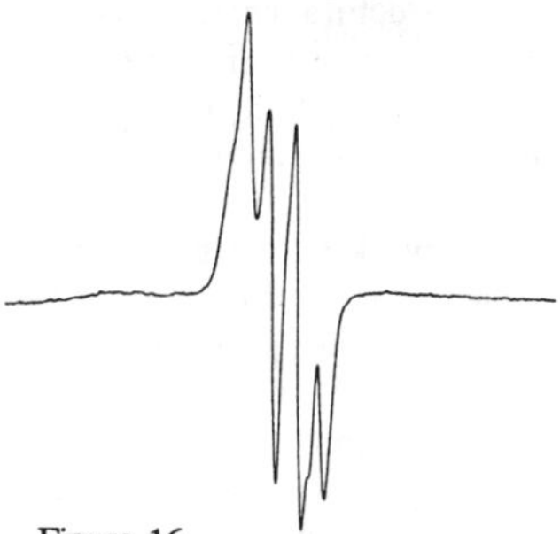

Figure 16

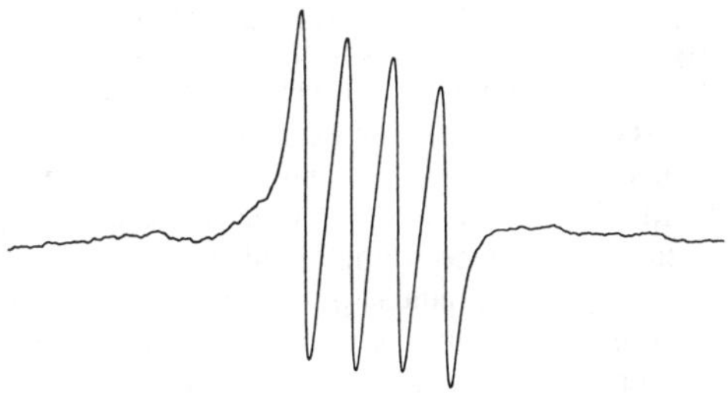

Figure 17

The mononuclear Au(II) complexes can also be prepared by reaction of Au(III) salts with crown thioethers under reducing conditions, although the yields using this method can be variable. Confirmation that these products are genuine mononuclear Au(II) complexes comes from the single crystal X-ray structures of $[Au([9]aneS_3)_2]^{2+}$ and of the Au(III) species $[Au([9]aneS_3)_2]^{3+}$. The single crystal X-ray structure of $[Au([9]aneS_3)_2]^{2+}$ shows (Figure 18) a distorted octahedral cation with all six thioether donors of the two crowns bound to the metal centre, Au-S = 2.452(5), 2.462(5), 2.839(5)Å. These structural parameters are entirely consistent with a Jahn-Teller distorted d^9 Au(II) species.[14] Oxidation of the Au(II) complex affords $[Au([9]aneS_3)_2]^{3+}$. The single crystal X-ray structure of $[Au([9]aneS_3)_2]^{3+}$ shows (Figure 19) square planar co-ordination of the two $[9]aneS_3$ crowns, Au-S = 2.348(4), 2.354(4)Å, with additional long-range interactions to two apical S-donors, Au···S = 2.926(4)Å consistent with the formation of a genuine d^8 Au(III) complex. The related Pd(II) species $[Pd([9]aneS_3)_2]^{2+}$ shows[16] a very similar stereochemistry and bond-lengths, the ionic radius of d^8 Pd(II) and d^8 Au(III) being very similar.

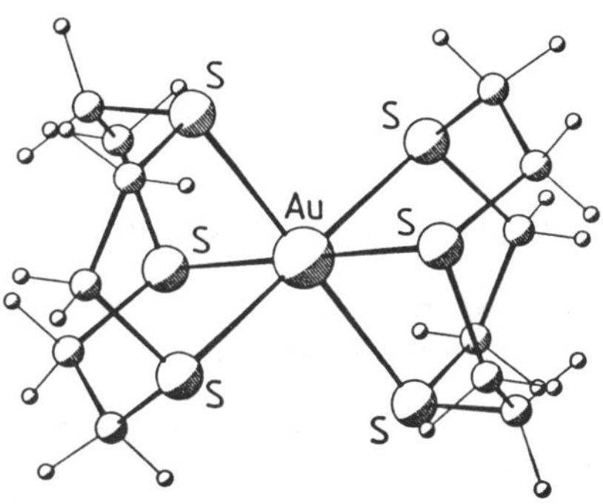

Figure 18

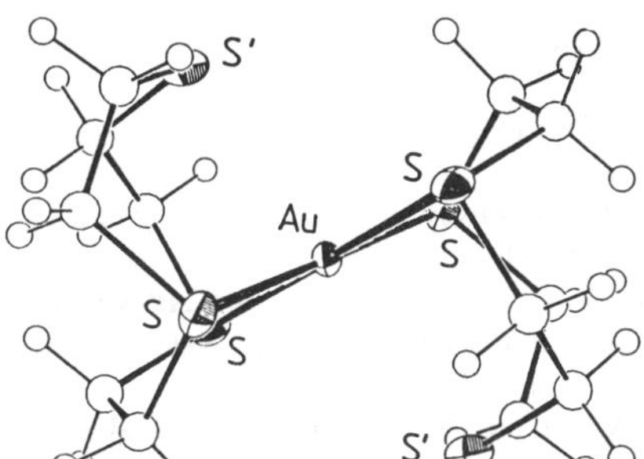

Figure 19

4. Conclusions

The work described herein confirms further the ability of polydentate thioether macrocycles to stabilise unusual metal radical centres. This stems from the co-ordinative flexibility of these ligands and the imposition of thioether co-ordination. Thioethers in general are recognised as weak σ-donors to metal centres.[17] In the above systems, this is counter-balanced by macrocyclic co-ordination and the potential ability of thioethers to act as either π-acceptors (*via* vacant d-orbitals on S, or S-C σ^{*}-orbitals) or π-donors (*via* lone-pair donation).[2] This aspect of covalency in metal-thioether bonding becomes potentially very important with soft metal ions of the second and third row transition series.

The redox interconversion of $[Au([9]aneS_3)_2]^{+/2+/3+}$ and the structural characterisation of each metal oxidation state illustrates the correlation between redox and structure in these systems. The normal oxidation states for Au are +1 and +3; in the complexes of $[9]aneS_3$ it is precisely these oxidation states of Au that are destabilised. $[Au([9]aneS_3)_2]^+$ incorporates a mono-dentate thioether donor which is therefore labile and readily lost to form complexes of type $[Au([9]aneS_3)(L)]^+$ (L = PR_3, NCMe).[18] $[Au([9]aneS_3)_2]^{3+}$ is unstable and very difficult to work with since this 3+ cation appears to readily undergo ring-opening reactions at the thioether crown;[12] this is supported by the observation that $[Au([9]aneS_3)_2]^{3+}$ is stabilised in acidic solution. In contrast, $[Au([9]aneS_3)_2]^{2+}$ is air-stable and relatively stable in solution, although conversion to Au(I) and Au(III) is relatively facile in the presence of redox agents. The stability of the $[Au([9]aneS_3)_2]^{2+}$ is inevitably due to the synergy between the Au(II) and the ligand donor set. The d^9 Au(II) centre is accommodated by octahedral S_6 co-ordination which can distort sufficiently to cope with the required Jahn-Teller distortion at the metal centre. Parallel to this, the co-ordinative requirements of the thioether crowns are also accommodated; it is in $[Au([9]aneS_3)_2]^{2+}$ that the co-ordinated $[9]aneS_3$ ligands are bound ***facially*** and are therefore configurationally stable. Thus, it is in $[Au([9]aneS_3)_2]^{2+}$ that the metal ion and ligand donors show optimum fit compared to $[Au([9]aneS_3)_2]^{+}$ and $[Au([9]aneS_3)_2]^{3+}$.

Assignment of the above oxidation products to mononuclear Ag(II) and Au(II) species is based on the observed redox potentials, electronic and esr spectroscopy, and in the case of Au(II), on a crystal structure determination. The unambiguous assignment of the silver complexes to Ag(II) species must await a crystal structure determination, although it should be stressed that none of the macrocyclic ligands show redox activity in the potential range discussed; these ligands show irreversible oxidations at much higher anodic potentials. This notwithstanding, a degree of delocalisation of charge from the metal centres to the S-donors, reflecting the covalency of M-S bonding, is undoubtedly important in stabilising these radical products.

5. Acknowledgements

We thank the SERC for support, Johnson Matthey for the loan of precious metals, The Royal Society of London for support (to DC), and the Royal Society of Edinburgh for a Support Research Fellowship (to MS).

6. References

1. *The Chemistry of Macrocyclic Ligand Complexes*, L.F. Lindoy, Cambridge University Press, Cambridge, UK, 1989.
2. A.J. Blake and M. Schröder, *Advances in Inorganic Chemistry*, 1990, **35**, 1; G. Reid and M. Schröder, *Chemical Society Reviews*, 1990, **19**, 43; M. Schröder, *Pure and Applied Chemistry*, 1990, **60**, 517.
3. A.J. Blake, R.O. Gould , M.A. Halcrow, A.J. Holder and M. Schröder, unpublished results; A.J. Holder, PhD Thesis, University of Edinburgh, 1987.
4. A.J. Blake, A.J. Holder, T.I. Hyde and M. Schröder, *J. Chem. Soc., Chem. Commun.*, 1987, 987.

5. A.J. Blake, R.O. Gould, A.J. Holder, T.I. Hyde, M.O. Odulate, A.J. Lavery and M. Schröder, *J. Chem. Soc., Chem. Commun.*, 1987, 118.

6. S.C. Rawle, R. Yagbasan, K. Prout and S.R. Cooper, *J. Am. Chem. Soc.*, 1987, **109**, 6181; A.J. Blake, R.O. Gould, A.J. Holder, T.I. Hyde and M. Schröder, *J. Chem. Soc., Dalton Trans.*, 1988, 1861; S.R. Cooper, S.C. Rawle, R. Yagbasan and D.J. Watkin, *J. Am. Chem. Soc.*, 1991, **113**, 1600.

7. A.J. Blake, R.O. Gould, A.J. Holder, T.I. Hyde and M. Schröder, *J. Chem. Soc., Dalton Trans.*, 1990, 1759.

8. A.J. Blake, R.O. Gould, A.J. Holder, T.I. Hyde and M. Schröder, *Polyhedron*, 1989, **8**, 513 and references therein.

9. A.J. Blake, G. Reid and M. Schröder, *J. Chem. Soc., Dalton Trans.*, 1990, 615.

10. H-J. Küppers, K. Wieghardt, Y-H. Tsay, C. Krüger, B. Nuber and J. Weiss, *Angew. Chem.*, 1987, **99**, 583; *Angew. Chem., Int. Ed. Engl.*, 1987, **27**, 575; J. Clarkson, R. Yagbasan, P.J. Blower, S.C. Rawle and S.R. Cooper, *J. Chem. Soc., Chem. Commun.*, 1987, 950.

11. A.J. Blake, R.O Gould, G. Reid and M. Schröder, *J. Chem. Soc., Chem. Commun.*, 1990, 974.

12. A.J. Blake, A.J. Holder, T.I Hyde, H-J. Küppers, M. Schröder and K. Wieghardt, *J. Chem. Soc., Chem. Commun.*, 1989, 1600.

13. R.S. Glass, G.S. Wilson and W.N. Setzer, *J. Am. Chem. Soc.*, 1980, **102**, 5068; R. Blom, D.W.H. Rankin, H.E. Robertson, M. Schröder and A. Taylor, *J. Chem. Soc., Perkin II*, 1991, 773.

14. A.J. Blake, R.O. Gould, J.A. Greig, A.J. Holder, T.I. Hyde and M. Schröder, *J. Chem. Soc., Chem. Commun.*, 1989, 876; A.J. Blake, J.A. Greig, A.J. Holder, T.I. Hyde, A. Taylor and M. Schröder, *Angew. Chem.*, 1990, **102**, 203; *Angew. Chem., Int. Ed. Engl.* 1990, **29**, 197 and references therein.

15. A.J. Blake, A. Taylor and M. Schröder, *Polyhedron*, 1990, **9**, 2911.

16. K. Wieghardt, H-J. Küppers, E. Raabe and C. Krüger, *Angew. Chem.*, 1986, **98**, 1136; *Angew. Chem., Int. Ed. Engl.*, 1986, **25**, 1101; A.J. Blake, A.J. Holder, T.I. Hyde, Y.V. Roberts, A.J. Lavery and M. Schröder, *J. Organomet. Chem.*, 1987, **323**, 261.

17. S.G. Murray and F.R. Hartley, *Chem. Revs.*, 1981, **81**, 365.

18. A. Taylor, PhD Thesis, University of Edinburgh, 1991.

3.2 BIOINORGANIC COMPOUNDS

BIOELECTROCHEMISTRY: MAKING USE OF THE ELECTROCHEMICAL BEHAVIOUR OF PROTEINS

H. ALLEN O. HILL
Inorganic Chemistry Laboratory,
South Parks Road,
Oxford OX1 3QR

ABSTRACT. Methods for obtaining the electrochemistry of redox proteins are now available making use of a variety of electrode surfaces. They are of two kinds: the preparation of metal surfaces upon which are adsorbed a variety of *promoters*, which, whilst not redox-active themselves, cause the proteins to adsorb in such a way that rapid electron transfer ensues. The second makes use of materials, especially edge-plane pyrolytic graphite, which have surfaces that adsorb proteins with, or without, the addition of non-redox-active metal ions. An important point that has recently been realised is that at, under certain conditions, the diffusion of the redox-active material may be *radial* in kind, rather than, as is usual, linear. This electrochemistry has been applied to, e.g, the study of protein-protein complexes and even to reactions of the mitochondria. The direct electrochemistry of enzymes has been observed and it shows much promise for the investigation of protein-enzyme complexes and even for analytical purposes. The latter achieved considerable success *before* methods of achieving direct electrochemistry of enzymes was discovered making use of ferrocenes as mediators with enzymes. This methodology as result in a sensor for glucose to be followed by one for cholesterol.

1. Introduction

Electron transfer is essential to Life yet how can the movement of electrons be coupled to the other obligatory processes? How are the structures of the proteins arranged to be appropriate for electron transfer? What are the nature of protein-protein interactions, whether mediated by small molecules or not? What part do the structures of the membranes play in such interactions? Are the static structures of such systems at all sufficient to give rise to an explanation of the *dynamics* of the redox events?

It may seem far removed from such questions to ask the simple question: can the ability of such molecules to take part in electron transfer be expressed at an electrode surface? Prior to 1977, the general opinion, with one exception, was: no, it cannot. A paper (Betso et al., 1972) drew attention to the fact that the electrochemistry of cytochrome *c* at a mercury electrode, though observed, was extremely short-lived. In 1977, two papers appeared (Eddowes and Hill, 1977, Yeh and Kuwana, 1977) that showed that, by attention to detail of the

A. J. L. Pombeiro and J. A. McCleverty (eds.),
Molecular Electrochemistry of Inorganic, Bioinorganic and Organomettallic Compounds, 133–149.

electrode surface, electrochemistry of e.g., cytochrome *c* was well-behaved. There then followed many studies, (see: Armstrong et al., 1988), from the search for electrode surfaces that allowed the electrochemistry of *any* redox protein to be expressed to the development of analytical systems based thereon.

Figure 1. The structure of horse heart cytochrome *c* showing the *a*-carbon backbone and the haem residue.

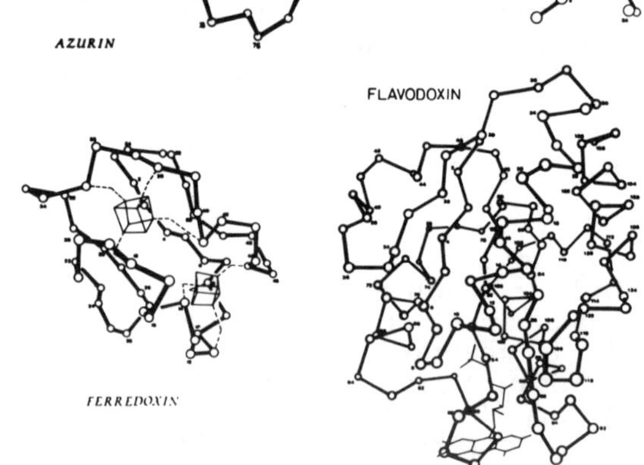

Figure 2. The structure of azurin, rubredoxin, ferredoxin and flavodoxin showing the *a*-carbon backbone.

2. Protein Electrochemistry

There now is **no** difficulty in achieving the electrochemistry of small redox proteins (see, for example, Armstrong et al., 1987, 1988, Frew and Hill, 1987, 1988) whether, e.g., cytochrome *c*, (Figure 1) ferredoxin, azurin, rubredoxin or flavodoxin (Figure 2). A variety of electrode surfaces exist at which the *direct* electrochemistry of proteins proceeds *without* the need for mediators. At some metal electrodes, what is required is the presence of some compound, called a *promoter* (Eddowes and Hill, 1977), which binds to the surface yet allows electron transfer to proceed (Figure 3). Over the years, many promoters have been reported (Taniguchi et al., 1982: Allen et al., 1984) and they are all bifunctional molecules, one part of which causes

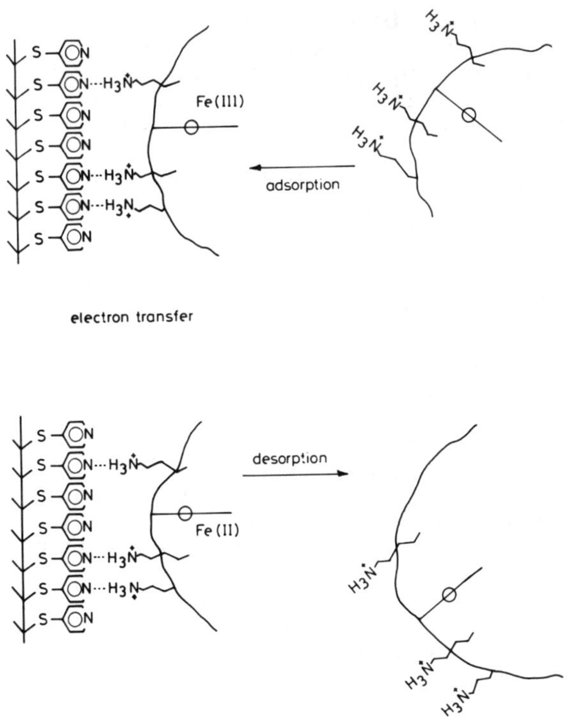

Figure 3. A schematic view of binding of cytochrome *c* of a gold electrode surface modified with 4,4'-dithiodipyridine.

binding to the electrode surface, the other interacts transiently with the protein. Of course, there is some evidence (Bowden et al., 1984) that *really* clean surfaces do not require promoters for the electrochemistry of some proteins to proceed but, in most laboratory environments, a trace of impurities is sufficient to lead to adsorption of the material of the surface and hence inhibition of electron transfer or of the protein to bind, or, indeed, both. This is included in a list of requirements, published some time ago (Allen et al., 1984):

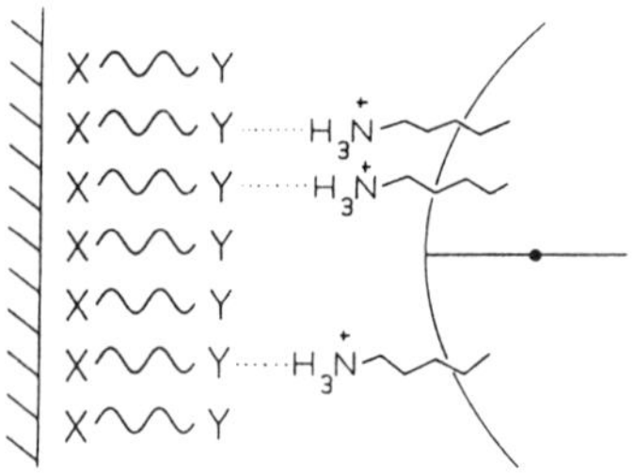

Figure 4. A schematic view of the interaction of, e.g., cytochrome *c* with an electrode surface modified with a promoter, **X~Y**.

(a) The compound should be bi- (or multi-) functional of the type **X~Y** (Figure 4) such that **X** is, or contains, a group, e.g., a pyridyl, phosphine, sulphydryl, thioether etc., which binds to a metal surface, and **Y** is a group, e.g., a carboxyl, pyridyl, etc., which interacts with a section of the proteins surface.

(b) The extension (~) should be such that it does **not** permit **Y** to bind to the electrode surface: it doesn't seem to matter whether it is aliphatic or aromatic in nature.

(c) Provided **X** is such that the molecule bind tightly to the surface, e.g., either a sulphydryl, a phosphine etc., a monolayer can be created, e.g., by dipping the gold or platinum electrode in a solution of the promoter, washing the electrode and then immersing in a solution of the protein.

(d) When **X** is a group, such as cysteine, in a peptide, **Y** can be chosen so that a carboxylate (as in a glutamate-containing peptide) or an amino group (as in a lysine-containing peptide), the former 'attracting' positively-charged proteins, such as cytochrome *c*, the latter, negatively-charged proteins, such as plastocyanin.

(e) By choosing **X** so that it can bind metal ions, such as Mg^{2+}, $Cr(NH_3)_6^{3+}$, or even $Pt(NH_3)_6^{4+}$, negatively-charged proteins such plastocyanin, ferredoxin or flavodoxin can be caused to bind to the modified electrode surface long enough for electron transfer to occur.

With this selection of promoters, it is possible to achieve the electrochemistry of *any* small redox protein. None fail to give good electrochemistry, often such that the half-wave potential corresponds to that assumed to be the thermodynamic potential of the protein. However, one must bear in mind that the experimental value reported is for the *protein-promoter-electrode* interaction: if this is significant, as, for example, when a metal ion is used as an ancillary promoter, then differences may be observed. These are not without interest: it may be that, e.g., the complex of magnesium (Armstrong et al., 1985) with plastocyanin, may be provide a more meaningful description of the complex of the protein in the photosynthetic unit than that of the 'free' protein. It should be noted that the behaviour of proteins at *absolutely* clean metal surfaces (a situation rarely encountered) is complex: transient short-lived electrochemistry may be observed in such systems, or irreversible adsorption on the metal surface with, or without, a marked change in potential, can occur. The purpose of promoters is to avoid such problems: the electrochemistry is well-behaved and reversible without the need of special conditions.

Other electrodes can be used in such studies. Perhaps the most extensive investigations have been made of graphite electrodes (Armstrong et al., 1984) where it was found that the electrochemistry at edge-plane pyrolytic graphite (Figure 5) with, and without, added metal ions, (depending on the charge of the protein), was well-behaved. Edge-plane pyrolytic graphite has, at its surface, a variety of functional groups including carboxylates. Depending on the state of oxidation of the freshly-prepared surface, the electrochemistry of cytochrome *c*, and, with added metal ions, plastocyanin etc., is quasi-reversible. In an attempt to make use of better-defined electrode surfaces, materials such as metallic ruthenium dioxide were employed, (Harmer and Hill, 1985) with some success. The results were similar to that reported at the same

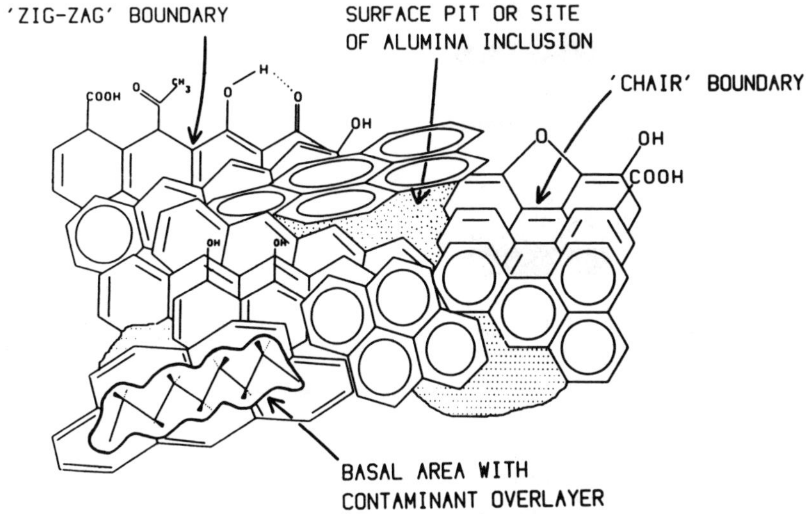

Figure 5. The hypothetical surface of pyrolytic graphite.

time as the first use of promoters, *viz.*, tin-doped indium oxide (Yeh and Kuwana, 1977) and followed up some time later by others (Bowden et al., 1984).

It would be misleading to give the impression that the results obtained in Oxford and elsewhere were straightforward. Of course, there were, and there still are, problems concerning the *exact* nature of the electrode surface. Attempts have been, and are being, made to examine the detailed nature of the surfaces, e.g., by ellipsometry (Elliott et al., 1986) and, more recently, by Scanning Tunneling Microscopy (Cataldi et al., 1990: an exact description in atomic terms still eludes us. No, the problems were, at one level, more mundane. Under a host of conditions, 'good' electrochemistry could not be obtained. Often, when a promoter was used, rather 'poor' electrochemistry resulted: when the electrode surface was modified in variety of ways, the electrochemical results indicated 'poor' electrochemistry when essentially no electrochemistry would have been expected. When the results were reassessed (Armstrong et al., 1989 (a) (b), Bond et al., 1990) in terms of the behaviour of the electroactive species *at a microelectrode*, i.e., one having dimensions of 1μm or less, they appeared consistent with rapid electron transfer rates. The difference was due to the effect of *radial* diffusion (Figure 6) of the electroactive species to the microelectrode as compared to the normally observed linear diffusion at macroelectrodes. The cyclic voltammogram of proteins whose diffusion to the small electrode is radial in kind resembles, but only superficially, poor electrochemistry at a macroelectrode. Detailed analysis of the experimental results showed

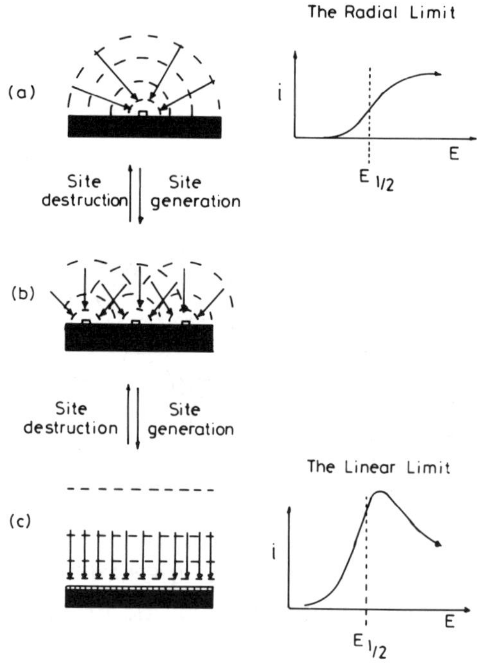

Figure 6. The effect of radial and linear diffusion on the resultant voltammogram.

that the electrochemistry corresponded, under all conditions, to situations in which the electron transfer rate was very fast. Another phenomenon which was elucidated by consideration of the radial diffusion was the *time* dependence of the electrochemistry, whether it was of situations corresponding to small amounts of promoter on the surface, or of mixtures of promoters or indeed of the mixture of a promoter with an inactive adsorbate. Even the gradual addition of increasing amounts of an adjunct promoter, e.g., $Cr(NH_3)_6^{3+}$, to a ferredoxin solution seemed to have a time dependence. Obviously this could be connected with the rearrangement of e.g., the $Cr(NH_3)_6^{3+}$/protein complex but many of the changes are now thought to be due to the rearrangement of the promoters *on the electrode surface*. This could, of course, occur by dissolution of the promoter followed by re-adsorption *but need not*: the promoters could rearrange whilst on the surface. In most cases, it is relatively easy to see how a mixture of promoter and inhibitor could, with time, rearrange to a set of islands, corresponding to microelectrodes. There are a few examples of situations where the reverse happens, i.e., where two different promoters rearrange from conditions corresponding to microelectrodes to that more normal.

Many, if not all, redox proteins, have no biological function *on their own*: they are associated with other redox proteins, assembled in electron transport systems in membranes or allied with redox enzymes in pursuit of some metabolic process. It was therefore of importance to see whether the electrochemical method of investigation could be applied to protein-protein complexes. The systems studied (Bagby et al., 1990) were the complexes formed between cytochrome *c* and plastocyanin, and cytochrome *c* and cytochrome b_5 and the availability of electrodes which were 'selective' towards the electrochemistry of given proteins, e.g., edge-plane pyrolytic graphite for cytochrome *c* or gold, coated with cys-lys-cys, for plastocyanin, enabled the behaviour of the complexes to be understood. It appears that the protein which binds to the electrode surface e.g., cytochrome *c* at edge-plane pyrolytic graphite, acts as an adjunct promoter holding e.g., plastocyanin in such an orientation that electron transfer to the latter can occur. This is consistent with **zinc** cytochrome *c*, or indeed many proteins which have an overall charge opposite to that of plastocyanin, acting in the same manner. There is one important point that emerged from these, and related, studies. *The structures of the complexes must be considered as* ***dynamic***: they move with respect, not only, to each other, but to the electrode surface. The initial description of electron transfer at the electrode envisaged a static arrangement of the protein, or proteins, at the surface but now one imagines a more mobile array of the protein(s) essentially moving in a lateral manner over the electrode surface until they reach a configuration where electron transfer to, or from, the surface is rapid. Such a dynamic view of the electrode/protein structure must even involve motions *within* the protein. Indeed, recent studies on the electrochemistry of genetically engineered (Burrows et al., in press) variants of cytochrome *c* and of its complexes with cytochrome b_5, indicate that subtle changes in the protein's structure affect its electrochemistry.

3. Enzyme Electrochemistry.

For a considerable time, there has been much interest in achieving the *direct* electrochemistry of enzymes, i.e., with no mediators, either simple redox proteins or small electroactive molecules, present. Of course, one expects their electrochemistry to be difficult to observe under most conditions: not only are most enzymes larger than simple redox proteins but one suspects them to be more idiosyncratic about the interaction with the electrode surface. For example, the charge on a site close to that involved in electron transfer might differ from the overall charge on the molecule. Indeed, such was the concern over the mechanism of electron transfer in redox enzymes, that they were divided (Armstrong et al., 1989) into two classes: **intrinsic** and **extrinsic**. In the former, electron transfer takes place between the substrates, e.g., glucose and dioxygen, and the prosthetic group in a well-ordered sequence *that does not involve electron transfer through the protein*. Of course, there may be enzymes in which the prosthetic group is situated so close to the surface of the protein that electron transfer can occur in any case but, one suspects, they will be exceptional. Extrinsic enzymes are those in which there is an electron transfer pathway through the enzyme primarily because they are associated with a redox protein when catalysing their normal reaction, e.g., cytochrome P-450 and putidaredoxin. This means that there is a part of the surface of the enzyme molecule arranged for binding of another molecule: if such exists, then it should be possible to model the surface of the electrode so that it mimics the partner protein.

Most of the apparently successful reports of such behaviour concerned flavoenzymes but it is more likely that the electro-chemistry is due to the free flavin which either yields its own electrochemistry or else acts as a mediator to intact enzyme. There were only two reports that went some way to proving that it was possible: the electro-chemistry of fungal laccase was mentioned in 1977 (Eddowes and Hill 1977) and latter fully reported (Lee et al., 1984): electron transfer to the blue copper component was observed. There had also been a brief report (Garindarajai et al., 1987) of the electrochemistry of a lysine oxidase. A few years ago, the electrochemistry of cytochrome *c* peroxidase was reported (Assefa and Bowden 1986, Armstrong and Lannon, 1987). This had been described earlier (Cass et al., 1985) using either cytochrome *c* or an organometallic compound, ferrocene, as mediators but in this case, *no mediators were required*, simply a variety of aminoglycosides which acted as promoters, adjunct or otherwise. Electrochemical behaviour was only observed in the presence of the substrate, H_2O_2, which is not suprising given that the simple iron(II)/iron(III) couple is not likely to be rapidly established. Direct electrochemistry *was* (Guo et al., 1989) reported for the enzyme, *p*-cresolmethylhydroxylase, a flavocytochrome, though for the cytochrome sub-unit only: presumably the electrochemistry of the other subunit is too slow to be directly observed. In the presence of the substrate, a quite different cyclic voltammogram is observed, a so-called catalytic wave. The height of this wave is directly related to the substrate concentration until, depending of the conditions employed, saturation behaviour is observed. The electrochemistry of the enzyme (Guo et al., 1990), sulphide:cytochrome *c* oxidoreductase behaved similarly to the cytochrome *c* peroxidase. i.e., it was dependent on the sulphide concentration though the enzyme itself did not show any electrochemistry. It was reassuring when it was found that the enzyme, methylamine dehydrogenase, which is *not* a flavocytochrome, indeed is not even a metalloprotein but has a co-factor which is related to pyrollequinolinequinone (PPQ), gave (Burrows et al., 1991) good electrochemistry, not only with the aminoglycosides referred to earlier but with a variety of cationic species. Once again, a linear dependence on the substrate concentration was observed. Investigations (Hill et al., to be published) of sulphite oxidase, ascorbate oxidase, a nitrate dehydrogenase and components of methane dehydrogenase show that, under different conditions for each enzyme, electrochemical observations can be observed.

4. Analytical Electrochemistry

Once the electrochemistry of enzymes has been observed, it is a simple matter to imagine a system or device making use of the analytical power inherent therein. However, when the direct electrochemistry of enzymes was first investigated, success was not immediately achieved. Of course, using proteins as mediators, enzymes could readily be used in an analytical mode. For example, as already referred to (Cass et al., 1985), cytochrome *c* peroxidase could be coupled to the electrode through the electrochemistry of cytochrome *c* and so could, without much difficulty, be adapted (Frew et al.,1986) into an analytical system for H_2O_2. Indeed the substrate for *any* enzyme which coupled to cytochrome *c* could be analysed in this way. For example, carbon monoxide oxidoreductase gives rise (Turner et al., 1984) to an efficient analytical system (Figure 7) for its substrate when coupled in this way. Similar observations were made for other redox proteins: azurin acted (Hill et al., 1985) as an excellent mediator to *p*-cresolmethylhydroxylase: however, the world is not crying out for a sensor for its substrate, *p*-cresol! Perhaps the most adventurous use of the electrochemistry of cytochrome *c* was (Coleman et al., 1983: Hill et al., 1985, Barker and Hill, 1988) as a

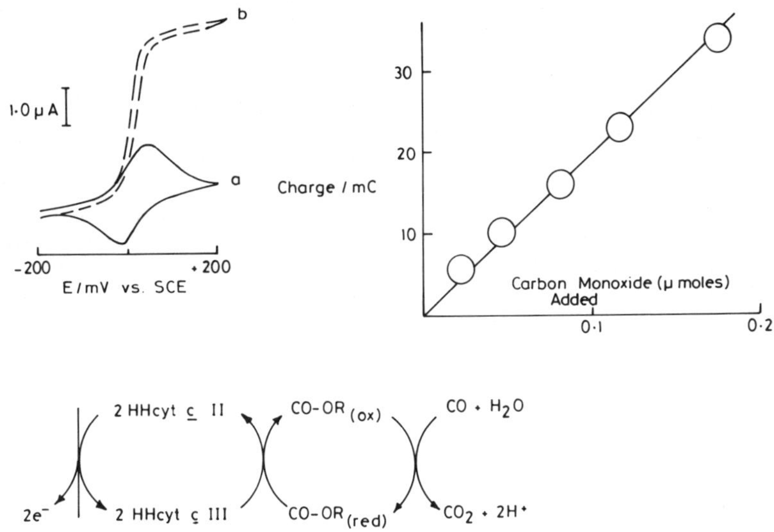

Figure 7. The cyclic voltammogram of a solution containing cytochrome *c* and carbon monoxide (a); as in (a) but with the enzyme, carbon monoxide oxidoreductase added.

mediator to mitochondria. At first, it was thought that the mitochondria were truly intact but latter studies suggested that diffusion of the cytochrome *c* took place through slight tears in the outer membrane. It was also shown that the electron transfer process increased markedly when, first mitoplasts the the solubilised cytochrome *c* oxidase where used. In a closed system, there was a remarkable consistency between the current passed and the reduction of dioxygen (Figure 8). More recent studies have made use of cytochrome *c* as a mediator to other enzymes in the mitochondria, e.g., sulphite oxidase.

Ten years ago, all enzyme electrodes (see, for example, Frew and Hill, 1987) made use of either the electrochemistry of the substrate or

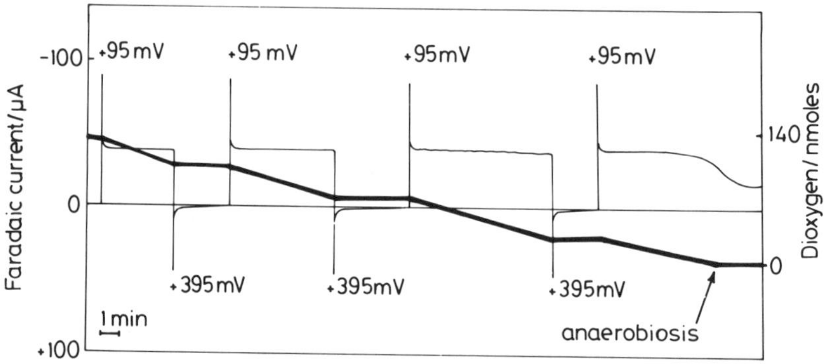

Figure 8. The use of dioxygen by a solution containing rat liver mitochondria and cytochrome *c* on its electrochemistry: the current passed through the solution is also shown.

product, usually O_2 or H_2O_2 or else employed a electroactive mediator such as tetramethylphenylenediamine to transport electrons to, or from, the enzyme. Though these provided good systems for laboratory analyses, they obviously required dioxygen, the mediator, itself, was often sensitive to oxidation. For various reasons, not least its *relative* insensitivity of the low oxidation state form to molecular oxygen, **ferrocene**, (Figure 9) was chosen as a mediator. (One can draw many parallels between ferrocene and cytochrome *c*, some just a little forced: the slowness of oxidation of the Fe(II) forms is the crucial one). It worked wonderfully well (see, for example, Green and Hill,

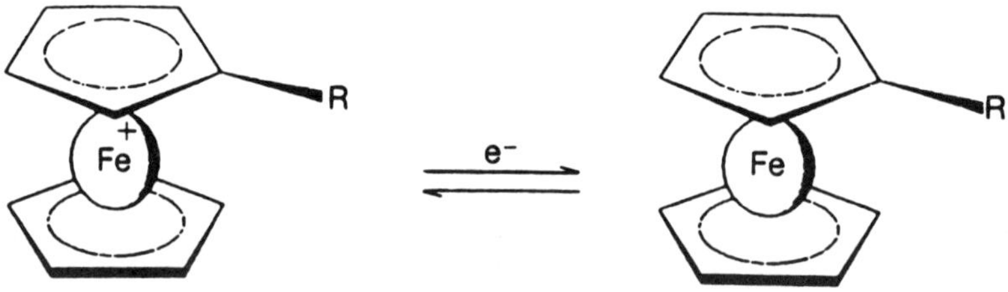

Figure 9. Ferrocene and its one-electron oxidation product, the ferricinium ion.

1986). Because our organometallic colleagues had not been idle over the years, it was possible to synthesize derivatives which had fast rates of reaction with all manner of enzymes. These species have been utilised in many forms, even to the extent of exploiting the electrochemistry of ferrocene-labelled enzymes.

Many enzymes, or collections of enzymes, were coupled in this way. However, the absence of any significant sensitivity to molecular oxygen, led to the exploitation of the ferrocene-linked glucose electrode (Figure 10) (Cass et al., 1984) as the basis for a glucose sensor, one that could be used 'at home' by diabetics.

After a relatively short period of time, this was launched on to the market by MediSense Inc., as the ExacTech™ sensor where it has

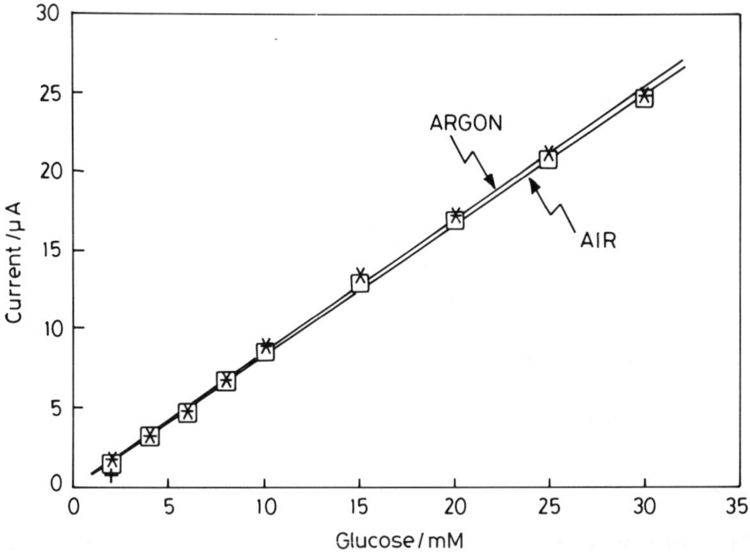

Figure 10. The correlation between the current passed and the glucose concentration of a solution containing a ferrocene, glucose oxidase and, of course, glucose.

been successful. The next device to reach the market was (Ball et al., 1986) a cholesterol sensor (Figure 11). This was necessarily more complicated, since it had to include a cholesterol esterase, and, after much investigation, it was decided that the system employing both a cholesterol oxidase, which gave H_2O_2 as the product, and a peroxidase, to deal with the latter, was the best for manufacture.

From a collection of enzymes, and the appropriate substrates, systems could be devised to arrange for the analysis of e.g., adenosine triphosphate or even the enzyme creatine kinase (Green et al., 1984) (Figure 12). Indeed, there would appear to be no limit to the complexity of such arrangements.

However, these, and related methods which employ amperometric analyses, suffer from a major weakness: they are *relatively insensitive*. There have been, and will, no doubt, be many approaches to resolving

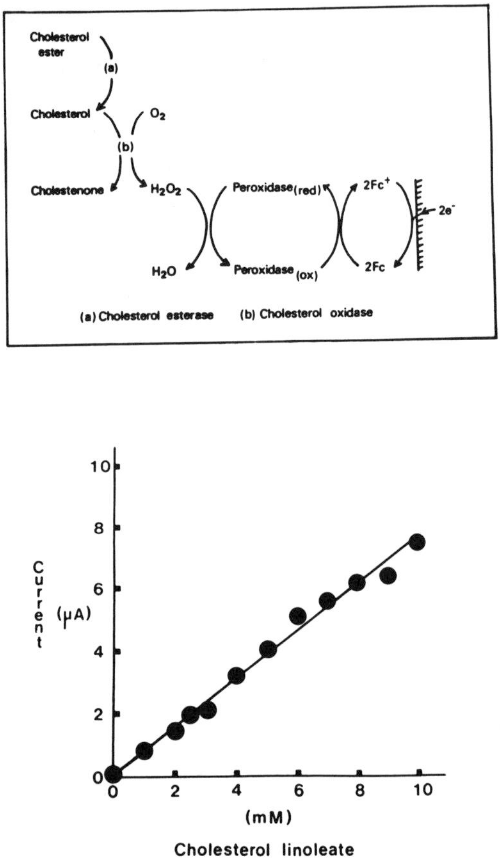

Figure 11. The outline scheme of a cholesterol sensor and the resultant correlation between the current observed and the amount of cholesterol linoleate.

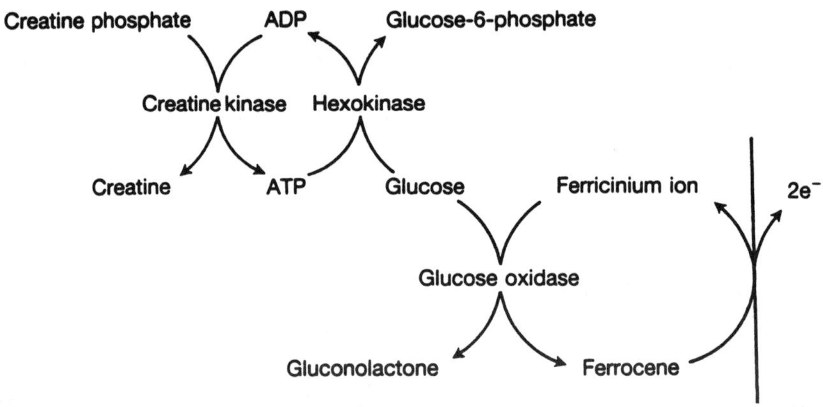

Figure 12. The procedure by which the ferrocene/glucose oxidase/glucose system can be linked to other enzymes and substrates.

this problem. Some exploit the power of enzymes to produce electroactive materials, others seek to make use of inhibitors, e.g., cyanide or sulphide, of enzymes which can be coupled electrochemically. One system (di Gleria, 1989) makes use of a 'cascade' of enzymes, the final one of which, trypsin, cleaves a ferrocenyl-peptide from an insoluble support to yield a mediator for the glucose electrode. It was exploited in an analysis for biotin but it could be applied to any antigen-antibody interaction *provided sufficient sensitivity was available*. That is the major remaining problem of amperometric analyses: the others, e.g., the proper coupling of enzyme to electrode whether directly or *via* mediator or through the electroactivity of substrate or product, the nature of the electrode material, the availability of enzymes or immunoglobulins, are in the process of solution. Whether electrodes suitable for such purposes can be made *really* small, e.g., 0.1μm or less by e.g., lithographic methods, remains to be seen.

5. References

Allen, P.M., Hill, H.A.O. and Walton, N.J. (1984) "Surface Modifiers for the Promotion of Direct Electrochemistry of Cytochrome *c*", J. Electroanal. Chem., 178, 69-86.

Armstrong, F. A., Hill, H.A.O. and Oliver, N.B. (1984) "Surface Selectivity in the Direct Electrochemistry of Redox Proteins. Contrasting Behaviour at Edge and Basal Planes of Graphite", J. Chem. Soc., Chem. Commun., 976-7.

Armstrong, F.A., Hill, H.A.O., Oliver, B.N. and Walton, N.J. (1984) "Direct Electrochemistry of Redox Proteins at Pyrolytic Graphite Electrodes", J. Amer. Chem. Soc., 106, 921-3.

Armstrong, F.A., Hill, H.A.O., Oliver, B.N. and Whitford, D. (1985) "Direct Electrochemistry of the Photosynthetic Blue Copper Protein, Plastocyanin. Electrostatic Promotion of Rapid Charge Transfer at an Edge-oriented Pyrolytic Graphite Electrode", J. Am. Chem. Soc., 107, 1473-6.

Armstrong, F.A., Hill, H.A.O. and Walton, N. J. (1986) "Electron-transfer Reactions of Redox Proteins at Electrodes", Quart. Rev. Biophys., 18, 261-322.

Armstrong, F.A., Cox, P.A., Lowe, V.J. and Oliver, B.N. (1987) "Metal Ions and Complexes as Modulators of Protein-Interfacial Electron Transport at Graphite Electrodes", J. Electroanal. Chem., 217, 331-366.

Armstrong, F.A., Hill, H.A.O. and Walton, N.J. (1988) "Direct Electrochemistry of Redox Proteins", Acc. Chem. Res., 21, 407-13.

Armstrong, F.A., Bond, A.M., Hill, H.A.O., Psalti, I.S.M. and Zoski, C.G. (1989) "A Microscopic Model of Electron Transfer at Electroactive Site of Molecular Dimensions for Reduction of Cytochrome *c* at Basal- and Edge-plane Graphite Electrodes", J. Phys. Chem., 93, 6485-93.

Armstrong, F.A., Bond, A.M., Hill, H.A.O., Oliver, B.N. and Psalti, I.S.M. (1989) "The Electrochemistry of Cytochrome *c*, Plastocyanin, Azurin, Rubredoxin and Ferredoxin at Edge- and Basal-plane Graphite Electrodes Interpreted *via* a Model Based on Electron Transfer at Electroactive Sites Approaching Molecular Dimensions in Size", J. Am. Chem. Soc., 111, 9185-9.

Armstrong, F.A. and Lannon, A.M. (1987) "Fast interfacial Electron Transfer Reactions of Cytochrome *c* Peroxidase and Graphite Electrodes Promoted by Aminoglycosides: Novel Electroenzymatic Catalysis of H_2O_2 Reduction", J. Am. Chem. Soc., 109, 7211-2.

Assefa, H. and Bowden, E.F. (1986) "Electron Transfer Reactions of Cytochrome *c*" Biochem. Biophys. Res. Commun., 139, 1003-8.

Bagby, S., Barker, P.D., Guo, L.-H. and Hill, H.A.O. (1990) "The Direct Electrochemistry of Protein-Protein Complexes Involving Cytochrome *c*, Cytochrome b_5 and Plastocyanin, Biochemistry, 29, 3213-9.

Ball, M.R., Frew, J.E., Green, M.J. and Hill, H.A.O. (1986) "Three Bioelectrochemical Assays for Total Cholesterol", in "Electrochemical Sensors for Biomedical Applications" Ed., C. K. N. Li, The Electrochemical Soc., (Proceedings volume), Vol. 86-14, pp. 16-25.

Barker, P.D., and Hill, H.A.O. (1988) "Direct Electrochemical Probes of Redox Protein and Redox Enzyme Structure and Function", in "Oxidases and Related Redox Systems" *Prog. Clinical Biological Research* Vol. 274 Ed. T. E. King, H. Mason and M. Morrison Alan Liss Inc., New York, N.Y., pp. 419-432

Betso, S.R., Klapper, M.H. and Anderson, L.B. (1972) "Electrochemical Studies of Heme Proteins: Coulometric, Polarographic and Combined Spectroelectrochemical Methods for the Reduction of the Heme Prosthetic Group in Cytochrome *c*", J. Amer. Chem. Soc., 94, 8197-8204

Bond, A.M., Hill, H.A.O., Page, D.J., Psalti, I.S.M. and Walton, N. J. (1990) "Evidence that the Rate of Electron Transfer for Reduction of Cytochrome *c*, Cytochrome b_5, Plastocyanin, Azurin and Ferredoxin is Exceptionally Fast and Assymetrically Disposed at Modified Gold Electrodes", Eur. J. Biochem., 191, 737-42.

Bowden, E.F., Hawkridge, F.M. and Blount, H.N.(1984) "Interfacial Electrochemistry of Cytochrome *c* at Tin Oxide, Iridium Oxide, Gold and Platinum Electrodes" J. Electroanal. Chem., 161, 355-76.

Burrows, A., Hill, H.A.O., Leese, T., McIntire, W., Nakayama, N.A. and Sanghera, G.S. (1991) "The Electrochemistry of Methylamine Dehydrogenase", Eur. J. Biochem. 199, 73-78

Burrows, A., Guo, L-H., Hill, H.A.O., McLendon, G. and Sherman, F. (1991) "The Electrochemistry of Genetically-substituted Cytochrome *c*", Eur. J. Biochem., 1991, *202*, 543-549.

Cass, A.E.G., Davis, G., Hill, H.A.O. and Nancarrow, D.J. (1985) "The Reaction of Flavocytochrome b_2 with Cytochrome *c* and Ferricinium Carboxylate. Comparative Kinetics by Cyclic Voltammetry and Chronoamperometry", Biochim. Biophys. Acta, 828, 51-7.

Cass, A.E.G., Davis, G., Francis, G.D., Hill, H.A.O., Aston, W.J., Higgins, I.J., Plotkin, E.V., Scott, L.D. and Turner, A.P.F. (1984) "Ferrocene-mediated Enzyme Electrode for Amperometric Determination of Glucose", Anal. Chem., 56, 667-71.

Cataldi, T.R.I., Blackham, I.G., Briggs, G.A.D., Hill, H.A.O. and Pethica, J.B. (1990) "*In Situ* Scanning Tunneling Microscopy:

New Insight for Electrochemical Electrode-surface Investigations", J. Electroanal. Chem., 290, 1-20.

Coleman, J.O.D., Hill, H.A.O., Walton, N.J. and Whatley, F.R. (1983) "Electrochemically-driven Respiration in Mitochondria and *Paracoccus denitrificans*", FEBS Lett., 154, 319-322

Coleman, J.O.D., Hill, H.A.O., Walton, N.J. and Whatley, F.R. (1984) "Electrochemically-driven electron Transport in Biological Systems: The Exploitation of Direct Redox-protein Electrochemistry". In: "Charge and Field Effects in Biosystems". Eds. M. J. Allen and P. N. R. Usherwood, Abacus Press, London, 1984, pp. 61-68

Di Gleria, K., Hill, H.A.O. and Chambers, J. (1989) "Homogeneous Amperometric Ligand-binding Assay Amplified by a Proteolytic Enzyme Cascade", J. Electroanal. Chem., 267, 83-91.

Eddowes, M.J. and Hill, H.A.O. (1977) "A Novel Method for the Investigation of the Electrochemistry of Metalloproteins: Cytochrome *c*", J. Chem. Soc., Chem. Commun. 771-2.

Elliott, D., Hamnett, A., Lettington, O.C., Hill, H.A.O. and Walton, N. J. (1986) "The Determination, by Ellipsometry, of the Mode of Adsorption on Gold of some Promoters of Redox-protein Direct Electrochemistry", J. Electroanal. Chem., 242, 303-314.

Frew, J.E., Harmer, M.A., Hill, H.A.O. and Libor, S.I. (1986) "A Method of Estimation of Hydrogen Peroxide Based on Mediated Electron Transfer Reactions of Peroxidases at Electrodes", J. Electroanal. Chem., 201, 1-12.

Frew, J.E. and Hill, H.A.O. (1987) "Electron Transfer Biosensors", Phil. Trans. R. Soc. Lond., 316, 95-106.

Frew, J.E. and Hill, H.A.O. (1987) "Biosensors", Anal. Chem., 59, 933A-944A.

Frew, J.E. and Hill, H.A.O. (1988) "Direct and Indirect Electron Transfer between Redox Proteins and Electrodes", Eur. J. Biochem., 172, 261-9.

Garindarajai, K., Unni Nair, Ramasami, N and Ramaswamy, D. (1987) "Active Site Chemistry of Lysyl Oxidase" J. Inorg. Biochem., 29 111-18.

Green, M.J., Davis G. and Hill, H.A.O. (1984) "Creatine Kinase Assay Using an Enzyme Electrode", J. Biomed. Eng., 6, 176-177.

Green, M.J. and Hill, H.A.O. (1986) "Amperometric Enzyme Electrodes", J. Chem. Soc. Farad. I 82, 1237-1243.

Guo, L.-H., Hill, H. A. O., Hopper, D. J., Lawrance, G. A. and Sanghera, G. S. (1989) "Direct Un-mediated Electrochemistry of the Enzyme, *p*-Cresolmethylhydroxylase", J. Electroanal. Chem., 266, 379-390.

Guo, L.-H., Hill, H. A. O., Hopper, D. J., Lawrance, G. A. and Sanghera, G. S. (1990) "Direct Un-mediated Voltammetry of the *Chromatium vinosum* Enzyme, Flavocytochrome c_{552}, J. Biol. Chem., 265, 1958-63.

Harmer, M.A. and Hill, H.A.O. (1985) "The Direct Electrochemistry of Redox Proteins at Metal Oxide Electrodes", J. Electroanal. Chem., 189, 229-246

Hill, H.A.O., Oliver, B.N., Page, D.J. and Hopper, D. (1985) "The Enzyme-catalysed Electrochemical Conversion of *p*-Cresol to *p*-Hydroxybenzaldehyde", J. Chem. Soc. Chem. Commun., 1469-70.

Hill, H.A.O., Walton, N.J., Whitford D. and Coleman, J.O.D. (1985) "The Measurement of Changes in pH Associated with Electrochemically-driven Respiration in Rat Liver Mitochondria", J. Inorganic Biochem., 23, 303-309.

Lee, C.-W., Gray, H.B., Anson, F.C. and Malmström, B.G. (1984) "Catalysis of the Reduction of Dioxygen at Graphite Electrodes Coated with Fungal Laccase", J. Electroanal. Chem., 172, 289-300.

Taniguchi, I., Iseki, M., Yamaguchi, H., Yasukouchi, K. (1984) "Surface Enhanced Raman Scattering Study of Horse Heart Cytochrome *c* at a Silver Electrode in the presence of *bis*(4-pyrdidyl) disulfide and Purine", J. Electroanal. Chem., 175, 341-8.

Turner, A.P.F., Aston, W.J., Higgins, I.J., Bell, J.M., Colby, J., Davis, G. and Hill, H.A.O. (1984) "CO:Acceptor Oxidoreductase from *Pseudomonas thermocarbooxydovorans* Strain C2 and its use in a Carbon Monoxide Sensor", Anal. Chim. Acta 163, 161-174

Yeh, P. and Kuwana, T. (1977) "Reversible Electrode Reaction of Cytochrome *c*" Chem. Lett., 1145-8.

ELECTROCHEMICAL STUDIES OF SYNTHETIC MODELS OF COPPER PROTEINS

J. O. CABRAL
Laboratório Ferreira da Silva, Departamento de Química
Faculdade de Ciências, 4000 Porto, Portugal

ABSTRACT. After a brief survey of the different types of natural copper proteins, and their metal-centred main functions, the characteristics of "good" synthetic models are outlined. Electrochemical studies, by cyclic voltammetry and/or differential pulse polarography, of some representative models are overviewed. Recent, unpublished, results on several copper complexes intended to mimick the Type I copper sites are described and briefly discussed.

1. COPPER PROTEINS AND MODELS

1.1. General aspects

Copper plays a most important role in living systems, copper-containing proteins being widely distributed both in plants and animals, including man. The biological functions of these proteins include, for instance, dioxygen reduction and electron transport, and are closely connected with the oxidation state and coordination sphere of copper in the metalloproteins.

Several books, including those originated from the famous, so-called "Copper Conferences", and reviews on copper proteins and related subjects have been published in recent years [1 - 5].

The generally accepted classification of copper proteins is based on their biological functions; each class comprises several types, each one of them with similar copper active centres but with some differences in the amino acid sequences of the proteins, for instance.

On the basis of spectroscopic properties, copper in proteins are classified into four main types, three involving copper(II) and one involving copper(I), with the proviso that any given type does not necessarily imply "identical coordination spheres or even similar properties" [6].

Type I copper sites are known to have one copper(II) ion and are characterized by a very intense absorption band around 600 mm; for this reason, proteins with such a copper site are called "blue copper pro-

A. J. L. Pombeiro and J. A. McCleverty (eds.),
Molecular Electrochemistry of Inorganic, Bioinorganic and Organometallic Compounds, 151–170.

teins". In their EPR spectra an extremely small hyperfine splitting is observed, and they have relatively high redox potentials (+300 mV to + 800 mV vs. NHE).

Type II copper sites, with one copper(II) ion in a practically square planar environment, possess no distinctive spectroscopic properties,which are comparable to those of "normal" copper compounds. They also "show considerable variability in redox behavior" [6].

Type III copper sites are considered particularly interesting because they are EPR-silent, in spite of having two copper(II) ions. This dicopper(II) centre may be synchronously reduced to a dicopper(I) one by accepting two electrons at one step. A mixed-valence intermediate cannot, however, be ruled out.

Type IV copper sites are those that do not fit any of the other three types; normally copper is in the copper(I) state.

Examples of different common copper proteins, and their main biological functions and properties, are given in TABLE 1.

TABLE 1. Properties and functions of some copper proteins[(a)]

Protein	Mol.wt.[(b)]	E^o [(b)] (mV)	Type I	Type II	Type III	Function
Single copper proteins:						Electron carriers
Plastocyanine	10500	+370	1			
Azurin	14000	+330	1			
Stellacyanine	20000	+184	1			
Umecyanine	14600	+283	1			
Multicopper oxidases:						O_2 reduction
Laccase	62000	+780	1	1	2	
Ascorbate oxidase	140000	+580	2	2	4	
Ceruplasmine	130000	+390	1	1	4	

(a) Adapted from Refs. 7, 8, 9 and 10.
(b) Values may vary with the origin of the protein.

1.2. Electrochemistry of natural proteins

Even accepting that an isolated metalloprotein is not different from what it was in its natural environment, the feasibility of direct electrochemistry is subject to three main kinds of limitations:

(i) The metal redox centres in proteins may be buried, which might require a mediator for fast electron transfer at the electrode interface;

(ii) Given their size, currents will be limited by the small values of diffusion coefficients of the proteins;

(iii) Proteins have a tendency to absorb strongly and eventually denature at the surface of electrodes.

In his excellent review, Armstrong [11] authoratively discussed the subject and emphasized the useful information that can be obtained from direct electrochemistry of metalloproteins.

Very recently, the differential pulse polarography of some copper proteins, made possible by promotion, was studied and discussed by Kovā *et*. *al*. [12].

Since voltammetric methods are dynamic in nature, they constitute the obvious choice to study the main oxidation-reduction processes

Cu(II)—— Cu(I)—— Cu(II)

Cu(II)...Cu(II)—— Cu(II)...Cu(I)—— Cu(I)...Cu(I)

with many turnovers, in which the copper proteins are engaged. One obvious caution will be to distinguish between copper-centered and protein-centered processes.

1.3. Synthetic models of copper proteins

Natural copper proteins are very complex molecules, the greatest contribution to the molecular weight coming from their proteic parts. In order to avoid this complexity and focus the attention on their copper active centres, a vast amount of synthetic models of these centres have been prepared and characterized. Nevertheless, it is not to be expected that such low-molecular weight compounds would reproduce the properties of the natural proteins because they lack the protein backbone [13].

Less than two years ago it was reported [14] that reduction of copper (II) to copper(I) in azurin, isolated from *Alcaligenes denitrificans*, involved minimal structural changes in the copper protein, one of the requirements for fast electron transfer.

What then would be hypothetical, ideal, synthetic models? Possibly copper complexes obeying the following criteria: the structural features of the model complex must be very similar to those of the natural copper protein; the ligand system must be such that only very small structural changes would result from redox processes; cycles of oxidation-reduction processes must not be strongly reflected on the life-time of the complex; the associated redox potentials of the model copper complex, as well as its spectral and magnetic properties, must be very close to those of the copper protein to be mimicked.

It is probably exceedingly ambitious to expect that any synthetic model would meet all the above-mentioned criteria. Urbach, for instance, considers that a suitable model only needs to "mimic some physical or chemical property of the protein site and thereby provide an improved understanding of the biological analog"[9].

A somewhat different approach is that of Reedijk *et*. *al*. [15].

They consider three types of models for different aims of the study of low-molecular weight compounds as analogs for metalloproteins:

a) *Speculative models*, "to help in discovering the metal chromophore and deducing the structure of the active site, by comparing the spectroscopic properties of the protein with the analog (of known structure);

b) *Corroborative models*, "to contribute to a better understanding of the structure and reactivity of the active sites in metalloproteins";

c) Applicative models, "to apply the knowledge of the structure and properties of the active site in metalloproteins", to, e.g., metal detoxification.

The electrochemical behaviour of the synthetic analogs is not included in speculative models but comes under "reactivity" or "properties" for corroborative and applicative models, respectively. Electrochemical studies may suffer from the disadvantage of having to be undertaken in solution. It all depends on the solvent, on how it affects the complex, because the ideal solvent, that is, one that is chemically inert towards the solute may not be found. If this inertness is lacking, the species in solution differ from the solid to a smaller or greater extent which may be difficult to detect and assess (see, for instance, Ref.16).

A limited literature survey of articles on models of copper proteins revealed some curious aspects, as expressed in the numbers shown in TABLE 2.

TABLE 2. Number of surveyed articles on copper protein models

	Without electrochemistry	With electrochemistry	TOTAL
Type I	24(*) (Refs.: 17-40)	34(*) (Refs.: 13,41-73)	58
Type II	4(*) (Refs.:25,74-77)	1(*) (Ref. 78)	5
Type III	79(*) (Refs.:77,78-156)	54(*) (Refs.:157-210)	133
			196

(*) References in approximate chronological order.

These numbers make clear the great interest on the models of Type I and Type III centres, and the almost complete lack of interest on the Type II, certainly owing to the typical properties of the first two types and the "normality" of the Type II. Exclusively on models of the Type II centres only three articles, in almost two hundred, were found, the other two, refs. 25 and 77, also contemplate models of Type I or Type III copper centres. Thus, in this overview Type II centres will not be considered.

2. MODELS AND VOLTAMMETRY OF TYPE I COPPER CENTRES

2.1. Representative models

With the idea of modelling the Type I copper centres a very large number of chelating ligands, including macrocyclic, and their copper complexes have been synthesized, characterized, and reported, with various measures of success in reproducing the spectral properties characteristic of the protein.

Since the coordination environment in the natural active sites consists of two imidazole-like nitrogen atoms, one thiolate sulfur and one thioether sulfur, ligands possessing the N_2S_2 set of donor atoms would probably be the most hopeful. In this respect the work of Reedijk and co-workers is outstanding, with dozens of different compounds to their credit. The main ligands they used are shown in Figure 1.

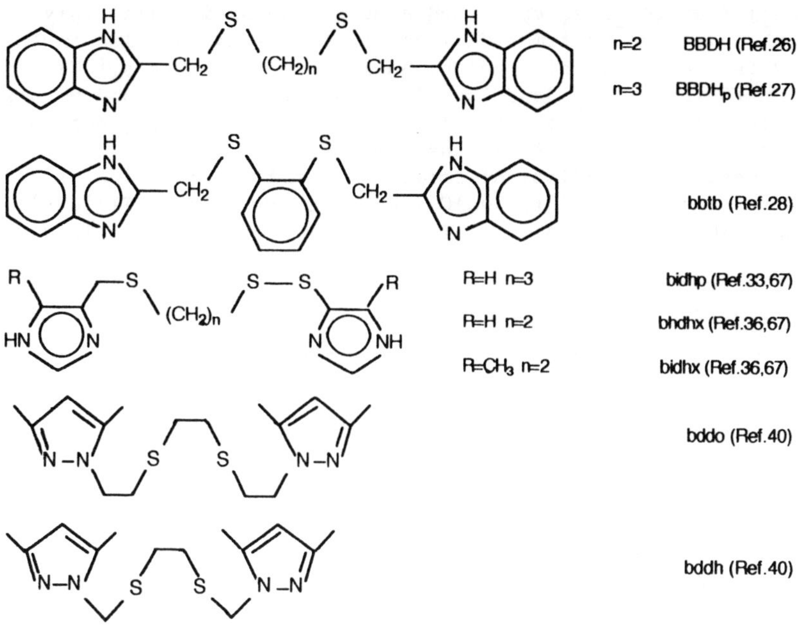

Figure 1. Schematic drawings of ligands with N_2S_2 donor atoms.

Much of this work was recently reviewed [67], and both the review and the original articles are readily available to anyone interested in the subject.

2.2. Electrochemistry of models

One of the characteristic properties of the Type I copper active sites is the relatively high potential for their Cu(II)/Cu(I) processes, in comparison with that of the Cu^{2+}(aq.)/Cu^{+}(aq.) couple. Consequently, a reasonable synthetic model will necessarily have a high potential, within the range found for natural proteins (see Table 1).

Some of the copper complexes studied by voltammetric methods have potentials of the right order of magnitude but are not reasonable models, because N_2S_2 set is absent; in other cases, this set is present but the values of the potentials are far from satisfactory and, therefore, are not suitable as models.

Possibly the first comprehensive study by cyclic voltammetry was that carried out by Rorabacher et al. [42], but only 3 of the 17 ligands they used had the N_2S_2 set of donor atoms, with potentials from +312 mV to

+361 mV vs. SHE.

Work on redox properties of copper complexes containing the N_2S_2 donor set has been extensively reviewed [61], until the end of 1987. Later work can be found, for instance, in references 66-70.

With commendable exceptions (e.g., Refs. 13, 50), authors tend to neglect the influence of solvent, and even of supporting electrolyte, not only on the cyclic voltammetric results themseves but also on the lifetime of the complex, which can be as low as $t_{1/2} \approx 0.5$ s [13]. In our work [68] on [Cu(bidhx)Cl_2] and [Cu(bidhx)Cl]Cl$(H_2O)_2$ we found both kinds of influences which we had to take into account for carrying out the experiments as well as to interpret the results.

More recently, we studied the cyclic voltammetric behaviour [72] of the following complexes: [Cu(bidhx)Cl(BF_4)] (A), [Cu(bhdhx)Cl][BF_4](B), [Cu(bhdhx)Cl_2](C), [Cu(bidhp)Cl_2]. The results are summarized in TABLE 3.

TABLE 3. Electrochemical data for the copper(II) complexes A, B, C and D in DMSO solution, 0.1 M in TEAP(a)

	Scan rate ($mV\ s^{-1}$)	$E_{1/2}$(b) (mV)	$i_{p,a}$ / $i_{p,c}$
[Cu(bidhx)Cl(BF_4)] (A)	10	+ 90 (80)	0.92
	20	+ 90 (80)	0.87
	50	+ 95 (90)	0.79
	100	+ 95 (110)	0.79
[Cu(bhdhx)Cl][BF_4] (B)	10	+ 110 (80)	0.81
	20	+ 110 (80)	0.72
	50	+ 115 (90)	0.62
	100	+ 120 (100)	0.61
[Cu(bhdhx)Cl_2] (C)	10	+ 8 (125)	0.71
	20	+ 10 (140)	0.68
	50	+ 13 (205)	0.61
	100	+ 40 (320)	0.50
[Cu(bidhp)Cl_2](c)	-	-	-

(a) Potentials vs. SCE, uncertainty ± 10 mV; all solutions freshly prepared in situ, about 1mM (A and B) or 0.1 mM (C) in complex.
(b) Peak separation in parentheses.
(c) Not surprinsingly (see Ref.33) it decomposes on dissolution.

Complexes A, B and C were affected by solvation, which proceeded practically to completion within hours. Thus, this decomposition is sufficiently slow to allow the recording of cyclic voltammograms from freshly prepared solutions, except in the case of complex C. Its low solubility in DMSO required a relatively long time to bring it into solution, which would favour decomposition by solvation, not only before starting the voltage scans but also during the few minutes involved in running successive voltammograms at four scan rates.

Whenever peak currents could be reasonably measured, it was found ney increased linearly with the square root of scan rates. By comparison ith the behaviour of the couple Fc^+/Fc under the same experimental con- itions, and taking into account the values of peak separations and peak urrents ratios, we suggest that complexes A and B undergo diffusion-con- rolled, one-electron, quasi-reversible redox processes.

The electrochemical behaviour of $[Cu(bidhx)Cl(BF_4)]$ requires some com- ents, since it is very different from that of $[Cu(bidhx)Cl_2]$ (Ref.68), lthough the two complexes only differ in one anionic ligand. To interpret he voltammetric results we assumed that a fast, practically complete irst solvation step would change

$$[Cu(bidhx)Cl(BF_4)]$$

nto

$$[Cu(bidhx)(BF_4)(DMSO)]^+$$

hich would be responsible for the observed wave.

We also studied the cyclic voltammetric behaviour of three copper(II) omplexes of the ligand bddo (see Figure 1), viz., $Cu(bddo)(H_2O)(BF_4)_2$, $u(bddo)(H_2O)(ClO_4)_2$, $Cu(bddo)Cl_2$, and compared it with that of $u(H_2O)_6(BF_4)_2$ and $Cu(H_2O)_6(ClO_4)_2$. The results are summarized in TABLE 4.

ABLE 4. Electrochemical data for copper(II) complexes of bddo and its etrafluoroborate and perchlorate salts in acetonitrile solution, 0.1M in EAP[(a)]

Compound	Scan rate (mV s^{-1})	$E_{1/2}$[(b)] (mV)	$i_{p,a}/i_{p,c}$
$u(bddo)(H_2O)(BF_4)_2$	50	+ 755 (70)	1.00
	100	+ 755 (70)	0.97
	200	+ 755 (70)	0.97
$u(bddo)(H_2O)(ClO_4)_2$	50	+ 755 (70)	1.00
	100	+ 755 (70)	0.98
	200	+ 750 (80)	0.98
$u(bddo)Cl_2$	50	+ 513 (155)	0.80
	100	+ 513 (155)	0.81
	200	+ 513 (175)	0.87
$u(H_2O)_6(BF_4)_2$	50	+ 935 (90)	0.95
	100	+ 935 (90)	0.80
	200	+ 945 (110)	0.77
$u(H_2O)_6(ClO_4)_2$	50	+ 938 (135)	0.97
	100	+ 940 (160)	0.89
	200	+ 940 (200)	0.74

a) Potentials vs. SCE, uncertainty about 10 mV; all solutions freshly prepared in situ, approximately 1 mM in complex.

b) Peak separation in parentheses.

The remarkable electrochemistry of the complexes with non-coordinated anions, BF^{-}_{4} and ClO_4^{-}, is very unusual, not only for being really reversible but also for the very high redox potentials, the highest we are aware of for this kind of complexes. Rorabacher et al. [71] reported a value of +0.83 V vs SHE (≈ 0.59 V vs. SCE), some 0.17 V lower than those.

The behaviour of Cu(bddo)Cl_2 has nothing unusual worth mentioning.

Another aspecto to consider is the decrease in $E_{1/2}$ on complexation of copper(II) with bddo, about 190 mV, as is to be expected on formation of stable complexes.

The reversibility of the redox process suggested that the reduced species must be stable in solution. However, no solid copper(I) complex could be isolated, despite several attempts.

3. MODELS AND VOLTAMMETRY OF TYPE III COPPER CENTRES

3.1. Representative models

Most of the synthetic models we found in our literature survey are concerned with hemocyanine, the oxygen transport protein in arthropods and molluscs, and tyrosinase, which catalyzes the two-electron oxidation of phenolic compounds. Both proteins contain a coupled binuclear copper active site, a Type III copper centre,"which reversibly binds dioxygen as peroxide bridging between the two copper ions" [153]. The Cu-Cu distance is of the order of 300-400 pm, and a tetragonal coordination is achieved with donor nitrogen atoms of imidazole ligands from histidine [142].

According to Nelson et al. [211], the probable requirements of worthwhile models of Type III copper centres will include:

(i) a binucleating synthetic ligand which permits the relatively close approach of the two copper ions, necessary for the strong antiferromagnetic interaction characteristic of the Type III centres, preferably with retention of the bimetallic unit in solution;

(ii) accessibility of the Cu(I) and the Cu(II) oxidation states;

(iii) coordinative unsaturation in the di-Cu(I) state (sic);

(iv) a coordination sphere for each metal ion with two (or three) nitrogen donors;

(v) the provision of a bridging group (cf. the endogenous bridge of the natural system) which mediates antiferromagnetic coupling between the metal ions in the oxidised state and which accounts for the ESR silence of the native proteins (sic).

In conclusion, a successful model should reproduce one or more of the following properties: reversible dioxygen binding; strong antiferromagnetic interaction, resulting in EPR-silence; reversible change from the oxy, di-Cu(II), to the deoxy, di-Cu(I), forms. Thus, the success of the model depends on the binucleating ligand.

Crane and Fenton [212] cite the two major strategies that have been used for ligand-based control of suitable dinuclear complexes. The first strategy is based on the size of the ligand, which should be large enough, contain sufficient donor atoms to coordinate both metal ions, and make allowance for any required exogenous bridging groups. The second strategy consists in the use of facially coordinating, tridentate ligands, which could give dinuclear complexes through additional coordination of potentially bridging groups.

Some of the ligands that have been used in the syntheses of prospective models of Type III copper centres are shown in Figure 2.

It is interesting to note that the first example of a complex which binds dioxygen reversibly was the mononuclear copper(I) complex of ligand **1** [79]. All the other ligands are binucleating, and their dicopper complexes are generally capable of coordinating a bridging group.

Many other ligands of different chemical natures have been used,

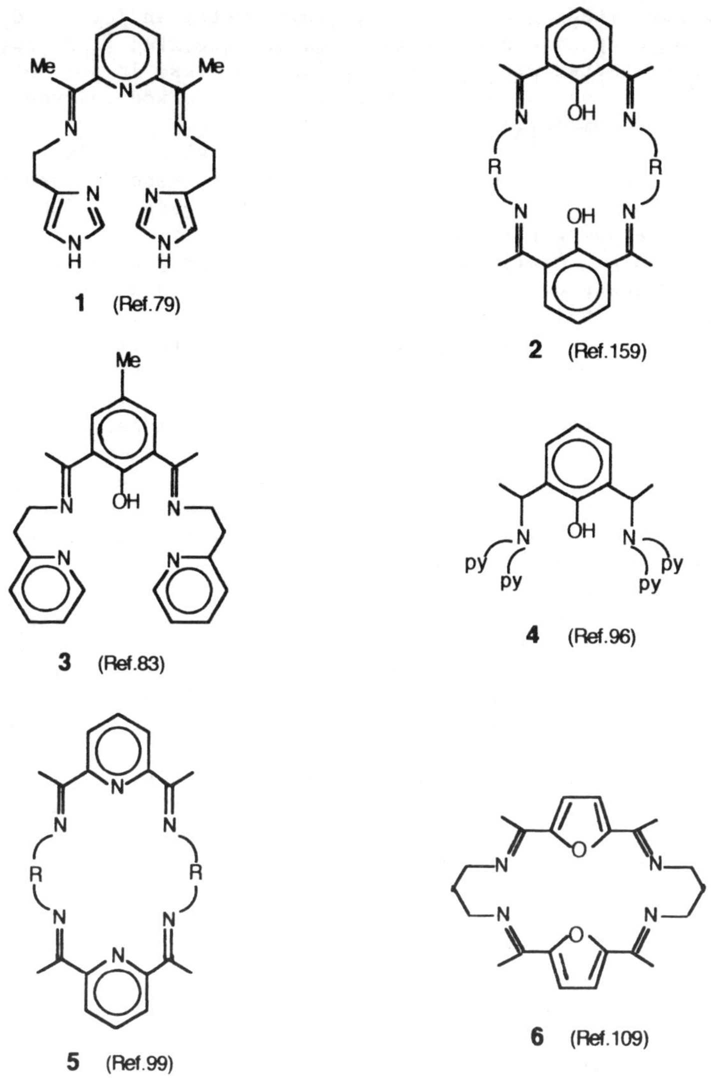

Figure 2. Schematic drawings of ligands for modelling Type III copper centres

in some cases having practically no similarity with natural Type III centres.

3.2. Electrochemistry of models

As mentioned before, a worthwhile model must possess a good accessibility of both the Cu(I) and Cu(II) oxidation states, and retain in solution not only the bimetallic unit but also any bridging group. To detect whether a proposed model of a Type III copper centre satisfies these requirements many studies on the cyclic voltammetry and/or the differential pulse polarography have been carried out and reported [157-210]. In some cases authors started with a di-copper(II) complex, in some other with a di-copper(I) complex and even eventually with a mixed-valence Cu(II)-Cu(I) dinuclear complex.

Two types of behaviour were observed: a single two-electron process, assumed to be the result of two one-electron processes at practically the same potentials; two distinct, one-electron processes, which allowed the estimation of a comproportionation constant.

Lintvedt et al. [161,162,177,182,191,209, e.g.] extensively studied di-copper complexes of acetylacetone derivatives, and were probably the first to report [161] the unprecedented "experimental observation of two sequential one-electron transfers at identical potentials" (sic), thus simulating an analogous property of the Type III centres.

The electrochemical investigations of di-copper(II) complexes of ligand 3 (see Figure 2), with hydroxo or chloro bridges, provided no evidence of the formation of either a stable mixed-valence, Cu(II)-Cu(I), or a dinuclear Cu(I) complex [163]. In contrast, from the somewhat similar ligand 4 (see Figure 2) a dinuclear copper(I) complex was prepared; cyclic voltammetric experiments in dimethylformamide revealed "a quasi-reversible one-electron (per copper) oxidation wave" [175].

It is curious to note that, although using very similar ligands of the general formula 2 (see Figure 2), Addison [157] observed only one cyclovoltammetric wave, but Gagné et al [159, 164], as well as Long and Hendrickson [178], observed two, both by cyclic voltammetry and differential pulse polarography. Addison used acetonitrile as solvent and the other authors used dimethylformamide. Will the difference in behaviour be the result of the difference in the solvents?

Ligands of the general formula 5 were obtained by template synthesis in the presence of Ba(II) or Sr(II), and, by transmetallation, gave di-Cu (II) complexes, which could accept anionic bridges (hydroxide or imidazolate, e.g.), and also a mononuclear Cu(II) complex. With three of those ligands

L^1 R = $(CH_2)_2$-O-$(CH_2)_2$-O-$(CH_2)_2$
L^2 R = $(CH_2)_2$-S-$(CH_2)_2$
L^3 R = $(CH_2)_3$

the following six complexes were prepared and characterized

(1) $[CuL^1][ClO_4]_2 \cdot H_2O$
(2) $Cu_2L^1(ClO_4)_4 \cdot 2H_2O$
(3) $Cu_2L^1(OH)(ClO_4)_3 \cdot H_2O$
(4) $Cu_2L^1(im)(ClO_4)_3$
(5) $Cu_2L^2(im)(ClO_4)_3 \cdot H_2O$
(6) $Cu_2L^3(OH)(ClO_4)_3 \cdot H_2O$

which we studied by cyclic voltammetry [181], with the results summarized in TABLE 5.

TABLE 5. Electrochemical data for mono and di-copper(II) complexes of ligands 5 in dimethylacetamide as solvent, 0.1M in TBAP[(a)]

Complex	Scan rate ($mV\ s^{-1}$)	$E_{1/2}$[(b)] (mV)		Comments
$[CuL^1][ClO_4]_2.H_2O$	50	- 390	(110)	Quasi-reversible
(1)	100	- 390	(110)	
$Cu_2L^1(ClO_4)_4.2H_2O$	50	- 33	(115)	Quasi-reversible
(2)		- 435	(70)	Reversible
$Cu_2L^1(OH)(ClO_4)_3.H_2O$	50	- 190[(c)]		Irreversible
(3)		- 390	(110)	Quasi-reversible
	100	- 200[(c)]		Irreversible
		- 395	(90)	Quasi-reversible
$Cu_2L^1(im)(ClO_4)_3$	50	- 250[(c)]		Irreversible
(4)		- 395	(70)	Reversible
$CuL^2(im)(ClO_4)_3.H_2O$	20	- 235	(90)	Quasi-reversible
(5)	50	- 235	(110)	Quasi-reversible
$CuL^3(OH)(ClO_4)_3.H_2O$	100	- 268	(65)	Reversible
(6)		- 600[(c)]		Irreversible

(a) TBAP=tetrabutylammonium perchlorate; potentials vs. SCE, uncertainty about 10mV; all solutions 10^{-3}-10^{-4} M in complex.
(b) Peak separation in parentheses.
(c) Cathodic peak potential, no anodic counterpart.

These results show that only in the case of the di-Cu(II) tetraperchlorate complex(2), without anionic bridge, can a fairly stable Cu(II)... Cu(I) mixed valence species be detected, and a conproportionation constant of ~ 10^7 be estimated.

All other details and interpretation of results will be found in the original article.

The furan based, 20-membered macrocyclic ligand 6 (see Figure 2) was prepared as the barium complex by template synthesis [109]. A series of dicopper(II) and dicopper(I) complexes, with and without bridging groups, were obtained from the barium complex by transmetallation, and properly characterized. It was found that di-copper(I) complexes could take up dioxygen, sometimes acting as oxygenation catalysts, depending on the solvent. We studied the electrochemical behaviour of some of these complexes by cyclic voltammetry [207], which proved to be always irreversible. This irreversibility might probably reflect the occurrence of a chemical process, such as transfer of a hydroxo bridging ligand, at each redox stage. The pair of irreversible reduction waves observed in the cyclic voltammograms of the di-hydroxobridged dicopper(II) complex, $[Cu_2(6)(HO)_2(H_2O)][ClO_4]_2$, after several consecutive scans, might indicate two sequential one-electron reductions of the dicopper centre, the first of which would correspond to a mixed valence, Cu(II)-OH-Cu(I), intermediate.

4. CONCLUDING REMARKS

This limited survey of the literature, and arbitrary choice of topics, on copper proteins and some synthetic models of their active metal centres can hopefully be considered a representative sample of the enormously vast amount of work that has been published on the subject.

It appears that a really worthwhile model has yet to be made but there is no doubt that the quest for such a model was, and is, the motivation behind a surge of the most interesting developments in the chemistry of copper and its coordination compounds.

Fortunately, the enthusiasm on this subject is not vanishing, many more valuable findings will be forthcoming and our knowledge on the matter will certainly greatly increase.

5. ACKNOWLEDGEMENTS

J.O.C. gratefully acknowledges the sponsorship of NATO (Collaborative Research Grant No.870370), and of Instituto Nacional de Investigação Científica (Centro de Investigação em Química, Linha de Acção nº.3).

6. REFERENCES

1. Sigel, H. (ed.) (1981) "Metal Ions in Biological Systems. Volume 12. Properties of Copper", Marcel Dekker, Inc., New York.
2. Sigel, H. (ed.) (1981) "Metal Ions in Biological Systems, Volume 13. Copper Proteins", Marcel Dekker, New York.
3. Karlin, K.D. and Zubieta, J. (eds.) (1983) "Copper Coordination Chemistry: Biochemical and Inorganic Perspectives", Adenine Press, Guilderland, New York.
4. Karlin, K.D. and Zubieta, J. (eds.) (1986) "Biological and Inorganic Copper Chemistry", Volumes I and II, Adenine Press, Guilderland, New York.
5. Karlin, K. D.; Gultneh, Y. Progr. Inorg. Chem., **1987**, 35, 219.
6. Jameson, R. F. "Coordination Chemistry of Copper with Regard to Biological Chemistry", Chapter 1, Ref.1.
7. Frieden, E. "The Evolution of Copper Proteins", Chapter 1, Ref. 2.
8. Lappin, A. G. "Properties of Copper "Blue" Proteins", Chapter 2, Ref. 2.
9. Urbach, F. L. "The Properties of Binuclear Copper Centres in Model and Natural Compounds", Chapter 3, Ref. 2.
10. Malmström, B. G. (1987) "Some Aspects of Structure and Function in Copper Containing Oxidases", in R. J. P. Williams and J. J. R. F. da Silva (eds.), "New Trends in Bio-inorgqnic Chemistry", Academic Press, New York.
11. Armstrong, F. A. Struct. Bonding (Berlin), **1990**, 72, 137.
12. Studnicková, M.; Pitrincová, J.; Kovár, J. Bioelectrochem. Bioenerg., **1991**, 25, 109.

13. Casella, L.; Guillotti, M.; Suardi, E.; Sisti, M.; Pagliarin, R.; Zanello, P. J. Chem. Soc. Dalton Trans., 1990, 2843.
14. Shepard, W. E. B.; Anderson, B. F.; Lewandoski, D. A.; Norris, G.; Baker, E. N. J. Am. Chem. Soc., 1990, 112, 7817.
15. Reedijk, J.; Driessen, W. L.; Van Rijm, J., in Ref. 4, Vol. II, 143, and references therein.
16. Kadish, K. M.; Deng, Y.; Zaydoun, S.; Idrissi, M. S. J. Chem. Soc. Dalton Trans., 1990, 2809.
17. Braithwaite, A. C.; Richard, C. E. F.; Waters, T. N. Inorg. Chim. Acta, 1978, 26, 63.
18. Styka, M. C.; Smierciak, R. C.; Blinn, E. L.; DeSimone, R. E.; Passariello, J. V. Inorg. Chem., 1978, 17, 82.
19. Schugar, H. J.; Fawcett, T. G.; Hendrickson, D. N.; Felthouse, T. R. Inorg. Chem., 1978, 17, 2707.
20. Dagdigian, J. V.; Reed, C. A. Inorg. Chem., 1979, 18, 2623.
21. Hendriks, H. M. J.; Reedijk, J. Recl. Trav. Chim. Pays-Bas, 1979, 98, 95.
22. DeSimone, R. E.; Blinn, E. L.; Mucker, K. F. Inorg. Nucl. Chem. Letters, 1980, 16, 23.
23. Downes, J. M.; Whelan, J.; Bosnich, B. Inorg. Chem., 1981, 20, 1081.
24. Birker, P. J. M. W. L.; Hendriks, H. M. J.; Reedijk,J.; Verschoor, G. C. Inorg. Chem., 1981, 20, 2408.
25. Takahashi, K.; Ogawa, E.; Oishi, N.; Nishida, Y; Kida, S. Inorg. Chim. Acta, 1982, 66, 97.
26. Birker, P. J. M. W. L.; Helder, J.; Henkel, G.; Krebs, B.; Reedijk, J. Inorg. Chem., 1982, 21, 357.
27. Schilstra, M. J.; Birker, P. J. W. L.; Verschoor, G. C.; Reedijk, J. Inorg. Chem., 1982, 21, 2637.
28. Rietmeijer, F. J.; Birker, P. J. M. W. L.; Gorter, S.; Reedijk, J. J. Chem. Soc. Dalton Trans., 1982, 1191.
29. Yamauchi, O.; Seki, H.; Shida, T. Bull. Chem. Soc. Jpn., 1983, 56, 3258.
30. Verheijdt, P. L.; Haasnoot, J. G.; Reedijk, J. Inorg. Chim. Acta, 1983, 76, L43.
31. Bernarducci, E.; Bharadwaj, P. K.; Krogh-Jespersen K.; Potenza, J. A.; Schugar, H. J. J. Am. Chem. Soc., 1983, 105, 3860.
32. Casella, L.; Silver, M. E.; Ibers, J. A. Inorg. Chem., 1984, 23, 1409.
33. Van Rijn, J.; Driessen, W. L.; Reedijk, J.; Lehn, J.-M. Inorg. Chem., 1984, 23, 3584.
34. Canters, G. W.; Groeneveld, C. M. (1986) "Electron Self-Exchange Reactions of Blue Copper Proteins", in A. V. Xavier (ed.), "Frontiers in Bioinorganic Chemistry", VCH, Weinheim.
35. Knapp, S.; Keenan, T. P.; Zhang, X.; Fikar, R.; Potenza, J. A.; Schugar, H. J. J. Am. Chem. Soc., 1987, 109, 1882.
36. Bouwman, E.; Day, R.; Driessen, W. L.; Tremel, W.; Krebs, B.; Wood, J. S.; Reedijk, J. Inorg. Chem., 1988, 27, 4614.
37. Bailey, N. A.; Bastida, R.; Fenton, D. E.; Lockwood, S. J.; McLean, C. H. J. Chem. Soc. Dalton Trans., 1988, 839.
38. Bouwman, E.; Hove, J. C. T.; Driessen, W. L.; Reedijk, J. Polyhedron, 1988, 7, 2591.

39. Brader, M. L.; Dunn, M. F. J. Am. Chem. Soc., 1990, 112, 4585.
40. Haanstra, W. G.; Driessen, W. L.; de Graaff, R. A. G.; Reedijk Inorg. Chim. Acta, 1991, 186, 215.
41. Jones, T. E.; Rorabacher, D. B.; Ochrymowycz, L. A. J. Am. Chem. Soc., 1975, 97, 7485.
42. Dockal, E. R.; Jones, T. E.; Sokol, W. F.; Engerer, R. J.; Rorabacher, D. B.; Ochrynowycz, L. A. J. Am. Chem. Soc., 1976, 98, 4322.
43. Karlin, K. D.; Dahlstrom, P. L.; Hyde, J. R.; Zubieta, J. J. C. S. Chem. Comm., 1980, 906.
44. Augustin, M. A.; Yandell, J. K.; Addison, A. W.; Karlin, K. D. Inorg. Chim. Acta, 1981, 55, L35.
45. Addison, A. W.; Hendriks, H. M. J.; Reedijk, J.; Thompson, L. K. Inorg. Chem., 1981, 20, 103.
46. Gisselbrecht, J. P.; Gross, M. J. Electroanal. Chem., 1981, 127, 127.
47. Karlin, K. D.; Sherman, S. Inorg. Chim. Acta, 1982, 65, L39.
48. Sorrell, T. N.; Jameson, D. L. Inorg. Chem., 1982, 21, 1014.
49. Karlin, K. D.; Hayes, J. C.; Juen, S.; Hutchinson, J. P.; Zubieta, J. Inorg. Chem., 1982, 21, 4108.
50. Nikles, D. E.; Powers, M. J.; Urbach, F. L. Inorg. Chem., 1983, 22, 3210.
51. Aoi, N.; Matsubayashi, G.; Tanaka, T. J. Chem. Soc. Dalton Trans., 1983, 1059.
52. Siegfried, L.; Kaden, T. A. Helv. Chim. Acta, 1984, 67, 29.
53. Balakrishnan, K. P.; Kaden, T. A.; Siegfried, L.; Zuberbühler, A. D. Helv. Chim. Acta, 1984, 67, 1060.
54. Aoi, N.; Matsubayashi, G.; Tanaka, T.; Nokatsu, K. Inorg. Chim. Acta, 1984, 85, 123.
55. Karlin, K. D.; Yandell, J. K. Inorg. Chem., 1984, 23, 1184.
56. Addison, A. W.; Rao, T. N.; Sinn, E. Inorg. Chem., 1984, 23, 1957.
57. Gouteron, J; Jeannin, S.; Jeannin, Y.; Livage, J.; Sanchez, C. Inorg. Chem., 1984, 23, 3387.
58. Hartman, J. A. R.; Cooper, S. R. J. Am. Chem. Soc., 1986, 108, 1202.
59. Martin, M. J.; Endicott, J. F.; Ochrymowyca, L. A.; Rorabacher, D. B. Inorg. Chem., 1987, 26, 3012.
60. Hay, R. W.; Govan, N.; Pujari, M. P. J. Chem. Soc. Dalton Trans., 1987, 963.
61. Zanello, P. Comments Inorg. Chem., 1988, 8, 45.
62. Addison, A. W.; Palaniandavar, M.; Driessen, W. L.; Paap, F.; Reedijk, J. Inorg. Chim. Acta, 1988, 142, 95.
63. Addison, A. W. Inorg. Chim. Acta, 1989, 162, 217.
64. Oka, H.; Nakao, Y.; Susuki, S.; Mori, W.; Kinoshita, I.; Ooi, S.; Ichimura, A. Inorg. Chim. Acta, 1989, 165, 111.
65. Bernardo, M. M.; Robandt, P. V.; Schroeder, R. R.; Rorabacher, D. B. J. Am. Chem. Soc., 1989, 111, 1224.
66. Gullotti, M.; Casella, L.; Pintar, A.; Suardi, E.; Zanello, P.; Mangani, S. J. Chem. Soc. Dalton Trans., 1989, 1979.
67. Bouwman, E.; Driessen, W.L.; Reedijk, J. Coord. Chem. Reviews, 1990, 104, 143.
68. Cabral, M. F.; Cabral, J. de O.; Bouwman, E.; Driessen, W. L.; Reedijk, J. Inorg. Chem. Acta, 1990, 167, 205.
69. Martin, E.M.; Berman, R. D. Inorg. Chim. Acta, 1991, 188, 221.

70. Ibid, 233.
71. Bernardo, M. M.; Schroeder, R. R.; Rorabacher, D. B. Inorg. Chem., 1991, 30, 1241.
72. Cabral, M. F.; Cabral, J. de O.; Bouwman, E.; Driessen, W. L.; Reedijk, J.; Turpeinen, U.; Hämälaine, R. Unpublished results.
73. Haanstra, W. G.; Cabral, M. F.; Cabral, J. de O.; Driessen, W. L.; Reedijk, J. Unpublished results.
74. Sokurai, T.; Suzuki, S.; Nakahara, A. Bull. Chem. Soc. Jpn., 1981, 54, 2313.
75. Sakurai, T.; Nakahara, A. Inorg. Chim. Acta, 1981, 55, 157.
76. Kitajima, N.; Whang, K.; Moro-oka, Y.; Uchida, A.; Sasada, Y. J. Chem. Soc. Chem. Commun., 1986, 1504.
77. Casella, L.; Gullotti, M.; Pallanza, G.; Buga, M. Inorg. Chem., 1991, 30, 221.
78. Addison, A. W.; Carpenter, M.; Lau, L.K.-M.; Wicholas, M. Inorg. Chem., 1978, 17, 1545.
79. Simmons, M. G.; Wilson, L. J. J. Chem. Soc. Chem. Commun., 1978, 634.
80. Hendriks, H. M. J.; Reedijk, J. Inorg. Chim. Acta, 1979, 37, L509.
81. Gagné, R. R.; Gall, R. S.; Lisensky, G. C.; Marsh, R. E.; Speltz, L. M. Inorg. Chem., 1979, 18, 771.
82. Pasquali; M.; Marini, G.; Floriani, C.; Gaetani-Manfredotti, A.; Guastini, C. Inorg. Chem., 1980, 19, 2525.
83. Gorzybowski, J. J.; Urbach, F. L. Inorg. Chem., 1980, 19, 2604.
84. Katz, R. N.; Kolks, G.; Lippard, S. J. Inorg. Chem., 1980, 19, 3845.
85. Coughlin, P. K.; Lippard, S. J.; Martin, A. E.; Bulkowsky, J. E. J. Am. Chem. Soc., 1980, 102, 7616.
86. Burnett, M. G.; McKee, V.; Nelson, S. M.; Drew, M. G. B. J. chem. Soc. Chem. Commun., 1980, 829.
87. Drew, M. G. B.; Cairns, C.; Lavery, A.; Nelson, S. M. J. Chem. Soc. Chem. Commun., 1980, 1122.
88. Hay, R. W.; Bembi, R.; McLaren, F. Inorg. Chim. Acta, 1981, 54, L161.
89. Murase, I.; Hamada, K.; Kida, S. Inorg. Chim. Acta, 1981, 54, L171.
90. Birker, P. J. M. W. L.; Hendriks, H. M. J.; Reedijk, J. Inorg. Chim. Acta, 1981, 55, L17.
91. Kolks, G.; Frihart, C. R.; Coughlin, P. K.; Lippard, S. J. Inorg. Chem., 1981, 20, 2933.
92. Co, M. S.; Hodgson, K. O.; Eccles, T. K.; Lontie, R. J. Am. Chem. Soc., 1981, 103, 984.
93. Burk, P. L.; Osborn, J. A.; Youinou, M. T.; Agnus, Y.; Louis, R.; Weiss, R. J. Am. Chem. Soc., 1981, 103, 1273.
94. Coughlin, P. K.; Lippard, S. J. J. Am. Chem. Soc., 1981, 103, 3228.
95. McKee, V.; Dagdigian, J. V.; Bau, R.; Reed, C. A. J. Am. Chem. Soc., 1981, 103, 7000.
96. Karlin, K. D.; Dahlstrom, P.L.; Cozzette, S. N.; Scensny, P. M.; Zubieta, J. J. Chem. Soc. Chem. Commun., 1981, 881.
97. Korp, J. D.; Bernal, J.; Merrill, C. L.; Wilson, L. J. J. Chem. Soc. Dalton Trans., 1981, 1951.
98. Karlin, K. D.; Dahlstrom, P. L.; DiPierro, L. T.; Simon, R. A.; Zubieta, J. J. Coord. Chem., 1981, 11, 61.
99. Nelson, S. M. Inorg. Chim. Acta, 1982, 62, 39.
100. Drew, M. G. B.; Nelson, S. M.; Reedijk, J. Inorg. Chim. Acta, 1982, 64, L189.

101. Moore, K.; Vigee, G. S. Inorg. Chim. Acta, 1982, 66, 125.
102. Bolus, D.; Vigee, G. S. Inorg. Chim. Acta, 1982, 67, 19.
103. Hendriks, H. M. J.; Birker, P. J. M. W. L.; Van Rijn, J.; Verschoor, G. C.; Reedijk, J. J. Am. Chem. Soc., 1982, 104, 3607.
104. Patel, K. V.; Bhattacharya, P. K. J. Chem. Soc. Dalton Trans., 1982, 359.
105. Drew, M. G. B.; Nelson, J.; Esho, F.; McKee, V.; Nelson, S. M. J. Chem. Soc. Dalton Trans., 1982, 1837.
106. Thompson, L. K. Can. J. Chem., 1983, 61, 579.
107. Hartstock, F. W.; Thompson, L. K. Inorg. Chem. Acta, 1983, 72, 227.
108. Casella, L.; Ghelli, S. Inorg. Chem., 1983, 22, 2458.
109. Nelson, S. M.; Esho, F.; Lavery, A.; Drew, M. G. B. J. Am. Chem. Soc., 1983, 105, 5693.
110. Karlin, K. P.; Hayes, J. C.; Hutchinson, J. P.; Zubieta, J. J. Chem. Soc. Dalton Trans., 1983, 376.
111. McKee, V.; Smith, J. J. Chem. Soc. Chem. Commun., 1983, 1465.
112. Murase, I.; Hamada, K.; Ueno, S.; Kida, S. Synth. React. Inorg. Met.-Org. Chem., 1983, 13, 191.
113. Adams, H.; Bailey, N. A.; Fenton, D. E.; Moss, S.; Jones, G. Inorg. Chim. Acta, 1984, 83, L79.
114. Sorrell, T. N.; Jameson, D. L.; O'Connor, C. J. Inorg. Chem., 1984, 23, 190.
115. Karlin, K. D.; Gultneh, Y.; Hayes, J. C.; Zubieta, J. Inorg. Chem., 1984, 23, 519.
116. Karlin, K. D.; Cruse, R. W.; Gultneh, Y.; Hayes, J. C.; Zubieta, J J. Am. Chem. Soc., 1984, 106, 3372.
117. McKee, V.; Zvagulis, M.; Dagdigian, J. V.; Patch, M. G.; Reed, C. A. J. Am. Chem. Soc., 1984, 106, 4765.
118. Blackburn, N. J.; Karlin, K.D.; Concannon, M.; Hayes, J. C.; Gultneh, Y.; Zubieta, J. J. Chem. soc. Chem. Commun., 1984, 939.
119. Sorrell, T. N.; Borovik, A. S. J. Chem. Soc. Chem. Commun., 1984, 1489.
120. Fallon, G. D.; Murray, K. S.; Spethmann, B.; Yandell, J. K.; Hodgkin, J. H.; Loft, B. C. J. Chem. Soc. Chem. Commun., 1984, 1561.
121. Matsumoto, N.; Asakawa, M.; Nogami, H.; Ohyoshi, A. Inorg. Chim. Acta, 1985, 96, 87.
122. Thompson, L. K.; Hartstock, F. W.; Rosenberg, L.; Woon, T. C. Inorg. Chim. Acta, 1985, 97, 1.
123. McKee, V.; Zvagulis, M.; Reed, C. A. Inorg. Chem., 1985, 24, 2914.
124. Sorrell, T. N.; O'Conner, C. J.; Anderson, O. P.; Reibenspies, J. H. J. Am. Chem. Soc., 1985, 107, 4199.
125. Karlin, K. D.; Gultneh, Y. J. Chem. Educ., 1985, 62, 983.
126. Casella, L.; Rigoni, L. J. Chem. Soc. Chem. Commun., 1985, 1668.
127. Butcher, R. J.; Diven, G.; Erickson, G.; Mockler, G. M.; Sinn, E. Inorg. Chim. Acta, 1985, 123, L17.
128. Karlin, K. D.; Cruse, R. W.; Gultneh, Y. J. Chem. Soc. Chem. Commun., 1987, 599.
129. Karlin, K. D.; Cohen, B. I.; Hayes, J. C.; Farooq, A.; Zubieta, J Inorg. Chem, 1987, 26, 147.
130. Mallah, T.; Kahn, O.; Gouteron, J.; Jeannin, Y.; O'Connor, C. J. Inorg. Chem., 1987, 26, 1375.

131. Pate, J. E.; Cruse, R. W.; Karlin, K. D.; Solomon, E. I. J. Am. Chem. Soc., 1987, 109, 2624.
132. Karlin, K. D.; Cruse, R. W.; Gultneh, Y.; Farroq, A.; Hayes, J. C.; Zubieta, J.J. Am. Chem. Soc., 1987, 109, 2668.
133. Sorrell, T. N.; Borovik, A. S. J. Am. Chem. Soc., 1987, 109, 4255.
134. Karlin, K. D.; Cruse, R. W.; Gultneh, Y. J. Chem. Soc. Chem. Commun., 1987, 599.
135. Van Rijn, J.; Reedijk, J.; Dartmann, M.; Krebs, B. J. Chem. Soc. Dalton Trans., 1987, 2579.
136. Karlin, K. D.; Haka, M. S.; Cruse, R. W.; Meyer, G. J.; Farroq, A.; Gultneh, Y.; Hayes, J. C.; Zubieta, J. J. Am. Chem. Soc., 1988, 110, 1196.
137. Casella, L.; Gullotti, M.; Palanza, G.; Rigoni, L. J. Am. Chem. Soc., 1988, 110, 4221.
138. Karlin, K. D.; Ghosh, P.; Cruse, R. W.; Farooq, A.; Gultneh, Y.; Jacobson, R. R.; Blackburn, N. J.; Strange, R. W.; Zubieta, J. J. Am. Chem. Soc., 1988, 110, 6769.
139. Kitajima, N.; Koda, T.; Hashimoto, S.; Kitagawa, T.; Moro-oka, Y. J. Chem. Soc. Chem. Commun., 1988, 151.
140. Drew, M. G. B.; Yates, P. C.; Esho, F. S.; Trocha-Grimshwa, J.; Lavery, A.; McKillop, K. P.; Nelson, S. M.; Nelson, J. J. Chem. Soc. Dalton Trans., 1988, 2995.
141. Buffinger, D. R.; Ridgely, M. M.; Cummings, S. C. J. Coord. Chem., 1988, 19, 197.
142. Tyeklár, Z.; Karlin, K. D. Acc. Chem. Res., 1989, 22, 241.
143. Goodwin, J. A.; Bodager, G. A.; Wilson, L. J.; Stanbury, D. M.; Scheidt, W. R. Inorg. Chem., 1989, 28, 35.
144. Menif, R.; Martell, A. E. J. Chem. Soc. Chem. Commun., 1989, 1521.
145. Kamiusuki, T.; Okawa, H.; Kitaura, E.; Koikawa, M.; Matsumoto, N.; Kida, S.; Oshio, H. J. Chem. Soc. Dalton Trans., 1989, 2077.
146. Vigabo, P. A.; Tamburini, S.; Fenton, D. E. Coord. Chem. Reviews, 1990, 106, 25.
147. Gelling, O. J.; Feringa, B. L. J. Am. Chem. Soc., 1990, 112, 7599.
148. Cole, J. L.; Clark, P. A.; Solomon, E. I. J. Am. Chem. Soc., 1990, 112, 9534.
149. Réglier, M.; Jorand, C.; Waegell, B. J. Chem. Soc. Chem. Commun., 1990, 1752.
150. Kamiusuki, T.; Okawa, H.; Kitaura, E.; Inoue, K.; Kida, S. Inorg. Chim. Acta, 1991, 179, 139.
151. Jacobson, R. R.; Tyeklár, Z.; Karlin, K. D.; Zubieta, J. Inorg. Chem., 1991, 30, 2035.
152. Nasir, M. S.; Karlin, K. D.; McGowty, D.; Zubieta, J J. Am. Chem. Soc., 1991, 113, 698.
153. Ross, P. K.; Solomon, E. I. J. Am. Chem. Soc., 1991, 113, 3246.
154. Casella, L.; Gullotti, M.; Bartosek, M.; Pallanza, G.; Laurenti, E. J. Chem. Soc. Chem. Commun., 1991, 1235.
155. Adams, H.; Bailey, N. A.; Dwyer, M. J. S.; Fenton, D. E.; Hellier, P. C.; Hempstead, P. D. J. Chem. Soc. Chem. Commun., 1991, 1297.
156. Kitajima, N.; Fujisawa, K.; Fujimoto, C.; Moro-oka, Y.; Hashimoto, S.; Kitagawa, T.; Toriumi, K.; Tatsumi, K.; Nakamura, A. J. Am.

Chem. Soc., 1992, 114, 1277.
157. Addison, A. W. Inorg. Nucl. Chem. Letters, 1976, 12, 899.
158. Gagné, R. R.; Allison, J. L.; Gall, R. S.; Koval, C. A. J. Am. Chem. Soc., 1977, 99, 7170.
159. Gagné, R. R.; Koval, C. A.; Smith, T. J. J. Am. Chem. Soc., 1977, 99, 8367.
160. Alberts, A. H.; Annunziata, R.; Lehn, J.-M. J. Am. Chem. Soc., 1977, 99, 8502.
161. Fenton, D. E.; Schroeder, R. R.; Lintvedt, R. L. J. Am. Chem. Soc., 1978, 100, 1931.
162. Fenton, D. E.; Lintvedt, R. L. J. Am. Chem. Soc., 1978, 100, 6367.
163. Grzybowski, J. J.; Merrell, P. H.; Urbach, F. L. Inorg. Chem., 1978, 17, 3078.
164. Gagné, R. R.; Koval, C. A.; Smith, T. J.; Cimolino, M. C. J. Am. Chem. Soc., 1979, 101, 4571.
165. Gagné, R. R.; Kreh, R. P.; Dodge, J. A. J. Am. Chem. Soc., 1979, 101, 6917.
166. Gagné, R. R.; Spiro, C. L. J. Am. Chem. Soc., 1980, 102, 1443.
167. Simmons, M. G.; Merrill, C. L.; Wilson, L. J.; Bottomley, L. A.; Kadish, K. M. J. Chem. Soc. Dalton Trans., 1980, 1827.
168. Nakamura, M.; Mikuriya, M.; Okawa, H.; Kida, S. Bull. Chem. Soc. Jpn., 1981, 54, 1825.
169. Kimura, M.; Suzuki, S.; Mori, W.; Nakahara, A. Inorg. Chim. Acta, 1981, 56, 83.
170. Gagné, R. R.; Spiro, C. L.; Smith, J. J.; Hamann, C. A.; Thies, W. R.; Shiemke, A. K. J. Am. Chem. Soc., 1981, 103, 4073.
171. Izumitani, T.; Nakamura, M.; Okawa, H.; Kida, S. Bull. Chem. Soc. Jpn., 1982, 55, 2122.
172. Nakamura, M.; Okawa, H.; Kida, S. Inorg. Chim. Acta, 1982, 62, 201.
173. Sorrell, T. N.; Malachowski, M. R.; Jameson, D. L. Inorg. Chem., 1982, 21, 3250.
174. Nishida, Y.; Oishi, N.; Kuramoto, H.; Kida, S. Inorg. Chim. Acta, 1982, 57, 253.
175. Karlin, K. D.; Gultneh, Y.; Hutchinson, J. P.; Zubieta, J. J. Am. Chem. Soc., 1982, 104, 5240.
176. Kitagawa, S.; Munakata, M.; Miyaji, N. Bull. Chem. Soc. Jpn., 1983, 56, 2258.
177. Lintvedt, R. L.; Kramer, L. S. Inorg. Chem., 1983, 22, 796.
178. Long, R. C.; Hendrickson, D. N. J. Am. Chem. Soc., 1983, 105, 1513.
179. Mandal, S. K.; Nag, K. J. Chem. Soc. Dalton Trans., 1983, 2429.
180. Oishi, N.; Takevehi, M.; Nishida, Y.; Kida, S. Polyhedron, 1984, 3, 157.
181. Cabral, J. de O.; Cabral, M. F.; McCann, M.; Nelson, S. M. Inorg. Chim. Acta, 1984, 86, L15.
182. Lintvedt, R. L.; Ranger, G.; Schoenfelner, B. A. Inorg. Chem., 1984, 23, 688.
183. Coughlin, P. K.; Lippard, S. J. Inorg. Chem., 1984, 23, 1446.
184. Mandal, S. K.; Nag, K. J. Chem. Soc. Dalton Trans., 1984, 2141.
185. Merrill, C. L.; Wilson, L. J.; Thamann, T. J.; Loehr, T. M.; Ferris, N. S.; Woodruff, W. H. J. Chem. Soc. Dalton Trans., 1984, 2207.

186. Agnus, Y.; Louis, R.; Gisselbrecht, J.-P.; Weiss, R. J. Am. Chem. Soc., 1984, 106, 93.
187. Mazurek, W.; Bond, A. M.; Murray, K. S.; O'Connor, M. J.; Wedd, A. G. Inorg. Chem., 1985, 24, 2484.
188. Latour, J.-M.; Limosin, D.; Rey, P. J. Chem. Soc. Chem. Commun., 1985, 464.
189. Mandal, S. K.; Thompson, L. K.; Hanson, A. W. J. Chem. Soc. Chem. Commun., 1985, 1709.
190. Mazurek, W.; Bond, A. M.; O'Connor, M. J.; Wedd, A. G. Inorg. Chem., 1986, 25, 906.
191. Lintvedt, R. L.; Schoenfelner, B. A.; Rupp, K. A. Inorg. Chem., 1986, 25, 2704.
192. Thompson, L. K.; Mandal, S. K.; Gabe, E. J.; Charland, J.-P. J. Chem. Soc. Chem. Commun., 1986, 1537.
193. Mandal, S. K.; Adhikary, B.; Nag, K. J. Chem. Soc. Dalton Trans., 1986, 1175.
194. Bailey, N. A.; Fenton, D. E.; Lay, J.; Roberts, P. B.; Latour, J.-M.; Limosin, D. J. Chem. Soc. Dalton Trans., 1986, 2681.
195. Thompson, L. K.; Mandal, S. K.; Gabe, E. J.; Lee, F. L.; Addison, A. W. Inorg. Chem., 1987, 26, 657.
196. Berends, H. P.; Stephan, D. W. Inorg. Chem., 1987, 26, 749.
197. Mandal, S. K.; Thompson, L. K.; Nag, K.; Charland, J.-P.; Gabe, E. J. Inorg. Chem., 1987, 26, 1391.
198. Basu, A.; Bhaduri, S.; Sapre, N. Y.; Jones, P. G. J. Chem. Soc. Chem. Commun., 1987, 1724.
199. Bailey, N. A.; Fenton, D. E.; Moody, R.; Barbarin, C. O. R.; Sciambarella, I. N.; Latour, J.-M.; Limosin, D.; McKee, V. J. Chem. Soc. Dalton Trans., 1987, 2519.
200. Lintvedt, R. L.; Rupp, K. A.; Hegg, M. J. Inorg. Chem., 1988, 27, 331.
201. Bartoncello, K.; Fallon, G. D.; Hodgkin, J. H.; Murray, K. S. Inorg. Chem., 1988, 27, 4750.
202. Benzebri, A.; Dubourdeaux, P.; Latour, J.-M.; Laugier, J.; Rey, P. J. Chem. Soc. Chem. Commun., 1987, 1564.
203. Barley, M.; Constable, E. C.; Corr, S. A.; Mc Queen, R. C. S.; Nutkins, J. C.; Ward, M. D.; Drew, M. G. B. J. Chem. Soc. Dalton Trans., 1988, 2655.
204. Collin, J.-P.; Osborn, J. A.; Youinou, M.-T. Inorg. Chim. Acta, 1989, 165, 41.
205. Bond, A. M.; Haga, M.; Creece, I. S.; Robson, R.; Wilson, J. C. Inorg. Chem., 1989, 28, 559.
206. Mandal, S. K.; Thompson, L. K.; Newlands, M. J.; Gabe, E. J. Inorg. Chem., 1989, 28, 3707.
207. Cabral, M. F.; Cabral, J.; Trocha-Grimshaw, J.; McKillop, K. P.; Nelson, S. M.; Nelson, J. J. Chem. Soc. Dalton Trans., 1989, 1351.
208. Spodine, E.; Atria, A. M.; Peña, O.; L'Helgouach, H.; Garland, M. T. J. Chem. Soc. Dalton Trans., 1989, 1435.
209. Zehetmair, J. K.; Lintvedt, R. L. Inorg. Chem., 1990, 29, 2201.
210. Salata, C. A.; Youinou, M.-T.; Burrows, C. J. Inorg. Chem., 1991, 30, 3454.

211. Nelson, S. M.; Trocha-Grimshaw, J.; Lavery, A.; McKillop, K. P.; Drew, M. G. B. "Substrate Binding and Dioxygen Activation at Dicopper sites", Ref. 4, Vol. II, 27-40.
212. Crane, J. D.; Fenton, D. E. Polyhedron, 1991, 10, 1809.

CYCLIC VOLTAMMETRIC BEHAVIOUR OF SOME COPPER COMPLEXES WITH A BIS(PYRAZOLE)BIS(THIOETHER)LIGAND

M. F. CABRAL & J. O. CABRAL
Laboratório Ferreira da Silva, Faculdade de Ciências
4000 Porto, Portugal
W. G. HAANSTRA, W. L. DRIESSEN & J. REEDIJK
Gorlaeus Laboratories, Leiden University
P.O. Box 9502, 2300 RA Leiden, The Netherlands

ABSTRACT. The cyclic voltammetric behaviour of four mononuclear copper(II) complexes and one binuclear copper(I) complex with the N_2S_2 ligand 1,6-bis(3,5-dimethyl-1-pyrazolyl)-2,5-dithiahexane (bddh) was studied. Their oxidation-reduction properties will be discussed in terms of the possible relevance of these complexes as models of the Type 1 site of the blue-copper proteins.

1. INTRODUCTION

It is known, from the crystal structures of some natural proteins, such as plastocyanin [1] and azurin [2], that the so-called blue-copper proteins, Type 1, have one copper(II) ion, with two Cu-S and two Cu-N coordinate bonds, in their active sites.

This copper(II) mononuclear complex is responsible for some remarkable properties, one of which is the high redox potentials of the copper(II) ions in these environments. These blue-copper proteins are engaged in electron transport, which involves a switching of the oxidation state of copper between +2 and +1.

A large variety of copper(II) complexes have been synthesized in attempts to model these natural proteins, among which several complexes with ligands possessing a set of N_2S_2 donor atoms. The successful synthesis of bddh provided a promising ligand (Figure 1)

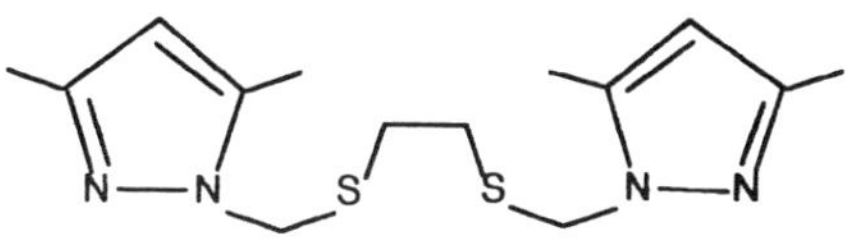

Figure 1. Schematic drawing of the ligand bddh

A. J. L. Pombeiro and J. A. McCleverty (eds.),
Molecular Electrochemistry of Inorganic, Bioinorganic and Organometallic Compounds, 171–176.

from which four mononuclear copper(II) complexes

(A) $[Cu(bddh)(NO_3)_2]$ (B) $[Cu(bddh)(H_2O)](BF_4)_2$

(C) $[Cu(bddh)Cl]_2[CuCl_4]$

(D) $[Cu(bddh)(Hdmpz)](BF_4)_2$ (dmpz= 3,5-dimethylpyrazole)

and one binuclear copper(I) complex

(E) $[Cu(bddh)]_2(BF_4)_2$

were prepared and characterized [3].

In this article we describe the electrochemical behaviour of these five complexes as studied by cyclic voltammetry in non-aqueous solution.

2. EXPERIMENTAL

The synthesis and characterization of these five complexes in the solid state, as well as the X-ray structures of (C), (D) and (E), have been reported [3].

All reagents were of the highest commercially available quality and used as received.

Cyclic voltammograms were obtained, as previously described [4], on freshly prepared solutions of the complexes directly in the cell, which already contained the solvent and supporting electrolyte (0.1 M TEAP).

All potentials are quoted versus the SCE, with the salt bridge, and are uncorrected for liquid junction potentials and iR drop. Peak currents were corrected for background currents, whenever relevant.

Electronic spectra of solution were obtained with a Hitachi U-2000 spectrophotometer.

3. RESULTS AND DISCUSSION

The experimental results are summarized in Table 1, and typical voltammograms are shown in Figures 2 and 3.

Two solvents were used, acetonitrile (MeCN) and dimethylsulfoxide (DMSO), because two of the mononuclear copper(II) complexes, viz., $[Cu(bddh)(NO_3)_2]$ and $[Cu(bddh)(Hdmpz)](BF_4)_2$, were soluble in MeCN, but not in DMSO, and the other two, $[Cu(bddh)(H_2O)](BF_4)_2$ and $[Cu(bddh)Cl]_2[CuCl_4]$, were soluble in DMSO, but not in MeCN. Only the dicopper(I) complex, $[Cu(bddh)]_2(BF_4)_2$, was soluble both in MeCN and, with decomposition, in DMSO.

The extent to which the species in solution differ from the solid depends on the degree of solvation, which may exert a strong influence on the results of cyclic voltammetry, as we had already noticed [4,5,6].

To check the degree of solvation, we compared the electronic spectra of the solid compounds with their spectra in solution, and these in the course of time. In the case of the complexes soluble in MeCN the spectral features of the solids and the solutions were not significantly different and remained the same over a period of three hours. We can, therefore, dismiss any appreciable solvation effects.

TABLE 1. Electrochemical data for the complexes[a]

Complex	Solvent[b]	S.R.[c]	$E_{p,c}$	$E_{p,a}$	ΔE_p	$E_{1/2}$	$i_{p,a}/i_{p,c}$
$[Cu(bddh)(NO_3)_2]$ (A)	MeCN	20	+500	+610	110	+555	0.73
		50	+500	+610	110	+555	0.73
		100	+500	+610	110	+555	0.78
		200	+480	+620	140	+550	0.86
$[Cu(bddh)(H_2O)](BF_4)_2$ (B)	DMSO	10	+ 40	+200	160	+120	-
		20	+ 30	+220	190	+125	-
		50	+ 15	+245	230	+130	-
$[Cu(bddh)Cl]_2[CuCl_4]$ (C)	DMSO	20	+150	+260	110	+215	0.91
		50	+120	+260	140	+190	0.91
		100	+110	+270	160	+190	0.93
		200	+100	+290	190	+195	0.98
$[Cu(bddh)(Hdmpz)](BF_4)_2$ (D)	MeCN	20	+445	+605	160	+525	0.98
		50	+435	+610	175	+523	0.93
		100	+425	+620	195	+523	0.80
		200	+400	+650	250	+525	0.75
$[Cu(bddh)]_2(BF_4)_2$ (E)	MeCN	20	+520	+650[d]	130	+585	0.90
		50	+510	+650[d]	140	+580	0.99
		100	+490	+700[d]	210	+595	1.0
		200	+475	+730[d]	255	+600	0.99

(a) Potentials in mV (uncertainty $\pm$5mV).

(b) MeCN=acetonitrile; DMSO=dimethylsulfoxide.

(c) Scan rates (S.R.) in mV s^{-1}.

(d) Broad; broadness increases with scan rate.

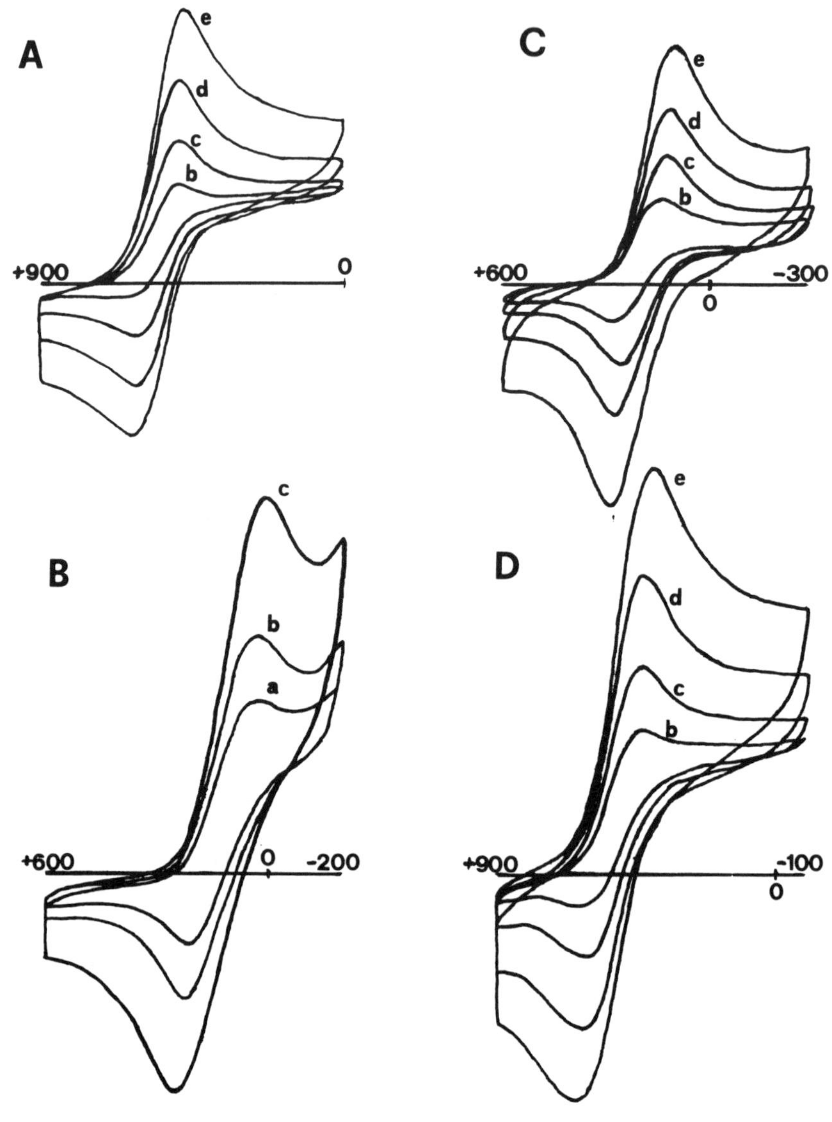

Figure 2. Cyclic voltammograms of $[Cu(bddh)(NO_3)_2]$ (A) in MeCN, $[Cu(bddh)(H_2O)](BF_4)_2$ (B) in DMSO, $[Cu(bddh)Cl]_2[CuCl_4]$ (C) in DMSO, and $[Cu(bddh)(Hdmpz)](BF_4)_2$ (D) in MeCN. Scan rates: a, 10 mV s^{-1}; b, 20 mV s^{-1}; c, 50 mV s^{-1}; d, 100 mV s^{-1}; e, 200 mV s^{-1}.

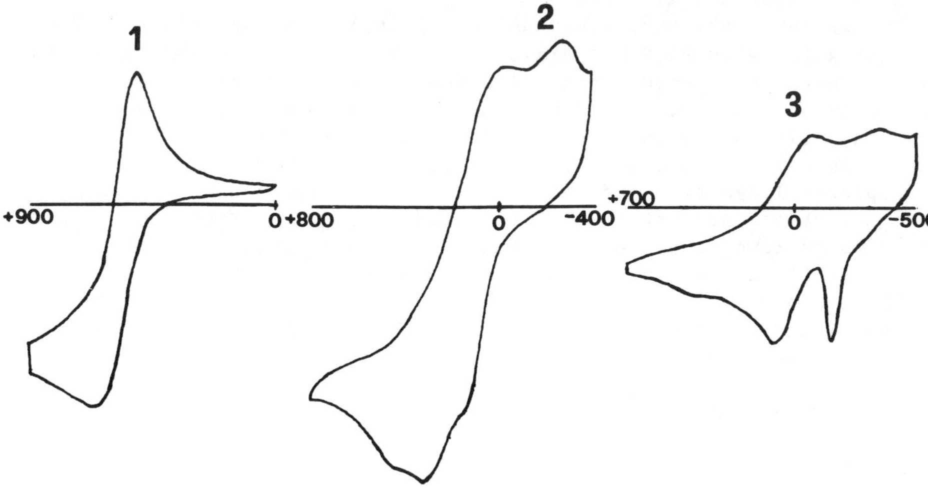

Figure 3. Cyclic voltammograms of $[Cu(bddh)]_2(BF_4)_2$ in acetonitrile (1) and in dimethylsulfoxide (2, immediately after dissolution; 3, 15 minutes later) at 50 mV s^{-1}.

The case of the complexes soluble in DMSO is completely different: the complexes are green in the solid state but the solutions are colourless, with the corresponding difference in the spectra. It appears, therefore, that copper(II) is somehow reduced by DMSO to copper(1).

It is clear from Table 1 and Figures 2 and 3 how solvation affects the cyclic voltammetric results. The behaviour of $[Cu(bddh)(NO_3)_2]$ and $[Cu(bddh)(Hdmpz)](BF_4)_2$, both soluble in MeCN with no detectable decomposition, can be considered quasi-reversible, diffusion controlled (currents varied linearly with the square root of the scan rates), one-electron redox processes, involving a sequence Cu(II)-Cu(I)-Cu(II). Their half-wave potentials are practically independent of the scan rate; their values, +555 mV vs. SCE (ca.+795 vs. NHE) and +525 mV vs. SCE (ca.+765 vs. NHE), respectively, are of the order of magnitude of the highest redox potentials found for the blue copper proteins [7].

The electrochemical behaviour in MeCN of the dicopper(I) complex, $[Cu(bddh)]_2(BF_4)_2$, deserves some comments. Since the solid is white and the solution is colourless. the degree of solvation could not be assessed by spectrophotometry. By cyclic voltammetry some signs of slow decomposition were detected: changes in peak potentials within a few hours at any scan rate. The cyclic voltammograms of this complex show sharp cathodic and broad anodic peaks, the broadness of which increases with scan rate. This suggests the possible formation of two copper(II) species, by oxidation at close potentials, with subsequent reduction to the same copper(I) species. Whether this is true or not, whether the dinuclear copper(I) complex breaks into two mononuclear copper(II) species on oxidation, or whether something else happens, cannot be said from the available experimental data.

As already mentioned, [Cu(bddh)(H_2O)](BF_4)$_2$ and [Cu(bddh)Cl]$_2$[$CuCl_4$] are not soluble in MeCN but dissolve in DMSO with visible decomposition. Cyclic voltammetry shows that solvation is less pronounced in the case of the chlorocomplex, possibly because H_2O is more easily displaced by DMSO than Cl^- and because Cu(I), if formed on dissolution, is more stable when Cl^- is present, in some agreement with the known tendency of copper halides to coordinate with DMSO [8]. The values of the half-wave potentials are not greatly affected by the scan rates, but are more than 300 mV lower than those of the other two copper(II) complexes, in spite of the similarity of these four complexes in the solid state.

The dicopper(I) complex, [Cu(bddh)]$_2$(BF)$_4$, also dissolve in DMSO with decomposition, as shown by the difference in cyclic voltammograms on successive scans at the same scan rate.

4. CONCLUDING REMARKS

The results we obtained in this work strongly emphasize the enormous influence solvation may have on electrochemical studies of models of copper proteins and the care that should be taken on extrapolating solid state structures into structures that may be pesent in solution.

As far as cyclic voltammetric behaviour is concerned, only that of [Cu(bddh)(NO_3)$_2$], and of [Cu(bddh)(Hdmpz)](BF_4)$_2$, have some relevance as models for Type 1 copper proteins, since they seem not to be seriously affected by solvation with MeCN, the cyclic voltammograms are reasonably well behaved, and the values of the half-wave potentials are close to the potentials of some natural proteins.

5. ACKNOWLEDGEMENTS

The financial support of NATO, Collaborative Research Grant Nº.870370, is gratefully acknowledged.

M. F. C. and J. O. C. also thank Instituto Nacional de Investigação Científica, Lisboa (Centro de Investigação em Química, Linha 3)

6. REFERENCES

1. Guss, J. M.; Freeman, H. C. J. Mol. Biol., 1983, 169, 521.
2. Baker, E. N. J. Mol. Biol, 1988, 203, 1071.
3. Haanstra, W. G.; Driessen, W. L.; de Graaff, R. A. G.; Reedijk, J.; Wang, Y. F.; Stam, C. H. Inorg. Chim. Acta, 1991, 186, 215.
4. Cabral, M. F.; Cabral, J. O.; Bouwman, E.; Driessen, W. L.; Reedijk, J. Inorg. Chim. Acta, 1990, 167, 205.
5. Cabral, M. F.; Cabral, J. de O.; Bouwman, E.; Driessen, W. L.; Reedijk, J.; Turpeinen, U.; Hämälainen, R.. Unpublished results.
6. Haanstra, W. G.; Cabral, M. F.; Cabral, J. de O.; Driessen, W. L.; Reedijk, J.. Unpublished results.
7. Beinert, H. Coord. Chem. Rev., 1980, 33, 55.
8. Cotton, F. A.; Francis, R. J. Am. Chem. Soc., 1960, 82, 2986.

PART 4. **DYNAMICS of ELECTRON TRANSFER**

CORRELATION BETWEEN STRUCTURE AND HETEROGENEOUS ELECTRON TRANSFER RATES OF COORDINATION COMPOUNDS

FRANKLIN A. SCHULTZ
Department of Chemistry
Indiana University Purdue University at Indianapolis
1125 East 38th Street
Indianapolis, Indiana 46205-2810
U.S.A.

ABSTRACT. Marcus-Hush theory predicts that a large change in molecular structure accompanying an electrode reaction will result in a large inner-shell barrier and a slow rate of electron transfer. Described here is an attempt to establish relationships between molecular structural features and electron transfer rates for two systems of bioinorganic interest: metalloporphyrins and model complexes for molybdenum-containing enzymes. For these compounds structural change occurs when an electron is added to an orbital that interacts strongly with donor atoms in the coordination environment of the metal. For the porphyrins, this results in increases in the metal atom out-of-plane distance and axial ligand bond length. For the molybdenum complexes, it results in cleavage of metal-ligand bonds in conjunction with electron transfer.

1. Introduction

Relationships between structure and electrochemical reactivity underlie participation of the transition elements in many important processes including catalysis, energy conversion and biological electron transfer. Considerable progress has been made in recent years on the theoretical and experimental aspects of electron transfer kinetics, but many facets of the topic remain unexplored. A particularly interesting circumstance arises when a large change in molecular structure accompanies an electrode reaction. Marcus-Hush theory [1] predicts that this condition will result in a large inner-shell activation barrier and limit the rate of electron transfer. Large structural change usually occurs when an electron is added to or removed from an orbital that interacts strongly with the donor atoms in the coordination environment of the metal. Thus, considerable information on the molecular features controlling reactivity can be derived from reactions of this type. This contribution reports recent efforts in our laboratory to investigate such transition metal electrode reactions.

The relationship between the standard heterogeneous electron transfer rate constant, $k_{s,h}$, and structural factors is expressed in terms of the Marcus-Hush theory applied to electrode reactions [2]:

. J. L. Pombeiro and J. A. McCleverty (eds.),
Molecular Electrochemistry of Inorganic, Bioinorganic and Organomettallic Compounds, 179–191.

$$k_{s,h} = A_e \cdot \exp^{-(\Delta G_{is}^* + \Delta G^*_{os})/RT} \tag{1}$$

$$A_e = K_o \cdot \Gamma_n \cdot \kappa_{el} \cdot \nu_n \tag{2}$$

$$\Delta G^*_{is} = 0.5 \sum f_s \left(\frac{\Delta a}{2}\right)^2 + 0.5 \sum f_b \left(\frac{\Delta \theta}{2}\right)^2 \tag{3}$$

$$\Delta G^*_{os} = 1/8(e^2) \left[1/r - 1/R_e\right] \left[1/\varepsilon_{op} - 1/\varepsilon_s\right] \tag{4}$$

In equations 1-4, A_e is the pre-exponential factor, and ΔG^*_{is} and ΔG^*_{os} are the inner- and outer-shell free energies of activation. A_e is defined as the product of κ_{el}, the electron tunneling probability, Γ_n, the nuclear tunneling factor, ν_n, the nuclear frequency factor and K_o, the distance from the surface over which efficient tunneling between the electrode and reactant occurs. Nominally, $A_e = 10^4$ cm s^{-1} at room temperature. ΔG^*_{is} is defined in equation 3 as the sum of harmonic metal-ligand bond length (Δa) and bond angle ($\Delta\theta$) displacements that accompany electron transfer; fs and f_b are the stretching and bending force constants, respectively. The outer-shell barrier is given by the dielectric continuum model of equation 4, where r is the radius of the reacting particle, R_e is the distance from its center to the image plane within the electrode, and ε_{op} and ε_s are the optical and static dielectric constants of the solvent.

In recent years, our laboratory has sought to apply these relationships in a qualitative to semi-quantitative way to transition metal electrode reactions characterized by large structural change. We also have sought to study compounds that are representative of transition elements in biological systems, since inner-shell effects often have important consequences on reactivity in such systems. In the following sections we describe results for a number of metalloporphyrins [3] and for several complexes of Mo that model the behavior of this element in molybdenum-containing enzymes [4].

2. Metalloporphyrins

2.1 ELECTRON TRANSFER SITE AND SPIN STATE

An important question in metalloporphyrin redox chemistry is whether a given electron transfer occurs at a metal or porphyrin site. The issue frequently is addressed by spectroscopic methods or by measurement of formal redox potentials. However, the rate of electron transfer also can be indicative of the site of charge transfer in a metalloporphyrin [5].

Figure 1 shows cyclic voltammograms recorded for oxidation and reduction of the five-coordinate Mn (III) porphyrin, Mn (TPP) Cl [6]. The left-hand side of the figure contains voltammograms corresponding to the one-electron porphyrin-centered oxidation:

$$Mn^{III}\,(TPP)\,Cl \;\Longleftrightarrow\; Mn^{III}\,(TPP\cdot)\,Cl^+ + e^- \tag{5}$$

The reaction is Nernstian under the experimental conditions; ΔE_p equals ca. 60 mV and is independent of sweep rate. We have obtained equivalent results for ring-centered oxidations and reductions of a number of metalloporphyrins [5]. In all cases, the standard heterogeneous rate constant is larger than can be measured experimentally ($k_{s,h} > 0.1$ cm s^{-1}). A large rate of electron transfer is anticipated, because the charge transferred to or from the porphyrin is distributed over a 24-atom framework necessitating little nuclear rearrangement.

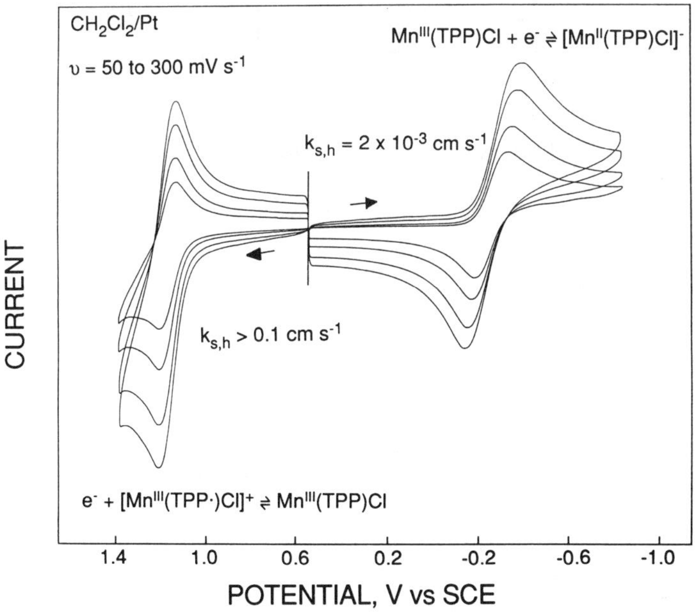

Figure 1. Cyclic voltammograms for reduction and oxidation of Mn (TPP) Cl as a function of sweep rate at a 0.02 cm^2 Pt disk electrode in 0.1 M Bu_4 NBF_4/CH_2Cl_2.

The right-hand side of Figure 1 contains voltammograms depicting the metal-centered reduction:

$$Mn^{III} (TPP) Cl + e- <===> Mn^{II} (TPP) Cl^- \qquad (6)$$

The rate of reaction 6 clearly is slower than that of reaction 5. The peak potential separation is greater than the Nernstian value and increases with sweep rate. A value of $k_{s,h} = 2.1 \times 10^{-3}$ cm s^{-1} is determined from the variation of ΔE_p with v [7].

The reason for the difference in rates is the inner-shell barrier associated with reaction 6. The inner-shell barrier arises because the electron that is added to the metal center as Mn (III) is reduced to Mn (II) enters the d_{x2-y2} orbital as illustrated in Figure 2. In this orbital the electron interacts strongly with the porphyrin N atoms and causes the Mn atom to move from a distance of 0.27 Å to 0.64 Å above the porphyrin plane. This out-of-plane displacement and other structural changes attending reduction of Mn (TPP) Cl to Mn (TPP) Cl^- is available from x-ray crystal structures of the isolated compounds [8].

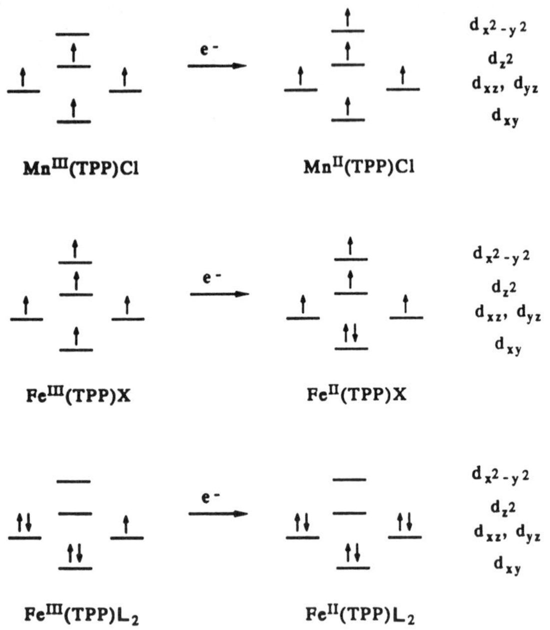

Figure 2. Changes in electronic configuration accompanying metal-centered reduction of metalloporphyrins.

Slow electron transfer also is observed for the metal-centered reduction of a series of five-coordinate, high-spin iron (III) porphyrins [9].

$$Fe^{III}\,(TPP)X + e^- \;<===>\; Fe^{II}\,(TPP)\,X^- \tag{7}$$

As shown in Figure 2 this electron enters the d_{xy} orbital, which has a degree of anti-bonding character. X-ray crystal structures of representative five-coordinate, high-spin Fe (III) and Fe (II) porphyrins [10] suggest that increases of ca. 0.2 Å in metal out-of-plane distance and ca. 0.1 Å in metal-axial ligand bond distance accompany reaction 7.

By contrast, low-spin Fe (III) porphyrins are six-coordinate and have the metal atom located essentially in the porphyrin plane. The orbital occupied by metal-centered reduction of these species:

$$Fe^{III}\,(TPP)\,L_2^+ + e^- \;<===>\; Fe^{II}\,(TPP)\,L_2 \tag{8}$$

is directed at neither the in-plane porphyrin N nor axial ligand donor atoms (Figure 2). Consequently, little change in nuclear coordinates accompanies electron transfer.

Table 1 contains heterogeneous electron transfer rate data for reduction of these three classes of metalloporphyrins. Clearly, metal-centered reduction of high-spin Mn (TPP) Cl and Fe (TPP) X is slower than that of low-spin Fe (TPP) L_2^+. In addition, the rate of Fe

(TPP) X reduction shows a dependence on axial ligand. Combination of the data in Table 1 with the Nernstian behavior observed for ring-centered oxidation of Mn (TPP) Cl (Figure 1) suggests the following sequence of rates for metalloporphyrin electrode reactions: ring-centered > LS metal-centered > HS metal-centered.

TABLE 1. Heterogeneous Rate Constants for Metal-Centered Reduction of Metalloporphyrins [a]

Complex	Spin State	$E_{1/2}$ (V vs. SCE)	$k_{s,h}$ (cm s^{-1})
Mn (TPP) Cl	HS	-0.29	2.1×10^{-3}
Fe (TPP) Br	HS	-0.15	1.0×10^{-2}
Fe (TPP) Cl	HS	-0.29	6.9×10^{-3}
Fe (TPP) F	HS	-0.51	2.4×10^{-3}
Fe (TPP) OAc	HS	-0.36	1.8×10^{-3}
Fe (TPP) OCH_3	HS	-0.76	1.1×10^{-3}
Fe (TPP) $(py)_2^+$	LS	0.09	2.0×10^{-2}
Fe (TPP) $(Im)_2^+$	LS	-0.20	1.6×10^{-2}
Fe (TPP) $(CN)_2^-$	LS	-0.46	3.1×10^{-2}

a Determined by cyclic voltammetry at Pt or Au electrodes in CH_2Cl_2 containing 0.1 M Bu_4NClO_4 or Bu_4NBF_4.

2.2. OUT-OF-PLANE DISPLACEMENT AND AXIAL LIGAND CONTRIBUTIONS TO THE INNER-SHELL BARRIER

The inner-shell barrier that causes metal-centered reduction of Mn (TPP) Cl to be slow relative to its ring-centered oxidation is movement of the metal atom with respect to the porphyrin plane. X-ray crystallography [8] indicates that the Mn atom moves from a distance of 0.27 to 0.64 Å above the porphyrin plane as Mn (TPP) Cl is reduced to Mn (TPP) Cl^- and that the Mn-Cl distance does not change. The Mn (TPP) $Cl^{\circ/-}$ couple is a good choice for assessing inner-shell effects on electron transfer kinetics, because structural information is available for both oxidation states and nuclear reorganization is limited almost entirely to out-of-plane displacement of the metal atom (distortion of the porphyrin core is neglected).

We have attempted to support this explanation for slow electron transfer kinetics by comparing experimental and calculated inner-shell barriers for Mn (TPP) Cl reduction. Table 2 contains data from temperature-dependent electrochemical measurements [11] in several non-aqueous solvents. The experimental barrier is obtained by determining ΔH^*_{real} from the temperature dependence of the standard heterogeneous rate constant [12] and subtracting from this quantity the value of ΔG^*_{os} calculated by equation 4. This difference, $\Delta G^*_{is, E}$, is to be compared with ΔG^*_{is} calculated by equation 3. For the purposes of the latter, values of $f_s = 1.32$ mdyne A^{-1} and $f_b = 0.26$ mdyne A rad^{-2} were estimated from spectroscopic data [13]. The bending and stretching components of the inner-shell motion contribute 7 and 9 kJ mol^{-1}, respectively, to the calculated barrier. The most energetic elements of these motions are elongation of the Mn-N bonds ($\Delta a = 0.175$ Å) and closing of the N-Mn-N angles ($\Delta\,\theta = 19°$) across the porphyrin core. The sum of these calculated quantities is in fair agreement with the range of values of 12 to 21 kJ mol^{-1} obtained for $\Delta G^*_{is,\,E}$. The results in Table 2 also indicate that despite a sizeable inner-shell contribution, the outer-shell component is a major part of the overall barrier. A total barrier height of ca. 30 kJ mol^{-1} and a pre-exponential factor of nearly 10^4 cm s^{-1} satisfactorily explain the heterogeneous kinetics of Mn (TPP) Cl reduction at room temperature.

TABLE 2. Activation Parameters for Mn (TPP) Cl Reduction [a]

Solvent	$k_{s,h}$ at 23° C (cm s^{-1})	A_e (cm s^{-1})	ΔH^*_{real} (kJ mol^{-1})	ΔG^*_{os} (kJ mol^{-1})	$\Delta G^*_{is,E}$ (kJ mol^{-1})
CH_3CN	0.089	2.4×10^4	26.2	13.8	12.4
CH_3NO_2	0.080	5.0×10^3	27.0	13.0	14.0
CH_3CH_2CN	0.042	3.7×10^3	28.4	12.0	16.4
CH_3CH_2	0.026	6.9×10^3	30.9	11.7	19.2
$C_6H_5NO_2$	0.014	9.5×10^2	27.4	10.1	17.3
PC	0.010	8.1×10^3	33.5	12.5	21.0

[a] Determined by cyclic voltammetry at a 127-μm diameter Hg (Au) electrode in 0.1 M Bu_4NPF_6.

The inner-shell barrier for reduction of high-spin iron (III) porphyrins consists of contributions from both metal atom out-of-plane displacement and axial ligand bond stretching. The influence of the latter term can be assessed by comparing the rate of Fe (III) /Fe (II) reduction with the strength of the Fe-X bond. The force constant for stretching of the axial ligand bond can be estimated from the relationship, $f_s = 4\pi^2\mu c^2\upsilon_s^2$, where υ_s is the stretching frequency of the Fe-X vibration. A linear relationship between -log ($k_{s,h}$) and υ_s^2 is anticipated if A_e, ΔG^*_{os} and other factors contributing to ΔG^*_{is} remain constant. For the five (TPP) X complexes in Table 1 values of $\upsilon_s = 270$, 357, 606, 530 and 540 cm^{-1} are obtained from literature data [9] for $X^- = Br^-$, Cl^-, OAc^- and CH_3O, respectively. Figure 3 shows the corresponding plot of -log ($k_{s,h}$) versus υ_s^2. The heterogeneous rate constant decreases as the electron donor strength of the axial ligand increases, the trend expected if Fe-X bond stretching were to contribute to the inner-shell barrier for Fe (TPP) X reduction. Furthermore, a value of $\Delta a = 0.14$ Å is derived from the slope of the line. This value is consistent with the extent of Fe-X bond lengthening accompanying Fe (III) to Fe (II) reduction [10].

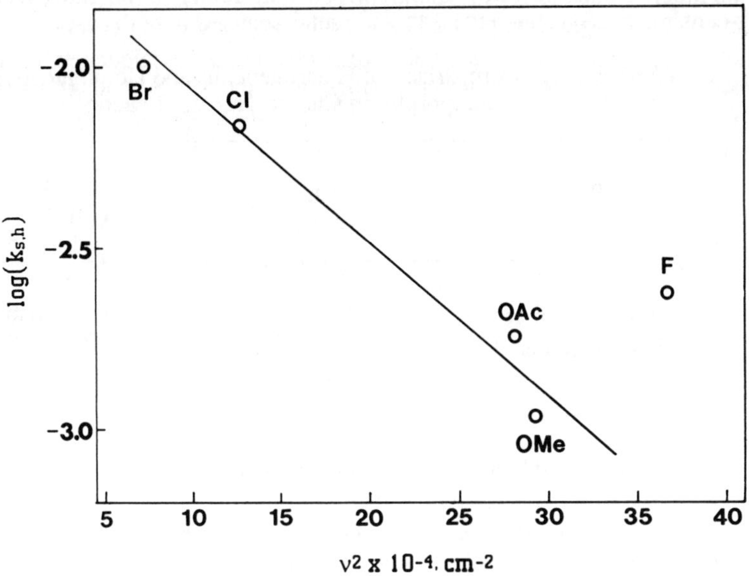

Figure 3. Variation of heterogeneous electron transfer rate constant with Fe-X stretching frequency for Fe (TPP) X reduction.

Data for the metal-centered reduction of several low-spin iron (III) porphyrins also are presented in Table 1. Values of $k_{s,h}$ for this electrode reaction (eq 8) are larger than for the high-spin reduction and do not show a substantial dependence on ligand type. This result is anticipated because little structural change is associated with metal-centered reduction of the low-spin species. However, the heterogeneous rate constants for Fe^{III} (TPP) L_2 reduction are significantly smaller than for ring-centered charge transfer reactions of metalloporphyrins. A possible explanation of this fact is that the more highly localized charge resulting from electron transfer at a metal site leads to a larger outer-shell barrier for this process.

2.3 RELEVANCE TO BIOLOGICAL SYSTEMS

The results described above have relevance to the electron transfer reactivity of metalloporphyrins in biological environments. First, it is apparent that inner-shell barriers can modulate the rate of electron transfer to metalloporphyrin sites. Electrochemically, we note that electron transfer is faster to the ring than to either high- or low-spin metal centers, and that the rate is faster for low-spin than high-spin species. Table 3 summarizes the results of our electrochemical measurements and compares them with representative values of homogeneous self-exchange rates taken form the literature. The rates follow the sequence ring-centered > LS metal-centered > HS metal-centered in both cases.

TABLE 3. Comparison of Heterogeneous and Homogeneous Rate Constants for Metalloporphyrin Charge Transfer Reactions

Reaction	$k_{s,h}$ (cm s^{-1})	k_{ex} [a] ($M^{-1}s^{-1}$)
HS metal-centered	.001 - .01	10^3 to 10^6
LS metal-centered	.02	10^7 to 10^8
Ring	>.1	10^8

a Taken from literature sources cited in [14].

A second observation is that the frequently cited correlation between electron transfer rate and spin state of metal porphyrins [15] is not truly a property of the spin state. Rather, it is a reflection of the amount of structural change that accompanies electron transfer within each spin state. Because high-spin reactions tend to involve orbitals that interact strongly with porphyrin N and axial ligand donor atoms, these reactions have larger inner-shell barriers and slower rates of electron transfer.

Finally, axial ligands also influence the rate of electron transfer. The trend is consistent with the explanation that stronger metal-ligand bonds produce larger inner-shell barriers and slower rates of electron transfer. The effect is most apparent at high-spin metal centers which experience a significant change in metal-axial ligand bond distance upon electron transfer. Hoffman et al. [16] have reported that variation of the exogenous ligand in a series of hybrid mixed-metal hemoglobins modulates the rate of interprotein electron transfer for these species. It has not been determined if this change is the result of inner-shell effects.

3. Oxomolybdenum Complexes

3.1 PERSPECTIVE

Oxomolybdenum centers are present in a series of enzymes known as molybdenum hydroxylases [4] which catalyze oxygen atom transfer reactions of a variety of small molecules. Examples include the oxidation of xanthine to uric acid and the reduction of NO_3^- to NO_2^-. The reactions are equivalent to two-electron transfers, and the molybdenum atom cycles between oxidation states VI and IV during these processes. Molybdenum (VI) typically contains two multiply bonded oxo (or sulfido) groups, and molybdenum (IV) contains one. Therefore, it is believed that MoO_2^{2+} and MoO^{2+}centers intercovert by coupled transfer of protons and electrons over a narrow potential interval during turnover of these enzymes. However, other explanations are possible.

To explore the influence of coordination environment on rates and mechanisms of electron transfer reactions at molybdenum centers, we have investigated [17-20] the electrochemical behavior of several families of oxomolybdenum (VI) complexes. The compounds are illustrated in Figure 4. The first three examples are seven-coordinate complexes and exhibit pentagonal bipyramidal geometry. These compounds contain a single, multiply bonded oxo group in one axial position. Four of the five coordination sites in the equatorial plane are occupied by the S atoms of two bidentate dithiocarbamate ligands. The remaining coordination sites may be occupied by two monodentate ligands, by a third dithiocarbamate ligand or by a bidentate catecholate ligand. The fourth example is the six-coordinate complex, MoO (Cat) (Sap), which contains a tridentate Schiff base ligand (Sap^{2-}) and a bidentate catecholate bound to the MoO^{4+} center.

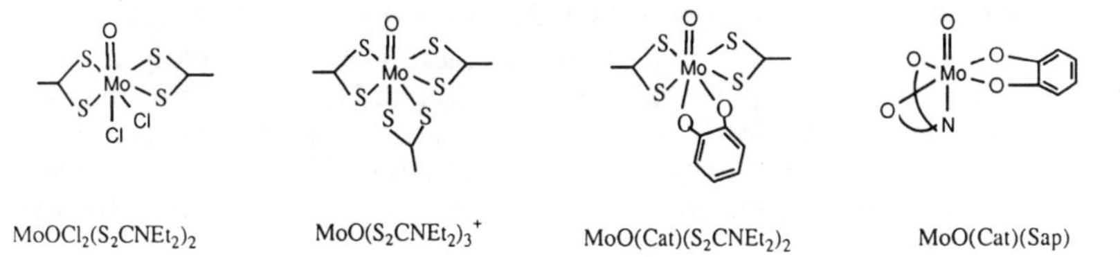

Figure 4. Structures of oxomolybdenum (VI) complexes.

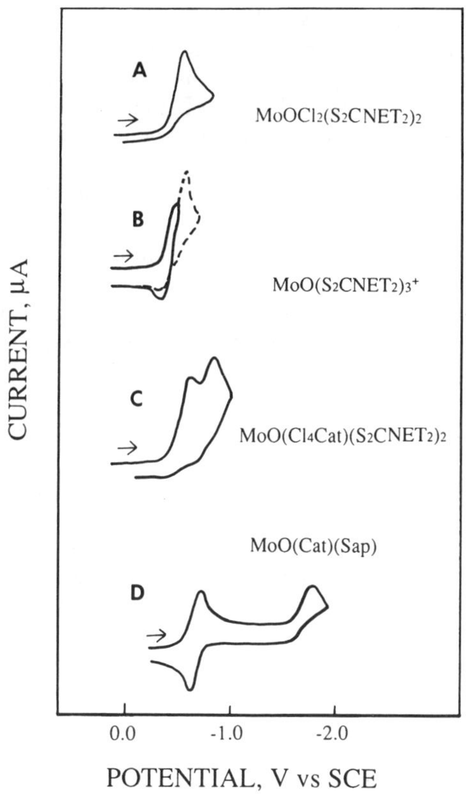

Figure 5. Cyclic voltammetric traces for reduction of oxomolybdenum (VI) complexes in non-aqueous solvents. (A) $MoOCl_2$ $(Et_2dtc)_2$, v =0.2 V s^{-1}; (B) MoO $(Et_2dtc)_3^+$, v = 25 V s^{-1}; (C) MoO (Cl_4cat) $(Et_2dtc)_2$, v = 0.2 Vs^{-1}; (D) MoO (Cat) (Sap), v = 0.2 V s^{-1}.

3.2 REDUCTION OF SEVEN-COORDINATE COMPLEXES

Electrochemical reduction of $MoOCl_2(Et_2dtc)_2$ in non-aqueous solvents proceeds by an irreversible two-electron transfer (Figure 5A) at -0.4 V versus SCE [17].

$$MoOCl_2(Et_2dtc)_2 + 2e^- \Longleftrightarrow MoO(Et_2dtc)_2 + 2\,Cl^- \qquad (9)$$

The Mo (IV) electrode product, $MoO(Et_2dtc)_2$, is reoxidized to $MoOCl_2(Et_2dtc)_2$ at +0.4 V in the presence of excess Cl^-. Equation 9 is therefore a chemically reversible two-electron process, although the potentials for reduction and reoxidation of the metal center are separated by 0.8 V. Reaction 9 is thought to proceed by a square scheme with loss or gain of a chloride ion coupled to each electron transfer. However, these reactions are rapid, and the individual steps in the mechanism cannot be isolated by use of low temperatures or fast sweep rates.

Electrochemical reduction of the tris (bidentate) $MoO(Et_2dtc)_3^+$ complex also proceeds by irreversible transfer of two electrons at slow sweep rates. However, if the sweep rate is increased to ca. 25 V s^{-1} (Figure 5B), the two-electron reaction becomes resolved into its one-electron components, and reversal of the scan after the first of these reductions shows the reaction to be a reversible process with $E^{o'} = -0.31$ V. If the reverse scan is continued in the positive direction after the two-electron reduction, $MoO(Et_2dtc)_2$ and Et_2dtc^- are identified as the electrode reaction products by their voltammetric peak potentials. The electrochemical behavior illustrated in Figure 5B therefore can be represented by the following mechanism consisting of closely spaced reversible and irreversible reactions.

$$MoO(Et_2dtc)_3^+ + e^- \Longleftrightarrow MoO(Et_2dtc)_3 \qquad (10)$$

$$MoO(Et_2dtc)_3 + e^- \longrightarrow MoO(Et_2dtc)_2 + Et_2dtc^- \qquad (11)$$

If one of the bidentate Et_2dtc^- ligands is replaced by catecholate, similar behavior is observed (Figure 5C) but there is a greater separation in potential between the two one-electron steps.

$$MoO(Cat)(Et_2dtc)_2 + e^- \Longleftrightarrow MoO(Cat)(Et_2dtc)_2^- \qquad (12)$$

$$MoO(Cat)(Et_2dtc)_2^- + e^- \longrightarrow MoO(Cat)(Et_2dtc)^- + Et_2dtc^- \qquad (13)$$

Also, the kinetics of the Mo (VI) /Mo (V) electrode reaction become dependent on the composition of the Mo coordination environment. Table 4 contains values of the apparent standard heterogeneous rate constant obtained from the sweep rate dependence of the peak potential separation of reactions 10 and 12 [18]. The values of $k_{s,h}$ for the catecholate complexes are smaller than that of $MoO(Et_2dtc)_3^+$ and decrease as the electron donating strengths of the catechol substituents increase.

The mechanism of electrochemical reduction of seven-coordinate MoO^{4+} complexes is illustrated in Figure 6. Each electron transfer is accompanied by a change in coordination number of the metal. This alters the energy of the redox orbital in such a way that transfer of two electrons becomes possible at nearly identical potentials. The mechanism in Figure 6 may not be relevant to the behavior of molybdenum hydroxylases, but it does illustrate that coordination reactions can sustain multielectron redox events at transition metal centers.

TABLE 4. Apparent Standard Heterogeneous Rate Constants for Mo (VI) /Mo (V) Reduction of MoO (Et_2dtc) 3+ and MoO (Cat) (Et_2dtc)2 Complexes.

Complex	E°' (V vs SCE)	α	ks,h (cm s-1)
MoO $(Et_2dtc)_3^+$	-0.31	0.64	10.0×10^{-2}
MoO (Cl_4cat) $(Et_2dtc)_2$	-0.44	0.45	2.0×10^{-3}
MoO (NO_2cat) $(Et_2dtc)_2$	-0.44	0.53	5.4×10^{-4}
MoO (Cat) $(Et_2dtc)_2$	-0.61	0.35	1.5×10^{-4}
MoO (DTBcat) $(Et_2dtc)_2$	-0.70	0.32	1.2×10^{-4}

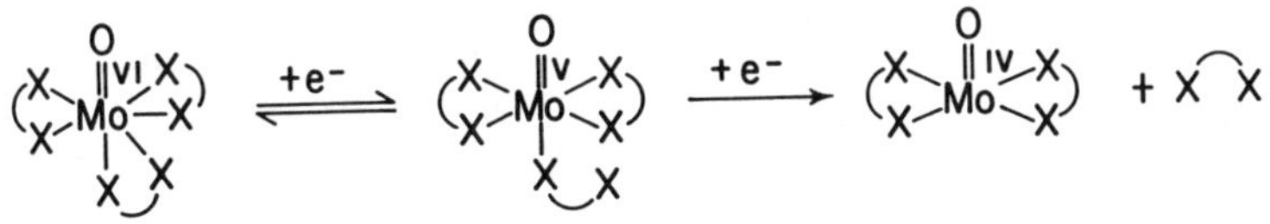

Figure 6. Electrode reaction mechanism for reduction of tris (bidentate) MoO^{4+} complexes.

When three bidentate ligands are present in the MoO^{4+} coordination sphere, Mo (VI) /Mo(V) reduction is chemically reversible. This property is retained because the bidentate ligand remains attached to the metal after one of its coordinate bonds is broken, and its chelated form is readily reestablished as the metal center is reoxidized. As illustrated in Figure 6 this reaction involves cleavage of a metal-ligand bond and changes in geometry about Mo as reduction proceeds. We believe the changes in apparent heterogeneous rate constant with coordination environment in Table 4 are a reflection of these processes, because the greater strengths of Mo-O versus Mo-S bonds should result in larger activation barriers for any structural rearrangement associated with electron transfer. A more quantitative interpretation of results is not possible at this time, because the changes in geometry that accompany Mo (VI) /Mo (V) reduction are complicated and structural information is not available for the Mo (V) intermediates.

3.3 REDUCTION OF SIX-COORDINATE COMPLEXES

Quasi-reversible to irreversible electron transfer is observed for seven-coordinate MoO^{4+} complexes because the electron that is added to these compounds interacts strongly with the coordination environment. Reduction of Mo (VI) to Mo (V) involves addition of an electron to the d_{xy} orbital which is located in the equatorial plane about the metal. Figure 7 shows that, when singly or fully occupied, this orbital can not be oriented in a way that will avoid unfavorable interaction with the five equatorial bonds to Mo that are present in pentagonal bipyramidal geometry.

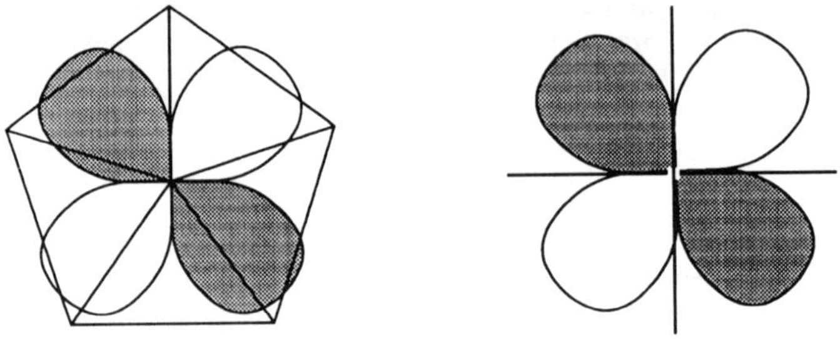

Pentagonal Bipyramidal Octahedral

Figure 7. Diagram illustrating orientation of in-plane d_{xy} orbital with five coordinate bonds in pentagonal bipyramidal geometry and four coordinate bonds in octahedral geometry.

We anticipated that if an octahedral complex with only four bonds in the equatorial plane could be examined, reversible electrochemistry would be observed at the MoO^{4+} center. Compounds of this type were not known at the time we initiated this study; consequently, we synthesized and structurally characterized the MoO (Cat) (Sap) complex illustrated in Figure 4 [19]. Consistent with these expectations, Figure 5D shows that MoO (Cat) (Sap) undergoes a fully Nernstian one-electron reduction [20], and we anticipate that this occurs with little or no structural reorganization. Figure 5D also illustrates that a large separation between the potentials of Mo (VI) /Mo(V) and Mo (V) /Mo (IV) reduction is observed for the six-coordinate complex. The range of electrochemical behavior exhibited in Figure 5 illustrates the extent to which coordination environment can influence electrochemical properties of MoO^{4+} complexes.

4. Acknowledgments

The author wishes to acknowledge the contributions of his coworkers who are cited in the references to this paper. In particular, Drs. Diwei Feng and Xi Hai Mu contributed to the work on metalloporphyrins and Drs. Julie Bradbury and Jalal Mondal to the work on Mo complexes. This research has been supported by the National Science Foundation.

5. References

[1a] Marcus, R.A. (1965) J. Chem. Phys. 43, 679-701.

[1b] Hush, N.S. (1968) Electrochim. Acta 13, 1005-1023.

[1c] Sutin, N. (1983) Prog. Inorg. Chem. 30, 441-498.

[2] Weaver, M.J. (1987) in R.G. Compton (ed.), Comprehensive Chemical Kinetics, Volume 27, Elsevier, Amsterdam, pp.1-60.

[3] Lever, A.B.P. and Gray, H.B. (eds.) (1982), Iron-Porphyrins, Addison-Wesley, New York.

[4] Spiro, T.G. (ed.) (1985), Molybdenum Enzymes, Wiley, New York.

[5] Mu, X.H.and Schultz, F.A. (1990) Inorg. Chem. 29, 2877-2879.

[6] Abbreviations: TPP = tetraphenylporphyrin; py = pyridine; Im = imidazole; Cat = catecholate; Cl_4cat = tetrachlorocatecholate; NO_2cat = 4-nitrocatecholate; DTB cat = 3,5-di-tert-butylcatecholate; Sap = N-salicylidene-2-aminophenolate; Et_2dtc = N,N-diethyldithiocarbamate; PC = propylene carbonate

[7] Nicholson, R.S. (1965) Anal. Chem. 37, 1351-1355.

[8a] Tulinsky, A. and Chen, B.M.L. (1977) J. Am. Chem. Soc. 99, 3647-3651.

[8b] Van Atta, R.B.; Strouse, C.E.; Hanson, L.K. and Valentine, J.S. (1987) J. Am. Chem. Soc. 109, 1425-1434.

[9] Feng, D. and Schultz, F.A. (1988) Inorg. Chem. 27, 2144-2149.

[10] Scheidt, W.R. and Reed, C.A. (1981) Chem. Rev. 81, 543-555.

[11] Mu, X.H. and Schultz, F.A. (1992) Unpublished work, IUPUI

[12] Weaver, M.J. (1976) J. Phys. Chem. 80, 2645-2651.

[13] Nakamoto, K. (1986) Infrared and Raman Specta of Inorganic and Coordination Compounds, 4th ed., Wiley, New York.

[14] Pasternack, R.F.; Spiro, E.G. (1978) J. Am. Chem. Soc. 100, 968-972.

[14b] Worthington, P. and Hambright, P. (1980) J. Inorg. Nucl. Chem. 42, 1651-1654.

[14c] Closs, G.L. and Sitzmann, E.V. (1981) J. Am. Chem. Soc. 103, 3217-3219.

[14d] Chapman, R.D. and Fleischer, E.B. (1982) J. Am. Chem. Soc. 104, 1575-1582.

[15] Kadish, K.M. and Su, C.H. (1983) J. Am. Chem. Soc. 105, 177-180.

[16] Natan, M.J.; Kuila, D.; Baxter, W.W.; King, B.C.; Hawkridge, F.M.; Hoffman, B.M. (1990) J. Am. Chem. Soc. 112, 4081-4082.

[17] Bradbury, J.R. and Schultz, F.A. (1986) Inorg. Chem. 25,4408-4416.

[18] Schultz, F.A. (1986) J. Electroanal. Chem. 213, 169-174.

[19] Mondal, J.U.; Schultz, F.A.; Brennan, T.D. and Scheidt, W.R. (1988) Inorg. Chem. 27, 3950-3956.

[20] Mondal, J.U. and Schultz, F.A. (1989) Inorg. Chim. Acta 157, 5-7.

SOME ELECTRONIC AND DYNAMICAL FACTORS IN ELECTRON-TRANSFER CHEMISTRY: METAL COMPLEXES, CLUSTERS, AND SURFACES

Michael J. WEAVER*
Fritz-Haber-Institut der Max-Planck-Gesellschaft
Faradayweg 4 - 6, D-1000 Berlin 33, Germany

ABSTRACT. Some selected concepts involving electronic and dynamical factors in the electron-transfer chemistry of inorganic and organometallic systems are outlined and illustrated by recent results primarily from the author's laboratory, with some emphasis placed on the redox properties of electrochemical interfaces in comparison with molecular reagents. Three interrelated topics are considered. The first concerns the control of electron-transfer rates by an interplay between donor-acceptor electronic coupling and nuclear dynamics. The role of the solvating medium in limiting the barrier-crossing frequency ("solvent friction" effects) is described, and the diagnostic capabilities (and limitations) of this phenomenon for probing the degree of electronic coupling by "tuning" the reaction dynamics are noted. Secondly, the behavioral differences anticipated between the kinetic properties of molecular redox reagents and electrode surfaces are discussed. A formalism is described for this purpose which intercompares homogeneous-phase and heterogeneous rate data on a unified basis, and is utilized in some illustrative experimental comparisons. Thirdly, the infrared spectroelectrochemical properties of high-nuclearity platinum carbonyl clusters in nonaqueous solvents are outlined in comparison with the potential-dependent properties of monocrystalline platinum electrode analogs. The results prompt a distinction between charge- and potential-dependent surface properties. The former factor controls localized surface bonding, which is apparently insensitive to the surface geometry; however, the surface charge-potential relationship (i.e. the surface capacitance) is appreciably different between the metal clusters and the planar metal interfaces.

*Humboldt Senior Scientist, 1992. Permanent address: Dept. of Chemistry, Purdue University, West Lafayette, Indiana 47907, U.S.A.

1. Introduction

The exploration of electron-transfer (ET) phenomena in chemistry has long been intertwined with inorganic and organometallic systems, and with

A. J. L. Pombeiro and J. A. McCleverty (eds.),
Molecular Electrochemistry of Inorganic, Bioinorganic and Organomettallic Compounds, 193–206.

electrode surfaces, both separately and in tandem. The former emphasis arose from the early recognition that such compounds offer a rich variety of single-electron redox couples having well-defined molecular structures and widely differing thermodynamics. Besides their practical utility, these systems have been utilized extensively for testing theoretical models of electron transfer; such strong interplay between theory and experiment is a hallmark of redox chemistry in general. Aside from the well-known utilization of electrodes for evaluating redox thermodynamics, electrochemical kinetics has long been a fertile field of investigation, including heterogeneous electron transfer involving inorganic and organometallic redox couples. Perhaps surprisingly, however, research into electrochemical kinetics has tended to develop in distinctly different directions to ET kinetics in homogeneous media. The overlap between these disciplines is nevertheless naturally broad.

The aim of this conference paper is to outline selected phenomena, of recent interest to the author, that illustrate some ET properties of inorganic and organometallic systems in relation to the behavior of metal-solution interfaces themselves. Three interrelated topics will be considered. The first concerns the limitations placed on ET reaction rates by the interplay of donor-acceptor site electronic coupling and nuclear reaction dynamics, and considers the role of the solvent medium in controlling the latter factor. Secondly, an anticipated relationship between the ET kinetic properties of molecular redox reagents and metal electrode surfaces will be outlined, and the predictions compared with experimental data. A third, closely related, issue concerns comparisons between the solution-phase redox properties of large metal clusters and metal-solution interfaces, involving specifically the infrared spectroelectrochemical properties of high-nuclearity platinum carbonyls and CO-saturated monocrystalline platinum electrodes. The, we believe novel, aim here is to compare the electronic and charge-dependent coordinative properties of structurally related metal clusters and metal electrodes.

2. The Interplay of Electronic and Nuclear Dynamical Factors in Electron Transfer

An obvious requirement for effective electron transfer between a pair of redox sites, either contained within molecules (or pairs of molecules) in solution or between a molecular species and a metal surface, is the occurrence of significant overlap between the donor and acceptor molecular orbitals. Electron-transfer reactions are unique in chemistry in proceeding even when there is extremely weak ($\ll 1$ kJ mol^{-1}) interaction between the reacting centers; indeed electron transfer is believed to occur effectively over 20 - 30 Å in some systems, such as metalloproteins [1]. Nevertheless, for bimolecular reactions between conventional inorganic or organometallic species, or for electrochemical processes, the reacting centers are commonly expected to be able to approach sufficiently closely (say within 4 - 8 Å) prior to electron transfer so to yield substantial orbital overlap. In general, the influence of such electronic interactions upon the efficiency of electron transfer is prescribed by the so-called *electronic coupling matrix element*, H_{12} [2,3]. Even for outer-sphere reactions, i.e. where no formal chemical bond exists between the redox centers, H_{12} is anticipated to vary substantially (say from ca. 0.1 to 10 kJ mol^{-1}) for inorganic and organometallic systems [3].

The magnitude of the electronic matrix coupling element can influence the kinetics of electron transfer in two ways. First, for weak coupling (small H_{12}), the frequency of electron tunneling within the transition state (and hence the preexponential component of the rate) is given by [2]:

$$\nu_{el} = H_{12}^2 (\pi^3/\Delta G^* h^2 k_B T)^{1/2} \quad (1)$$

where ΔG^* is the activation free energy, h is the Planck constant and k_B is the Boltzmann constant. Equation (1) displays a clear sensitivity of ν_{el}, and hence the reaction rate, to the magnitude of H_{12}. As the electronic coupling (H_{12}) is increased, however, the ET rate will eventually become limited by the frequency of passage along the nuclear reaction coordinate at the barrier top, ν_n, as prescribed by the dynamics of the various nuclear modes (reactant vibrations, solvent polarization, etc.) contributing to the activation barrier (vide infra).

More generally, then, it is useful to consider a *net barrier-crossing frequency*, $A_{et} = \kappa_{el}\nu_n$, with the *electronic transmission coefficient* κ_{el} describing the (fractional) probability of electron transfer occuring each time the system reaches the barrier top. An illustrative, albeit simplified, expression for κ_{el} is [2]:

$$\kappa_{el} = 2[1-\exp(-\nu_{el}/2\nu_n)]/[2-\exp(-\nu_{el}/2\nu_n)] \quad (2)$$

Two limiting cases of Eq (2) are noteworthy. Firstly, for sufficiently weak electronic coupling (i.e. small H_{12}) such that $\nu_{el} \ll \nu_n$, Eq (2) reduces to $\kappa_{el} \approx \nu_{el}/\nu_n$, so that the net preexponential factor $A_{et} = \nu_{el}$. Processes within this limit are designated as occurring via *nonadiabatic pathways*, whereupon the barrier-crossing frequency is proportional to $(H_{12})^2$, and *independent* of the nuclear dynamics (ν_n). Secondly, for sufficiently strong electronic coupling (i.e., large H_{12}) such that $\nu_{el} \gg \nu_n$, Eq (2) reduces simply to $\kappa_{el} \to 1$, so that $A_{et} \approx \nu_n$. Reactions fulfilling this condition are denoted as proceeding via *adiabatic pathways*, whereupon the barrier-crossing frequency is limited solely by the nuclear reaction dynamics.

Nevertheless, the electronic coupling can still influence the rate of such adiabatic pathways by diminishing the barrier height. A general expression for the observed rate constant of bimolecular homogeneous-phase or electrochemical reactions is [2,4,5]:

$$k_{ob} = K_p \kappa_{el} \nu_n \exp(-\Delta G^*/k_B T) \quad (3)$$

where K_p is a precursor equilibrium constant, describing the probability of finding the reactant(s) in a spatial configuration suitable for electron transfer.

To a first approximation [6], the barrier height ΔG^* is diminished below the cusp value ΔG^*_c which would apply in the absence of electronic coupling, by

$$\Delta G^* = \Delta G^*_c - H_{12} \tag{4}$$

Consequently, then, reaction rates are generally anticipated to be sensitive to the nature of the donor-acceptor electronic coupling.

For intramolecular ET reactions having relatively strong electronic coupling, H_{12} can be evaluated experimentally in suitable cases from the intensity of the near-infrared optical ET transition [9]. For most homogeneous-phase outer-sphere, or electrochemical reactions, however, such direct assessments of H_{12} are unavailable. Nonetheless, as already mentioned, a distinct behavioral difference between nonadiabatic and adiabatic ET processes is that the rates for the latter, but not the former, should be sensitive to the nuclear dynamics as embodied in ν_n. Provided that some means is available by which ν_n can be altered systematically, an experimental distinction between adiabatic and nonadiabatic pathways, and perhaps the diagnosis of intermediate cases, might be forged from the sensitivity of k_{ob} to ν_n. While certainly not straightforward, an interesting tactic along these lines has recently become apparent from considerations of dynamical solvent effects on ET processes. Since we have recently reviewed this overall topic in detail elsewhere [10], only a brief discussion of some pertinent points is given here.

The solvent has long been known to exert important influences upon the activation barrier to electron transfer; such *energetic* factors form a mainstay of the classic theoretical treatments due to Marcus and others. Over the last ten years or so, it has become apparent that ET rates can also be affected importantly by the solvent *dynamical* properties. In essence, the frequency of net progress along the reaction coordinate for an ET barrier composed primarily of outer-shell (solvent) reorganization is often predicted to be dominated by so-called *overdamped* solvent motion, associated with the collective repolarization of solvent dipoles necessary to reach the transition state and hence consummate electron transfer. The notion of "overdamped" dynamics refers to the commonly anticipated situation where the motion of individual dipoles is impeded by irreversible energy transfer to surrounding molecules (dissipative relaxation), so that the *net* solvent repolarization dynamics as described by an effective dielectric relaxation time, τ_{eff}, can be considerably slower than that for rotation of individual solvent dipoles, τ_{rot}. This situation, where $\tau_{eff} < \tau_{rot}$, corresponds to a breakdown of conventional transition-state theory (TST), and provides a number of interesting ramifications [11].

For the present purposes, it is sufficient to note that τ_{eff} in a number of simple polar solvents can be approximated by the *longitudinal relaxation time*, τ_L, extracted from solvent dielectric loss spectra. Interestingly, substantial (up to ca 50 fold) variations in τ_L are observed in such media, from ca. 0.25 to 10 ps [12]. Since analytic theories predict commonly that $\nu_n \sim \tau_L^{-1}$, correspondingly large variations in ν_n are anticipated to be achieved by suitable alterations in the solvent medium. The apparent ability to "tune" the nuclear dynamics in this manner is clearly of substantial value.

A complication, however, is that changing the solvent medium will generally alter the reaction energetics (ΔG^*) as well as dynamics. The former effect can be especially large for energetically nonsymmetric reactions, since the driving force ΔG° is then usually solvent dependent. A simpler situation applies to electron exchange processes, for which $\Delta G^\circ = 0$, since ΔG^* will now equal (aside from work terms) the *intrinsic barrier* ΔG^*_{int}. The solvent dependence of the latter can be predicted to a first approximation from dielectric-continuum theory. A substantial number of studies of both electrochemical exchange and homogeneous self-exchange reactions in various solvents have been published since 1985, with solvent-dependent kinetic analyses aimed at unraveling the sensitivity of the rate constant for electron exchange, k_{ex}, to the solvent dynamics [10]. The usual strategies involved are embodied conveniently in the relation:

$$\log k_{ex} = \log \kappa_{el}\nu_n + \log K_p - C(\varepsilon_{op}^{-1} - \varepsilon_o^{-1}) \qquad (5)$$

This equation follows from Eq (3) by assuming that ΔG^* equals the outer-shell intrinsic barrier as predicted by the usual dielectric-continuum model, contained in the last term in Eq (5); ε_{op} and ε_o are the solvent optical and zero-frequency (i.e. static) dielectric constants, respectively, and C is a constant that depends on the precursor-state geometry.

Provided that K_p is indeeed solvent insensitive, distinctly different solvent dependencies of k_{ex} can often be anticipated from Eq (5) for reactions following largely adiabatic and nonadiabatic pathways [10]. In the latter case, as noted above the preexponential term $\kappa_{el}\nu_n$ will be independent of ν_n and hence the solvent dynamics, so that log k_{ex} should correlate with $(\varepsilon_{op}^{-1} - \varepsilon_o^{-1})$. On the other hand, substantial departures from this behavior should be seen for adiabatic pathways since $\kappa_{el}\nu_n$ can then be strongly solvent dependent. For polar Debye solvents, log τ_L^{-1} tends to increase along with $(\varepsilon_{op}^{-1} - \varepsilon_o^{-1})$; this correlation can yield log k_{ex} - $(\varepsilon_{op}^{-1} - \varepsilon_o^{-1})$ slopes of *opposite sign* to those anticipated for nonadiabatic processes. Illustrative examples of reactions following both limiting cases have been described [13]. Several other related solvent-dependent analyses have been discussed [10b,14,15]. All are limited in various ways by the assumed applicability of the dielectric-continuum model of the reaction energetics and dynamics, and are therefore valid at best only on a semiquantitative level [10b].

In a few cases, however, a more quantitative analysis can be undertaken whereby the desired solvent-dependent ΔG^*_{int} values are obtained experimentally from optical ET reorganization energies for related binuclear systems. An example of this approach, involving self exchanges of metallocenium-metallocene ($Cp_2M^{+/o}$) redox couples in various polar Debye solvents is worth highlighting here since ET rate-solvent dependencies are observed that span the limits of nonadiabatic and adiabatic behavior [12]. For ferrocenium-ferrocene couples, the rate constants k'_{ex} (corrected for ΔG^*-solvent variations) are almost independent of the solvent dynamics, as discerned by τ_L^{-1}. For the more facile cobaltocenium-cobaltocene couples, log k'_{ex} is seen to be significantly dependent upon log τ_L^{-1}; for the fastest systems, the log k'_{ex} - log τ_L^{-1} slope approaches unity towards smaller τ_L^{-1} values, as expected for adiabatic pathways. By comparing this spectrum of k'_{ex} - τ_L^{-1} behavior with corresponding plots calculated for different degrees of electronic coupling, approximate estimates of H_{12} could be obtained [12]. These values, varying between ca. 0.4 and 5 kJ mol^{-1}, are in approximate agreement with theoretical H_{12} estimates (for $Cp_2Fe^{+/o}$ and $Cp_2Co^{+/o}$) [3,16]. The greater electronic coupling obtained for cobaltocenes versus ferrocenes can be understood qualitatively in terms of the ligand- and metal-centered molecular orbitals, respectively, involved in these ET processes [17].

This sensitivity of the log k'_{ex} - log τ_L^{-1} dependence, as well as the ET self-exchange rates themselves, to the metallocene electronic structure is in contrast to the electrochemical exchange kinetics, which exhibit similarly facile rates with k_{ex} - τ_L^{-1} dependencies indicating the presence of largely adiabatic pathways [17,18]. Indeed, the large majority of electrochemical exchange processes subjected to solvent-dependent analyses exhibit k_{ex} - τ_L^{-1} behavior consistent with the occurrence of adiabatic pathways, although exceptions have been noted [13]. This suggests that even outer-sphere electrochemical processes commonly proceed via pathways involving at least moderate electronic coupling (say, $\gtrsim 2$ kJ mol^{-1}). While consistent with some theoretical expectations, such apparent dependencies of the ET rates on the solvent dynamics, however, can also result from experimental artifacts [10b].

Finally, it is worth noting that the nature of the solvent-dependent ET dynamics is also predicted to be affected by the presence of inner-shell (reactant vibrational) contributions to the activation barrier [19]. As might be expected, the presence of higher-frequency vibrational contributions to the activation barrier can yield a marked attenuation in the degree to which overdamped solvent dynamics control the adiabatic barrier-crossing frequency [19]. The experimental exploration of such effects is limited in part by the paucity of redox couples suitable for solvent-dependent studies that exhibit known vibrational barriers, and complicated by the qualitatively similar behavior expected for nonadiabatic pathways. Nevertheless, there is some evidence that vibrational activation can indeed attenuate the role of solvent dynamics, although the theoretical predictions appear to overestimate the magnitude of this effect [10b,20].

3. Kinetic Properties of Metal Electrodes versus Molecular Redox Reagents

A fundamental topic in electron-transfer chemistry involves understanding reaction rates in terms of the intrinsic, as distinct from thermodynamic, properties of the redox couples involved, and the environment in which the reaction occurs. The intercomparison of homogeneous-phase rates involving related pairs of redox couples by means of treatments based on Marcus theory is well known, especially in inorganic chemistry [2]. An equally intriguing, but less extensively explored, issue of this type involves comparisons of the kinetics of related electrochemical and homogeneous-phase reactions. The most well-known formula for this purpose, based on Marcus theory, relates the rate constants for corresponding electrochemical exchange and homogeneous self exchange processes, k^e_{ex} and k^h_{ex}, respectively, by [21].

$$k^e_{ex}/A^e \simeq (k^h_{ex}/A^h)^{1/2} \tag{6}$$

where A^e and A^h are the appropriate overall preexponential factors. An additional, perhaps more useful, equation relates electrochemical rate constants, k^e_1 and k^e_2, for a pair of reactions at the same electrode potential, E, with the homogeneous-phase rate constants for the same two redox couples, k^h_1 and k^h_2, reacting with a common reagent R [21]:

$$(k^e_1/k^e_2)_E = (k^h_1/k^h_2)_R \tag{7}$$

This latter relation is applicable to chemically irreversible as well as reversible processes.

While these relationships can be valuable for collating data and for understanding as well as predicting kinetic trends, it is enlightening to consider more specifically the manner in which the ET properties of metal surfaces can be compared with molecular (i.e. solution-phase) redox reagents. The following is extracted from a recent discussion by the present author [22]. For this purpose, it is useful to regard electrochemical (metal-solution) interfaces as a special type of redox reagent, having a continuously variable redox potential (i.e. the electrode potential, E), and which requires no nuclear reorganization to transfer the electron(s). By contrast, molecular redox reagents have a fixed redox potential (E°) under a given set of chemical and physical conditions. Electron transfer requires some nuclear reorganization associated with the surrounding solvent and often additionally from alterations in reactant bond distances (inner-shell activation) associated with the change in ionic charge. On this basis, then, one would expect electrode surfaces to commonly yield more facile ET reaction pathways than those provided by molecular redox reagents under the same thermodynamic conditions.

The order to compare directly the rates of corresponding electochemical and homogeneous-phase processes along these lines, it is first necessary to make allowance for the difference in rate units, cm s^{-1} and M^{-1} s^{-1}, respectively, for these two types of reaction, arising from the two- and three-dimensional nature of the reaction environments. A simple means of achieving this entails shrinking (hypothetically) the metal electrode radius from infinity (i.e. a plane surface) down to a sphere of a same size as the molecular reagent without altering its properties. This allows electrochemical rate constants, k^e(cm s^{-1}), to be transformed into "equivalent second-order" rate constants, k^{e2} ($M^{-1}s^{-1}$), that would be observed if the metal surface offered the same geometric environment to the reactant as the molecular reagent. The required transformation from linear to spherical coordinates is

$$k^{e2} = 4\pi N r_h^2 k^e \qquad (8)$$

where N is Avogadro's number and r_h is the radius of the molecular reagent [22].

The comparison between k^{e2} and the corresponding rate constant, k^h, for reduction (or oxidation) of the same species by a molecular reagent is most straightforward when the former is evaluated at an electrode potential equal to the standard (or formal) potential for the latter reagent, since the thermodynamic (driving force) terms thereby approximately cancel. Under these conditions, and in the specific cases where: (a) the inner-shell barrier associated with the molecular reagent ($\Delta G^*_{i,h2}$) is negligible and (b) in the absence of stabilizing reactant-electrode image interactions, we expect that:

$$k^{e2} = k^h (\kappa^e_{el} \delta r_e / \kappa^h_{el} \delta r_h) \qquad (9)$$

Here κ^e_{el} and κ^h_{el} are the electronic transmission coefficients associated with the electrochemical and homogeneous reagent pathways, and δr_e and δr_h are the corresponding "reaction-zone thickness"; i.e., the range of inter-reactant distances within the precursor states that contributes effectively to the observed reaction rate. When assumption (a) above does not apply, i.e. when the molecular reagent contributes a significant inner-shell barrier, the observed rate constants k^{h2} can be corrected readily for this effect if $\Delta G^*_{i,h2}$ can be calculated, yielding "inner-shell corrected" rate constants k^{h2}. Fortunately, $\Delta G^*_{i,h2}$ for many redox reagents is small and/or can be estimated reliably.

Comparisons between corresponding k^{e2} and k^{h2} values obtained in this manner can yield interesting insights into electrochemical versus homogeneous-phase ET reactivity [22]. Table I contains an abbreviated illustration of some reactivity comparisons along these lines, taken from ref. 22. Three types of comparative reactions are represented, all referring to one-electron electroreductions at mercury electrodes. The first variant (entries 1 - 5) involves the outer-sphere reductions of cationic transition-metal complexes by $Ru(NH_3)_6^{2+}$ in aqueous media. This reductant is of particular interest since it obliges the reactions to follow outer-sphere ET pathways. The penultimate right-hand column lists the "equivalent second-order" electrochemical rate contents, k^{e2}, obtained from the work-corrected values measured at the formal potential, E_f=-0.185 V vs SCE, for the reducing agent $Ru(NH_3)_6^{2+}$, by means of Eq (8). The far right-hand column contains corresponding work-corrected rate constants for the homogeneous-phase reduction by $Ru(NH_3)_6^{2+}$, corrected for the small (1.5 kJ mol^{-1}) inner-shell barrier contributed by this reductant (see ref. 22 for details).

Comparison between these corresponding k^{e2} and k^{h2} values in Table I shows that uniformly $k^{e2} > k^{h2}$, the electrochemical rates ranging between ca 10 - 10^3 fold faster than the homogeneous-phase values. The largest inequalities are seen for the metal ammine reductions. As discussed in ref. 22, the origins of these notably higher reactivities in the electrochemical environment probably lie at least partly in the greater efficiency of electron tunneling for the electro-

TABLE I Illustrative Comparison between "Equivalent Second-Order" Rate Constants for One-Electron Reductions at Mercury Electrodes and Rate Constants involving Homogeneous-Phase Molecular Reagents (abstracted from ref. 22)

No.	Oxidant	Reductant	Environment[a]	$k^{e\,b}$ cm s^{-1}	$k^{e2\,c}$ M^{-1} s^{-1}	$k^{h2\,d}$ M^{-1} s^{-1}
1	$Co(NH_3)_6^{3+}$	$Ru(NH_3)_6^{2+}$	Aqueous	$4x10^{-5}$	$1.3x10^{3}$	0.45
2	$Co(NH_3)_5F^{2+}$	$Ru(NH_3)_6^{2+}$	"	$6x10^{-5}$	$2.0x10^{3}$	2.7
3	$Ru(NH_3)_6^{3+}$	$Ru(NH_3)_6^{2+}$	"	2.5	$8.5x10^{7}$	$3.5x10^{5}$
4	$Fe(OH_2)_6^{3+}$	$Ru(NH_3)_6^{2+}$	"	~5	$1.7x10^{8}$	$2.7x10^{7}$
5	$V(OH_2)_6^{3+}$	$Ru(NH_3)_6^{2+}$	"	$5.5x10^{-7}$	20	1.4
6	O_2	$Ru(NH_3)_6^{2+}$	"	0.06	$2.0x10^{6}$	$1.1x10^{2}$
7	Cp_2Co^+	Cp_2Co	DMF	2.0	$9x10^{7}$	$7x10^{7}$
8	Cp_2Co^+	Cp_2Co	TMU	0.35	$1.5x10^{7}$	$3.5x10^{7}$

[a]$Ru(NH_3)_6^{2+}$ reductions refer to aqueous media; Cp_2Co reductions (Cp=cyclopentadianyl) refer to dimethylformamide (DMF) and tetramethylurea (TMU) solvents, as indicated.
[b]Work-corrected rate constant for reduction at mercury electrode, at electrode potential equal to E_f of reductant [-0.185 V vs SCE for $Ru(NH_3)_6^{3+/2+}$] (see ref. 22 for data sources, etc).
[c]"Equivalent second-order rate constant" for electrochemical reduction, obtained from k^e value in adjacent column by using Eq (8) (see ref. 22).
[d]Work-corrected rate constant for corresponding reduction by homogeneous-phase reducing agent, also corrected for inner-shell barrier contribution from the reductant (see ref. 22 for data sources)

chemical versus the homogeneous-phase processes, i.e., $(\kappa^e_{el}\delta r_e/\kappa^h_{el}\delta r_h) \gg 1$ in Eq (9). Thus evidence has been mashalled to indicate that at least the Co(III) ammine-$Ru(NH_3)_6^{2+}$ reactions are strongly nonadiabatic [23], whereas the Co(III) ammine and (to a lesser extent) the metal aquo electroreductions appear to follow largely adiabatic pathways [24]. The latter is consistent with the close approach of metal ammine reactants to the metal surface deduced on the basis of the sensitivity of the ET rates to variations in the double-layer potential profile [25]. An additional factor favouring the electrochemical processes may be the presence of stabilizing reactant-electrode image interactions, which are not considered in Eq (9). Evidence against at least the predominant importance of this factor, however, can be gleaned on both experimental and theoretical grounds, as outlined in ref. 22.

The second type of reaction in Table I, represented by entry 6, concerns the irreversible reduction of a non-metal species, dioxygen. Comparison of the k^{e2} and k^{h2} values shows again that the former is substantially faster, by 2×10^4 fold. This result is of significance since it demonstrates that a reaction pathway for dioxygen reduction that is notably more facile than those engendered by the outer-sphere reductant $Ru(NH_3)_6^{2+}$ can be engendered even at the relatively "noncatalytic" mercury-aqueous interface. Such a rate acceleration may again be due partly to greater electronic coupling, but appears more likely to result from specific solvation and other short-range environmental factors [22].

The last type of example in Table I, entries 7 and 8, concerns $Cp_2Co^{+/o}$ electron exchange in a pair of solvents, dimethylformamide (DMF) and tetramethylurea (TMU). These media were selected since they offer significantly different overdamped solvent dynamics, with τ_L^{-1} values of 7.5×10^{11} and $1.7 \times 10^{11} s^{-1}$, respectively. This type of reaction is included partly to emphasize that the present treatment is applicable to electron-exchange, as well as nonsymmetric ET processes. In the former case, one merely compares directly rate data for electrochemical exchange and homogeneous self-exchange since the former necessarily refers to an electrode potential equalling E_f for the molecular reductant (or oxidant). In contrast to the above examples, both the corresponding k^{e2} and k^{h2} values are seen to be closely similar, within ca. twofold of each other. It therefore appears that the electrochemical and homogeneous-phase reductants offer similarly facile reaction environments in this case. Given the discussion in the last section, it is worth noting that the variation in k^{e2} with τ_L^{-1} is distinctly greater than that observed for k^{h2} (Table I). This difference, which is observed more generally, most likely reflects the occurrence of stronger electronic coupling at the metal surface than for the homogeneous-phase bimolecular process, thereby yielding a larger dependence of the ET rate on the solvent dynamics, as expected for a more adiabatic pathway.

4. Charge-Dependent Spectroelectrochemical Properties of High-Nuclearity Pt Carbonyl Clusters in relation to Pt-CO Electrodes

Such behavioral differences between the redox properties of metal surface and metal complex reagents raises the more general issue of the manner in which the electronic nature of electrode surfaces might usefully be intercompared with molecular solutes. It has long been recognized that metal clusters represent a state of matter that, being intermediate between small metal complexes and metal surfaces, enable the differences between the chemistry of such systems to be bridged in an experimental fashion [26]. Besides the preparation and examination of naked metal clusters in the gas phase, which has attracted considerable recent attention [27], a variety of ligand-stabilized metal clusters have been synthesized and structurally characterized, especially in the solid state [28]. Nevertheless, there has been surprisingly little attention devoted so far to the electrochemical redox properties of the latter systems. In particular, one might anticipate that larger solution-soluble clusters would exhibit a rich redox chemistry as a result of the ready availability of unfilled low-lying molecular orbitals.

In collaboration with the group of L. F. Dahl (University of Wisconsin-Madison), we have recently explored the infrared spectroelectrochemical properties of several high-nuclearity platinum carbonyl clusters in nonaqueous media [29-31]. The dianions $[Pt_{24}(CO)_{30}]^{2-}$, $[Pt_{26}(CO)_{32}]^{2-}$, and $[Pt_{38}(CO)_{44}]^{2-}$ have previously undergone detailed x-ray structural examination by Dahl *et al.*; they feature hexagonal close-packed "surface facets" with CO coordinated in both terminal and twofold bridging geometries in an analogous fashion to monocrystalline platinum surfaces [28]. Intriguingly, all three Pt clusters exhibit a rich sequence of reversible redox transitions involving net charges, n, from 0 to -10 [29,30]. For $[Pt_{24}(CO)_{30}]^{n}$ and $[Pt_{38}(CO)_{44}]^{n}$, uniformly one-electron voltammetric steps are observed, although the electrode-potential regions over which the odd-charge states are stable (i.e. the spacings between successive E_f values) are markedly smaller than for the even-charge states. Indeed, for $[Pt_{26}(CO)_{32}]^{n}$ two-electron steps involving even-charge states are chiefly observed. The clusters therefore provide effectively "solubilized microscopic chunks" of Pt metal that can be charged progressively more negative in a fashion closely akin to Pt electrodes at potentials negative of the potential of zero charge (pzc) [30].

Moreover, infrared spectroelectochemistry shows that the C-O stretching frequencies for the binding and especially the terminal (ν^{t}_{CO}) coordinated ligands decrease systematically as n becomes more negative; for example, ν^{t}_{CO} for $[Pt_{24}(CO)_{30}]^{n}$ diminishes by 15 - 20 cm^{-1} per added electron. These ν_{CO} frequency-cluster charge dependencies are closely analogous to the corresponding ν_{CO} frequency-potential dependencies commonly observed for CO adlayers at platinum electrodes, including monocrystalline surfaces in aqueous and nonaqueous media under similar conditions (electrode potentials, solvents) as for the metal clusters [32,33]. These latter ν_{CO} frequency-electrode potential dependencies are usually ascribed to increased $d\pi$ (Pt) → $2\pi^*$ (CO) backbonding as the surface charge becomes more negative, or to a first-order Stark (electric field) effect [34].

Two related comparisons between these charge-dependent metal-cluster and analogous metal-surface properties [30] are worth noting specifically here. The first involves examining the effective ν^{t}_{CO}-E slopes (commonly, if imprecisely, termed "Stark-tuning rates") for corresponding Pt cluster-solvent and Pt electrode/CO-solvent interfaces. While the surface charge-potential behavior is necessarily quantized for the former systems, approximate ν^{t}_{CO}-E slopes can nonetheless be extracted from the changes in ν^{t}_{CO} brought about over a potential span which is sufficiently large (say 2 V) so to encompass several sequential redox steps. Significantly, the ν^{t}_{CO}-E slopes observed for the clusters, ca 40 - 50 $cm^{-1}V^{-1}$, are significantly (2 - 4 fold) larger than those obtained for the corresponding Pt(111)/CO-nonaqueous interfaces.

These differences were found to be due chiefly to larger "effective surface" capacitances (i.e. charge-E_f dependencies) for the cluster solutes than for the electrode-solution interfaces. By estimating the surface area of the clusters, approximate values of the effective cluster surface capacitance, $C_S \approx 15$ - 20 μF cm^{-2}, were obtained. These values are significantly higher than those measured for the corresponding Pt(111)/CO- and Pt(110)/CO-acetonitrile interfaces, 6(±1)μF cm^{-2} [30]. Such differing capacitances for effectively spherical and planar conducting surfaces can be accounted for largely by simple geometric electrostatic considerations [30]. This finding is noteworthy: given that stepwise cluster reduction results in a progressive filling of molecular orbitals, the increasing energies of these orbitals might be expected to yield lower effective C_s values in comparison with metal surfaces which feature an energetically continuous electronic band structure.

One possible point of confusion regarding the notion of "cluster surface capacitance" is also worthy of comment here. For electrode surfaces, capacitance charging is usually denoted as a nonfaradaic process to distinguish it from faradaic events that involve necessarily electron transfer across the metal-solution interface. From the cluster solutes, however, all charge build-up must necessarily occur via electron transfer from a source present in , or in contact with, the same solution (such as the gold electrode utilized in the spectroelectrochemical measurements). Of course, charging of plane metal surfaces may also occur by this faradaic mechanism, but is more conveniently achieved by connection to an external electrical source, thereby acting as a polarizable electrode.

The second comparison of note here concerns examining instead the dependence of ν^t_{co} on the electronic charge *per surface Pt atom*, σ_{Pt} ("surface charge density"). Significantly, when examined in this manner all the cluster-solvent systems yield almost coinicident linear ν^t_{co} - σ_{Pt} plots, with a slope of 400 (±20) cm^{-1} (e^- per Pt atom)$^{-1}$ and an intercept close to 2080 cm^{-1}. A similar ν^t_{co} - σ_{Pt} dependence was also extracted for the Pt(111)/CO-acetonitile interface from the combined ν^t_{co} - E and surface capacitance data. This remarkably uniform ν^t_{co} - σ_{Pt} behavior infers that the Pt-CO surface bonding has a localized character in that it is dependent largely on the local (albeit averaged) electron density rather on the longer-range potential distribution. For a given surface charge density, the latter quantity is markedly dependent on the geometry of the metal-solvent interface, accounting for the appreciably larger ν^t_{co} - E (and σ_{Pt} - E) dependencies observed for the microscopically spherical clusters in comparison with the planar electrodes.

An interesting consequence of these combined effects is that remarkably large negative surface charge densities can be established on the Pt clusters. Thus the $[Pt_{26}(CO)_{32}]^{-10}$ state, formed by a final two-electron reduction at -2.3 V (vs. ferrocenium-ferrocene) in acetonitile, corresponds to a σ_{Pt} value of nearly 0.5 e^- per surface Pt atom. Such charge densities are markedly higher than can generally be achieved on planar electrode surfaces, for which usually $\sigma_M \leqq 0.1$ e^- per surface metal atom. While larger σ_M values could in principle be established, solvent reduction (or oxidation) commonly acts to limit the degree of electrode charging (i.e. polarization).

Overall, then, the examination of metal cluster electrochemistry (and spectroelectrochemistry) can offer useful insight into the electronic properties of such molecular solutes in relation to electrochemical interfaces in addition to the fundamental features of the clusters themselves. Recalling the foregoing discussion, it would clearly be of interest to compare the electron-transfer *kinetic* properties of such high-nuclearity metal clusters in comparison with both small molecule (mononuclear) redox reagents and with metal electrodes. Considerations paralleling those noted here for the platinum carbonyl clusters could also be developed, for example, for the fullerenes given their similarly rich electrochemical properties [35]. In general, therefore, exploration of the relationships between the redox properties of interfacial electrochemical and homogeneous molecular-based systems promises to become an increasingly broadbased, as well as enticing, endeavour in the future.

Acknowledgements

It is a pleasure to acknowledge the continued support of the research outlined in this article by the U.S. Office of Naval Research, with additional funding from the National Science Foundation. The author is also grateful to Alexander von Humboldt-Stiftung for a Senior Research Award.

References and Notes

(1) For example, see: S. L. Mayo, W. R. Ellis, Jr., R. J. Crutchley, and H. B. Gray, Science **233**, 948 (1986)

(2) For a general overview, see: N. Sutin, Prog. Inorg. Chem. **30**, 441 (1983)

(3) M. D. Newton, Chem. Revs. **91**, 767 (1991)

(4) J. T. Hupp and M. J. Weaver, J. Electroanal Chem. **152**, 1 (1983)

(5) M. J. Weaver, in "Comprehensive Chemical Kinetics", Vol. 27, R. G. Compton, ed., Elesevier, Amsterdam, 1987, Chapter 1

(6) Strictly speaking, Eq (4) applies only to reactions involving small (or zero) driving forces. In addition, a more general limitation of Eq (4) has recently been described, whereupon the effective barrier lowering by electronic coupling is predicted to be smaller than given by this simple relation [7]. It presently appears, however, that Eq (4) nevertheless represents a viable approximation under most circumstances [8].

(7) H. J. Kim and J. T. Hynes, J. Chem. Phys. **93**, 5144, 5211 (1990)

(8) H. J. Kim and J. T. Hynes, J. Chem. Phys., in press; J. N. Gehlen, D. Chandler, H. J. Kim, and J. T. Hynes, J. Phys. Chem., in press

(9) For a review, see: Creutz, C., Prog. Inorg. Chem. **30** 1(1983)

(10) (a) M. J. Weaver and G. E. McManis, Acc. Chem. Res. **23**, 294 (1990); (b) M. J. Weaver, Chem. Revs., to appear May 1992

(11) For explanative overviews see: (a) J. T. Hynes, in "The Theory of Chemical Reactions", Vol. 4, M. Baer, ed., CRC Press, Boca Raton, FL, 1985, p. 171; (b) J. T. Hynes, J. Stat. Phys. **42**, 149 (1986)
(12) G. E. McManis, R. M. Nielson, A. Gochev, and M. J. Weaver, J. Am. Chem. Soc. 111, 5533 (1989)
(13) For example, see: M. J. Weaver, D. K. Phelps, R. M. Nielson, M. N. Golovin, and G. E. McManis, J. Phys. Chem. **44**, 2949 (1990)
(14) G. Grampp and W. Jaenicke, Ber. Bunsenges. Phys. Chem. **95**, 904 (1991)
(15) W. R. Fawcett and C. Foss, J. Electroanal Chem. **270**, 103 (1989)
(16) M. D. Newton, K. Ohta, and E. Zhang, J. Phys. Chem. **95**, 2317 (1991)
(17) R. M. Nielson, M.N. Golovin, G. E. McManis, and M. J. Weaver, J. Am. Chem. Soc. **110**, 1745 (1988)
(18) T. Gennett, D. F. Milner, and M. J. Weaver, J. Phys. Chem. **89**, 2787 (1985)
(19) (a) H. Sumi and R. A. Marcus, J. Chem. Phys. **84**, 4894 (1986); (b) W. Nadler and R. A. Marcus, J. Chem. Phys. **86**, 3906 (1987)
(20) D. K. Phelps and M. J. Weaver, submitted for publication
(21) R. A. Marcus, J. Phys. Chem. **67**, 853 (1963)
(22) M. J. Weaver, J. Phys. Chem. **94**, 8608 (1990)
(23) J. F. Endicott and T. Ramasami, J. Phys. Chem. **90**, 3740 (1986)
(24) J. T. Hupp and M. J. Weaver, J. Phys. Chem. **88**, 1463 (1984)
(25) M. J. Weaver and T. L. Satterberg, J. Phys. Chem. **81**, 1772 (1977); T. L. Satterberg and M. J. Weaver, J. Phys. Chem. **82**, 1784 (1978); M. J. Weaver, H. Y. Liu, and Y. Kim, Can. J. Chem. **59**, 1944 (1981)
(26) For example, see: E. L. Mutterties, R. M. Wexler, Survey Prog. Chem. **10**, 61 (1983); E. L. Mutterties, T. N. Rhodin, E. Band, C. F. Brucker, and W. R. Pretzer, Chem. Revs. **79**, 91 (1979)
(27) For example, see: M. M. Kappes, Chem. Rev. **88**, 369 (1988); M. L. Cohen and W. D. Knight, Physics Today, December 1990, p. 42
(28) K. C. C. Kharas and L. F. Dahl, Adv. Chem. Phys. **70** (part 2), 1 (1988)
(29) G. J. Lewis, J. D. Roth, R. A. Montag, L. K. Safford, X. Gao, S.-C. Chang, L. F. Dahl, and M. J. Weaver, J. Am. Chem. Soc. **112**, 2831 (1990)
(30) J. D. Roth, G. J. Lewis, L. K. Safford, X. Jiang, L. F. Dahl, and M. J. Weaver, J. Am. Chem. Soc., in press
(31) J. D. Roth, G. J. Lewis, X. Jiang, L. F. Dahl and M. J. Weaver, J. Phys. Chem., submitted
(32) For an overview, see: S.-C. Chang and M. J. Weaver, J. Phys. Chem. **95**, 5391 (1991)
(33) S.-C. Chang, X. Jiang, J. D. Roth, and M. J. Weaver, J. Phys. Chem. **95**, 5378 (1991)
(34) For example, see: (a) D. K. Lambert, J. Chem. Phys. **89**, 3847 (1988); (b) A. B. Anderson, J. Electroanal Chem. **280**, 37 (1990)
(35) D. Dubois, K. M. Kadish, S. Flanagan, and L. J. Wilson, J. Am. Chem. Soc. 113, 7773 (1991)

THEORETICAL STUDY OF ELECTRON TRANSFER REACTIONS IN PORPHYRINS AND CYTOCHROMES

L. G. ARNAUT
Departamento de Química
Universidade de Coimbra
3049 Coimbra Codex
Portugal

ABSTRACT. A theoretical study of the electron self-exchange in porphyrins and in cytochrome c, and of the free-energy dependence of porphyrin-cytochrome c systems, shows that these reactions are not easily amenable to an explanation in the framework of the theory of Marcus. On the other hand, the intersecting-state model can be used to calculate the self-exchange rates and provide an useful rationalization of the free-energy relationships observed in these systems.

1. Introduction

The transfer of an electron between two chemical entities is one of the most fundamental processes in Chemistry and in Biochemistry. It is also the simplest chemical reaction that can be studied theoretically. The first quantitative attempt to calculate absolute rates of electron transfer reactions from first principles, plus assumptions considered reasonable on *a priori* grounds, was made by Marcus nearly 40 years ago.[1, 2] The fact that no bonds are broken or formed in the course of an electron exchange, certainly influenced this author to assign an important kinetic role to the reorganization of the solvation shell. Indeed this is considered to be the only contribution to the activation free energy in the cases where the oxidized and reduced species have the same geometry.

In large bioinorganic molecules like cytochrome c, where the redox centers are surrounded by the protein structure, the difference in equilibrium protein structure between the reduced and oxidized species can be seen as the solvent reorganization.[3] The theoretical advantage of this situation over the currently considered case of an electron exchange in solution, is that the detailed X-ray structure of systems like ferro and ferricytochrome c is available,[4, 5] and it is possible to use this experimental information to calculate the protein reorganization energy resulting from the dielectric relaxation upon the oxidation of the heme. Warshel carried out this calculation and concluded that when the cytochrome c self electron exchange occurs at a distance of 800 pm, the protein reorganization energy should be 27 kJ/mol.[3] Using the crystallographic data for

A. J. L. Pombeiro and J. A. McCleverty (eds.),
Molecular Electrochemistry of Inorganic, Bioinorganic and Organomettallic Compounds, 207–214.

cytochrome c, assuming the force constants for the Fe-N bonds are similar to those found in $Fe(bpy)_3^{2+/3+}$,[6] and using the force constant for the Fe-S bond obtained by Kincaid,[7, 8] we calculate by the theory of Marcus an internal reorganization energy of 5 kJ/mol, similar to the one assigned to other metalloproteins.[9] However, the experimentally derived reorganization energy at the heme edge-to-heme edge distance of r=890 pm is 69.5 kJ/mol.[10, 11] Thus, this model overestimates the self-exchange rate in cytochrome c by a factor of 3.8x10^6. Part of this factor is due to the distance dependence of the electronic coupling

$$V_r^2 = V_0^2 \exp[-\beta(r - r_0)] \tag{1}$$

and can be corrected with the parameters typical of proteins, β=0.014 pm^{-1}, $V_0 \approx 1$ and r_0=360 pm.[12] However, this correction still leaves a discrepancy of 2x10^3.

It is also surprising that the reorganization energies obtained for the thermal backward ET of electrostatically associated complexes between metallocytochrome c and metallouroporphyrins are similar to those obtained for the systems with two hemes.[13] These systems are very similar in terms of their reaction centers, but the environment of the heme porphyrin is different from that of the uroporphyrin moiety. The latter is more exposed to bulk solvent and should have a higher outer-sphere contribution to the solvent reorganization. Actually, diporphyrin synthetic models react ca. 10^4 times faster in methanol than the cytochrome systems at similar ΔG^0,[14] although conventional wisdom would expect the synthetic models to exhibit a larger solvent reorganization energy. The anomalous features of this system go even further. The photoinduced forward and thermal backward electron transfers extend over a similar range of free energies, but the former exhibit the Rehm-Weller plateau for the highly exothermic reactions,[15] while the latter shows the inverted region theoretically predicted by Marcus.[16] Finally, the maximum rate observed for these electron transfers between two redox centers 1.2 nm apart, is only 3.7x10^6 s^{-1}, in sharp contrast with the maximum rates larger than 2x10^9 s^{-1} observed in intramolecular ET between donor and acceptor groups separated 1.7 nm by rigid spacers, in isooctane.[17]

The detailed structural information on the cytochrome c system, its biological relevance and the inability to offer an adequate explanation of its features with the current theories of electron transfer, prompted us to apply an alternative model to predict and rationalize the electron transfer rates observed. In the following, we present the intersecting-state model and apply it to the calculation of the self-exchange rates of a synthetic analogue of a cytochrome and to cytochrome c itself. The model is then applied to rationalize the photochemical and thermal ET reactions between metallocytochrome c and metallouroporphyrins. Our calculations show that these systems are amenable to a quantitative explanation by the theoretical model we employ.

2. Theory

The intersecting-state model, ISM, was developed to estimate the energy of activation of atom or group transfer reactions, based on the structural information about the reactants and products and a geometric criterion for the localization of the transition state.[18] Latter, Formosinho extended his model to other systems, including weak interaction processes like electron transfer reactions.[19] The key feature of ISM is the calculation of the energy increase associated with a single bond extension of the reactant until its configuration in the transition state is reached, with no separation between the internal and the solvent coordinates. In ET there are two reactants and at least two coordinates for the reactants' potential energy surfaces, and there should be at least two extensions. Actually there are many coordinates for each reactant, but only those coordinates of the reaction center whose distortion leads to large energy variations need to be considered.

Let us take, for example, the electron self-exchange between two octahedral metal complexes, where the central metal ion is coordinated with 6 identical ligands, c_n=6. Some combination of vibrations will be active in bringing the reactants in a multidimensional potential energy surface to a nuclear configuration where they may exchange an electron. The most intuitive coordinate system to describe this process is made of the 12 interatomic metal-to-ligand separations and associated bond angles, plus the intermetal separation and the angles describing the relative orientation of the two reactants. It is obvious that this is not a convenient coordinate system for ET. In our search for a simple representation of the ET process, our first approximation is to neglect the bond angle variations, which lead to smaller energy changes than the bond extensions. We will treat the effect of the intermetal distance separately and assume its effect on the rate constants can be represented by a factor with the form of eq 1. We will also assume the complexes are spherically symmetric and neglect the orientation dependence of the ET rates in this treatment, although it may be added latter via a separate factor. Furthermore, we wish to describe the energy changes of the two reactants, oxidized and reduced species, by one effective force constant, and do identically for the two products. The criterium to obtain the force constant for each species has to be the same employed to determine the single reaction coordinate that connects the equilibrium configuration of that species to its transition state configuration, TS.

The reaction coordinate given by ISM is the sum of the reduced bond extensions from the reactant's or product's equilibrium configurations to the transition state one, $d = \eta\,(l_{ox}+l_{red})$, with the reduced bond extension given by[19]

$$\eta = \frac{a'}{2\,n^{\ddagger}} \ln \frac{1 + \exp\left(\sqrt{2\,n^{\ddagger}}\;\Delta G^{0}/\lambda\right)}{1 - \left[1 + \exp\left(\sqrt{2\,n^{\ddagger}}\;\Delta G^{0}/\lambda\right)\right]^{-1}} \tag{2}$$

where a' is a constant (a'=0.156), ΔG^0 is the reaction free energy, λ is an energy term related to the dynamics of the reaction and $n^{\ddagger}$ is the transition state bond order. In reactions involving weak interaction of the reactants, $n^{\ddagger}$ is identical to the chemical bond order of each bond and does not vary along the reaction coordinate. Exceptions to this rule occur when excited electronic configurations of low energy are present and can be accessed in the transition state. Then, $n^{\ddagger}$ will increase if these configurations have bonding character or decrease if they have antibonding character. When each reactant is modelled by an harmonic oscillator, the displacement x from the equilibrium to the TS configuration and the effective force constant f_{eff} of the oscillator, determine the TS energy, $V^{\ddagger}=(1/2)\,f_{eff}\,x^2$.

The bond order of a reactant does not vary in any combination of its ML vibrations. Thus, a possible reaction coordinate may be one where each ML bond of the complex gives an equal contribution to the TS configuration, $x = x_i$ and $f_{eff} = c_n f_i$, where x_i is the displacement of the bond i and f_i its stretching force constant. This is the approach followed in Marcus theory. However, if the same energy is placed in only one ML bond, another extreme situation must be considered, where $x=\sqrt{c_n}\,x_i$ and $f_{eff} = f_i$. In general, if we restrict the distribution of the TS energy to multiples of $V^{\ddagger}/c_n$, the total number of combinations of ML bonds that lead to $V^{\ddagger}$ is

$$W_i = \frac{c_n!}{(c_n - b_i)! \prod_j (b_{ij}!)} \tag{3}$$

where b_i is the number of bonds in the distribution i with an energy multiple of $V^{\ddagger}/c_n$, and b_{ij} is the number of bonds in this distribution with the same multiple j of $V^{\ddagger}/c_n$. A bond in its equilibrium position does not contribute to the force constant of the distribution W_i. Thus, the effective force constant of the oxidized species is, if all the ML bonds are identical,

$$f_{eff,ox} = \frac{\sum_i W_i\, b_i}{\sum_i W_i} f_{ox} \tag{4}$$

A similar expression applies for $f_{eff,red}$. In a random distribution of energy among the vibrational modes of the oxidized species, a relation of the type $f_{eff,ox}=\sqrt{c_n}\, f_{ox}$ was obtained.[6] These two approaches are compared in Figure 1. Unless c_n and the energy barrier are high, the results are similar. There is evidence that the vibrational modes with lower frequency are more probable for a given vibrational energy of the molecule above the lowest vibrational state.[20] Thus, the distributions that contribute with smaller force constants to $f_{eff,ox}$, should be given most weight and the dependence of $f_{eff,ox}$ on $\sqrt{c_n}\, f_{ox}$ should be approached. This is the dependence we will use to calculate the reactant´s force constant. Assuming that both reactants undergo similar bond extensions to the transition state, their effective force constant is $f_r=(f_{eff,ox}+f_{eff,red})/2$. A similar expression can be obtained for f_p. The reactant and product minimum extensions to a common equipotential curve, x, are related by

$$1/2\ f_r\, x^2 = 1/2\ f_p\, (d - x)^2 + \Delta G^0 \tag{5}$$

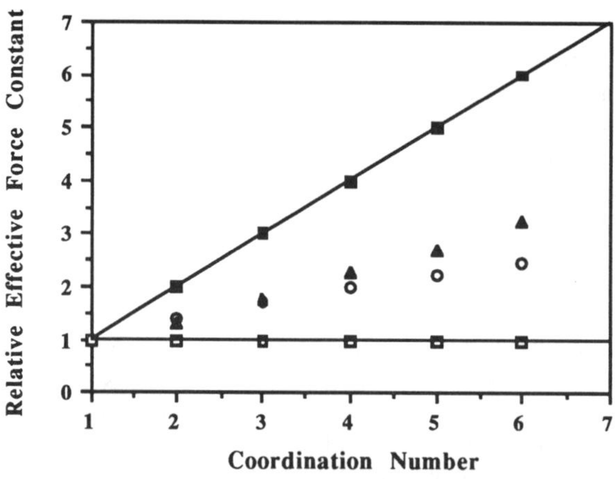

Figure 1. Effective force constants obtained according to Marcus theory (■), eq 4 (triangles), random energy distribution (circles) and placing all the energy in only one mode (□).

which can be used to estimate the activation free energy, $\Delta G^{\ddagger}=(1/2)\, f_r\, x^2$.

In this approach we neglected the explicit effect of the solvent in the reaction coordinate. In fact, this effect can be incorporated in the parameters λ or $n^{\ddagger}$ and, to a smaller extent, in the force constants. ISM predicts that the effect of the solvent on the ET rates can be small, as apparent from the similar energy barriers for the electron exchange of alkylhydrazines$^{0/+}$ in acetonitrile and in the gas phase.[21]

Although this model was initially developed for reactions occurring in an adiabatic potential energy surface, it can be extended to weak interaction systems under the assumptions mentioned above, because the reaction path is obtained in terms of independent stretches of the two reactants. This contrasts with the Agmon-Levine approach,[22, 23] which is restricted to atom or group transfer reactions, where reactant and product are connected via a family of parallel curves each defined by the same positive bond order, maintained constant through the compensation between the decrease in the reactant bond order and the increase in the product one. One caution must be observed in our analysis: the extensions x are not directly related to the intersection of the reactants curves, although it is convenient to represent them that way.[24] Therefore, we can use the transition state expression to estimate the rates of ET reactions

$$k = \chi \frac{k_B T}{h} c_0^{1-m} \exp\left(-\Delta G^{\ddagger}/RT\right) \qquad (6)$$

where χ is an electronic transmission factor whose distance dependence has the form of eq. 1, c_0 is the standard concentration (1M) and m is the molecularity of the reaction

3. Electron Self Exchanges

It is convenient to consider, in the first place, the application of ISM to the electron self-exchange in synthetic models of cytochrome c, in order to avoid the difficulties introduced by the ΔG^0 dependence and the structural complexity. For the couple iron tetraphenylporphyrin bis-1-methylimidazole, $FeTPP(1\text{-}MeIm)_2^{0/+}$, the electron self-exchange rate of 8.1×10^7 M^{-1} s^{-1} was measured directly in CD_2Cl_2 at 252 K.[25] Using the bond lengths and force constants of $Fe(phen)_3^{2+/3+}$,[6] which are similar low spin iron complexes, we calculate 9.9×10^7 M^{-1} s^{-1}, with $n^{\ddagger}=2$. This value of $n^{\ddagger}$ is typical of phenanthroline and bipyridine complexes with metals carrying empty d-orbitals, as discussed in detail elsewhere.[6]

In the calculation of the electron self-exchange rate in cytochrome c according to ISM, we utilize the data previously mentioned in the context of Marcus theory. Now we make $n^{\ddagger}=1.83$ because it is the average of five Fe-N bonds with $n^{\ddagger}=2$ and one Fe-S bond with $n^{\ddagger}=1$. The calculated rate is 3.6×10^7 M^{-1} s^{-1}, a factor of 7×10^3 higher than the experimental. In order to reproduce the observed rate, the electronic coupling distance dependence must have a coefficient $\beta=0.0167$ pm^{-1}, very close to the value typical of proteins. In a previous study of the distance dependence of electron transfers between organic donors and acceptors separated by a rigid bridge, we had obtained $\beta=0.0063$ pm^{-1},[26] while the average value for this type of systems is $\beta=0.007$ pm^{-1}.[12] The remarkable agreement between the calculated and the experimental rates in self-exchange reactions, gives us confidence to extend the model to the study of free-energy relationships in similar systems.

4. Free-Energy Relationships

Simple calculations with asymmetric systems can be performed averaging the force constants for the two reactants, and doing the same with the bond lengths and transitions state bond orders. The relevant parameters of cytc and Mn-cytc are assumed to be very similar. For $FeUp(OH_2)_2$, $MnUp(OH_2)_2$ and $FeUp(CN)_2$ the porphyrin part was assumed to be identical to cytc and the axial ligands to the correspondent hexaaquo or hexacyano complexes.[6] The average $n^{\ddagger}$ for these complexes is 1.67. Zn^{2+} has a preference for square-pyramidal five-coordinate geometries in porphyrin complexes.[27] We use the crystallographic data of a Zn-porphyrin complex, assume that Zn-cytc has five identical Zn-N bonds,[28] and estimate the force constant for these bonds from the ratio of the vibrational frequencies of $Fe(bpy)_3^{2+}$ and $Zn(bpy)_3^{2+}$ complexes.[29] A similar procedure is followed for the complex ZnUp, but here only four bonds are considered. The electronic configuration of Zn^{2+} is d^{10} and there are no low-lying empty orbitals to increase the bonding order at the transition state. Therefore, for the Zn complexes, we assume $n^{\ddagger}=1$. This leads to the interesting prediction that the self-exchange rate of $Zn(bpy)_3^{2+/3+}$ must be 100 times slower than of $Fe(phen)_3^{2+/3+}$. The same is predicted and verified for $Cu^{+/2+}$ complexes.[30] Finally, we have shown that aromatic molecules may be characterized by l=200 pm, f=155 J/(mol pm^2) and $n^{\ddagger}=1$.[26] The electronic transmission factor of $\chi=10^{-3}$ can be estimated from the value of β obtained for cytc and the distance between the reaction centers in the Up-cytc complex.[31]

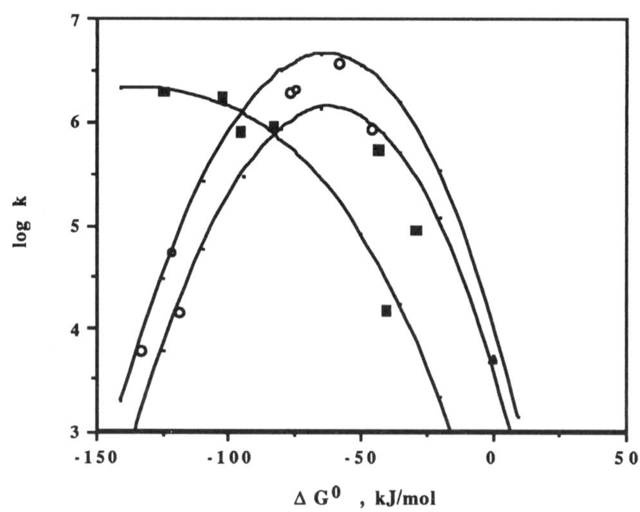

Figure 2. ET in uroporphyrin-cytochrome c complexes at 298 K. The circles indicate thermal backward ET reactions and two theoretical bell-shaped curves are associated with them: the lower curve was calculated with the parameters of the system $Zn\text{-}cytc^{\cdot+}$ + $Fe(II)Up(OH_2)_2$, and the upper curve with those of the system $ZnUp^{\cdot+}$ + cytc(II); in both these curves $\chi=10^{-3}$ and $\lambda=110$ kJ/mol. The squares represent photoinduced forward ET reactions and the correspondent curve was calculated with the same values as the upper curve, but using $\chi=5\times10^{-6}$ and $\lambda=180$ kJ/mol. The triangle indicates the cytochrome c self exchange.

Figure 2 shows the results of the application of ISM to the photoinduced forward ET of, in order of decreasing exothermicity, Zn-cytc$\cdot^{+}$ + Fe(II)Up$(CN)_2$, Zn-cytc$\cdot^{+}$ + Mn(II)Up$(OH_2)_2$, Zn-cytc$\cdot^{+}$ + Fe(II)Up$(OH_2)_2$, Up$\cdot^{+}$ + Mn(II)cytc, ZnUp$\cdot^{+}$ + Mn(II)cytc, Up$\cdot^{+}$ + cytc(II), ZnUp$\cdot^{+}$ + cytc(II) and to the thermal backward transfers, whose order of decreasing exothermicity is reversed. According to our analysis, the set of reactants is not homogeneous. For the thermal transfers we present the two curves obtained with the sets of data we obtained that lead to larger differences in the calculated rates. The lower bell-shaped curve refers to the couple Zn-cytc$\cdot^{+}$ + Fe(II)Up$(OH_2)_2$, which is characterized by f_r=244 J/(mol pm^2), f_p=261 J/(mol pm^2), l_r+l_p=413 pm, $n^{\ddagger}$=1.34, and the upper bell-shaped curve to the couple ZnUp$\cdot^{+}$ + cytc(II), for which f_r=256 J/(mol pm^2), f_p=254 J/(mol pm^2), l_r+l_p=416 pm, $n^{\ddagger}$=1.41. The other parameters used in the calculation of all the thermal reactions were $\chi=10^{-3}$ and λ=110 kJ/mol. The curve calculated for the photoinduced ET employed the same parameters as for ZnUp$\cdot^{+}$ + Mn(II)cytc, but the adjustment to the experimental data required $\chi=5\text{x}10^{-6}$ and λ=180 kJ/mol. The 200 times lower χ value of the photochemical reaction may be assigned to the nature of the excited state, which is a triplet. The lower λ value obtained for the thermal ET is marginally the lowest λ value we have obtained for ET reactions. Such a low value reveals a more relaxed or poorly solvated activation complex relative to the corresponding reactants,[26] in agreement with the expectation that the protein environment is a poor solvent. The difference between the λ values of the thermal and photochemical reactions can be rationalized noting that the electronic excitation of Zn(II)-cytc or Zn(II)Up may place the electron in one of the ligands, like in *Ru$(bpy)_3^{2+}$,[32] making the structure of the photochemical reaction center resemble more the one of the product, and demanding less protein reorganization in course of the reaction.

5. Conclusion

The electron self-exchange in iron porphyrins follows a pattern close to the one of iron-trisphenanthroline or bipyridine complexes. The self-exchange in cytochrome c has a higher $\Delta G^{\ddagger}$ in part due to the nature of the Fe-S bond in the transition state, but its distance dependence is similar to that observed in other proteins. The free-energy relationships observed in porphyrin-cytochrome c systems depend on the nature of the reactant electronic state and, relative to the reactants, may exhibit a more relaxed transition state than the reactions that occur in solution.

6. References

(1) Marcus, R. A. *J. Chem. Phys.* **1956**, *24*, 966.
(2) Marcus, R. A. *J. Chem. Phys.* **1957**, *26*, 867.
(3) Churg, A. K.; Weiss, R. M.; Warshel, A.; Takano, T. *J. Phys. Chem.* **1983**, *87*, 1683.
(4) Takano, T.; Dickerson, R. E. *J. Mol. Biol.* **1981**, *153*, 95.
(5) Takano, T.; Dickerson, R. E. *J. Mol. Biol.* **1981**, *153*, 79.
(6) Formosinho, S. J; Arnaut, L. G. *in preparation.*
(7) Hu, S.; Kincaid, J. R. *J. Am. Chem. Soc.* **1991**, *113*, 2843.
(8) Hu, S.; Kincaid, J. R. *J. Am. Chem. Soc.* **1991**, *113*, 9760.
(9) Gray, H. B.; Malmström, B. G. *Biochem.* **1989**, *28*, 7499.
(10) Dixon, D. W.; Hong, X.; Woehler, S. E. *Biophys. J.* **1989**, *56*, 339.

(11) Dixon, D. W.; Hong, X.; Woehler, S. E.; Mauk, A. G.; Sishta, B. P. *J. Am. Chem. Soc.* **1990**, *112*, 1082.

(12) Moser, C. C.; Keske, J. M.; Warncke, K.; Farid, R. S.; Dutton, P. L. *Nature* **1992**, *355*, 796.

(13) Zhou, J. S.; Rodgers, M. A. J. *J. Am. Chem. Soc.* **1991**, *113*, 7728.

(14) Heiler, D.; McLendon, G.; Rogalskyi, P. *J. Am. Chem. Soc.* **1987**, *109*, 604.

(15) Rehm, D.; Weller, A. *Isr. J. Chem.* **1970**, *8*, 259.

(16) Marcus, R. A. *Faraday Discuss. Chem. Soc.* **1960**, *29*, 21.

(17) Closs, G. L.; Calcaterra, L. T.; Green, N. J.; Penfield, K. W.; Miller, J. R. *J. Phys. Chem.* **1986**, *90*, 3673.

(18) Varandas, A. J. C.; Formosinho, S. J. *J. Chem. Soc. Faraday Trans.* **1986**, *82*, 953.

(19) Formosinho, S. J. in *Understanding Chemical Reactivity Through the Intersecting-State Model*; Formosinho, S. J.; Csizmadia, I. G. Arnaut, L. G. (Eds.); NATO ASI, Kluwer, Dordrecht, 1991; pp 159.

(20) Henry, B. R. *Acc. Chem. Res.* **1977**, *10*, 207.

(21) Nelsen, S. F.; Rumack, D. T.; Meot-Ner (Mautner), M. *J. Am. Chem. Soc.*. **1987**, *109*, 1373.

(22) Agmon, N.; Levine, R. D. *J. Chem. Phys.* **1979**, *71*, 3034.

(23) Agmon, N. *Int. J. Chem. Kinetics* **1981**, *13*, 333.

(24) Arnaut, L. G. *J. Phys. Org. Chem.* **1991**, *4*, 726.

(25) Dixon, D. W.; Barbush, M.; Shirazi, A. *J. Am. Chem. Soc.* **1984**, *106*, 4638.

(26) Arnaut, L. G.; Formosinho, S. J. *J. Mol. Struct. (Theochem)* **1991**, *233*, 209.

(27) Spaulding, L. D.; Eller, P. G.; Bertrand, J. A.; Felton, R. H. *J. Am. Chem. Soc.* **1974**, *96*, 982.

(28) McLendon, G.; Miller, J. R. *J. Am. Chem. Soc.* **1985**, *107*, 7811.

(29) Saito, Y.; Takemoto, J.; Hutchinson, B.; Nakamoto, K. *Inorg. Chem.* **1972**, *11*, 2003.

(30) Endicott, J. F.; Kumar, K.; Ramasami, T.; Rotzinger, F. P *Prog. Inorg. Chem.* **1983**, *30*, 141. Alternatively, we may keep $n^{\ddagger}=2$ and make $\chi<<1$ for the transmission of a σ^*-d electron along the bpy ligand. These approaches are equivalent in terms of ISM because they both reflect a decrease in the rate of the reaction due to the intervention of an unfavourable electronic factor.

(31) Zhou, J. S.; Granada, E. S. V.; Leontis, N. B.; Rodgers, M. A. J. *J. Am. Chem. Soc.* **1990**, *112*, 5074.

(32) Bradley, P. G.; Kress, N.; Hornberger, B. A.; Dallinger, R. F.; Woodruff, W. H. *J. Am. Chem. Soc.* **1981**, *103*, 7441.

PART 5. **POLYNUCLEAR and SUPRAMOLECULAR COORDINATION COMPOUNDS**

SUPRAMOLECULAR ELECTROCHEMISTRY: POLYNUCLEAR METAL COMPOUNDS.

S.ROFFIA AND C.PARADISI
Dipartimento di Chimica "G.Ciamician", Università di Bologna, Via Selmi 2, 40126 BOLOGNA, Italy.

C.A.BIGNOZZI
Dipartimento di Chimica, Centro di Fotochimica del CNR, Università di Ferrara, Via L.Borsari 46,44100 FERRARA, Italy

ABSTRACT. The electrochemical behaviour of the binuclear complex NC-$Ru^{II}(bpy)_2$-CN-$Ru^{II}(bpy)_2$-CN^{+}(**1**) and of the trinuclear complexes NC-$Ru^{II}(bpy)_2$-CN-$Ru^{II}(bpy)_2$-NC-$Ru^{II}(bpy)_2$-CN^{2+} (**2**), NC-$Ru^{II}(bpy)_2$-CN-Ru^{II}(DCE-bpy$)_2$-NC-$Ru^{II}(bpy)_2$-CN^{2+} (**3**) is described. These complexes give rise to extended redox series. Metal localized one-electron oxidation processes are assigned on the basis of the comparison with the model $Ru(bpy)_2(CN)_2$ complex. Ligand localized reductions are fully reversible at -54°C. The assignment of the redox sites, which are involved in such processes, is given taking into account the different electronic perturbation introduced at metal centres by the asymmetric bridging ligand as well as the energy difference between π^* accepting orbital of the different polypyridine ligands. The experimentally determined redox series consist of nine, thirteen and fourtheen redox steps for the binuclear complex **1** and for the trinuclear complexes **2** and **3** respectively.

1. Introduction

There is a growing and wide interest today in the electrochemical and photophysical properties of supramolecular systems constituted by appropriate assemblies of simple molecular compounds [1]. An important goal of this ongoing research is to obtain molecules capable of accomplishing functions through the concerted or simultaneous action of different subunits. Notable examples are polynuclear metal complexes in which photoinduced intramolecular electron and energy transfer processes may occur [2,3]. Such processes are relevant not only in the field of fundamental but also in applied research. In fact, interesting potential applications such as conversion of light into chemical and electrochemical energy, treatment of electrical signals at molecular level and multielectron photocatalysis may be expected from the use of such systems.

The search for molecular components performing as energy or electron transfer photosensitizers is directed toward the identification of species having reversible redox behaviour, suitable ground and excited state redox potentials, stability towards thermal and

A. J. L. Pombeiro and J. A. McCleverty (eds.),
Molecular Electrochemistry of Inorganic, Bioinorganic and Organometallic Compounds, 217–228.

photochemical decomposition reactions, a sufficiently large extinction coefficient in the visible region and a high lifetime of the triplet excited state. The fragment $Ru(bpy)_2^{2+}$ (bpy=2,2'-bipyridine) seems indeed to satisfy most of the necessary requirements, provided strong field ligands are incorporated in the two vacant coordination sites to make the appropriate links with other molecular units. Cyanide ligands appear to be a natural choice from this point of view and in this respect a number of polynuclear cyano-bridged bipyridine ruthenium complexes have been studied [4-7].

Along this line of research the following complexes

$NC\text{-}Ru^{II}(bpy)_2\text{-}CN\text{-}Ru^{II}(bpy)_2\text{-}CN^{+}$ (1)

$NC\text{-}Ru^{II}(bpy)_2\text{-}CN\text{-}Ru^{II}(bpy)_2\text{-}NC\text{-}Ru^{II}(bpy)_2\text{-}CN^{2+}$ (2)

$NC\text{-}Ru^{II}(bpy)_2\text{-}CN\text{-}Ru^{II}(DCE\text{-}bpy)_2\text{-}NC\text{-}Ru^{II}(bpy)_2\text{-}CN^{2+}$ (3)

(DCE-bpy = 5,5'-dicarboxyethyl-bpy)

have been synthesized and the results obtained [8, 9] are summarized and discussed here, with particular attention to the electrochemical behaviour.

The interest in the electrochemical study of these systems is mainly related to the identification and characterization of the redox sites and to understanding of the their mutual interactions.

2. Electrochemical behaviour of binuclear complex 1 and trinuclear complexes 2 and 3

In view of the high number of oxidizable and reducible centres these species give rise to very extended redox series, i.e. set of compounds having identical stoichiometric composition and differing only in the overall number of electrons [10].

A general feature of the electrochemical behaviour of these species is that they present oxidation processes that can be considered essentially localized on Ru(II) centres and reduction processes localized on bpy and DCE-bpy ligands.

2.1. METAL CENTERED REDOX SERIES AND SPECTROSCOPIC PROPERTIES

In cyclic voltammetry (CV) the binuclear complex **1** and trinuclear complexes **2** and **3**, show two and three one-electron reversible processes respectively, that can be attributed confidently to the oxidation of Ru(II) to Ru(III). The results, obtained at 25°C, are summarized in Fig.1 in terms of the half-wave potentials $E_{1/2}$ for the various steps. The $E_{1/2}$ of the $Ru(bpy)_2(CN)_2$ complex is also reported for comparison.

For complexes **1** and **2** the identification of the redox sites involved in the different oxidation steps can be made by taking into account the electronic asymmetry of the CN bridge. In particular, since the nitrogen end of cyanide behaves as an amine-type ligand, the negative shift of the first oxidation potential in going from mononuclear to trinuclear species can be explained by considering an higher electron density at N-bonded Ru(II) centre. As a consequence, the ease of oxidation increases in the order NC-Ru-CN, NC-Ru-NC, CN-Ru-NC.

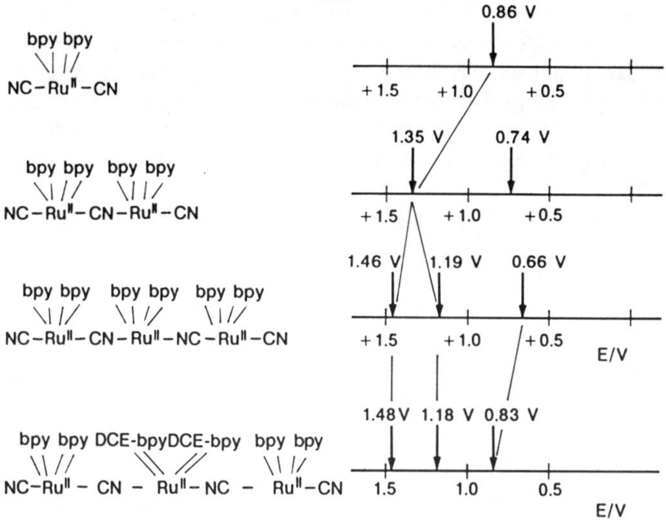

Figure 1. Comparison of $E_{1/2}$ values relative to the oxidation of Ru(II) centres for $5x10^{-4}$ M ACN solution in 0.1 M $(C_2H_5)_4NBF_4$ of $Ru(bpy)_2(CN)_2$, binuclear complex **1** and trinuclear complexes **2** and **3**. T=25°C.

The comparison, between the oxidation potentials of the $Ru(bpy)_2(CN)_2$ complex (0.86 V) with the first oxidation potential of complex **1** (0.66 V), indicates that the dπ orbitals, of the N-bonded Ru(II), are 0.2 V higher in energy than those of the mononuclear complex.

In complex **3**, the lower energy of the π* orbitals of the DCE-bpy ligand, compared to bpy, enhances dπ-π* back-bonding thus stabilizing the dπ orbitals at the N-bonded Ru(II) centre. The higher stabilization induced by changes in the polypyridine ligand, however, should not overcome the effect of the electronic asimmetry of the bridge since the first oxidation in complex **3** is only 0.17 V more positive than that of complex **2**. In this hypothesis, the first oxidation process observed for complex **3** should involve the ruthenium of the - Ru^{II}(DCE-bpy)$_2$- moiety.

The comparison between the electronic absorption spectrum of complex **3** and of its one-electron oxidized form allows to confirm this assignment. These spectra are shown in Fig.2 for D_2O solutions of the two species. Complex **3** exhibits an intense absorption at 287 nm (34.8 kK) due to intraligand π-π* (bpy) transition. The shoulder at 310 nm (32.3 kK) can be assigned to π-π* (DCE-bpy) transitions, based on the lower intensity (two DCE-bpy vs four bpy ligands) and on the lower energy of the π-π* bands in the model Ru(DCE-bpy)$_2$(Cl)$_2$ complex compared to $Ru(bpy)_2(Cl)_2$ (λ_{max}, in CH_3OH solutions, are at 300 and 294 nm respectively). The fact that this shoulder shifts to 325 nm (30.8 kK) in the one-electron oxidized form of complex **3**, is a consequence of the increased positive charge at the ruthenium coordinated to the DCE-bpy ligand. Comparable energy decreases of π-π* transitions are, in fact, observed upon going from bivalent to tervalent polypyridine complexes of iron, ruthenium and osmium [11, 12].

On the basis of these results, the one-electron oxidized form of complex **3** can be formulated

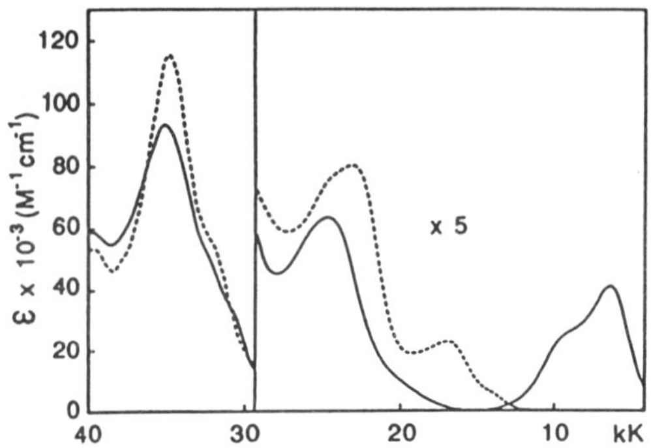

Figure 2. Absorption spectra of NC-Ru^{II}(bpy)$_2$-CN-Ru^{II}(DCE-bpy)$_2$-NC-Ru^{II}(bpy)$_2$CN^{2+}(-----) and of NC-Ru^{II}(bpy)$_2$-CN-Ru^{III}(DCE-bpy)$_2$-NC-Ru^{II}(bpy)$_2$CN^{3+}(——) in D_2O solutions.

as NC-Ru^{II}(bpy)$_2$-CN-Ru^{III}(DCE-bpy)$_2$-NC-Ru^{II}(bpy)$_2$-CN^{3+}.

The absorption spectra of complex **3** and of its mixed-valence form, in the vis/NIR region, are similar to those observed for complex **2** [7] and can be rationalized on the basis of the electrochemical results. For the unoxidized complex, the band at 590 nm is assigned to metal-to ligand charge transfer (MLCT) transitions localized on the CN-Ru^{II}(DCE-bpy)$_2$-NC-unit, while the band at 434 nm is due to MLCT transitions localized on the external -NC-Ru^{II}(bpy)$_2$-CN units.In the mixed valence complex, the absorption band at 1550 nm is assigned to metal - to metal charge transfer transitions between Ru(II)/Ru(III) centres [7]. This assignment is supported by infrared and resonance raman results that have recently obtained for the mixed valence NC-Ru^{II}(bpy)$_2$-CN-Ru^{III}(bpy)$_2$-CN^{2+} complex [12] and by the energy difference between the optical energy in the mixed valence form of complexes **2** and **3** (0.17 eV) which is coincident with the difference between the first oxidation potentials of the two complexes (Fig.1).

As far as the other oxidation processes are concerned, the second oxidation process in the binuclear complex corresponds to the oxidation of the Ru(II) centre in the electrochemically generated species NC-Ru^{II}(bpy)$_2$-CN-Ru^{III}(bpy)$_2$-CN^{2+}. The considerable splitting between the redox potentials of the first and second oxidation process ($\Delta E_{1/2}$=0.6 V) is mainly attributed to the electrostatic perturbation that the Ru(III) center exerts on Ru(II) and to partial delocalization between the metal centres of the odd electron.

The same considerations can be made for the second oxidation processes of the trinuclear complexes **2** and **3**. These oxidations involve one of the two lateral Ru(II) of the NC-Ru^{II}-CN-Ru^{III}-NC-Ru^{II}-CN species and the $E_{1/2}$ for these processes are therefore more positive than that of the mononuclear complex. The lower $E_{1/2}$ values, with respect to the binuclear complex, are consistent with a lower electron withdrawing effect of the central Ru(III), being shared between the two side Ru(II) centres.

The third process in the trinuclear species are due to the oxidation of the remaining lateral Ru(II) centres. The considerable separation from the second oxidation step, $\Delta E_{1/2}$=270 mV, instead of the 36 mV expected on the the basis of the statistical factor for two non interacting equivalent redox centres [14] is indicative of the effectiveness of the CN-Ru(III)-NC bridge in mediating the electronic interaction between the two identical side units.

2.2. LIGAND BASED REDOX SERIES. REDUCTION OF BIPYRIDINE LIGANDS

The complete redox series relative to reduction of the bipyridine ligands, in the binuclear and trinuclear complexes, should be constituted by eigth and twelve one-electron redox processes, respectively, since each bpy carries one redox orbital [10]. At room temperature and in non strictly aprotic conditions, chemical reactions, involving one of the members of the redox series, can take place, causing the termination of the redox series. The reduction processes for these species were therefore studied in strictly anhydrous DMF solutions at -54°C [8].

2.2.1 *Binuclear complex 1.* Fig.3a shows the low temperature voltammogram for the binuclear complex recorded for a sweep rate of 50 V/s. The various peaks can be grouped in two sets separated by a gap of about 0.5 V.

The analysis of the cyclic voltammetric curve (cvc) indicates that the first apparently single peak of the first set corresponds to two consecutive one-electron reversible processes with closely spaced standard potentials. The same obtains for the two successively closely spaced peaks and for the first peak of the second set. The last peak instead corresponds to the exchange of one electron. The experimental ligand based redox series for the molecule is thus constituted by seven one-electron transfer steps. Since the maximum possible size of the ligand based redox series for this molecule is eight, one redox step is missing. Except for the last process for which the $E_{1/2}$ value was obtained by averaging the cathodic and anodic peak potential, the $E_{1/2}$ values of all other processes were deduced by means of convolutive analysis of the variuos doublets according to a procedure described by Ammar and Savéant [14] and are reported in Tab.1.

The simulated curve obtained utilizing the above values of $E_{1/2}$ (fig.3b) is in agreement with the experimental one.

As to the failure to observe the last reduction step, it is probable that the relative electrode process is masked by the discharge of the background electrolyte. The suggestion to this effect comes from the consideration of the dependence of redox potentials on the number of electrons introduced into the system. Indeed, if the redox potentials for the various steps are plotted against this number the redox potential, expected for the exchange of a hypotetical eighth electron, is more negative than the discharge of the background electrolyte.

The comparison with the behaviour of the bpy ligand and of the $Ru(bpy)_2(CN)_2$ and $Ru(bpy)_3^{2+}$ complexes allows a reasonable identification of the redox sites relative to the various electrons introduced into the molecule.

The first set of peaks can be attributed to the introduction of electrons into redox orbitals localized on each bpy ligand. The second set of peaks is consistent with an energy pairing of about 0.5 V, as estimated from the electrochemistry of the free ligand and of the mononuclear complexes.

Considering the higher electronic density at the N-bonded Ru(II) centre, with respect to the C-bonded moiety, the first electron is assigned to the bpy bound to the C-bonded moiety. The second bpy to be reduced belongs to N-bonded $Ru(bpy)_2CN$ unit. This ensues from the consideration that the $E_{1/2}$ values of the first two reduction steps, separated by 60 mV, should correspond to reduction of two bpy ligands of different metal centres (for the reduction of two bpy ligand of the same metal centre, larger potential differences are found in monuclear complexes). For the other reduction processes similar considerations can be made and the proposed assignment is shown schematically in fig.3a.

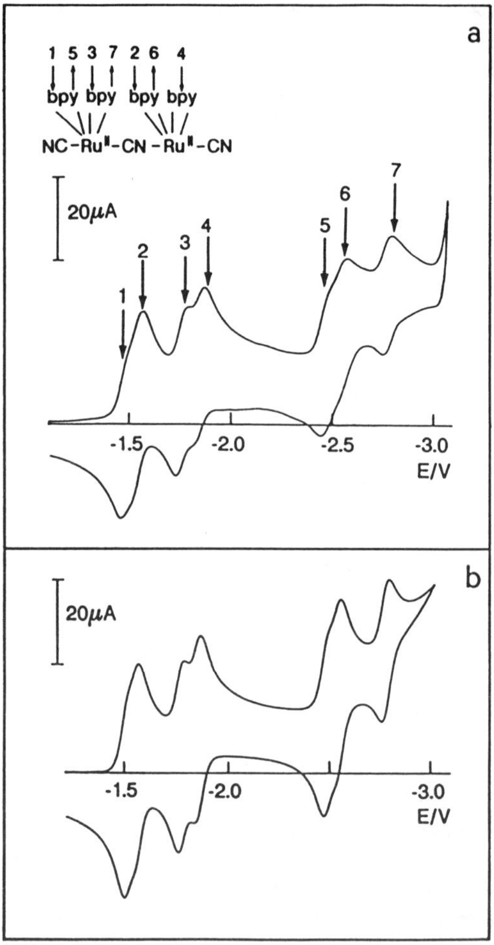

Figure 3.(a) Cyclic voltammetric curve for a $5x10^{-4}$ M DMF solution of the binuclear complex **1** in 0.1 M $(C_2H_5)_4NBF_4$. Working electrode:Pt.T=-54°C.Sweep rate: 50 V/s.The arrows refer to the identification of the redox sites relative to the various electrons introduced into the molecule.(b) Simulated curve for the system and conditions reported in (a).

2.2.2. *Trinuclear complexes 2 and 3.* Fig. 4a shows the low temperature cvc of the trinuclear complex 2. The peaks observed can be grouped in two sets separated by about 0.5 V. This separation, on the basis of the previous discussion, suggests that the first set corresponds to the introduction of one electron into a redox orbital localized on each bpy, while the second set involves electrons which pair with those of the preceding set. For the first set of peaks, it was concluded that the apparently single first and second peaks correspond to the reversible

TABLE 1. Reduction potentials in DMF at - 54 °C vs. SCE*

Species	$E_{1/2}$/V										
	(1)	(2)	(3)	(4)	(5)	(6)	(7)	(8)	(9)	(10)	(11)
bpy	- 2.09	- 2.69									
DCE-bpy	- 1.37	- 1.78									
$Ru(bpy)_2(CN)_2$	- 1.50	- 1.72	- 2.36	- 2.58							
$Ru(bpy)_3^{2+}$	- 1.33	- 1.48	- 1.70	- 2.32	- 2.55	- 2.85					
$Ru(DCE\text{-}bpy)_2(Cl)_2$	- 0.94	- 1.16	- 1.56	- 1.73							
Binuclear 1	- 1.47	- 1.53	- 1.74	- 1.83	- 2.44	- 2.52	- 2.78				
Trinuclear 2	- 1.52	- 1.57	- 1.63	- 1.80	- 1.84	- 2.02	- 2.57	- 2.62	- 2.83	- 2.90	
Trinuclear 3	- 0.98	- 1.15	- 1.55	- 1.73	- 1.78	- 1.83	- 2.08	- 2.13	- 2.75	- 2.79	- 2.99

* Numbers 1-11 refer to $E_{1/2}$ of the different one-electron redox steps.

exchange of three and two electrons respectively, while the third peak involve the reversible transfer of one electron. This interpretation is confirmed by the voltammogram, recorded with a Pt electrode with periodical renewal of the diffusion layer, where the heights of the first three waves are in the ratio 3:2:1.

As far as the second set of peaks is concerned its first member corresponds to two successive one-electron reversible processes, with closely spaced standard potentials.

The more negative peak corresponds to the exchange of two electrons. This interpretation is tentative since, as will be shown later, the curve simulated under the hypothesis that the last two electrons are exchanged reversibly does not superimpose on the experimental one as closely as for the preceding processes.

The values of $E_{1/2}$ for the various steps, deduced by simulation technique [15], are reported in Table 1. The simulated curve, obtained utilizing the above values of $E_{1/2}$,is reported in Fig.4b. As one can see, the agreement, between the experimental and simulated curve, is rather good for the first eight processes, while it is not so satisfactory for the last two.

As to the sequence and localization of the redox sites, in which the electrons are introduced, considering that, in the context of the first three electron transfers, the first two one-electron processes are characterized by very closely spaced half-wave potentials ($\Delta E_{1/2}$= 50mV), the corresponding two electrons should be localized essentially on two slightly interacting chemically equivalent redox sites. These two redox centres should correspond to two bpys bound to the two lateral Ru atoms. The third electron should instead go into a bpy coordinated to the central Ru atom, these bpys being less easily reducible than the preceding ones. As to the couple of electrons, which follow the first three electrons, the corresponding processes being characterized by very closely spaced $E_{1/2}$s, they should go into two chemically equivalent weakly interacting redox sites which can be identified confidently with two bpys coordinated to the side Ru atoms. This assignment is also substantiated by the fact that $\Delta E_{1/2}$ for the first and fourth processes (0.28 V) and the second and the fifth ones (0.27 V) practically coincide. Furthermore, the difference in $E_{1/2}$ for the peaks of these two doublets, each of them corresponding to the successive introduction of two electrons in two bpys coordinated to the

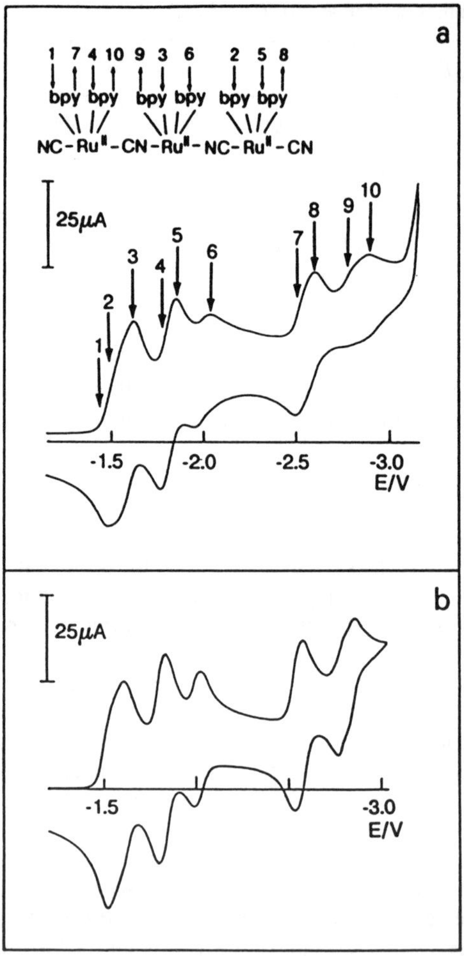

Figure 4.(a) Cyclic voltammetric curve for a $5x10^{-4}$ M DMF solution of the trinuclear complex **2** in 0.1 M $(C_2H_5)_4NBF_4$. Working electrode:Pt.T=-54°C.Sweep rate: 10 V/s.The arrows refer to the identification of the redox sites relative to the various electrons introduced into the molecule.(b) Simulated curve for the system and conditions reported in (a).

same Ru atom, compares favourably with the difference in $E_{1/2}$ for the peaks of the two doublets for the binuclear species (0.27 and 0.30 V). The sixth process observed should correspond to the introduction of one electron into the second bpy bound to the central atom.

For the second part of the cvc, the two apparently single peaks with shoulders, correspond to the successive exchange of two couples of electrons which pair with those of the preceding set.

As to the first two electrons of the second set, since the difference in $E_{1/2}$ of the corresponding processes is about 50 mV, they should enter into weakly interacting, chemically equivalent redox sites which can be identified with two bpys coordinated to the two ruthenium side atoms. As to the last two electrons, the first of these should enter into a bpy coordinated to the central Ru atom, while the second one should enter into a bpy coordinated to the Ru side atom. The assignment for these last two electrons is, however, tentative, since the nature of the corresponding electrode processes is not as clear cut as the preceding ones.

The localization of the redox sites for the various electrons introduced into the molecule is illustrated in Fig.4a.

The electrochemical behaviour of compound **3** allows to confirm the preceding assignment. Fig.5a shows the low temperature cvc for complex **3**. The first three peaks correspond to three one-electron reversible processes, the fourth peak to three one-electron processes with closely spaced standard potentials, both fifth and sixth peak to two one-electron reversible processes with very closely standard potentials, and finally, the last peak to a one-electron reversible process. On the overall, the total number of electrons exchanged is eleven.

A confirmation of the number of electrons exchanged in the first five peaks comes from the voltammogram, recorded with a Pt electrode with periodical renewal of the diffusion layer, where the heights of the first five waves are in the ratio 1:1:1:3:2. Such coincidence is not verified for the part of the cvc following the fifth peak due to the chemical instability of the complex at very negative potentials. This does not happen in CV due to the shorter time of the experiment.

The values of $E_{1/2}$ for the various steps are reported in Tab.1. The curve simulated utilizing the above values of $E_{1/2}$ is in agreement with the experimental one (Fig.5b).

In view of the presence of the electron-withdrawing -COOEt groups, which causes the reduction of the DCE-bpy ligand to occur at potentials circa 0.8 V more positive than those corresponding to bpy, one expects DCE-bpy's in species **3** to be reduced first. As a consequence, the redox processes 3, 6 and 9 of the central unit of complex **2** should be shifted in complex **3** towards more positive potentials. Besides, a fourth peak, unlike in **2**, corresponding to the last electron introduced into the DCE-bpy's bonded to the central Ru atom, should be observed. In particular, considering a spin pairing energy for the DCE-bpy ligand of 0.4 V, as deduced by the potential difference between the first and the second peak of the free ligand and between the second and third peak of the mononuclear $Ru(DCE\text{-}bpy)_2Cl_2$ complex (Tab.1), the cvc curve for complex **3** is expected to exhibit, in the less negative region, a system of two doublets, separated by a potential difference of about 0.4 V. The rest of the cvc, corresponding to the reduction of the bpys, should also exhibit a pattern with two doublets, corresponding to the processes indicated by the numbers 1, 2 and 4, 5 respectively. At a distance of about 0.6 V (the potential difference corresponding to the spin pairing energy for the bpy ligand) these processes should be followed by one doublet and one singlet corresponding to the electrons 7 , 8 and 10 respectively.

The results obtained confirm the pattern expected and support the previous assignment for the localization of the redox sites in the trinuclear species **2**.

The identification of the redox sites relative to the various electrons, introduced into the species **3**, is shown schematically in Fig.5a.

As to the failure to observe the last reduction peak, which would make the ligand based redox series complete, the considerations made previously, with regard to species **2**, are still valid, that is the redox potential, expected for the last electron, is more negative than the discharge of the background electrolyte.

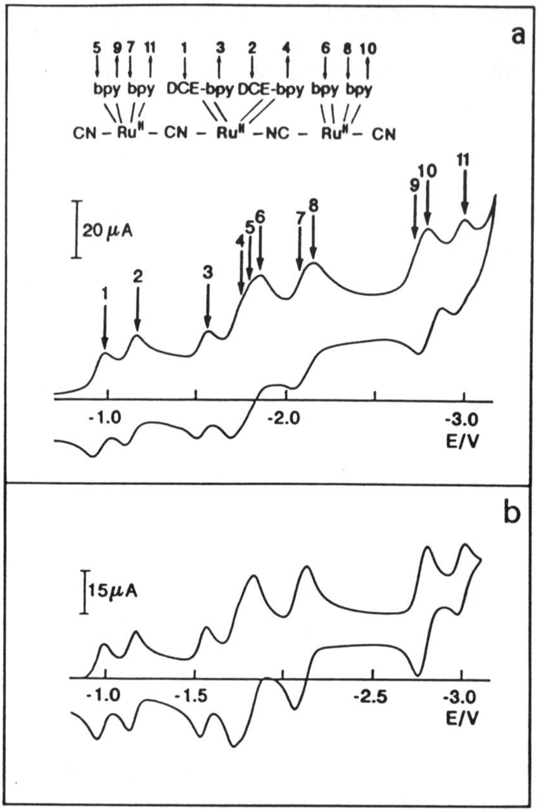

Figure 5.(a) Cyclic voltammetric curve for a $5x10^{-4}$ M DMF solution of the trinuclear complex **3** in 0.1 M $(C_2H_5)_4NBF_4$. Working electrode:Pt.T=-54°C.Sweep rate: 10 V/s.The arrows refer to the identification of the redox sites relative to the various electrons introduced into the molecule.(b) Simulated curve for the system and conditions reported in (a).

3. Conclusions

The polynuclear complexes, studied in this work, give rise to redox series which are typical of systems with multiple interacting redox centres.

The redox series are formed by nine, thirteen and fourteen redox steps for the binuclear complex **1** and for trinuclear complexes **2** and **3** respectively.

The oxidation processes can be considered essentially localized on Ru(II) centres, while reduction processes are largely localized on the bipyridine ligands.

Comparison between the behaviour of the polynuclear complexes and $Ru(bpy)_2(CN)_2$, bpy, DCE-bpy and $Ru(bpy)_3^{2+}$, has allowed a satisfactory assignment for the localization of the electrons removed or introduced into the molecules. Furthermore, information hasbeen obtained on how the assembly of single molecular components in a supramolecular structure affects their redox properties.

For these systems, rather strong metal-metal interactions via CN bridges appear to be operative, as deduced from the metal localized redox series, while, remote ligand-ligand interactions (e.g. L-M-CN-M-L) are weaker, as shown by the study of the characteristics of ligand based redox series.

Furthermore, the behaviour of these complexes, which can be considered as made up of identical or similar building blocks connected by asymmetric bridges, appears to be influenced deeply by the asymmetry of the bridge. As a matter of fact, the more easily oxidizable Ru(II) centres in the polynuclear complexes are those that contain the greater number of N-bonded cyanides.

The extension of the redox series both for the binuclear and trinuclear complexes, does not reach, under the conditions utilized, the maximum number of redox steps predicted, probably due to the high negative potentials expected for the last steps. Although one or two steps of the ligand based redox series are missing, the redox series, reported in this work, are among the most extended ever observed.

From the whole of the electrochemical, spectroscopic and photophysical information, achieved from these and previous studies, it has been possible to designed an antenna sensitizer trinuclear complex NC-$Ru^{II}(bpy)_2$-CN-$Ru^{II}(bpy(COO)_2)_2$-NC-$Ru^{II}(bpy)_2$-CN^{2-}, which has been shown to perform as an efficient molecular device for the sensitization of semiconductor electrodes in the visible region [16-19].

4. Acknowledgments

Financial support by Ministero della Università e della Ricerca Scientifica e Tecnologica and by Consiglio Nazionale delle Ricerche is gratefully acknowledged.

5. References

1. V.Balzani and F.Scandola, Supramolecular Photochemistry, Ellis Horwood, London, 1990
2. V.Balzani (Ed.), Supramolecular Photochemistry, NATO ASI Series, Reidel, Dordrecht, 1987.
3. F.Scandola, M.T.Indelli, C.Chiorboli and C.A.Bignozzi, Top.Curr.Chem., (1990) 158.
4. C.A.Bignozzi, S.Roffia and F.Scandola, J.Am.Chem.Soc., (1985) **107**, 1644.
5. S.Roffia, C.Paradisi and C.A.Bignozzi, J.Electroanal.Chem., (1986) **100**, 105.
6. C.A.Bignozzi, C.Paradisi, S.Roffia and F.Scandola, Inorg.Chem., (1988) **27**, 408.
7. C.A.Bignozzi, S.Roffia, C.Chiorboli, S.Davila, M.T.Indelli and F.Scandola, Inorg.Chem., (1989) **28**, 4350.
8. S.Roffia, R.Casadei, F.Paolucci, C.Paradisi, C.A.Bignozzi and F. Scandola, J.Electroanal. Chem., (1991) **302**, 157.
9. M.G.Teixeira, S.Roffia, C.A.Bignozzi, C.Paradisi and F.Paolucci, J. Electroanal. Chem., submitted.
10. A.A.Vlcek, Coord. Chem. Rev.,(1982) **43**, 39; Rev.Chim.Min., (1983) **20**, 612.

11. G.M.Bryant, S.E.Ferguson,and H.K.S.Powell, Aust.J.Chem., (1971) **24**, 257.
12. G.M.Bryant and S.E.Ferguson, Aust.J.Chem., (1971) **24**, 275.
13. C.A.Bignozzi, R.Argazzi, J.R.Schoonover, K.C.Gordon, R.B.Dyer and F.Scandola, Inorg. Chem., submitted.
14. F.Ammar and J.M.Savéant, J.Electroanal.Chem., (1973) **47**, 215.
15. K.Niki, J.Kobayashi and H.Matsuda, J.Electroanal.Chem., (1984) **178**, 333.
16. R.Amadelli, R.Argazzi, C.A.Bignozzi and F.Scandola, J.Am.Chem.Soc., (1990) **112**, 7099.
17. M.K.Nazeeruddin, P.Liska, S.Moser, N.Vlachopoulos and M.Graetzel, Helv.Chim.Acta., (1990) **73**, 1788.
18. B.O'Regan and M.Graetzel, Nature., (1991) **353**, 737.
19. T.E.Mallouk, Nature., (1991) **353**, 698.

THE REDOX PROPENSITY OF HIGH NUCLEARITY METAL CLUSTERS

P.ZANELLO
*Dipartimento di Chimica dell'Università di Siena,
Piano dei Mantellini, 44, 53100 Siena, Italy*

ABSTRACT. The electrochemical behaviour of a number of metal-sulfur and metal-carbonyl clusters of increasing nuclearity is presented. As much as possible, the structural consequences of the electron addition/removal processes are discussed on the basis of the solid-state X-ray structure of the involved redox couples.

1. Introduction

Metal clusters are commonly thought to be able to act as electron sponges, even if it is often reported that their LUMO level is metal-metal antibonding. In order to clear up such contrasting generalizations, we carried out a systematic survey of the electron-transfer processes exhibited by homometal and heterometal cluster compounds with both sulfur[1,2] and carbonyl[3,4] ligands, giving also evidence to the stereochemical reorganizations accompanying such redox changes.

We present here some illustrative examples of the redox propensity of metal-sulfur and metal-carbonyl clusters, in order of their increasing nuclearity.

2. Results and discussion

2.1. *The redox behaviour of some metal-sulfur clusters*

2.1.1. $[Fe_4(\mu_3\text{-}S)_3(NO)_7]^-$

The monoanion $[Fe_4S_3(NO)_7]^-$ was prepared by Roussin in 1858 and its X-ray structure has been solved about twenty years ago.[5,6] As illustrated in Figure 1a, it is constituted by a flattened tetrahedron of iron atoms, three faces of which are capped by triply-bridging sulfur atoms. Each one of the three basal irons binds two nitrosyl groups, whereas the apical iron bears only one nitrosyl ligand.

*A. J. L. Pombeiro and J. A. McCleverty (eds.),
Molecular Electrochemistry of Inorganic, Bioinorganic and Organomettallic Compounds*, 229–245.

In confirmation of previous brief reports,[7,8] Figure 1b shows that $[Fe_4S_3(NO)_7]^-$ undergoes three successive, chemically reversible, one-electron reductions (E°′ = -0.68 V, -1.26 V, -1.75 V, *vs.* S.C.E., respectively).

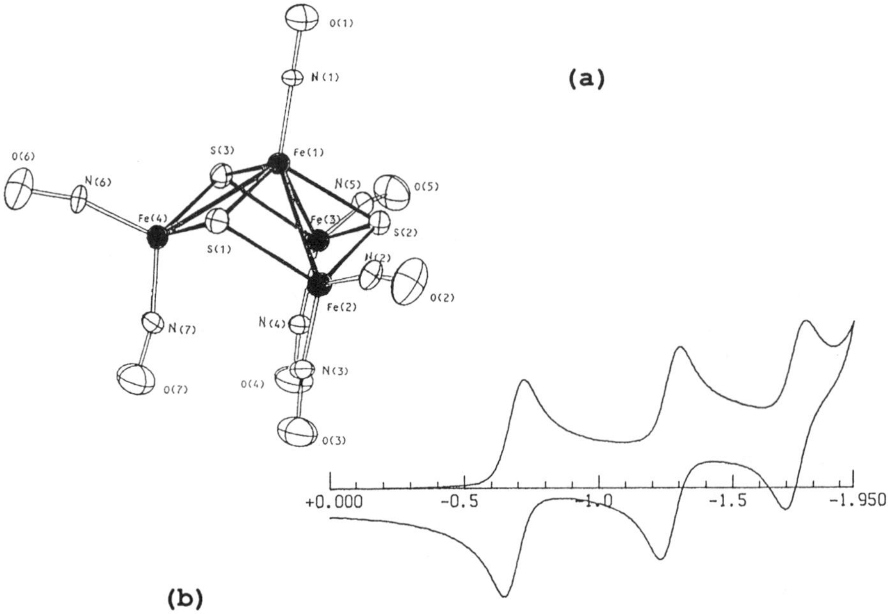

Figure 1. *(a) Perspective view of the monoanion $[Fe_4S_3(NO)_7]^-$. (b) Cyclic voltammetric response recorded at a platinum electrode on a MeCN solution of $[Fe_4S_3(NO)_7]^-$. $[NEt_4][ClO_4]$ (0.2 mol dm^{-3}) supporting electrolyte. Scan rate 0.2 Vs^{-1}.*

The peak-to-peak separation relative to each electron transfer, approaching the theoretical value of 59 mV expected for an electrochemically reversible one-electron step, allows one to foresee that such electron additions do not cause large structural reorganizations with respect to the starting geometry. Really, the crystal structure of the dianion $[Fe_4(\mu_3\text{-}S)_3(NO)_7]^{2-}$ shows that the main rearrangements are simply constituted by an increase of both the Fe_{apical}-Fe_{basal} (mean ≈ 0.06 Å) and the Fe_{apical}-S distances (mean ≈ 0.02 Å), as well as a decrease of the Fe-N distances (mean ≈ 0.02 Å).[9]

2.1.2. $[M_6(\mu_3\text{-}S)_8(PEt_3)_6]^{n+}$ (M = Fe, n = 2; M = Co, n = 1)

The two isostructural species $[Fe_6(\mu_3\text{-}S)_8(PEt_3)_6]^{2+}$ and $[Co_6(\mu_3\text{-}S)_8(PEt_3)_6]^{+}$ are constituted by an octahedron of metal atoms, each face of which is symmetrically capped by one triply-bridging sulfur ligand. In addition, each metal atom is linked to a triethylfosphine group.

The high redox ability of these complexes is illustrated in Figure 2.[10, 11]

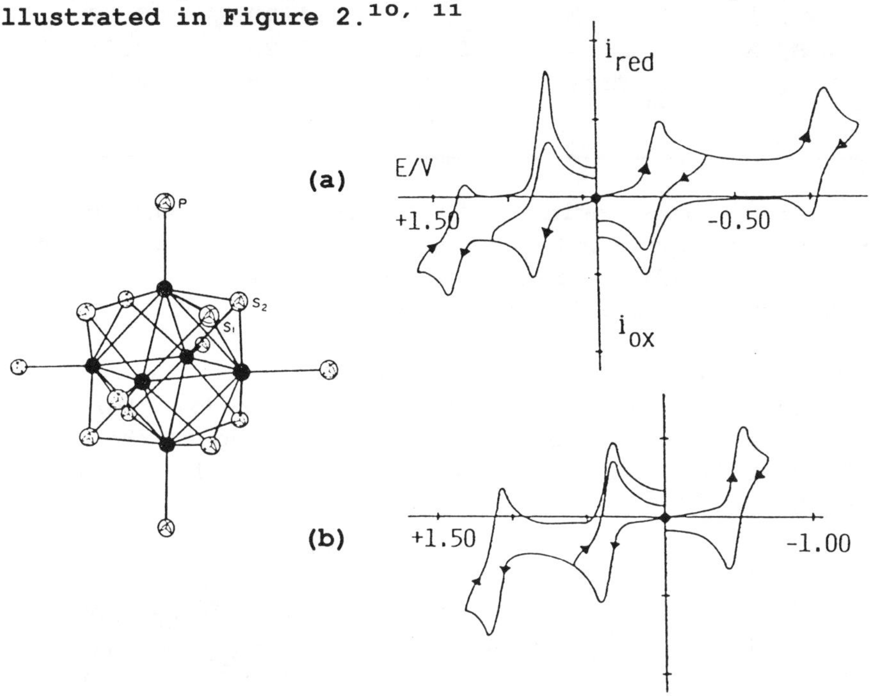

Figure 2. *Cyclic voltammograms recorded at a platinum electrode on CH_2Cl_2 solutions of: (a) $[Fe_6S_8(PEt_3)_6]^{2+}$; (b) $[Co_6S_8(PEt_3)_6]^{+}$. $[NBu_4][ClO_4]$ (0.1 mol dm^{-3}) supporting electrolyte. Scan rate 0.2 Vs^{-1}.*

Also in this case, the peak-to-peak separation of the different one-electron transfers, approaching the theoretical value of 59 mV, preludes to a substantial maintaining of the starting molecular frameworks. As a matter of fact, the redox congeners we succeeded in crystallizing showed to be isostructural with the parent precursors. Table 1 summarizes the relevant minimal structural rearrangements.

Table 1. Main structural reorganizations observed in the two redox couples: $[Fe_6S_8(PEt_3)_6]^{2+/+}$, $[Co_6S_8(PEt_3)_6]^{+/0}$

structural parameters	$[Fe_6S_8]^{2+}$	$[Fe_6S_8]^{+}$	$[Co_6S_8]^{+}$	$[Co_6S_8]^{0}$
M-M, Å	2.62	2.64	2.79	2.82
M-S, Å	2.25	2.25	2.23	2.23
M-P, Å	2.29	2.26	2.16	2.14
M-S-M, deg	71.2	71.7	77.4	78.2

It results evident that the LUMO level of both $[Fe_6S_8(PEt_3)_6]^{2+}$ and $[Co_6S_8(PEt_3)_6]^{+}$ is weakly metal-metal antibonding, metal-sulfur nonbonding, and metal-phosphorus bonding in character.

2.1.3. $[Ni_9(\mu_4\text{-}S)_3(\mu_3\text{-}S)_6(PEt_3)_6]^{2+}$

In confirmation of the remarkable redox flexibility of metal-sulfur clusters, Figure 3 shows the cyclic voltammetric response exhibited by $[Ni_9(\mu_4\text{-}S)_3(\mu_3\text{-}S)_6(PEt_3)_6]^{2+}$.[1]

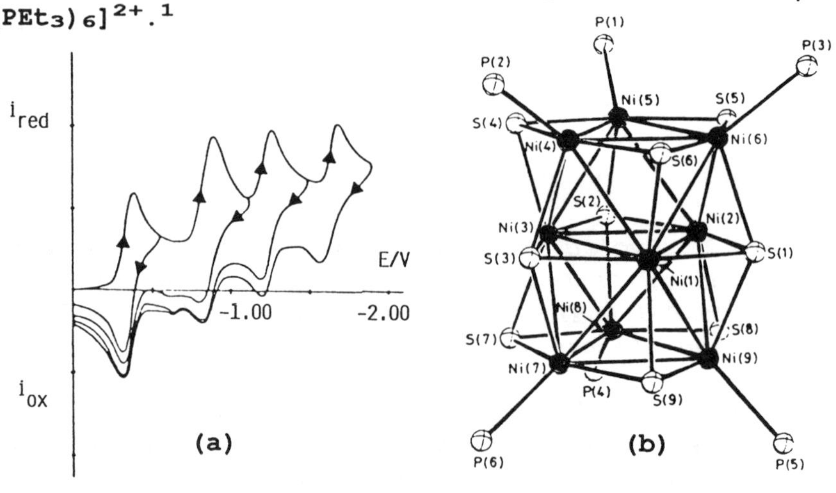

Figure 3. *(a) Cyclic voltammogram recorded at a platinum electrode on a MeCN solution of $[Ni_9S_9(PEt_3)_6]^{2+}$. $[NEt_4][ClO_4]$ (0.1 mol dm^{-3}) supporting electrolyte. Scan rate 0.2 Vs^{-1}. (b) Perspective view of $[Ni_9S_9(PEt_3)_6]^{2+}$.*

Even if we did not yet succeed in obtaining crystals of the one-electron reduced species, which proved to be the only stable congener in the longer times of macroelectrolysis, it seems not ventured to predict, on the basis of the substantial electrochemical reversibility of the first cathodic step, that the monocation $[Ni_9S_9(PEt_3)_6]^+$ maintains the metallic cofacial bioctahedral geometry of the corresponding dication.

2.2. *The redox behaviour of some metal-carbonyl clusters*

2.2.1. $[M_6(CO)_{15}]^{2-}$ (M = Co, Ir)

As illustrated in Figure 4, both the octahedral dianions $[Co_6(CO)_{15}]^{2-}$ and $[Ir_6(CO)_{15}]^{2-}$ undergo two subsequent one-electron oxidations.[12]

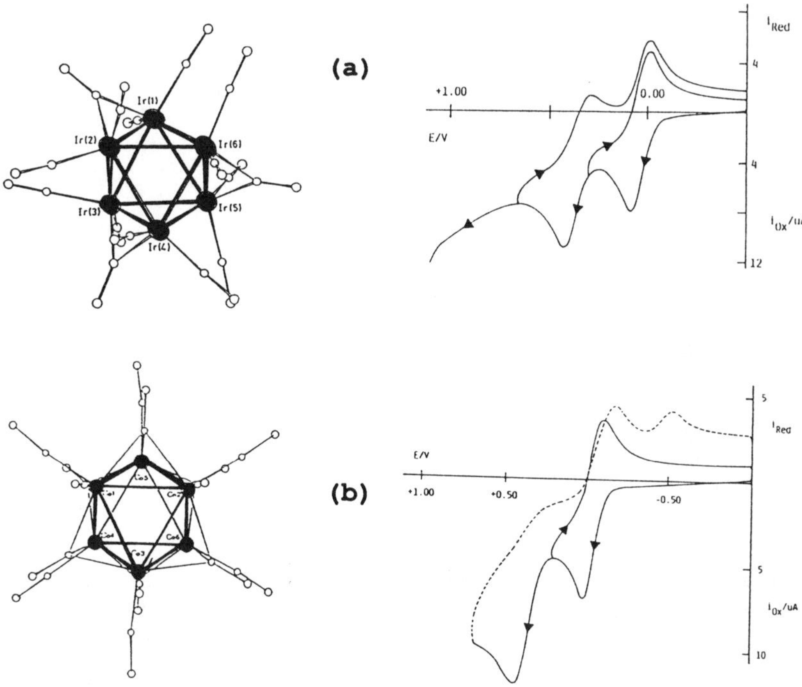

Figure 4. *Cyclic voltammograms recorded at a platinum electrode on THF solutions of: (a)* $[Ir_6(CO)_{15}][N(PPh_3)_2]_2$*; (b)* $[Co_6(CO)_{15}][NEt_4]_2$*.* $[NBu_4][ClO_4]$ *(0.1 mol* dm^{-3}*) supporting electrolyte. Scan rate 0.2* Vs^{-1}.

In both cases, the monoanion congener results long-lived. In contrast, the neutral $Ir_6(CO)_{15}$ is short-lived, whereas $Co_6(CO)_{15}$ is quite unstable.

An interesting aspect of the reactivity of the $[Ir_6(CO)_{15}]^{n-}$ congeners is deducible from Figure 5, which compares the redox pattern of $[Ir_6(CO)_{15}]^{2-}$ in the presence of chloride ions (a, b) with that of $[Ir_6(CO)_{15}Cl]^-$ (c).

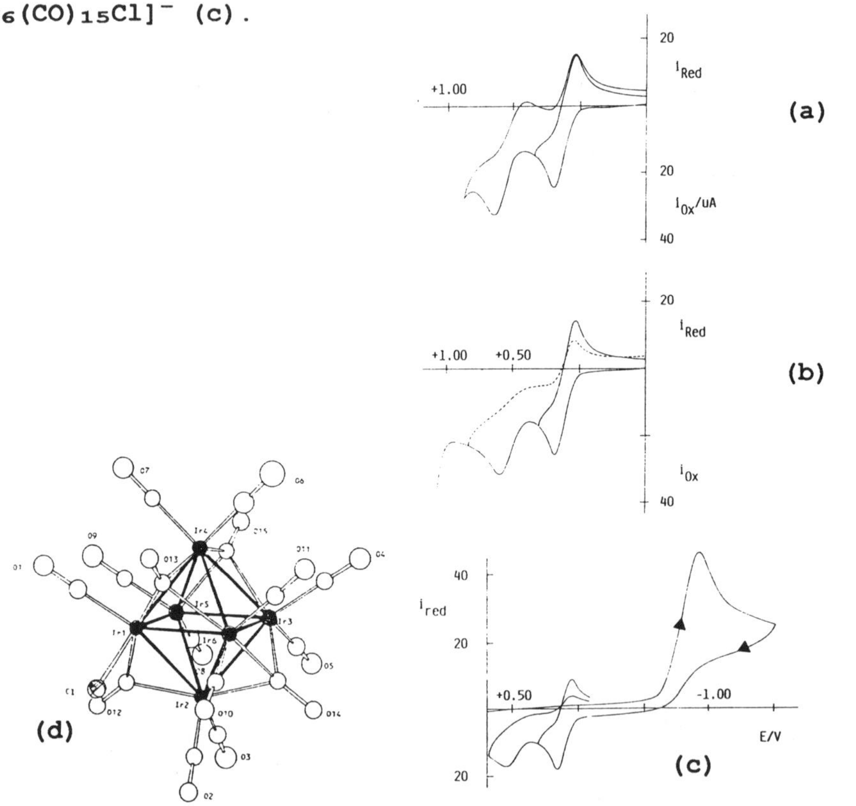

Figure 5. *Cyclic voltammograms recorded at a platinum electrode on THF solutions containing: (a) $[Ir_6(CO)_{15}]^{2-}$; (b) after the addition of an excess of Cl^- ions; (c) $[Ir_6(CO)_{15}Cl][NEt_4]$. $[NBu_4][ClO_4]$ (0.1 mol dm^{-3}) supporting electrolyte. Scan rate 0.2 Vs^{-1}. (d) Perspective view of $[Ir_6(CO)_{15}Cl]^-$.*

The lack of the backward response associated to the second anodic step in Figure 5b, as well as the appearance of the redox fingerprint of $[Ir_6(CO)_{15}]^{2-}$ upon two-electron reduction of $[Ir_6(CO)_{15}Cl]^-$ in Figure

5c well support the following redox cycle:

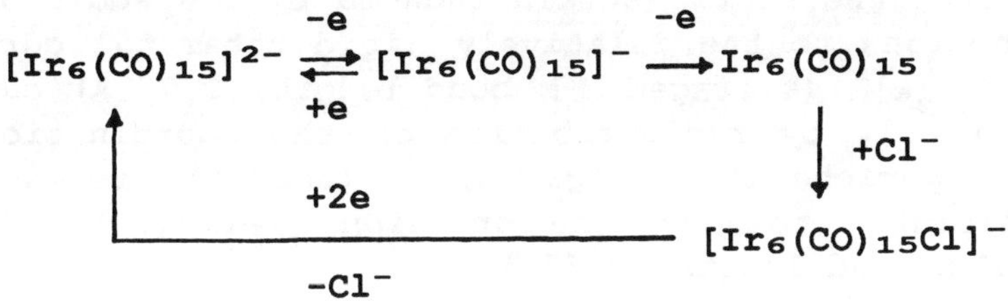

2.2.2. $[FeIr_5(CO)_{16}]^-$/$[FeIr_5(CO)_{15}]^{3-}$

Figure 6 shows the redox patterns exhibited by the two octahedral clusters $[FeIr_5(CO)_{15}]^{3-}$ and $[FeIr_5(CO)_{16}]^-$, respectively.[13]

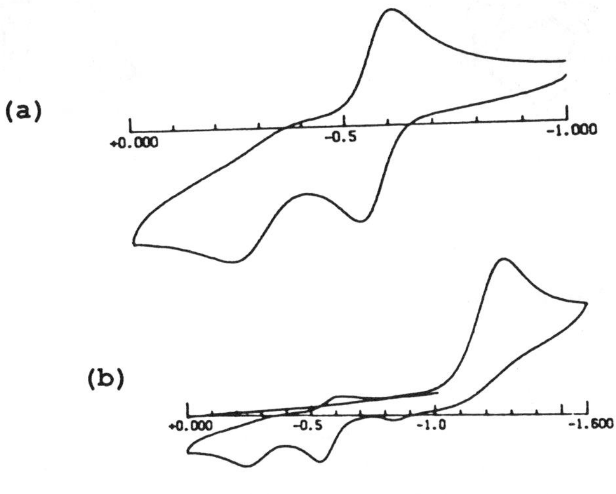

Figure 6. *Cyclic voltammetric responses recorded at a platinum electrode on MeCN solutions containing: (a) $[FeIr_5(CO)_{15}][NEt_4]_3$; (b) $[FeIr_5(CO)_{16}][N(PPh_3)_2]$. $[NEt_4][ClO_4]$ (0.2 mol dm^{-3}) supporting electrolyte. Scan rate 0.2 Vs^{-1}.*

It is well evident that the trianion $[FeIr_5(CO)_{15}]^{3-}$ undergoes, reversibly, an easy one-electron removal, whereas the more carbonylated monoanion $[FeIr_5(CO)_{16}]^-$ undergoes a fragmentating two-electron addition. Nevertheless, the relevant backward response reveals that $[FeIr_5(CO)_{16}]^-$, upon reduction, converts to $[FeIr_5(CO)_{15}]^{3-}$ simply by loss of one CO molecule.

From the crystal structures shown in Figure 7, one can infer that the steric strain induced by the addition of two electrons to the relatively rigid octametal core of $[FeIr_5(CO)_{16}]^-$ (averaged M-M bond length, 2.77 Å) can be relieved only by redistribution of the coordination mode of the peripheral CO ligands, included the release of one molecule, to give the trianion $[FeIr_5(CO)_{15}]^{3-}$ (averaged M-M bond length, 2.77 Å).

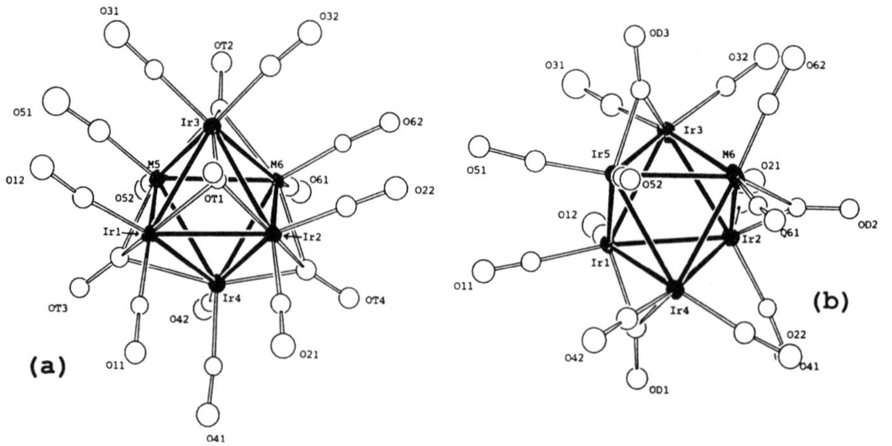

Figure 7. *The molecular structure of: (a) $[FeIr_5(CO)_{16}]^-$; (b) $[FeIr_5(CO)_{15}]^{3-}$.*

2.2.3. $[Fe_3Pt_3(CO)_{15}]^{2-}$

The dianion $[Fe_3Pt_3(CO)_{15}]^{2-}$ is the first structurally characterized cluster of the three-membered family $[Fe_3Pt_3(CO)_{15}]^{2-/-/0}$.[14,15]

As schematized in Figure 8a, the dianion is composed by a two-dimensional metallic assembly in which an equilateral Pt_3 triangle is inscribed into an equilateral Fe_3 triangle. Figure 8b shows that $[Fe_3Pt_3(CO)_{15}]^{2-}$ undergoes two subsequent one-electron removals, chemically and electrochemically reversible in character.[16]

In agreement with the electrochemical reversibility, which suggests the lack of important structural changes accompanying such oxidation steps, Table 2 shows that the successive electron removals, while maintaining intact the primary raft-like geometry, simply strenghten the Pt-Pt bonds. These results put in evidence that the HOMO level of the dianion is predomi-

nantly antibonding among the three Pt atoms, as well as substantially nonbonding with respect to the Fe-Pt interactions.

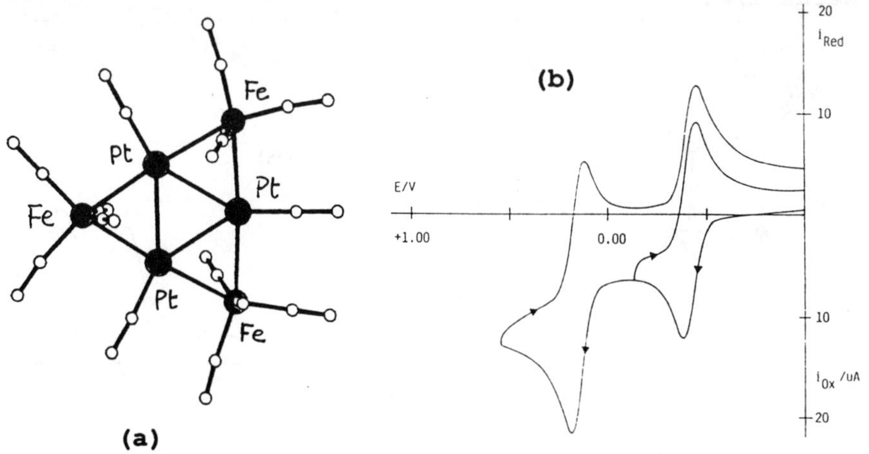

Figure 8. *(a) Perspective view of the dianion $[Fe_3Pt_3(CO)_{15}]^{2-}$. (b) Cyclic voltammogram recorded at a platinum electrode on a CH_2Cl_2 solution of $[Fe_3Pt_3(CO)_{15}]^{2-}$. $[NBu_4][ClO_4]$ (0.1 mol dm^{-3}) supporting electrolyte. Scan rate 0.2 Vs^{-1}.*

Table 2. Main structural changes observed in the series $[Fe_3Pt_3(CO)_{15}]^{2-/-/0}$

bond lengths (in Å)	$[Fe_3Pt_3]^{2-}$	$[Fe_3Pt_3]^{-}$	$[Fe_3Pt_3]^{0}$
Pt-Pt	2.75	2.66	2.59
Pt-Fe	2.60	2.59	2.58

2.2.4. $[Co_6Ni_2(CO)_{16}(C)_2]^{2-}$

As illustrated in Figure 9a, the octametal core of the dianion $[Co_6Ni_2(CO)_{16}(C)_2]^{2-}$ can be described as deriving from two trigonal prisms fused along a square face.[17] Figure 9b shows that such a dianion is the only stable member of its potential redox family, in that it

undergoes two separated one-electron reduction steps, both complicated by decomposition reactions, as well as an oxidation step, which involves an ECECE type mechanism.

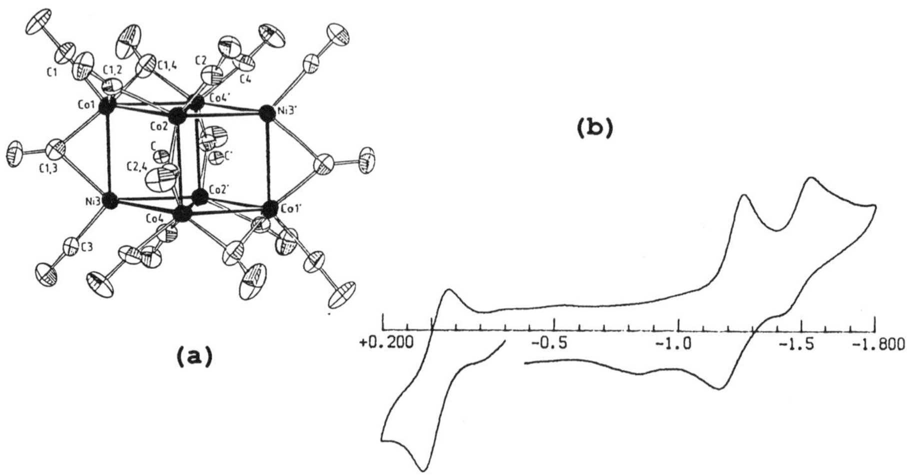

Figure 9. *(a) Perspective view of* $[Co_6Ni_2(CO)_{16}(C)_2]^{2-}$. *(b) Cyclic voltammetric response recorded at a platinum electrode on a THF solution of* $[Co_6Ni_2(CO)_{16}(C)_2]^{2-}$. $[NBu_4][ClO_4]$ *(0.2 mol* dm^{-3}*) supporting electrolyte. Scan rate 0.2* Vs^{-1}.

2.2.5. $[Pd_6Ru_6(CO)_{24}]^{2-}$

The molecular structure of the dianion $[Pd_6Ru_6(CO)_{24}]^{2-}$ is shown in Figure 10a.[18] It is composed by a trigonal antiprismatic assembly of Pd atoms, the lateral faces of which are capped by $Ru(CO)_4$ groups.

Figure 10. *X-ray structure of* $[Pd_6Ru_6(CO)_{24}]^{2-}$.

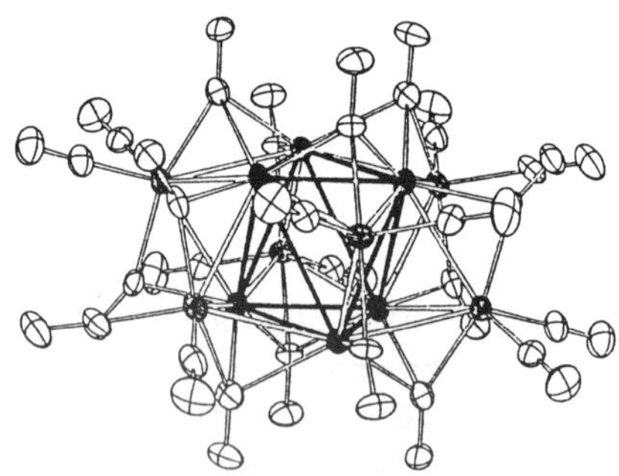

As illustrated in Figure 11, the dianion undergoes both a chemically reversible one-electron oxidation and a single-stepped two-electron reduction, complicated by slow decomposition of the primarily electrogenerated tetranion.

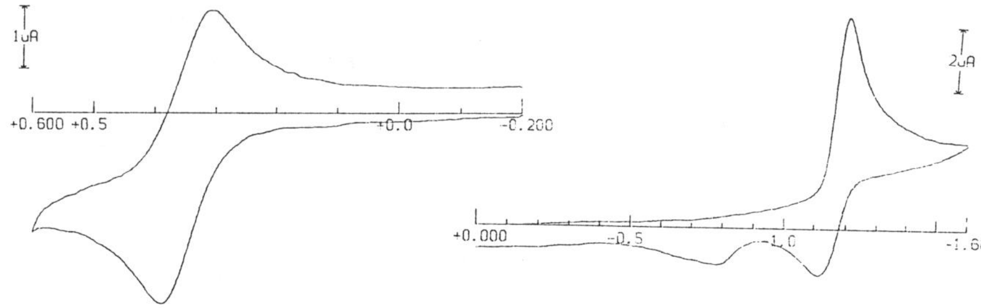

Figure 11. *Cyclic voltammograms recorded at a platinum electrode on a THF solution of $[Pd_6Ru_6(CO)_{24}][NEt_4]_2$. $[NBu_4][ClO_4]$ (0.2 mol dm^{-3}) supporting electrolyte. Scan rate 0.05 Vs^{-1}.*

Even if we have not yet characterized by X-ray diffraction the monoanion $[Pd_6Ru_6(CO)_{24}]^-$, it is not ventured to presume that, based on the substantial electrochemical reversibility of the $[Pd_6Ru_6(CO)_{24}]^{2-/-}$ electron transfer, it likely maintains the geometry of the dianion precursor.

2.2.6. $[Co_{13}(CO)_{24}(C)_2]^{4-}$

The metal core of the tetranion $[Co_{13}(CO)_{24}(C)_2]^{4-}$ can be described as two, mutually rotated, prismatic Co_6C units sharing one common vertex (Co1), capped, at opposite sides, by two five-bridging cobalt atoms (Co12, Co13), Figure 12.[19]

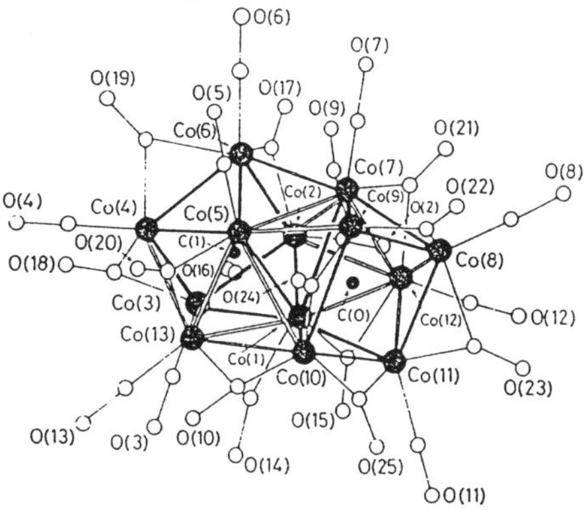

Figure 12. *Perspective view of $[Co_{13}(CO)_{24}(C)_2]^{4-}$.*

As illustrated in Figure 13, it undergoes, reversibly, both two subsequent one-electron reductions and a one-electron oxidation.[3]

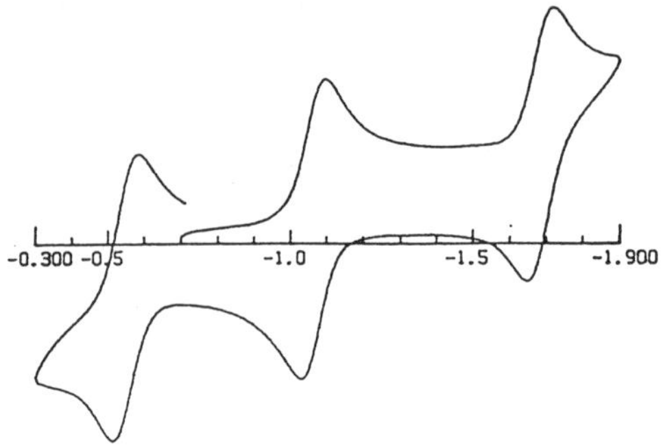

Figure 13. *Cyclic voltammogram recorded at a platinum electrode on a MeCN solution of* $[Co_{13}(CO)_{24}(C)_2]^{4-}$. $[NEt_4][ClO_4]$ *(0.1 mol* dm^{-3}*) supporting electrolyte. Scan rate 0.2* Vs^{-1}.

The inference that such electrochemically reversible redox changes do not induce significant structural rearrangements in the starting framework is supported by the crystal structure of the corresponding trianion $[Co_{13}(CO)_{24}(C)_2]^{3-}$.[20] It resulted quite isostructural with the parent tetranion. The averaged Co-Co distance remains unaltered (2.57 Å), even if a slight shortening of the two longest Co1-Co5 and Co1-Co7 bonds (≈ 0.02 Å) as well as of the mean $Co\text{-}C_{carbide}$ distance (≈ 0.01 Å) occurs.

2.2.7. $[Ni_{11}Bi_2(CO)_{18}]^{3-}$

Figure 14 shows the molecular structure of the trianion $[Ni_{10}Bi_2(\mu_{12}\text{-}Ni)(CO)_{18}]^{3-}$.[21] It is constituted by a Ni-centred pentagonal antiprism of Ni atoms, capped, at opposite sides, by two Bi atoms.

As illustrated in Figure 15, $[Ni_{11}Bi_2(CO)_{18}]^{3-}$ undergoes both a one-electron oxidation and a one-electron reduction. Even if we were unable to obtain crystals of the relevant redox congeners, the electrochemical reversibility of such electron transfers strongly preludes to the maintaining of the starting molecular framework.

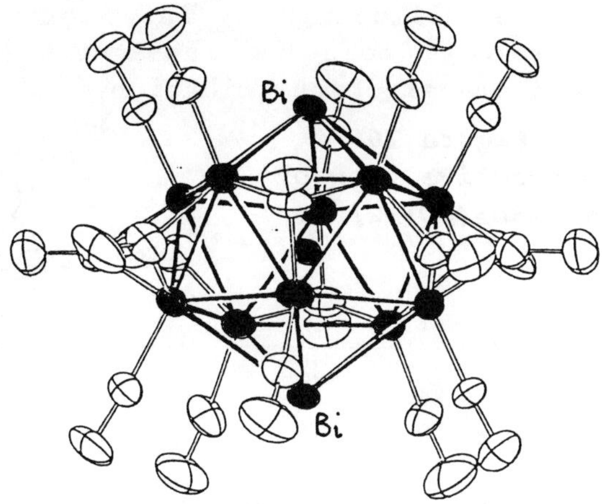

Figure 14. *Perspective view of $[Ni_{11}Bi_2(CO)_{18}]^{3-}$.*

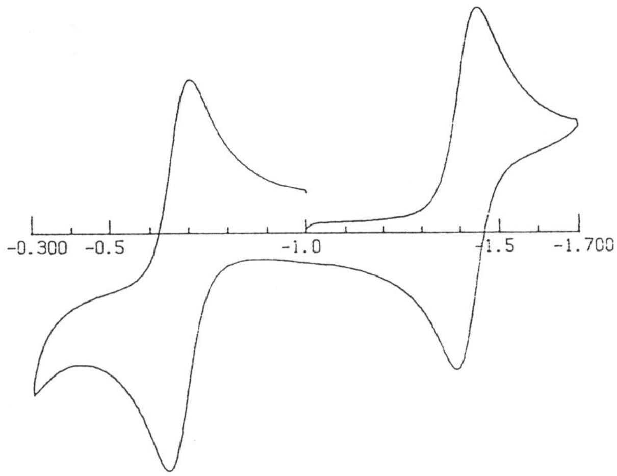

Figure 15. *Cyclic voltammogram recorded at a platinum electrode on a MeCN solution of $[Ni_{11}Bi_2(CO)_{18}]^{3-}$. $[NEt_4][ClO_4]$ (0.2 mol dm^{-3}) supporting electrolyte. Scan rate 0.2 Vs^{-1}.*

2.2.8. **$[Ni_{13}Sb_2(CO)_{24}]^{3-}$**

As shown in Figure 16, the structure of the trianion $[Ni_{10}Sb_2(\mu_{12}\text{-}Ni)\{Ni(CO)_3\}_2(CO)_{18}]^{3-}$ parallels that of the preceding trianion $[Ni_{10}Bi_2(\mu_{12}\text{-}Ni)(CO)_{18}]^{3-}$. It is in fact constituted by a similar Ni-centred icosahedral assembly of metal atoms, bearing, unexpectedly, two dangling $Ni(CO)_3$ fragments.[22]

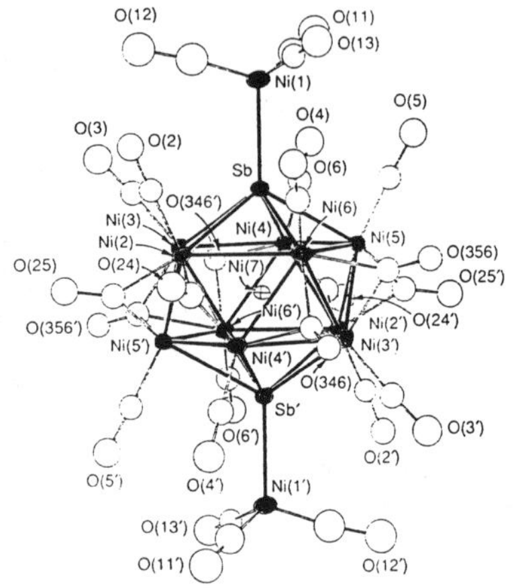

Figure 16. ***Perspective view of $[Ni_{13}Sb_2(CO)_{24}]^{3-}$.***

The similarity between $[Ni_{11}Bi_2(CO)_{18}]^{3-}$ and $[Ni_{13}Sb_2(CO)_{24}]^{3-}$ is not limited to their structural features, but also encompasses their redox propensity. In fact, as illustrated in Figure 17, $[Ni_{13}Sb_2(CO)_{24}]^{3-}$ undergoes, reversibly, both a one-electron oxidation and a one-electron reduction.

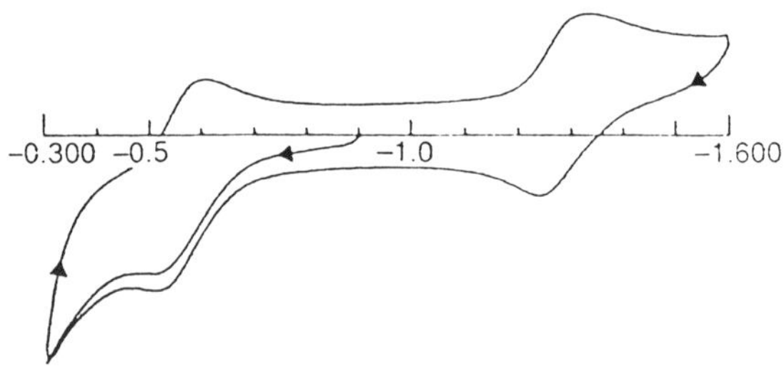

Figure 17. ***Cyclic voltammogram recorded at a platinum electrode on a MeCN solution of $[Ni_{13}Sb_2(CO)_{24}]^{3-}$. $[NEt_4][ClO_4]$ (0.1 mol dm^{-3}) supporting electrolyte. Scan rate 0.2 Vs^{-1}.***

As expected on the basis of the electrochemical reversibility of the $[Ni_{13}Sb_2(CO)_{24}]^{3-/2-}$ electron transfer, the dianion congener resulted isostructural with the trianion, and the minimal structural changes induced by the removal of one electron are summarized in Table 3.

Table 3. Main structural variations observed in the redox couple $[Ni_{13}Sb_2(CO)_{24}]^{3-}/[Ni_{13}Sb_2(CO)_{24}]^{2-}$

structural parameters	bonding distances (in Å) $[Ni_{13}Sb_2]^{3-}$	$[Ni_{13}Sb_2]^{2-}$
$Ni_{interstitial}$-Sb	2.40	2.40
Ni-Ni(intrapentagonal)	2.78	2.77
Ni-Ni(interpentagonal)	2.52	2.51
$Ni_{interstitial}$-$Ni_{pentagonal}$	2.58	2.57

2.2.9. $[Pt_{19}(CO)_{22}]^{4-}$

As a final example of the redox ability of high nuclearity metal clusters, Figure 18a shows the cyclic voltammetric response exhibited by the tetranion $[Pt_{19}(CO)_{22}]^{4-}$.[3]

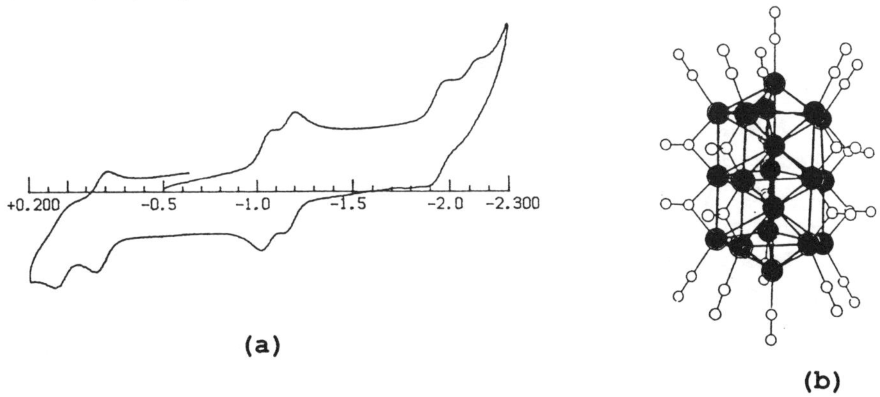

Figure 18. (a) Cyclic voltammogram recorded at a platinum electrode on a MeCN solution of $[Pt_{19}(CO)_{22}]^{4-}$. $[NEt_4][ClO_4]$ (0.1 mol dm^{-3}) supporting electrolyte. Scan rate 0.2 Vs^{-1}. (b) Perspective view of $[Pt_{19}(CO)_{22}]^{4-}$.

This layered pentagonal assembly of Pt atoms[23] is able to undergo two consecutive pairs of closely spaced one-electron reductions, as well as one pair of one-electron oxidations. Except for the two most extreme one-electron steps, all the involved redox changes seem reversible.

3. Conclusion

The few examples of the redox aptitude of metal clusters here reported are sufficient to put in evidence that electrochemistry must become a routine technique for their characterization, just like the different spectroscopic techniques do. As a matter of fact, the electrochemical investigation not only provides with an unequivocal fingerprint of the cluster, but also gives a prompt information about the existence and the relative stability of redox congeners, which otherwise requires laborious chemical manipulations.

4. References

1. P.Zanello, *Coord.Chem.Rev.*, 1988, *83*, 199.
2. P.Zanello, *Coord.Chem.Rev.*, 1988, *87*, 1.
3. P.Zanello, in *Stereochemistry of Organometallic and Inorganic Compounds*, I.Bernal, ed; Elsevier, Amsterdam, Vol. 5, in the press.
4. P.Zanello, *Struct.Bonding* (Berlin), in the press.
5. G.Johansson and W.N.Lipscomb, *Acta Cryst.*, 1958, *11*, 599.
6. C.T.-W. Chu and L.F.Dahl, *Inorg.Chem.*, 1977, *16*, 3245.
7. E.J.Wharton and J.A.McCleverty, *Inorg.Nucl.Chem.Letters*, 1970, *6*, 549.
8. J.A.Crayston, C.Glidewell, and R.J.Lambert, *Polyhedron*, 1990, *9*, 1741.
9. G.Longoni, F.Demartin, and P.Zanello, work in progress.
10. F.Cecconi, C.A.Ghilardi, S.Midollini, A.Orlandini and P.Zanello, *J.Chem.Soc.,Dalton Trans.*, 1987,831.
11. F.Cecconi, C.A.Ghilardi, S.Midollini, A.Orlandini and P.Zanello, *Polyhedron*, 1986, *5*, 2021.
12. A.Cinquantini, P.Zanello, R.Della Pergola, L.Garlaschelli, and S.Martinengo, *J.Organomet.Chem.*, 1991, *412*, 215 .
13. A.Ceriotti, R.Della Pergola, L.Garlaschelli, F.Laschi, M.Manassero, N.Masciocchi, M.Sansoni, and P.Zanello, *Inorg.Chem.*, 1991, *30*, 3349.
14. G.Longoni, M.Manassero, and M.Sansoni, *J.Am.Chem.Soc.*, 1980, *102*, 7974.
15. R.C.Adams, G.Chen., and J.-G.Wang, *Polyhedron*, 1989, *8*, 2521.

16. R.Della Pergola, L.Garlaschelli, C.Mealli, D.M.Proserpio, and P.Zanello, *J.Cluster Science*, 1990, *1*, 93.
17. A.Ceriotti, R.Della Pergola, L.Garlaschelli, G.Longoni, M.Manassero, N.Masciocchi, M.Sansoni, and P.Zanello, *Gazz.Chim.Ital.*, in the press.
18. E.Brivio, A.Ceriotti, R.Della Pergola, L.Garlaschelli, F.Demartin, M.Manassero, and M.Sansoni, *Proceedings of CISCI 91*, p. 682.
19. V.G.Albano, D.Braga, P.Chini, G.Ciani, and S.Martinengo, *J.Chem.Soc.,Dalton Trans.*, 1982, 645.
20. V.G.Albano, D.Braga, A.Fumagalli, and S.Martinengo, *J.Chem.Soc., Dalton Trans.*, 1985, 1137.
21. V.G.Albano, F.Demartin, M.C.Iapalucci, G.Longoni, M.Monari, and P.Zanello, *J.Chem.Soc., Dalton Trans.*, in the press.
22. V.G.Albano, F.Demartin, M.C.Iapalucci, F.Laschi, G.Longoni, A.Sironi, and P.Zanello, *J.Chem.Soc., Dalton Trans.*, 1991, 739.
23. D.M.Washecheck, E.J.Wucherer, L.F.Dahl, A.Ceriotti, G.Longoni, M.Manassero, M.Sansoni, and P.Chini, *J.Am.Chem.Soc.*, 1979, 101, 6110.

REDOX-ACTIVE TETRAPALLADIA UNITS $[Pd_4]^{n+}$ (n = 0, 1, 2, 3, 4) FROM DOUBLE OXIDATIVE CARBON-CARBON COUPLING

P. ESPINET [a] and G. GARCÍA-HERBOSA [b]

a) Departamento de Química Inorgánica. Universidad de Valladolid. 47005 Spain.
b) Departamento de Química Inorgánica. Colegio Universitario. 09002 BURGOS. Spain.

ABSTRACT. Chemical or electrochemical oxidation of the orthometallated amido dimeric complex $[Pd\text{-}o\text{-}C_6H_4\text{-}C(Me)=N\text{-}NPh)L]_2$ (L= $P(OMe)_3$, 1) leads to double carbon-carbon coupling on the *para* carbons of the phenyl groups giving tetrapalladia units $[Pd4]^{n+}$ (n = 0,1,2,3,4) related by fully reversible monoelectronic electron transfers. The species $[Pd_4]^{4+}$ undergoes splitting with Cl^- giving dimeric units $[Pd_2]$ where a 4,4′-biphenylene quinone diimine is trapped and stabilised between two palladia atoms; reduction of $[Pd_2]$ or $[Pd4]^{4+}$ with $CoCp_2$ affords $[Pd_4]^0$.

Introduction

Mixed-valence polynuclear complexes are of current interest[1] as they exhibit unusual structural, bonding, reactivity spectroscopic, and magnetic properties[2] which can find applications in areas such as the study of electron transfer processes[3] and the design of molecular electronics.[4]

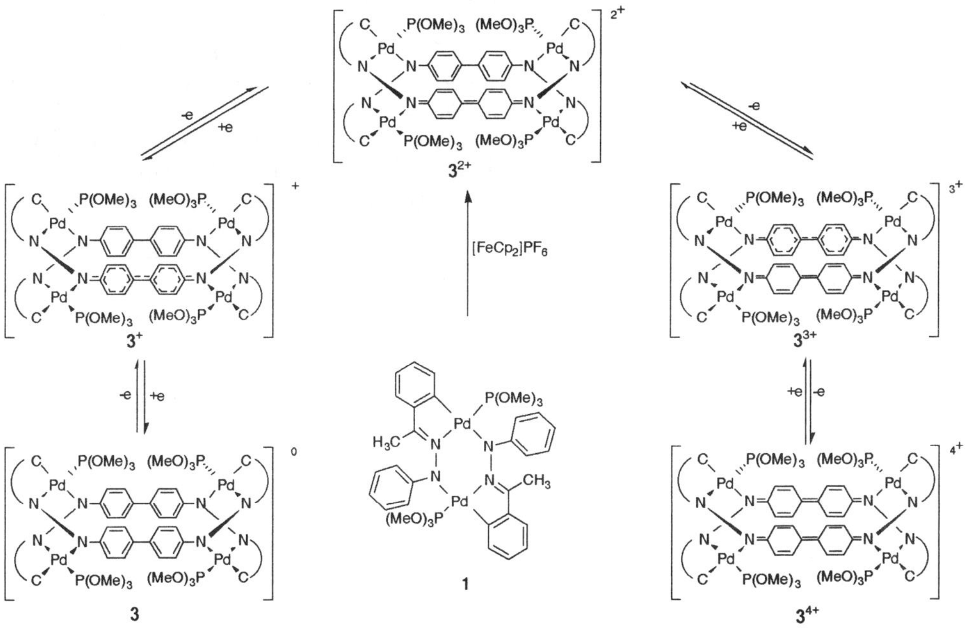

Scheme 1

A. J. L. Pombeiro and J. A. McCleverty (eds.),
Molecular Electrochemistry of Inorganic, Bioinorganic and Organomettallic Compounds, 247–251.

We have previously prepared and described[5] binuclear orthometallated complexes of palladium(II) which could be used as precursors to prepare potentially mixed-valence compounds. The presence of unusual Pd-N amido bonds in these complexes adds further interest based on the reactivity shown by amides of the platinum group metals.[6]

Formation of carbon-carbon bonds is quite common in electron transfer reactions[7-10] sometimes leading to polymers with special electrical properties;[11-13] chemical oxidative C-C coupling in amido complexes of platinum has been described[14,15] and we have shown[16] that the use of electrochemical techniques to characterise the processes allows a better understanding of the nature of the reaction and the products formed. Here we describe the double oxidative coupling of a binuclear orthometallated Pd(II) amido complex leading to a family of tetranuclear compounds (Scheme 1) which are represented in the following as $[Pd_4]^{n+}$, where n= 0,1,2,3,4. The chemical and electrochemical behaviour of these complexes is presented and discussed below.

Results

Chemical oxidation of $[Pd\text{-}o\text{-}C_6H_4\text{-}C(Me){=}N\text{-}NPh)L]_2$ (L= $P(OMe)_3$,**1**) with $[Cp_2Fe]PF_6$ in CH_2Cl_2, molar ratio 1:1, afforded complex $\mathbf{3}^{2+}$ which gives the cyclic voltammogram shown in Figure 1. The RE (rotating electrode) experiment reveals that $\mathbf{3}^{2+}$ undergoes two one-electron oxidations to $\mathbf{3}^{4+}$ through $\mathbf{3}^{3+}$, and two one-electron reductions to **3** through $\mathbf{3}^{+}$ as shown in Scheme 1. All the redox processes are diffusion controlled ($i/v^{1/2}$ constant for scan rates 50 to 500 mVs^{-1}) and fully reversible (i.e. for an oxidation process, $i_{red}/i_{ox} = 1$ and *vice versa*, for all scan rates used).

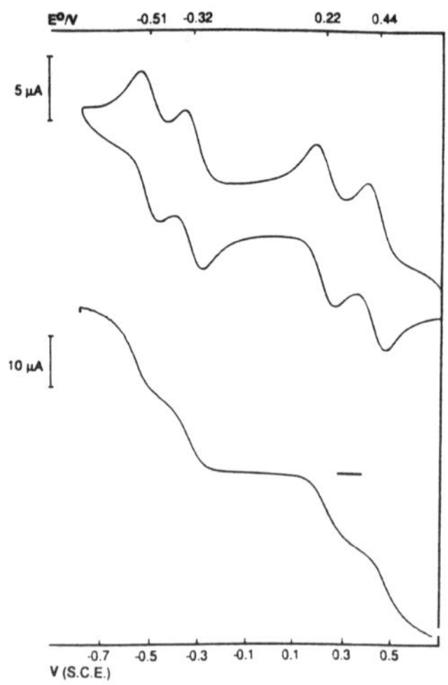

FIGURE 1. Cyclic Voltammetry and Rotating Electrode curves of complex 3^{2+} in CH_2Cl_2, 0.1 M $(TBA)PF_6$ at a Pt-bead electrode. Scan rate (CV) 200 mVs^{-1} and (RE) 20 mVs^{-1}.

CPE (controlled potential electrolysis) experiments on **1** show that the $[Pd_4]^{n+}$ system is produced only if the applied potential is above 0.5 V; in these conditions a solution containing $\mathbf{3}^{4+}$ is produced. This suggests that the formation of $\mathbf{3}^{2+}$ by chemical oxidation is a complex process involving intermediates in a higher oxidation state.

Stoicheiometric oxidation of $\mathbf{3}^{2+}$ with $[Cp_2Fe]PF_6$ in dichloromethane affords $\mathbf{3}^{3+}$ or $\mathbf{3}^{4+}$ respectively. Similarly, stoicheiometric reduction of $\mathbf{3}^{2+}$, $\mathbf{3}^{3+}$ or $\mathbf{3}^{4+}$ with $[CoCp_2]$ in dichloromethane leads to the neutral complex **3**, which can be extracted in toluene. Conproportionation of $\mathbf{3}^{2+}$ and **3** gives $\mathbf{3}^{+}$ but we could not isolate it with satisfactory analytical results.

All attempts to grow suitable crystals of any $\mathbf{3}^{n+}$ for X-ray determination were unsuccessful. Although nice, apparently well shaped, metallic like bright-blue crystals of $\mathbf{3}^{4+}$ were obtained occasionally by diffusion or electrochemical methods,[17] in each occasion diffraction experiments proved them to be compact powders. However, the tetranuclear nature of the complexes is supported by mass spectrometry (FAB) of complex **3** : A peak at m/e = 1750 with the expected pattern for the isotopic abundance of palladium was found. This is

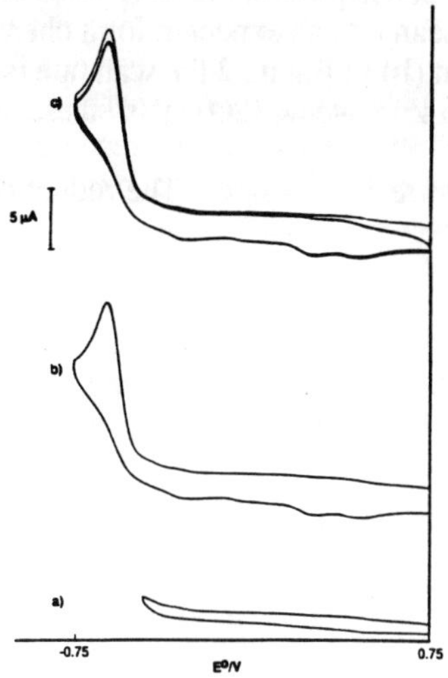

FIGURE 2. Cyclic Voltammograms of complex 6 in CH_2Cl_2, 0.1 M $(TBA)PF_6$ at a Pt-bead electrode. Scan rates 100 mVs^{-1} in **a** and **b** and 200 mVs^{-1} in **c**.

consistent with an oxidative coupling of two dimeric units of **1** involving the loss of one H_2 per dimer.

Magnetic measurements with a Gouy balance reveal diamagnetic behaviour for **3**, $\mathbf{3}^{2+}$, and $\mathbf{3}^{4+}$, whereas complex $\mathbf{3}^{3+}$ exhibits e.s.r. response in a frozen glass (CH_2Cl_2-THF) at 105 K. The spectrum consists of a broad line with a g value of 1.997 (width 100 G, no hyperfine splitting observed).

$\mathbf{3}^{2+}$ and $\mathbf{3}^{4+}$ show broad signals in their ^1H n.m.r. spectra which remain unresolved and uninformative even at low temperature (-40°C); we are not sure whether this broadening is due to unresolved non-rigid behaviour of the complexes in solution or to traces of paramagnetic impurities below the level detectable in the Gouy balance. By contrast, clean ^1H and ^{13}C n.m.r. spectra were obtained for complex **3**. The ^1H, ^{13}C and ^{13}C DEPT2 n.m.r. spectra support the proposed oxidative coupling.

Treatment of $\mathbf{3}^{4+}$ with $[(PhCH_2)Et_3N]Cl$ produces bridge splitting affording **6** where a

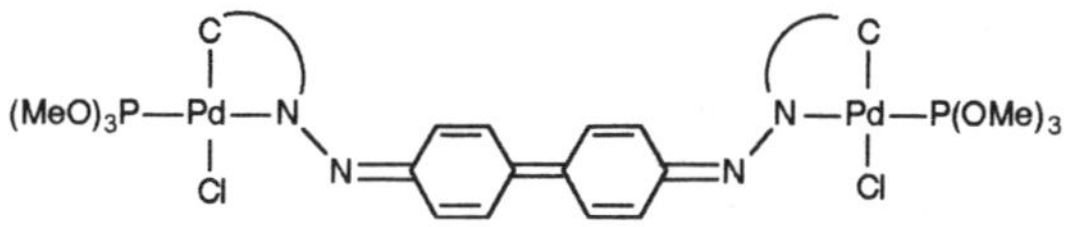

6

quinonediimine ligand is trapped and stabilised between two palladium atoms. Reduction of **6** with cobaltocene in non-aqueous media affords **3**.

Discussion

From the electrochemical data in CV's of $\mathbf{3}^{n+}$ conproportionation constants for the processes $\mathbf{3}^{(n+1)+} + \mathbf{3}^{(n-1)+} \rightleftharpoons \mathbf{3}^{n+}$ can be calculated ($Kc = 10^{\Delta E/0.059}$). The K_c values for $\mathbf{3}^{+}$ ($1.7x10^3$), $\mathbf{3}^{2+}$ ($1.4x10^9$), and $\mathbf{3}^{3+}$ ($1.9x10^4$) show that these complexes are stable towards disproportionation.

The cyclic voltammogram of complex **6** is shown in Figure 2. Scanning from +0.75V to -0.45 V (**a** in Figure 2) gives a flat voltammogram (i.e. no waves are present). Further scanning to more negative potentials (**b** in Figure 2) produces an irreversible wave at -0.6 V; in the reverse scan small anodic waves appear at -0.28, +0.26 and +0.45 V. The irreversible reduction at -0.6 V is assigned to the two electron process **6**/$\mathbf{6}^{2-}$. At this point, two units of $\mathbf{6}^{2-}$ undergo coupling according to equation 1.

$$2\,[\mathbf{6}]^{2-} \longrightarrow [\mathbf{3}] + 4\,Cl^- \qquad (1)$$

In the reverse scan **3** shows oxidations to 3^{+}, 3^{2+}, 3^{3+} and 3^{4+}, but 3^{4+} is able to react with the Cl^{-} liberated in the coupling step, giving back **6**. In consequence, the intensities of the cathodic waves due to reductions of 3^{4+} depend on the scan rate as expected for a chemical step (reaction with Cl^{-}) following the electron transfer. In (**b**) of Figure 2 the scan rate is 100 mVs^{-1} and cathodic peaks are not seen, but in (**c**) where the scan rate was 200 mVs^{-1} these peaks begin to be observed.

At scan rates of 1700 mVs^{-1} the CV shown in Figure 3 is obtained: The reduction of

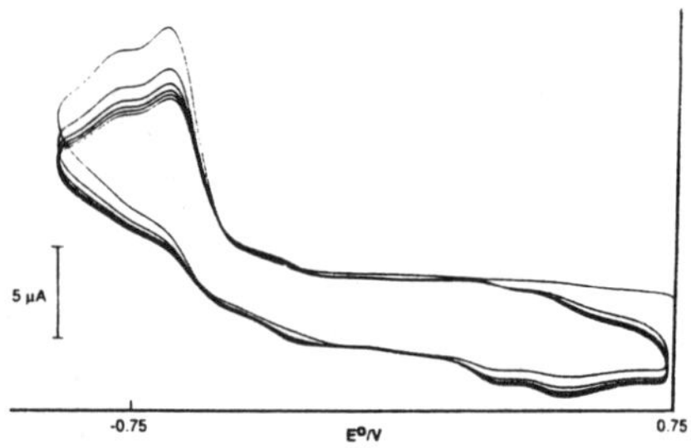

FIGURE 3. Cyclic Voltammograms of complex **6** in CH_2Cl_2, 0.1 M (TBA)PF_6 at a Pt-bead electrode. Scan rate 1700 mVs^{-1}.

6 appears now as two peaks (two electrons are involved) and the $3^{4+}/3^{3+}$ and $3^{3+}/3^{2+}$ electron transfers have become reversible meaning that the chemical reaction of 3^{4+} with Cl^{-} has been suppressed. The overall process observed in the voltammograms of **6**, can be regarded as a case of an EECEEEEC mechanism.

The cyclic voltammograms shown in Figure 4 were recorded on a solution obtained by controlled potential electrolysis of **6** at -0.75 V: A scanning from -0.75 V to -0.15 V shows reversible waves for the processes **3**/3^{+} and $3^{+}/3^{2+}$; however, if the potential +0.55 V is reached formation of **6** (by reaction in the electrode of 3^{4+} with chloride) occurs, being noticeable by the increased intensity of the cathodic wave at -0.6 V in the reverse scan.

Benzidine diamido ligands turn promising as bridging ligands capable to stabilise mixed oxidation states of a molecule, which is a goal of current research,[18-20] and in the design of molecules of interest as photoreductors in photochemical electron transfer processes.[21,22]

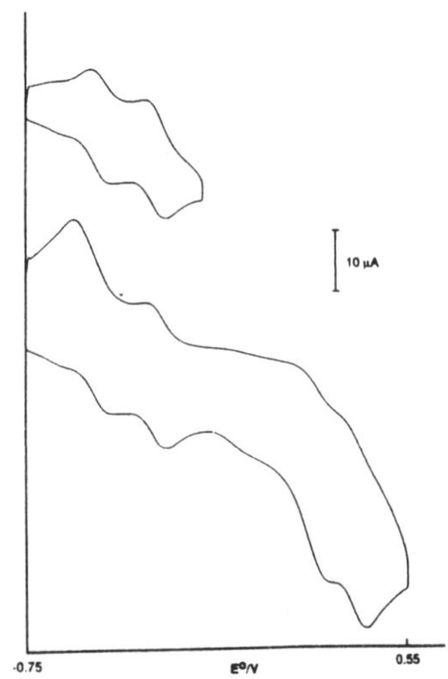

FIGURE 4.Cyclic Voltammograms after Controlled Potential Electrolisis at -0.9 V of complex **6** in CH_2Cl_2, 0.1 M

References

(1) For recent literature on "mixed valence" compounds see for instance Haga, M.A.; Bond, A. M. *Inorg. Chem.* **1991**, *30*, 475 and references therein.

(2) Creutz, C. *Prog. Inorg. Chem.*, **1983**, *30*, 1.

(3) Taube, H. *Angew. Chem., Int. Ed. Engl.* **1984**, *23*, 329.

(4) Mikkelsen, K. V.; Ratner, M. A. *Chem. Rev.* **1987**, *87*, 113.

(5) Espinet, P.; García-Herbosa, G.; Herrero, F. J.; Jeannin, Y.; Philoche-Levisalles, M. *Inorg. Chem.* **1989**, *28*, 4207.

(6) Fryzuk, M. D.; Montgomery, C. D. *Coord. Chem. Rev.* **1989**, *95*, 1.

(7) Adams, H.; Bailey, N. A.; Willett, D. G.; Winter, M. J. *J. Organomet. Chem.* **1987**, *333*, 61.

(8) Aggarwal, R. P.; Connelly, N. G.; Dunne, B. J.; Gilbert, M.; Orpen, A. G. *J. Chem. Soc. Dalton Trans.* **1991**, 1.

(9) Filippou, A. C.; Grünleitner, W. *J. Organomet. Chem.* **1990**, *393*, C10.

(10) Klein, H. F.; Beck-Menetsberger, H.; Reitzel, L.; Rodenhaeuser, B.; Cordier, G. *Chem. Ber.* **1989**, *122*, 43.

(11) Caslbore-Miceli, G.; Beggiato, G. S.; Emmi, S. S.; Geri, A. *J. Appl. Electrochem.* **1988**, *18*, 885.

(12) Casalbore-Miceli, G.; Beggiato, G.; Daolio, S.; Marco, P. G. D.; Emmi, S. S.; Giro, G. *J. Appl. Electrochem.* **1987**, *17*, 1111.

(13) Yue, J.; Epstein, A. J. *J. Am. Chem. Soc.* **1990**, *112*, 2800.

(14) O'Sullivan, R. D.; Parkins, A. W.; Alcock, N. W. *J. Chem. Soc. Dalton Trans.* **1986**, 571.

(15) Alcock, N. W.; O'Sullivan, R. D.; Parkins, A. W. *J. Chem. Soc. Chem. Commun.* **1980**, 1286.

(16) Espinet, P.; García-Herbosa, G.; Ramos, J. M. *J. Chem. Soc. Dalton Trans.* **1990**, 2931.

(17) Stephens, D. A.; Rehan, A. E.; Compton, S. J.; Barkhan, R. A.; Williams, J. M. *Inorg. Syn.*, **1986**, *24*, 135.

(18) Ernst, S.; Kasack, V.; Kaim, W. *Inorg. Chem.* **1988**, *27*, 1146.

(19) Kaim, W.; Kasack, V. *Inorg. Chem.* **1990**, *29*, 4696.

(20) Richardson, D. E.; Taube, H. *J. Am. Chem. Soc.* **1983**, *105*, 40.

(21) Benbouazza, A.; Buntinx, G.; Poizat, O.; Valat, P.; Wintgens, V. *New J. Chem.* **1990**, *14*, 629.

(22) Guichard, V.; Bourkba, A.; Poizat, O.; Buntinx, G. *J. Phys. Chem.* **1989**, *93*, 4429.

PART 6. MOLECULAR ELECTROACTIVATION and ELECTROCATALYSIS

USING ELECTROCHEMICAL INFORMATION FOR ORGANOMETALLIC COORDINATION CHEMISTRY: STOICHIOMETRIC AND CATALYTIC REACTIONS

W. BRUNS, W. KAIM*, M. LADWIG, B. OLBRICH-DEUSSNER,
T. ROTH and B. SCHWEDERSKI
Institut für Anorganische Chemie der Universität Stuttgart
Pfaffenwaldring 55
D-7000 Stuttgart 80
Germany

ABSTRACT. Following a brief comment on the particular electrochemical behaviour of transition and main group organometallics we illustrate several synthetically relevant applications of cyclic voltammetry, mainly by example of carbonyl complexes of group 6 and 7 metals with π acceptor ligands. Such applications include the discovery of stable oxidation states and guidelines to their chemical generation as well the determination of electronic structures. Synthetically useful consequences of electron transfer such as the tendency towards coordinative saturation, isomerization reactions, and (catalytic) substitutional activation can be evident from cyclic voltammetry and may be correlated with the electrochemical data in ways that allow the rational planning of non-electrochemical reactions.

1. Is Organometallic Electrochemistry Special ?

Per definitionem, organometallic compounds are distinguished by an often reactive metal-carbon bond. Typical organic ligands are carbon monoxide CO, polyhapto-coordinated π systems such as olefins or aromatics, and η^1-coordinated alkyl or aryl ligands [1].

Electrochemically, complexes with the latter type of ligands can be problematic because of the possible exposure of a polar, reactive metal-carbon single bond to the electrode surface. The frequently poor responses in electrochemical experiments [2-4] with such compounds can be overcome, however, by using bulky alkyl or aryl groups instead of e.g. a methyl ligand, i.e. by steric protection of the metal-carbon bonds. Such protection is particularly necessary in the little studied area of main group organometallic electrochemistry where polar metal carbon bonds are common [1]; examples that were shown to behave rather well towards glassy carbon working electrodes - in contrast to related systems [2-4] - include the triisopropylsilyl [5], triisopropylaluminium [5], di-tert.-butylzinc [6] and mesitylboryl groups [7] in compounds (1).

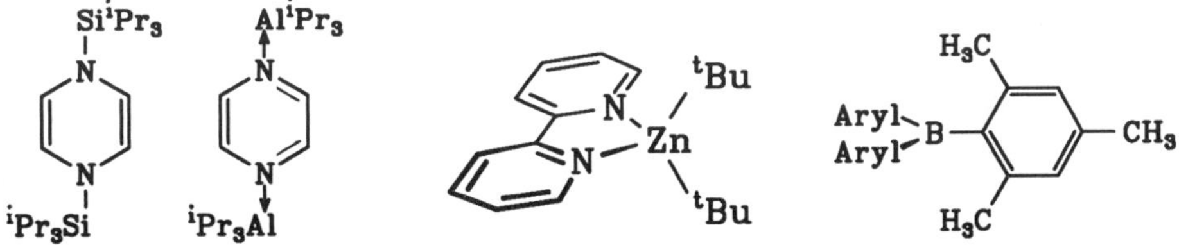

(1)

A. J. L. Pombeiro and J. A. McCleverty (eds.),
Molecular Electrochemistry of Inorganic, Bioinorganic and Organomettallic Compounds, 255–268.

In the transition metal series, coordinatively unsaturated complexes, e.g. with d^8 configuration, can pose similar problems; steric shielding of axial positions by η^1-mesityl or η^5-pentamethylcyclopentadienyl can to some extent stabilize reactive oxidized states such as Pt(III) [8] or highly reduced intermediates such as electron-rich and reactive Rh(I)/''Rh(0)'' [9] or Ir(I)/''Ir(0)'' (2, cf. Section 3).

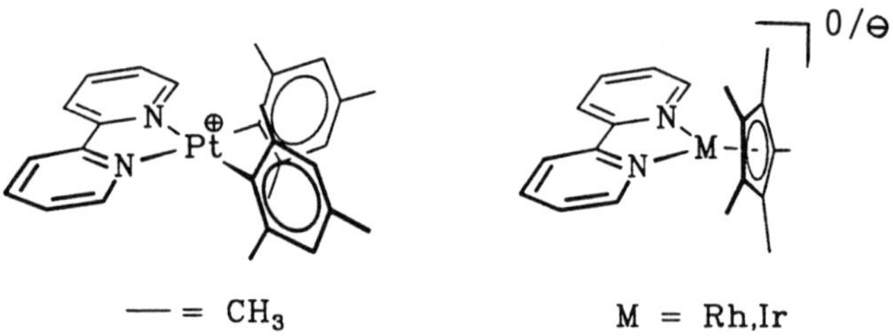

(2)

Another difficulty which is related to the potentially catalytic use of organometallics concerns the often enormous substitutional labilization of such systems after heterogeneous or homogeneous electron transfer. Typical textbook cases are the ligand exchange reactions of 18 valence electron (VE) complexes which can be accelerated by many orders of magnitude via one-electron oxidation (17 VE intermediates) or reduction (19 VE intermediates) [10,11]. It is shown below (Section 6) how even partial intramolecular electron transfer from ligands to metals can activate these for CO substitution.

Undesired reactions of this kind with the electro*chemical* medium can be suppressed by particularly careful preparation of electrodes and by proper modification of the chemical environment. In organometallic electrochemistry the solvents and electrolytes cannot only be selected on the basis of available ''potential windows'' [12]; in addition, their coordinating properties such as the basicity [13], the type of potentially ligating atoms and the susceptibility towards transition metal catalyzed activation towards substitution or polymerization must also be considered (cf. reactions of halogenated hydrocarbons [14]). For an example, cyclic voltammetric studies [15] of the very reactive, coordinatively unsaturated organometallic complex fragment $W(CO)_3(PR_3)_2$, R = ^{i}Pr, and of its complex with η^2-H_2 have to be carried out using tetrabutylammonium *tetraphenylborate* n-Bu_4NBPh_4 in THF solution [15]. An even less coordinating electrolyte for the study of labile organometallics such as certain metallocenes is tetrakis[3,5-bis(trifluoromethyl)phenyl]borate [16].

2. A Simple Application of Electrochemistry: Discovering Stable Redox States

If the tendency towards redox comproportionation as determined from the potential range ΔE between two reversible or even irreversible waves for oxidation and reduction of that state is rather large and if this intermediate exists within about ±1 V vs. conventional reference electrodes (SCE or Ag/AgCl), then the chemical synthetis of such redox states can be straightforward even if their existence or stability might not have been anticipated. Moreover, analytical electrochemistry such as cyclic voltammetry may provide guidelines concerning the conditions which lead to the selective generation of such intermediates, concerning e.g. the choice of oxidants, reductants or solvents.

Two curiosely related organometallic and organic examples include the discovery [17,18] of the first stable organometallic analogue (Fig. 1) of the mixed-valent Creutz-Taube ion and the isolation in crystalline form of a simple monocyclic radical cation (Fig. 2) [19,20].

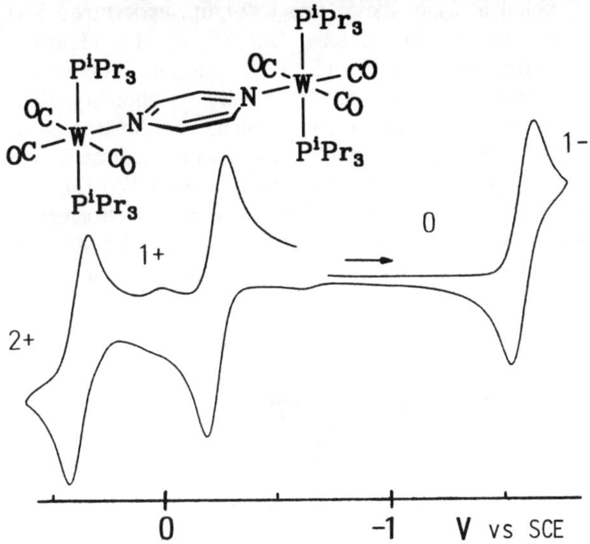

Figure 1. The cyclic voltammogram of a pyrazine-bridged W(0)/W(0) complex in dichloromethane/0.1 M Bu_4NClO_4 showed an electrochemically stable mixed-valent W(0)/W(I) oxidation state ($K_c = 10^{10.3}$ [17]) which was subsequently synthesized by oxidation with $AgPF_6$, isolated and characterized in the form of a pure tetraphenylborate salt [18]. A caveat: The seemingly reversible one-electron reduction to an anion radical complex at rather negative potentials results in the EPR-detectable loss of one phosphine ligand at each metal center [18].

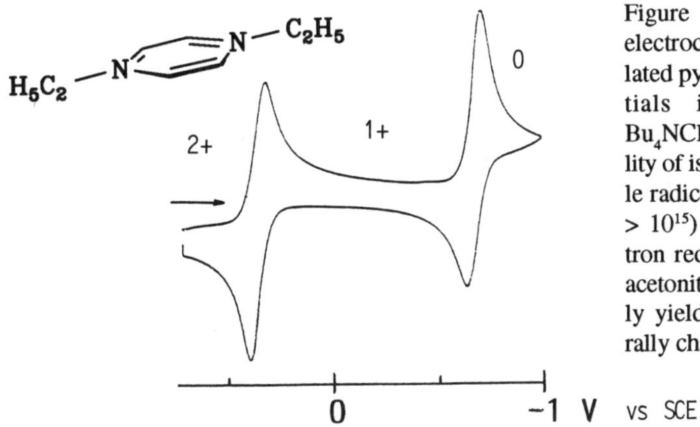

Figure 2. Two-step reversible electrochemistry of N,N'-dialkylated pyrazine at moderate potentials in acetonitrile/0.1 M Bu_4NClO_4 suggested the possibility of isolating the unusually stable radical cation intermediate ($K_c > 10^{15}$) [19]. Selective one-electron reduction of the dication in acetonitrile using iodide eventually yielded a crystalline, structurally characterized material [20].

3. Clues to Electronic Structures from Comparative Electrochemistry

While electrochemical methods in itself, i.e. without the help of accompanying spectroscopic methods (spectroelectrochemistry) are not very well suited to obtain detailed information on electronic structures such as molecular orbital characteristics, a comparison within a class of compounds can sometimes provide valuable pertinent information. Two examples shall illustrate this.

3.1 The molecule 3,6-bis(4-pyridyl)-1,2,4,5-tetrazine (cf. 3) was proposed by us as an unusual new conjugated bridging ligand system capable of topologically and electronically connecting two metal centers [21]. A molecular orbital analysis shows two low-lying unoccupied MOs which are rather close in energy but very different in character. One MO (a_u) has almost no participation of the metal-binding pyridyl rings whereas the other (b_{1u}) represents a delocalized MO with reasonable coefficients at the pyridyl nitrogen centers. Cyclic voltammetry of the bis(pentacarbonyltungsten) complex (3) showed very little effect of metal coordination on the ligand-based reduction, the anodic shift of only 0.14 V clearly points to an occupation of a_u as "redox orbital" as do the results from EPR spectroscopy [21]. A series of other $W(CO)_5$ complexes with ambidentate nitrogen ligands was studied in a similar way; the electrochemical potential shifts E(complex) vs. E(ligand) were found to correlate with calculated LUMO coefficients at the (thus established) coordinating centers of the ambidentate conjugated systems [21].

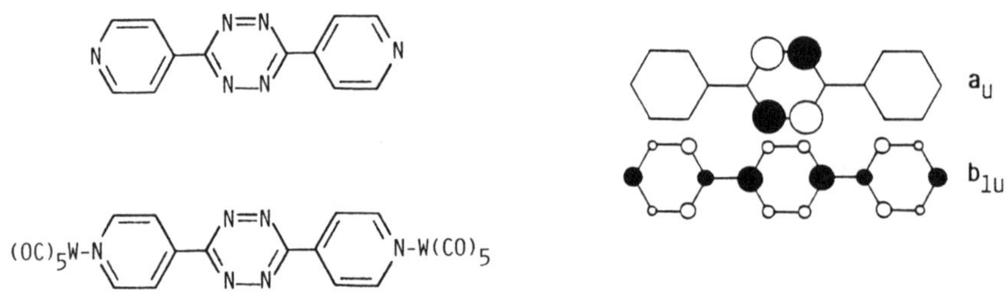

(3)

3.2 The general effect of facilitated ligand reduction upon metal coordination is due to the charge shift ligand-to-metal, associated with the formation of a coordinative bond. This effect may be reversed, however, if there is very strong ''back donation'' or even full electron transfer from the metal to the ligand which can more than compensate for the primary polarization metal$^{\delta-}$/ligand$^{\delta+}$. Such phenomena are observed when π electron rich organometallic 16 VE complex fragments such as $(C_5R_5)(CO)_2Mn$ or $(C_6R_6)(CO)_2Cr$ are combined with the very strongly π accepting TCNE and TCNQ ligands which are reduced at positive potentials vs. SCE (cf. Section 6 and Table 2) [22]. Less expected is such an effect, however, when a common ligand such as 2,2'-bipyridine (bpy) is employed because it is reduced at potentials below -2 V vs. SCE. Accordingly, almost all complexes of bpy are reduced more easily than the free ligand, even if the metal centers are fairly electron rich (Table 1), the reason being the charge shift from *two basic* N lone pairs (pK_a 4.4) to the chelated metal center.

With $(C_5Me_5)Rh$ [9] or $(C_5Me_5)Ir$ [28] as very π electron rich metal complex fragments, however, the complexes of bpy and of the related bidiazines [9] become *more difficult to reduce* than the free ligands (Table 1), suggesting an extremely strong degree of π back donation from the metal centers to the unsaturated ligands. While such an interpretation is also supported by results from NMR, UV/Vis and EPR spectroscopy, the energetic proximity and ligand-dependent [9] overlap of metal d and ligand π* orbitals as deduced from electrochemistry at very negative potentials is probably responsible for the excellent "two-electron reservoir" behaviour [15,29] of the complexes $(C_5Me_5)M(bpy)$ which can thus serve as catalytic intermediates in homogeneous hydride transfer [30], including H_2 production [31].

Table 1. Potentials for the reduction of 2,2'-bipyridine and of some of its complexes in acetonitrile

compound	E_{red} [V vs. SCE]	reference
(bpy)Re(CO)$_3$Cl	-1.24[a]	[23,24]
[(bpy)Cu(PPh$_3$)$_2$]$^+$	-1.40[a]	[24]
(bpy)Mo(CO)$_4$	-1.40[a]	[25]
(bpy)Zn(tert.-Bu)$_2$	-1.40	[6]
(bpy)PtPh$_2$	-1.48	[24,26]
(bpy)Mo(CO)$_2$(PBu$_3$)$_2$	-1.75[a]	[27]
bpy	-2.13	[9]
(bpy)Rh(C$_5$Me$_5$)	-2.15	[9]
(bpy)Ir(C$_5$Me$_5$)	-2.36	[28]

[a]In DMF.

The catalytic cycle for the two-electron reduction of 2H$^+$ to H$_2$ [31] involves the two-electron reductive elimination of a ligand L from the Rh(III) precursor (4), a rapid oxidative addition of the first H$^+$ to form a monohydride (5), and the protonation of this hydride to yield dihydrogen in exchange for the eliminated group L (6) [31].

$$\text{L-[Rh}^{\text{III}}]^+ + 2\,e^- \longrightarrow [\text{Rh}^{\text{I}}] + \text{L}^- \qquad (4)$$

$$[\text{Rh}^{\text{I}}] + \text{H}^+ \longrightarrow \text{H-[Rh}^{\text{III}}]^+ \qquad (5)$$

$$\text{H-[Rh}^{\text{III}}]^+ + \text{H}^+ + \text{L}^- \longrightarrow \text{L-[Rh}^{\text{III}}]^+ + \text{H}_2 \qquad (6)$$

$\text{L}^- = \text{Cl}^-$; $[\text{Rh}] = (\text{C}_5\text{Me}_5)\text{Rh}(\alpha\text{-diimine})$

The further reducible (Table 1) intermediate [9] can be formulated as

$$(\text{bpy})\text{Rh}^{\text{I}} \longleftrightarrow (\text{bpy}^{\cdot-})\text{Rh}^{\text{II}} \qquad (7)$$

and seems to be able to act as a single chromophor and two-electron reservoir with a catalytic metal site.

4. Electrochemical Evidence for the Facile Formation of Polynuclear Complexes

Electron transfer to a ligand in a coordination compound may strongly alter its characteristics and make it much more basic [32,33] solely because of the net charge effect. In a situation where labile coordination equilibria exist, e.g. in the presence of a polar electrolyte, the electron transfer to a complex may thus produce via EC [12] spectroscopically detectable and possibly even isolable coordination compounds of the reduced ligand with higher nuclearity [34,35]. The reoxidation potential of such species may be shifted anodically to a very large extent [35,36], producing non-reduced polynuclear systems in an extended ECEC [12], "double-square" or "ladder" [37] scheme (8).

$$\begin{array}{ccccc} \mathrm{L} + 2\,\mathrm{M} & \rightleftharpoons & \mathrm{L\text{-}M} + \mathrm{M} & \rightleftharpoons & \mathrm{LM_2} \\ \updownarrow E_1 & & \updownarrow E_2 & & \updownarrow E_3 \\ \mathrm{L}^{-\cdot} + 2\,\mathrm{M} & \rightleftharpoons & (\mathrm{L\text{-}M})^{-\cdot} + \mathrm{M} & \rightleftharpoons & (\mathrm{LM_2})^{-\cdot} \end{array} \qquad (8)$$

When the electrophile is a highly labile alkali metal cation [35,36], the dissociation of the non-reduced polynuclear system (top right in (8)) is often so rapid that there is no reduction peak visible for this compound [35]. In other instances, the reactions may be slower and the polynuclear systems may be persistent enough to show a reversible behaviour which then suggests ways to generate and isolate such complexes, stabilized by internal charge shifts, on a preparative scale [38].

The series of complexes (bcd)M(CO)$_5$ (9) of the potentially dinucleating 2,1,3-benzochalcogenadiazoles (bcd) shows a quantitatively very variable response towards electron transfer-induced dinucleation. The most rapid reaction according to (9) was observed (Fig. 3) with the combination of the lightest partners, X = O and M = Cr, wheras the combination of the heaviest species, X = Se and M = W, yielded a rather long-lived Se-coordinated [39] mononuclear anion which converts to the dinuclear anion radical species only after several hours at 290 K in THF [38]. Taking advantage of these results, the relatively labile *neutral* dinuclear complexes (bcd)[M(CO)$_5$]$_2$, M = Cr,Mo,W, could be prepared and studied, e.g. with respect to their optical properties [38].

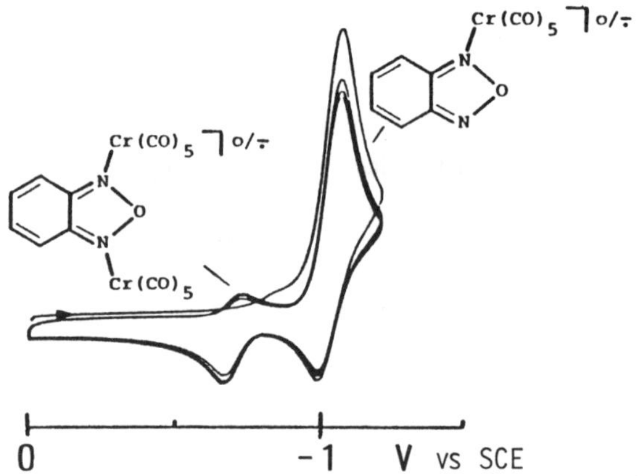

Figure 3. Multiscan cyclovoltammogram of (bod)Cr(CO)$_5$ in CH_3CN/0.1 M Bu_4NClO_4, illustrating the appearance of anodic *and* cathodic peaks of the dinuclear complex after initial reduction of the mononuclear system (bod: 2,1,3-benzoxadiazole) [38].

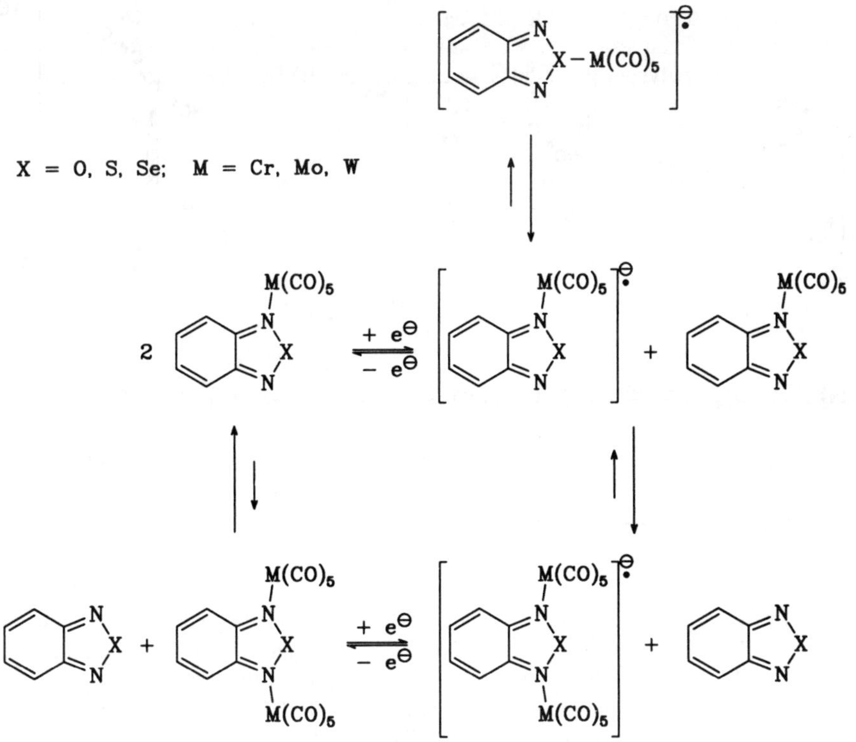

(9)

Although the ability to detect such new species depends on the rate of their formation in relation to the scan rate of the electrochemical experiment, the example given underscores the synthetic opportunities which lie in the following up on initially disturbing irreversible processes.

5. Electrochemical Evidence for Electron Transfer Induced Linkage Isomerization: An Organometallic Example of "Molecular Hysteresis"

The electrochemical response to frequently occurring structural changes after electron transfer can be very different, depending on the relation of reaction rates and scan rates. Unusual cyclic voltammetric behaviour is observed for slowly interconverting conformations of sterically restricted organic molecules [40]; in the organometallic field [41] it is frequently the variable, redox state-dependent hapticity of unsaturated organic or other ligands which lends itself to studies by cyclic voltammetry [42-46].

A relatively simple case based on ligand-centered electron transfer involves the olefin complex $(\eta^2\text{-TCNE})W(CO)_5$, TCNE = tetracyanoethylene (10) [46]. This compound undergoes a $\eta^2(C{=}C) \rightarrow \eta^1(N)$, i.e. a $\pi \rightarrow \sigma$ rearrangement on ligand-based reduction to the anion and, in particular, to the dianion.

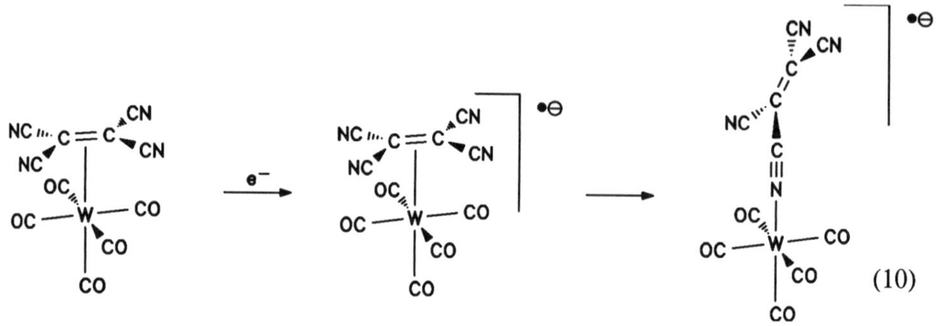

(10)

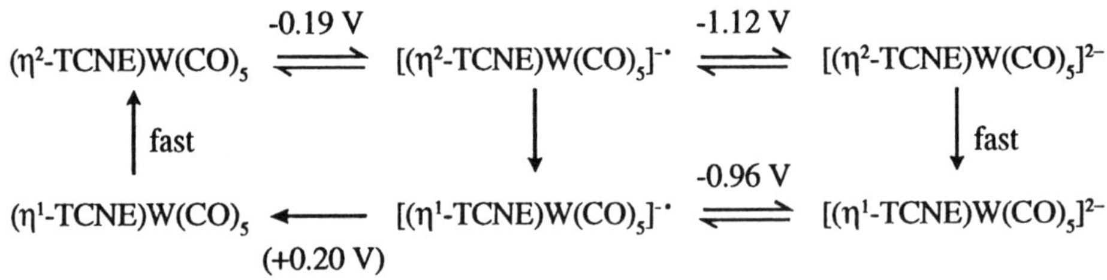

The first part of the underlying ''double square'' [46] scheme (10) contains a single electron transfer to the π(C=C) bonded tetracyanoethylene (TCNE) ligand in the blue complex with $W(CO)_5$ which leads - in a relatively slow chemical reaction on the cyclovoltammetric time scale - to a yellow isomerized anion radical product in which $TCNE^{-\bullet}$ coordinates via one nitrile nitrogen lone pair. The high resolution EPR, UV/Vis and infrared data of the singly reduced complex clearly illustrate the lowered symmetry and diminished π back donation in the secondary (η^1) monoanionic form [46].

In CH_2Cl_2/0.1 M Bu_4NPF_6, the neutral complex (η^2-TCNE)$W(CO)_5$ is reduced reversibly at -0.19 V vs Fc/Fc$^+$ (Fig. 4). This potential is virtually identical with that of free TCNE (-0.20 V), indicating considerable π back donation in the ground state (cf. Section 3.2). The position of the first reduction wave remains unchanged between 20 and 50000 mV/s.

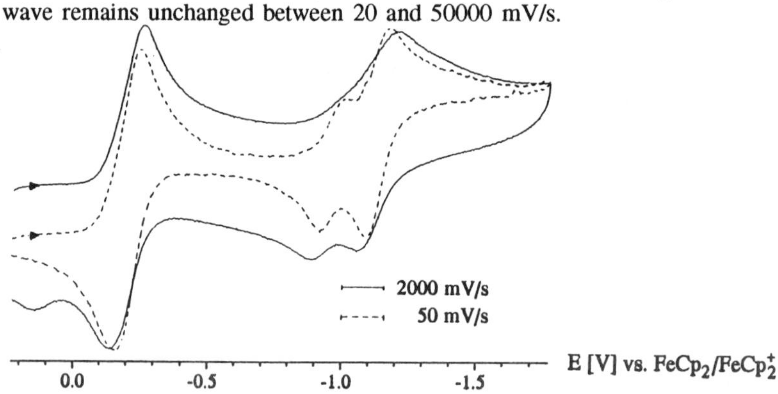

Figure 4. Sweep rate-dependence of cyclic voltammograms for the reduction of (TCNE)$W(CO)_5$ in CH_2Cl_2/0.1 M Bu_4NPF_6 (different peak current scales).

The second reduction to the dianion is usually a reversible process for TCNE and its complexes [22], however, there are now two such waves in the cyclovoltammograms, their ratio depending on the scan rate (Fig. 4). At fast scan rates, one major wave is observed at a potential of -1.12 V which is the expected [22] 0.9-1.0 V more negative than the potential for the first reduction. There are two small additional reoxidation peaks at -0.85 V for the dianion and at +0.20 V for the anionic form (Fig. 4), shifted positively relative to the main features. At slower scan rates, a secondary reversible wave appears for the second reduction at a potential of -0.96 V which clearly grows at the expense of the primary second reduction wave (Fig. 4). The small additional reoxidation peak for the anion at +0.2 V disappears at slower scan rates.

According to the interpretation in (10), the first reduction produces a primary anionic species (η^2) which is converted in a relatively slow chemical step ($t_{1/2} > 10$ s) to a secondary monoanionic product (η^1) as evident from two competing second reduction processes to two different dianions. Whereas the reoxidation peaks of the quite rapidly converting ($t_{1/2} < 1$ s) primary and secondary dianions are always observable within the range of scan rates employed, the reoxidation of the secondary monoanionic species (η^1) to the neutral form occurs predominantly at the "η^2"-potential; only at rather high scan rates is a small reoxidation peak detectable for the oxidation to the rapidly ($t_{1/2} < 0.1$ s) isomerizing secondary (η^1) neutral species at +0.20 V (Fig. 4). Previous studies [22] have focused on the first reduction process at moderate scan rates and could thus not reveal the more complicated reaction pattern in (10) [46]; the term "molecular hysteresis" has recently been coined for such schemes by Sano and Taube [47].

The driving force behind the isomerization in (10) lies in the requirement for the bonding of $W(CO)_5$. While a (weakly) nucleophilic *and* π accepting site is needed for π coordination of olefins according to the Dewar-Chatt-Duncanson model [1], the C=C-centered π^* acceptor orbital becomes singly occupied on reduction and the effectiveness of d(metal)-to-π^*(olefin) back bonding is diminished. On the other hand, the negative charge acquired on reduction turns $TCNE^{\cdot -}$ into an acceptable ligand of different character; the relatively long range of electrostatic interactions makes a coordination via the nitrile N lone pair more attractive. In fact, all known complexes containing $TCNE^{\cdot -}$ have the metal coordinated through the nitrile-N atoms [22,48,49]. The positive shifts of both reduction potentials for the η^1-bonded isomers (10) are in agreement with diminished π back donation in these persistent σ(N) coordinated forms.

The example (10) for electron transfer induced isomerization is distinguished by showing both a change of hapticity [42,43] and a change of the type of donor atoms [38,39,46]. Such changes of coordination preference on electron transfer to organometallic compounds may be relevant in homogeneous catalysis. Although this isomerization represents a qualitatively significant structural change, it did not become apparent from cyclic voltammetric studies at conventional scan rates. Extension of the potential range, variation of scan rates, and use of several spectroscopic methods proved to be necessary [46] to detect and clarify these processes.

6. Electrochemical Requirements for Electron Transfer (Auto-)Catalyzed Substitution

Transition metal carbonyl complexes containing 17 oder 19 valence electron metals [10,11] or an anion radical ligand at an "18+δ" VE metal center [50] display a strong tendency towards substitutional activation. While "normal" ET-induced substitution reactions such as $Cl^- \rightarrow PR_3$ in Re(I) halide complexes of singly reduced α-diimines (11a) are stoichiometric albeit with characteristically variable rates [24], the CO $\rightarrow$ PR_3 substitution in related complexes (11b) shows evidence for an electron transfer catalyzed chain reaction, based on the ratio of redox potentials [50-52].

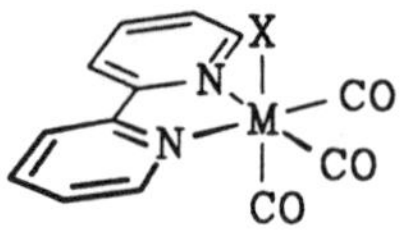

a) M = Re
X = $Cl^- \rightarrow PR_3$

b) M = Mo,W
X = $CO \rightarrow PR_3$

(11)

Using Scheme (12), we have previously shown that a rather small delocalization of spin [51,53] and negative charge to the metal fragment yields what we introduced as ''18+δ'' electron species [50,51] which are sufficiently labilized for CO substitution in cis-position to the spin-bearing π system. No further substitution was observed after single ligand exchange at one metal center because of the resulting charge delocalization primarily to the incoming ligand.

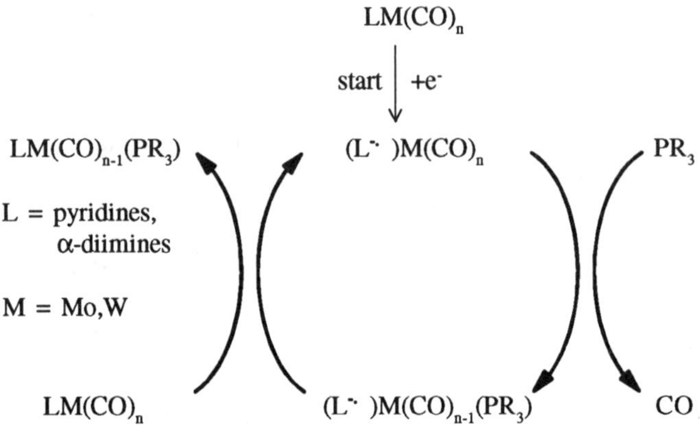

(12)

$$E_{red}[LM(CO)_{n-1}(PR_3)] < E_{red}[LM(CO)_n]$$

A chemically or electrochemically generated carbonylmetal complex of an anion radical ligand $L^{\cdot -}$ undergoes one cis-CO substitution, e.g. by a phosphine, in the rate determining step. The electron transfer catalytic cycle is closed if the the substituted anion radical complex can reduce the starting material in a homogenous electron transfer step, a condition (12) which requires a simple comparison of redox potentials and which is fulfilled, sometimes only to a small extent, for several such systems (pyridines, α-diimines) [51-53]. A less than stoichiometric amount of electrons, introduced either cathodically [52] or chemically [51,53] can thus effect substitution of CO by PR_3 under mild conditions, especially when the usual routes fail because starting materials or products are thermolabile or light-sensitive. The chemical method, employing potassium metal in THF solution, has been thus used in the synthesis of several new carbonylmetal complexes with mixed donor *and* acceptor ligands, such as tricarbonylphosphane-molybdenum(0) and -tungsten(0) complexes of α-diimines (11b) and tetracarbonylphosphanetungsten(0) complexes of electron deficient pyridines [51,53].

While process (12) has several advantages with respect to synthetic convenience, selectivity, usefulness and mechanistic clarity [50], the substitution is fairly slow in comparison to metal centered ETC processes [10,11,54,55], requiring a relatively high amount of catalyst (10-20%) and a relatively long time (>1 hour) for completion. This is due to the only partial character of electron transfer to the 18+δ VE metal center via σ^*/π^* hyperconjugation.

Much more rapid electron transfer catalyses occur when oxidation or reduction of 18 VE species is metal centered. While such reactions chains involving 17 or 19 VE intermediates are typically started by electron transfer from an electrode or from external chemical sources [11,54,55], the most economical process would involve the substrate (here: the substituting ligand) as chain starter: ''electron transfer autocatalysis'' [22].

For a variety of 18 VE carbonylmetal precursors which contain a labile 2 e-donor ligand such as THF, the differences between anodic peak potentials for oxidation and the reduction potential of the strong π acceptor TCNE vary considerably (Table 2). A thus preset electron transfer between the two reaction partners, the organometallic precursor and TCNE, leads to various amounts of substitutionally labile 17 valence electron precursor complexes as essential intermediates in the catalytic chain (13) [22,56].

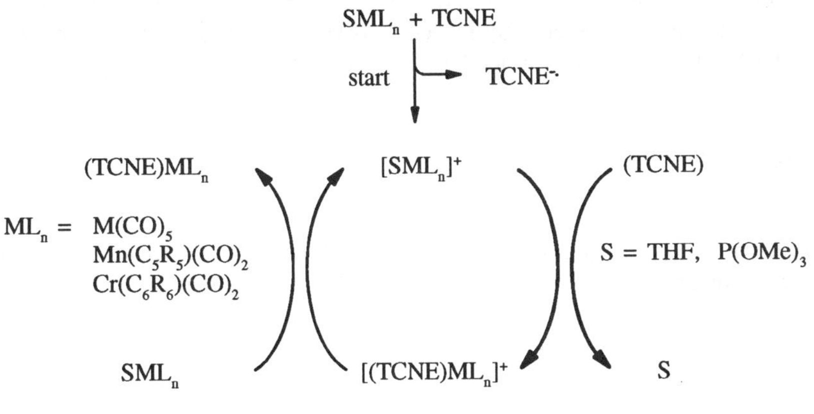

(13)

$E_{ox}[SML_n] < E_{ox}[(TCNE)ML_n]$

Table 2. Anodic peak potentials E_{ox} (V vs. Fc) and kinetic data (M^{-1} s^{-1}) for the electron transfer autocatalyzed reaction L_nMS + TCNE --> $L_nM(TCNE)$ + S (TCNE has E_{red} = -0.24 V vs. Fc)

L_nM	S	E_{ox} (L_nMS)	E_{ox} (L_nM-TCNE)	k_{subst}
$W(CO)_5$	THF	+0.48	+0.9	0.043
$Cr(CO)_5$	THF	+0.20	+0.7	0.39
$(C_5H_4\ Me)(CO)_2Mn$	THF	+0.08	+0.5	> 14000
$(C_5Me_5)(CO)_2Mn$	THF	-0.22	+0.4	1900
$(C_6Me_6)(CO)_2Cr$	$P(OR)_3$	-0.22	+0.3	1.59

Although the peak potentials of the organometallic precursors cannot be directly compared with the half-wave potential for TCNE reduction, it is safe to assume that the amount of 17 VE intermediate produced increases with decreasing oxidation peak potential of the organometallic precursor. In fact, we have shown that complete electron transfer (14) in the ground state occurs between the widely used acceptor TCNE [22] and the organometallic fragments $(C_5R_5)(CO)_2Mn$ and $(C_6R_6)(CO)_2Cr$ while $(CO)_5Cr$ and, in particular, $(CO)_5W$ exhibit metal-to-ligand charge transfer predominantly in the MLCT excited states [25,57]. Most characteristically, the TCNE-substituted product complexes of the carbonyl/carbocycle coordinated "16 VE" fragments (cf. 14) are reduced at more negative potentials than the free TCNE ligand [22], a relation which is opposite to that found in "normal" complexes (cf. Section 3.2).

$$\text{TCNE} + (C_5R_5)(CO)_2Mn^{I}(\text{THF}) \xrightarrow[\text{-THF}]{} (C_5R_5)(CO)_2Mn^{II}(\eta^1\text{-TCNE}^{\cdot-}) \qquad (14)$$

Table 2 illustrates and scheme (13) explains the very variable rates of (electron-transfer catalyzed) substitution by TCNE in seemingly related carbonylmetal compounds, the manganese systems reacting faster by a factor of at least 10^5 than the pentacarbonyltungsten complex.

7. Acknowledgements

Support from Deutsche Forschungsgemeinschaft (SFB 270), Volkswagenstiftung (program "Organometallics for Organic Synthesis") and Fonds der Chemischen Industrie is gratefully acknowledged.

8. References

(1) Elschenbroich, C.; Salzer, A. *Organometallics - A Concise Introduction;* VCH: Weinheim, 1989.
(2) Bock, H.; Lechner-Knoblauch, U. *J. Organomet. Chem.* **1985**, *294,* 295.
(3) El-Shazly, M.F. *Inorg. Chim. Acta* **1978**, *26,* 173.
(4) DuPont, T.J.; Mills, J.L. *J. Am. Chem. Soc.* **1975**, *97,* 6375.
(5) Baumgarten, J.; Bessenbacher, C.; Kaim, W.; Stahl, T. *J. Am. Chem. Soc.* **1989**, *111,* 2126 and 5017.
(6) Stahl, T. *Ph.D. Thesis,* University of Stuttgart, 1992.
(7) Schulz, A.; Kaim, W. *Chem. Ber.* **1989**, *122,* 1863.
(8) Klein, A.; Hausen, H.-D.; Kaim, W., unpublished.
(9) Ladwig, M.; Kaim, W. *J. Organomet. Chem.* **1991**, *415,* 233.
(10) Chanon. M., Ed. *Importance of Paramagnetic Organometallic Species in Activation, Selectivity and Catalysis;* Kluwer Academic Publishers: Dordrecht, 1989.
(11) Trogler, W.C., Ed. *Organometallic Radical Processes;* Elsevier: Amsterdam, 1990.
(12) Heinze, J. *Angew. Chem.* **1984**, *96,* 823; *Angew. Chem. Int. Ed. Engl.* **1984**, *23,* 831.
(13) Schulz, A.; Kaim, W. *Chem. Ber.* **1991**, *124,* 129.
(14) Crabtree, R.H.; Hamilton, D.G. *Adv. Organomet. Chem.* **1988**, *28,* 299.
(15) (a) Bruns, W.; Kaim, W.; Ladwig, M.; Matheis, W.; Roth, T. in *Energieträger Wasserstoff;* VDI-Verlag: Düsseldorf, 1991; p 126.

(16) Hill, M.G.; Lamanna, W.M.; Mann, K.R. *Inorg. Chem.* **1991**, *30,* 4687.
(17) Bruns, W.; Kaim, W. *J. Organomet. Chem.* **1990**, *390,* C45.
(18) Bruns, W.; Kaim, W. in *Mixed Valency Systems - Applications in Chemistry, Physics and Biology;* Prassides, K., Ed.; Kluwer Academic Publishers: Dordrecht, 1991; p 365. Cf. also Bruns, W. *Ph.D. Thesis,* University of Stuttgart, 1992.
(19) Kaim, W. *Heterocycles* **1985**, *23,* 1363.
(20) Hausen, H.-D.; Schulz, A.; Kaim, W. *Chem. Ber.* **1988**, *121,* 2059.
(21) Kaim, W.; Kohlmann, S. *Inorg. Chem.* **1990**, *29,* 1898.
(22) Olbrich-Deussner, B.; Kaim, W.; Gross-Lannert, R. *Inorg. Chem.* **1989**, *28,* 3113.
(23) Cf. Sullivan, B.P.; Bolinger, C.M.; Conrad, D.; Vining, W.J.; Meyer, T.J. *J. Chem. Soc., Chem. Commun.* **1985**, 1414.
(24) Vogler, C. *Ph.D. Thesis,* University of Stuttgart, 1990.
(25) Ernst, S.; Kaim, W. *J. Am. Chem. Soc.* **1986**, *108,* 3578.
(26) Vogler, C.; Schwederski, B.; Klein, A.; Kaim, W. *J. Organomet. Chem.*, in print.
(27) Ernst, S.; Kohlmann, S.; Kaim, W. *J. Organomet. Chem.* **1988**, *354,* 177.
(28) Ladwig, M.; Kaim, W., unpublished results.
(29) Pugh, J.R.; Bruce, M.R.M.; Sullivan, B.P.; Meyer, T.J. *Inorg. Chem.* **1991**, *30,* 86.
(30) Steckhan, E.; Herrmann, S.; Ruppert, R.; Thömmes, J.; Wandrey, C. *Angew. Chem.* **1990**, *102,* 445; *Angew. Chem. Int. Ed. Engl.* **1990**, *29,* 388.
(31) (a) Kölle, U.; Grätzel, M. *Angew. Chem.* **1987**, *99,* 572; *Angew. Chem. Int. Ed. Engl.* **1987**, *26,* 568. (b) Ziessel, R. *Angew. Chem.* **1991**, *103,* 863; *Angew. Chem. Int. Ed. Engl.* **1991**,*30,* 844.
(32) Kölle, U.; Kang, B.-S.; Infelta, P.; Compte, P.; Grätzel, M. *Chem. Ber.* **1989**, *122,* 1869.
(33) Cf. Krishnan, C.V.; Creutz, C.; Schwarz, H.A.; Sutin, N. *J. Am. Chem. Soc.* **1983**, *105,* 5617.
(34) Gross, R.; Kaim, W. *J. Organomet. Chem.* **1987**, *333,* 347.
(35) Kaim, W. in *Electron and Proton Transfer in Chemistry and Biology;* Müller, A.; Diemann,
E.; Junge, W.; Ratajczak, H.; Eds.; Elsevier: Amsterdam, in print.
(36) Kaim, W. *Acc. Chem. Res.* **1985**, *18,* 160.
(37) Evans, D.H. *Chem. Rev.* **1990**, *90,* 739.
(38) Kaim, W.; Kohlmann, S.; Lees, A.J.; Zulu, M.M. *Z. Anorg. Allg. Chem.* **1989**, *575,* 97.
(39) Kaim, W.; Kasack, V. *Angew. Chem.* **1982**, *94,* 712; *Angew. Chem. Int. Ed. Engl.* **1982**, *21,* 700.
(40) Dietrich, M.; Heinze, J.; Fischer, H.; Neugebauer, F.A. *Angew. Chem.* **1986**, *98,* 999; *Angew. Chem. Int. Ed. Engl.* **1986**, *25,* 1021.
(41) Connelly, N.G.; Geiger, W.E. *Adv. Organomet. Chem.* **1984**, *23,* 1 and **1985**, *24,* 1.
(42) Geiger, W.E. *Prog. Inorg. Chem.* **1985**, *33,* 275. Potential applications of such systems in the field of molecular electronics are described in Kölle, U. *Angew. Chem.* **1991**, *103,* 970; *Angew. Chem. Int. Ed. Engl.* **1991**, *30,* 956.
(43) Geiger, W.E.; Salzer, A.; Edwin, J.; von Philipsborn, W.; Piantini, U.; Rheingold, A.L. *J. Am. Chem. Soc.* **1990**, *112,* 7113.
(44) Gaudiello, J.G.; Wright, T.C.; Jones, R.A.; Bard, A.J. *J. Am. Chem. Soc.* **1985**, *107,* 888.
(45) Bond, A.M.; Colton, R.; Feldberg, S.W.; Mahon, P.J.; Whyte, T. *Organometallics* **1991**, *10,* 3320.
(46) Roth, T.; Kaim, W. *Inorg. Chem.*, in print.

(47) Sano, M.; Taube, H. *J. Am. Chem. Soc.* **1991**, *113*, 2327.

(48) (a) Rettig, M.F.; Wing, R.M. *Inorg. Chem.* **1969**, *8*, 2685. (b) Krusic, P.J.; Stoklasa, H.; Manzer, L.E.; Meakin, P. *J. Am. Chem. Soc.* **1975**, *97*, 667. (c) K.L. Amos and N.G. Connelly, *J. Organomet. Chem.* **1980**, *194*, C57.

(49) Gross-Lannert, R.; Kaim, W.; Olbrich-Deussner, B. *Inorg. Chem.* **1990**, *29*, 5046.

(50) Kaim, W.; Olbrich-Deussner, B.; Gross, R.; Ernst, S.; Kohlmann, S.; Bessenbacher, C. in ref. 10, p 283.

(51) Olbrich-Deussner, B.; Kaim, W. *J. Organomet. Chem.* **1988**, *340*, 71.

(52) Miholova, D.; Vlcek, A.A., *J. Organomet. Chem.* **1985**, *279*, 317.

(53) Olbrich-Deussner, B.; Kaim, W. *J. Organomet. Chem.* **1989**, *361*, 335.

(54) (a) Chanon, M.; Tobe, M.L. *Angew. Chem.* **1982**, *94*, 27; *Angew. Chem. Int. Ed. Engl.* **1982**, *21*, 1. (b) Astruc, D. *Angew. Chem.* **1988**, *100*, 662; *Angew. Chem. Int. Ed. Engl.* **1988**, *27*, 643.

(55) Kochi, J.K. *J. Organomet. Chem.* **1986**, *300*, 139.

(56) Schwederski, B.; Olbrich-Deussner, B.; Roth, T.; Kaim, W., unpublished.

(57) Stufkens, D.J.; Snoeck, T.L.; Kaim, W.; Roth, T.; Olbrich-Deussner, B. *J. Organomet. Chem.* **1991**, *409*, 189.

OXIDATIVELY INDUCED, ELECTROCATALYTIC LIGAND SUBSTITUTION IN 17-ELECTRON MANGANESE COMPLEXES: AN INVESTIGATION OF THE KINETICS BY DERIVATIVE CYCLIC VOLTAMMETRY

Mats Tilset,*[a] Vidar Skagestad[a] and Vernon D. Parker[b]

[a] *Department of Chemistry, University of Oslo, P. O. Box 1033 Blindern, N-0315 Oslo, NORWAY*
[b] *Department of Chemistry and Biochemistry, Utah State University, Logan, Utah 84322-0300, USA*

ABSTRACT. The electrocatalytic ligand substitutions of Cp'Mn(CO)$_2$(NCMe) and Cp'Mn(CO)$_2$(pyr) with phosphines and phosphites have been investigated by derivative cyclic voltammetry (DCV) at sweep rates that are sufficiently fast to preclude high catalytic turnovers on the experimental time scale. Digital simulations have established that a single working curve may be used to analyze the DCV data, provided that the substrate is more readily oxidized than the product by 250 mV or more. Rate constants and activation parameters were obtained for the reactions between Cp'Mn(CO)$_2$(NCMe)$^{\cdot+}$ and PPh$_3$ and P(OMe)$_3$, and between Cp'Mn(CO)$_2$(pyr)$^{\cdot+}$ and PEt$_3$ and PBu$_3$. Excellent agreement is observed between the kinetic data obtained with this technique and previously reported values that were obtained under conditions that allowed for high catalytic turnover numbers.

1. Introduction

Recent advances in the field of organotransition-metal radical chemistry[1] has established that much of odd-electron organometallic reactivity may be described in terms of 17/19-electron cycles that parallel the 18/16-electron rule that describes much of the reactivity of conventional, closed-shell systems. In a classic study, Kochi and co-workers investigated the details of the oxidatively induced ligand substitution reactions of Cp'Mn(CO)$_2$L (Cp' = η^5-C$_5$H$_4$Me; L = 2-electron donor) and other complexes.[2] A highly efficient electrocatalytic process was unravelled, in which up to several hundred substitutions ensued per electron removed. A detailed mechanistic and kinetic investigation that to a great extent was based on comparison of experimental cyclic voltammograms with voltammograms obtained by digital simulation established the mechanism shown in Scheme I. For L, L' pairs that fulfilled the condition E_{ox}(MnL') >> E_{ox}(MnL), catalytic amounts of oxidant or oxidizing current induced the complete transformation of the substrate Cp'Mn(CO)$_2$L to the product Cp'Mn(CO)$_2$L' due to highly efficient ligand substitution, (a) in Scheme I, and cross electron transfer (b) reactions.

The kinetic parameters for the reaction between Cp'Mn(CO)$_2$(NCMe)$^{\cdot+}$ and PPh$_3$ were elucidated by comparison of experimental and simulated voltammograms over a wide temperature range and reaction conditions and resulted in k(25 °C) = 1.3 x 10^4 M^{-1}s^{-1}, $\Delta H^{\#}$ = 4.4 kcal/mol, and $\Delta S^{\#}$

Scheme I

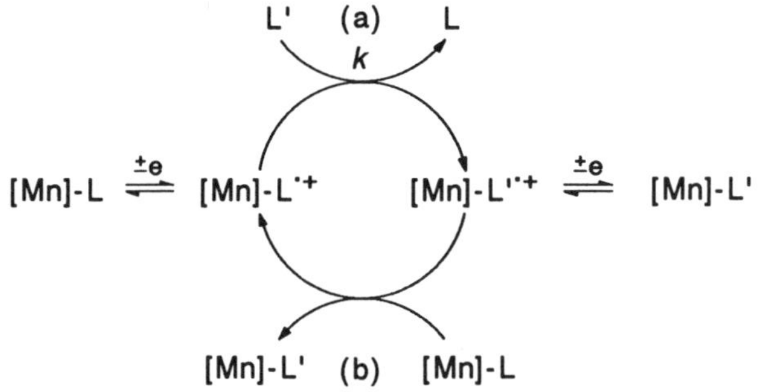

A. J. L. Pombeiro and J. A. McCleverty (eds.),
Molecular Electrochemistry of Inorganic, Bioinorganic and Organomettallic Compounds, 269–275.

= -25 cal/(K·mol),[2] consistent with an extremely facile associative mechanism involving a 19-electron intermediate or transition state.

Derivative cyclic voltammetry (DCV) has been shown to offer great advantages in the detailed investigation of electrode reactions.[3] The DCV reaction-order analysis conveniently gives access to overall kinetic rate laws. Arrhenius activation energies (E_a) are readily obtained, even without a detailed knowledge about the reaction mechanism. Once the overall mechanism has been established, comparisons with theoretical working curves provide rate constants and the derived quantities $\Delta H^{\#}$ and $\Delta S^{\#}$.

Here, we report the results of a DCV investigation of the substitution reactions described above. Included are experimental data which in part have been previously reported.[4] The details of a digital simulation analysis of the DCV response for the reaction are presented. It is shown that the kinetics of the substitution reaction is amenable to analysis by DCV with the use of only one working curve if E_{ox}(MnL′) - E_{ox}(MnL) > 250 mV.

2. Cyclic Voltammetry Analysis of the Oxidation of Cp′Mn(CO)$_2$(NCMe) in the Presence of PPh$_3$

Figure 1 displays derivative cyclic voltammograms for the oxidation of Cp′Mn(CO)$_2$(NCMe) (**1**; left) and Cp′Mn(CO)$_2$(PPh$_3$) (**2**; right) at a voltage sweep rate v = 0.2 V/s. Cyclic voltammetry peak potentials correspond to the intersections between the DCV curves and the base line. The reversible potentials for the oxidation of **1** and **2**, taken as the midpoints between the anodic and cathodic CV peaks, are located at -0.12 and +0.22 V vs the ferrocene/ferricinium couple (Fc), respectively. The unity ratio of the cathodic to anodic derivative peak currents demonstrates the chemical reversibility of the **1**/**1**$^{\cdot +}$ and **2**/**2**$^{\cdot +}$ couples.

Figure 2 shows DCV curves for the oxidation of **1** in the presence of 10 equiv PPh$_3$ at v = 0.2 and 100 V/s. At v = 100 V/s, the extent of reaction of **1**$^{\cdot +}$ is somewhat limited. The DCV trace shows significant chemical reversibility for the oxidation of **1**, and the derivative peak current corresponding to the oxidation of the product **2** is of low intensity. The time scale does not allow for high catalytic turnovers, and **1** is largely consumed due to the anodic process. At v = 0.2 V/s, the time scale makes possible high catalytic turnover numbers. Therefore, **1** is mostly consumed due to the catalytic cycle in which *no current is consumed*, and the resulting waves for the **1**/**1**$^{\cdot +}$ couple are of low intensity. The features of the cyclic voltammograms have been excellently described by Kochi[2] and will not be discussed in further detail here.

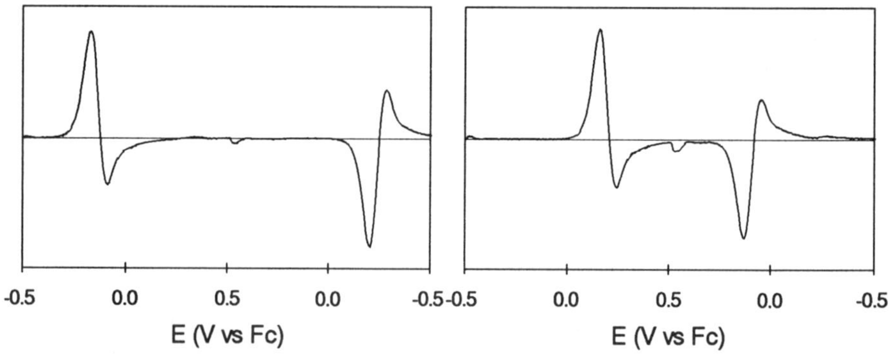

Figure 1. Derivative cyclic voltammograms for the oxidation of Cp′Mn(CO)$_2$(NCMe) (**1**; left) and Cp′Mn(CO)$_2$(PPh$_3$) (**2**; right). 1.0 mM substrate, acetonitrile/0.1 M $Bu_4N^+PF_6^-$, 20 °C, 0.6 mm Pt disk electrode, voltage sweep rate v = 0.2 V/s.

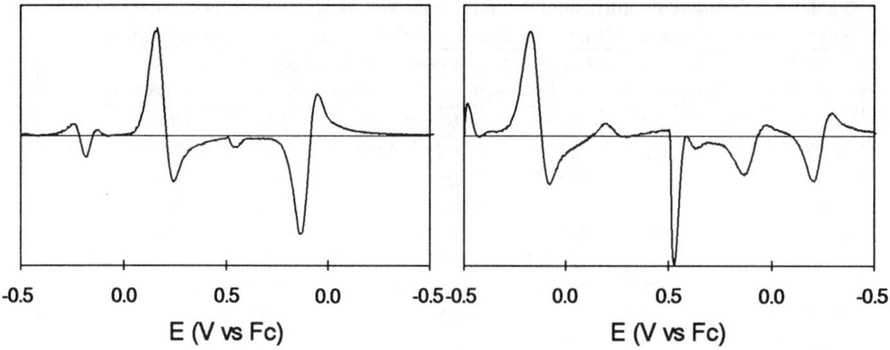

Figure 2. Derivative cyclic voltammograms for the oxidation of Cp'Mn(CO)$_2$(NCMe) (1.0 mM) in the presence of PPh_3 (10 mM) at v = 0.2 (left) and 100 (right) V/s. Conditions as in Figure 1.

The fact that high voltage sweep rates give access to time windows in which the consumption of **1** is essentially due to the anodic process suggested that the ligand substitution kinetics could be amenable to a DCV analysis. The kinetic information for the substitution should be contained in the partially reversible wave due to the **1**/**1**$^{\cdot+}$ couple. This is reasonable because the chain-carrying cross electron transfer would be expected to be fast and irreversible when E_{ox}(product) >> E_{ox}(substrate). In the following, we will describe results from digital simulations of this mechanism that indeed substantiate this assumption.

3. Digital Simulation of the DCV Response

Kochi[2] established that the predicted theoretical response for the mechanism in Scheme I is rather insensitive to several parameters, including the relative magnitudes of the diffusion coefficients of the species involved, transfer coefficients, cross electron transfer rates and heterogeneous electron transfer rates. Therefore, we focused only on variations in the parameters k (ligand substitution rate constant) and ΔE = E_{ox}(product) - E_{ox}(substrate). Feldberg's method of finite differences[5] was employed. The cross electron transfer was assumed to be in equilibrium for all values of ΔE. The transfer and diffusion coefficients were taken to be identical for all species, and Nernstian electrode processes were assumed. Simulation voltage step increments were 0.1 mV or less per pass through the simulation loop. Pseudo first-order ligand substitution conditions were assumed in the simulations. The simulation rate constants were chosen such as to ensure that no more than 5% conversion of the substrate cation radical resulted per pass. The quantitative results were unaffected by a 10-fold simultaneous decrease in potential step size and simulation rate constant. For all simulations, the distance between the CV switching potential E_{switch} and the reversible oxidation potential for the substrate was fixed at 200 mV. The simulated product oxidation potentials were 200, 250, 300, and 350 mV anodic of the substrate oxidation potential. Consequently, the sweep reversal occurs *before* significant product oxidation takes place (except for E_{switch} - E_{ox}(substrate) = 200 mV). In a DCV analysis,[3] the observable parameter of primary interest is the cathodic to anodic derivative peak current ratio, R'_i, for the substrate wave. The simulation output yielded R'_i as a function of the dimensionless rate constant kC/a, where C is the attacking ligand concentration and $a = Fv/RT$. During experimental measurements, the voltage sweep rate v is fine-tuned to yield a predetermined value for the observable R'_i, to be held constant throughout a measurement series.

Table I lists the theoretical DCV response, as measured by the derivative peak current ratio R'_i, as a function of the dimensionless rate constant kC/a, for E_{ox}(product) - E_{ox}(substrate) = 200, 250, 300,

Table I. Simulated dimensionless rate constant $(kC/a)^a$ and normalized cyclic voltammetry peak current $(I/I_0)^b$ as a function of the derivative peak current ratio R'_i

R'_i	kC/a for E_{ox}(product) - E_{ox}(substrate) = 200 mV		250 mV		300 mV		350 mV	
	kC/a	I/I_0	kC/a	I/I_0	kC/a	I/I_0	kC/a	I/I_0
0.10	0.2243	0.807	0.2615	0.785	0.2706	0.780	0.2723	0.779
0.20	0.1728	0.843	0.1957	0.828	0.2023	0.824	0.2034	0.823
0.30	0.1376	0.871	0.1518	0.861	0.1555	0.858	0.1562	0.857
0.40	0.1099	0.894	0.1191	0.887	0.1213	0.885	0.1216	0.885
0.50	0.0868	0.914	0.0928	0.909	0.0940	0.908	0.0942	0.908
0.60	0.0667	0.932	0.0705	0.929	0.0713	0.929	0.0714	0.928
0.70	0.0488	0.949	0.0512	0.948	0.0516	0.947	0.0516	0.947
0.80	0.0327	0.965	0.0339	0.964	0.0341	0.964	0.0341	0.964
0.90	0.0178	0.981	0.0183	0.980	0.0184	0.980	0.0184	0.980

^{a} $a = F\nu/RT$. ^{b} I^0 = CV peak current in the absence of nucleophile.

and 350 mV. Also included are the normalized CV peak currents, defined as the substrate peak current relative to the current simulated for the chemically reversible system. The I/I_0 data show that even for R'_i as low as 0.10 (for kC/a = 0.27 at ΔE = 300 mV), the CV peak current amounts to 78% of that obtained for the no-reaction case, attesting that the catalytic turnover number is modest under these conditions.

The data in Table I show that for a given value of R'_i, the corresponding rate constant is quite insensitive to the magnitude of the oxidation potential difference between substrate and product. For any value of R'_i in the interval listed (except ΔE = 200 mV), the calculated rate constant never differs from the corresponding value for ΔE = 300 mV by more than 3%. This small deviation is well within normal experimental uncertainties, and the results suggest that the 300 mV data may be employed as a single working curve for all values of ΔE greater than 250 mV. Even when ΔE = 200 mV, under typical experimental conditions involving measurements at R'_i = 0.50, the deviation is only 8%. The advantages involved in using this approach are obvious because the need to perform digital simulations for all possible values of ΔE is averted. We will present experimental results that demonstrate that the DCV approach may be of great value for the kinetic investigation of these ligand substitutions.

4. Experimental Results

The reactions that have been investigated are those between Cp'Mn(CO)$_2$(NCMe)$^{\cdot+}$ and PPh$_3$ and P(OMe)$_3$, and between Cp'Mn(CO)$_2$(pyr)$^{\cdot+}$ and PEt$_3$ and PBu$_3$. The oxidation potentials (vs Fc) for the species Cp'Mn(CO)$_2$L that were involved, are -0.14 (pyr), -0.12 V (L = MeCN), +0.15 (PBu$_3$), +0.16 (PEt$_3$), +0.22 (PPh$_3$), and +0.40 (P(OMe)$_3$). Thus, for all the reactions that were investigated, E_{ox}(product) - E_{ox}(substrate) exceeds 290 mV.

A DCV reaction-order analysis of the selected reactions showed the expected first-order behavior with respect to the substrate cation radical concentrations and with respect to the attacking nucleophile concentrations. The temperature dependence of $\nu_{0.5}$ (1.0 mM substrate, 10 mM nucleophile) was investigated by DCV. The data obtained for a representative measurement series for Cp'Mn(CO)$_2$(NCMe)$^{\cdot+}$ and PPh$_3$ in acetonitrile are depicted in Figure 3. The slope of the Arrhenius-type plot of ln($\nu_{0.5}/T$) vs 1000/T yields the activation energy E_a = 4.3 kcal/mol. Rate constants were obtained by comparison with digital simulation data and yielded k(25 °C) = 16.8 x 10^3 M^{-1}s^{-1}, $\Delta H^\#$ = 3.7 $\pm$ 0.2 kcal/mol, and $\Delta S^\#$ = -26.7 $\pm$ 0.5 cal/(K·mol). The agreement between these data[4] and Kochi's values[2] (k(25 °C) = 13 x 10^3 M^{-1}s^{-1}, $\Delta H^\#$ = 4.4 kcal/mol, and $\Delta S^\#$ = -25 cal/(K·mol)) is excellent.

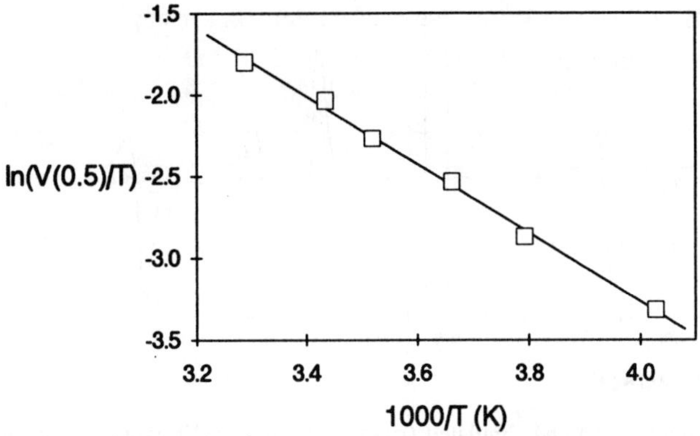

Figure 3. Arrhenius plot of $\ln(v_{0.5}/T)$ vs $1000/T$ (E_{switch} - E_{rev} = 200 mV) for the reaction between $Cp'Mn(CO)_2(NCMe)^{\cdot +}$ (1.0 mM substrate) and PPh_3 (10 mM) in acetonitrile/$Bu_4N^+PF_6^-$ (0.1 M) at a Pt disk electrode (d = 0.4 mm) in the temperature range -25 - +32 °C.

The sweep rate dependence of R'_i was investigated for the same reaction in the interval $15 < v < 160$ V/s at 20 °C. Figure 4 shows theoretical (line; based on data in Table I) and experimental (squares) data for R'_i vs kC/a assuming $k = 12.8 \times 10^3$ $M^{-1}s^{-1}$. Again, the agreement between theory and experiment is good.

The consistency of the method was checked by one more method. Figure 5 (left) shows a DCV trace for the oxidation of $Cp'Mn(CO)_2(NCMe)$ in the presence of 10 equiv PPh_3 in acetonitrile/0.1 M $Bu_4N^+PF_6^-$ at 18 °C and v = 35 V/s. These are conditions similar to those used for kinetics measurements, except that the sweep has been extended in the anodic direction in order to include the waves that are due to the accumulated product $Cp'Mn(CO)_2(PPh_3)$. The right half of Figure 5 displays a simulated DCV curve for the same process. The simulation rate constant corresponds to a macroscopic rate constant $k = 12.2 \times 10^3$ $M^{-1}s^{-1}$ for the substitution process and represents the case for which the agreement between simulated and experimental voltammograms was best, as judged from a visual

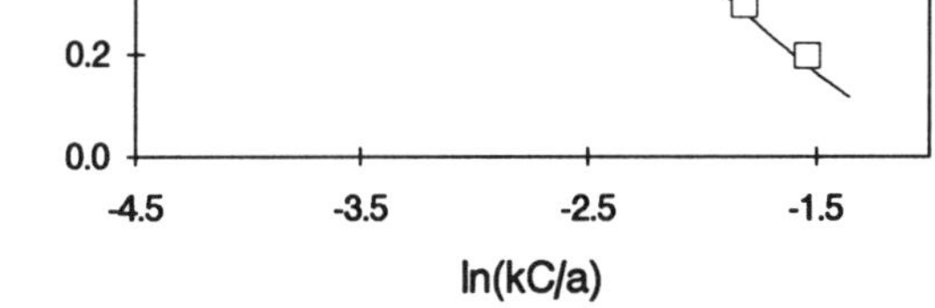

Figure 4. Comparison of experimental (squares) and theoretical (line) data for the reaction between $Cp'Mn(CO)_2(NCMe)^{\cdot +}$ (1.0 mM) and PPh_3 (10 mM) at 20 °C, using $k = 12.8 \times 10^3$ $M^{-1}s^{-1}$ to convert values of v to kC/a.

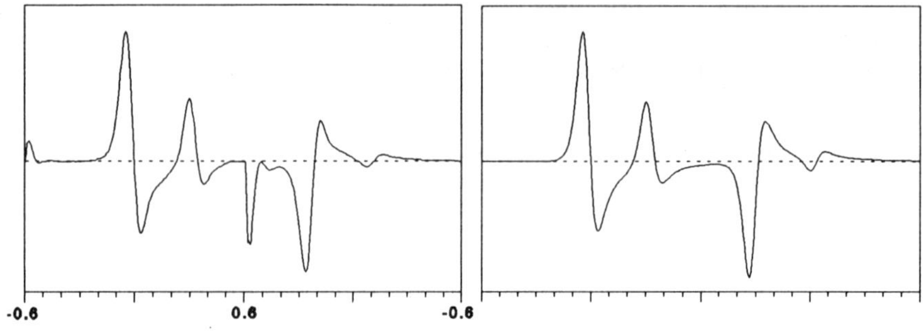

Figure 5. Experimental (left) and simulated (right) derivative cyclic voltammograms for the oxidation of Cp'Mn(CO)$_2$(NCMe) (1 mM) in the presence of PPh_3 (10 mM). Experimental parameters: 18 °C, 0.1 M $Bu_4N^+PF_6^-$, 0.4 mm Pt disk electrode, sweep rate v = 35 V/s. Simulated rate constant, k = 12.2 x 10^3 $M^{-1}s^{-1}$. (Reproduced with permission from ref 4).

comparison of the two. The interpolated k(18 °C) (based on the estimated enthalpy and entropy of activation obtained from the variable-temperature run described above) is 14.2 x 10^3 $M^{-1}s^{-1}$.

Table II summarizes the results of the kinetics measurements that were performed for reactions between Cp'Mn(CO)$_2$(NCMe) and PPh_3 (in acetonitrile and acetone) or P(OMe)$_3$ (acetonitrile), and between Cp'Mn(CO)$_2$(pyr) and PEt_3 and PBu_3 (acetone). Kochi's data, when available,[2] are included in parentheses. The agreement is excellent in all cases except for the reaction of Cp'Mn(CO)$_2$(NCMe) with PPh_3 in acetone. We found no significant solvent effect when changing from acetone to acetonitrile for this system, whereas Kochi observed a rate enhancement of a factor of 4 in acetonitrile. The reason for this discrepancy is not clear.

All our DCV-based results provide consistent kinetic data for the substitution processes. The close agreement between our kinetic parameters and Kochi's values is especially appreciated when taking into account the great difference in the time scales involved in the measurements, and reflects the fact that these systems are extremely well-behaved. Kochi's cyclic voltammetry potential sweep rates were relatively slow, typically 0.20 V/s. The CV response therefore reflected the intervention of several catalytic turnovers. In contrast, our DCV measurements were much faster, in the range 10-50 V/s for the Cp'Mn(CO)$_2$(NCMe)/PPh_3 system. This is fast enough to effectively outrun the catalytic process, and therefore the response mainly reflects the direct cation radical/nucleophile reaction.

Table II. Kinetics for Ligand Substitution Reactions of Manganese complexes[a]

Substrate	Nucleophile	10^{-3} x k (25 °C), $M^{-1}s^{-1}$ [b]		$\Delta H^\#$, kcal/mol	$\Delta S^\#$, cal/(K·mol)
Cp'Mn(CO)$_2$(NCMe)[c]	PPh_3	16.8	(13)	3.7 ± 0.2	-26.7 ± 0.5
Cp'Mn(CO)$_2$(NCMe)[d]	PPh_3	12.0	(3.3)	4.2 ± 0.3	-25.6 ± 0.9
Cp'Mn(CO)$_2$(NCMe)[c]	P(OMe)$_3$	2.52		4.7 ± 0.6	-27.1 ± 2.0
Cp'Mn(CO)$_2$(pyr)[d]	PEt_3	1.18	(1.3)	2.8 ± 0.2	-35.2 ± 0.6
Cp'Mn(CO)$_2$(pyr)[d]	PBu_3	0.85		2.8 ± 0.2	-36.2 ± 0.6

[a] From ref 4. [b] From the Eyring plot of the rate constants obtained by comparison of experimental DCV $v_{0.5}$ data (E_{switch} - E_{rev} = 200 mV) with digital simulation data. Literature values[2] in parentheses. [c] In acetonitrile/0.1 M $Bu_4N^+PF_6^-$. [d] In acetone/- 0.1 M $Bu_4N^+PF_6^-$.

The results show that the DCV approach offers great advantages for kinetic measurements on this and related systems. It should be pointed out that the DCV reaction-order and variable-temperature measurements alone are not sufficient to establish the occurrence of the electrocatalytic mechanism, and that supporting evidence must be obtained by other means. It is when the presence of the electrocatalytic mechanism has been reliably established by alternative methods that the DCV analysis, in conjunction with the simulated working curves, shows its true power in providing easy access to the relevant kinetic parameters without the necessity of performing additional simulations.

Acknowledgment.

We gratefully acknowledge support from Statoil under the VISTA program, administered by the Norwegian Academy of Science and Letters (stipend to V.S.), and from the Norwegian Council for Science and the Humanities, NAVF.

References

1. (a) *Paramagnetic Species in Activation/Selectivity, Catalysis*; Chanon, M., Julliard, M., Poite, J. C., Eds.; Kluwer Academic: Dordrecht, 1989. (b) *Organometallic Radical Processes*; Trogler, W. C., Ed.; Elsevier: Amsterdam, 1990. (c) Tyler, D. R. *Acc. Chem. Res.* **1991**, *24*, 325. (d) Connelly, N. G. *Chem. Soc. Rev.* **1989**, *18*, 153. (e) Astruc, D. *Chem. Rev.* **1988**, *88*, 1189. (f) Baird, M. C. *Chem. Rev.* **1988**, *88*, 1217.
2. Herschberger, J. W.; Klingler, R. J.; Kochi, J. K. *J. Am. Chem. Soc.* **1983**, *105*, 61.
3. (a) Ahlberg, E.; Parker, V. D. *J. Electroanal. Chem. Interfacial Electrochem.* **1981**, *121*, 73. (b) Parker, V. D. *Electroanal. Chem.* **1986**, *14*, 1.
4. Tilset, M. In *Energetics of Organometallic Species*; Simões, J. A. M.; Beauchamp, J. L.; Connor, J. A., Eds.; Kluwer Academic: Dordrecht, The Netherlands, 1992 (in print).
5. Feldberg, S. W. *Electroanal. Chem.* **1969**, *3*, 199.

ELECTRON-TRANSFER-CHAIN CATALYSIS: ROLE OF TRANSITION-METAL RADICALS, SIDE REACTIONS AND COUPLING WITH ORGANOMETALLIC CATALYSIS

D. ASTRUC , M.-H. DELVILLE and J. RUIZ
Laboratoire de Chimie Organic et Organométallique, URA CNRS N°35,
Université Bordeaux I, 351 Cours de la Libération,
33405 Talance cedex , France

ABSTRACT. Most of the working Electron-Transfer-Chain (ETC) catalyzed reactions need an exergonic cross ET step (see a former review, ref 2d). Three examples of organometallic ETC reactions illustrate this principle: i) For the one-decade old ligand exchange in $[FeCp(arene)]^+$ by two-electron donors, high efficiency is observed. Spectroscopic techniques as well as kinetics allow providing a reasonnably complete mechanistic picture. Emphasis is given on the 17e↔19e flexibility of the intermediate iron radicals. ii) The chelation of the dithiocarbamate ligand Fe^{II} complexes can be initiated either by oxidation or by reduction, both chain pathways being subject to considerable counter-ion and ion-pairing effects. iii) The appropriate application of the above principle allows designing a system in which both ETC catalysis and organometallic are coupled. The first efficient example concerns the polymerization of terminal alkynes by $[W(CO)_3(NCMe)_3]$.

Introduction

Electron-transfer-chain catalysis[1] has found considerable success in organo-transition metal chemistry[2]. It applies indeed to the single ligand exchange reaction[2] as well as to the more mechanistically sophisticated insertion[3] (and its reverse : extrusion[2b,d]). Note that both types are involved in organometallic catalysis. However despite the large number of examples of ETC catalyzed organo-transition metal reactions, there was no example of coupling with organometallic catalysis. We describe the first example a such a coupling process which was designed to work efficiently[4]. Nevertheless there is a substantial amount of prerequisite mechanistic knowledge of organo-transition metal radicals as a background[2]. Thus we will start with a pioneering, simple example of ETC catalyzed ligand exchange working with a high coulombic efficiency[5,6], then turn to the question of whether a given ligand exchange reaction should be catalyzed by oxidation or reduction[7]. In other terms, we examine to which extent the side reactions of the radicals formed in the catalytic cycles compete with the desired efficient reactions[8,9]. We also discuss the medium effects, especially the role of ion-pairing and salt effect[10].

A. J. L. Pombeiro and J. A. McCleverty (eds.),
Molecular Electrochemistry of Inorganic, Bioinorganic and Organometallic Compounds, 277–294.

Electron Transfer-Chain Catalysis in Organo-Transition Metal Chemistry

The process of ETC catalysis involves two parts : the initiation and the propagation. The initiation can be oxidation or reduction and this choice is crucial. It is dictated by the nature of the propagation cycle. The latter is decomposed in a "chemical" step and a "cross ET" step. The overall propagation cycle must be exergonic overall ($\Delta G° < 0$) and the mechanism of the propagation (oxidative vs reductive initiation) should not matter. However the efficient propagation reactions must compete favorably with the side reactions of the radicals formed in the propagation cycle. Very little is usually known concerning the chemical propagation step.

On the other hand, it is easy to have access to the knowledge of the thermodynamics of the cross ET step[12]. Usually suffices to record the cyclic voltammograms of the starting material and final products for a good estimation (if the waves are at least partly chemically reversible, the estimation of E° is excellent)[13]. Rewieving the many literature examples[2d] leads to the conclusion that ETC reactions working efficiently are indeed those which have an exergonic cross ET propagation step ($\Delta G° < 0$). It results that there is most often only one efficient way, oxidation or reduction, to insure initiation[2d].

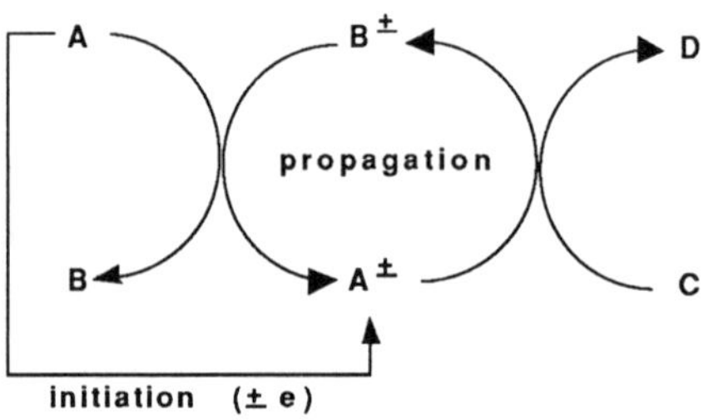

overall : $A + C \rightarrow B + D$

mechanism :

initiation $A \rightarrow A^{\pm}$

propagation $A^{\pm} + C \rightarrow B^{\pm} + D$ (chemical step

$B^{\pm} + A \rightarrow B + A^{\pm}$ (cross ET step)

However, even with a judicious choice of the mean of initiation, the ETC catalyzed process can be inefficient if a radical of the propagation cycle has a side reaction which is faster than the "chemical" propagation step. For instance, the oxidatively ETC ligand exchange of CO in mononuclear transition metal complexes usually does not work efficiently because of the fast dimerization of the radical anions generated[14,15].

Arene Exchange by 2-Electron Ligands in Cationic Fe^{II} Complexes : Comparison between Electrocatalytic and Stoichiometric Reactions

In the initial work, it was found that a catalytic amount of cathodic current at - 1.7 V vs SCE decomposes $[FeCp(C_6H_6)]^+PF_6^-$ in 95 % ethanol on Hg cathode in basic media. Cyclopentadiene, benzene and $Fe(OH)_2$ were characterized and the electron-transfer chain mechanism was proposed to account for the benzene exchange for solvent molecules. The driving force of the propagation cycle is the exergonic cross ET step since the three EtOH ligands do not allow back bonding whereas benzene does[5].

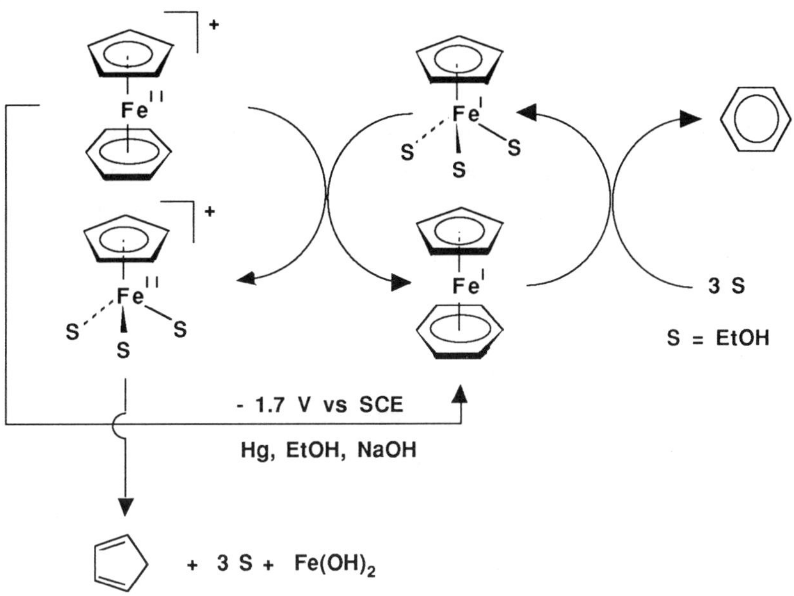

Later the system was extended to other ligands with high coulombic efficiencies[8,16]. The method is an alternative to decomplex coordinated aromatics although it has not yet really been used in this way (the visible light photolysis is probably even simpler)[17]. On the other hand we believe this electrocatalyzed reaction is the best route to the complexes $[Fe^{II}CpL_3]^+$ with a large variety of ligands L. Synthetically it is convenient to introduce a 1 % amount of $[Fe^ICp(C_6H_5R)]$, R = H or CH_3 to a Schlenk flask containing a THF suspension of $[Fe^{II}Cp(C_6H_5R)]^+PF_6^-$, R = H or CH_3. In the case of a diphos ligand, the acetonitrile ligand easily allows to obtain $[Fe^{II}Cp(diphos)(CH_3CN)]^+PF_6^-$. The yields are quantitative, the coulombic efficiency is at least 100 and the reaction can be carried out on a large scale.

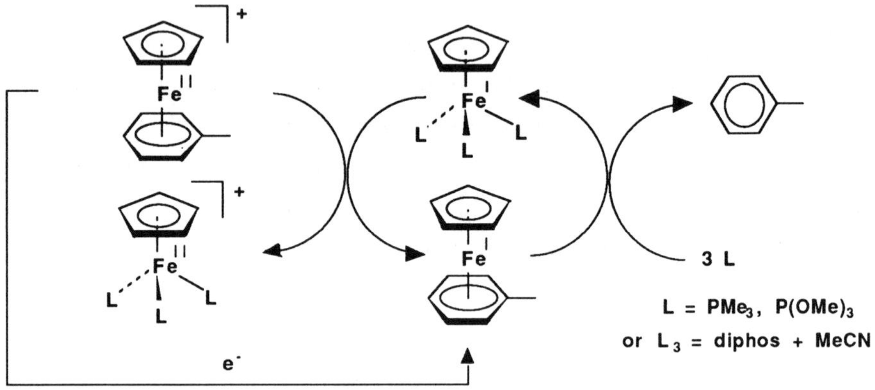

Again, the driving force for this efficient electrocatalytic cycle is the cross ET step. An estimation of this driving force can be obtained by recording the low-temperature cyclic voltammograms of $[Fe^{II}CpL_3]^+$ in order to observe a fraction of chemical reversibility. The mean values $1/2(E_{pc} + E_{pa})$ gives the approximate values of E° = - 1.76 V vs SCE (L = $P(OMe)_3$ and E° = - 2.05 V vs SCE (L = PMe_3) in DMF compared to E° = - 1.36 V vs SCE for $[FeCp(C_6H_6)]^+PF_6^-$ and E° = -1.41 V for $[FeCp(C_6H_5CH_3)]^+PF_6^-$; thus the driving forces change from 0.35 V (34 kJ $Mole^{-1}$) to 0.69 V (66 kJ $Mole^{-1}$) and ET is very fast according to Marcus theory[18]. On the other hand with L = $P(OMe)_3$, E° $FeCpL_3^{+/o}$ is only - 1.2 V vs SCE and the cross ET is thus endergonic in every case. It was indeed found that, consistently, the ETC reaction does not proceed and only the stoichiometric reaction is observed[6].

The stoichiometric reactions of 19e Fe^I complexes with P donors are of interest because they mimic the chemical step of the propagation cycle in the electrocatalytic process. Moreover their study gives hints concerning the side reactions of the electrocatalytic process. Note that these reactions of the Fe^I complexes with P donors occur at low temperature (typically - 20°C) whereas the ETC reaction proceeds conveniently at room temperature.

The Fe^{I} complexes react with L = PMe_3, PPh_3, dppm and dppe and give the hydrides $[Fe^{II}CpL_2(H)]$ in THF without addtional H atom source. This general reaction, proceeding usually with around 60 % isolated yield on a mmole scale, is the most convenient route to this class of hydride complexes. If the reaction is carried out in CH_2Cl_2 instead of THF, the chloro complex $[Fe^{II}CpL_2(Cl)]$ is isolated. With PPh_2H and PPH_2D, the hydride (resp. deuteride) and phosphido complexes are isolated in equal amounts. Finally the radical Arbuzov rearrangement is found with $P(OMe)_3$ an P-O cleavage is found with $P(OPh)_3$. These reactions are summarized below (for details see ref 6).

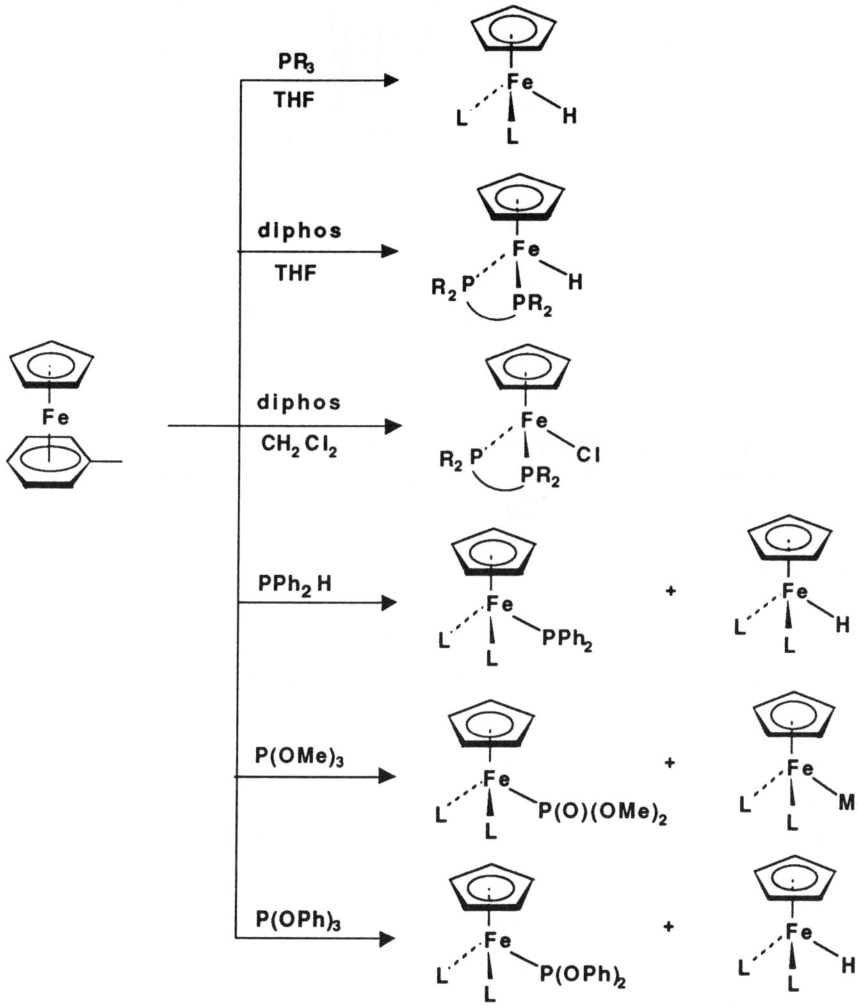

It is possible to characterize by ESR the 17-electron $Fe^{I}CpL_2$ intermediate (L = PPh_3) by its triplet indicating the presence of the two phosphorus ligands.

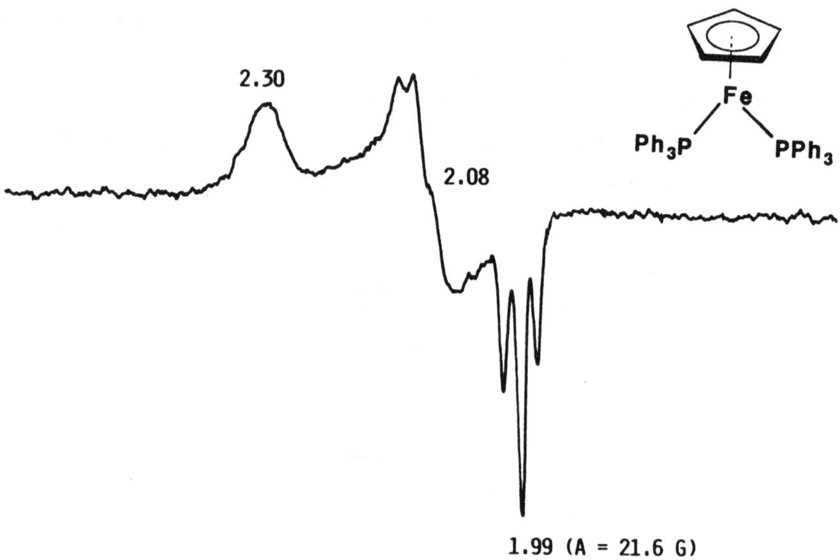

ESR spectrum of the proposed 17e species $Fe^{I}Cp(PPh_3)_2$ observed from a frozen THF solution at 77K by reaction of $Fe^{I}Cp(C_6H_5Me)$ with PPh_3 in THF at - 20 °C after 10 mn.

A remarkable feature concerning these reactions is that they are inhibited in THF in the presence of one equiv. [Na^+,PF_6^-] which induces the disproportionation of Fe^{I} to Fe^{0} and Fe^{II}. Indeed the 17e radicals are in fast equilibria with the 19e radicals resulting from the addition of 2e ligand present in the medium. This 19e complex is extremely electron-rich and can reduce $Fe^{I}Cp(C_6H_6)$ to its 20 e anionic counterpart. In the absence of a salt the large organometallic ion pair formed is stabilized to a sufficient extent so that its formation is reversible. This is no longer the case in the presence of [Na^+,PF_6^-] provoking double ion-pair exchange with the ions of the organometallic ion pair which drives the chemistry by further reactions[9]. Similarly in the presence of polyaromatic acceptors A, the 19 e form is trapped by ET[6] :

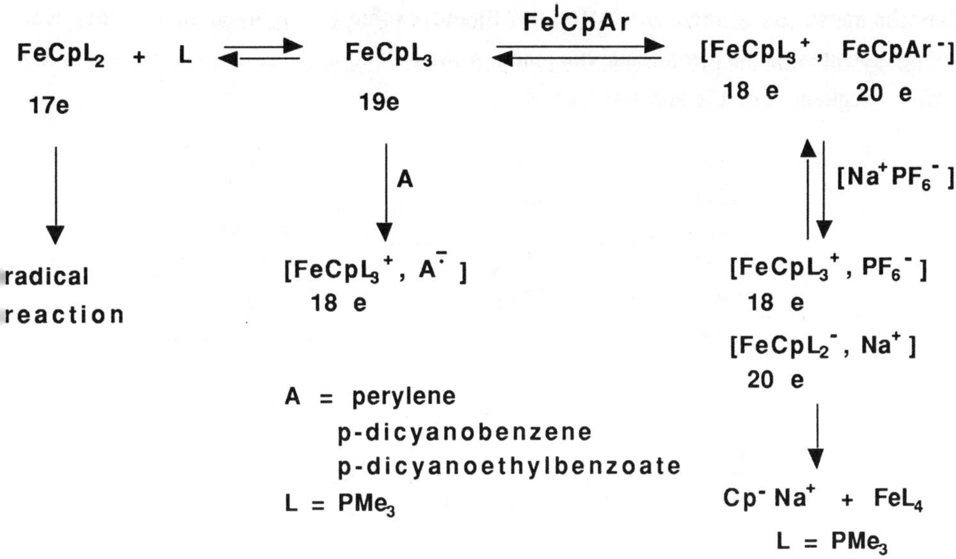

The E° value for $FeCp(C_6H_6)^{o/-}$ is about - 2 V vs SCE[19]. Thus with L = PMe_3, the ET responsible for the disproportionation found in the presence of $NaPF_6$ is about isoergonic; but with $P(OMe)_3$, it is endergonic by about 0.25 V (24 kJ $Mole^{-1}$) and nevertheless driven by the follow-up double ion-pair exchange and ensuing irreversible reactions[6].

The rate limiting step of the ligand exchange reactions of the Fe^{I} complexes with P donors is found to follow a second-order rate constant which depends on both the nature of the arene and of that of the P donor.

$$- d[FeCp(toluene)] / dt = k [FeCp(toluene)][P\ donor]$$

For $[Fe^{I}Cp(toluene)]$, the associative mechanism is confirmed by the large entropy of activation ($\Delta S^{\neq} = - 22 \pm 3$ cal $Mole^{-1}K^{-1}$ for $P(OMe)_3$). In order to avoid the high energy 21-electron transition state, we proposed a fast pre-equilibrium 19e $\rightleftharpoons$ 17e providing a 17e form responsible for the attack of L.

In passing, note that the form responsible for the reactivity of transition metal radicals is not the major one, but of slighly higher energy (similarly the responsible form for the dimerization is a minor 18e Fe^{II} form with a pentahaptocyclohexadienyl radical)[20,21].

Thus the most likely mechanism for the ligand exchange involved in the chemical propagation step of the electrocatalytic reaction (as well as in the soichiometric reactions) involves sequences of 17e and 19e species :

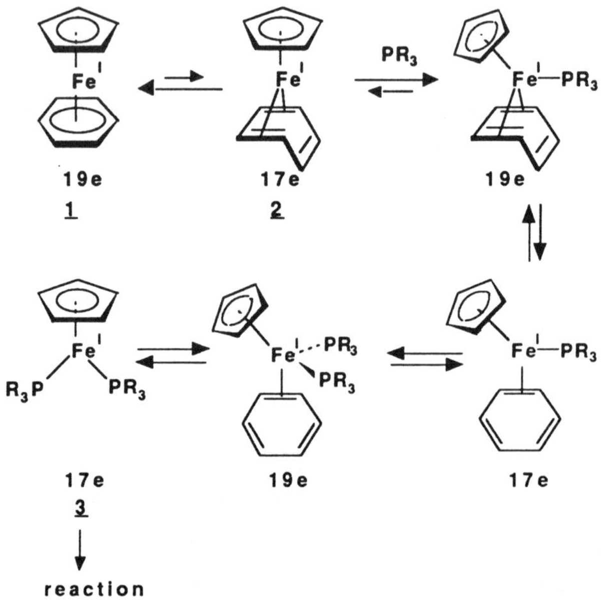

First Reaction which can be Electrocatalyzed either by Oxidative and Reductive Initiation : Chelation of Dithiocarbamate (dtc) and its Side Reactions.

Catalytic amounts of a ferricinium salt or of anodic current induce the chelation of $[Fe(C_5R_5)(CO)_2(\eta^1\text{-dtc})]$ with liberation of CO in THF[22]. The cross ET is endergonic but the E° of the oxidation could not be obtained because the partial reversibility of the oxidation was unobservable even after scanning at 5000 V / s[22d]. The electrocatalytic system is thus driven by the chemical propagation step (fast chelation). The coulombic efficiency of the ferricinium catalyzed chelation was found to be low but, interestingly, depends on the nature of the ferricinium counter anion (Table I). Unchanged starting material is found after the reaction when the optimum chelation yield is obtained (Table II) ;

15 % of $FeCp_2^+PF_6^-$ is necessary in order to obtain the optimum chelation yield of 45 %. Under these conditions 38 % of starting material is recovered and the stoichiometric oxidation by ferricinium gives the corresponding amount of the 17e complex $\underline{2b}^+PF_6^-$.

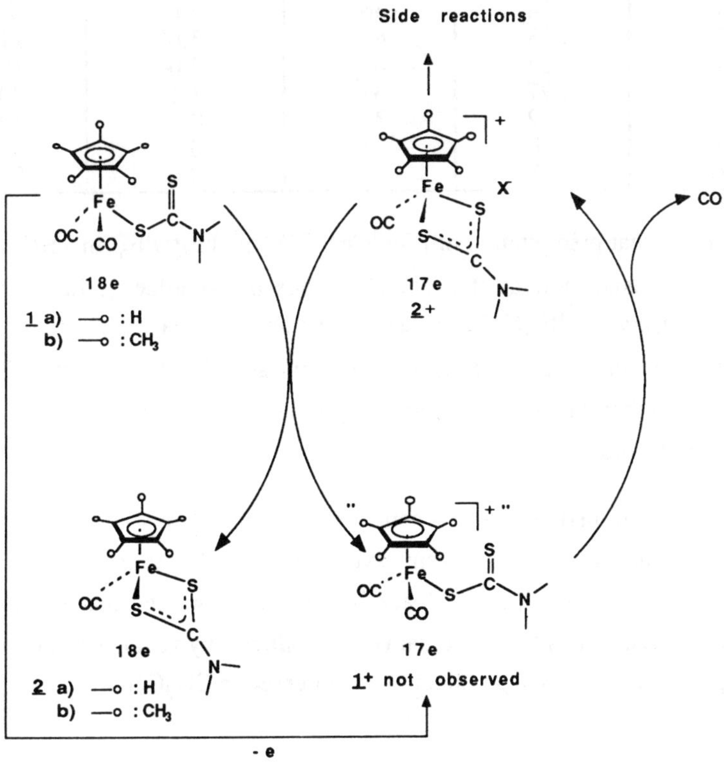

Table I : Influence of the nature of the counter anion of the catalyst on the chelation yields : $[Fe(\eta^5\text{-}C_5R_5)(\eta^1\text{-dtc})(CO)_2]$ + 10 % $[FeCp_2]^+X^-$ $\rightarrow$ $[Fe(\eta^5\text{-}C_5R_5)(\eta^2\text{-dtc})(CO)]$ R = H or Me		
X^-	% (R = H)	% (R = Me)
BF_4^-	47	23
PF_6^-	52	36
$SbCl_6^-$	63	38

Table II : Influence of the catalyst quantity on the evolution of the chelation reaction of complex $[Fe^{II}Cp^*(CO)_2(\eta^1\text{-dtc})]$

% of added $[FeCp_2]^+PF_6^-$	% of unreacted complex	% of chelation	catalytic activity	% $\underline{2b}^+PF_6^-$ formed
5	67	20	4.0	4
7	60	31	4;4	6
10	54	36	3.6	7
15	38	45	3.0	11
20	33	41	2.0	19
30	27	33	1.1	30
40	22	31	0.8	39
50	7	33	0.7	49
100	0	0	0.0	95

Thus it appears that precipitation of $[Fe^{III}Cp^*(CO)(\eta^2\text{-dtc})]^+PF_6^-$ in THF is a dead end since the THF suspension of this complex does not oxidize 1. In the presence of supporting electrolyte $n\text{-}Bu_4N^+BF_4^-$, $\underline{2a}^+$ is not stable. We have identified the dead-end complex found in the electrochemical experiment as being $[FeCp^*(dtc)(THF)]^+$. This distinction shows that the medium, ion-pairing and counter-anion effects are important in the electrocatalytic process.

We have also examined the side reaction of $\underline{2b}^+$ in other solvents[23] as attempted electrocatalytic reactions in other solvent gave poor results. We found that CO is extremely labile in the 17e Fe^{II} complex $\underline{2b}^+$ (back-bonding is poor) and that CO is exchanged rapidly. The kinetics of this ligand exchange was followed by visible spectroscopy and two isosbestic points at λ = 488 and 600 nm were observed in CH_3CN.

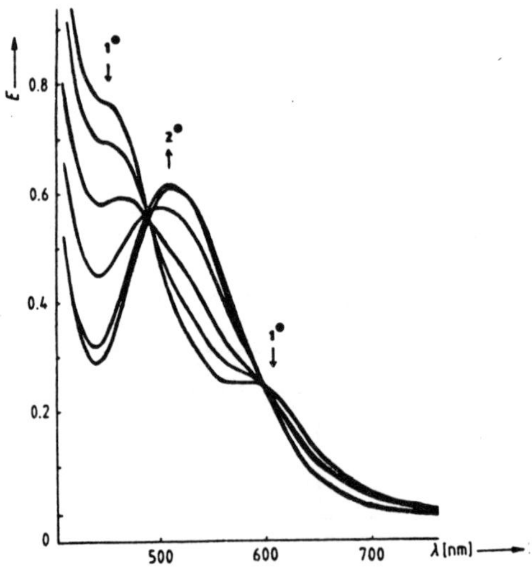

Visible spectroscopy of the conversion of $\underline{1}^+$ (c = 3.2 10^{-3} mol L^{-1}) to $\underline{2}$ in CH_3CN at 5 °C. E = Extinction coefficient

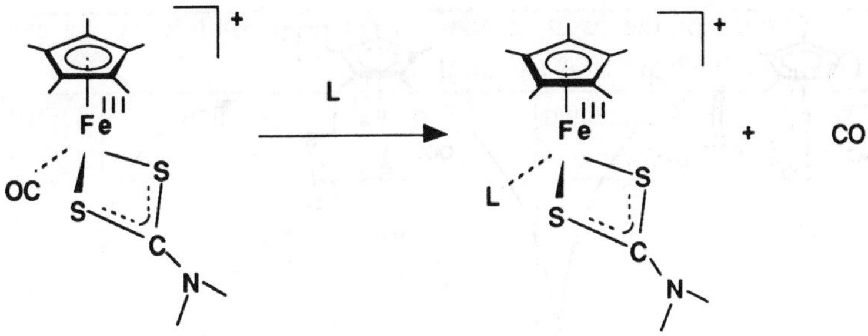

L = MeCN, Me_2CO, CH_2Cl_2

This ligand exchange reaction was found to follow a pseudo first-order kinetics depending on the incoming ligand ($t_{1/2}$ = 57 mn in CH_2Cl_2, 25 °C ; 13.5 mn in Me_2CO, 5°C ; 5 mn in CH_3CN, 5 °C) suggesting an associative mechanism via 19e intermediates or transition states. Clearly, this reaction dramatically competes with the endergonic (thus slow) cross ET step of propagation cycle. This study has led us to isolate and characterize a variety of 17e $[Fe^{III}Cp(dtc)L]^+$ complexes such as those involved in this propagation cycle and its side reations. In these radicals, the Mössbauer and ESR spectra indicate that the radicals are metal centered, as confirmed by theoretical studies[23b].

We turned to reductive initiation of the electrocatalytic reaction since it was the appropriate way to generate an exergonic redox step[7]. Indeed under these conditions, no starting material was found under optimum conditions. The cyclic voltammograms show that the reduction of 1 is irreversible and that the CV wave of 2 appears as a consequence of a cathodic reduction of 1. However, the CV wave of $[FeCp^*(CO)_2]_2$ is also observed. The chelation yields and efficiency of the electrocatalytic cycles are better than upon oxidative initiation but limited by the formation of the side product $[FeCp^*(CO)_2]_2$ resulting from the cleavage of the dtc ligand in the radical anion $[FeCp^*(\eta^1\text{-dtc})(CO)_2]^-$. In the latter, chelation with CO loss and dtc cleavage compete since the dtc ligand does not tolerate the low Fe^I oxidation state well. The extent of the side reaction also depends on the ion-pairing. The Na^+ counter cation seems to optimize the chelation (66% yield using naphthyl sodium) possibly because its binding with organometallic anions allows decreasing the electronic density in the iron coordination sphere (for instance possible binding to a carbonyl oxygen would assist the CO cleavage concomitant with the dtc chelation).

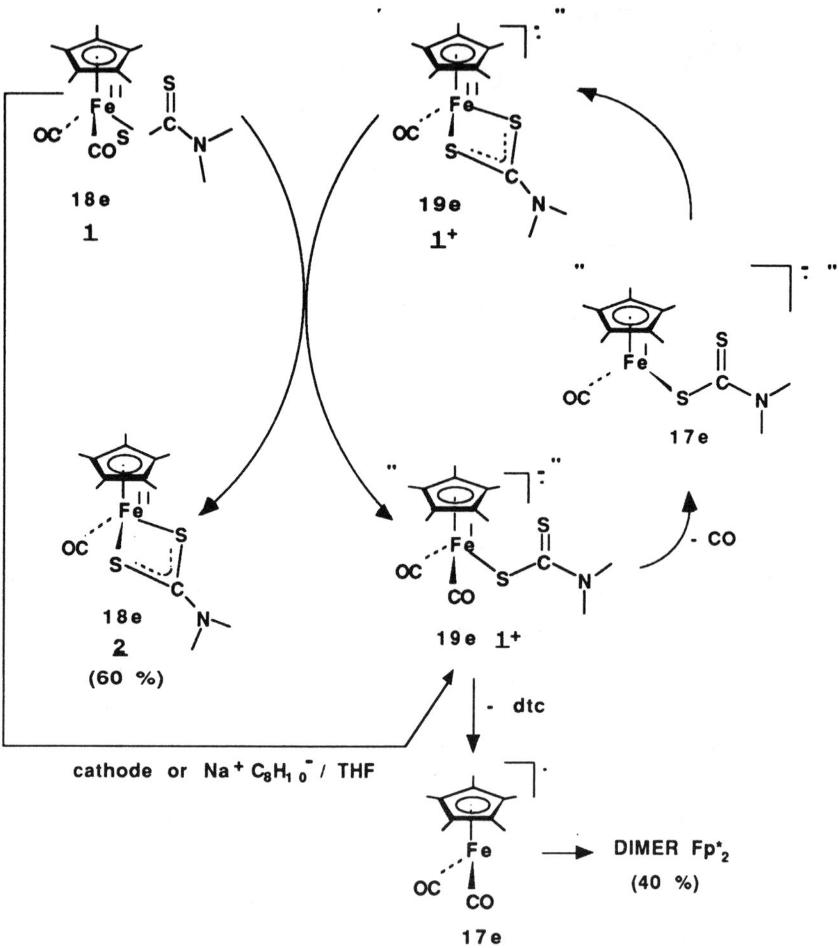

In conclusion, the comparison between reductive and oxidative initiation and examination of the side reactions allows determining which propagation step works well and which one is marred by problems of side reactions in each case. The knowledge of the ion-pairing effects is always important to minimize side reactions. These mechanistic investigations are useful in order to design systems in which the coulombic efficiency is expected to be high, for instance when ETC chain catalysis will be coupled to organometallic catalysis.

Coupling ETC Catalysis with Organometallic Catalysis : Carbenic Polymerization of Terminal Alkynes.

Organometallic catalysis can be significantly improved if ligand exchange on the active metal center is accelerated. This step is indeed often the rate-limiting step in catalytic processes. Since ligand exchange is the most efficient an extensively studied ETC catalyzed reaction in inorganic and organometallic chemistry, it is possible to design an ETC system in which the correct mode of initiation (oxidation vs reduction) is used. If this system is applied to a transition metal precatalyst in the presence of the substrate, the catalytically active species is formed at a very high rate. As a first example, we have chosen the polymerization of terminal alkynes which is known to be catalyzed by W^0 complexes at temperatures around 100 °C for several hours.

$$\equiv\!\!-R \xrightarrow{\text{cata}} \text{polyacetylenes}$$

The complex $W(CO)_3(CH_3CN)_3$ is a known starting material for ligand exchange[24]. Yet its use for alkyne polymerization still requires a week to reach completion with modest yields[25]. ETC catalyzed ligand exchange of CH_3CN by P donors is known[26]. We have applied CH_3CN exchange by alkynes in $W(CO)_3(CH_3CN)_3$ although there was no precedent for ETC catalyzed introduction of an alkyne onto a transition metal center. If a catalytic amount of the W complex slowly catalyzes the alkyne polymerization, the latter is considerably accelerated by the use of catalytic amounts of $FeCp_2^+PF_6^-$ vs the W complex. Thus this amount of Fe^{III} initiator must be very small (of the order of 1 mg). Under these conditions, much better yields of polymer are obtained in 10 mn. It is important to note that the maximum yield is reached for a proportion $FeCp_2^+ / W = 0.1$ to 0.2 and that when this ratio is increased the yields drop. In particular the yields are much weaker when this ratio reaches 1. This indicates that W^0 is the operating catalyst (not W^I). Finally, excess MeCN (solvent) inhibits the process as expected by competing with the alkyne coordination and $FeCp_2^+PF_6^-$ without the W catalyst is inefficient.

$$\equiv\!\!-R \xrightarrow[\text{FeCp}_2^+\text{PF}_6^-]{\text{cata W(CO)}_3\text{(NCMe)}_3} \text{polyacetylenes}$$

R = Ph or alkyl

Table III : Solvent effect on the polymerization of PhC_2H by $W(CO)_3(NCMe)_3$ and 0.1 eq. of $[FeCp_2]^+PF_6^-$

Solvent	THF	CH_2Cl_2	CH_3CN	toluene	bulk
Polymer yield (%)[a]	45(15)[b]	42(45)[b]	14(12.6)	36(13)	13(7)
M_W[c]	32000 (21350)	30200 (22250)	2070 (1530)	7000 (30600)	21400 (16800)
Mn[c]	13200	11600	800	3400	6400
Polydispersity index[c]	2.42 (200)	2.61 (3.56)	2.57 (2.63)	2.06 4.37)	3.34 (2.99)

[a] polymerized at RT, in the dark, under argon for 10 mn. ; 50 eq. of alkyne per W complex was used ; [b] as indication : numbers in brakets are the yields and characterizations obtained without addition of the cocatalyst $[FeCp_2]^+PF_6^-$ and after one week reaction times ; [c] determined by gel permeation chromatography. The polydispersity index being close to 2, indicates either a slow initiation or a random disruption of chain growth.

Table IV : Polymerization of PhC_2H by $W(CO)_3(NCMe)_3$ and $[FeCp_2]^+PF_6^-$: influence of the amount of $[FeCp_2]^+PF_6^-$ per eq. of W complex (in THF, at room temperature, in the dark, under argon, 10 mn). Results are average vaalues of a series of 5 experiments.

$[FeCp_2]^+PF_6^-$	0	0.1	0.2	0.3	0.5	0.7	1
Polymer yield (%)[a]	1	15.7	26.7	21.44	14.10	12.30	5.97

Table V : Polymerization of alkynes (50 eq.) RC_2R' by $W(CO)_3(NCMe)_3$ and 0.1 eq. of $[FeCp_2]^+PF_6^-$ per W (in 10 mL of THF at room temperature, in the dark and for 10 mn).

R	R'	Yields	M_W[a]	M_n[a]	Polydispersity index[a]
Ph	H	45 (15)	32000 (21300)	13200 (10600)	2.42 2.00
nBu	H	35.5 (12)	31350 (1800)	14150 (12000)	2.21 (1.55)
C_3H_7[b]	CH_3[b]	6(0)	6100	2500	2.45

[a] determined by gel permeation chromatography. Numbers in brakets are yields and characterizations of polymers obtained without addition of $[FeCp_2]^+PF_6^-$; [b] most of the product consisted in oligomers of Mw around 600.

Table VI : Evolution of the yields of polymerization when increasing the ratio alkyne / W ; the ratio $[FeCp_2]^+PF_6^-$ / W = 10 remaining constant (solvent : THF (10 mL)). Results are average values of a series of 5 experiments (reactions were performed et room temperature).

|alkyne| / |W|	50	62	70	100
yieds (%)	35.2	30.1	21.2	15.0
turnover	17.6	18.6	14.8	15

Katz[27] has proposed a carbenic Chauvin-type metathesis mechanism for the polymerization of alkynes. This mechanism involves two alkynes coordinated to W^0 in the catalytically active species. The ETC catalyzed ligand exchange process is adequate to rapidly introduce two or three alkyne ligands on W^0 in such a way that the Katz mechanism is then operating as shown on the following scheme.

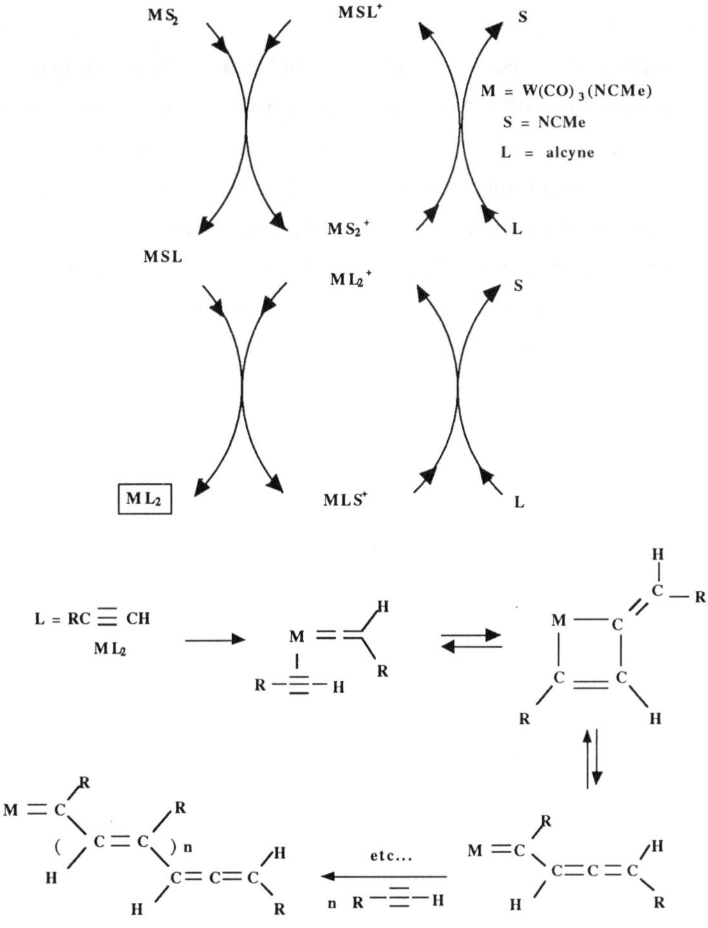

We are now trying to extend this type of coupling to other important catalytic processes involving transition metal homogeneous catalysts.

Conclusion and Prospects

ETC catalysis works with a large number of ligand exchange reactions in transition metal chemistry. The mode of initiation (oxidation vs reduction) can now be forecasted to design efficient processes. Application in organometallic syntheses is possible in every case where an efficient coulombic efficiency is occuring. It is then most practical to use an electron-reservoir sandwich complex (Fe^{III} ox ; Fe^{I} red) in Schlenk flasks for high-scale syntheses. The first ETC catalyzed reaction where either oxidative or reductive initiation can be used has been disclosed. Comparison of both modes brings about essential mechanistic features involving side reactions and ion-pairing. These informations are very useful in order to design and forecast multi-catalytic processes with ETC. The first case of ETC catalysis coupled with organometallic catalysis was shown to be efficient : the carbenic polymerization of terminal alkyne was considerably accelerated and its yields improved. This powerful principle should be extended.

Acknowledgments

We ow much to the French electrochemical school, for instance to the laboratories of Dijon, Rennes and ENS (Paris) for their enriching collaborations cited in the references. We also acknowledge the excellent pioneering work in our Rennes group on the $FeCp_2^+$ or anodically induced chelation of $Fe(C_5R_5)(CO)_2(dtc)$ by Drs D. Catheline and A. Madonik. Finally we thank the CNRS, the NSF, the Humboldt Fundation, the University Bordeaux I and the Région Aquitaine for financial support.

References

1 - **a)** Savéant, J.-M. *Acc. Chem. Res.* 1980, *13*, 323 ;
b) Bunnett, J.F. *Ibid* 1978, *11*, 413 ;
c) Russell, G.A. *Spec. Publ. Chem. Soc.* 1970, *24*, 271 ;

d) Kornblum, N. *Angew. Chem. Int. Ed. Engl.* 1975, *14*, 734.

2 - **a)** Feldberg, S.W. *J. Chem. Phys.* 1971, *75*, 2377 ;
b) Kochi, J.K. *J. Organomet. Chem.* 1986, *300*, 139 ;
c) Chanon, M. *Acc. Chem. Res.* 1987, *20*, 214 ;
d) Astruc, D. *Angew. Chem. Int. Ed. Engl.* 1988, *27*, 643 ;
e) For excellent reviews of the redox processes of organometallic complexes see : Geiger, W.E. ; Connelly, N.G. *Advan. Organomet. Chem.* 1985, *24*, 87 ; Connelly, N.G. ; Geiger, W.E. *Ibid.* 1984, *23*, 1.

3 - **a)** Miholova, D. ; Vlcek, A.A. *J. Organomet. Chem.* 1982, *240*, 413 ;
b) Magnuson, R.P. ; Meirowitz, R. ; Zulu, S.J. ; Giering, W.P. *Organometallics*, 1983, *2*, 460.

4- Desbois M.-H. ; Astruc, D. *J. Chem. Soc, Chem. Commun* 1988, 472 ; *New J. Chem.* 1989, *13*, 595.

5- **a)** Moinet, C. ; Román, E. ; Astruc, D. *J. Electroanal. Interfac. Chem.* 1981, *121*, 241 ;
b) For $[Fe(arene)_2]^{2+}$, see Karpinski, Z.I. ; Kochi, J.K. *J. Organomet. Chem.* in press.

6- Ruiz, J. ; Lacoste, M. ; Astruc, D. *J. Am. Chem. Soc.* 1990, *112*, 5471.

7- Desbois, M.-H. ; Astruc, D. *J. Chem. Soc., Chem. Commun.* 1990, 943.

8- Ruiz, J. ; Astruc, D. *J. Chem. Soc., Chem. Commun.* 1989, 815.

9- Ruiz, J. ; Lacoste, M. ; Astruc, D. *J. Chem. Soc., Chem. Commun.* 1989, 813.

10- Loupy, A. ; Tchoubar, B. ; Astruc, D. submitted to *Chem. Rev.*

11- **a)** Hershberger, J.W. ; Kochi, J.K. *J. Chem. Soc, Chem. Commun* 1982, 212 ;
b) Hershberger, J.W. ; Klinger, R.J. ; Kochi, J.K. *J. Am. Chem. Soc.* 1983, *105*, 61.

12- Rehm, D. ; Weller, A. *Isr. J. Chem.* 1970, *8*, 259.

13- Bard, A.J. ; Faulkner, L.R. "*Electrochemical Methods*", Wiley, New York, 1980, chapter 6.

14- Bezems, G.J. ; Rieger, P.H. ; Visco, S. *J. Chem. Soc, Chem. Commun* 1981, 265.

15- However for a few successful cases in complexes bearing electron-sink ligands see :
a) Kaim, W. ; Olbrich-Deussner, B. ; Gross, R. ; Ernst, S. ; Kolhmann, S. ; Bessenbacher, C. In "*Paramagnetic Species in Activation, Selectivity and Catalysis*" Chanon, M. ; Juillard, M. ; Poite, J.-C. Eds ; Kluwer, Dordrecht, 1989, p.283 ;
b) Lahuerta, P. ; Estenau, F.R. ; Latorre, J. *Inorg. Chim. Acta* 1986, *116*, L33.

16- Darchen, A. *J. Chem. Soc, Chem. Commun* 1983, 768.

17- **a)** Hamon, J.-R. ; Astruc, D. ; Michaud, P. *J. Am. Chem. Soc.* 1981, *103*, 758 ; **b)** Catheline, D. ; Astruc, D. *J. Organometal. Chem.* 1984, *272*, 417.

18- Marcus, R.A. *J. Chem. Phys.* 1956, *24*, 966.

19- **a)** Astruc, D. ; Dabard, R. ; Laviron, E. C.*R. Acad. Sci.* Paris, ser. C 1969, *269*, 608 ;
b) Gubin, S.P. *Pure Appl. Chem.* 1970, *23*, 463.

20- Astruc, D. *Acc. Chem. Res.* 1991, *24*, 36.

21- Astruc, D. In "*Mechanisms and Processes in Molecular Chemistry*" Astruc, D. Ed., *New J. Chem.*, 1992, *16*, 000.

22- **a)** Catheline, D. ; Astruc, D. *Coord. Chem.* 1982, *23*, F41 ;
b) Astruc, D. *Acc. Chem. Res.* 1986, *19*, 377 ;
c) Astruc, D. *Comments Inorg. Chem.*, 1987, *6*, 61. This paper briefly reviewed various electrocatalytic organometallic reactions including the oxidatively induced electrocatalytic dithiocarbamate chelation in $Fe^{II}Cp(CO)_2(dtc)$;
d) Verpeaux, J.-N. ; Desbois, M.-H. ; Madonik, A. ; Amatore, C. ; Astruc, D. *Organometallics* 1990, *9*, 630 ;
f) Desbois, M.-H. ; Nunn, C.M. ; Cowley, A.H. ; Astruc, D. *Organometallics* 1990, *9*, 640 ;
g) For an interesting similar example see Morán, M. ; Cuadrado, J.R. ; Masaguer, J.R. ; Losada, J. *J. Organomet. Chem.* 1987, *335*, 255.

23- **a)** Desbois, M.-H. ; Astruc, D. *Angew. Chem. Int. Ed. Engl.* 1989, *28,* 460
b) Delville, M.-H. ; Mross, S. ; Astruc, D. ; Cowley, A.H. ; Rabaa, H. ; Saillard, J.-Y. work in progress.

24- Tate, D.P. ; Knipple, W.R. ; Augl. J. M. *Inorg. Chem.* 1962, *1,* 433 ; Tate, D.P. ; Buss, A. ; Augl. J.M. *Ibid* 1963, *2*, 427.

25- For catalysis of alkyne polymerization using W complexes, see :
a) Landon, S. J. ; Shulman, P.M. ; Geoffroy, G. *J. Am. Chem. Soc.* 1985, *107*, 6739 and ref cited ;
b) Parker, A. ; Rudler, H. *J. Chem. Soc., Chem. Commun.*, 1986, 514 and ref cited.

26- Hershberger, J.W. ; Klinger, R.J. ; Kochi, J.K. *J. Am. Chem. Soc.* 1982, *104*, 3034.

27- **a)** Katz, T.J. ; Lee, S. *J. Am. Chem. Soc.* 1980, *102*, 422 ;
b) Katz, T.J. ; Ho, T.H. ; Shih, N.Y. ; Ying, V.C. ; Stuart, Y.I.W. *J. Am. Chem. Soc.* 1984, *106*, 2659.

MOLECULAR CATALYSIS OF ELECTROCHEMICAL REACTIONS BY LOW VALENT TRANSITION METAL COMPLEXES: METALLOPORPHYRINS.

D. LEXA and J.-M. SAVEANT
Laboratoire d'Electrochimie Moleculaire, UA 438
Université de Paris 7, 2 Place Jussieu,
75251 Paris cedex 05, France

ABSTRACT.Low valent transition metalloporphyrins are shown to react by an innersphere mechanism in the catalytic electrochemical reduction of electrophilic molecules like vicinal dihalides and CO_2. For vicinal dibromo alcanes the comparison of the rate constants of the key reaction with catalysts reacting by an outer-sphere mechanism (ET) like aromatic or heteroaromatic radical anion having the same standard potential, leads to the conclusion that the chemical mechanism is an E2 elimination rather than an SN2 mechanism. Not only is the catalytic efficiency much better, but also selectivity is drammatically improved. Iron(0) porphyrins react with CO_2 leading to CO and $CO_3{}^{2-}$ .But the porphyrin is rapidly consumed after a progressive carboxylation on the ring and the catalytic yields decrease. The addition of $Mg^{2+}C1O^{4-}$ triggers a spectacular increase of the catalytic efficiency and of the stability of the catalyst. The Mg^{2+} play the role of a co-substrat. Large temperature effects are discussed.

Introduction

Very numerous electrochemical reactions require a significant overpotential to proceed at appreciable rates owing to the slowness and irreversibility of the electron transfer and/of chemical steps they involve. Homogeneous catalysis of such electrochemical processes may be carried out along two conceptually different types of mechanism. In one case, "redox catalysis" [1], the active form of the catalyst, generated at the electrode, exchanges electrons with the reactant in an outer-sphere manner yielding the reaction products and regenerating the starting form of the catalyst.

On the other hand, the use of metal complexes as metalloporphyrins acting by a inner-sphere "chemical catalysis"[2] involves the transient formation of an adduct between the active form of the catalyst and the reactant. The chemical bond thus formed has to be cleaved successively or after the exchange of additional electrons for eventually yield the products and regenerating the starting form of the catalyst.

The catalytic efficiency is related to the nature of the catalyst through its standard potential in redox catalysis. The resulting kinetic law have been amply illustrated with examples in the reduction of aliphatic and aromatic halides in aprotic solvents. The catalytic efficiency is then an increasing function of the corresponding rate constant k_1 which is itself related to the difference between the standard potential of the redox catalyst $E°_{cat}$, and the redox potential concerning the substrate $E°_{sub}$, according to a Brönstedt-Marcus activation driving force free energy relationship.

A. J. L. Pombeiro and J. A. McCleverty (eds.),
Molecular Electrochemistry of Inorganic, Bioinorganic and Organometallic Compounds, 295–306.

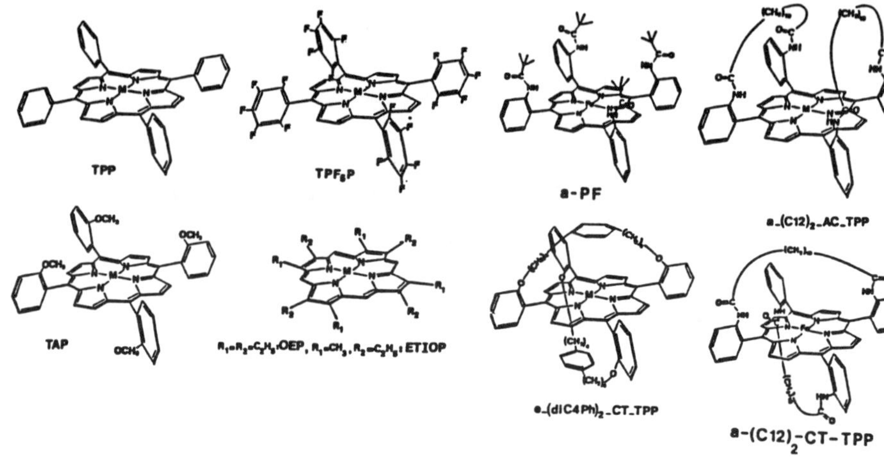

Figure 1. Porphyrins used in the present work with their conventional designations.

Identification of a chemical catalysis process is more difficult. Thus far, the only approach has been to characterise the intermediate adducts formed by the reaction of the active form of the catalyst with the substrate or to calculate the driving forces relative to each proposed mechanisms and to make the more suitable choice according to the experimental results.

In the following we describe an another approach to the problem which consists in comparing chemical and redox catalysis of the same reduction process, in terms of catalytic efficiency and thus of rate constants of the rate determining step, as a function of the driving force offered by the catalyst couple, itself measured by its standard potential.

Two examples illustrating this approach, will be discussed: one is the electrochemical reduction of vicinal dihalides leading to the corresponding olefins[2e] and the second one is the electrochemical reduction of carbon dioxide[3]. In both cases, two advantages are expected from chemical (inner-sphere) catalysis over redox (outer-sphere) catalysis, namely the enhancement of efficiency for catalysts having the same standard potential and better opportunities to achieve selective reactions.

The chemical catalysts will be electrochemically reduced iron(I) and iron(0) porphyrins and other low valent metalloporphyrins as Co(I) porphyrins which have been shown previously to react on aliphatic halides along an inner-sphere SN2 mechanism leading to the corresponding alkyl-σ-complexes[4] and Ni(I) porphyrins[5] (the various porphyrins used in this work are presented in Figure 1). The outer-sphere electron donors will be chosen among aromatic and heteroaromatic anion radicals of known increasing standard potentials .

Materials and Methods

All the cyclic voltammetric experiments have been conducted in DMF + 0.1M NEt_4Cl_4 on glassy carbon electrodes. The potentials are mesured versus an SCE (saturated with NaCl).
Detailed experimental conditions and preparation of the materials have been published[2e,3b].

Results and discussions

CATALYTIC REDUCTION OF VICINAL DIBROMOALCANES[2e].

The discussion will deal with three main possible mechanisms involving the following rate determining steps:

-outer-sphere electron transfer (ET): $OlX_2 + D^-(D^{\cdot -}) \rightarrow OlX^{\cdot} + D^{\cdot}(D) + X^-$

-inner-sphere mechanisms:

halogen-atom transfer (HA): $OlX_2 + D^-(D^{\cdot -}) \rightarrow OlX\cdot + DX^{\cdot -}(DX^-)$

halonium ion transfer concerted with halide expulsion (E2 elimination):

$$OlX_2 + D^-(D^{\cdot -}) \rightarrow Ol + X^- + DX(DX^{\cdot})$$

The rate constants for the reaction of four acyclic vic-dibromides and three dibromocyclohexanes with aromatic anion radicals (including reduced zinc(II) and copper(II) porphyrins) as well as with iron(I) and cobalt(I) porphyrins have been determined. They were derived from the catalytic current obtained at the initially reversible wave of the parent compound upon addition of the dibromoalkane to the solution. In all cases, the catalytic currents that corresponds to two electrons transfered on the overall from the donor to the dibromide were observed. Derivation of the rate constant for the reaction of the electrogenerated electron donor to the dibromoalkane was carried out along known procedures.

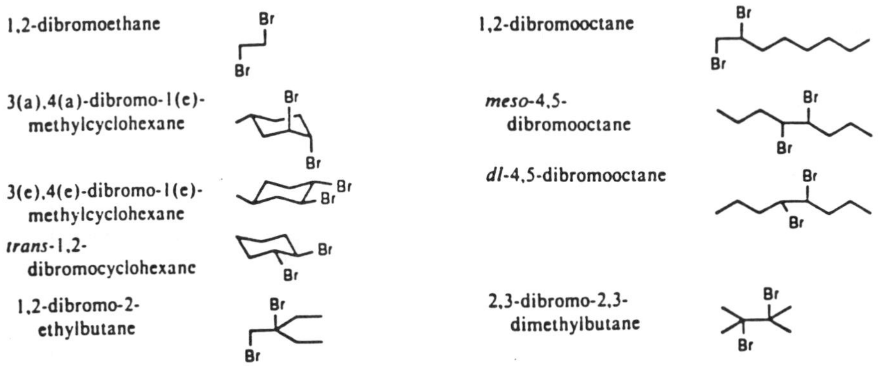

Figure 2. Vicinal dibromoalcanes used in this work

On Figures 3 and 4 are representated the variations of the rate constants with the standard potential of the electron donor couple.

As noted earlier[2d], reduced zinc(II) and copper(II) porphyrins behave as aromatic anion radicals in the sense that the representative points follow the same activation *vs.* driving force line as the aromatic anion radicals.

Let us now examine the reaction of vic-dibromoalkanes with the low oxidation state iron and cobalt porphyrins. A first indication of their inner-sphere character is the fact that they react

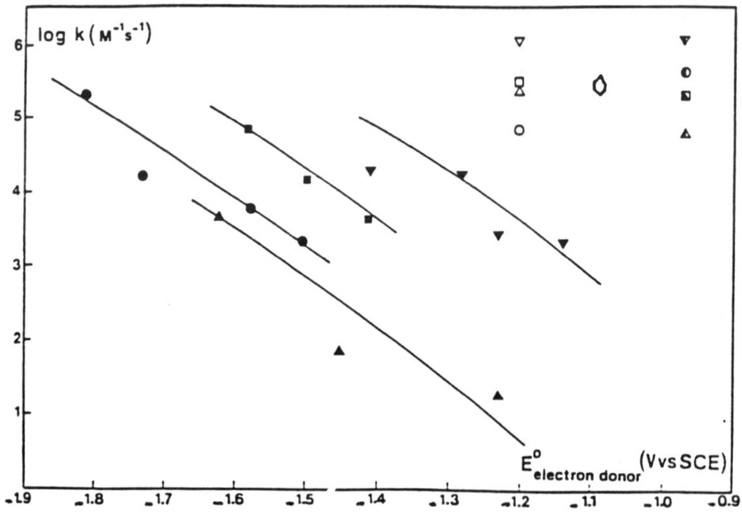

Figure 3. Reaction of electrogenerated anion radicals (solid symbols), OEPFe(I) (open symbols), an EtioPCo(I) (half-open symbols) with 2,3-dibromo-2,3-dimethylbutane (▼ , ▼ , ▽), 1,2-dibromo-2 ethylbutane (■ , ◩ , □), 1,2-dibromooctane (● , ◐ , ○), and 1,2-dibromoethane (▲, ▲ , △) (◇ represents the reaction of 1,2-dibromoethane with e-(di C_4 Ph)$_2$-CT-TPPFe(I). Variation of the rate constar with the standard potential of the electron donor.

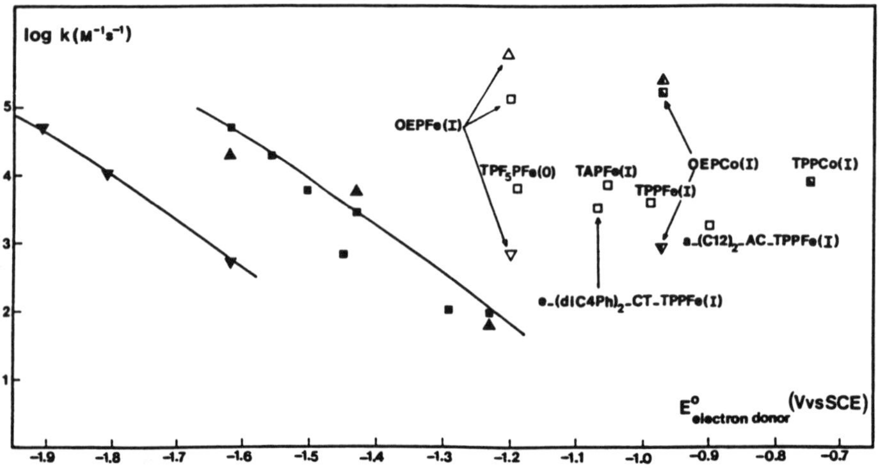

Figure 4. Reaction of electrogenerated anion radicals (solid symbols), iron(I) and iron(0)porphyrins (open symbols), and cobalt(I) porphyrins(half-open symbols) with *trans*-1,2-dibromocyclohexane (■,◩,□)axial axial-*trans*-dibromo-3-methylcyclohexane (▲ , ▲ , △)equatorial,equatorial-*trans*-dibromo-3-methylcyclohexane (▼,▼,▽).Variation of the rate constant with the standard potential of the electror donor.

much faster, by a factor of 10^3-10^4 for the iron porphyrins and 10^5-10^6 for cobalt porphyrins, than the aromatic anion radicals that would have the same standard potential (Figures 3). The same effect is found in the particular case of OEPFe(I) and EtioPCo(I) in their reaction with *trans*-dibromocyclohexane[1,2], and the blocked isomers such as 3(a),4(a)-dibromo-l(e)-methylcyclohexane and 3(e),4(e)-dibromo-l(e)methylcyclohexane (Figure 4).

For these last compounds we have shown previously stereochimical effect due to an antiperiplanar preference of the β-bromoradical produced after the first reaction step, the stability of which in the partial double bond C-C-Br increases for the a-a isomer vis-à-vis the e-e isomer of about 0.20 eV. Thus the last one is more difficult to reduce. For the two series of catalysts this antiperiplanar preference is maintained but is slightly larger by a factor 10^3 instead of 10^2 for the iron and cobalt porphyrins. Alkyl substitution at functional carbon does not seem, however to have a large effect.

The stereoselectivity obtained during the preparation of the corresponding olefin (Ol) by preparative scale electrolysis mediated by different catalysts have been checked. In Table 1 are gathered the ratio of the E/Z isomers obtained after electrolysis of meso and dl-4,5 dibromooctanes using different kind of mediators. Using 1,4-diacetylbenzene there is not a good selectivity, even if the racemic is not obtained. On the other hand, using the innersphere mediators leads to a very good selectivity according to the starting material.

4,5-dibromooctane	mediator (concn in mM)	cathode potential V vs SCE	E/Z 4-octene
	1,4-diacetylbenzene		
3	(57)	-1.41	87/13
	(5.7)	-1.41	87/13
4	(57)	-1.41	36/64
	(11,4)	-1.41	41/59
	(5.7)	-1.41	38/62
	OEPFe		
3	(1.4)	-1.23	100/0
4	(1.4)	-1.23	0/100
	ETIOCo		
3	(1.4)	-1.03	100/0
4	(1.4)	-1.03	O/100

Table 1. Yields of (*E*)- and (*Z*)-4-Octenes upon Mediated Preparative Electrolysis of *meso*-(**3**) and *dl*-4,5-Dibromooctanes(**4**). In DMF + 0.1M Bu_4NBF_4; average of three determinations.

Among the various conceivable inner-sphere mechanisms we can rule out an SN2 attack at carbon followed by elimination of Fe(III)Br or Co(III)Br in the case of Fe(I) and Co(I) and Fe(II)Br in the case of Fe"0". This follows from the antiperiplanar preference which would imply strong steric hindrance by the second bromine atom in the transition state. It is confirmed by the fact that sterically congested e-(diC4Ph)2-CT-TPPFe(I) complex reacts at almost the same rate that unencumbered TAPFe(I) complex which has about the same standard potential (Figure 3 and 4).

Two other mechanisms remain, namely halogen atom (HA) and halonium ion (E2) abstraction. For these two mechanisms no significant influence of steric hindrance at the electron donor is likewise expected. The fact that total stereoselectivity is obtained with both iron(I) and cobalt(I) porphyrins (Table 1) demonstrates that the reaction follows an E2 mechanism rather than an HA mechanism. In the later case indeed a partial lost of stereoselectivity should occur as with aromatic anion radicals since the reaction goes entirely through the same $OlX^{\cdot-}$ radical.

We have calculated from data obtained from direct reduction of the substrate at a glassy electrode, thermochimica data and the stabilities constants values of the complexes of Fe(III), Fe(II) and Co(III),Co(II) with bromide, the driving forces of the rate determining step in the ET, HA, and E2 mechanisms[2e]. The E2 mechanism has a strong driving force advantage above the others (about 1 to 1.9 eV) depending of the metalloporphyrin concerned. The halonium abstraction is related to the strong affinity of the iron(III) and cobalt(III) toward halide ions and thus is the mechanism involved in the inner-sphere catalysis.

REDUCTION OF CO_2[3]

The electrochemical reduction of carbon dioxide has attracted considerable attention as a possible source of carbon for the synthesis of organic molecules and as a possible means of energy storage. CO_2 is a low energy molecule and its electrochemical reduction into its anion radical requires a quite negative potential[6]. The products of the direct electrochemical reduction depend both on the electrode and the reaction medium. In low acidity solvents, such as N,N'-dimethylformamide (DMF), and with inert electrodes, such as mercury and lead, the reduction takes place at potential slightly negative to the standard potential of the $CO_2/CO_2^{\cdot-}$ couple.

To orient the reaction selectively, various heterogeneous catalysts[7], mostly metals, have been used but also homogeneous molecular catalysts[6e]. Concerning the latter, a distinction should be made between redox catalysts and chemical catalysts. Aromatic and heteroaromatic anion radicals often react as outersphere electron donors. However in the case of CO_2 reduction, the $CO_2^{\cdot-}$ anion radical couples with the aromatic anion radical leading to an hydrocarboxylation reaction that destroys the catalyst. It appears however that benzonitrile, benzoic esters, their substituted analogues as well as phthalic esters escape this rule and are thus able to act as redox catalyst precursors[6e]. Oxalate is then the main reaction product as in the direct reduction of CO_2 on inert electrodes.

In this context, we used iron(0) porphyrins (TPP, a-PF, and a-$(C12)_2$-CT-TPP) wHich are able to develop iron-centered chemistry involving bonded interactions and eventually bond formation with electrophilic centers. There were thus some hope that iron(0) porphyrins may serve as chemical catalysts of the reduction of carbon dioxide.

The main conclusions emerging from the results of extensive cyclic voltammetry and preparative-scale electrolysis investigations are the following. Iron("O") porphyrins do catalyse the electrochemical reduction of carbon dioxide. The main reduction product is carbon monoxide. However, the porphyrin is rapidly consumed and the catalysis stops after a small number of cycles (Figure 5). The degradation of the porphyrin appears to be the result of progressive saturation of the ring through carboxylation and/or hydrogenation.

The addition of Mg^{2+}, under the form of a non-hydrated salt with a non-nucleophilic (or weakly nucleophilic) counter-anion such as perchlorate, and by dissolution of the Mg anode[8] during preparative scale electrolysis, as shown on Figure 6, triggers a spectacular increase of the catalytic efficiency and of the stability of the catalyst. These improvements are best achieved with the simple TPP whereas the superstructured porphyrins investigated continue to be consumed in the catalytic reaction albeit to a lesser extent than whithout Mg^{2+} ions in the solution. Lewis acids such as LiCl, $BaCl_2$, $AlCl_3$ also improve the catalytic efficiency and the stability of the catalyst but to a much lesser extent than "free" (although complexed by solvent molecules) Mg^{2+} ions.

Decreasing the temperature increases the catalytic efficiency mostly because CO_2 is more soluble. The fact that this effect is not compensated by a decrease of the rate determining kinetic constant indicates that the overall reaction possesses a small activation energy. The degradation of the porphyrin also appears to be lowered upon decreasing the temperature.

The comparison with homogeneous catalysis by outersphere electron donors[6e] (benzonitrile anion radicals and similar nitriles or esters) clearly indicates that iron("O") porphyrins act as chemical rather than redox catalysts in the reduction of CO_2: the main product is different (CO instead of oxalate) and the catalytic efficiency is higher for the same standard potential of the catalyst couple.

At low temperature (-40°C) as well as at room temperature (20°C), the catalytic process begins with the introduction of one molecule of CO_2 in the coordination sphere of the iron atom (Schemes I and II) forming a carbene-type complex in which electrons have been transferred from the iron porphyrin to the CO_2 moiety. At low temperature, the second step, in the absence as well as in the presence of Mg^{2+} ions, is the addition of a second molecule of CO_2 in an acid-base type manner.The C-O bond of the first CO_2 molecule is then broken with formation of the Fe(II)CO complex and of a carbonate ion.

This reaction is accelerated by the presence of Mg^{2+} ions, presumably through the intermediary formation of a magnesium complex. At room temperature, the Mg^{2+} ions participate earlier to the breaking of the C-O, namely at the level of the first iron-CO_2 adduct. This variation of the reaction pathway with temperature falls in line with the expected differences in the entropy-enthalpy balance for the formation of an ion-pair with a Mg^{2+} ion on one hand and the addition of a CO_2 molecule on the other, taking account of the interactions of these two species with the solvent.

At room temperature the coordination of CO with the iron(I) porphyrin is weak and the oxidized form of the catalyst (*i.e.*, the iron(I) porphyrin) is regenerated without difficulty. This is not the case at low temperature where the dissociation of the $Fe(I)CO^-$ complex is more difficult. The regeneration of the catalyst thus follows the above pathway only in part, the remainder of the $Fe(I)CO^-$ porphyrin being reduced at a slightly more negative potentials thus regenerating the catalyst in its reduced form.

The role played by the Mg^{2+} ions in the catalytic reaction offers a noteworthy example of a bimetallic chemical catalysis of an electrochemical reduction where an electron-rich center, here the iron("O") porphyrin, commences the reduction process in an innersphere fashion and an

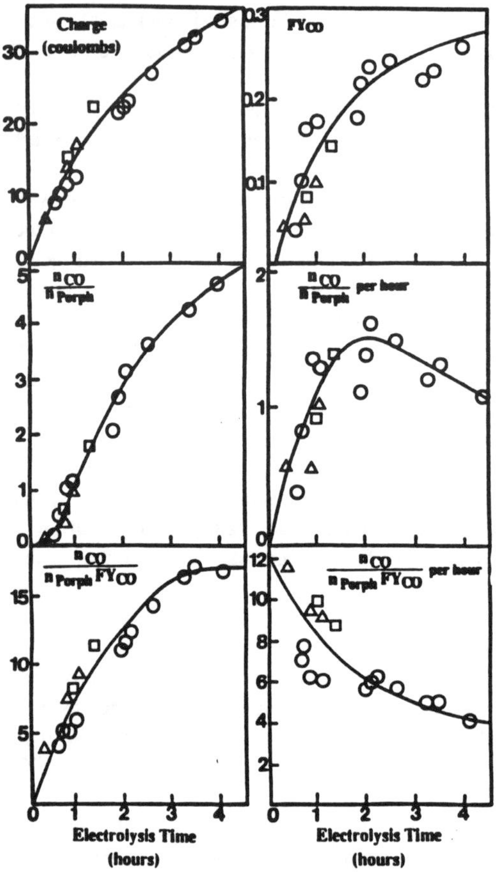

Fig.5. Electrolysis of CO_2 (1 atm) in the presence of TPPFe(I) (0.5 mM) in DMF + 0.1 M NEt_4ClO_4 on a mercury pool electrode in a two-compartment cell. Charge, faradaic yield in CO (FYco), turnover number, (n_{co}/n_{Porph}) turnover number per hour, turnover number for the production of CO ($n_{co}/n_{Porph}\ FY_{co}$), turnover number for the production of CO per hour as functions of the electrolysis time. Electrolysis potential : -1.8 V vs SCE. Temp.: 20°C. The three types of points (Δ, □, O) correspond to three different runs.

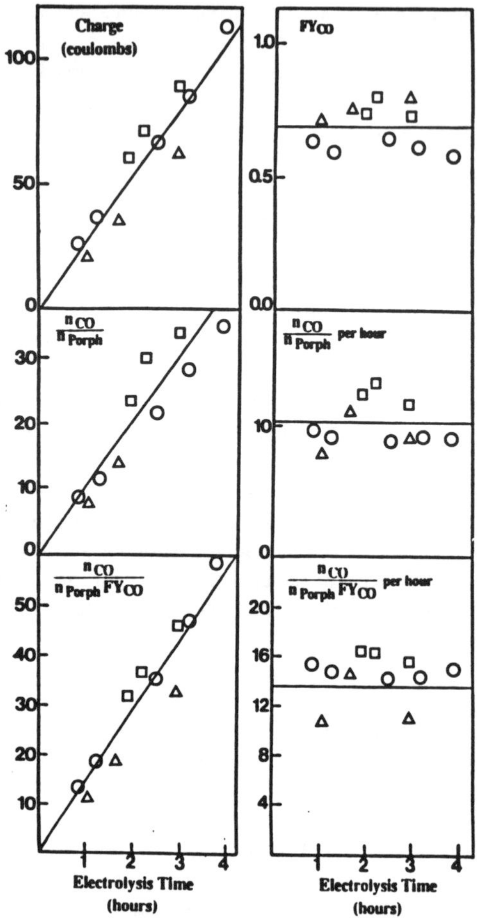

Fig. 6. Electrolysis of CO_2 (1 atm.) in the presence of TPPFe(I) (0.5 mM) and Mg^{2+} ions (24mM) in DMF + 0.1 M NEt_4ClO_4 on a mercury pool electrode in an undivided cell with a soluble magnesium anode. Charge, faradaic yield in CO (FY_{co}), turnover number (n_{co}/n_{Porph}), turnover number per hour, turnover number for the production of CO ($n_{co}/n_{Porph}FY_{co}$), turnover number for the production of CO per hour as functions of the electrolysis time. Electrolysis potential : -1.7 V vs SCE. Temperature : 20°C. The three types of points (Δ, □, O) correspond to three different runs.

SCHEME I

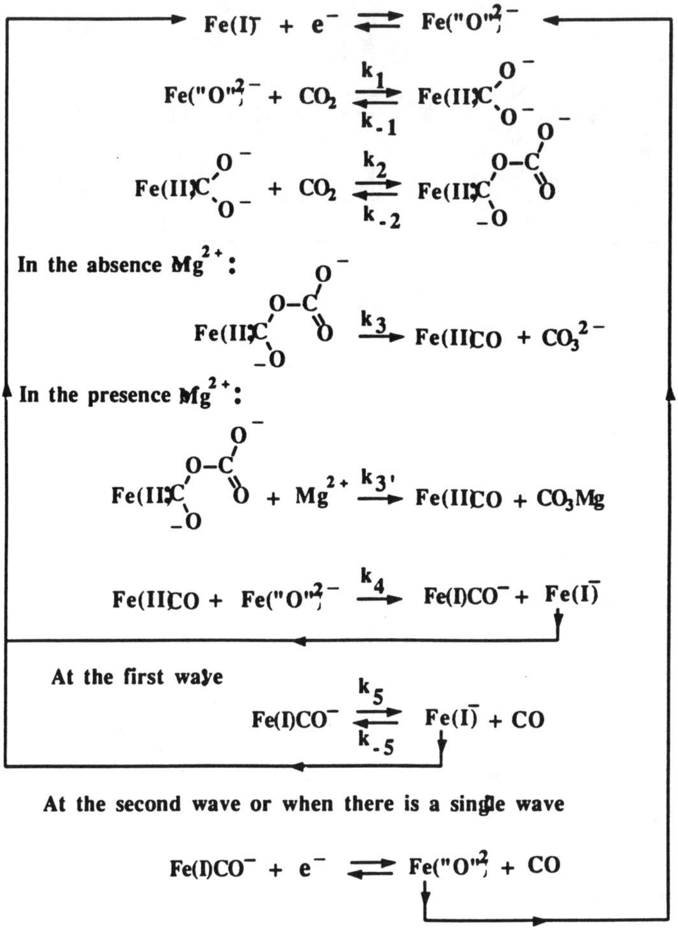

(1)

(2)

(3)

(3')

(4)

(5)

(6)

SCHEME II

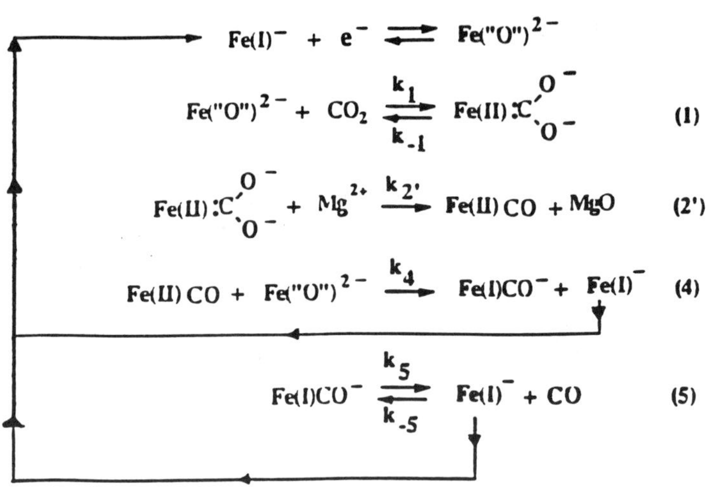

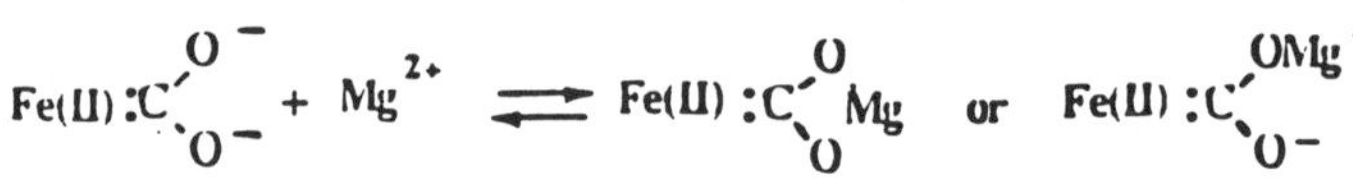

(2'')

electron-poor center, here a Mg^{2+} ion, completes the bond transformation process in a manner that is reminiscent of electrophilic assistance in nucleophilic substitution. It should however be noted that Mg^{2+} ions play the role of a cosubstrate rather than that of a co-catalyst.

Conclusion

These two examples show how electrochemically mediated catalysis depends on the mechanisms involved in the rate determining reaction with the substrate. The investigated systems underline the superiority of inner-sphere(chemical) catalysis over outer-sphere(redox) catalysis of electrochemical reactions. Not only the catalytic efficiency is much better since the rate constants of the key step are larger, given the standard potential of the catalyst, but also selectivity is dramatically improved.

Acknowledgements

I would like to acknowledge my various coworkers in the above described studies, whose names can be found in the reference list work. The collaboration with Dr. M. Momenteau, Pr. H.J. Shafer and B. Vering, who synthesized the basket-handle porphyrins and the substituted vicinal dibromoalkanes was essential. This investigation was supported by the Centre National de la Recherche Scientifique.

References.

1. Andrieux, C. P., Dumas-Bouchiat, J-M., and Savéant, J-M.(1978), J. Electroanal. Chem. , *87*, 39

2 (a) Lexa, D., Savéant, J-M., and Wang, D. L.(1986), Organometallics, *5*, 1428. (b) Lexa D., Savéant, J-M. , Su, K. B., and Wang, D. L.(1988), J. Am. Chem. Soc., *110*, 7617 . (c) Gueutin,C., Lexa, D., Savéant, J-M. and Wang, D. L.(1989). Organometallics, 8, 1607. (d) Lexa, D., Savéant, J-M., Su, K. B. and Wang, D. L., J. Am. Chem. Soc., (1987).*109*, 6464 (e) Lexa, D., Savéant, J-M., Schäfer, H. J., Su, K. B., Vering, B., and D. L. Wang, D. L., J. Am. Chem. Soc., (1990).*112*, 6162

3.(a) Hammouche, M., Lexa, D., Momenteau, M., and Savéant, J-M., J. Electroanal.,(1988) *249*, 347 .(b) Hammouche, M., Lexa, D., M. Momenteau, Savéant, J-M., (1991), J.Am.Chem.Soc. ,*113*, 8455, .

4 .Lexa,.D., Mispelter, J., Savéant, J-M., J. Am. Chem. Soc. (1981),*103*, 6806.

5. Lexa, D., Momenteau, M., Mispelter, J., Savéant, J-M., Inorg. Chem. (1989), *28*, 30.

6. (a): the standard potential of the $CO_2/CO_2^{\bullet-}$ couple in N,N'-dimethylformamide (DMF) with 0.1 M NEt_4ClO_4 as supporting electrolyte has been estimated to be -2.2 V vs SCE by cyclic voltammetry on a mercury electrode[6b]. More recent results obtained by means of redox catalysis[6c,d] indicate a somewhat more negative value that depends upon the nature of the solvent, the nature and concentration of the supporting electrolyte[6e]. It may also depends upon the CO_2 concentration in view of possible acid-base interactions between CO_2 and $CO_2^{\bullet-}$. (b) Lamy, E.; Nadjo, L.; Savéant, J-M., J. Electroanal. Chem. (1977), *78*, 403. (c) Andrieux, C. P.; Savéant, J-M., "Electrochemical Reactions" in Investigations of Rates and Mechanisms ; Bernasconi, C. F., Ed.; Wiley: New York, (1986); Vol. 6, 4/E, Part 2, pp. 305. (d) Andrieux, C. P.; Hapiot, P.; Savéant, J-M., Chem. Rev. (1990), *90*, 723.(e) Vianello, E., private communication, (1990). (f) Kaiser, U.; Heitz, E. Ber. Bunsenges. Phys. Chem. (1973), *77*, 818. (g) Gressin, J. C.; Michelet, D.; Nadjo, L.; Savéant, J-M. Nouv. J. Chim. (1979), *3*, 545. (h) Amatore, C.; Savéant, J-M. J. Am. Chem. Soc. (1981), *103*, 5021. (i) Amatore, C.; Savéant, J-M. J. Electroanal. Chem. (1981), *125*, 22. (j) Amatore, C.; Nadjo, L.; Savéant, J-M. Nouv. J. Chim. (1984), *8*, 565. (k) Fisher, J.; Lehmann, T.; Heitz, E., J. Appl. Electrochem. (1981), *11*, 743.

7. (a) : for recent reviews see refs 7b-d. (b) Silvestri, G. in "Carbon Dioxide as a Source Carbon", Aresta, M.; Forti, G. Eds., NATO ASI Series; Ser. C, Reidel, Dordrecht, (198 pp. 339. (c) Collin, J. P.; Sauvage, J. P. Coord. Chemistry Rev. (1989), *93*, 245. Silvestri, G.; Gambino, S.; Filardo, G. in "Enzymatic and Model Carboxylation a Reduction Reactions for CO_2 Utilisation," NATO ASI Series, Reidel, Dordrecht.
8. (a) Silvestri, G.; Gambino, S.; Filardo, G.; Guainazzi, M.; Ercoli, R. Gazz. Chim. I (1972), 102, 818. (b) Sibille, S; Coulombeix, J.; Perichon, J; Fuch, J.M; Mortreux, Petit, F., J. Mol. Catal. (1985), *32,* 239. (c) Chaussard, J.; Folest, J.C.; Nédélec J. Périchon, J.; Sibille, S.; Troupel, M., Synthesis. (1990), 369.

ONE TWO-ELECTRON *vs* TWO ONE-ELECTRON MECHANISMS IN THE CATALYSIS OF ELECTROCARBOXYLATION OF ARYL HALIDES BY NICKEL AND PALLADIUM COMPLEXES

Christian AMATORE, Anny JUTAND and Merete Folmer NIELSEN
Ecole Normale Supérieure
Département de Chimie. URA CNRS 1110
24 rue Lhomond.
75231 Paris Cedex 05
France.

ABSTRACT.

Investigation of the mechanism of the electrocarboxylation of aryl halides catalyzed by palladium or nickel complexes clearly evidences the crucial role played not by the metal, but by the ligand (*e.g.* monodentate *vs* bidentate) on the propagation of the catalytic cycle. Both mechanisms involve a succession of chemical and electron tranfer reactions. But for bidentate ligand (illustrated by $NiCl_2(dppe)$) one observes only distinct one electron tranfer reactions and the catalyst activates both aryl halide and carbon dioxide, whereas when the catalyst is ligated by a monodentate ligand (illustrated by $PdCl_2(PPh_3)_2$) overall two-electron transfer reactions are involved and the catalyst only activates the aryl halide but not carbon dioxide, the carboxylation step occuring then between CO_2 and a free aryl anion.

1. Introduction

The electrosynthesis of aromatic carboxylic acids from aryl halides and carbon dioxide is either catalyzed by nickel[1-4] or palladium[5] complexes.

$$ArX + CO_2 + 2\,e \xrightarrow{M^{II}X_2L_2} ArCO_2^- + X^- \qquad M = Ni, Pd$$

The most efficient catalysts reported so far are $NiCl_2(dppe)$ (dppe = $PPh_2CH_2CH_2PPh_2$) in which the nickel is ligated by a bidentate ligand and $PdCl_2(PPh_3)_2$ in which the palladium is ligated by a monodentate ligand. Whereas the electrochemical reduction of divalent complexes of nickel and palladium coordinated by monodentate ligands involves one overall two-electron process[6-8], the reduction of divalent complexes coordinated by bidentate ligands involves two one-electron processes.[6,9-11] This difference due to the nature of the ligand, led us to investigate the mechanism of the electrocarboxylation of aryl halides catalyzed by $NiCl_2(dppe)$ and $PdCl_2(PPh_3)_2$ and to examine the influence of the ligand on the carbon dioxide activation process.

A. J. L. Pombeiro and J. A. McCleverty (eds.),
Molecular Electrochemistry of Inorganic, Bioinorganic and Organomettallic Compounds, 307–310.

2. Mechanism of the electrocarboxylation of aryl halides catalysed by $NiCl_2$(dppe).

The catalytic cycle of the carboxylation of aryl halides catalyzed by the bivalent complex $NiCl_2$(dppe) is initiated by a reduction of this complex by a two one-electron process[9] at the very beginning of the electrolysis (see Scheme 1) which affords a zerovalent nickel complex Ni°(dppe). This low ligated complex can undergo a fast oxidative addition with the aryl halide to form a σ-arylnickel complex $ArNi^{II}X$(dppe). This latter can be reduced, more easily than ArX and its reduction involves one electron providing a paramagnetic species $ArNi^{I}$(dppe). The electrosynthesis yields the carboxylic acid only when it is performed at the reduction potential of $ArNi^{II}X$(dppe)[2] and the resulting complex $ArNi^{I}$(dppe) is the efficient species which can activate carbon dioxide. This activation proceeds in two steps: fixation of CO_2 by $ArNi^{I}$(dppe) to form a new paramagnetic species $ArNi^{III}(CO_2)$(dppe) and then insertion of carbon dioxide into the C-Ni bond providing a monovalent nickel carboxylate $ArCO_2Ni^{I}$(dppe) whose one-electron reduction to the initial zerovalent nickel allows the catalytic cycle to propagate.

Scheme 1[3]

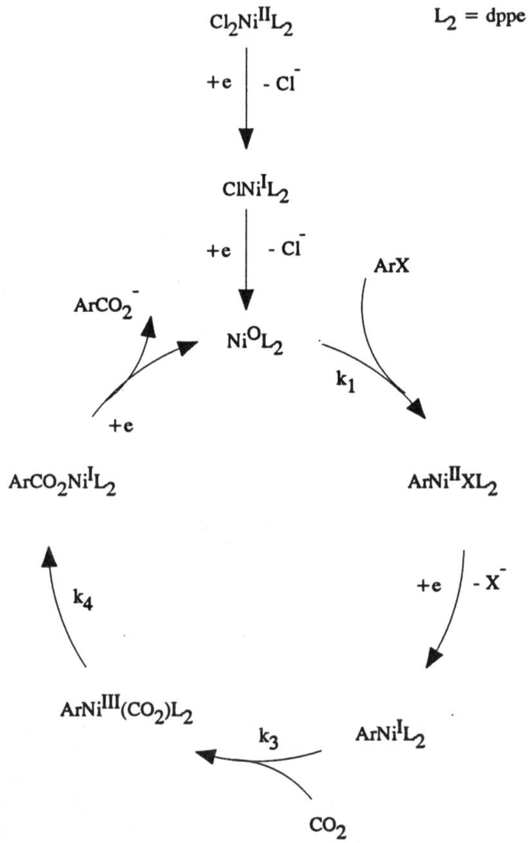

In the case of PhBr

$k_1 = 10^5 M^{-1} s^{-1}$

$k_3 = 10^5 M^{-1} s^{-1}$

$k_4 = 100 s^{-1}$

Therefore, the catalytic cycle is found to proceed with a succession of elementary steps involving paramagnetic and diamagnetic nickel complexes. The catalytic cycle is a succession of single electron transfer reactions (whose half-wave potentials have been characterized) and chemical steps (whose rate constants have been determined by transient electrochemical technics)[3,4]. The first oxidative addition step and the fixation of CO_2 by $ArNi^I(dppe)$ are fast reactions while the last chemical step, *e.g.* the insertion step of CO_2 is the rate determining step of the catalytic process.
In this catalytic reaction the nickel plays a dual role: first, it activates the aryl halide and makes its reduction more easy, and then it activates carbon dioxide. In such reactions, the carboxylation steps are found to occur within the coordination shell of the metal.

3. Mechanism of the electrocarboxylation of aryl halides catalysed by $PdCl_2(PPh_3)_2$.

The catalytic cycle of the carboxylation of aryl halides catalyzed by the bivalent complex $PdCl_2(PPh_3)_2$ is initiated by one overall two-electron reduction (see Scheme 2) which affords a zerovalent palladium complex $Pd°(PPh_3)_2$ This low ligated complex can undergo an oxidative addition with the aryl halide to form a σ-arylpalladium complex $ArPd^{II}X(PPh_3)_2$.[8] which can be reduced, more easily than ArX and its reduction involves one two-electron process providing an anionic species $ArPd^{\circ}(PPh_3)_2]^-$. This latter is in fast equilibrium with $Pd^{\circ}(PPh_3)_2$ and the anion Ar^- which reacts with carbon dioxide by a nucleophilic attack.

Scheme 2

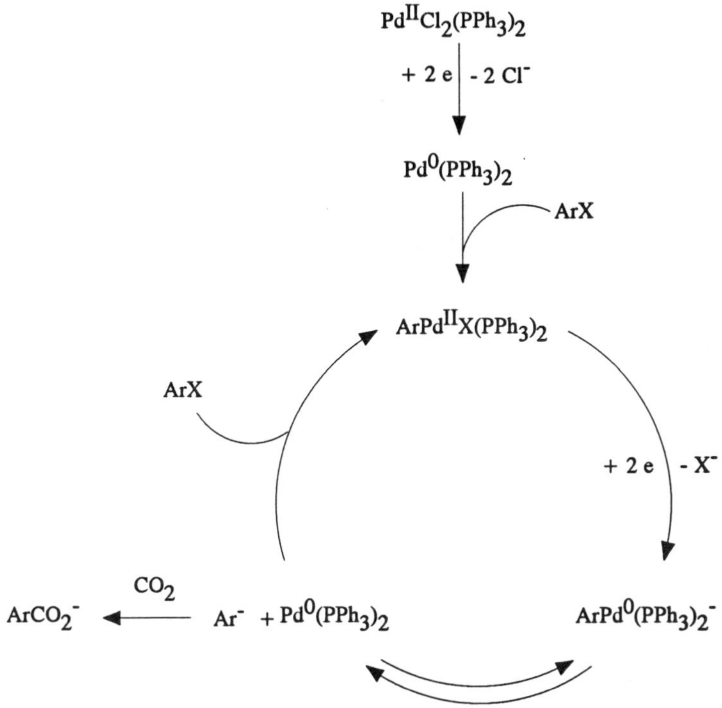

In contrast with the above reported mechanism of the $NiCl_2$(dppe) catalysis, the $PdCl_2(PPh_3)_2$ catalysis is found to proceed with a succession of elementary steps involving only paramagnetic palladium complexes. The catalytic cycle is a succession of two-electron transfer steps (which have been characterized by their potential) and chemical steps. The oxidative addition is the rate determining step of the catalytic cycle while the nucleophilic attack of CO_2 by the anion Ar^- is a fast reaction.
In this catalytic reaction the palladium plays only one role: it activates the aryl halide and makes its reduction more easy (Ar^- is formed at a less negative potential) but the metal does not play any role in the activation of carbon dioxide. In such reactions, the carboxylation step does not occur within the coordination shell of the metal. [12]

4. Conclusion.

The electrocarboxylation of aryl halides catalyzed by either $NiCl_2$(dppe) or $PdCl_2(PPh_3)_2$ proceeds via two different mechanisms as a function of the catalyst. This difference is not due to the metal but to the ligand. When the metal is ligated by a bidentate ligand ($NiCl_2$(dppe)), only distinct one-electron transfer reactions are involved and the catalytic cycle proceeds with a succession of steps involving diamagnetic and paramagnetic nickel complexes. Whereas when the metal is ligated by a monodentate ligand ($PdCl_2(PPh_3)_2$) overall two-electron transfer reactions are involved and the catalytic cycle proceeds with a succession of steps involving only diamagnetic palladium complexes. The main consequence is that in the former case, the carboxylation reaction occurs in the coordination shell of a nickel complex whereas in the latter case, the carboxylic acid is formed by a nucleophilic attack of CO_2 by a free anion Ar^-.

5. References.

1. M. Troupel, Y. Rollin, J. Périchon, J.F. Fauvarque, *New J. Chem.* **1981,** *5,* 26.
2. J.F. Fauvarque, C. Chevrot, A. Jutand, M. Francois, J. Périchon, *J. Organomet. Chem.*, **1984,** *264* , 273.
3. C. Amatore, A. Jutand, *J. Am. Chem. Soc.*, **1991,** *113,* 2819.
4. C. Amatore, A. Jutand, *J. Electroanal. Chem.*, **1991,** *306,* 141.
5. S. Torii, H. Tanaka, T. Hamatani, K. Morisaki, A. Jutand, F. Pflüger, J.F. Fauvarque, *Chem. Lett.*, **1986,** 169.
6. C. Amatore, M. Azzabi, P. Calas, A. Jutand, C. Lefrou, Y. Rollin, *J. Electroanal. Chem.*, **1990,** *288,* 45.
7. C. Amatore, M. Azzabi, A. Jutand, *J. Am. Chem. Soc.*, **1991,** *113,* 1670.
8. C. Amatore, M. Azzabi, A. Jutand, *J. Am. Chem. Soc.*, **1991,** *113,* 8375.
9. C. Amatore, M. Azzabi, A. Jutand, *Organometallics.*, **1988,** *7,* 2203.
10. C. Amatore, A. Jutand, L. Mottier *J. Electroanal. Chem.*, **1991,** *306,* 125.
11. C. Amatore, S Aziz, A. Jutand, F. Kahlil, M.A. M'Barki, L. Mottier *unpublished results.* **1992.**
12. C. Amatore, A. Jutand, M.F. Nielsen, *J. Am. Chem. Soc.*, **1992,** submitted.

ELECTROCHEMICAL REDUCTION OF CO_2 CATALYSED BY TRANSITION METAL COMPLEXES

Armando GENNARO, Abdirisak A. ISSE and Elio VIANELLO
Department of Physical Chemistry, Padua University,
Via Loredan 2, 35131 Padua
Italy

ABSTRACT. Cobalt and nickel complexes of N,N'–1,2–phenylenebis(salicylideneiminato) (salophen = L) have been investigated by cyclic voltammetry and controlled potential electrolysis as potential catalysts for the electrochemical activation of CO_2. In the case of [Ni(II)L], no catalysis was observed. The first electron uptake is ligand–based, yielding the radical anion $[Ni(II)L]^{\cdot -}$ which rapidly dimerizes. On the contrary, the electron transfer to [Co(II)L] is metal–centered giving the $[Co(I)L]^-$ complex which in AN containing $LiClO_4$ as background electrolyte appears to catalyze the CO_2 reduction to CO and CO_3^{2-}, in a potential region where direct CO_2 reduction does not occur. Mechanistic investigations point to the interaction of CO_2 with [Co(I)(L)Li] to give a precursor complex containing a head–to–tail CO_2 dimer. Further reduction of the latter yields a Co–carbonyl complex which releases quite slowly the final CO_2 reduction product CO.

Introduction

The electrochemical activation of carbon dioxide is intensely investigated as a possible means of converting this rather inert molecule into useful chemicals. Direct electrochemical reduction on most metal electrodes suffers from a large overpotential and an extreme dependance of product distribution both on the nature of the electrode material and of the reaction medium [1]. Efficient reduction requires catalysis. A large number of transition metal complexes has been successfully used as catalysts for CO_2 reduction, mainly to CO and formic acid [2]. Electrogenerated organic anion radicals have also been shown to give efficient catalysis of CO_2 reduction to oxalate [3].

In this paper, the electrochemical reduction of cobalt and nickel complexes of the ligand N,N'–1,2–phenylenebis(salicylideneiminato) (salophen = L) and its relation to the electrochemical activation of CO_2 is discussed. These complexes have been investigated as potential electrocatalysts of CO_2 reduction. Indeed, cobalt and nickel complexes containing tetraazamacrocycles or tetradentate Schiff base ligands have been recognized as powerful catalysts in the electrochemical reduction of CO_2 [4,5]. Bifunctional fixation of CO_2 by nucleophilic Co(I)–Schiff base complexes assisted by alkali cations coordinated to the same ligand has been reported [6].

A. J. L. Pombeiro and J. A. McCleverty (eds.),
Molecular Electrochemistry of Inorganic, Bioinorganic and Organometallic Compounds, 311–316.

Experimental

The complexes were prepared by procedures described in the literature [6,7]. Dimethylformamide (DMF) was fractionally distilled under reduced pressure and stored in a dark bottle under N_2 atmosphere. Reagent–grade acetonitrile (AN) was used without further purification. Anhydrous $LiClO_4$ was further dried in a vacuum oven at 100 °C. CO_2 (> 99.998%) was supplied by SIAD (Italy).

The electrochemical setup was already described [5]. All potentials are reported versus saturated calomel electrode *sce*. CO_2 reduction products were analyzed by chromatographic techniques as previously described [5].

Results and discussion

ELECTROCHEMICAL REDUCTION OF THE COMPLEXES

Both complexes exhibit a very rich redox chemistry, both metal and ligand being involved in the reduction processes. We shall describe here only the primary reduction processes which appear to be relevant to the electrocatalytic reduction of CO_2.

Figure 1 shows voltammetric curves of Ni(II)L recorded on a Hg electrode in DMF. On the cathodic scan, three well defined reduction peaks A, B and C are formed whereas on the reverse sweep, four anodic peaks D, E, F and G can be observed. At

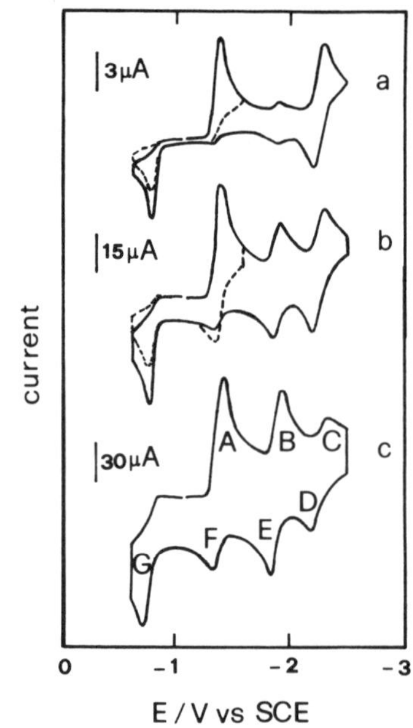

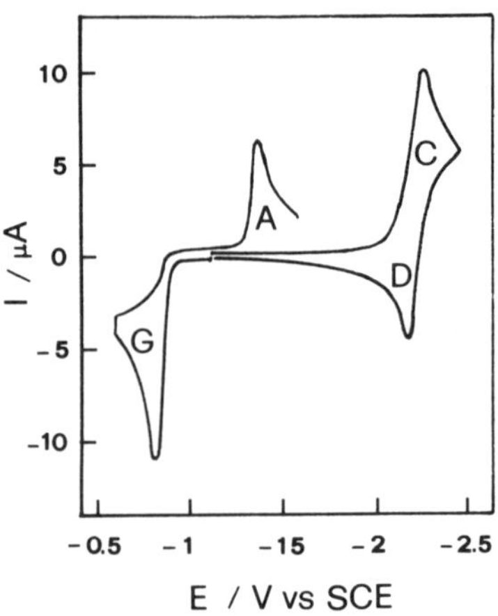

Fig. 1. Cyclic voltammograms of 0.9 mM Ni(L) in DMF + 0.1 M TBAP on a Hg electrode at 25 °C. sweep rate: (a) 1 Vs^{-1}, (b) 50 Vs^{-1}, (c) 200 Vs^{-1}.

Fig. 2. Cyclic voltammograms of 1 mM Ni(L) in DMF + 0.1 M TBAP solution recorded on a Hg electrode after an exhaustive electrolysis at —1.5 V *vs. sce*.

slow scan rates v peak A is chemically irreversible but as v increases an anodic peak F appears and increases in height until a ratio $i_p(F)/i_p(A)$ close to the unity is reached. The standard potential of the process, estimated as the mid point between peaks A and F, is –1.39 V *vs. sce*. The anodic peak G appears to be associated with the reduction process A. Variations of the voltammetric peak potentials $E_p(A)$ and $E_p(G)$ were examined both as a function of v and concentration c. Logarithmic analysis of $E_p(A)$ as a function of v were performed at several concentrations. An average cathodic shift of *ca* 19 mV for each 10–fold increase in v and an anodic shift of 19 mV for each 10–fold increase in c were observed. $E_p(G)$ was found to be dependent only on the sweep rate with a variation of *ca* 30 mV per log v unit.

Bulk electrolysis of Ni(II)L at –1.50 V consumed one electron per molecule of complex while the solution changed colour from red to green. A cyclic voltammogram recorded after electrolysis (Fig. 2) showed, on an initially positive–going scan, a well formed oxidation process at a potential corresponding to that of peak G and a coupled reduction process at the potential of peak A. On an initially negative–going scan, only one peak couple occurring at the potential of the peak couple C/D could be observed. Back–electrolysis of the green solution at –0.6 V consumed again one electron per molecule of Ni(L) and regenerated the original complex with a quantitative yield.

On the basis of these results, process A can be attributed to a one–electron reversible charge transfer followed by a fast chemical reaction yielding a stable product which can be reoxidized to the original Ni(L) complex. The observed variations of $E_p(A)$ with log v and log c indicate that the chemical reaction is a bimolecular reaction involving the primary reduction product [8]. The same results also suggest direct radical–radical coupling as the most probable dimerization reaction. The dimeric nature of the electrolytic product has also been confirmed by a comparison of its voltammetric behaviour with that of a well characterized dimer having two salophen units joined through a C–C bond, prepared by reaction of metallic sodium with Ni(II)L [7]. We therefore suggest the following reaction mechanism:

$$[\mathrm{Ni(II)L}] + e^- \rightleftharpoons [\mathrm{Ni(II)L}]^{\cdot -} \qquad \text{peaks A/F} \qquad (1)$$

$$2[\mathrm{Ni(II)L}]^{\cdot -} \longrightarrow [\mathrm{Ni(II)L}]_2^{2-} \qquad (2)$$

$$[\mathrm{Ni(II)L}]_2^{2-} - 2e^- \longrightarrow 2[\mathrm{Ni(II)L}] \qquad \text{peak G} \qquad (3)$$

We ascribe [9] both processes B/E and C/D (Fig. 1) to nickel–centered reversible charge transfer processes. The former is observed only at high scan rates when the dimerization reaction is hampered and is thus assigned to the reduction of the Ni(II) anion radical $[\mathrm{Ni(II)L}]^{\cdot -}$, whereas the latter corresponds to the reduction of the dimer $[\mathrm{Ni(II)L}]_2^{2-}$ (Fig. 2).

$$[\mathrm{Ni(II)L}]^{\cdot -} + e^- \rightleftharpoons [\mathrm{Ni(I)L}]^{2-} \qquad \text{peaks B/E} \qquad (4)$$

$$[\mathrm{Ni(II)L}]_2^{2-} + 2e^- \longrightarrow [\mathrm{Ni(I)L}]_2^{4-} \qquad \text{peaks C/D} \qquad (5)$$

The electrochemical reduction mechanism of Co(II)L is quite complex. In DMF containing TBAP as background electrolyte, up to four reduction processes can be observed. If, however, the potential scan is confined to *ca* –2.2 V *vs. sce*, only two reduction processes, which appear to be diffusion controlled one–electron reversible charge transfers, as indicated by anodic to cathodic peak seperation of *ca* 60 mV, unit ratio of the anodic to cathodic peak currents and constancy of the peak current function $i_p/v^{1/2}$, are observed. The first cathodic process can be unambiguously assigned to the reduction of Co(II) to Co(I), as has been previously reported for electrochemical and ESR studies [10].

$$[Co(II)L] + e^- \rightleftharpoons [Co(I)L]^- \quad (6)$$

As to the second reduction process, even though reduction to Co(0) has been previously suggested [10] our results [11] are more indicative of a ligand–based electron transfer leading to the formation of a stable Co(I)–ligand anion radical.

$$[Co(I)L]^- + e^- \rightleftharpoons [Co(I)L]^{2-} \quad (7)$$

ELECTROCATALYTIC REDUCTION OF CO_2

Ni(salophen) shows no catalytic activity toward CO_2 reduction. This seems to be quite reasonable in view of the reduction mechanism we discussed above. In fact in the vast majority of transition metal catalyzed CO_2 reduction processes, catalysis is attributed to an interaction between the metal complex, with the metal reduced to a lower oxidation state and CO_2. Some polypyridyl metal complexes, which give rise to ligand-based anion radicals as primary reduction products, make an exception [12]. But even in those cases metal–stabilized anion radicals which react slowly with CO_2 are formed.

Unlike the nickel complex, Co(salophen) is an effective elctrocatalyst for CO_2 reduction. The electrocatalytic process was studied in AN containing $LiClO_4$ as background electrolyte. Under these conditions cyclic voltammetry of Co(II)L shows only one reversible peak couple (Fig. 3a) with $E^0 = -1.018$ V *vs. sce*, corresponding to the reduction of Co(II) to Co(I).

$$[Co(II)L] + e^- \overset{Li^+}{\rightleftharpoons} [Co(I)(L)Li] \quad (8)$$

Saturation of an AN + 0.1 M $LiClO_4$ solution with CO_2 at 1 atm, corresponding to a 0.28 M concentration of the latter [13], does not produce any appreciable reduction current before the background electrolyte discharge. Nor does CO_2 addition to AN solution of Co(II)L modify the voltammetric pattern of the complex, provided that the potential is not swept beyond –1.30 V. However, when the potential scan is extended to more negative values, a new reduction peak appears with $E_p = -1.53$ V *vs. sce*, and a substantial decrease of the anodic peak current corresponding to the oxidation of the Co(I) complex is observed (Fig. 3b). Removal of CO_2 by a stream of argon restores the voltammetric pattern of Fig. 4a. Controlled potential electrolysis of a CO_2–saturated AN solution of Co(II)L, performed at the potential of the Co(II)/Co(I) couple, resulted in the formation of a new cobalt complex, with the consumption of one electron per molecule of Co(L). After electrolysis, the reversible peak couple at –1.018 V disappeared while a new irreversible reduction peak appeared at –1.26 V (Fig. 4a). Again, however, removal of CO_2 by a stream of argon restored the voltammetric pattern characteristic of the Co(II)/Co(I) redox system.

These results indicate a reversible reaction between electrogenerated Co(I) and CO_2. It is of particular interest to note that, depending on the time scale of the experiment, two different Co–CO_2 complexes with reduction potentials –1.26 V and –1.53 V *vs. sce* may be formed. Catalytic reduction of CO_2 to CO and carbonate was observed with high selectivity and satisfactory current efficiency, when 1–2 mM Co(II)L solutions saturated with CO_2 were electrolyzed at potentials beyond –1.26 V. Note that sustained electrocatalysis was obtained indifferently whether the elctrolysis potential was beyond the reduction of the first or second CO_2 complex. This result possibly

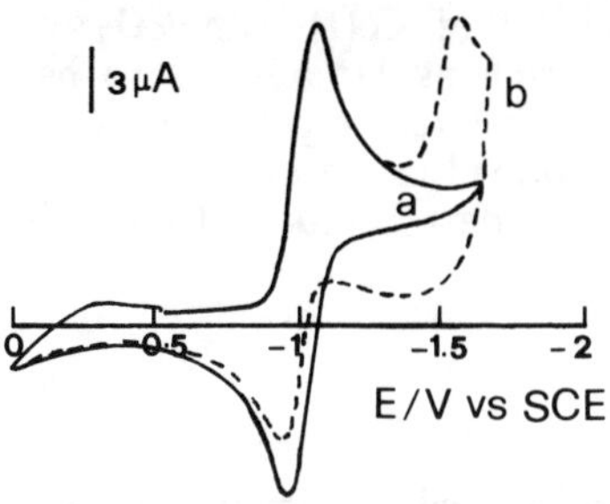

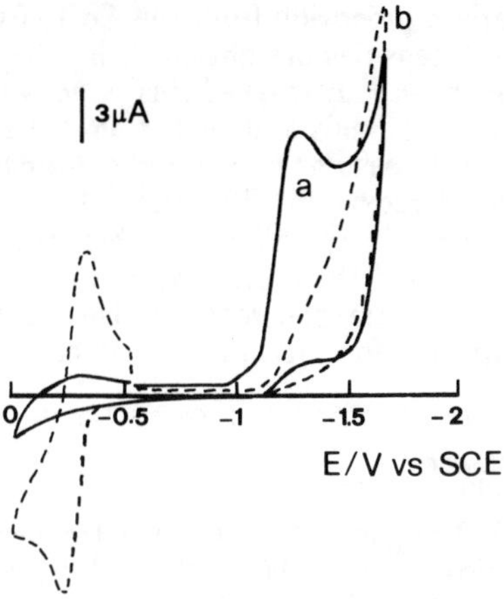

Fig. 3. Cyclic voltammograms of 1.32 mM Co(L) in AN + 0.1 M $LiClO_4$ under Ar (curve a) or CO_2 atmosphere (curve b), using a Hg electrode at a sweep rate of 200 mVs^{-1}.

Fig. 4. Cyclic voltammograms of 1.32 mM Co(L) in CO_2–saturated AN + 0.1 M $LiClO_4$ recorded on a Hg electrode after controlled potential electrolysis at (a) –1.0 and (b) –1.50 V. $v = 200$ mVs^{-1}.

implies that both CO_2 complexes are intermediates in the electrocatalytic cycle. Successive incorporation of two CO_2 molecules by Co(I)L leading to the formation of a complex containing a head–to–tail CO_2 dimer $C(O)OCO_2$, is a plauxible explanation, as has been reported for other transition metal complexes [14]. The cathodic peaks at –1.26 V and –1.53 V may be attributed to the reduction of the two CO_2 complexes, the one containing the dimer being more easily reducible than the other, owing to the better delocalization of the negative charge by the $C(O)OCO_2$ structure respective to CO_2.

The catalytic cycle involves the intermediate formation of a Co(I)–carbonyl complex, as evidenced by its reversible oxidation at –0.29 V (Fig. 4b). This was confirmed by electrochemical investigations on Co(II)L performed under CO atmosphere [5]. Release

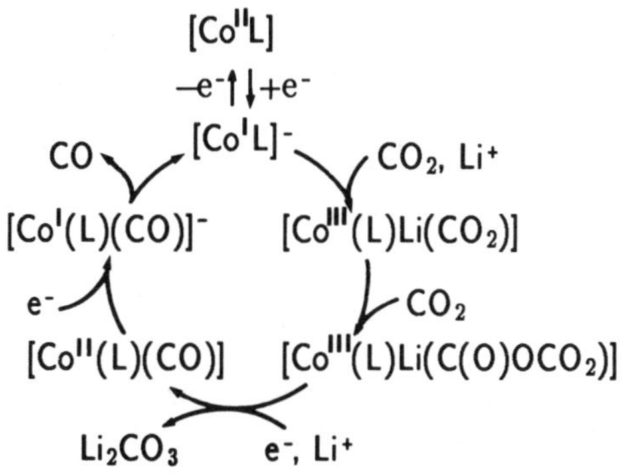

Scheme 1

of carbon monoxide from the Co(I)(CO)L complex is rather slow, and consequently low current densities are obtained, but high turnover numbers (>20) of the catalyst can be attained without marked loss of activity.

The mechanism depicted in Scheme 1 may be proposed. The first step of the process is assumed to be the formation, by the reaction of Co(I) with CO_2, of a cobalt–carboxylate. The presence of an alkali cation such as Li^+ appears to be of fundamental importance for this step. The CO_2 complex first reacts with another CO_2 molecule and then, after being further reduced, is transformed into a carbonyl complex by an oxide transfer reaction. Here CO_2 itself acts as an oxide acceptor and this indeed explains the formation of CO_3^{2-} concomitantly with CO.

References

1 I. Taniguchi (1989) 'Electrochemical and photochemical reduction of carbon dioxide', in J. M. Bockris, R. E. White and B. E. Conway (eds), Modern Aspects of Electrochemistry, Plenum Publishing Corporation, New York, pp. 327–400.
2 J. P. Collin and J. P. Sauvage (1989) 'Electrochemical reduction of CO_2 mediated by molecular catalysts', Coord. Chem. Rev. 93, 245–268.
3 G. Filardo, S. Gambino, G. Silvestri, A. Gennaro and E. Vianello (1984) 'Electrocarboxylation of styrene through homogeneous redox catalysis', J. Electroanal. Chem. 177, 303–309.
4 B. Fischer and R. Eisenberg (1980) 'Electrocatalytic reduction of CO_2 by using macrocycles of nickel and cobalt', J. Am. Chem. Soc. 102, 7361–7363.
5 A. A. Isse, A. Gennaro, E. Vianello and C. Floriani (1991) 'Electrochemical reduction of carbon dioxide catalyzed by [Co(I)(salophen)Li]', J. Mol. Catal. 70, 197–208.
6 S. Gambarotta, F. Arena, C. Floriani and P. F. Zanazzi (1982) 'Carbon dioxide fixation: Bifunctional complexes containing acidic and basic sites working as reversible carriers', J. Am. Chem. Soc. 104, 5082–5092.
7 S. Gambarotta, F. Urso, C. Floriani, A. Chiesi–Villa and C. Guastini (1983) 'Carbon–carbon bond forming and breaking by a metal–assisted redox process in a nickel(II)–Schiff base complex', Inorg. Chem. 22, 3966–3972.
8 J. M. Saveant and E. Vianello (1967) 'Potential–sweep voltammetry: Theoretical analysis of monomerization and dimerization' Electrochim. Acta 12, 1545–1561.
9 A. A. Isse, A. Gennaro and E. Vianello (1992) 'A study of the electrochemical reduction mechanism of Ni(salophen) in DMF', Electrochim. Acta 37, 113–118.
10 A. M. Van Den Berger, D. J. Brockway and B. O. West (1983) 'The synthesis and electrochemistry of some perfluoroalkyl organometallic compounds', J. Organomet. Chem. 249, 205–225.
11 A. A. Isse, A. Gennaro and E. Vianello, to be published.
12 C. M. Bolinger, N. Storey, B. P. Sullivan and T. J. Meyer (1988) 'Electrocatalytic reduction of carbon dioxide by 2,2'–bipyridine complexes of rhodium and iridium', Inorg. Chem. 27, 4582–4587.
13 A. Gennaro, A. A. Isse and E. Vianello (1990) 'Solubility and electrochemical determination of carbon dioxide in some dipolar aprotic solvents', J. Electroanal. Chem. 289, 203–215.
14 G. R. Lee, J. M. Maher and N. J. Cooper (1987) 'Reductive disproportionation of CO_2 by dianionic carbonylmetallates of the transition metals', J. Am. Chem. Soc. 109, 2956–2962.

ORGANOMETALLIC ELECTROCHEMISTRY OF METAL ALKYNE AND RELATED COMPLEXES

N.G. CONNELLY
School of Chemistry
University of Bristol
Bristol BS8 1TS
U.K.

ABSTRACT. The wide applicability of electron-transfer reactions in organotransition metal chemistry, in studies of both bonding and reactivity, is illustrated with reference to complexes of alkynes and their oligomers. The complexes $[Cr(CO)_2(\eta\text{-}RC_2R)(\eta\text{-}C_6Me_6)]$ are reversibly oxidised to monocations having substantial unpaired electron density on the alkyne ligand; the cation $[Cr(CO)_2(\eta\text{-}Me_3SiC_2SiMe_3)(\eta\text{-}C_6Me_6)]^+$ is an intermediate in oxidatively induced vinylidene-alkyne isomerisation. Examples are given of other C_2-ligand rearrangements induced by electron-transfer. On one-electron oxidation, the alkyne ligand of $[Co_2(CO)_2(\mu\text{-dppm})_2(\mu\text{-}MeC_2Me)]$ rotates with respect to the metal-metal bond, in accord with the results of m.o. calculations. The reductive cleavage of the metal-metal bond in $[Co_2(CO)_6(\mu\text{-}RC_2R)]$, and the oxidative cleavage of $[Pd_2(\mu\text{-}PhC_2Ph)(\eta\text{-}C_5Ph_5)_2]$, give routes to novel mono- and binuclear alkyne complexes and to unusual complexes of Pd(I) and Pd(III). The reduction of bis(alkyne) complexes such as $[Mo(NCMe)(\mu\text{-}MeC_2Me)_2(\eta\text{-}L)]^+$ (L = C_5H_5 or indenyl) results in alkyne oligomerisation to give $[Mo_2(\mu\text{-}C_8Me_8)(\eta\text{-}C_5H_5)_2]$ and $[Mo_2(\mu\text{-}C_6Me_6)(\eta^5\text{-indenyl})_2]$ respectively; the former undergoes C-H bond activation by a double-oxidation deprotonation mechanism. Finally, 'flyover' dicobalt, dirhodium, and diruthenium complexes related to $[Mo_2(\mu\text{-}C_8Me_8)(\eta\text{-}C_5H_5)_2]$ undergo reversible C-C bond formation and cleavage reactions by overall two-electron transfer reactions.

1. Introduction

In recent years there has been an enormous increase in interest[1] in the redox reactions of organometallic complexes, such reactions providing not only new synthetic pathways to otherwise inaccessible compounds but also a means of probing electronic structure.

The bonding in redox-active organometallics may be investigated by two methods. X-ray crystallographic studies on redox-related pairs of complexes (*i.e.* complexes differing only by one electron) reveal changes in bond legth and angle which are related to the composition of the frontier orbitals[2]. In addition, the nature of the s.o.m.o. of the paramagnetic member of the redox pair can often be deduced from its e.s.r. spectrum[3].

A. J. L. Pombeiro and J. A. McCleverty (eds.),
Molecular Electrochemistry of Inorganic, Bioinorganic and Organomettallic Compounds, 317–329.

Considerable changes in the reactivity of organometallic complexes can be brought about by one-electron oxidation or reduction. For example, neutral 18-electron complexes give 17-electron[4] cations or 19-electron[5] anions which can be more reactive towards nucleophiles and electrophiles respectively. Enormous increases in reaction rates can be observed, *e.g.* of the nucleophilic substitution[6] of $[Fe(CO)_3(PPh_3)_2]$ on one-electron oxidation; in some cases, reactions are electron-transfer chain (ETC) catalysed[7].

Although most redox reactions seem to be largely metal-based, the C-H and C-C bonds of co-ordinated hydrocarbons can also be activated[8]. For example, the one-electron oxidation of cyclo-octatetraene (cot) complexes of the iron and cobalt groups[9] results in the stereo-and regiospecific formation and cleavage of C-C bonds. This chemistry has helped to establish the general principles involved in the activation of hydrocarbon ligands. For example, if the hydrocarbon is able to change hapticity, metal-based oxidation can be followed by ligand rearrangement, delocalisation of unpaired electron density onto the hydrocarbon, and radical-like C-C coupling.

The oxidation of co-ordinated cot gives novel polycyclic C_{16}-hydrocarbons which are, however, complex in structure and likely to have little practical utility. We have therefore turned our attention to the activation of simpler ligands, and in particular to alkynes and related unsaturated C_2 hydrocarbons which are important precursors in both organometallic and organic chemistry. Alkynes can bond to transition metals as 2-, 3-, or 4-electron donors and it seemed likely therefore that these different bonding modes, and therefore the reactivity of the ligand itself, should be interconvertible by single electron transfer reactions. Relatively little mention has been made of the redox chemistry of co-ordinated alkynes[1,10] but we show here that they, their oligomers, and other C_2-ligands do indeed have a rich and diverse redox-based chemistry. We also describe how metal-alkyne bonding, in both mono- and polynuclear complexes, can be investigated by X-ray structural and e.s.r. spectroscopic methods.

2. The One-Electron Oxidation of $[Cr(CO)_2(\eta\text{-}RC_2R)(\eta\text{-}C_6Me_6)]$ and Related C_2 Complexes

2.1. THE BONDING IN $[Cr(CO)_2(\eta\text{-}RC_2R)(\eta\text{-}C_6Me_6)]^+$

In the mid-1970's, we showed that the reaction between $[NO]^+$ and $[Cr(CO)_2(\eta\text{-}PhC_2Ph)(\eta\text{-}C_6Me_6)]$, and that between $[Cr(CO)_2(PPh_3)(\eta\text{-}C_6Me_6)]$ and $[N_2C_6H_4F\text{-}p]^+$ gave the 17-electron cations $[Cr(CO)_2L(\eta\text{-}C_6Me_6)]^+$ (L = PhC_2Ph[11] and PPh_3[12] respectively). On oxidation, large shifts to higher energy in the carbonyl stretching frequencies were observed, suggesting that in both cases the cations might be regarded as metal-based radicals. Further studies[13] have shown, however, that the electronic structures of the two cations are very different.

Figure 1 shows a plot of E^o, the potential for the reversible one-electron oxidation of $[Cr(CO)_2L(\eta\text{-}C_6Me_6)]$, against $\nu(CO)$, the average of the carbonyl stretching frequencies for the dicarbonyl complexes. Linear correlations between E^o values and other physical quantities have been noted previously[14]. In many cases there is no obvious quantitative basis for the observed linear correlations. However, such correlations can be qualitatively useful and in the example reported here the very different relationships between E^o and $\nu(CO)$ for the PR_3 and alkyne complexes imply rather different electronic structures.

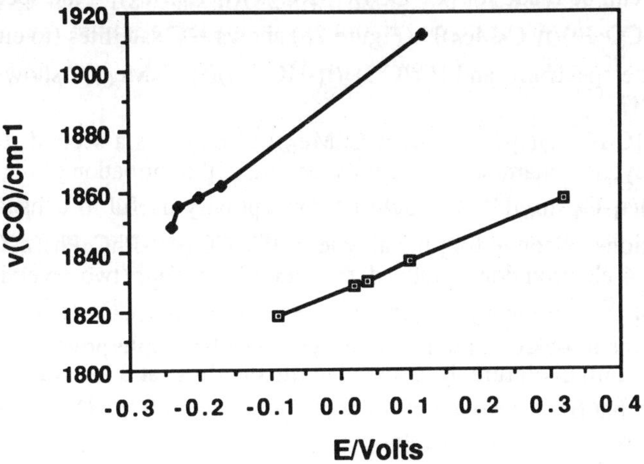

Figure 1. One-electron oxidation potential (E/Volts) *vs.* average carbonyl stretching frequency [$\nu(CO)/cm^{-1}$] for [$Cr(CO)_2L(\eta$-$C_6Me_6)$] (open square, L = phoshine or phosphite; filled square, L = alkyne).

The difference is more clearly shown by e.s.r. spectroscopy, as in Figure 2. Thus, the PR_3 complexes give spectra only in frozen solution whereas those of the alkyne complexes are observed in fluid solution at room temperature. In addition, g_{ave} for the former are *ca.* 2.034 whereas g_{iso} for the latter (*ca.* 1.998) are much closer to the free electron value.

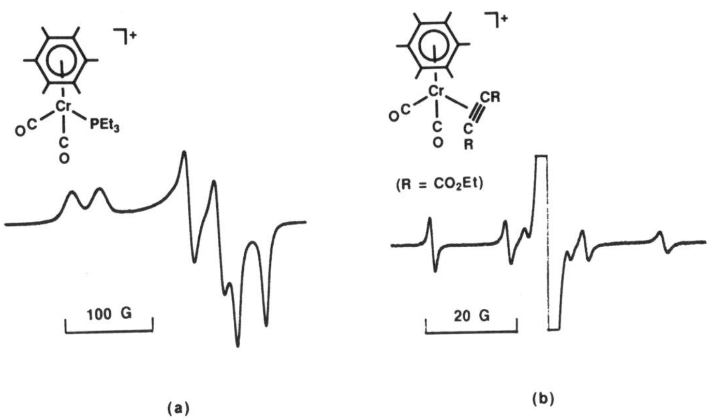

Figure 2. The e.s.r. spectra, in CH_2Cl_2:thf (1:2), of (a) [$Cr(CO)_2(PEt_3)(\eta$-$C_6Me_6)]^+$ at 90K, and (b) [$Cr(CO)_2(EtCO_2MeC_2CO_2Et)(\eta$-$C_6Me_6)]^+$ at 260K.

A more quantitative analysis[13] of the spectra shows that the PR_3 complex cations are best described as metal-based radicals. By contrast, an estimate of *ca.* 30-40% spin density on the alkyne can be made for $[Cr(CO)_2(\eta\text{-}RC_2R)(\eta\text{-}C_6Me_6)]^+$; the spectrum of $[Cr(CO)_2(\eta\text{-}EtCO_2C_2CO_2Et)(\eta\text{-}C_6Me_6)]^+$ (Figure 2b) shows ^{13}C satellites (to either side of the main peak of the spectrum) and $[Cr(CO)_2(\eta\text{-}HC_2Ph)(\eta\text{-}C_6Me_6)]^+$ shows alkyne proton coupling of 4.2G.

The complex $[Cr(CO)_2(\eta\text{-}PhC_2Ph)(\eta\text{-}C_6Me_6)]$ also shows a second reversible oxidation wave in the cyclic voltammogram, corresponding to the formation of the dication $[Cr(CO)_2(\eta\text{-}PhC_2Ph)(\eta\text{-}C_6Me_6)]^{2+}$. It would be conceptually useful to consider these electron-transfer reactions as converting the alkyne in $[Cr(CO)_2(\eta\text{-}PhC_2Ph)(\eta\text{-}C_6Me_6)]$ from a 2-, to a 3-, to a 4-electron donor. Indeed, the redox behaviour (two reversible one-electron reductions) of the tetraphenylcyclobutadiene complex $[Cr(CO)_2(\eta\text{-}PhC_2Ph)(\eta\text{-}C_4Ph_4)]$ (which contains a 4-electron donating alkyne) has been interpreted[15] in such a way. However, this simple picture is somewhat misleading and preliminary m.o. calculations suggest[16] that from the point of view of the alkyne in $[Cr(CO)_2(\eta\text{-}RC_2R)(\eta\text{-}C_6Me_6)]^+$ it is closer to an electronic excited state than to a 3-electron cation radical. Whatever the description of the bonding, the results described below confirm that the chemical reactivity of co-oordinated alkynes is considerably altered on electron-transfer.

2.2 THE REDOX ACTIVATION OF C_2 FRAGMENTS

Whereas the alkyne complexes $[Cr(CO)_2(\eta\text{-}RC_2R)(\eta\text{-}C_6Me_6)]$ show reversible one-electron oxidation, the orange crystalline complex (**1**) prepared from $[Cr(CO)_3(\eta\text{-}C_6Me_6)]$ and $Me_3SiC_2SiMe_3$, having the formula $[Cr(CO)_2L(\eta\text{-}C_6Me_6)]$ ($L = Me_3SiC_2SiMe_3$), has a very different cyclic voltammogram; an irreversible oxidation wave (peak potential 0.18V) is accompanied by a reversible product wave centred at -0.24V≠. Complex (**1**) is oxidised with $[Fe(\eta\text{-}C_5H_5)_2]^+$ in CH_2Cl_2 to give a yellow solution with an e.s.r. spectrum fully consistent with the presence of the alkyne cation $[Cr(CO)_2(\eta\text{-}Me_3SiC_2SiMe_3)(\eta\text{-}C_6Me_6)]^+$; treatment of the yellow solution with $[BH_4]^-$ gave a second neutral species which, in a few minutes, reverted at room temperature to (**1**). The X-ray structure of (**1**) has not been determined but the ^{13}C n.m.r. spectrum is consistent with the vinylidene complex $[Cr(CO)_2\{\eta\text{-}C{=}C(SiMe_3)_2\}(\eta\text{-}C_6Me_6)]$; the second neutral species, which has an i.r. carbonyl spectrum and oxidation potential ($E^o = -0.24V$) consistent with those of other members of the series (Figure 1), is undoubtedly the alkyne complex $[Cr(CO)_2(\eta\text{-}Me_3SiC_2SiMe_3)(\eta\text{-}C_6Me_6)]$. The redox-based chemistry of (**1**) can therefore be summarised by a 'square-scheme' (Scheme 1) in which there is an unprecedented vinylidene-alkyne rearrangement at the 17-electron stage. The reverse process, at the 18-

≠All potential data quoted are relative to the saturated calomel electrode and are taken from cyclic voltammetric measurements at a platinum bead electrode in CH_2Cl_2 containing $[NBu_4^n][PF_6]$ as base electrolyte. Under these conditions the E^o value for the ferrocene-ferrocenium couple is 0.47V.

electron level, is known[17] to provide a synthetic pathway to metal vinylidene complexes.

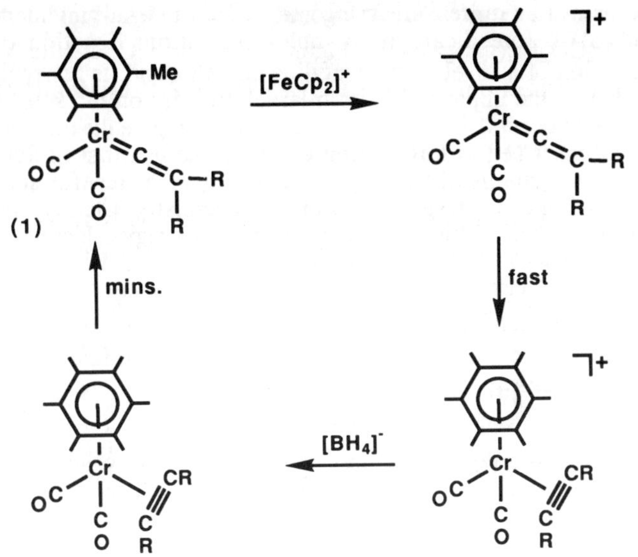

Scheme 1. A 'square scheme' for the oxidative isomerisation of the vinylidene complex $[Cr(CO)_2\{\eta\text{-}C{=}C(SiMe_3)_2\}(\eta\text{-}C_6Me_6)]$ to the alkyne complex $[Cr(CO)_2(\eta\text{-}Me_3SiC_2SiMe_3)(\eta\text{-}C_6Me_6)]$

Redox-induced rearrangements of co-ordinated C_2-ligands are undoubtedly involved in reactions (i) - (iii) (Scheme 2)[18-20] and may also be implicated in the formation of $[Os(\sigma,\eta^2\text{-}PhC{\equiv}CC{=}CHPh)(PMe_3)_4]^+$ from $[Os(C{\equiv}CPh)_2(PMe_3)_4]$ and Ag^+ [21], and of $[(\eta\text{-}C_5H_5)(CO)_2Cr{\equiv}CC(Ph){=}C(Ph)C{\equiv}Cr(CO)_2(\eta\text{-}C_5H_5)]$ from $[Cr(C_2Ph)(CO)_2(\eta\text{-}C_5H_5)]$ and sodium-potassium alloy[22]. Both reactions (i) and (ii) (Scheme 2) involve the

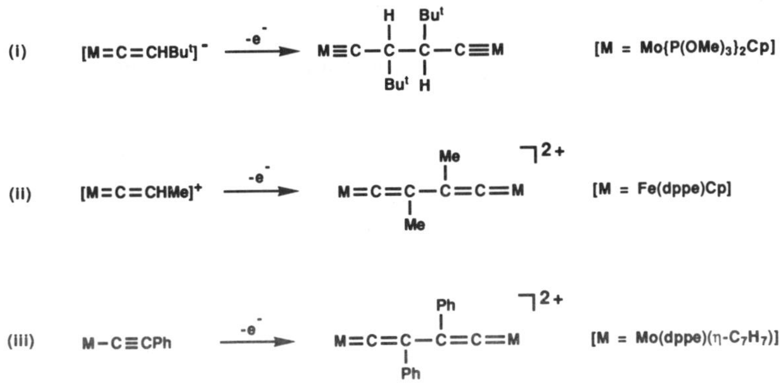

Scheme 2. Oxidative transformations of C_2-ligand complexes.

oxidation of a vinylidene complex and a subsequent dimerisation step leading to C-C bond formation. However, they differ from one another in that (ii) also involves[18] proton loss to give a dicarbene whereas (i) requires[19] rearrangement (from a metal-vinylidene radical to a metal carbyne radical) before dicarbyne complex formation. Reaction (iii) certainly involves[20] the formation of a 17-electron acetylide complex and its rearrangement to an alkylidenyl cation where the unpaired electron largely resides on the substituted carbon atom. Dimerisation again occurs, this time regiospecifically, to give the dicarbene complex.

The reactions described in this section are collected together in Scheme 3 (with adjustments for overall charge differences, *etc.*) to give a general scheme for the rearrangement and linking of C_2-fragments *via* redox chemistry. Clearly, there is a great deal of scope for the extension of this chemistry in stereo- and regioselective synthesis.

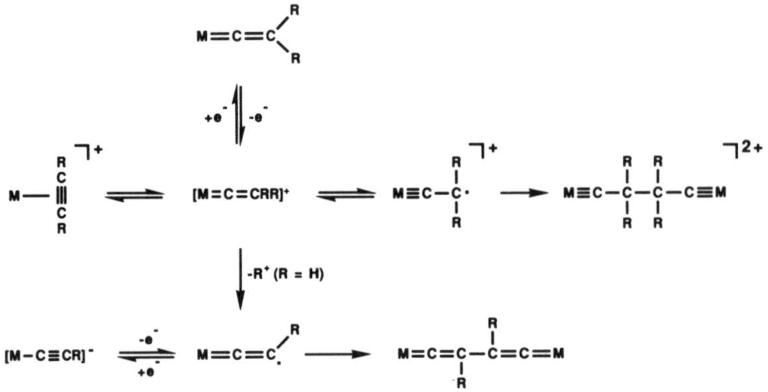

Scheme 3. The interconversion and oligomerisation of C_2-fragments *via* redox chemistry.

3. The One-Electron Redox Reactions of Alkyne-Bridged Bimetallics

When more than one metal is present in an organometallic complex rather different electrochemical behaviour is possible. For example, metal-metal bonds can function as 'electron sinks' so that extended redox series, linked by sequential one-electron transfer steps, may become available. In addition, the paramagnetic members of such series may show unusual properties associated with mixed valency. Here we describe not only the reactions of binuclear species which undergo two or more sequential electron-transfer reactions but also a study of the bonding in alkyne-bridged binuclear species.

3.1. THE BONDING IN $[Co_2(CO)_6(\mu\text{-}RC_2R)]$

In independent m.o. studies, Anderson[23], and Thorn and Hoffmann[24], showed that the replacement of cobalt by iron (effectively a two-electron reduction) in the alkyne-bridged complexes $[Co_2(CO)_6(\mu\text{-}RC_2R)]$ should result in a 20 - 30° rotation of the alkyne relative to the metal-metal bond. However, the X-ray structure of $[Fe_2(CO)_6(\mu\text{-}Bu^tC_2Bu^t)]$ showed[25] instead that the molecule distorted in a different way, namely *via* the relative twisting of the two $M(CO)_3$ groups away from the 'sawhorse' structure of the dicobalt complex. We have shown that, provided the sawhorse structure is retained, the expected distortion is observed on one-electron oxidation.

Though the hexacarbonyls $[Co_2(CO)_6(\mu\text{-}RC_2R)]$ are oxidised at very positive potentials, carbonyl substitution with PR_3 ligands gives species such as $[Co_2(CO)_{6-n}L_n(\mu\text{-}RC_2R)]$ which are easier to oxidise. The monocations $[Co_2(CO)_{6-n}\{P(OMe)_3\}_n(\mu\text{-}RC_2R)]^+$ (R = Ph or CF_3, n = 3 or 4) have been briefly mentioned[26] but our attempts to crystallise similar species for comparitive X-ray studies failed. However, the complex $[Co_2(CO)_2(\mu\text{-dppm})_2(\mu\text{-}MeC_2Me)]$ is readily oxidised by $[FeCp_2]^+$ to the stable monocation $[Co_2(CO)_2(\mu\text{-dppm})_2(\mu\text{-}MeC_2Me)]^+$ which was successfully structurally characterised[27]. Both complexes are constrained to the 'sawhorse' structure by the two bridging dppm ligands. On oxidation, the metal-metal bond is shortened, from 2.513 to 2.418Å , in accord with the removal of an electron from an orbital with metal-metal antibonding character. However, the most significant change is in the geometry of the $Co_2(\mu\text{-}C_2)$ core where the Co-C bonds vary from 1.950 and 1.960Å in the neutral molecule to 2.103 and 1.941Å in the cation. This change corresponds almost exactly to that predicted by theory, namely a rotation of *ca.* 12°.

As noted above, bi- and polynuclear species may undergo more than one electron-transfer reaction. Accordingly, $[Co_2(CO)_6(\mu\text{-}RC_2R)]$ is reduced to a stable radical anion[28]. Though not isolable, such radicals are readily generated in solution, and detailed e.s.r. spectroscopic studies have shown that the s.o.m.o. of the anion is σ^* with respect to the cobalt-cobalt bond. Thus, the combination of X-ray and e.s.r. spectroscopic studies has provided a detailed picture of the bonding in a three-membered redox series.

It is interesting to note that $[Fe_3(CO)_9(\mu^3\text{-}EtC_2Et)]$, which has the alkyne orthogonal to one metal-metal bond of the trinuclear cluster, also undergoes redox-induced rearrangement; two-electron reduction gives the dianion $[Fe_3(CO)_9(\mu^3\text{-}EtC_2Et)]^{2-}$ in which the alkyne is parallel to the same metal-metal bond[29].

3.2. OXIDATIVE AND REDUCTIVE METAL-METAL BOND CLEAVAGE

The alkyne-bridged radical anions $[Co_2(CO)_6(\mu\text{-}RC_2R)]^-$ are not only of interest because of their electronic structure but also for their chemical reactivity. First, they are intermediates in the ETC catalysed carbonyl substitution reactions[26,30] of $[Co_2(CO)_6(\mu\text{-}F_3CC_2CF_3)]$ with phosphine ligands (population of the σ^* Co-Co orbital leads to metal-metal bond cleavage, the formation of a reactive 17-electron metal centre, and rapid nucleophilic substitution). However, where the alkyne bridge is less robust, as in $[Co_2(CO)_6(\mu\text{-}PhC_2Ph)]^-$ complete cleavage occurs to give a novel series of mononuclear paramagnetic alkyne complexes, namely $[Co(CO)_{2-n}L_n(RC_2R)]$ (n = 1 or 2) [31]. These species show novel isomerism as indicated by their e.s.r. spectra. The chemical reactivity of the co-ordinated alkyne has not yet been fully explored.

The complex $[Pd_2(\mu\text{-}PhC_2Ph)(\eta\text{-}C_5Ph_5)_2]$ (**2**) shows[32] extensive electrochemistry, with reversible oxidation waves centred at 0.54 and 1.16V and a near reversible reduction wave centred at -1.11V [a small reversible product wave is observed at -0.30V which corresponds to the formation of the free $[C_5Ph_5]^-$ anion after reduction of (**2**)]. Chemical oxidation of (**2**) with one equivalent of Ag^+ affords good yields of the air-stable paramagnetic monocation $[Pd_2(\mu\text{-}PhC_2Ph)(\eta\text{-}C_5Ph_5)_2]^+$ (**2**$^+$). Unlike the neutral precursor

(**2**) or [$Co_2(CO)_2(\mu$-dppm$)_2(\mu$-$MeC_2Me)]^+$, (**2**$^+$) reacts with nucleophiles[33,34] to give either mononuclear products *via* metal-metal bond cleavage or novel asymmetric alkyne-bridged binuclear species *via* metal-cyclopentadienyl bond rupture. With phosphines or dienes[33], (**2**$^+$) gives [$PdL_2(\eta$-$C_5Ph_5)]^+$ which are reversibly reduced and oxidised to Pd(I) and Pd(III) radicals stabilised by the bulky pentaphenylcyclopentadienyl ring[35]. In the case of [$Pd(\eta^4$-dbcot$)(\eta$-$C_5Ph_5)]^+$ (dbcot = dibenzocyclo-octatetraene) chemical or electrolytic reduction gave the neutral Pd(I) radical [$Pd(\eta^4$-dbcot$)(\eta$-$C_5Ph_5)$] which reacts with dibenzoylperoxide to yield the Pd(II) complex [$Pd\{\sigma,\eta^2$-$C_8H_{12}OC(O)C_6H_5\}(\eta$-$C_5Ph_5)$]. The last is related to [$Pd(\sigma,\eta^2$-$C_8H_{12}OR)(\eta$-$C_5H_5)$] which is formed by nucleophilic addition of alkoxide to [$Pd(\eta^4$-cod$)(\eta$-$C_5H_5)]^+$ and which is regarded as a model for the first intermediate in the Pd(II) catalysed oxidation of ethylene to acetaldehyde (*i.e.* the Wacker process). While the Wacker process itself is unlikely to involve paramagnetic intermediates, radical mechanisms for other reactions formally involving the Pd(II)/Pd(0) couple should not be discounted.

The reactions of (**2**$^+$) with donor solvents such as acetone and acetonitrile[34] are very different from those with other nucleophiles in that $Pd(\eta$-$C_5Ph_5)$ bond cleavage occurs (Scheme 4); the products, [$Pd_2(\mu$-$PhC_2Ph)L_3(\eta$-$C_5Ph_5)]^+$ (L = acetone or MeCN) are precursors to novel asymmetric alkyne-bridged dipalladium complexes such as [$Pd_2(\mu$-$PhC_2Ph)(PPh_3)(bipy)(\eta$-$C_5Ph_5)]^+$.

It is not easy to propose a mechanism which accounts for the formation of both types of product when (**2**$^+$) reacts with nucleophiles. The first step in the reaction sequence

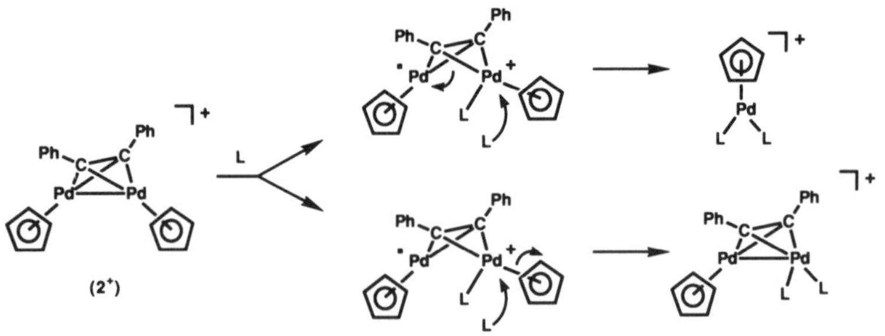

Scheme 4. The oxidative cleavage of [$Pd_2(\mu$-$PhC_2Ph)(\eta$-$C_5Ph_5)_2]^+$ (**2**$^+$).

is no doubt nucleophilic attack of a ligand (L in CH_2Cl_2, or the donor solvent itself) as shown in Scheme 4; at this stage metal-metal bond cleavage would generate one 17-electron centre and one cationic 18-electron centre bound to the incoming nucleophile. The next step might then involve cleavage of one alkyne-metal bond and coordination of the second ligand L, to give [$PdL_2(\eta$-$C_5Ph_5)]^+$, or loss of C_5Ph_5 and reformation of the metal-metal bond to give [Pd_2(L-L)$(\mu$-$PhC_2Ph)(\eta$-$C_5Ph_5)]^+$. Where the incoming ligand is a π-acceptor, for example PPh_3 or dienes, the first possibility is likely to be favoured; σ-donor ligands such as acetone and acetonitrile are more likely to result in the second pathway.

4. The Redox-Induced Oligomerisation of Co-ordinated Alkynes

The electron-transfer reactions of bis(alkyne) complexes are rather different from those of monoalkyne complexes. Thus, although $[Mo(CO)(PhC_2Ph)_2(\eta\text{-}C_5Me_5)]^+$ is said[36] to undergo two reversible one-electron reductions, Green and co-workers have shown[37] that $[Mo(NCMe)(\eta\text{-}MeC_2Me)_2(\eta\text{-}C_5H_5)]^+$ and reducing agents such as $[Fe(CO)_2(\eta\text{-}C_5H_5)]^-$ give a binuclear complex containing an alkyne tetramer, namely (3) (Scheme 5). The C_8

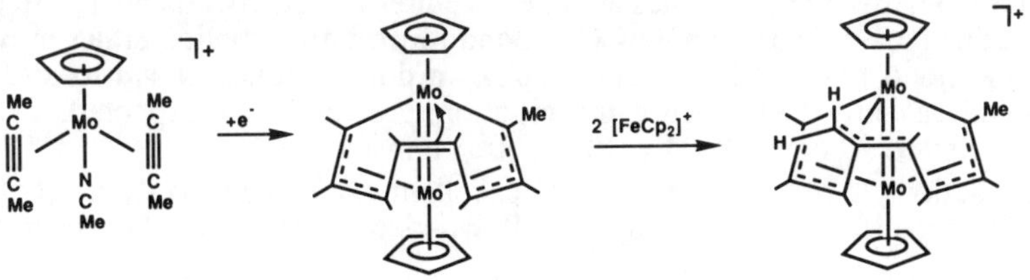

Scheme 5. The reductive oligomerisation of $[Mo(NCMe)(\eta\text{-}MeC_2Me)_2(\eta\text{-}C_5H_5)]^+$ and the oxidative activation of $[Mo_2(\mu\text{-}C_8Me_8)(\eta\text{-}C_5H_5)_2]$ (3).

chain bridges two Mo atoms, with both termini bound to the same molybdenum atom. Electrochemical studies[38] revealed that one-electron reduction to $[Mo(NCMe)(\eta\text{-}MeC_2Me)_2(\eta\text{-}C_5H_5)]$, followed by rapid loss of the nitrile ligand, precedes oligomerisation but the route by which $[Mo(\eta\text{-}MeC_2Me)_2(\eta\text{-}C_5H_5)]$ then yields the final product (3) is unknown. Further work is necessary to understand this potentially important reaction but it is clear that the ancillary ligands are important; the reduction of $[Mo(NCMe)(\eta\text{-}MeC_2Me)_2(\eta^5\text{-indenyl})]^+$ gives a 'flyover' complex, $[Mo_2(\mu\text{-}C_6Me_6)(\eta^5\text{-indenyl})_2]$, in which an alkyne trimer bridges the two metal atoms; in this case the C_6 chain begins at one metal and ends at the second.

5. The Redox Activation of Co-ordinated Alkyne Oligomers

One-electron transfer reactions are not only involved in the formation of co-ordinated alkyne oligomers, as described above, but also in their reactions. Such reactions can involve both C-H and C-C bond activation.

5.1. OXIDATIVE ACTIVATION OF C-H BONDS

The reaction of complex (3) with $[CPh_3]^+$ results[39] in the isolation of cation (4) (Scheme 5) apparently by hydride ion abstraction from one of the C_8Me_8 methyl groups. However, electrochemical studies show that H^- loss does not occur. The cyclic voltammogram of (3) has a reversible one-electron oxidation wave at -0.11V followed by a second, irreversible, oxidation wave at 0.64V. We have shown previously[40] that $[Ru_2(\mu\text{-}CH_2)(\mu\text{-}CO)(\mu\text{-}dppm)(\eta\text{-}C_5H_5)_2]$ is converted into $[Ru_2(\mu\text{-}CH)(\mu\text{-}CO)(\mu\text{-}dppm)(\eta\text{-}C_5H_5)_2]^+$ by a double-oxidation deprotonation mechanism and it is likely that a similar process occurs to give (**4**) from (**3**) *i.e* there is loss of H^+ from ($\mathbf{3}^{2+}$) rather than H^- from (**3**). However, ($\mathbf{3}^+$) has not yet been isolated and the possibility of hydrogen atom loss after one-electron oxidation of (**3**), as suggested[41] for the formation of $[Fe_2(\mu\text{-}CH)(\mu\text{-}CO)(CO)_2(\eta\text{-}C_5H_5)_2]^+$ from $[Fe_2(\mu\text{-}CH_2)(\mu\text{-}CO)(CO)_2(\eta\text{-}C_5H_5)_2]$, cannot yet be ruled out.

5.2. C-C BOND FORMATION AND CLEAVAGE

The thermal oligomerisation of alkynes at dimetal centres is well documented[42]. In general, the two electrons required for each new C-C bond formed are supplied either by a metal-metal multiple bond (*e.g.* a triple bond becomes a double bond) or *via* loss of a two-electron ligand (usually CO) during the reaction. As might be expected, C-C bond formation can also be accomplished by two-electron transfer.

The reactions in Scheme 6 provide[43] excellent examples of fully reversible C-C bond formation and cleavage induced by overall two-electron processes. However, C-C

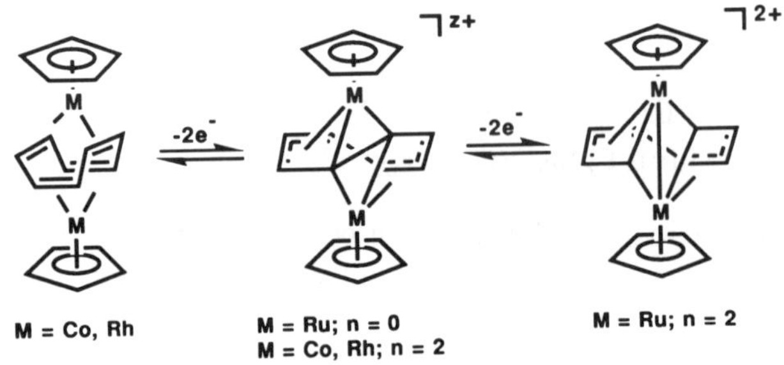

Scheme 6. Reversible redox-induced cleavage of C-C bonds in μ-cot complexes.

bond formation can also occur after *one*-electron reduction, *e.g.* arenes are formed[44] from the flyover complexes $[Co_2(CO)_4(\eta\text{-}C_6R_6)]$ (*via* the arene radical anions), and by two-electron oxidation, *e.g.* when the free arene 1,2-$C_6H_4Bu^t_2$ is formed on treatment[45] of the flyover complex $[Co_2(CO)_4(\eta\text{-}C_6H_4Bu^t_2)]$ with bromine (a strong but non-selective oxidant). Much work remains to be carried out to understand these fascinating, and potentially important, C-C bond forming reactions.

6. Conclusions and Future Developments.

The redox chemistry of alkyne and related complexes provides excellent examples of the wide applicability of electron-transfer reactions in organometallic chemistry. Electrochemical, spectroscopic, and structural studies provide an overall picture of the bonding of alkynes in mono-, bi-, and trinuclear centres, and ligand isomerisation, oligomerisation, C-C and C-H bond activation can be brought about by both oxidation and reduction. The determination and control of the factors which can lead to stereo- and regiospecificity in redox-induced organic transformations will play a large part in future developments in this area. Similarly, redox-catalysed processes may be developed. In this context, the coupling of ETC catalysis and 'normal' catalysis in the production[46] of alkyne polymers from terminal alkynes, $[W(CO)_3(NCMe)_3]$ and $[Fe(\eta-C_5H_5)_2]^+$ is particularly relevant.

7. Acknowledgement

It is a pleasure to acknowledge the efforts of the many students and post-doctoral workers who have contributed to the chemistry outlined here, and the invaluable collaborations I have had with Guy Orpen (University of Bristol, U.K.), Bill Geiger (University of Vermont, U.S.A.), and Phil and Nancy Rieger (Brown University, U.S.A.).

8. References

1. N.G. Connelly and W.E. Geiger, *Adv. in Organometal. Chem.,* 1984, **23**, 1; W.E. Geiger and N.G. Connelly, *Adv. in Organometal. Chem.,* 1985, **24**, 87; J.C. Kotz, in "Topics in Organic Electrochemistry" Plenum Press, 1986; J.K. Kochi, *J. Organomet. Chem.,* 1986, **300**, 139; P. Lemoine, *Coord. Chem. Revs.*, 1988, **83**, 169; C.J. Pickett, in "Electrochemistry", A Specialist Periodical Report, Ed. D. Pletcher, The Royal Society of Chemistry, Vol. 8, p. 81, 1983 and Vol. 9, p. 162, 1984.
2. see, for example, A.G. Orpen and N.G. Connelly, *Organometallics*, 1990, **9**, 1206; A.G. Orpen, N.G. Connelly, M.W. Whiteley, and P. Woodward, *J. Chem. Soc., Dalton Trans.,* 1989, 1751; N.G. Connelly, W.E. Geiger, G.A. Lane, S.J. Raven, and P.H. Rieger, *J. Am. Chem. Soc.*, 1986, **108**, 6219; N.G. Connelly, M.J. Freeman, A.G. Orpen, A.R. Sheehan, J.B. Sheridan, and D.A. Sweigart, *J. Chem. Soc., Dalton Trans.,* 1985, 1019; M.J. Freeman, A.G. Orpen, N.G. Connelly, I. Manners, and S.J. Raven, *J. Chem. Soc., Dalton Trans.,* 1985, 2283.
3. P.H. Rieger, p. 375 in "Paramagnetic Organometallic Species in Activation/Selectivity, Catalysis", Ed. M. Chanon, M. Julliard, and J.C. Poite, Kluwer Academic Publishers, Holland, 1989; P.H. Rieger, in "Organometallic Radical Processes", Ed. W.C. Trogler, Elsevier, Amsterdam, in press.
4. M.C. Baird, *Chem. Revs.*, 1988, **88**, 1217.
5. D.R. Tyler and F. Mao, *Coord. Chem. Revs.*, 1990, **97**, 119; D. Astruc, *Chem. Revs.*, 1988, **88**, 1189.
6. M.J. Therien, C.-L. Ni, F.C. Anson, J.G. Osteryoung, and W.C. Trogler, *J. Am. Chem. Soc.*, 1986, **108**, 4037.
7. D. Astruc, *Angew. Chem., Int. Ed. Engl.,* 1988, **27**, 643; M. Chanon, *Acc. Chem. Res.,* 1987, **20**, 214.
8. N.G. Connelly, *Chem. Soc. Revs.,* 1989, **18**, 153; N.G. Connelly, p. 71 in "Paramagnetic Organometallic Species in Activation/Selectivity, Catalysis", Ed. M. Chanon, M. Julliard, and J.C. Poite, Kluwer Academic Publishers, Holland, 1989.
9. R.P. Aggarwal, N.G. Connelly, B.J. Dunne, M. Gilbert, and A.G. Orpen, *J. Chem. Soc., Dalton Trans.,* 1991, 1; N.G. Connelly, M.J. Freeman, A.G. Orpen, J.B. Sheridan, A.N.D. Symonds, and M.W. Whiteley, *J. Chem. Soc., Dalton Trans.,* 1985, 1027.
10. J.L. Templeton, *Adv. in Organometal. Chem.,* 1989, **29**, 1; M.J. McGeary, A.S. Gamble and J.L. Templeton. *Organometallics*, 1988, **7**, 271; P.B. Winston, S.J.N. Burgmayer, T.L. Tonker, and J.L. Templeton, *Organometallics*, 1986, **5**,

1707; J.L. Templeton, R.S. Herrick, and J.R. Morrow, *Organometallics*, 1984, **3**, 535.
11. N.G. Connelly and G.A. Johnson, *J. Organomet. Chem.*, 1974, **77**, 341.
12. N.G. Connelly, Z. Demidowicz, and R.L. Kelly, *J. Chem. Soc., Dalton Trans.*, 1975, 2335.
13. N.G. Connelly, A.L. Rieger, P.H. Rieger, and C.J. Scott, unpublished results.
14. A.B.P. Lever, *Inorg. Chem.*, 1990, **29**, 1271; B.E. Bursten and M.R. Green, *Progr. in Inorg. Chem.*, 1988, **36**, 393; R.H. Morris, K.A. Earl, R.L. Luck, N.J. Lazarowych, and A. Sella, *Inorg. Chem.*, 1987, **26**, 2674, and refs. therein.
15. D.J. Wink, J.R. Fox, and N.J. Cooper, *J. Am. Chem. Soc.*, 1985, **107**, 5012.
16. N.G. Connelly, A.G. Orpen, P.H. Rieger, and G.M. Rosair, unpublished results.
17. M.I. Bruce, *Chem. Revs.* 1991, **91**, 197.
18. R.G. Beevor, M.J. Freeman, M. Green, C.E. Morton, and A.G. Orpen, *J. Chem. Soc., Chem. Comm.*, 1985, 68.
19. R.S. Iyer and J.P. Selegue, *J. Am. Chem. Soc.*, 1987, **109**, 910.
20. R.L. Beddoes, C. Bitcon, A. Ricalton, and M.W. Whiteley, *J. Organomet. Chem*, 1989, **367**, C21; J.S. Adams, C. Bitcon, J.R. Brown, D. Collison, M. Cunningham, and M.W. Whiteley, *J. Chem. Soc., Dalton Trans.*, 1987, 3049.
21. J. Gotzig, H. Otto, and H. Werner, *J. Organomet. Chem.*, 1985, **287**, 247.
22. N.A. Ustynyuk, V.N. Vinogradova, V.G. Andrianov, and Yu. T. Struchkov, *J. Organomet. Chem.*, 1984, **268**, 73.
23. A.B. Anderson, *Inorg. Chem.*, 1976, **15**, 2598.
24. D.L. Thorn and R. Hoffmann, *Inorg. Chem.*, 1978, **17**, 126.
25. F.A. Cotton, J.D. Jamerson, and B.R. Stults, *J. Am. Chem. Soc.*, 1976, **98**, 1774.
26. see ref. 19 in M. Arewgoda, P.H. Rieger, B.H. Robinson, J. Simpson, and S.J. Visco, *J. Am. Chem. Soc.*, 1982, **104**, 5633.
27. R.P. Aggarwal, N.G. Connelly, M.C. Crespo, B.J. Dunne, P.M. Hopkins, and A.G. Orpen, *J. Chem. Soc., Dalton Trans.*, 1992, 655.
28. B.M. Peake, P.H. Rieger, B.H. Robinson, and J. Simpson, *J. Am. Chem. Soc.*, 1980, **102**, 156.
29. D. Osella, R. Gobetto, P. Montangero, P. Zanello, and A. Cinquantini *Organometallics*, 1986, **5**, 1247
30. M. Arewgoda, B.H. Robinson, and J. Simpson, *J. Am. Chem. Soc.*, 1983, **105**, 1893; S.D. Jensen, B.H. Robinson, and J. Simpson, *Organometallics*, 1987, **6**, 1479, and refs. therein.
31. J.A. DeGray, Q. Meng, and P.H. Rieger, *J. Chem. Soc., Faraday Trans. I,* 1987, **83**, 3565; L.V. Casagrande, T. Chen, P.H. Rieger, B.H. Robinson, J. Simpson, and S.J. Visco, *Inorg. Chem.*, 1984, **23**, 2019.
32. K. Broadley, N.G. Connelly, G.A. Lane, and W.E. Geiger, *J. Chem. Soc., Dalton Trans.*, 1986, 373.
33. G.A. Lane, W.E. Geiger, and N.G. Connelly, *J. Am. Chem. Soc.*, 1987, **109**, 402.
34. N.G. Connelly, W.E. Geiger, A.G. Orpen, J.J. Orsini, jun., and K.E. Richardson, *J. Chem. Soc., Dalton Trans.*, 1991, 2967.
35. N.G. Connelly and I. Manners, *J. Chem. Soc., Dalton Trans.*, 1989, 283, and refs. therein.
36. P. Leoni, F. Marchetti, M. Pasquali, and P. Zanello, *J. Chem. Soc., Dalton Trans.*, 1988, 635.
37. M. Green, *Polyhedron*, 1986, **5**, 427.
38. D. Pufahl, W.E. Geiger, and N.G. Connelly, *Organometallics*, 1989, **8**, 412.

39. S.G. Bott, L. Brammer, N.G. Connelly, M. Green, A.G. Orpen, J.F. Paxton, C.J. Schaverien, S.Bristow and N.C. Norman, *J. Chem. Soc., Dalton Trans.*, 1990, 1957.
40. N.G. Connelly, N.J. Forrow, B.P. Gracey, S.A.R. Knox, and A.G. Orpen, *J. Chem. Soc., Chem. Comm.,* 1985, 14.
41. ref. 29 in B.E. Bursten and R.H. Cayton, *J. Am. Chem. Soc.*, 1986, **108**, 8241.
42. S.A.R. Knox, R.F.D. Stansfield, F.G.A. Stone, M.J. Winter, and P. Woodward, *J. Chem. Soc., Dalton Trans.,* 1982, 173, and refs. therein.
43. J. Edwin and W.E. Geiger, *J. Am. Chem. Soc.*, 1990, **112**, 7104; W.E. Geiger, A. Salzer, W. von Philipsborn, U. Piantini, and A.L. Rheingold, *J. Am. Chem. Soc.*, 1990, **112**, 7113.
44. C.M. Arewgoda, A.M. Bond, R.S. Dickson, T.F. Mann, J.E. Moir, P.H. Rieger, B.H. Robinson, and J. Simpson, *Organometallics*, 1985, **4**, 1077.
45. R.S. Dickson and P.J. Fraser, *Adv. in Organometal. Chem.,* 1974, **12**, 323.
46. M.-H. Desbois and D. Astruc, *New J. Chem.*, 1989, **13**, 595.

ELECTROCHEMICAL BEHAVIOUR OF COMPLEXES DERIVED FROM THE ACTIVATION OF ALKYNES, ISOCYANIDES AND NITRILES

ARMANDO J.L. POMBEIRO
Centro de Química Estrutural, Complexo I,
Instituto Superior Tecnico, Av. Rovisco Pais
1096 Lisboa codex, Portugal

ABSTRACT. The redox properties of a variety of isocyanide, nitrile and alkyne-derived vinylidene and phenylallene complexes, commonly with an electron-rich Re(I) or Fe(II)-phosphinic centre, as well as of the derived protonated species (with aminocarbyne, methyleneamido, carbyne, η^2-vinyl or hydride ligands), as studied by cyclic voltammetry and controlled potential electrolysis in aprotic media, are discussed in terms of ligand and metal site effects on the redox potential, and the values of the electrochemical P_L parameter which measures the ligand net electron-donor/acceptor character are also presented for these ligands. Expressions are indicated which allow to predict the oxidation potential for the octahedral-type 18-electron complexes $[M_sL_n]$ (n=2 or 3) or $[M_sLL']$. Moreover, the electrode processes of the protonated complexes are shown to involve the following types of electroinduced reactions, which are also discussed: anodic N-H, C-H or metal-H heterolytic bond cleavage; cathodic N-H, C-H or metal-H homolytic bond rupture; anodic nucleophilic attack at the metal; anodic ligand isomerization.

1. INTRODUCTION

In pursuit of our interest on the activation of small molecules, particularly C-N and C-C unsaturated substrates of nitrogenases (the N_2-reducing enzymes in the biological nitrogen fixation process), we have been investigating [1] the coordination chemistry of isocyanides (C≡NR; R = alkyl, aryl or H), nitriles or cyanides (N≡CR, including cyanamide ,N≡C-NH_2) and alkynes, commonly at electron-rich dinitrogen-binding d^6 metal centres of the types {M(dppe)$_2$} (M = Mo or W, dppe = $Ph_2PCH_2CH_2PPh_2$), {ReCl(dppe)$_2$} and derived sites, or {FeH(dppe)$_2$}$^+$.

Series of isocyanide, nitrile (cyanide) or alkyne-derived vinylidene (=C=CHR), allene (η^2-CH_2=C=CHPh) or alkynyl (-C≡CR) complexes have been obtained, usually by replacement of dinitrogen and/or chloro ligand at a parent N_2-complex, *e.g.*, *trans*-[M(N_2)$_2$(dppe)$_2$] or *trans*-[ReCl(N_2)(dppe)$_2$], with involvement, in the case of the alkynes, of hydrogen-migration processes along their carbon-framework. As typical examples of such complexes we may cite the

A. J. L. Pombeiro and J. A. McCleverty (eds.),
Molecular Electrochemistry of Inorganic, Bioinorganic and Organomettallic Compounds, 331–344.

following 18-electron octahedral-type species: *trans*-$[M(CNR)_2(dppe)_2]$ (M = Mo or W) [2], *trans*-$[ReCl(L)(dppe)_2]$ { L = CNR [3,4], NCR [5], C=CHR [6], η^2-CH_2=C=CHPh [7] }, $[ReL_2(dppe)_2][BF_4]$ (L = CNR [8], CO [8] or NCR [9]), *trans*-$[Re(CNMe)L(dppe)_2]X$ (L = CNR, CO or NCMe; X = Cl, BF_4 or PF_6) [10], *trans*-$[ReL(CNR)(dppe)_2]^+$ (L = $NCNH_2$ or $NCNH^-$) [11], $[Re(N_2)(NCR)(dppe)_2][BF_4]$ [9,12], *mer*-$[ReCl(N_2)(CNR)\{P(OMe)_3\}_3]$ and $[ReCl(N_2)(CNMe)(PPh_3)\{P(OEt)_3\}_2]$ [13].

Moreover, such unsaturated ligands are often activated by the binding metal centre towards β-electrophilic attack (usually protonation) to give, *e.g.*, a variety of multiple metal-carbon bonded species, such as: the aminocarbynes *trans*-$[ReCl(\equiv C \cdots NHR)(dppe)_2]^+$ from the parent isocyanide complexes [14,4], the carbynes *trans*-$[ReCl(\equiv C\text{-}CH_2R)(dppe)_2]^+$ from the vinylidenes [15] and the η^2-vinyl *trans*-$[ReCl\{\eta^2\text{-}C(CH_2Ph)CH_2\}(dppe)_2]^+$ from the η^2-phenylallene complex [16]. In addition, the methyleneamido compounds $[ReCl(N{=}CHR)(dppe)_2]^+$ have been obtained by β-protonation of the parent nitrile species [17]. These reactions, summarized by the overall eqs. 1-4, constituted (with the exception of the vinylidene protonation — eq. 3) novel routes for such types of organic or organonitrogenated products; they have been reviewed recently [1] and will not be discussed herein.

$$M\text{-}C{\equiv}NR \quad + \quad H^+ \quad \rightarrow \quad M{\equiv}C{\cdots}NHR^{\,+} \tag{1}$$

$$M\text{-}N{\equiv}CR \quad + \quad H^+ \quad \rightarrow \quad M\text{ - }N{=}CHR^{\,+} \tag{2}$$

$$M = C{=}CHR \quad + \quad H^+ \quad \rightarrow \quad M{\equiv}C\text{-}CH_2R^{\,+} \tag{3}$$

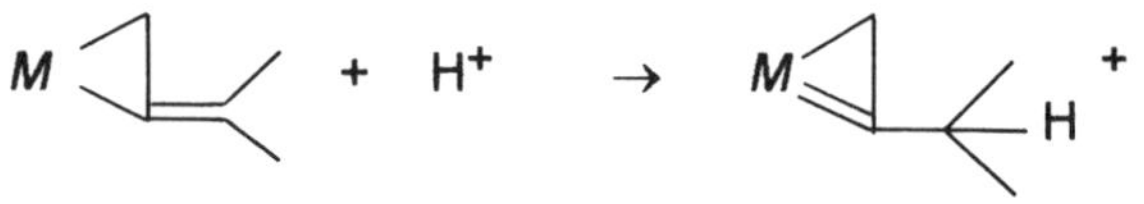

(4)

In addition, we have also been interested in the application of electrochemical methods to the study of the properties of those coordination compounds, namely of the redox and electronic type; ligand and structural effects on their redox potential have been investigated and expressions developed for electrochemical quantification or prediction of the redox properties and the electron donor/acceptor character of metal sites and ligands. Electroinduced reactivity (for both the binding metal centres and the ligated substrates, often involving the electroactivation of metal-H, N-H or C-H bonds) has also been studied, as well as, by electrochemical techniques, the mechanisms of some of these processes.

Such types of electrochemical studies are briefly discussed in the present review.

2. SINGLE-ELECTRON REVERSIBLE OXIDATIONS, REDOX POTENTIAL, METAL / LIGAND AND STRUCTURAL EFFECTS

2.1. REDOX PROPERTIES

Most of the abovementioned isocyanide, nitrile, vinylidene, allene and dinitrogen complexes, in an aprotic medium (such as THF or NCMe/0.2 M $[Bu_4N][BF_4]$, as the normal conditions used throughout this work), can undergo, by cyclic voltammetry (CV), at a Pt electrode, two successive single-electron oxidations, the first one commonly being reversible and corresponding to the Mo(0/I) or Re(I/II) redox couples. It occurs, for the neutral complexes, at a considerable low half-wave oxidation potential, $E_{1/2}^{ox}$, in agreement with the high electron-richness of their neutral d^6 metal centres, *e.g.*, *trans*-$[M(CNR)_2(dppe)_2]$ ($E_{1/2}^{ox}$ = -1.1 to -0.33 V *vs.* SCE) [2], $[ReCl(NCR)(dppe)_2]$ [5,9] and $[Re(NCNH)(CNR)(dppe)_2]$ [11] (-0.4 to 0.0 V), *trans*-$[ReCl(C{=}CHBu^t)(dppe)_2]$ (-0.26 V) [18], *trans*-[ReCl $\{\eta^2\text{-}CH_2{=}C{=}CHPh\}(dppe)_2]$ (-0.03 V) [18] or *trans*-$[ReCl(CNR)(dppe)_2]$ (+0.1 to +0.4 V) [3]. As expected, the cationic complexes are oxidized at more anodic potentials [8-11,18].

An isomeric effect can be detected in some cases and, *e.g.*, *cis*-$[Re(NCR)_2(dppe)_2]^+$ are oxidized at considerably higher potentials (by *ca.* 0.1-0.2 V) than the corresponding *trans*-isomers [9], thus showing a greater stabilization of the HOMO in the former species in agreement with some theoretical calculations [19] which indicate that the *cis* isomers of related complexes should be thermodynamically more stable than the *trans* ones.

The first reversible anodic wave, in some cases,may be conveniently monitored by CV along the course of a reaction. This, together with NMR spectroscopy, has allowed the investigation of the mechanism for the formation of some of the Mo-isocyanide complexes, such as *cis*-$[Mo(CNMe)_2L_4]$ ($L{=}PMe_2Ph$) formed by reacting CNMe with *cis*-$[Mo(N_2)_2L_4]$ [20]. The data were consistent with the first and the second dinitrogen losses being the rate limiting steps, according to eqs. 5 and 6 (where *Mo* stands for MoL_4).

$$cis\text{-}[Mo(N_2)_2] \xrightarrow[k_1]{-N_2} [Mo(N_2)] \xrightarrow{CNMe} cis\text{-}[Mo(N_2)(CNMe)] \quad (5)$$

$$cis\text{-}[Mo(N_2)(CNMe)] \xrightarrow[k_2]{-N_2} [Mo(CNMe)] \xrightarrow{CNMe} [Mo(CNMe)_2] \quad (6)$$

By comparing the obtained values of the k_2/k_1 ratio for the replacement of N_2 by CNR or NCR it was proposed [20] a *cis*-N_2-labilising order (CNMe > N_2 > NCR) which follows the π-accepting ability of these ligands and reflects the importance of π-effects in the electronic properties of these electron-rich complexes.

2.2. CORRELATIONS of the REDOX POTENTIAL with HAMMETT'S CONSTANTS and with ELECTRONIC and IR SPECTROSCOPIC DATA.

The aryl-isocyanide [3] or -cyanide [5,9] complexes of our study are invariably harder to oxidize than their alkyl homologues and, within any aryl series (with a variable phenyl substituent), a gross linear correlation is observed between $E_{1/2}^{ox}$ and the Hammett's σ_p or σ_p^+ constant, the former increasing with the latter, in accord with the stabilization of the HOMO as a result of an increase of the electron-acceptor character of a ligand; moreover, the HOMO can be conjugated to the aryl substituent ($E_{1/2}^{ox}$ *vs.* σ_p^+ correlation) [3], again reflecting the important role of π-effects.

From the linear correlation between $E_{1/2}^{ox}$ and σ_p or the Taft polar σ^* substituent constants recognized for a number of ferrocene derivatives, it was possible to estimate such constants for various isocyanide groups and the (high) values obtained (*e.g.*, σ^* *ca.* 2.8 for -N≡C) substantiate the strong electron withdrawing ability of the isocyano group [21].

Moreover, a coarse linear inverse correlation between $E_{1/2}^{ox}$ and the energy of the metal to ligand charge transfer band has been observed in some of the isocyanide [22,23] or nitrile series [9] of complexes, which has been interpreted by considering that the LUMO presents a greater dependence than the HOMO on the substituent, in agreement with some simplified π-MO schemes [24] which indicate that the former orbital, in contrast to the latter, is mainly based on orbitals of the unsaturated ligand.

In addition, gross linear relationships have been recognized between $E_{1/2}^{ox}$ and the IR ν(C≡N) or ν(N≡N), for series of related isocyanide and/or dinitrogen complexes [25], a lowering in the IR frequency corresponding to a decrease of $E_{1/2}^{ox}$, thus following the expected increase of the π-electron releasing ability of the metal centre to the π^* orbitals of such ligands.

2.3. ELECTROCHEMICAL LIGAND and METAL SITE PARAMETERS

In order to get a better insight into the electronic properties of the ligands and binding metal sites of this study, from a coordination point of view, we have followed the systematic approach proposed by others [26] based on the linear relationship expressed by eq. 7 which relates $E_{1/2}^{ox}$ of 18-electron octahedral-type complexes [M_SL] (M_S = 16-electron square pyramid metal site) with the so-called P_L ligand parameter, defined by eq. 8 and which constitutes a measure of the net π-electron acceptor minus σ-donor character of the ligand L: the stronger this character, the higher (*i.e.*, usually the less negative) is P_L.

The parameter E_S, defined by eq. 9, is proposed as a measure of the "electron-richness" of the metal site — the lower this parameter, the higher is the electron-richness — whereas β is the polarisability of the site.

$$E_{1/2}^{ox}\ [M_sL] = E_s + \beta . P_L \quad (7)$$

$$P_L = E_{1/2}^{ox}\ [Cr(CO)_5L] - E_{1/2}^{ox}\ [Cr(CO)_6] \quad (8)$$

$$E_s = E_{1/2}^{ox}\ [M_s(CO)] \quad (9)$$

Initially, P_L was considered [26] to be characteristic of the ligand and independent of the binding metal centre. However, at least for ligands such as isocyanides which present a high ability to accommodate their electron donor/acceptor ability to the electronic properties of the binding metal site, the P_L ligand parameter can also depend on this site, as we have recognized [3] first for *trans*-[ReCl(CNR)(dppe)$_2$]. In order to account for the validity of eq. 7 applied to these and related complexes with the common electron-rich metal site {ReCl(dppe)$_2$}, the isocyanide P_L values, estimated from eq. 8, have to be corrected by adding 0.26 V (eq. 10, where the superscript *l* stands for *low* value).

$$P_L(CNR) = 0.26 + P_L{}^l(CNR) \quad (10)$$

Therefore, at such an electron-rich centre (E_s = 0.68 V), the isocyanides behave as stronger net electron acceptors than when they bind {Cr(CO)$_5$} (E_s = 1.50 V [26]) which was used to define that parameter. The former isocyanide behaviour is also expected for other N_2-binding metal centres with a high electron-richness, such as {Mo(CO)(dppe)$_2$} (E_s = -0.11V [26]), whereas the low P_L values ($P_L{}^l$) have been reported [8,10] for the less electron-rich sites, {ReL(dppe)$_2$}$^+$ [*e.g.*,L = CNMe (E_s = 1.15 V) or CO (E_s = 1.42 V)] [8,10], apart from {Cr(CO)$_5$}.

The P_L dependence on the metal site has also been found for nitrile ligands, for which the P_L values increase with the increase of the electron-richness of the binding metal site, *e.g.*, in the order {Re(CO)(dppe)$_2$}$^+$ < {Re(NCR)(dppe)$_2$}$^+$ < {ReCl(dppe)$_2$} [9].

The P_L values for the unsaturated ligands of our study, usually at {ReCl(dppe)$_2$}, have been estimated by using eq.7 or derived ones, the measured $E_{1/2}^{ox}$ or $E_{p/2}^{ox}$ values (the latter for the irreversible anodic processes), and the knowledge [3] of E_s (0.68 V) and β (3.4) for this centre. This has allowed to order them as follows (with the P_L values in parentheses), according to their net π-electron acceptor / σ-donor character:

Carbynes (*ca.* 0.27) [18] > η^2-vinyl (0.22) [18] > aminocarbynes (*ca.* 0.1) [18] > methyleneamides (*ca.* 0.07) [27] > CO (0) > isocyanides (-0.1 to -0.2) [3,8, 10] > η^2-allene (-0.21) [18] > vinylidenes (*ca.*-0.27) [18] > nitriles (-0.29 to -0.33) [9] > cyanamide (-0.85) [11] > alkynyl (-1.22) [28] > hydrogen cyanamide (-1.34 V) [11].

All the protonated ligands (carbynes, η^2-vinyl, aminocarbynes and methyleneamides) behave as better net electron acceptors than CO, whereas hydrogen cyanamide (-NCNH) is the strongest net electron donor, followed by the alkynyl ligand (-C≡CPh).

Within each type of ligands, P_L also reflects the (electronic) properties of their particular moieties and, *e.g.*, for the isocyanides (CNR), it increases with the net electron acceptor ability of R [3]; moreover, a P_L increase of *ca.*0.2 V results from the introduction of a positive charge at this group, in some ferrocenyl-isocyanides (R = ferrocene or a derived group), as a consequence of the oxidation of the iron(II) centre [29].

2.4. ESTIMATE of ELECTROCHEMICAL METAL SITE PARAMETERS

Electrochemical metal site parameters can be estimated by using data obtained with related centres and expressions derived from the abovementioned basic equations.

Hence, *e.g.*, the electron-richness (E_S) of $\{ReL(dppe)_2\}^+$ (L = CNR or N_2) can easily be estimated from the knowledge of E_S (1.42 V) and β (0.62) of the related CO centre $\{Re(CO)(dppe)_2\}^+$, by considering the definition of E_S (eq. 9) followed by the application of eq. 7 to the latter site (eqs. 11 and 12) [8].

$$E_S\{ReL(dppe)_2\}^+ = E_{1/2}{}^{ox}[ReL(CO)\,(dppe)_2]^+ \tag{11}$$

$$= E_S\{Re(CO)\,(dppe)_2\}^+ + \beta\{Re(CO)\,(dppe)_2\}^+.P_L(L) \tag{12}$$

Moreover, the polarisability β of the $\{ReL(dppe)_2\}^+$ site is simply given by an expression obtained through the application of eq. 7 to $[ReCl(L)(dppe)_2]$ (with known redox potential) [8].

As expected, the electron-richness of the metal site decreases (*i.e.*, E_S increases) with the increase of the net electron acceptor character of L, from Cl^- (E_S = 0.68 V, β= 3.4) to CNR (E_S = 1.15 -1.19 V, β = 0.92 - 0.75), N_2 (E_S = 1.38 V, β = 0.92) or CO (E_S = 1.42 V, β = 0.62), whereas the polarisability tends to decrease in the same order; therefore, a strong net electron acceptor, such as isocyanide or CO, presents a high ability to buffer changes in electron density at the metal (*i.e.*, corresponding to low β values) [8,10].

2.5 ESTIMATE of the OXIDATION POTENTIAL for the MIXED LIGAND COMPLEXES $[M_SLL']$ and for $[M_SL_2]$

The oxidation potential of the closed shell mixed ligand complexes $[M_SLL']$ (M_S= 14-electron square planar metal centre) can be estimated as the average of the oxidation potentials of $[M_SL_2]$ and $[M_SL'_2]$, according to eq. 13, derived [10] from the application of eq. 7 to the metal sites involved, $\{M_SL\}$ and $\{M_SL'\}$. It is expected to be valid if these centres present similar values of β (and this implies that the ligands L and L' have comparable electron donor/acceptor properties) and if the

P_L value, for each ligand, is the same for the two metal sites; its validity has been tested successfully to predict $E_{1/2}{}^{ox}$ for *trans*-$[Re(CNMe)L(dppe)_2]^+$ (L = CNR or CO) [10].

$$E_{1/2}{}^{ox}[M_sLL'] = 1/2\ \{E_{1/2}{}^{ox}[M_sL_2] + E_{1/2}{}^{ox}[M_sL'_2]\} \tag{13}$$

Moreover, the oxidation potential of $[M_sL_2]$ can be predicted by using eq. 14, also derived [8] from the abovementioned basic eqs. 7-9 and derived ones (*e.g.*, eq. 12), provided one knows the oxidation potential of one member of the series, the dicarbonyl complex $[M_s(CO)_2]$, as well as E_s and β of a suitable auxiliary metal site, $\{M_sL'\}$, which should include the basic $\{M_s\}$ centre of the series. It is also necessary to know the values of P_L for L and the auxiliary L' ligand.

If P_L depends on the metal site, a more complex expression should be used [8].

$$E_{1/2}{}^{ox}[M_sL_2] = \left[1 - \frac{P_L(L)}{P_L(L')}\right]^2 E_{1/2}{}^{ox}[M_s(CO)_2] + \left[\frac{2P_L(L)}{P_L(L')} - \frac{P_L{}^2(L)}{P_L{}^2(L')}\right] E_s\{M_sL'\} + \frac{P_L{}^2(L)}{P_L(L')}\,\beta\{M_sL'\} \tag{14}$$

The validity of this expression has been proved for *trans*-$[Re(CNR)_2(dppe)_2]^+$, by considering the auxiliary $\{ReCl(dppe)_2\}$ site (L' = Cl) [8].

2.6. GENERALIZATION of the ELECTROCHEMICAL P_L LIGAND and METAL SITE PARAMETERS

The abovementioned definitions of the P_L, E_s and β parameters have been extended [30] to octahedral 18-electron complexes of the types $[M_sL_n]$ with 14-electron (n=2) or 12-electron (n=3) metal sites, by defining a new overall P_{nL} ligand parameter (eq.15) and considering the general linear expression 16, in which the electron-richness E_s of the metal site $\{M_s\}$ is defined (eq. 17) as $E_{1/2}{}^{ox}$ of the di- or tri-carbonyl complex (n=2 or 3, respectively) with this centre. These expressions have been successfully tested for the metal centres $\{Re(dppe)_2\}^+$, $\{Mo(dppe)_2\}$, $\{cis\text{-}M(CO)_4\}$ (M=Cr, Mo or W) and $\{fac\text{-}M(CO)_3\}$ [30].

$$P_{nL} = E_{1/2}{}^{ox}[Cr(CO)_{6-n}L_n] - E_{1/2}{}^{ox}[Cr(CO)_6] \tag{15}$$

$$E_{1/2}{}^{ox}[M_sL_n] = E_s + \beta.P_{nL} \tag{16}$$

$$E_S = E_{1/2}^{ox}\,[M_S(CO)_n] \tag{17}$$

Moreover, the P_L ligand parameter appears to be additive, at least for *cis* ligands which are not too strong electron donors or acceptors (therefore without an extensive interaction), and the overall P_{nL} parameter is then equal to the sum of the n individual P_L values (eq. 18) [30].

$$P_{nL} = n.P_L \tag{18}$$

3. ELECTRON-TRANSFER WITH COUPLED CHEMICAL REACTION PROCESSES

We have been mainly discussing the redox properties of compounds which commonly undergo single-electron reversible or *quasi*-reversible oxidations. However, for some types of complexes, irreversible redox processes with coupled chemical reactions have been detected and examples are considered below.

3.1. AMINOCARBYNE, METHYLENEAMIDE, CYANAMIDE and η^2-VINYL COMPLEXES

The aminocarbyne complexes *trans*-$[ReCl(CNHR)(dppe)_2]^+$, formed by β-protonation of the corresponding isocyanide compounds, undergo, in a slightly basic solvent, such as THF, a partial dissociation to the parent species which are also formed by cathodic reduction or, in an oxidized form, by anodic oxidation with proton loss which has been demonstrated by acid-base titration of the electrolysed solution. These results are summarized by eqs. 19, where *Re* = *trans*-$\{ReCl(dppe)_2\}$,for the case of the simplest aminocarbyne species *trans*-$[ReCl(CNH_2)(dppe)_2]^+$ [18,31].

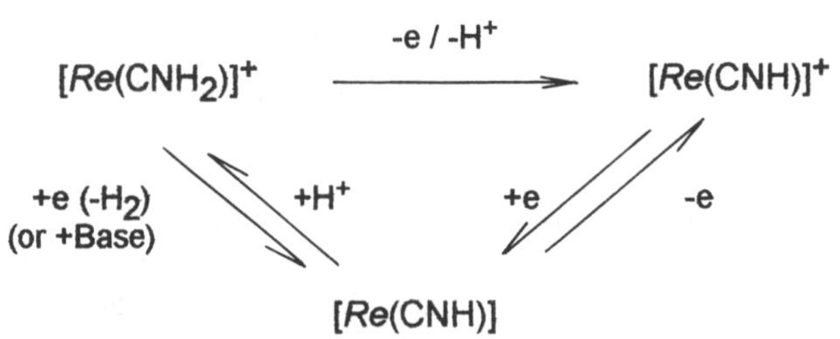

(19)

The methylaminocarbyne compound *trans*-$[ReCl(CNHMe)(dppe)_2]^+$ exhibits a related electrochemical behaviour, although the anodic process is more complex involving an overall two-electron oxidation to afford the di-oxidized, rather than the mono-oxidized, isocyanide compound (eqs. 20) [32]. A similar reaction profile upon electrochemical oxidation is followed by the dinuclear adducts $[ReCl\{CN(M)R\}(dppe)_2]$ [R = Me or Bu^t; *M* = $CoCl_2(THF)$ or $ReOCl_3(PPh_3)$] formed upon addition of the appropriate Lewis acid to the isocyanide complex *trans*-$[ReCl(CNR)(dppe)_2]$ [33]. In fact, they undergo dissociation with liberation of the Lewis acid upon two-electron anodic oxidation, in agreement with some extended Hückel calculations [33] which indicate the appearance of a positive charge on the nitrogen atom as a result of their oxidation.

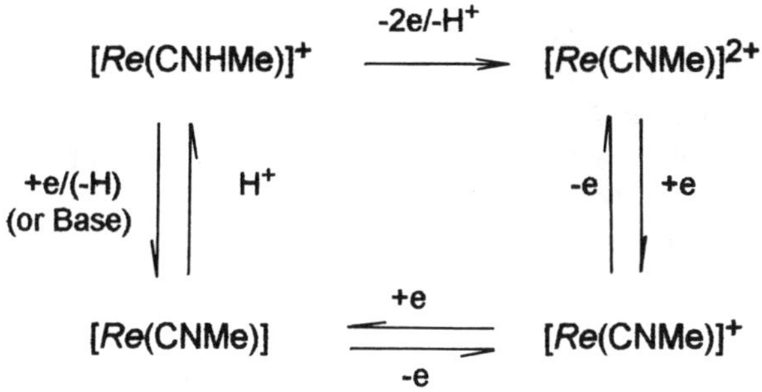

(20)

The electroinduced formation of the oxidized or the neutral isocyanide compounds from the corresponding aminocarbyne complexes has been clearly established by the detection of their characteristic CV waves formed upon oxidation or reduction, respectively, of the latter complexes.

An analogous anodic behaviour (eq. 21) is observed for the methyleneamido complexes $[ReCl(NCHR)(dppe)_2]^+$ (R = phenyl or substituted phenyl), derived from β-proton attack at the parent nitrile compounds [27].

$$[ReCl(NCHR)(dppe)_2]^+ \xrightarrow{-2e/-H^+} [ReCl(NCR)(dppe)_2]^{2+} \qquad (21)$$

The anodically induced deprotonation of those ligands (involving the heterolytic N-H or C-H bond cleavage) is consistent with the expected increase of acidity of the amino or the methyleneamido group upon oxidation of the metal site.

Moreover, related anodically-induced N-H and C-H bond rupture processes with proton loss (in the latter case possibly from methylene groups activated by an adjacent amino- or oxy-carbene moiety) have been reported [34] for a series of diamino- or aminooxy-carbene complexes of Pd or Pt, as discussed separately.

The one-electron cathodic reduction of the aminocarbyne complexes leads to dehydrogenation, by homolytic N-H bond cleavage at the CNHR ligand; H_2 evolution has been detected by gas chromatography in the case of the CNH_2 complex [18] but, for the other aminocarbynes, the fate of hydrogen has not been ascertained. The same cathodic behaviour has been recognized [11] for the cyanamide complexes *trans*-$[Re(CNR)(NCNH_2)(dppe)_2][BF_4]$ (R = Me or Bu^t) which, upon reduction, afford the hydrogen cyanamide species *trans*-$[Re(NCNH)(CNR)(dppe)_2]$.

The abovementioned allene-derived η^2-vinyl species *trans*-$[ReCl\{\eta^2\text{-}C(CH_2Ph)CH_2\}(dppe)_2]^+$ exhibits an identical cathodic process leading to the parent allene compound, whereas the described type of anodic behaviour, with β-proton elimination, conceivably also occurs, although this could not be confirmed in view of the higher anodic potential of the η^2-vinyl complex ($E_{p/2}{}^{ox}$ = 1.34 V) relative to that of the oxidation process (which is irreversible) of the possible oxidized allene product [18].

The oxidation of all these cationic complexes occurs at considerably higher potentials than those observed for the parent neutral compounds, in accord with the higher π-acceptance of the protonated ligands relative to the parent ones, as indicated also by their P_L values, discussed above.

3.2. HYDRIDE COMPLEXES

The 7-coordinated hydride-nitrile complexes $[ReClH(NCR)(dppe)_2]^+$, obtained by metal protonation at the parent nitrile compounds [9,35], undergo, by preparative electrolysis, two-electron anodically-induced deprotonation (at 0.7 - 0.9 V) to afford $[ReCl(NCR)(dppe)_2]^{2+}$ (as observed for the methyleneamido complexes) (eq. 22) and one-electron cathodically-induced dehydrogenation (at -1.0 ~ -1.2 V) to give the corresponding neutral nitrile complexes (eq. 23) [9,36] (following similar processes to those discussed above for the aminocarbyne, η^2-vinyl or cyanamide compounds).

$$[ReClH(NCR)(dppe)_2]^+ \xrightarrow{-2e/-H^+} [ReCl(NCR)(dppe)_2]^{2+} \quad (22)$$

$$[ReClH(NCR)(dppe)_2]^+ \xrightarrow{+e/(-H^\cdot)} [ReCl(NCR)(dppe)_2] \quad (23)$$

The detailed mechanisms of these reactions have been investigated by CV simulation techniques and shown [37] to involve bimolecular EC processes [eqs. 24 and 25 for the anodic process, where *Re* = $\{ReCl(NCR)(dppe)_2\}$], which are discussed separately.

$$ReH^+ \underset{}{\overset{-e}{\rightleftharpoons}} ReH^{2+} \rightleftharpoons Re^+ + H^+ \quad (24)$$

$$Re^+ + ReH^{2+} \rightarrow Re^{2+} + ReH^+ \quad (25)$$

A distinct anodic process is observed [38] for the hydride-isocyanide complexes of iron(II) *trans*-$[FeH(CNR)(dppe)_2]^+$ (R = alkyl or aryl) which follow an ECECE-type reaction mechanism [eqs. 26, where *Fe* = *trans*-$\{Fe(CNR)(dppe)_2\}$] as proved [39] by digital simulation.

$$FeH^+ \overset{-e}{\rightleftharpoons} FeH^{2+} \overset{-H^+}{\rightarrow} Fe^+ \overset{-e}{\rightleftharpoons} Fe^{2+} \overset{F^-}{\rightarrow} FeF^+ \overset{-e}{\rightleftharpoons} FeF^{2+} \quad (26)$$

Proton elimination, by metal-hydride bond cleavage, results from the oxidation of the original complex (E_p^{ox} = 0.92 to 1.1 V) to give an unstable Fe(I) species which is oxidized to Fe(II), an unsaturated dicationic metal site. This is activated towards nucleophilic attack by BF_4^-, yielding, upon fluorination, *trans*-$[FeF(CNR)(dppe)_2]^+$ which is reversibly oxidized at a slightly higher potential ($E_{1/2}^{ox}$ = 1.0 to 1.2 V).

The homogeneous rate constant for the anodically-induced proton loss has been estimated and shown to correlate with the electron donor/acceptor ability of the *trans* isocyanide ligand, as indicated by P_L, the Taft σ^* polar (for the alkyl isocyanides) or the Hammett's σ_p+ constant (for the aromatic ones) [39].

It is also noteworthy to mention that the hydride complex of the series with the simplest isocyanide, *i.e.*, *trans*-$[FeH(CNH)(dppe)_2]^+$, undergoes a dominant anodic process which is very different from that we have described. In fact, as a result of a single electron oxidation, the isocyanide appears to isomerize to hydrogen cyanide, a stronger net electron donor, giving $[FeH(NCH)(dppe)_2]^{2+}$ [40].

4. FINAL REMARKS

This study provided a few examples which illustrate the application of molecular electrochemistry to the investigation of coordination compounds with unsaturated-C or -N ligands, in a number of perspectives, namely directed towards the elucidation of reaction mechanisms (either in their synthesis or in further reactions), the rationalization and prediction of their redox or electronic properties (for both the unsaturated ligands and their binding metal centres), the recognition of correlations between electrochemical and other physico-chemical parameters, and the electroinduction of chemical reactivity (electroactivation of those ligands or their coordination metal sites).

Electron-transfer at aminocarbyne (CNH_2, CNHR or CN*M*R where *M* = Lewis acid metal centre), methyleneamido (NCHR), cyanamide ($NCNH_2$),

η^2-vinyl [η^2-C(CH_2Ph)CH_2] or hydride complexes is often followed by chemical steps, and examples of redox induced proton or hydrogen elimination (involving heterolytic or homolytic N-H, C-H or metal-H bond cleavage), nucleophilic attack or isomerization reactions have been presented.

This work also gave an insight into the electrochemical behaviour of complexes with multiple metal-carbon bonds (presenting aminocarbyne, carbyne, η^2-vinyl, vinylidene or carbene ligands), a field which has not yet been explored in spite of the rich chemical reactivity which has already been developed for such a type of species. Further developments are also expected, particularly in the fields of the electroactivation of the C- or N-unsaturated species and of the mechanistic investigation of their redox processes.

ACKNOWLEDGEMENTS

Thanks are due to the co-authors indicated in the citations, particularly Mrs. M.A N.D.A. Lemos, Miss M.F.C. Guedes da Silva and Dr. M.F.N.N. Carvalho (Lisbon), as well as Dr. C.J. Pickett, Dr. R.L. Richards (Sussex, U.K.) and Dr. C. Amatore (Paris). The author also gratefully acknowledges Prof. J.J.R. Fraústo da Silva for laboratory facilities and general support. The work has been mainly supported by INIC and JNICT (Portugal).

REFERENCES

[1] For reviews, see, *e.g.*: Pombeiro, A.J.L. (1989), *in* 'Advances in Metal Carbene Chemistry', Schubert, U. (ed.), Kluwer Academic Publ., Dordrecht, p. 79-99; Pombeiro A.J.L. (1988) J. Organometal. Chem. 358, 273; Pombeiro, A.J.L. (1988), *in* 'Coordination Chemistry and Catalysis', Ziolkowski, J.J. (ed.), World Scientific Publ. Co., Singapore, p.100-124; Pombeiro A.J.L. (1992), Inorg. Chim. Acta 198-200, 179; Pombeiro A.J.L., and Richards R.L. (1990), Coord. Chem. Rev. 104, 13.

[2] Chatt J., Elson C.M., Pombeiro A.J.L., Richards R.L., and Royston G.H.D. (1978), J. Chem. Soc ., Dalton Trans ., 165 .

[3] Pombeiro A.J.L., Pickett C.J., and Richards R.L. (1982), J.Organometal. Chem. 224, 285.

[4] Pombeiro A.J.L., Hughes D.L ., Pickett C .J., and Richards R.L . (1986), J. Chem. Soc., Chem. Commun., 246.

[5] Pombeiro A.J.L., Silva M.F.C.G., Hughes D.L., and Richards R.L. (1989), Polyhedron 8, 1872; Silva M.F.C.G., and Pombeiro A.J.L. (1989) , Portugaliæ Electrochim. Acta 7 , 95 .

[6] Pombeiro A.J.L., Almeida S.S.P.R., Silva M.F.C.G., Jeffery J.C., and Richards R.L. (1989), J. Chem. Soc., Dalton Trans., 2381

[7] Hughes D.L., Pombeiro A.J.L., Pickett C.J., and Richards R.L. (1984), J. Chem. Soc., Chem. Commun., 992.

[8] Pombeiro A.J.L. (1985), Inorg. Chim. Acta 103, 95.

[9] Silva M.F.C.G., and Pombeiro A.J.L., unpublished.

[10] Carvalho M.F.N.N., and Pombeiro A.J.L. (1989), J. Chem. Soc., Dalton Trans., 1209.
[11] Carvalho M. F. N. N., and Pombeiro A. J. L. (1991), J. Organometal. Chem. 410, 347.
[12] Leigh G.J., Morris R.H., Pickett C.J., Stanley D.R., and Chatt J. (1981), J. Chem. Soc., Dalton Trans., 800.
[13] Carvalho M.F.N.N., Pombeiro A.J.L., Orama O., Pickett C.J., and Richards R.L. (1985), J. Chem. Soc., Dalton Trans., 2079.
[14] Pombeiro A.J.L., Carvalho M.F.N.N., Hitchcock P.B., and Richards R.L. (1981), J. Chem. Soc., Dalton Trans., 1629
[15] Pombeiro A.J.L., Hills A., Hughes D.L., and Richards R.L. (1988), J. Organometal. Chem. 352, C5.
[16] Pombeiro A.J.L., Hughes D.L., Richards R.L., Silvestre J., and Hoffmann R. (1986), J. Chem. Soc., Chem. Commun., 1125.
[17] Pombeiro A.J.L., Hughes D.L., and Richards R.L. (1988), J. Chem. Soc., Chem. Commun., 1052.
[18] Lemos M.A.N.D.A., and Pombeiro A.J.L. (1988), J. Organometal. Chem. 356, C79; Almeida S.S.P.R., Lemos M.A.N.D.A., and Pombeiro A.J.L. (1989), Portugaliæ Electrochim. Acta 7, 91.
[19] Mingos D.M.P. (1979), J. Organometal. Chem. 179, C29; Dubois D.L., and Hoffmann R. (1977), Nouveau J. Chim. 1, 479; Murrell J.N., Al-Derzi A., Leigh G.J., and Guest M.F. (1980), J. Chem. Soc., Dalton Trans., 1425.
[20] Pombeiro A.J.L., Pickett C.J., Richards R.L., and Sangokoya S.A. (1980), J. Organometal. Chem. 202, C15.
[21] Silva M.E.N.P.R.A., Pombeiro A.J.L., Fraústo da Silva J.J.R, Herrmann R., Deus N., Castilho T., and Silva M.F.C.G. (1991) J. Organometal. Chem. 421, 75.
[22] Pombeiro A.J.L., and Richards R.L. (1979) J. Organometal. Chem. 179, 459.
[23] Pombeiro A.J.L. (1981), Rev. Port. Quim. 23, 179.
[24] Pombeiro A.J.L. (1979), Rev. Port. Quim. 21, 90; Chatt J., Leigh G.J., Pickett C.J., Pombeiro A.J.L., and Richards R.L. (1978), Nouv. J. Chim. 2, 541.
[25] Carvalho M.F.N.N., and Pombeiro A.J.L. (1989), Portugaliæ Electrochim. Acta 7, 101.
[26] Chatt J., Kan C.T., Leigh G.J., Pickett C.J., and Stanley D.R. (1980), J. Chem. Soc., Dalton Trans., 2032.
[27] Silva M.F.C.G., and Pombeiro A.J.L. (1991), Portugaliæ Electrochim. Acta 9, 189.
[28] Hills A., Hughes D.L., Kashef N., Richards R.L., Lemos M.A.N.D.A., and Pombeiro, A.J.L. (1988), J. Organometal. Chem. 350, C4.
[29] El-Shihi T., Siglmuller F., Herrmann R., Carvalho M.F.N.N., and Pombeiro A.J.L. (1987), J. Organometal. Chem. 335, 239.
[30] Pombeiro A.J.L. (1985), Portugaliæ Electrochim. Acta 3, 41
[31] Pombeiro A.J.L. (1989), XI Reunion del Grupo de Electroquímica de la Real Sociedad Española de Química, Valladolid, CP-2.

[32] Pombeiro A.J.L., Lemos M.A.N.D.A., and Carvalho M.F.N.N. (1990), IX Congresso Iberoamericano de Electroquimica, La Laguna, Tenerife, 4-12, p.453.

[33] Carvalho M.F.N.N., Pombeiro A.J.L., Bakalbassis E.G., and Tsipsis C.A. (1989), J. Organometal. Chem. 371, C26.

[34] Bertani R., Mozzon M., Michelin R.A., Benetollo F., Bombieri G., Castilho T.J., and Pombeiro A.J.L. (1991), Inorg. Chim. Acta 189, 175; Castilho T.J., Guedes da Silva M.F.C., Pombeiro A.J.L.,Bertani R., Mozzon M., and Michelin R.A. (1993) *in* 'Molecular Electrochemistry of Inorganic, Bioinorganic and Organometallic Complexes', Pombeiro A.J.L. and McCleverty J. (eds.), NATO ASI Series, Kluwer Academic Publishers, Dordrecht (this book).

[35] Silva M.F.C.G., Pombeiro A.J.L., and Richards R.L. (1991), Intern. Conf. Inorg. Chem., The Royal Society of Chemistry / Gesellschaft Deutscher Chemiker, Brighton, P.21

[36] Silva M.F.C.G., Pombeiro A.J.L. (1991), Journées d'Electrochimie 1991, Brest, CA-4.3.

[37] Guedes da Silva M.F.C., Fraústo da Silva J.J.R., Pombeiro A.J.L., Amatore C. and Verpeaux J.N. (1993), *in* 'Molecular Electrochemistry of Inorganic, Bioinorganic and Organometallic Complexes', Pombeiro A.J.L. and McCleverty J. (eds.), NATO ASI Series, Kluwer Academic Publishers, Dordrecht (this book).

[38] Lemos M.A.N.D.A., and Pombeiro A.J.L. (1987), J. Organometal. Chem. 332, C17.

[39] Lemos M.A.N.D.A., and Pombeiro A.J.L. (1991), Portugaliæ Electrochim. Acta 9,171; J. Organometal. Chem., in press.

[40] Almeida S.S.P.R., Fraústo da Silva J.J.R., and Pombeiro A.J.L. (1987), Journées d'Électrochimie 1987, Dijon, 3-26.

ELECTROCHEMICAL BEHAVIOUR OF AMINOOXY-, DIOXY- AND DIAMINOCARBENE COMPLEXES OF PALLADIUM(II) AND PLATINUM(II)

T.J. CASTILHO, M.F.C. GUEDES DA SILVA and A.J.L. POMBEIRO
Centro de Quimica Estrutural, Complexo I, Instituto Superior Tecnico, Av. Rovisco Pais, 1096 Lisboa Codex, Portugal.

R. BERTANI, M. MOZZON and R.A. MICHELIN
Centro di Chimica e Tecnologia dei Composti Metallorganici degli Elementi di Transizione del C.N.R. and Istituto di Chimica Industriale, Facoltà di Ingegneria, Università di Padova, Via F. Marzolo 9, 35131 Padova, Italy.

ABSTRACT. The electrochemical behaviour of a series of carbene complexes of Pd(II) and Pt(II), where the carbene ligand contains the N and/or O heteroatom directly bonded to the carbene carbon, has been studied by cyclic voltammetry and controlled potential electrolysis in aprotic media. The anodic behaviour has shown to present irreversible oxidation waves commonly with a multi-electron character and involving liberation of a similar number of protons, likely to result from N-H or C-H bond cleavage of the carbene ligand. The cathodic behaviour has shown to present, in some cases, an irreversible reduction wave,corresponding to a two-electron or to a single-electron cathodic process, possibly to form low-ligated Pd(0) species.

1. Introduction

Since only a small number of electrochemical studies have been reported on complexes with multiple metal-carbon bonds, and in view of our interest on such compounds, we have embarked upon the electrochemical investigation of series of carbene or carbyne complexes of Cr, Mo, W [1], Re [2], Pd or Pt [3,4]. In this work we summarize the results obtained previously [3,4] on a series of Fischer-type carbene complexes of Pd(II) and Pt(II) and report the extension of this study to further mono- and di-carbenes, particularly *cis*-$Cl_2Pt(PPh_3)[CN(C_6H_4OMe\text{-}4)CH_2CH_2NH]$, *cis*-$Cl_2M(PMe_2Ph)[CN(C_6H_4CH_2Cl\text{-}2)CH_2CH_2NH]$ (M = Pd or Pt), *cis*-$Cl_2Pt(PPh_3)[CN(C_6H_4OMe\text{-}4)CH_2CH_2O]$ and *trans*-$Cl_2Pd[CN(C_6H_4OMe\text{-}4)CH_2CH_2NH]_2$. The cyclic voltammetry (CV) and controlled potential electrolysis (CPE) studies have been undertaken in 0.2 mol dm^{-3} $[Bu_4N][BF_4]$/NCMe, at a Pt electrode (except for the cathodic CPE when a Hg-pool electrode was employed).

A. J. L. Pombeiro and J. A. McCleverty (eds.),
Molecular Electrochemistry of Inorganic, Bioinorganic and Organomettallic Compounds, 345–350.

2. Electrochemical Studies

2.1. ANODIC BEHAVIOUR

By CV the monocarbene complexes *cis*-$Cl_2M(L)[\overline{CN(R)CH_2CH_2NH}]$ (**1**; M = Pd or Pt; L = PMe_2Ph, PPh_3 or $CNBu^t$; R = C_6H_4OMe-4, $C_6H_4CH_2Cl$-2 or Bu^t) or *cis*-Cl_2Pt-$(PPh_3)[\overline{CN(C_6H_4OMe\text{-}4)CH_2CH_2O}]$ (**2**) undergo an irreversible oxidation at E_p^{ox} = 1.5-1.9 V *vs*. SCE (Table), which can be followed by a second one at a higher potential (1.8-2.0 V). By CPE the first wave normally involves *ca.* 2 electrons with extrusion of *ca.* $2H^+$ as measured by acid-base titration (either potentiometric or with an acid-base indicator), whereas a higher number of electrons and protons can be involved at the second anodic wave.

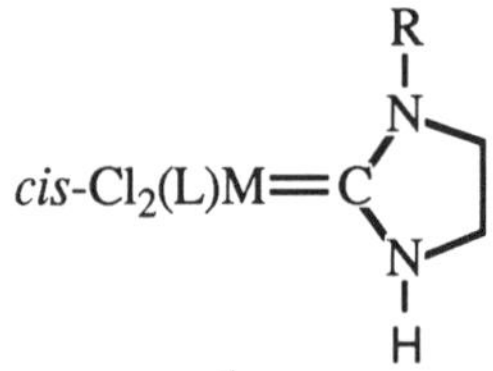

1

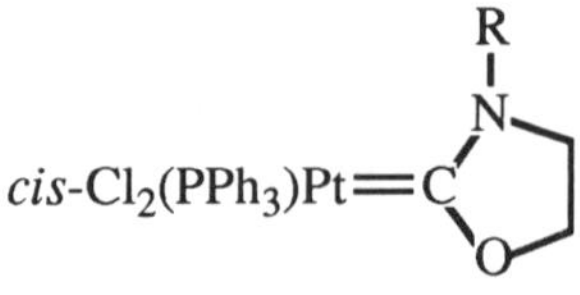

2

The dicarbene complexes *cis*-$Cl_2M[\overline{CN(R)CH_2CH_2NH}]_2$ (**3**; M = Pd or Pt; R = C_6H_4-OMe-4) or *cis*-$Cl_2Pt[\overline{CN(R)CH_2CH_2O}]_2$ (**4**) also exhibit a sequence of complex anodic processes, the first one occurring at lower potentials (E_p^{ox} = 1.0-1.5 V) than those observed for the monocarbene complexes and involving, by CPE, *ca.* 1 electron, for (**3**), or *ca.* 2 electrons, for (**4**), and a similar number of protons extruded. The second anodic wave corresponds to a multielectron and multiproton loss process [*ca.* 4 electrons and 4 protons for (**3**, M = Pt, R = C_6H_4OMe-4) and (**4**)].

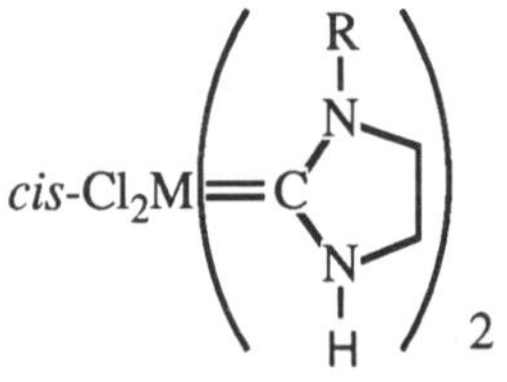

3

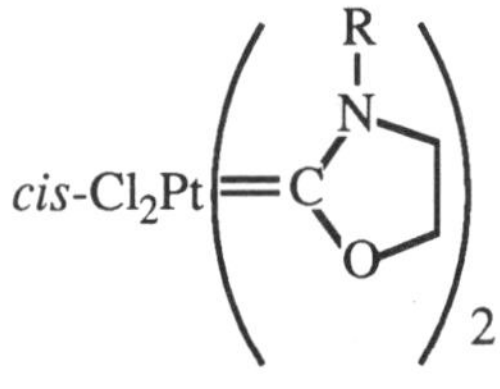

4

An anodic behaviour similar to that observed for the dicarbene complexes (**3**) or (**4**) is exhibited by the dinuclear diaminocarbene species $\{(PPh_3)ClM[\mu\text{-}\overline{CNCH_2CH_2X}]\}_2$ (**5**; M = Pd or Pt; X = NMe, NC_6H_4OMe-4 or O) [3].

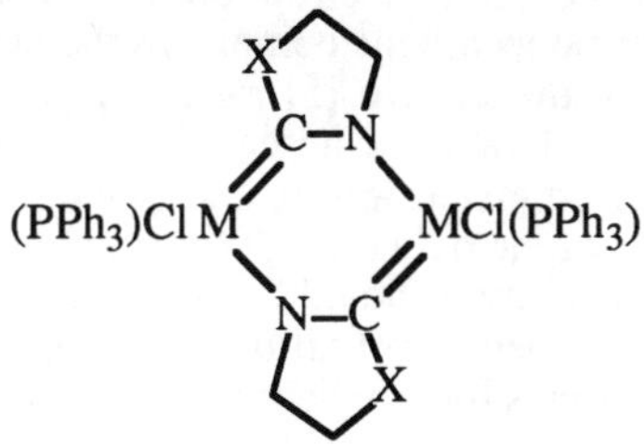

5

Table. Summary of Electrochemical[(a)] and Acid-Base Titration Data for *cis*-$Cl_2M(L)[\overline{CN(R)CH_2CH_2NH}]$ (**1**), *cis*-$Cl_2Pt(PPh_3)[\overline{CN(C_6H_4OMe\text{-}4)CH_2CH_2O}]$ (**2**), *cis*- or *trans*-$Cl_2M[\overline{CN(R)CH_2CH_2NH}]_2$ (**3**) and *cis*-$Cl_2Pt[\overline{CN(R)CH_2CH_2O}]_2$ (**4**).

Complex	E_p^{ox}/V *vs.* SCE		CPE		Ref.
	I	II	n(e⁻)	$n(H^+)$	
1, M = Pd, L = PMe_2Ph, R = $C_6H_4CH_2Cl$-2	1.52	1.80	2.1 (I)	1.7 (I)	*b*
1, M = Pt, L = PMe_2Ph, R = $C_6H_4CH_2Cl$-2	1.70	1.78	2.1 (I + II)	2.3 (I + II)	*b*
1, M = Pd or Pt, L = PPh_3, R = Bu^t	*ca.* 1.9		1.9	2.0	[3]
1, M = Pt, L = PPh_3, R = C_6H_4OMe-4	1.68	1.88	1.9 (I)	*ca.* 2 (I)	*b*
1, M = Pt, L = PPh_3, R = C_6H_4OMe-4	1.75	1.98	1.8 (I)	1.7 (I)	*b*
			ca. 5.0 (II)	4.9 (II)	
2, M = Pt, R = C_6H_4OMe-4	1.78		1.7	1.9	[3]
3, M = Pd, R = C_6H_4OMe-4 (*trans*)	1.00	1.46	0.6 (I)	3.7 (I + II)	*b*
		1.78	4.3 (II)		
3, M = Pt, R = C_6H_4OMe-4 (*cis*)	1.20	1.35	0.9 (I)	0.8 (I)	[3]
		1.46	3.9 (II)	3.6 (II)	
4, R = C_6H_4OMe-4	1.46	1.78	1.7 (I)	1.9 (I)	*b*, [3]
			> 4 (II)	> 4 (II)	

[a] In NCMe/0.2 M $[Bu_4N][BF_4]$, at a Pt wire (CV) or gauze (CPE) electrode; E_p^{ox} values in V *vs.* SCE, measured at 200 mV/s. [b] This study.

In these anodic processes, the observed proton evolution is consistent with the expected increase of the acidity character of the aminocarbene ligands and should involve anodically induced N-H and/or C-H (from a methylene group) heterolytic bond cleavage. Although we cannot ascertain the site of the initial electron removal, there is circumstantial evidence which suggests [3] that for the monocarbene complexes the oxidation can be metal centred (the oxidation potential lies in the known [5] range for the oxidation of related Pt(II) complexes without the carbene ligand), whereas for the dicarbenes and the dinuclear carbenes it can be based on the aminocarbene ligand (the oxidation potential values are in the range reported [6] for those of the oxidation of free aliphatic secondary amines).

The extent of the proton evolution should be promoted by the basicity of the medium and also depend on other experimental conditions such as the temperature and the scan rate. Figure 1 illustrates these effects for the dicarbene complex (**4**).

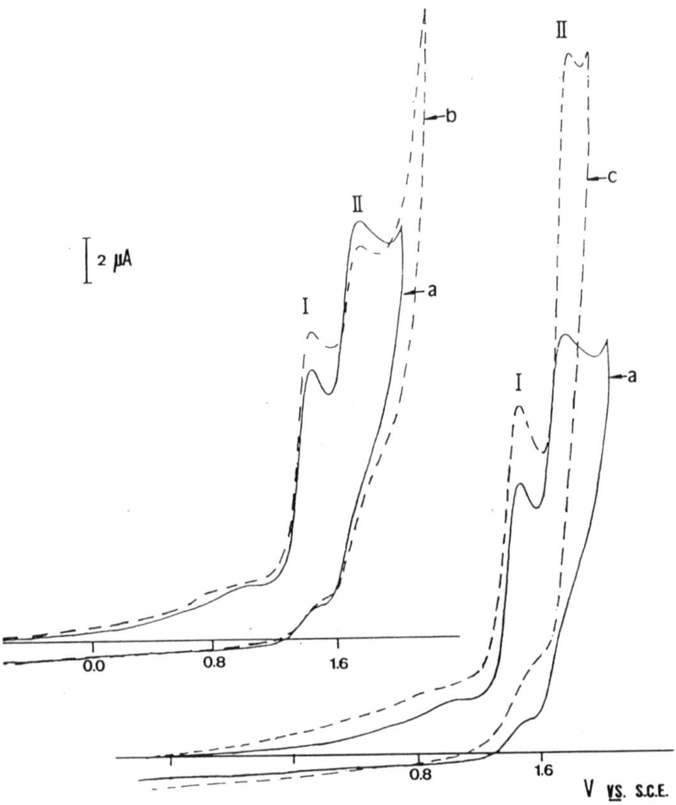

Fig. 1. Cyclic voltammograms of complex (**4**) in NCMe/0.2 M $[Bu_4N][BF_4]$, at a Pt electrode (effect of pyridine addition and of the temperature, T) (scan rate v = 200 mV s^{-1}). (a, solid line) T = -20°C. (b, dashed line) Same as (a) but upon addition of pyridine (1.2:1 molar ratio). (c, dashed line) T = 22°C.

The enhancement of the current-function for the anodic wave (I), which involves proton dissociation, upon addition of pyridine is evident [Fig. 1(a) and (b)], whereas the decrease of the current-function for wave (II) conceivably results from the reaction of pyridine with the complex product of the first oxidation (however, in other complexes, namely of the dinuclear dicarbene type [3], this current-function is promoted by base). The second anodic process, which possibly involves a sequence of electron transfer/deprotonation steps (≥ 4 electrons, by CPE), is more affected by the temperature and scan rate than the first one (which involves a smaller number of electrons): either a decrease of the temperature [Fig. 1(c), (a)] or an increase of the scan rate leads to a reduction of the current function of wave (II) relative to that of wave (I), by hampering the sequence of the chemical steps involved. Potentiometric titration of the exhaustively electrolyzed solution at the second wave indicates the presence of an acidic product (besides the liberated protons), possibly with an acidic methylene group.

2.2. CATHODIC BEHAVIOUR

The diaminocarbene complexes (**1**) undergo, by CV, an irreversible cathodic process (at E_p^{red} = -1.5 to -1.8 V), which, by CPE, involves *ca.* two electrons and appears to generate zerovalent species, possibly of the types $Cl_2ML(diaminocarbene)^{2-}$ or $ClML(diaminocarbene)^-$, which, however, have not yet been isolated [4].

Nevertheless, no cathodic wave has been detected for comparable dicarbene complexes, whereas an irreversible single-electron reduction has been observed for *trans*-$[(PPh_3)_2Pt(H)(\overline{COCH_2CH_2O})][BF_4]$ leading to the quantitative formation of CO_2 and $CH_2{=}CH_2$ (from fragmentation of the dioxycarbene ligand [4]) and possible generation of a low-coordinated zerovalent platinum species.

3. Acknowledgements

This work has been partially supported by INIC (Portugal)/C.N.R. (Italy) cooperative research program, as well as by JNICT (Portugal) and MURST (Italy).

4. References

[1] Ribeiro, M.T.A.; Pombeiro, A.J.L.; Facchin, G.; Mozzon, M.; and Michelin, R.A. (1992), in A.J.L. Pombeiro and J. McCleverty (eds.), Molecular Electrochemistry of Inorganic, Bioinorganic and Organometallic Complexes, Kluwer Academic Publ., Dordrecht (P7 of the NATO ARW under that title).

[2] Lemos, M.A.N.D.A.; and Pombeiro, A.J.L. (1988), J. Organometal. Chem. 356, C79.

[3] Bertani, R.; Mozzon, M.; Michelin, R.A.; Benetollo, F.; Bombieri, G.; Castilho, T.J.; and Pombeiro, A.J.L. (1991), Inorg. Chim. Acta 189, 175.

[4] Bertani, R.; Mozzon, M.; Michelin, R.A.; Castilho, T.J.; and Pombeiro, A.J.L., J. Organometal. Chem., in press.

[5] (a) Chen, L.; Davies, J.A.; and Staples, R. (1989), Inorg. Chim. Acta 163, 11. (b) Davies, J.A.; Uma, V.; and Mierzwiak, J.G. (1985), J. Electroanal. Chem. 196, 329 and refs. cited therein.

[6] Mann, C.K. (1964), Anal. Chem. 36, 2424.

ELECTROCHEMISTRY OF PLATINUM PHOSPHINE COMPLEXES: C-H AND C-X ACTIVATION BY HIGHLY REACTIVE INTERMEDIATES

J. A. DAVIES*, L. CHEN, C.T. EAGLE AND R.J. STAPLES
Department of Chemistry
University of Toledo
Toledo, Ohio 43606
U.S.A.

ABSTRACT. The electrochemical reduction of *cis*-[PtX_2L_2] (X = halide, L = tertiary phosphine) complexes in CH_3CN/C_6H_6/TBAP (Hg electrode) generates [PtL_2] equivalents. The reactivity of these complexes is determined by the nature of the monodentate phosphine ligands, *e.g.* when L = PPh_3 the complex does not react with benzonitrile but when L = PEt_3 the complex reacts with PhCN *via* oxidative addition of the rather inert C-CN bond. When the monodentate ligands are replaced by a bidentate ligand, electrochemical reduction leads to the generation of nonlinear [Pt(bidentate)] complexes. The reactivity is altered by the presence of the bidentate ligand, *e.g.* when L_2 = $(c\text{-Hx})_2P(CH_2)_3P(c\text{-Hx})_2$, generation of [Pt(bidentate)] in CH_3CN/C_6H_6/TBAP results in C-H oxidative addition of benzene to produce a phenylplatinum(II) hydride complex. This contrasts with the electrochemical generation of [$Pt(PEt_3)_2$] in the same medium where a subsequent acid/base reaction with the $N(n\text{-Bu})_4^+$ cation leads to the formation of *trans*-[$PtH(Cl)(PEt_3)_2$] with production of tri(*n*-butyl)amine. Electrochemical reduction of *trans*-[$PtH(Cl)(PEt_3)_2$] in CH_3CN/C_6H_6/TBAP (Hg electrode) results in H^- transfer to CH_3CN and then further reactions due to the cyanomethyl anion that is produced. Electrochemical oxidation of *trans*-[$PtH(Cl)(PEt_3)_2$] at a platinum mesh electrode does not lead to transformation of Pt(II) into Pt(IV) but rather induces the formal oxidation of H^- to H^+. Reduction of the platinum(II) aryl complex [$PtPh_2L_2$] (L_2 = $Ph_2PCH_2CH_2PPh_2$) in CH_3CN/TBAP (Hg electrode) leads to cleavage of the Pt-C bonds and formation of benzene (but not biphenyl) through scavenging of the organic fragments. Cleavage of Pt-C bonds can similarly be induced by oxidation in certain cases and this process is the main focus of the current report. Thus, although oxidative electrolysis of *cis*-[$PtPh_2(PEt_3)_2$] in CH_3CN produces the expected [$PtPh_2(CH_3CN)_2(PEt_3)_2$]$^{2+}$ without Pt-C bond cleavage, oxidation of the benzylplatinum(II) complexes *trans*-[$PtBz(Cl)(PEt_3)_2$] and *cis*-[$PtBz_2(PEt_3)_2$] generates benzyl alcohol and benzaldehyde *via* oxidation of the cleaved organic fragments. These results demonstrate not only that C-H and C-X bond cleavage, accompanied by Pt-H, Pt-X, and Pt-C bond formation, can be induced by electrochemical strategies but also that Pt-H and Pt-C cleavage processes, accompanied by the formation of useful organic products, can be achieved with the use of electrochemical methods.

1. Introduction

In 1986 Mortreux and Petit[1] provided a review of coordination catalysis by electrochemically

A. J. L. Pombeiro and J. A. McCleverty (eds.),
Molecular Electrochemistry of Inorganic, Bioinorganic and Organomettallic Compounds, 351–356.

generated transition metal complexes. These authors observed that the typical methods of producing low valent transition metal complexes that catalyze reactions of olefins and dienes involve reactions between transition metal complexes and Lewis acids/reducing agents such as organoaluminum compounds. The resulting low valent species are typically complicated mixtures that are hard to characterize. Electrochemistry provides a method of generating low valent species under controlled conditions and so may provide opportunities to investigate individual steps in potential catalytic cycles. There is considerable interest in catalysis involving platinum phosphine complexes and electrochemical methods have been used to generate reactive platinum species that catalyze reactions of unsaturated substrates. Paumard, Mortreux and Petit[2] have shown that reduction of $[PtCl_2L_2]$ (L_2 = *2,3-O*-isopropylidene-2,3-dihydroxy-1,4-bis(diphenylphosphino)butane, *i.e.* diop) at a tin or iron anode leads to the generation of precursors active for olefin hydroformylation and that these precursors exhibit remarkable regioselectivities for linear aldehydes. We have considered the individual steps of C-H and C-X bond cleavage, as well as M-H, M-X, and M-C bond formation, that are typically involved in the interaction of an electrochemically generated low valent platinum phosphine complex with an organic substrate and have summarized current work in this general area.[3] In the following section we briefly describe our own work on the interaction of low valent platinum complexes with organic substrates where C-X and/or C-H bonds are broken and M-H, M-X, and/or M-C bonds are formed. This work is described in detail elsewhere.[3-14] Following this brief review we discuss the question of M-H and M-C cleavage by electrochemical methods and present new work that focuses specifically on the topic of M-C cleavage by electrochemical oxidation.

1.1 C-H AND C-X BOND CLEAVAGE WITH M-H, M-X, AND M-C BOND FORMATION

The electrochemical reduction of *cis*-$[PtX_2L_2]$ (X = halide, L = tertiary phosphine) complexes in CH_3CN/C_6H_6/TBAP (TBAP = tetra(*n*-butyl)ammonium perchlorate) at a mercury pool electrode generates 14-electron, 2-coordinate $[PtL_2]$ complexes.[4-6] The reactivity of these complexes is determined by the nature of the monodentate phosphine ligands. With either PPh_3 or PEt_3 as ligands, the $[PtL_2]$ complex can be trapped by acetylenes such as PhCCPh or MeOOCCCCOOMe to generate the corresponding $[PtL_2(\text{acetylene})]$ complexes.[4,5] When L = PPh_3, the 2-coordinate, 14-electron complex does not react with Ph-CN but when L = PEt_3 the complex reacts with this substrate *via* oxidative addition of the typically rather inert C-CN bond. A similar dichotomy of reactivity is found when chlorobenzene is used as the organic substrate.[5] Other ligand effects are clearly manifested in cyclic voltammetric measurements of simple *cis*-$[PtX_2L_2]$ complexes[7-10] and in the chemical reactivities of the derived $[PtL_2]$ species.[4-6] Indeed, relationships exist between electrochemical parameters measured for the irreversible Pt(II)/Pt(0) reduction of *cis*-$[PtCl_2L_2]$ complexes and spectroscopic properties of the Pt(II) complexes as well as with parameters that describe the electronic and steric nature of the phosphine ligands.[10] These relationships are not as well defined nor as well understood as those for reversible electrochemical processes exhibited by other transition metal complexes.

When the monodentate ligands of a *cis*-$[PtX_2L_2]$ complex are replaced by a bidentate ligand, electrochemical reduction leads to the generation of a nonlinear [Pt(bidentate)] complex.[11,12] The reactivity is completely altered by the presence of the bidentate ligand. When the bidentate ligand is $(c\text{-Hx})_2P(CH_2)_3P(c\text{-Hx})_2$, generation of [Pt(bidentate)] in CH_3CN/C_6H_6/TBAP results in selective C-H oxidative addition of benzene to produce a phenylplatinum(II) hydride complex.[12] This type of reactivity contrasts with that observed for the electrochemical generation of $[Pt(PEt_3)_2]$ in the same medium where a subsequent acid/base reaction with the $N(n\text{-Bu})_4^+$ cation leads to the formation of *trans*-[$PtH(Cl)(PEt_3)_2$] with production of tri(*n*-butyl)amine. (*i.e.* an electrochemically initiated Hofmann elimination reaction).[6] These results demonstrate that C-H and C-X bond cleavage, accompanied by Pt-H, Pt-X, and Pt-C bond formation, can be induced by electrochemical reductive strategies. In the remainder of this article we discuss Pt-H and Pt-C bond cleavage processes, beginning with the cleavage of platinum hydrides.

1.2 M-H BOND CLEAVAGE

We have studied both the reductive[5] and the oxidative[14] cleavages of the Pt-H bond in *trans*-[$PtH(Cl)(PEt_3)_2$]. Prolonged reductive electrolysis of *trans*-[$PtH(Cl)(PEt_3)_2$] at a mercury pool electrode (-2.1 V vs. Ag/AgCl) in CH_3CN/C_6H_6/TBAP results in hydride transfer from platinum to the acetonitrile solvent generating the cyanomethyl anion. This anion can subsequently be trapped by *trans*-[$PtH(Cl)(PEt_3)_2$] (which is regenerated in solution *via* the reaction of [$Pt(PEt_3)_2$] with $N(n\text{-}Bu)_4^+$ and Cl^-) to produce *trans*-[$PtH(CH_2CN)(PEt_3)_2$].[5]

In contrast, oxidative electrolysis of *trans*-[$PtH(Cl)(PEt_3)_2$] in CH_3CN/TBAP at a platinum mesh electrode produces almost exclusively platinum(II) complexes as products, none of which contain Pt-H bonds.[14] The major oxidation product is *trans*-[$PtCl(solvent)(PEt_3)_2$]$^+$ and this product appears as the hydride ligand (formally H^-) is oxidized to H^+. Thus, the Pt-H bond of *trans*-[$PtH(Cl)(PEt_3)_2$] is a source of both H^- and of H^+ *via* electrochemical cleavage under differing conditions.

1.3 M-C BOND CLEAVAGE

Following any potential catalytic process, release of a transformed substrate as a product from a metal center will involve cleavage of a M-C bond. The electrochemical reduction[15] of [$PtPh_2L_2$] (L_2 = $Ph_2PCH_2CH_2PPh_2$) in CH_3CN/TBAP (Hg electrode) leads to Pt-C bond cleavage. Examination of the reaction medium by GC-MS and other techniques allowed identification of benzene (but not biphenyl) perhaps implying carbanion generation. No products from the attack of phenyl anion on CH_3CN could be identified, however, suggesting that the Pt-C cleavage process may be more complex than simple carbanion expulsion from a reduced metal center.

Here we describe new studies aimed at understanding the electrochemical oxidative cleavage of the M-C bonds of simple *cis*-[PtR_2L_2] and *trans*-[$PtR(Cl)L_2$] complexes. Our initial work involved complexes of triethylphosphine since we have routinely used this ligand in other studies[3] and since the voltammetric behavior of *cis*-[PtR_2L_2] complexes has been investigated by others[16] in the case where L = triethylphosphine.

2. Results and Discussion

2.1 OXIDATIVE ELECTROLYSIS OF *cis*-[$PtPh_2(PEt_3)_2$]

Controlled potential oxidative electrolysis of *cis*-[$PtPh_2(PEt_3)_2$] in CH_3CN/TBAP at +1.30 V (vs. Ag/AgCl; potential determined by cyclic voltammetry) at a Pt mesh electrode under dried Ar using the methods we have previously reported[14] proceeds with transfer of *ca* 2 electrons/Pt. Work-up of the solution and examination by $^{31}P\{^1H\}$ NMR spectroscopy allowed identification of [$PtPh_2(CH_3CN)_2(PEt_3)_2$]$^{2+}$ as the major (> 90% as estimated by peak heights) phosphorus-containing product (δP = 13.0 ppm; $^1J(Pt,P)$ = 1733 Hz). The identity of this complex was confirmed by addition of chloride ion which produced $[PtPh_2Cl_2(PEt_3)_2]$ (an independently prepared sample of $[PtPh_2Cl_2(PEt_3)_2]$ gave identical NMR data; δP = -6.2 ppm; $^1J(Pt,P)$ = 1830 Hz). Attempts to prepare pure samples of the solvated dication by halide abstraction from $[PtPh_2Cl_2(PEt_3)_2]$ with $AgPF_6$ were not fruitful since the dication catalyzed the hydrolysis of the PF_6^- ion by traces of moisture, producing $PO_2F_2^-$ (δP = -15.9 ppm; $^1J(P,F)$ = 965 Hz).

Oxidative electrolysis of *cis*-$[PtPh_2(PEt_3)_2]$ thus proceeds ***without*** cleavage of the Pt-Ph bonds. This inert character of platinum-aryl bonds is further demonstrated by an almost complete absence of aryl migration when *cis*-$[PtPh_2(PEt_3)_2]$ and *cis*-$[PtCl_2(PEt_3)_2]$ are stirred together in CH_3CN for 48 hours (only *ca* 5% *trans*-$[PtPh(Cl)(PEt_3)_2]$ is produced; *c.f.* *ca* 90% alkyl transfer in the case of the methyl analog).

2.2 OXIDATIVE ELECTROLYSES OF *trans*-$[PtBz(Cl)(PEt_3)_2]$ AND *cis*-$[PtBz_2(PEt_3)_2]$ (Bz = BENZYL)

Controlled potential oxidative electrolysis of the alkyl complex *trans*-$[PtBz(Cl)(PEt_3)_2]$ (**1**) in CH_3CN/TBAP at +1.60 V (vs. Ag/AgCl; potential determined by cyclic voltammetry) at a Pt mesh electrode under dried Ar using the methods we have previously reported[14] proceeds with transfer of only *ca* 0.5 electrons/Pt. Work-up of the solution obtained in one experiment and examination by $^{31}P\{^1H\}$ NMR spectroscopy allowed identification of **1** (δP = 13.4 ppm; $^1J(Pt,P)$ = 2865 Hz; 30%), *trans*-$[PtBz(CH_3CN)(PEt_3)_2]^+$ (**2**, δP = 14.1 ppm; $^1J(Pt,P)$ = 2762 Hz; 35%), *cis* and *trans*-$[PtCl_2(PEt_3)_2]$ (**3**, δP = 8.9 ppm; $^1J(Pt,P)$ = 3496 Hz and **4**, δP = 12.0 ppm; $^1J(Pt,P)$ = 2391 Hz; 25% total), and *trans*-$[PtCl(CH_3CN)(PEt_3)_2]^+$ (**5**, δP = 15.7 ppm; $^1J(Pt,P)$ = 2171 Hz; 10%). A second experiment, with a slightly longer electrolysis time, gave rise to the following product distribution: **1** (35%), **2** (45%), **3** and **4** (20% total). The cation **5** was not observed as a product from the second reaction. Our working hypothesis was that the electrochemical oxidation begins with Pt-C bond cleavage, producing **5**, and that this cation abstracts chloride ion from **1**, thus generating **2** and also the isomers **3** and **4**. When **5** was generated from **3** with $AgPF_6$ and mixed with **1**, halide transfer indeed occurred, producing **2**, **3**, and **4**. Generation of **3** and **4** also explains why the electrolysis stopped after transfer of only *ca* 0.5 electrons/Pt, even though **1** remained in solution. Compounds **3** and **4** are largely electroinactive under the experimental conditions employed and are known to coat the platinum electrode, hindering current flow.[13] Indeed, attempts to electrolyze a 1:1 mixture of **1** and **3** under the conditions employed here gave rise to a very low current flow and essentially no net chemical reaction.

The products resulting from the cleaved benzyl groups in the oxidative electrolysis of **1** were identified from the 400 MHz ^1H NMR spectrum of a concentrated sample of the reaction mixture. In addition to resonances attributable to the platinum complexes, the solvents, and TBAP, additional signals at 4.32 ppm and 9.62 ppm were observed. Addition of a small aliquot of benzyl alcohol caused enhancement of the signal at 4.32 ppm while addition of benzaldehyde caused the signal at 9.62 ppm to increase in size. Thus, the benzyl groups cleaved from platinum were found in the oxidized organic products $C_6H_5CH_2OH$ and C_6H_5CHO. Since benzyl radicals are oxidized[17] to benzyl cations at +0.40 V and the electrolysis of **1** is performed at +1.60 V, the oxidized products presumably have their origins in further chemistry of the benzyl cation.

Similar controlled potential oxidative electrolyses of *cis*-$[PtBz_2(PEt_3)_2]$ (**6**; δP = 7.3 ppm; $^1J(Pt,P)$ = 1925 Hz) in CH_3CN/TBAP at +1.40 V (vs. Ag/AgCl) produced solutions containing **2** along with $C_6H_5CH_2OH$ and C_6H_5CHO. Thus it seems that only one of the two benzyl groups is readily cleaved from **6** (*i.e.* oxidative cleavage of a cation is more difficult than oxidative cleavage of a neutral molecule).

3. Experimental

Acetonitrile[18] and TBAP[13] were purified and dried as previously described. Compounds **3** and **4** were prepared by reported methods.[13] Compounds **1** and **6** were prepared from **3** *via* the reaction with benzyl magnesium chloride and separated by column chromatography on alumina with benzene as the eluent.[18] *cis*-[$PtPh_2(PEt_3)_2$] was prepared from **3** by reaction with phenyllithium.[18] [$PtPh_2Cl_2(PEt_3)_2$] was prepared from *cis*-[$PtPh_2(PEt_3)_2$] by reaction with one equivalent of the molecular chlorine adduct of iodobenzene in chloroform at room temperature.[18] Cyclic voltammetric measurements and controlled potential bulk electrolyses were performed using the methods that we have previously described.[13, 14] NMR spectra were obtained with use of a JEOL FX90Q NMR spectrometer and a Varian VXR 400 NMR spectrometer. Chemical shifts were referenced to external 85% phosphoric acid (^{31}P) and internal tetramethylsilane (^1H). Product distributions were estimated from peak heights in the $^{31}P\{^1H\}$ NMR spectra and so represent only an approximate measure of the distribution of species present.

4. Conclusions

Our studies of *cis*-[$PtR_2(PEt_3)_2$] (R = Ph, Bz) and *trans*-[$PtR(Cl)(PEt_3)_2$] (R = Bz) show that oxidative electrolysis leads to cleavage of the Pt-C bonds where R = Bz but not where R = Ph. This result parallels observations made when *cis*-[$PtR_2(PEt_3)_2$] (R = Ph, Bz) complexes are chemically oxidized with chlorine (as its iodobenzene adduct) where [$PtPh_2Cl_2(PEt_3)_2$] is readily formed from *cis*-[$PtPh_2(PEt_3)_2$] but only [$PtCl_2(PEt_3)_2$] is obtained from *cis*-[$PtBz_2(PEt_3)_2$] under comparable conditions.[18] Thus it seems that the Pt-Ph bond is less susceptible to oxidative cleavage than the Pt-Bz bond and this may relate to the relative stabilities of the phenyl and benzyl cation equivalents that may be produced during electrolysis.

5. Acknowledgements

Thanks are expressed to the Petroleum Research Fund, administered by the American Chemical Society, for support of this work.

6. References

1. Mortreux, A.; Petit, F. *Applied Catalysis* **1986**, *24*, 1.
2. Paumard, E.; Mortreux, A.; Petit, F. *J. Chem. Soc., Chem. Commun.* **1989**, 1380.
3. Davies, J.A.; Staples, R.J. In "*Selective Hydrocarbon Activation: Principles and Progress*", Davies, J.A.; Watson, P.L.; Greenberg, A.; Liebman, J.F. (Eds.), VCH Publishers: New York, **1990**, 379.
4. Davies, J.A.; Eagle, C.T.; Otis, D.E.; Uma, V. *Organometallics* **1986**, *5*, 1264.
5. Davies, J.A.; Eagle, C.T.; Otis, D.E.; Uma, V. *Organometallics* **1989**, *8*, 1080.
6. Davies, J.A.; Eagle, C.T. *Organometallics* **1986**, *5*, 2149.
7. Davies, J.A.; Uma, V. *Inorg. Chim. Acta Lett.* **1983**, *76*, L291.
8. Davies, J.A.; Uma, V. *J. Electroanal. Chem.* **1983**, *158*, 13.

9. Davies, J.A.; Uma, V. *J. Electroanal. Chem.* **1984**, *179*, 273.
10. Davies, J.A.; Uma, V.; Mierzwiak, J. *J. Electroanal. Chem.* **1985**, *196*, 329.
11. Davies, J.A.; Staples, R.J. *Polyhedron* **1991**, *10*, 899.
12. Davies, J.A.; Staples, R.J. *Polyhedron* **1991**, *10*, 909.
13. Chen, L; Davies, J.A.; Staples, R.J. *Inorg. Chim. Acta* **1989**, *163*, 11.
14. Chen, L; Davies, J.A. *Inorg. Chim. Acta* **1990**, *175*, 41.
15. Davies, J.A.; Staples, R.J.; Wise, E.T. unpublished results.
16. Seeber, R.; Mazzocchin, G.A.; Minniti, D.; Romeo, R.; Uguagliati, P.; Belluco, U. *J. Organomet. Chem.* **1978**, *157*, 69.
17. Wayner, D.D.M.; Griller, D. *J. Am. Chem. Soc.* **1990**, *112*, 1577.
18. Chen, L. *Ph.D. Thesis* University of Toledo, **1991**.

AZAMETALLIC ELECTROCHEMISTRY

C J PICKETT
The Nitrogen Fixation Laboratory
AFRC Institute of Plant Science Research
University of Sussex
Brighton BN1 9RQ
United Kingdom.

ABSTRACT
Electrochemical reactions which lead to the formation, cleavage or transformation of transition metal - nitrogen ligand bonds are described. The ligands discussed are : dinitrogen (N_2), diazenides (NNR), diazoalkanes ($NNCR_2$), hydrazides (NNR_2), nitrides (N), imides (NR), amides (NR_2), methyleneamide (NCH_2), cyanide (CN), aminocarbide (CNH_2) and nitric oxide (NO) where R is usually an alkyl group or H. Metal centres considered include molybdenum tertiaryphosphine, dithiocarbamate, tetrathioether and dithiolate dithioether assemblies. Certain of the electrochemical reactions described provide routes from molecular nitrogen to ammonia and other N -products and some may be of relevance, at the functional group level, to biological nitrogen fixation.

1. Introduction

Azametallic electrochemistry is concerned with electrochemical reactions which involve the formation or cleavage of metal - nitrogen ligand bonds and the transformation of nitrogen ligands at metal centres. This paper describes some research in this area that has been carried out at the Nitrogen Fixation Laboratory. It focuses on azametallic electrochemistry that bears upon abiological and biological aspects of nitrogen fixation.

The electrochemical reactions discussed are primarily those which involve the binding and transformation of molecular nitrogen at robust molybdenum and tungsten tertiary diphosphine sites of the type *trans* - $\{M(Ph_2PCH_2CH_2PPh_2)_2\}$ (M = Mo or W) . Such sites provide a stable platform at which a ligand(s) in the axial positions can be bound, cleaved, or transformed by nucleophilic, radical, or electrophilic attack as a consequence of electrochemically adjusting the electron complement of the metal centre to a 19-, 18-, 17- or 16 - electron configuration.

Some recent work on the electron - transfer reactions of thioether and thiolate complexes of molybdenum is also introduced together with an approach to modification of electrodes with metallo-sulphur clusters - the first steps towards the assembly of artificial metalloenzyme sites for fixing dinitrogen or other small molecules electrochemically.

A. J. L. Pombeiro and J. A. McCleverty (eds.),
Molecular Electrochemistry of Inorganic, Bioinorganic and Organomettallic Compounds, 357–380.

2. Electrosynthesis of dinitrogen complexes from molecular nitrogen

Dinitrogen complexes of molybdenum and tungsten with mono- , di - and tri - tertiary phosphine co - ligands are electrosynthesised from molecular nitrogen and Mo^{II} or W^{II} precursors *via* cleavage of halide or thiolate bonds and from Mo^{IV} or W^{IV} oxide , imide or hydrazide complexes by reduction in the presence of a source of protons [1-10]. Examples of such reactions are illustrated by Schemes A and B. Generally current -yields are high but competing reactions can lead to poor yields. For example, a triphenylphosphine ligand can rearrange to give a π -arene complex Scheme B [9] and certain solvato - intermediates decarbonylate to give carbonyl species (see below).

The electrochemistry of *trans* - $[Mo(CN)X(Ph_2PCH_2CH_2PPh_2)_2]$ **I** (X = Cl or Br) under argon and dinitrogen is summarised by Scheme C and is illustrative of electron - transfer pathways by which dinitrogen complexes are formed from Mo^{II} or W^{II} precursors. The primary reduction of **I** involves a single electron - transfer to give a 17 - electron Mo^{I} anion. Loss of Br^- is rapid and the quasi - reversible reduction of a solvato-intermediate *trans* - $[Mo(CN)(solvent)(Ph_2PCH_2CH_2PPh_2)_2]$ is detected. Exchange of solvent by molecular nitrogen gives a seventeen - electron complex which is reduced to the stable anion at a potential *positive* of that for the parent species **I**, *ie* binding of the π^* electron acceptor N_2 in place of a donor solvent molecule promotes secondary electronation. When X = Cl, then loss of chloride from the anion of **I** is relatively slow and, at a more negative potential, the irreversible reduction of *trans* - $[Mo(CN)X(Ph_2PCH_2CH_2PPh_2)_2]^-$ is observed. This gives the 18 - electron anion *trans* - $[Mo(CN)(solvent)(Ph_2PCH_2CH_2PPh_2)_2]^-$ under argon and stable *trans* - $[Mo(N_2)(CN)(Ph_2PCH_2CH_2PPh_2)_2]^-$ under molecular nitrogen [7,8,10] Scheme C.

In dimethylformamide, dmf, the Mo^{I} solvato - intermediate *trans*- $[Mo(CN)(dmf)(Ph_2PCH_2CH_2PPh_2)_2]$ is unstable, the dmf ligand decarbonylates to give *trans* - $[Mo(CN)(CO)(Ph_2PCH_2CH_2PPh_2)_2]$ in a reaction that is competitive with formation of the dinitrogen complex, Scheme D [10].

3. Electron - transfer reactions of dinitrogen complexes: 17 - and 16 - electron species

Electrochemical oxidation of closed - shell, tertiary phosphine dinitrogen complexes of molybdenum and tungsten give successively 17- and 16 - electron species . The stability of the electron - deficient species is determined by the rate of loss of dinitrogen : removal of electrons weakens the metal - dinitrogen ligand bond by diminishing $d\pi$-π^* back - bonding interactions. Thus a 16 - electron dinitrogen complex is less stable than its 17 - electron counterpart, which in turn is less stable than the 18 - electron parent [11,12].

For a given electron complement the nature of the metal, the co -ligands and the geometry of the complex also influence the stability of the metal - dinitrogen bond [11-14].

Trans - $[W(N_2)_2(PMe_2Ph)_4]$ oxidises reversibly at 25 °C to give a stable 17 - electron cation, whereas the oxidation of the molybdenum analogue is only partially reversible under the same conditions [13,14]. This parallells the behaviour of the parent 18 - electron complexes and related bis (diphosphine) complexes, dinitrogen more readily dissociates from Mo^0 than from W^0 . It is likely that the W -N_2 bond is generally stronger than the Mo -N_2 bond because

SCHEME A

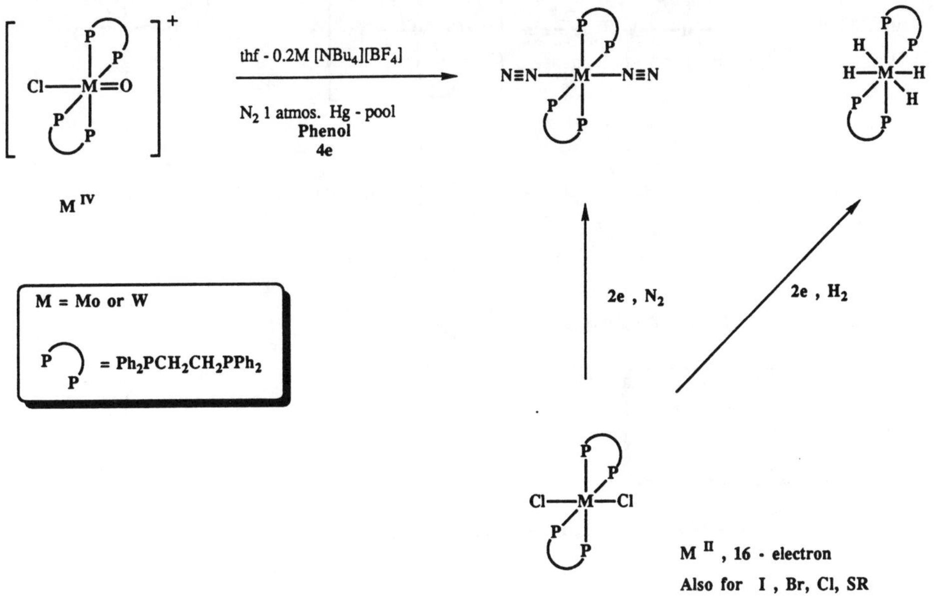

SCHEME B

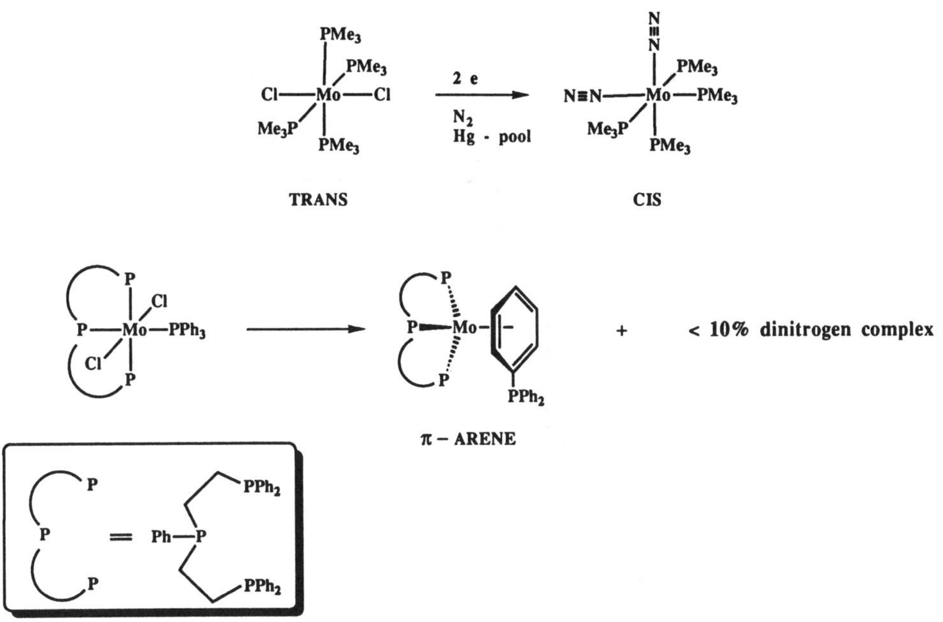

SCHEME C

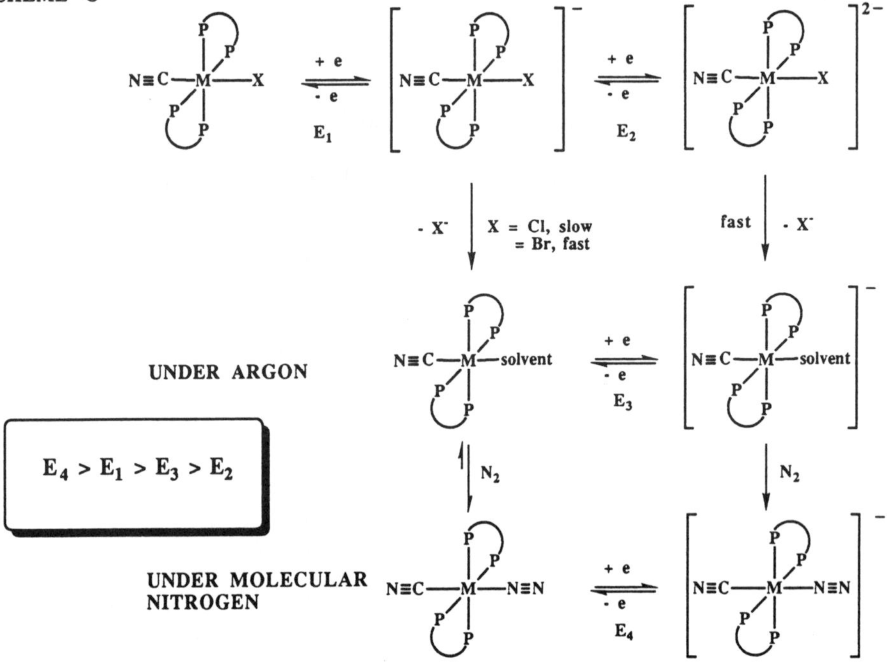

SCHEME D

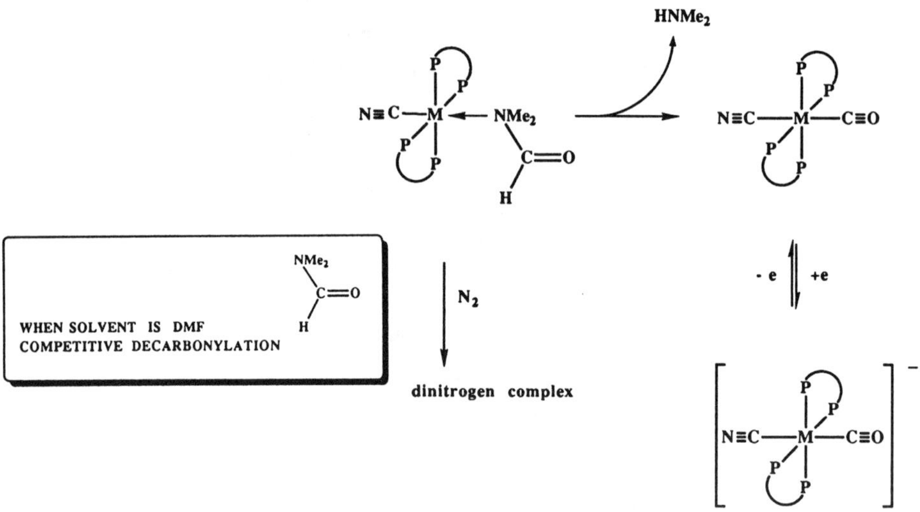

$d\pi$-π^* overlap is more effective for the former metal : enhanced synergy may also differentiate the relative σ - bonding contributions the metal - dinitrogen bond strength of the two metals. Notably, ν (N_2) is around 50 cm^{-1} lower in frequency in tungsten compounds than in the molybdenum analogues [12].

The arrangement of the tertiary phosphine co - ligands around the metal - centre has a substantial effect on the lability of the M - N_2 bond. Mono- tertiary phosphines in complexes of the type $[M(N_2)_2(PR_3)_4]$ (M = Mo or W ; R = aryl or alkyl) can adopt two geometries with the dinitrogen ligands in a *cis* - or *trans* -arrangement. Where steric demands are not restrictive, as for PMe_2Ph and PMe_3 ligands, the *cis* - geometry is favoured whereas for $PMePh_2$, the *trans* - complex is the stable isomer. Although the complexes *trans* - $[M(N_2)_2(PMe_2Ph)_4]$ isomerise, they do so only slowly and, for M = W, the *trans* - isomer can be isolated in a reasonable pure form. Cyclic voltammetry of *cis*- and *trans* - isomers of $[M(N_2)_2(PMe_2Ph)_4]$ shows that the latter undergoes reversible one - electron oxidation at room - temperature whereas the former oxidises irreversibly under the same conditions . It seems there is a rather delicate balance between the competing demands for metal $d\pi$ - electron density by the phosphine and dinitrogen ligands which leads to a greater dinitrogen lability in both M^0 and M^I oxidation states of the cis - compounds relative to the trans isomers [13,14].

There is no measurable difference in the primary oxidation potentials of the *cis* and *trans* isomers of the Mo or of W dinitrogen complexes . Moreover, the primary redox potentials of the complexes of these two different metals are also remarkably similar, as is generally observed for their 17-electron / 18-electron couples [12-14].

Structure, spectroscopic properties, reactivity and redox potentials have been correlated for octahedral dinitrogen complexes of Mo, W and Re bearing a wide range of ligands *trans* or *cis* to dinitrogen [11,12,15-17]. This work and earlier ideas [18] led to the suggestion that the oxidation potential $E^{o/}$ of an octahedral complex M_sL, in which a ligand L is bound to the site M_s , might be expressed in terms of a site constant E_s , a ligand constant P_L , and a site parameter β :

$$E^{o/}(M_sL) = E_s + \beta P_L$$

E_s is formally the oxidation potential of the $[M_s(CO)]$ complex and provides a measure of the electron - richness of the site, β provides a measure of the sensitivity of the HOMO to a change in the ligand L [17].

4. The role of 17 - electron intermediates in the formation of bonds between nitrogen and carbon.

Anionic dinitrogen complexes of Mo^0 and W^0 of the type *trans* - $[M(N_2)X(Ph_2PCH_2CH_2PPh_2)_2]^-$ (X = N_3, CN, SCN) are moderately strong reducing agents with oxidation potentials in the vicinity of -1 V versus SCE. They undergo an outer sphere electron - transfer with alkyl halides to produce M^I dinitrogen 17 - electron intermediates and alkyl radicals, Scheme E. The fates of both

SCHEME E

17- electron dinitrogen complexes and N - C bond formation

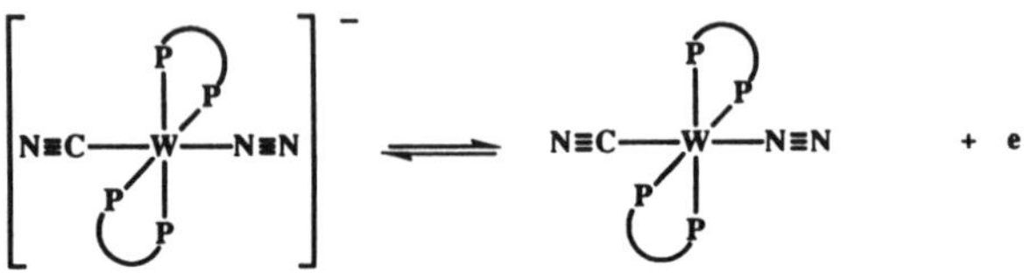

$E^{o\prime}$ = -1.59 V *versus* $fc^{+/0}$

$$RX + e \longrightarrow R^{\cdot} + X^{-}$$

$PhCH_2Cl$	$PhCH_2^{\cdot}$
MeI	$Me^{\cdot}$
Me_3SiCH_2I	$Me_3SiCH_2^{\cdot}$

SCHEME F

Coupling of 17-electron species with radicals

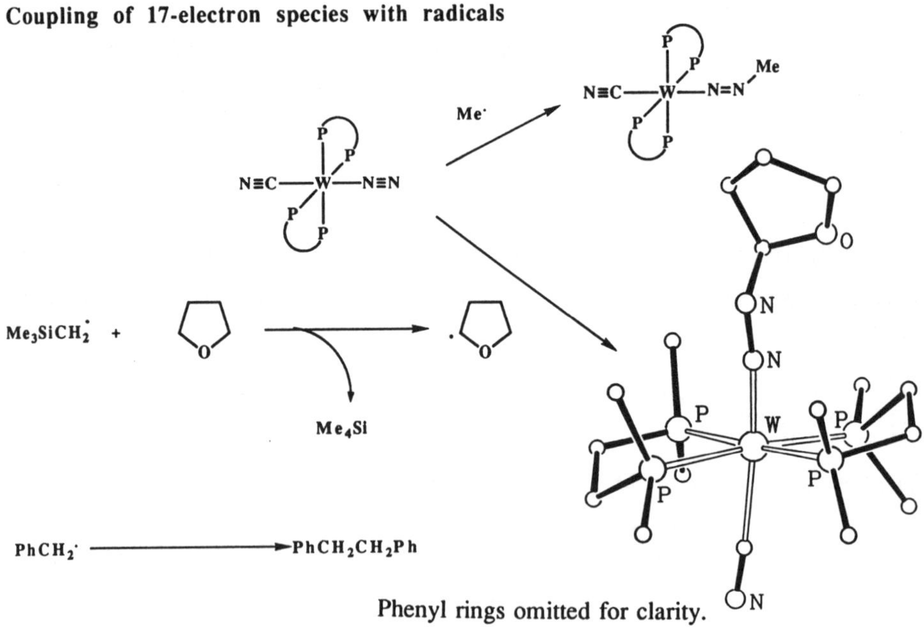

Phenyl rings omitted for clarity.

the metallo- and organo-radical so formed are interdependent. For example, the reaction of MeI with *trans* -$[Mo(N_2)(CN)(Ph_2PCH_2CH_2PPh_2)_2]^-$ in thf generates a methyl radical which subsequently attacks the β nitrogen of the metallo -radical to give a methyldiazenide. $PhCH_2Cl$ reacts with the same anion to produce a relatively unreactive radical which dimerises: the 17 - electron intermediate which is co-produced loses the dinitrogen ligand and thence reacts with further $PhCH_2Cl$ to give the stable Mo^{II} complex *trans* -$[MoCl(CN)(Ph_2PCH_2CH_2PPh_2)_2]$. The corresponding reaction with the tungsten analogue gives a relatively stable intermediate *trans* -$[W(N_2)(CN)(Ph_2PCH_2CH_2PPh_2)_2]$ [8, 10, 15]

An interesting situation arises with the reaction of Me_3SiCH_2I with *trans* -$[W(N_2)(CN)(Ph_2PCH_2CH_2PPh_2)_2]^-$ in thf. The Me_3SiCH_2 radical attacks thf to produce a new radical by H - abstraction from an α carbon atom. The tetrahydrofuranyl radical attacks the 17 - electron dinitrogen intermediate to give a diazenide, the structure of which we have recently established by X - ray crystallography, Scheme F [10].

Neutral dinitrogen complexes of the type *trans* -$[M(N_2)_2(Ph_2PCH_2CH_2PPh_2)_2]$ (M = Mo or W) oxidise at potentials positive of -0.2 V *versus* SCE and are incapable of reducing alkyl bromides or iodides at a significant rate by an outer - sphere electron - transfer mechanism. However, reduction of such organohalides does occur by an inner - sphere atom - transfer pathway . Loss of N_2, thermally (Mo) or photochemically (W), leads to the homolytic cleavage of the alkyl halide at the vacated metal centre , formation of M^I dinitrogen intermediates and organic radicals, and thence N-C bond formation, as discussed above. Thus neutral 17 - electron dinitrogen intermediates are involved in *both* outer - sphere and inner - sphere routes to alkyldiazenides [15, 19-21].

5. Alkyldiazenides and diazoalkanes: interconversion and coupling reactions

Alkyldiazenides of the type *trans* -$[M(N_2CH_2\ R)X(Ph_2PCH_2CH_2PPh_2)_2]$ (M = Mo or W, R = H, alkyl, or aryl) undergo reversible one - electron oxidation in the vicinity of 0V versus SCE to give moderately stable 17 - electron complexes [22,23]. However, their oxidation in the presence of a base can lead to the formation of diazoalkane compounds *trans* -$[M(N_2CH_2\ R)X(Ph_2PCH_2CH_2PPh_2)_2]$ in an irreversible two - electron process. Diazoakanes with R = alkyl or aryl undergo reversible one - electron reduction to give unstable stable 19- (or 17- electron) intermediates. These are attacked by electrophiles such as MeI or H^+ and, with a further electron - transfer, stable diazenides are formed. For example, controlled potential reduction of *trans* -$[M(N_2CH\ \{C_6H_4\text{-}p\ \text{-OMe}\})Br(Ph_2PCH_2CH_2PPh_2)_2]^+$ in the presence of phenol is an overall two-electron process which gives the diazenide *trans* -$[M(N_2CH_2C_6H_4\text{-}p\ \text{-OMe})Br(Ph_2PCH_2CH_2PPh_2)_2]$ in good yield, the same product is produced by hydride addition to the parent cation [24].

In contrast, the electroreduction of the diazomethane cation *trans* -$[Mo(N_2CH_2)Br(Ph_2PCH_2CH_2PPh_2)_2]^+$ involves an irreversible single - electron transfer which leads to a carbon - carbon coupling reaction and the formation of the bridging bis-diazenide *trans*, -

SCHEME G

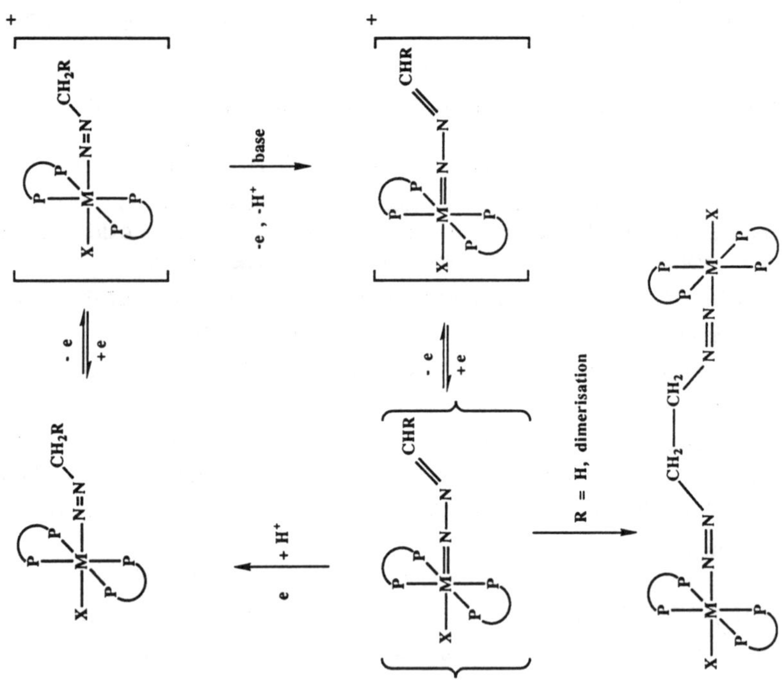

SCHEME H

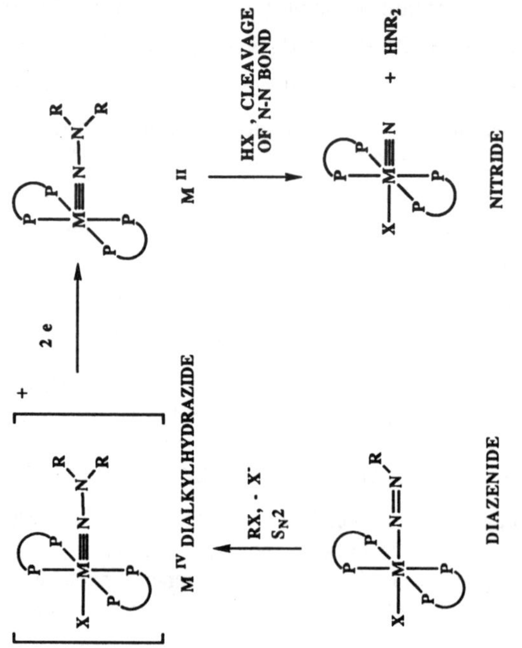

$[(Ph_2PCH_2CH_2PPh_2)_2BrMo(N_2CH_2CH_2N_2)MoBr(Ph_2PCH_2CH_2PPh_2)_2]$ [25]. Scheme G summarises the electrochemical reactions of the alkyldiazenides and diazoalkanes.

6. Dialkylhydrazides : cleavage of an N-N bond

Alkyldiazenides are nucleophilic. The β N - atom is sp^2 hybridised and N-H or N-C bonds are formed by protonation with acids or by reaction with alkyl halides ; cationic mono- or dialkylhydrazides of M^{IV} are produced respectively, Scheme H [19,22,26]. The rates of reaction of MeI and EtI with a wide range of ethyldiazenide complexes of Mo and W have been determined electrochemically and the mechanism of secondary alkylation shown to conform to an S_N2 process ; the β N - atom on W is considerably more nucleophilic than it is on Mo [22].

Dialkylhydrazides of the type *trans* -$[MX(N_2R_2)(Ph_2PCH_2CH_2PPh_2)_2]$ (M = Mo or W ; X = Br or I ; R = alkyl) are electroactive oxidatively and reductively. In a trifluoroacetic acid electrolyte they undergo reversible one - electron oxidation to give di-cations at potentials near +1V *versus* SCE [27] and, in aprotic electrolytes such as MeCN or thf, irreversible two electron reductions are observed near -1.8V versus SCE. Electrochemical reduction leads to the cleavage of the M - X bond and the formation of stable, five-coordinate 18 - electron M^{II} dialkylhydrazides , *trans* -$[M(N_2R_2)(Ph_2PCH_2CH_2PPh_2)_2]$ [28-30].

The M^{II} dialkylhydrazides undergo reversible one - electron oxidation to M^{III} cations in a thf electrolyte near -0.8V *versus* SCE. In the presence of the nucleophiles Br^- or MeCN the oxidation becomes an irreversible *ece* process. With Br^- *trans* -$[M(N_2R_2)Br(Ph_2PCH_2CH_2PPh_2)_2]^+$ is regenerated, with MeCN *trans* -$[M(N_2R_2)(MeCN)(Ph_2PCH_2CH_2PPh_2)_2]^{2+}$ is produced [28] .

Perhaps the most important reaction of M^{II} dialkylhydrazides is that with protic acids. The N-N bond is cleaved and an amine and a nitride (or imide) are formed Scheme H [29 ,30]

Such reactions provide the first examples of the of the stepwise cleavage of the N-N bond and suggest that a similar protonation -electronation - protonation sequence might operate during the early stages of reduction of molecular nitrogen by molybdo-nitrogenase [31]. The formation of nitrides from coordinated dinitrogen also provides the key to the synthesis of new products from molecular nitrogen, as described below.

7. Nitrides and Imides

Nitrides of the type *trans* -$[MNX(R_2PCH_2CH_2PR_2)_2]$ (M = W or W ; X = halide, N or OMe ; R = Ph or Et) undergo reversible one - electron oxidation in aprotic electrolytes to give unstable 17 - electron cations at potentials in the range 0.0 to + 0.3V *versus* SCE [32]. Products of oxidation have not been identified.

The nitride ligand is nucleophilic, protonation gives imides and alkylation gives alkylimides, Schemes I,J [26,32]. The imides *trans* - $[M(NH)X(Ph_2PCH_2CH_2PPh_2)_2]^+$ undergo reversible oxidation in a trifluoroacetic acid electrolyte at potentials in the region of +1V versus SCE to give unstable 17-electron dications. Under similar conditions, electrochemical oxidation of

SCHEME I

Reduction of imide to dinitrogen complex and ammonia : imide is sacrificial source of protons

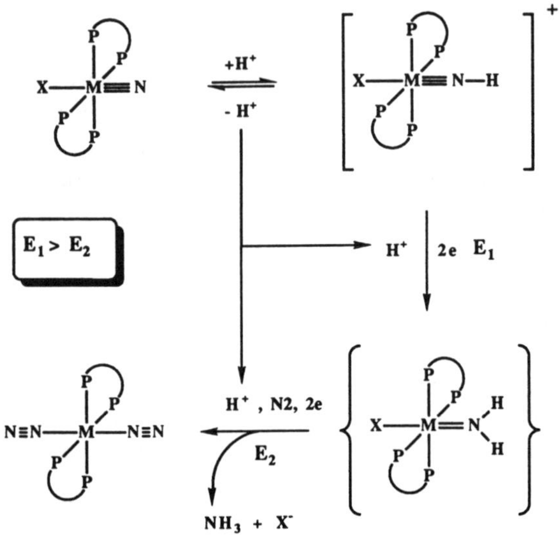

unstable, reverts to nitride.

SCHEME J

Conversion of methylimide to methyleneamide and to dinitrogen complex

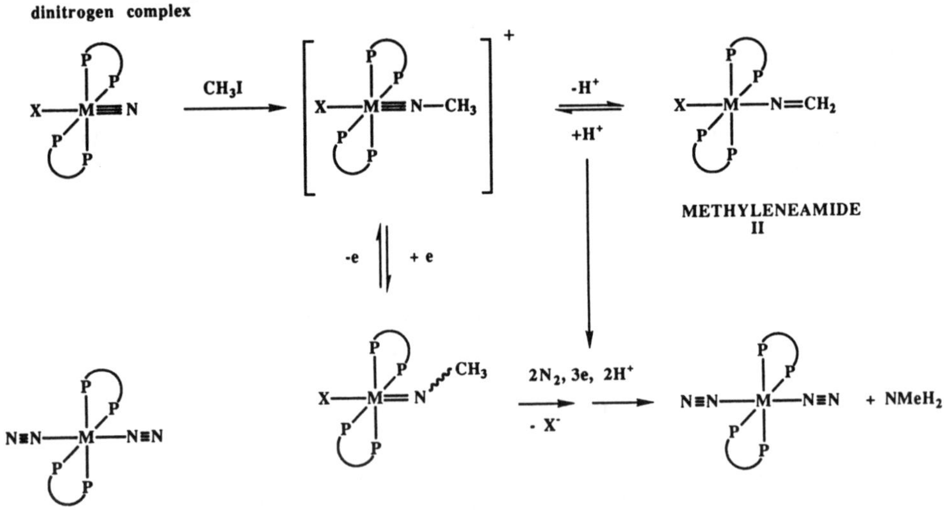

METHYLENEAMIDE
II

$[M(NMe)X(Ph_2PCH_2CH_2PPh_2)_2]^+$ gives relatively stable paramagnetic 17 - electron dications such as *trans* - $[Mo(NMe)Cl(Ph_2PCH_2CH_2PPh_2)_2]^{2+}$. X -ray crystallography, multinuclear EPR and ENDOR measurements show that the principle structural features of the parent 18 - electron cation are retained in the 17 - electron dication [27].

The imide complexes *trans* - $[M(NH)X(Ph_2PCH_2CH_2PPh_2)_2]^+$ undergo an irreversible electrochemical reduction at potentials in the range -1.8 to -2.1V *versus* SCE in a thf electrolyte. Under molecular nitrogen the electroreduction at mercury can lead to the formation of a dinitrogen complex, the nitride *trans* - $[M(N)X(Ph_2PCH_2CH_2PPh_2)_2]$ and to the release of ammonia. The parent imide serves as both the electroactive material and the source of protons for the formation of ammonia and this provides one of the pathways by which a nitride is formed by electrolysis Scheme I. The formation of an amido - intermediate is a key step on the pathway to ammonia. For example, if the electrolysis of *trans* - $[Mo(NH)Cl(Ph_2PCH_2CH_2PPh_2)_2]^+$ is performed at a potential which encompasses the reduction of the intermediate $[Mo(NH_2)Cl(Ph_2PCH_2CH_2PPh_2)_2]$ then ammonia, the dinitrogen complex *trans* - $[Mo(N_2)_2(Ph_2PCH_2CH_2PPh_2)_2]$ and a nitride are formed. However, if it is carried out at a less negative potential then only the nitride is produced on the electrolysis time - scale, presumably because the intermediate $[Mo(NH_2)Cl(Ph_2PCH_2CH_2PPh_2)_2]$ loses dihydrogen, Scheme I [4,6].

The methylimide *trans* - $[Mo(NCH_3)Cl(Ph_2PCH_2CH_2PPh_2)_2]^+$ reduces cleanly under molecular nitrogen in the presence of phenol as a source of protons to give methylamine and *trans* - $[Mo(N_2)_2(Ph_2PCH_2CH_2PPh_2)_2]$ in an overall four electron process. In the absence of a proton source, low temperature voltammetry shows that the reduction of the methylimide cation is a partially reversible one electron process which gives either a 19 - electron intermediate with a linear MoNMe group or, more likely, a 17 - electron species in which the N atom functions as a single electron donor and the group adopts a bent configuration [4,8].

Bulk electrolysis of the methylimide under molecular nitrogen in the absence of phenol gave a surprising result. Some dinitrogen complex was formed together with methylamine and a new metal product which was identified as trans -$[Mo(NCH_2)Cl(Ph_2PCH_2CH_2PPh_2)_2]$ **II** an hitherto unknown methyleneamide complex. This complex was subsequently prepared by an independent route: *trans* - $[Mo(NCH_3)Cl(Ph_2PCH_2CH_2PPh_2)_2]^+$ reacts with $K(OBu^t)$ to give isolable **II** [7,8]. It is probable that **II** is formed in the electrolysis by deprotonation of the parent methylimide cation by the primary reduction product $[Mo(NCH_3)Cl(Ph_2PCH_2CH_2PPh_2)_2]^0$ and by following step(s) which generate basic metallo - intermediates [4,6,8] Scheme J.

The methyleneamide **II** shows a remarkable chemical and electrochemical reactivity. It reacts directly with elemental oxygen, sulphur and selenium to give isocyanato-, isothiocyanato-, and isoselenocyanato- complexes, for example, *trans* - $[Mo(NCS)Cl(Ph_2PCH_2CH_2PPh_2)_2]$ which has been characterised by X -ray crystallography [8] Scheme K.

Electrochemical oxidation of **II** in the presence of a base gives the cyanide complex *trans* - $[Mo(CN)Cl(Ph_2PCH_2CH_2PPh_2)_2]$ the structure of which has been determined by X - ray crystallography. Isotopic studies show that the oxidative rearrangement of the MoNC framework of the methyleneamide to the MoCN unit of the cyanide occurs by an intramolecular process

SCHEME K

Reactions of methyleneamide (M = Mo, X = Cl)

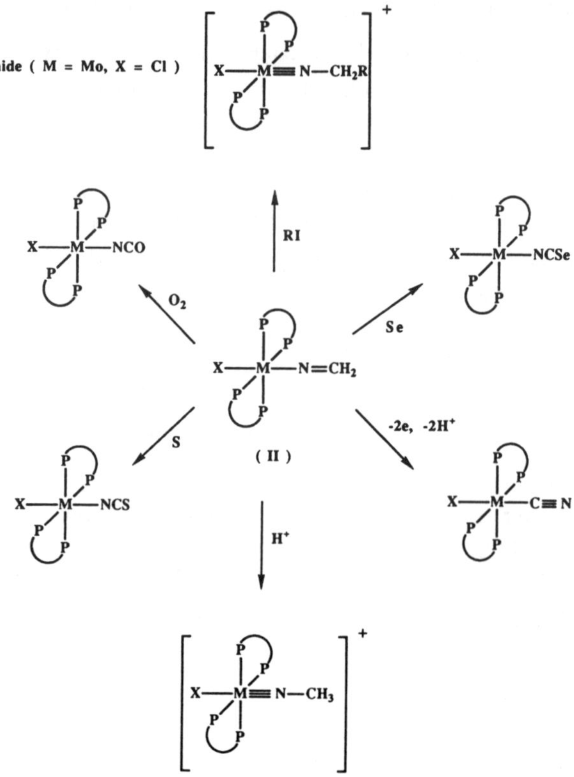

SCHEME L

Reduction of cyanide to aminocarbyne and pathway for isomerisation of methyleneamide to aminocarbyne.

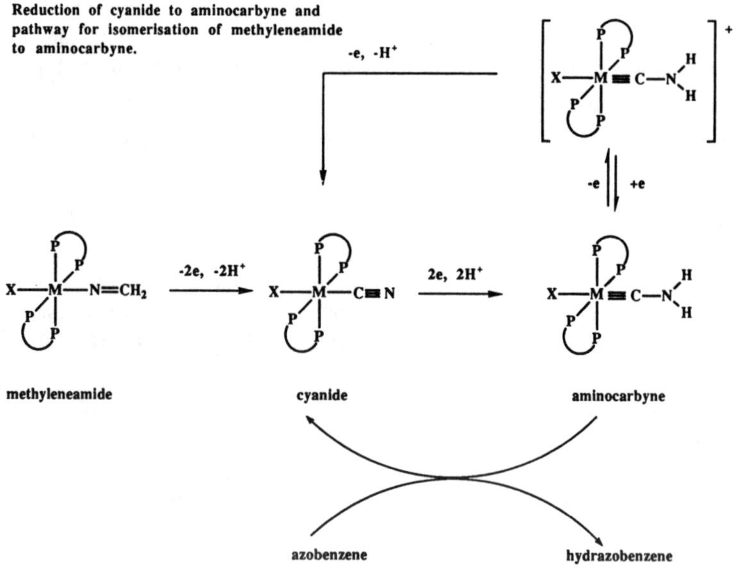

presumably involving a sideways bound CN or CNH group [8], Scheme L.

As discussed above, the reduction of *trans* - $[Mo(CN)Cl(Ph_2PCH_2CH_2PPh_2)_2]$ under molecular nitrogen gives a dinitrogen complex. However, if the electrochemical reduction of this species, or its tungsten analogue, is carried out in the presence of phenol as a source of protons then the cyanide ligand is reduced to an aminocarbyne and the halide is retained. X - ray structures of electrosynthesised *trans* - $[W(CNH_2)Cl(Ph_2PCH_2CH_2PPh_2)_2]$ III and of its one - electron oxidation product *trans* - $[W(CNH_2)Cl(Ph_2PCH_2CH_2PPh_2)_2][BF_4]$ have now been obtained and that of the later clearly shows that C and H adopt a trigonal planar geometry about the N atom of the aminocarbyne group [7,8,10] , Scheme L.

The reduction of coordinated cyanide to an aminocarbyne ligand is unique and there is only one other example of an authenticated CNH_2 complex, *trans* - $[Re(CNH_2)Cl(Ph_2PCH_2CH_2PPh_2)_2][BF_4]$ that we prepared indirectly by cleavage of the Si-C bond of the isocyanide ligand in *trans* - $[Re(CNSiMe_3)Cl(Ph_2PCH_2CH_2PPh_2)_2]$ with HBF_4 [33]. It is possible that a $MoCNH_2$ group lies on the pathway whereby cyanide is reduced by molybdonitrogenase to ammonia, methane and methylamine [7].

The oxidation of the methyleneamide to a cyanide and the reduction of the cyanide to aminocarbyne provides a pathway for isomerisation of the metal bound NCH_2 group to CNH_2 ; oxidation of *trans* - $[M(CNH_2)Cl(Ph_2PCH_2CH_2PPh_2)_2]$ in the presence of a base regenerates the cyanide , Scheme L. *Trans* - $[W(CNH_2)Cl(Ph_2PCH_2CH_2PPh_2)_2]$ acts as a 2H reductant : at 50^0C in thf azobenzene is reduced by III to hydrazobenzene and the cyanide is regenerated. So far we have been unable to condense the Mo or $W^{IV}CNH_2$ group with aldehydes or ketones, although this type of reaction takes place with isoelectronic hydrazides *trans* - $[M(NNH_2)Cl(Ph_2PCH_2CH_2PPh_2)_2]^+$

8. Electrosynthesis of ammonia from molecular nitrogen

Since the discovery that ligating dinitrogen in certain tetrakis monophosphine complexes of Mo and W is converted into ammonia upon protolysis [26] there existed the engaging possibility that a cyclic process for nitrogen fixation at ambient temperature and pressure might be based on $\{MP_4\}$ assemblies, M = Mo or W. Such a process required that the irreversible oxidative degradation of the MP_4 core be circumvented. We have shown that the robust $\{W(Ph_2PCH_2CH_2PPh_2)_2\}$ assembly can sustain the binding, protonation and electronation of molecular nitrogen without degradation and this provides the basis of a cyclic system for the electrosynthesis of ammonia [2,3,31]

The key steps involved in the conversion of dinitrogen to ammonia at the conserved bis(di-tertiary phosphine) metal - site are those outlined in Scheme M .Protonation of bis -dinitrogen complexes with acids HX give stable hydrazides of the type *trans* - $[M(NNH_2)X(Ph_2PCH_2CH_2PPh_2)_2]^+$ [26] but their electroreduction generally leads to unproductive N-H bond cleavage and the formation of a diazenide or to the regeneration of the parent dinitrogen complex [3]. However, if X is a good leaving group such as tosylate, methanesulphonate, or p-xylene sulphonate and M = W , then the primary product of two - electron reduction is the reactive W^{II} - hydrazide $[W(NNH_2)(Ph_2PCH_2CH_2PPh_2)_2]$ [2] . Subsequent protic attack by the parent (acidic) hydrazide leads to N - N bond cleavage with the release of

SCHEME M
Electrosynthesis of ammonia

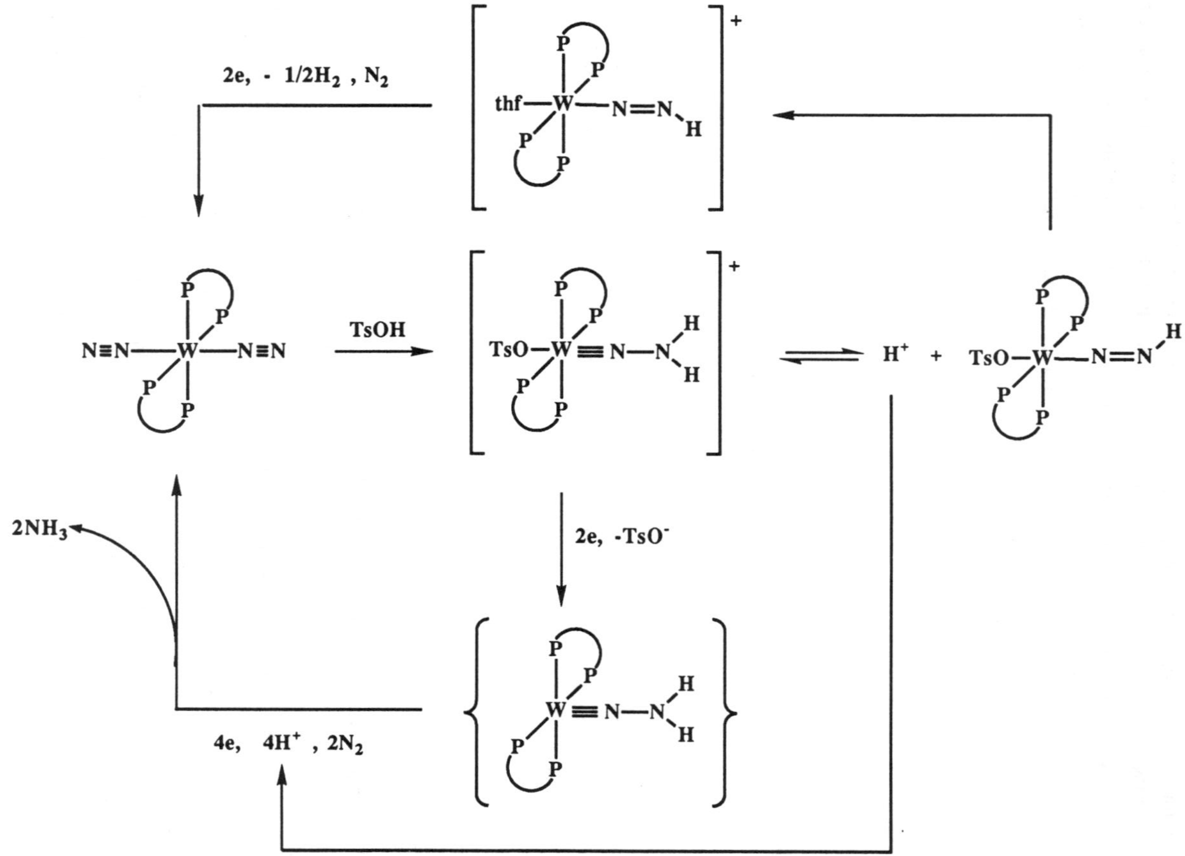

ammonia and the formation of an imide (cf. behaviour of diakylhydrazides). Reduction of the imide leads to the release of further ammonia and the regeneration of the parent dinitrogen complex.

In the final stage of electrolysis the electroinactive conjugate base *trans* - $[W(NNH)(OTs)(Ph_2PCH_2CH_2PPh_2)_2]$ solvates to give *trans* - $[W(NNH)(thf)(Ph_2PCH_2CH_2PPh_2)_2]^+$. This reduces under dinitrogen to reform the parent dinitrogen complex. . Protonation of the regenerated dinitrogen complex with tosylic acid completes the cycle [2]. In so far as the inputs N_2, H^+ and e lead to ammonia at ambient temperature and pressure, the electrosynthesis has some operational parallel with biological fixation [31].

9. Metallo - sulphur compounds

Studies of the reactions of dinitrogen at phosphine ligated transition metal centres have established an extensive functional group chemistry that bears on biological fixation and which has begun to open up new routes for the conversion of molecular nitrogen to products other than ammonia . However, molybdenum at the active site of the molybdo-protein of nitrogenase is ligated by sulphur, and is part of the framework of an iron-sulphur cluster assembly of indeterminate structure .

The electrochemistry of mononuclear metal - sulphur complexes and metallo-sulphur clusters which interact with dinitrogen (or other substrates and inhibitors of the enzyme) or which bear groups such as hydrazide, nitride, or amide are consequently of some relevance to biological nitrogen fixation.

The electron - transfer reactions of organodiazenide, organohydrazide and nitride groups bound at various molybdenum dithiocarbamate sites have been studied and the reductive release of an organohydrazine shown [34 - 37] but the 'parent' dinitrogen complexes are unknown [36,37].

Elegant chemistry by Yoshida's group has shown that dinitrogen binds to molybdenum in a macrocyclic tetrathioether complex and that some of the reactions of the coordinated dinitrogen parallel those of the Mo bis(diphosphine) systems [38]. Comparison of the reduction potentials for the $Mo^{II/I}$ dihalide couples of the sulphur macrocycle with those reported by us for corresponding tertiary phosphine dihalides [1] led the Japanese group to conclude that their dihalides complexes were 'extraordinarily' easy to reduce, that this was because the S atoms were very weak σ-donors, and that their compounds should therefore offer a facile route to low valent species.

We find that the potential of the $Mo^{III/II}$ -couple of their compounds has been incorrectly assigned the to that of the $Mo^{II/I}$ -couple, the latter process is observed at low temperature at a potential considerably *negative* of that for the related phosphine systems. This accords with the S atoms of the macrocycle acting as a strong donor ligands , as is consistent with the observed chemistry and spectroscopic data. Low temperature studies reveal that the $Mo^{II/III}$ dibromide system exhibits fluxionality [39] Scheme N.

Cyclic voltammetry of the Mo^{II} dibromide under molecular nitrogen (1 atmosphere) shows that reduction gives some dinitrogen complex but other species, also observed under argon, are co - produced. In contrast, electrochemical reduction under carbon monoxide is a relatively clean process, Scheme N.

Sellmann and co-workers have described the synthesis of tetradentate dithioether dithiolate

SCHEME N
Tetrathioether Complexes

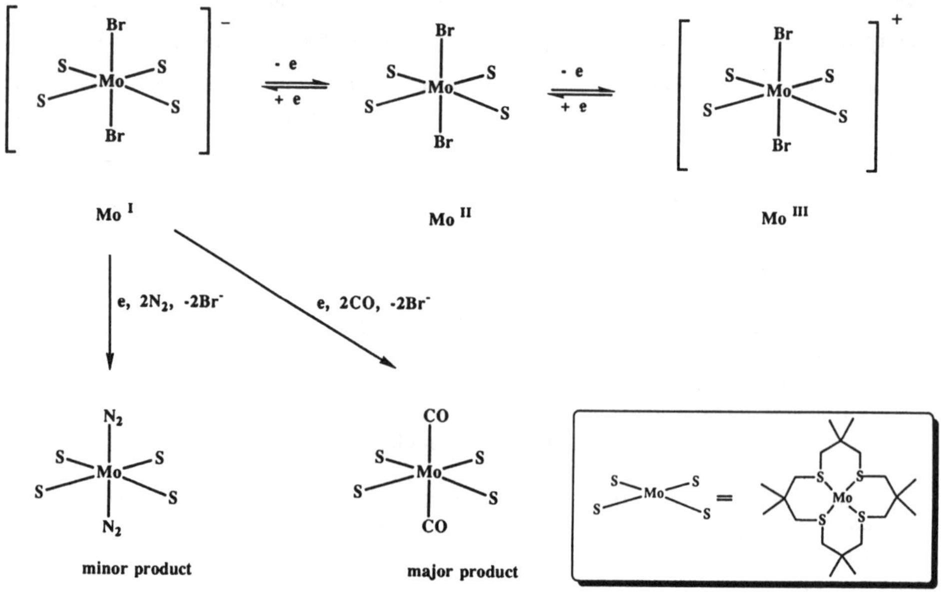

SCHEME O
Ethylene extrusion and conversion of NO to ammonia or nitrite at molybdenum - sulphur site

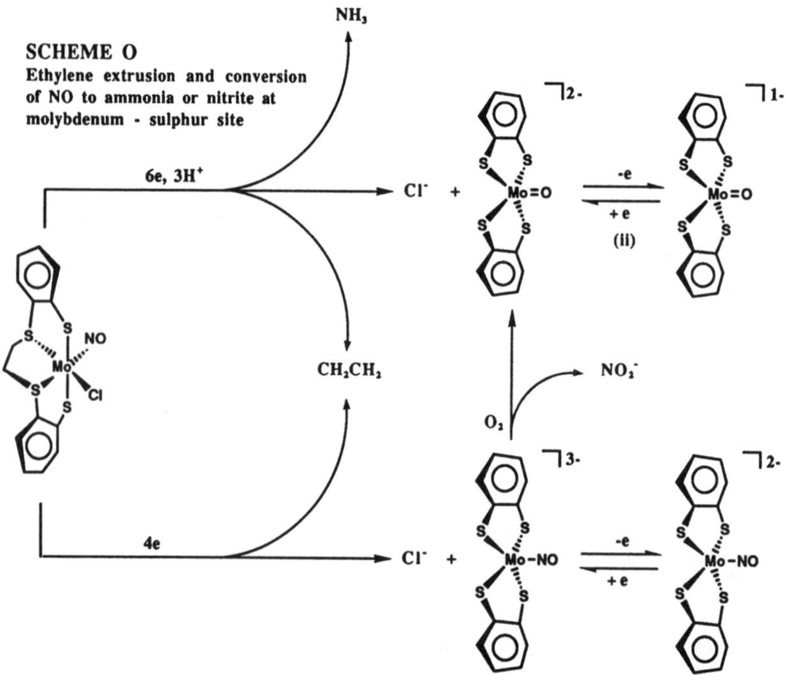

complexes of molybdenum bearing NO and Cl or NR_2 groups [40]. Electrochemical reduction of this type of compound leads to the extrusion of ethene from the backbone of the sulphur ligand and this unmasks a dithiolene centre at which ligated NO can be reduced in the presence of phenol to ammonia or oxygenated to nitrite, reactions which involve oxygen atom transfer to molybdenum [41] , Scheme O. The amide complexes reduce in a similar fashion except that an amine is released instead of chloride.

These reactions serve to illustrate that quite complex multi-electron reductions of coordinated groups can be effected at such molybdenum sulphur sites but the dithiolate dithioether ligand is also transformed.

10. Metallo - sulphur clusters

Much effort has been devoted to the synthesis of iron - sulphur and heterometallic clusters with the aim of modelling the spectroscopic, electron - transfer and catalytic properties of clusters within proteins such hydrogenase, the ferredoxins, the high potential iron proteins and, not least, the components of nitrogenase.

Whilst synthetic structural models of Fe_4S_4 centres are well developed for ferredoxins and the high potential iron - proteins, it is still unclear how the protein microenvironment can exert such a dramatic influence on redox potentials. For example, E^o 's of the $\{Fe_4S_4\}^{2+/3+}$ - couple of the ferredoxins spans about 400mV even though the primary ligation of the cluster by cysteinyl ligands *and* NH...S hydrogen bonding interactions around the cluster are conserved [44]. Access to and stability of, the four redox states [42] of the $\{Fe_4S_4\}$ - core is also subject to control by the protein environment of the cluster.

Metallo - sulphur clusters are contained within the two proteins which comprise the nitrogenase system. The smaller of these proteins contains an $\{Fe_4S_4\}^{2+}$ -centre and serves to transfer electrons to the larger with concomitant hydrolysis of APT. The larger protein, an $\alpha 2\beta 2$ dimer, contains a molybdenum iron cluster assembly of approximate stoichiometry $MoFe_{5-7}S_{4-9}$ and also unusual iron sulphur centre(s) . It is believed that molecular nitrogen is bound and reduced at the molybdenum cluster in molybdenum nitrogenase (or at corresponding vanadium or iron clusters in the less common V and Fe nitrogenases) [43].

Synthetic structural models for clusters within the nitrogenase proteins or of the molybdenum or vanadium iron - sulphur cofactors isolated from them, are at an early stage of development and there are as yet no well - defined systems which model electron - transfer chemistry involving the binding and reduction of molecular nitrogen or other substrates : dinitrogen binding to synthetic iron - sulphur clusters or heterometallic iron - sulphur clusters of any type has not been established structurally nor spectroscopically , [44- 46].

With this background in mind, we have begun to investigate ways by which synthetic metallo-clusters or co - factors extracted from nitrogenase proteins, might be incorporated within functionalised polymers on electrode surfaces. The aims are to model the diverse properties of clusters in ferredoxins, hydrogenases and nitrogenases by controlling the microenvironment of the cluster centres within the polymer and to construct artificial enzyme - electrode assemblies capable of reducing molecular nitrogen or other small molecules. Our first steps in this direction include the binding of a wide range of clusters in ion-exchange polymers [47] and $\{Fe_4S_4\}^{2+}$ - centres in novel cysteine and glycinyl - cysteine derivatised pyrrole co-polymers *via* S - ligation *ie* ferredoxin electrodes, Scheme P [48].

Scheme P

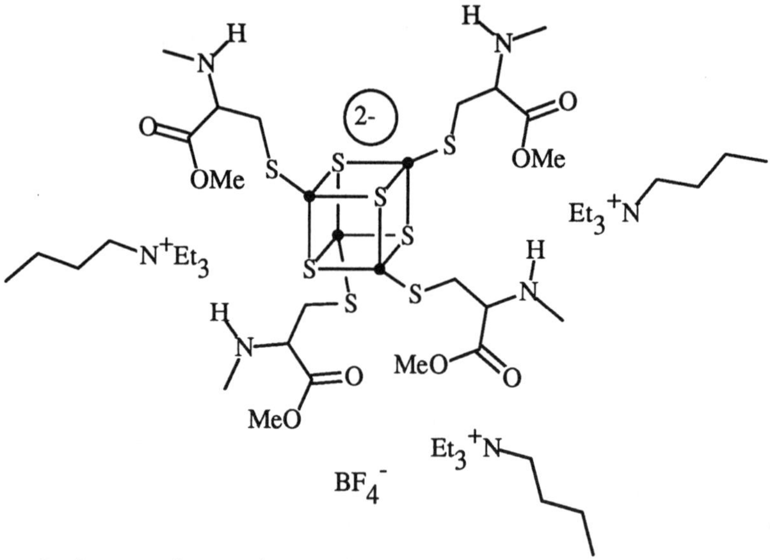

Ferredoxin centre in pyrrole co-polymer on electrode.

Acknowledgements

I would like to thank my students, colleagues and friends who have collaborated in this work and whose names appear in the references. Particular thanks go to David Hughes for crystallographic studies, and to Jean Talarmin (Brest) , Adrian George (Nebraska) Armando Pombeiro (Lisboa) Saad Ibrahim and Karl Ryder of the Nitrogen Fixation Laboratory.

REFERENCES

1. Al-Salih, T.I. and Pickett, C.J.
Electron-transfer reactions in nitrogen fixation. Part 1. The electrosynthesis of dinitrogen, hydride, isocyanide, and carbonyl complexes of molybdenum: intermediates, mechanisms, and energetics
J.Chem.Soc. Dalton Trans., (1985), 1255-1264

2. Pickett, C.J. and Talarmin, J.
Electrosynthesis of ammonia
Nature, (1985), 317, 652-653

3. Pickett, C.J., Ryder, K.S. and Talarmin, J.
Electron-transfer reactions in nitrogen fixation. Part 2. The electrosynthesis of ammonia: Identification and estimation of products
J.Chem.Soc. Dalton Trans., (1986), 1453-1457

4. Mohammed, M.Y. and Pickett, C.J.
From metal imides and molecular nitrogen to ammonia and dinitrogen complexes
J.C.S. Chem. Commun., (1988), 1119-1121

5. Hughes, D.L., Mohammed, M.Y. and Pickett, C.J.
Metallo-oxides in nitrogen fixation
J.Chem.Soc., Chem.Commun., (1988), 1481-1482

6. Hughes, D.L., Lowe, D.J., Mohammed, M.Y., Macdonald, C.J. and Pickett, C.J.
Electrochemistry of ligands multiply bonded to molybdenum and tungsten
Polyhedron, (1989), 8, 1653-1656

7. Hughes, D.L., Mohammed, M.Y. and Pickett, C.J.
Electroreduction of co-ordinated cyanide to the aminocarbyne ligand (CNH_2) and a pathway for isomerisation of ligating methyleneamide (NCH_2): Reactions at molybdenum of relevance to cyanide reduction by nitrogenase
J.Chem.Soc., Chem. Commun., (1989), 1399-1400

8. Hills, A., Hughes, D.L., Macdonald, C.J., Mohammed, M.Y. and Pickett, C.J.
Transformation of a Methyleneamide Ligand at Molybdenum: Electrochemical Oxidation to a Cyanide, Reactions with Elemental Oxygen, Sulphur or Selenium and X-Ray Crystal Structures of trans-$[Mo(CN)Cl(dppe)_2]\cdot MeOH$ and trans-$[Mo(NCS)Cl(dppe)_2]$; Electroreduction of the Cyanide to an Aminocarbyne, trans-$[Mo(CNH_2)Cl(dppe)_2]$ (dppe = $Ph_2PCH_2CH_2PPh_2$)
J. Chem. Soc., Dalton Trans., (1991), 121-129

9. Talarmin, J., Al Salih, T.I., Pickett, C.J., Bossard, G.E., George, T.A. and Noddings, C.M.
The electrochemical Generation of Low Valent Molybdenum Tris (Tertiary phosphine) Complexes: Interactions with Monophosphines and Molecular Nitrogen

J. Chem. Soc., Dalton Trans. (1992), in press.

10. Hughes, D.L., Mohd Ali, H. and Pickett, C.J. to be published.

11. Leigh, G.J. and Pickett, C.J.
Electrochemical behaviour of organonitrile dinitrogen complexes of molybdenum(0) and tungsten(0) and the anchoring of a dinitrogen complex to an electrode surface
J.Chem.Soc. Dalton Trans., (1977), 1797-1800

12. Hussain, W., Leigh, G.J., Modh-Ali, H., Pickett, C.J. and Rankin, D.A.
Dinitrogen binding and electrochemistry in complexes of molybdenum and tungsten
J.Chem.Soc. Dalton Trans., (1984), 1703-1708

13. George, T.A., Hayes, R.K., Mohammed, M.Y. and Pickett, C.J.
Cis and trans isomers of $Mo(N_2)_2(PMe_2Ph)_4$
Inorg.Chem., (1989), 28, 3269-3270

14. De Bord, J., George, T.A., Kaul, B.B., Pickett, C.J. and Rose, D.J.
Cis and trans isomers of $W(N_2)_2(PMe_2Ph)_4$
Inorg. Chem., (1992).

15. Chatt, J., Leigh, G.J. Neukomm, H., Pickett, C.J. and Stanley, D.R.
Redox potential-structure relationships in metal complexes. Part 2. The influence of trans-substituents upon the redox properties of certain dinitrogen complexes of molybdenum and tungsten and some carbonyl analogues: inner-sphere versus outer-sphere electron transfer in the alkylation of co-ordinated dinitrogen
J.Chem.Soc. Dalton Trans., (1980), 121-127

16. Leigh, G.J., Morris, R.H., Pickett, C.J., Stanley, D.R. and Chatt, J.
Redox-potential-structure relationships in metal complexes: Part 4. Electron-poor dinitrogen complexes of rhenium(I)
J.Chem.Soc. Dalton Trans., (1981), 800-804

17. Chatt, J., Kan, C.T., Leigh, G.J., Pickett, C.J. and Stanley, D.R.
Transition-metal binding sites and ligand parameters
J.Chem.Soc. Dalton Trans., (1980), 2032-2038

18. Pickett, C.J. and Pletcher, D.
Electrochemical measurement as a probe for the investigation of the structure of metal carbonyls
J.Organomet.Chem., (1975), 102, 327

19. Chatt, J., Head, R.A., Leigh, G.J. and Pickett, C.J.

Mechanism of alkylation of dinitrogen co-ordinated to molybdenum(0) and tungsten(0)
J.C.S. Chem.Comm., (1977), 299-300

20. Chatt, J., Leigh, G.J., Pickett, C.J., Richards, R.L. and Pombeiro, A.J.L.
A molecular-orbital rationale of reactions of N_2 in the complexes of trans-$[M(N_2)_2(Ph_2PCH_2CH_2PPh_2)_2]$ (M = Mo or W); A comment on the paper by Hoffmann and Dubois
Nouveau Journal de Chimie, (1978), 2, 541-542

21. Chatt, J., Head, R.A., Leigh, G.J. and Pickett, C.J.
Mechanism of alkylation and acylation of dinitrogen co-ordinated to molybdenum and tungsten
J.Chem.Soc. Dalton Trans., (1979), 1638-1647

22. Chatt, J., Hussain, W., Leigh, G.J. Neukomm, H., Pickett, C.J. and Rankin, D.A.
The mechanism of the secondary alkylation of dinitrogen and the relationship between structure and reactivity in the alkylation of co-ordinated dinitrogen
J.C.S. Chem.Comm., (1980), 1024-1025

23. Hussain, H., Leigh, G.J., Modh-Ali, H. and Pickett, C.J.
Mechanism of alkylation of alkyldiazenido-complexes of molybdenum(II) and tungsten(II): influence of metal and co-ligands on the nucleophilicity of a diazenido-group
J.Chem.Soc. Dalton Trans., (1986), 1473-1477

24. Ryder, K.S. and Pickett, C.J., to be published.

25. Pickett, C.J., Tolhurst, J.E., Copenhaver, A., George, T.A. and Lester, R.K.
Electrochemical carbon-carbon coupling of diazomethane ligands
J.C.S. Chem.Comm., (1982), 1071-1072

26. Henderson, R.A., Leigh, G.J. and Pickett, C.J.
The chemistry of nitrogen fixation and models for the reactions of nitrogenase
Advances in Inorganic Chemistry and Radiochemistry, (1983), 27, 197-292

27. Hughes, D.L., Lowe, D.J., Mohammed, M.Y., Pickett, C.J. and Pinhal, N.M.
Determination of Structural Features of Electrogenerated trans-$[MoCl(NMe)(Ph_2PCH_2CH_2PPh_2)_2]^{2+}$ by Multinuclear Electron Paramagnetic Resonance and Electron Nuclear Double Resonance Spectroscopy and Comparison of Interatomic Distances with those measured by X-Ray Analysis of the Parent Monocation
J. Chem. Soc., Dalton Trans., (1990), 2021-2027

28. Pickett, C.J. and Leigh, G.J.
Towards a nitrogen-fixing cycle: Electrochemical reduction

of a hydrazido-complex of molybdenum(IV) under N_2 to yield the dialkylhydrazine and a molybdenum(0) dinitrogen complex
J.C.S. Chem.Comm., (1981), 1033-1035

29. Hussain, W., Leigh, G.J. and Pickett, C.J.
Stepwise conversion of dinitrogen co-ordinated to molybdenum into an amine and an imido-complex. Relevance to the reactions of nitrogenase
J.C.S. Chem.Comm., (1982), 747-748

30. Henderson, R.A., Leigh, G.J. and Pickett, C.J.
Mechanistic studies on hydrazido(2-)-complexes. Cleavage of the nitrogen-nitrogen bond in the reaction of $[Mo\{NN(CH_2)_4CH_2\}(Ph_2PCH_2CH_2PPh_2)_2]$ with acid
J.Chem.Soc., Dalton Trans., (1989), 425-430

31. Pickett, C.J., Cate, K., Macdonald, C.J., Mohammed, M.Y., Ryder, K.S. and Talarmin, J.
Electron-transfer reactions in nitrogen fixation
Nitrogen Fixation: Hundred Years After, Eds. H. Bothe, F.J. de Bruijn and W.E. Newton: pub. Gustav Fischer, Stuttgart, New York (1988), 51-55.

32. Hughes, D.L., Mohammed, M.Y. and Pickett, C.J.
Synthesis, Reactivity and Electrochemistry of some New Nitrides of Molybdenum and Tungsten: Crystal Structure of Trinuclear
$[\{\mu\text{-MoN}(N_3)_2\}\{NMo(N_3)(Et_2PCH_2CH_2PEt_2)_2\}_2]$
J. Chem. Soc. Dalton Trans., (1990), 2013-2019.

33. Pombeiro, A.J.L., Hughes, D.L., Pickett, C.J. and Richards, R.L.
The aminocarbyne ligand CNH_2: Metal-centred synthesis from a cyanosilane, preparation and X-ray structure of trans-$[ReCl(CNH_2)(Ph_2PCH_2CH_2PPh_2)_2]BF_4$
J.C.S. Chem.Commun., (1986), 246-247

34. Butler, G., Chatt, J., Leigh, G.J. and Pickett, C.J.
Structure-redox potential relations of metal complexes Part 1. Organodiazenido-complexes of molybdenum: Neighbouring-group influence on redox potential
J.Chem.Soc. Dalton Trans., (1979), 113-116

35. Crichton, B.A.L., Dilworth, J.R., Pickett, C.J. and Chatt, J.
Redox potential-structure relationships in metal complexes. Part 3. The electrochemical activation of co-ordinated hydrazido(2-)-ligands
J.Chem.Soc. Dalton Trans., (1981), 419-424

36. Crichton, B.A.L., Dilworth, J.R., Pickett, C.J. and Chatt, J.
Open-shell molybdenum carbonyl complexes and the binding of π-acid ligands to higher-oxidation-state molybdenum

sites
J.Chem.Soc. Dalton Trans., (1981), 892-893

37. Dilworth, J.R., Neaves, B.D., Pickett, C.J., Chatt, J. and Zubieta, J.A.
Electrochemical generation of sulfur-ligated molybdenum(II) and molybdenum(III) substrate binding sites. Preparation and crystal structures of $[MoCl(S_2CNEt_2)_2(Ph_2PCH_2CH_2PPh_2][BF4]$ and $[MoCl(S_2CNEt_2)_2(PPh_2Me)_2][PF_6]$ and the mechanism of their electrochemical reduction
Inorg.Chem., (1983), 22, 3524-3529

38. Yoshida, T., Adachi, T., Kaminaka, M. and Veda, T.
A novel molybdenum(O) dinitrogen complex containing crown thioether as the sole auxillary ligand: trans-$[Mo(N_2)_2Me_8[16]aneS_4]$
J.Am.Chem.Soc., (1988), 110, 4872-4873.

39. Durrant, M.C., Hughes, D.L., Pickett, C.J., Richards, R.L. and Talarmin, J., to be published

40. Sellmann, D., Pohlmann, G., Knock, F. and Moll, M.
Transition metal complexes with sulphur ligands, XLII $[Mo(NO)(NR_2)dttd]$: metal amido complexes with rotationaly stable $M-NR_2$ bonds (R = H, Me, Et; dttd = 2,3,8,9-dibenzo-1,4,7,10-tetradecane(2-); synthesis, properties and X-ray structures
Z.Naturforsch, (1989), 44b, 312-322

41. Ibrahim, S.A. and Pickett, C.J.
Electrochemical Transformation of the Molybdenum Nitrosyl [Mo(NO)Cl(dttd)] dttd = 1,2-bis(2-mercaptophenylthio)ethane to give the Oxide $[MoO(1,2-C_6H_4S_2)_2]^{2-}$ and either Ammonia by Protic Attack or Nitrite by Oxygenation: New Reactions which involve Ethylene Extrusion from the Backbone of the Sulphur Ligand
J. Chem. Soc., Chem. Commun., (1991), 246-249.

42. Pickett, C.J.
A simple hydrocarbon electrolyte: completing the electron-transfer series $[Fe_4S_4(SPh)_4]^{1-/2-/3-/4-}$
J.C.S. Chem.Commun., (1985), 323-326

43. Smith, B.E. and Eady, R.R.
Metalloclusters of the nitrogenases
Eur.J.Biochem., (1992), 205, 1-15

44. Backes, G., Mino, Y., Loehr, T.M., Meyon, T.E., Cusanovich, M.A., Sweeney, W.V., Adman, E.T. and Sanders-Loehr, J.
J.Am.Chem.Soc., (1991), 113, 2055-2064

45. Holm, R.H., Ciurli, S. and Wiegel, J.A.
Subsite-specific structures and reactions in native and

synthetic [4Fe-4S] cubane-type clusters
Prog.Inorg.Chem.: Bioinorg.Chem., Ed. Lippard, S.J. (1990), 38, 1-74

46. Al-Ani, F.T., Hughes, D.L. and Pickett, C.J.
From an $\{Fe_4S_4\}$-cluster to $\{Fe_2S_2\}$- and $\{Fe_3S\}$-carbonyls. Crystal structure of $[Fe_3S(CO)_9]^{2-}$
J.Organomet.Chem., (1986), 307, C31-C34

47. Al-Ani, F.T. and Pickett, C.J.
On the reduction of iron-sulphur clusters under carbon monoxide
J.Chem.Soc., Dalton Trans., (1988), 2329-2334

48. Moutet, J-C. and Pickett, C.J.
Iron sulphur clusters in ionic polymers on electrodes
J.Chem.Soc., Chem.Commun., (1989), 188-190

49. Pickett, C.J., Ryder, K.S. and Moutet, J-C.
Synthesis and anodic polymerisationi of an L-cystine derivatised pyrrole; co-polymerisation with a tetraalkylammonium pyrrole allows reduction of the cystinyl film to a cysteinyl state that binds electroactive $\{Fe_4S_4\}^{2+}$ centres
J. Chem. Soc., Chem. Commun. (1992), in press

ELECTROCATALYSIS OF THE REDUCTION OF O_2 WITHIN POLYELECTROLYTE COATINGS ON ELECTRODES

C. SHI and F. C. ANSON
Division of Chemistry & Chemical Engineering, California Institute of Technology, Pasadena CA 91125, U.S.A

ABSTRACT
The reaction of $Ru(NH_3)_5OH_2^{2+}$ with $CoP(py)_4$ (cobalt meso-tetrakis(4-pyridyl)porphyrin) within Nafion coatings on graphite electrodes produces $[CoP(pyRu(NH_3)_5)_4]^{8+}$. This complex, immobilized within the polyelectrolyte coating, acts as a catalyst for the four-electron reduction of O_2 under conditions where mixtures of $CoP(py)_4$ with $Ru(NH_3)_6^{2+}$ or $Ru(NH_3)_5py^{2+}$ do not. The results provide evidence for the intramolecular delivery of four electrons from the four coordinated $Ru(NH_3)_5^{2+}$ groups to O_2 molecules associated with the Co(II) center of the porphyrin.

INTRODUCTION
Catalysis is required in order to achieve the direct, four-electron reduction of O_2 to H_2O at potentials near the thermodynamically allowed values and metalloporphyrins are frequently employed as catalysts for this purpose. Cobalt porphyrins, which are active at reasonably positive potentials, typically do not carry the reduction beyond the two-electron step to H_2O_2.

One strategy to achieve rapid, multiple-electron transfers (while avoiding undesired intermediates) involves the use of catalysts containing multiple centers which can serve as electron donors or acceptors. We have been exploring this strategy by coordinating the substitutionally inert $Ru^{II}(NH_3)_5$ center to cobalt tetrapyridylporphyrin, $CoP(py)_4$, to obtain a potential catalyst containing four electron donors (1). The preparative reactions were carried out with the reactants confined to coatings of the polyelectrolyte Nafion (2) on graphite electrode surfaces. The resulting multinuclear complex was demonstrated to be a catalyst for the four-electron reduction of O_2 by a mechanism involving multiple, intramolecular electron transfers.

EXPERIMENTAL
The preparation of the necessary reactants and the synthesis of the multinuclear catalyst have been described, as have the experimental apparatus and procedures (1).

RESULTS AND DISCUSSION
The cobalt tetrapyridylporphyrin pre-catalyst, $CoP(py)_4$, becomes soluble in aqueous acid because of the protonation of the four pyridine groups to form a tetra positive cation, $CoP(pyH)_4^{4+}$. This cation is slowly incorporated by Nafion coatings but it exhibits essentially no electrochemical response because the low mobility of the tetrapositive cation within the Nafion prevents it from diffusing to the electrode surface where the oxidation/reduction of the cobalt center could proceed (Figure 1A).

A. J. L. Pombeiro and J. A. McCleverty (eds.),
Molecular Electrochemistry of Inorganic, Bioinorganic and Organometallic Compounds, 381–389.

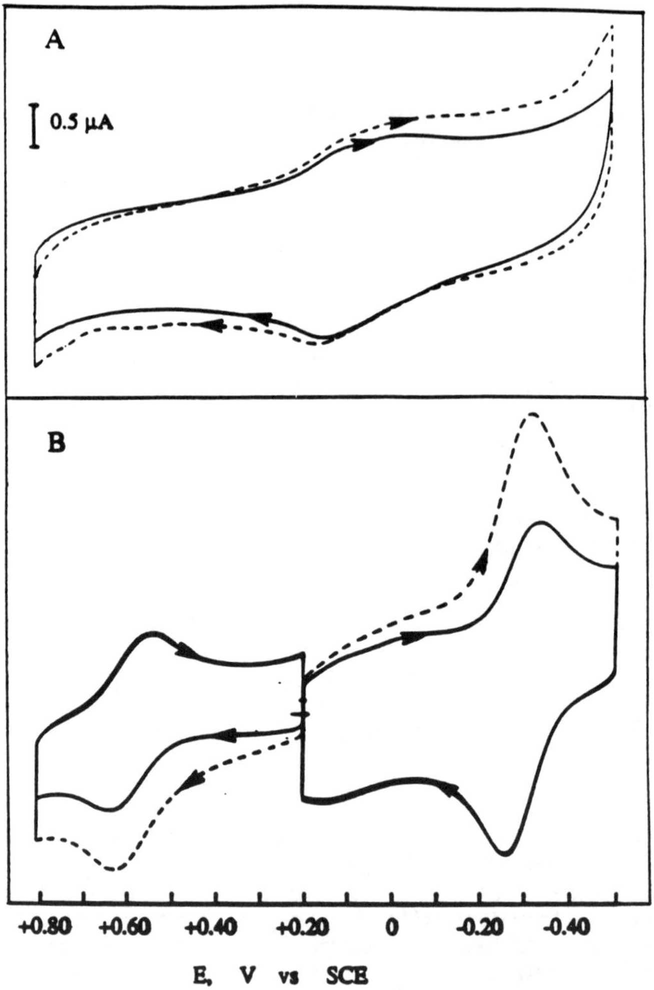

Figure 1. A. Cyclic voltammograms recorded with a Nafion-coated pyrolytic graphite electrode containing 1.1×10^{-7} mole cm^{-2} of sulfonate groups. The solid line was obtained with the Nafion coating alone; the dashed line resulted after incorporation of 2.2×10^{-10} mole cm^{-2} of $CoP(py)_4$ in the coating.
B. Voltammograms recorded with the coating from A after 3.9×10^{-10} mole cm^{-2} of $Os(bpy)_3{}^{2+}$ and 1.0×10^{-9} mole cm^{-2} of $Ru(NH_3)_6{}^{3+}$ were incorporated. The initial and final potential was 0.2 V for every scan. The dashed curves are the first scans after the cobalt center in the incorporated $CoP(py)_4$ had been oxidized or reduced. Supporting electrolyte: 0.1 M acetate buffer (pH 4.5) saturated with argon. Scan rate: 10 mV s^{-1}.

To provide electronic coupling between the incorporated $CoP(pyH)_4^{4+}$ counterions and the graphite electrode surface it was necessary to add to the coatings suitable redox mediators with much larger diffusion coefficients. In Figure 1B is shown the cyclic voltammetric response obtained from a Nafion coating which contained $Os(bpy)_3^{2+}$ (bpy = 2,2'-bipyridine) and $Ru(NH_3)_6^{3+}$ as well as $CoP(pyH)_4^{4+}$. The pair of current peaks near −0.25 V arises from the $Ru(NH_3)_6^{3+/2+}$ couple and the pair near 0.6 V from the $Os(bpy)_3^{3+/2+}$ couple. This pair of mediators was chosen because they both have reasonable diffusional rates in Nafion and their formal potentials lie on either side of the formal potential of the $CoP(pyH)_4^{5+/4+}$ couple (−0.1V) so that the cobalt center of the porphyrin could be repetitively cycled between its oxidized and reduced states by the mediator counterions which diffused to and from the electrode surface where they were oxidized or reduced.

The dashed curves with the larger peak currents in Figure 1B were obtained during the first voltammetric scan from 0.2 V with the cobalt center of the incorporated porphyrin in its oxidized state (enhanced cathodic peak for reduction of $Ru(NH_3)_6^{3+}$) or its reduced state (enhanced anodic peak for oxidation of $Os(bpy)_3^{2+}$). The solid curves show the steady responses obtained in the absence of incorporated porphyrin or when the voltammetric scans were restricted to the range between 0.2 and −0.5 V or between 0.2 and 0.8 V so that the cobalt center remained in its reduced or oxidized state, respectively. The areas between the dashed and solid curves in Figure 1B were used to measure the quantities of $CoP(pyH)_4^{4+}$ incorporated in the Nafion coatings.

In Table 1 are listed coulometric assays of the quantities of $CoP(pyH)_4^{4+}$ incorporated in these Nafion coatings. The agreement between the assays obtained using the $Os(bpy)_3^{2+}$ or the $Ru(NH)_6^{3+}$ mediators is good.

TABLE 1

Coulometric Assays of Quantities of $CoP(pyH)_4^{4+}$ Incorporated in Nafion Coatings [a]

Electrode No.	$10^{10}\Gamma_{Co}$ [b], mole cm^{-2}	$10^{10}\Gamma_{Co}$ [c], mole cm^{-2}
1	4.7	4.3
2	2.2	2.2
3	2.3	2.2

a. Each coating contained 1.1×10^{-7} mole cm^{-2} of sulfonate groups.
b. Measured from the difference in the areas under the cathodic peaks during the first and subsequent scans of the $Ru(NH_3)_6^{3+/2+}$ couple.
c. Measured from the difference in the areas under the anodic peaks during the first and subsequent scans of the $Os(bpy)_3^{3+/2+}$ couple.

Coordination of $Ru(NH_3)_5$ Centers to $CoP(py)_4$ in Nafion Coatings

To coordinate $Ru(NH_3)_5^{3+/2+}$ groups to the four pyridine sites present in the $CoP(py)_4$ molecules incorporated in the Nafion coatings, $Ru(NH_3)_5OH_2^{2+}$, with a moderately labile aqua ligand, was introduced into the coatings. A convenient procedure for doing so was to incorporate $Ru(NH_3)_5Cl^{2+}$ as a precursor complex. Electrochemical reduction of this complex to $Ru(NH_3)_5Cl^{+}$ causes it to aquate at a modest rate and the conversion to $Ru(NH_3)_5OH_2^{2+}$ is readily monitored electrochemically (3, 4).

The first cyclic voltammogram recorded after $Ru(NH_3)_5Cl^{2+}$ is incorporated in a Nafion coating exhibits the single wave corresponding to the reversible interconversion of the $Ru(NH_3)_5Cl^{2+/+}$ complexes. However, some of the $Ru(NH_3)_5Cl^{+}$ complex spontaneously aquates to yield $Ru(NH_3)_5OH_2^{2+}$. The chloro complex can be converted completely to the aqua complex by maintaining the electrode potential for several minutes at a value (−0.6 V) where the complex is maintained in the Ru(II) oxidation state. Following such a treatment, the voltammetry consists of a single wave at the potential corresponding to the $Ru(NH_3)_5OH_2^{3+/2+}$ couple.

With both $CoP(py)_4$ and $Ru(NH_3)_5OH_2^{2+}$ incorporated in the Nafion coating the reaction between them (equation 1) was monitored by cyclic voltammetry until it was completed.

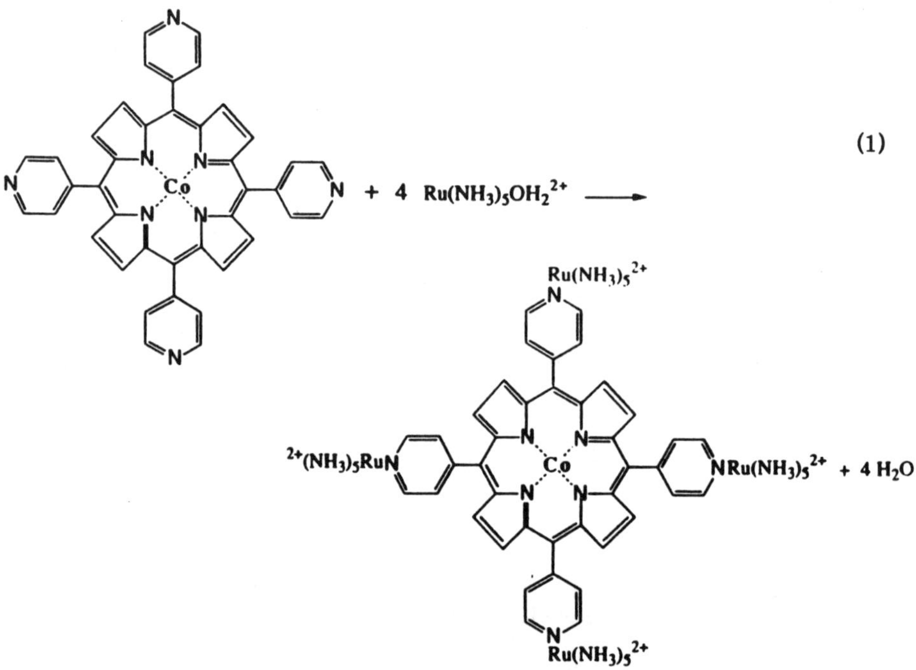

(1)

The quantities of $Ru(NH_3)_5OH_2^{2+}$ incorporated in the coatings were typically greater than four times the amount of $CoP(py)_4$ present in order to drive the reaction to completion which required about 2 hours. The excess $Ru(NH_3)_5OH_2^{2+}$ was easily removed by reverse ion-exchange. The resulting coatings exhibited a voltammogram with a single, reversible response at the potential expected for the $Ru(NH_3)_5py^{3+/2+}$ couple (Figure 2). The quantity of $Ru(NH_3)_5$ groups present in the coatings was determined from the area under voltammograms such as the one in Figure 2 but recorded at a scan rate of 10 mV s^{-1}. Combination of these values with the coulometric assays for $CoP(py)_4$ as described above provided a measure of the stoichiometric ratio of $Ru(NH_3)_5$ to $CoP(py)_4$. The results, shown in Table 2,demonstrate that $Ru(NH_3)_5$ groups were coordinated all four of the pyridine groups of the $CoP(py)_4$ molecule.

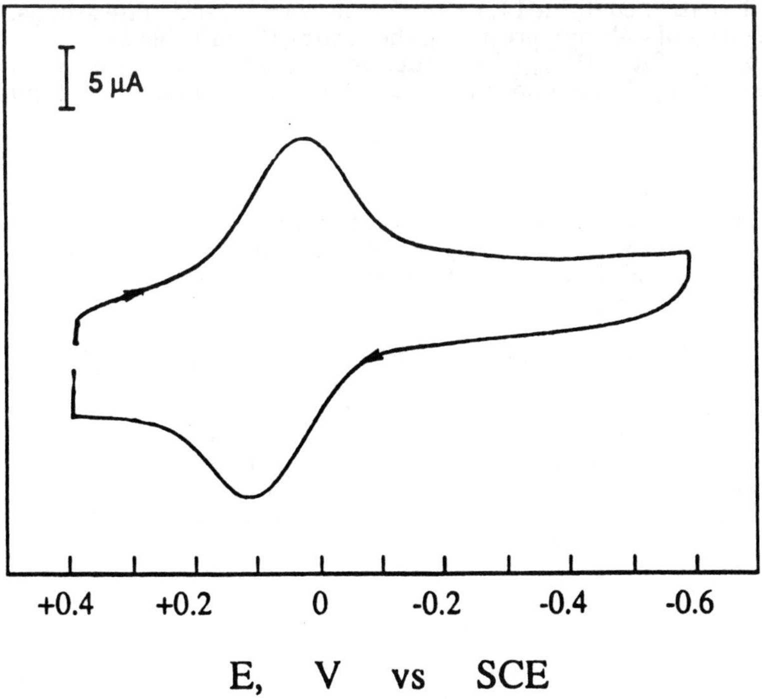

Figure 2. Cyclic voltammetry with a Nafion coating containing 3.8 x 10^{-10} mole cm^{-2} of $[CoP(pyRu(NH_3)_5)_4]^{8+}$. Supporting electrolyte: 0.25 M Na_2SO_4. Scan rate: 50 mV s^{-1}.

TABLE 2

Stoichiometric Ratio in the Reaction Between $CoP(py)_4$ and Excess $Ru(NH_3)_5OH_2^{2+}$ in Nafion Coatings [a]

Coating No.	$10^{10}\Gamma_{Co}$ [b], mole cm^{-2}	$10^{10}\Gamma_{Co\text{-}Ru}$ [c] mole cm^{-2}	$\Gamma_{Co\text{-}Ru}/\Gamma_{Co}$
1	4.5	18.1	4.0
2	2.2	9.3	4.2
3	2.25	8.5	3.8

a. Each coating contained 1.1×10^{-7} mole cm^{-2} of sulfonate groups.
b. Quantity of $CoP(py)_4$ present in the coating (from Table 1).
c. Quantity of $Ru(NH_3)_5py^{3+}$ groups present in the coating as measured from the cathodic charge under voltammograms such as the one in Figure 2.

The fact that the four $Ru(NH_3)_5py$ centers give rise to only a single cyclic voltammetric peak shows that the centers undergo reduction and oxidation independently and at essentially the same formal potential. This is a property which is essential if catalysts for the transfer of multiple electrons at the same potential are desired.

Catalysis of O_2 Reduction by $[CoP(pyRu(NH_3)_5)_4]^{8+}$ in Nafion Coatings
The electroreduction of O_2 at an electrode coated with Nafion in which $(CoP(pyRu(NH_3)_5)_4]^{8+}$ was prepared is shown in Figure 3. The voltammogram for the catalyst in the absence of O_2 is shown in the dashed curve. The solid curve resulted in the presence of O_2. The evident catalysis of the reduction of O_2 proceeds at the potential where the ruthenium centers are reduced which shows that the $Ru(NH_3)_5$ centers coordinated to the porphyrin ring serve both to shuttle electrons from the electrode to the immobilized catalyst molecules and to deliver the electrons to the Co-O_2 adduct that is presumed to be the species being reduced.

Results obtained with a rotating Nafion-coated graphite disk-platinum ring electrode are shown in Figure 4. When $CoP(py)_4$ and $Ru(NH_3)_6^{3+/2+}$ (as mediator) are present in the Nafion coating (Figure 4B) the reduction of O_2 yields a smaller disk current (upper curve) and a ring current (lower curve) which corresponds to essentially quantitative formation of H_2O_2. By contrast, with the $CoP(pyRu(NH_3)_5)_4^{8+}$ catalyst present in the Nafion, a larger disk current and smaller ring current result (Figure 4A). The magnitude of the currents shows that the multinuclear catalyst is able to divert over 70% of the O_2 into the four-electron reduction pathway.

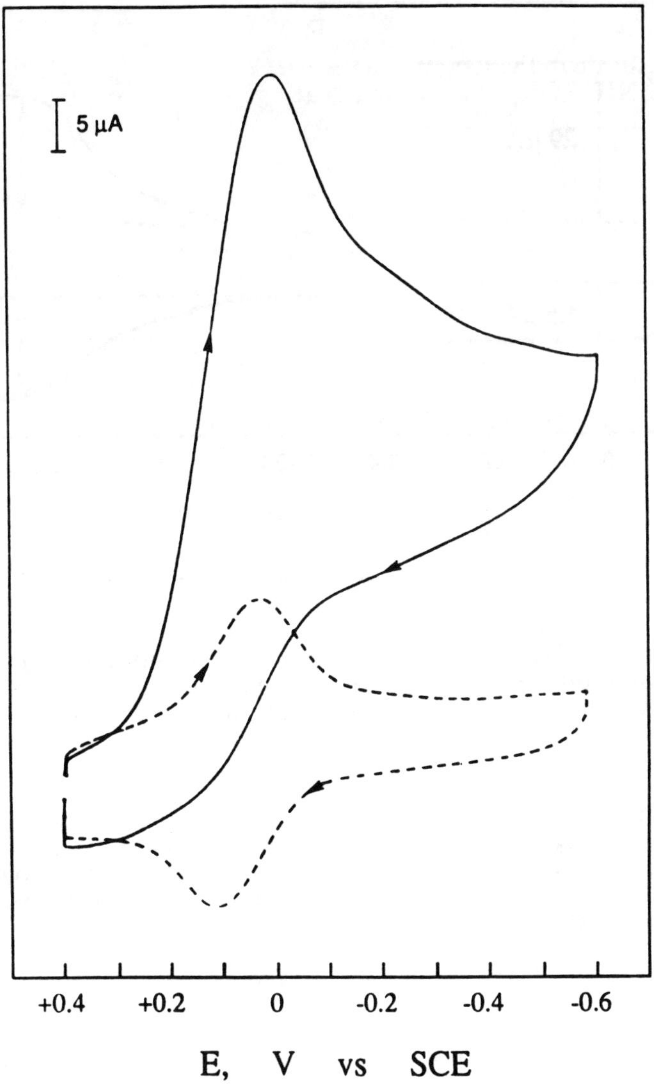

Figure 3. Voltammetric reduction of O_2 at a Nafion-coated electrode containing 3.8×10^{-10} mole cm^{-2} of $[CoP(pyRu(NH_3)_5)_4]^{8+}$. Supporting electrolyte: 0.5 M CF_3COOH saturated with air. The dashed curve was recorded in the absence of O_2. Scan rate: 50 mV s^{-1}.

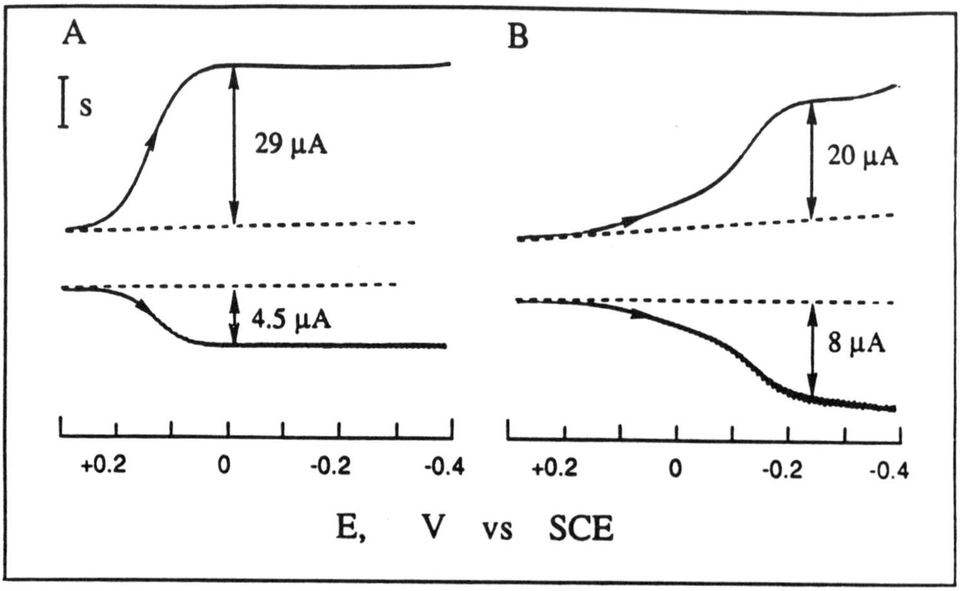

Figure 4. Reduction of O_2 at a rotating graphite disk-platinum ring electrode. The Nafion-coated disk contained 1.1 x 10^{-7} mole cm^{-2} of sulfonate groups (A) 7 x 10^{-10} mole cm^{-2} of $[CoP(pyRu(NH_3)_5)_4]^{8+}$ and (B) 7 x 10^{-10} mole cm^{-2} of $CoP(pyH)_4^{4+}$ and 1.2 x 10^{-9} mole cm^{-2} of $Ru(NH_3)_6^{3+}$; .

The coordination of the $Ru(NH_3)_5$ groups to the pyridine ligands attached to the meso position of the porphyrin ring is essential to obtain the four-electron reduction. Thus, Nafion coatings in which both $CoP(pyH)_4^{4+}$ and $Ru(NH_3)_5py^{2+}$ (as mediator) are incorporated yield only the two-electron reduction of O_2 and at a low rate.

To account for the difference in behavior it seems reasonable to surmise that the coordinated $Ru(NH_3)_5^{2+}$ centers are in more intimate electronic contact with the cobalt(II) center in the porphyrin ring and with the O_2 molecules that may coordinate to it than are uncoordinated Ru(II) complexes also present in the coatings. The production of H_2O_2 instead of H_2O during the catalysis of the electroreduction of O_2 by most cobalt porphyrins presumably results from the escape of partially reduced intermediates from the coordination sphere of the cobalt center before they can be further reduced. By pre-attaching the four electron-donating centers to the porphyrin ring in $CoP(pyRu(NH_3)_5)_4^{8+}$ the rate of intramolecular electron delivery to the cobalt center, and substrates coordinated to it, is evidently sufficiently enhanced to accomplish the complete reduction of O_2 to H_2O before the intermediate reduction products escape into solution. (It was established in separate experiments that the

$CoP(pyRu(NH_3)_5)_4^{8+}$ complex is not a catalyst for the electroreduction of H_2O_2.)

The results presented are encouraging to continuing efforts to prepare catalysts with the capacity to transfer multiple electrons at the same energy. Numerous analogs of the successful $CoP(pyRuNH_3)_5)_4^{8+}$ catalyst can be envisioned which could behave similarly, or even better, as multiple-electron transfer agents.

ACKNOWLEDGEMENTS

This work was supported by the National Science Foundation and DARPA/ONR. Discussions with Dr. Koiti Araki were helpful. Portions of Figures 1 to 4 and Tables 1 and 2 are reprinted with permission from reference 1. Copyright 1991 American Chemical Society.

REFERENCES

1. C. Shi and F. C. Anson, *J. Am. Chem. Soc.*, **1991**, 113, 9564.
2. Nafion is a trademark of E. I. duPont de Nemours and Co., Inc.
3. Lim, H. S.; Barclay, D. J. and Anson, F. C., *Inorg. Chem.*, **1972**, 11, 1460.
4. Lieber, C. M.; Schmidt, M. H. and Lewis, N. S., *J. Am. Chem. Soc.*, **1986**, 108, 6103.

ELECTROCHEMICAL STUDIES ON ACTIVATION OF DIOXYGEN AT COBALT MACROCYCLES DERIVED FROM COBALOXIME

G. COSTA and C. TAVAGNACCO
Department of Chemical Sciences - University of Trieste
Via Valerio 38
34127 - Trieste - Italy

ABSTRACT. The electrochemistry of a series of cobalt chelates derived from cobaloxime (bis dimethylglyoximato cobalt complexes) and the behaviour towards molecular oxygen are reviewed. Correlations between the structure of the equatorial ligand, the relative stability of Co(III), Co(II), Co(I) oxidation states, the extent of interaction with O_2, the formation of dioxygen adducts, mechanisms and products of the redox catalysis for the reduction of O_2 in the presence of protons, initiation of a radical mechanism of autoxidation reactions are pointed out and discussed.

1. INTRODUCTION

Transport of molecular oxygen and oxidation reactions with molecular oxygen are involved in the most important processes in the biosphere, where reduction of O_2 is utilized as a "fuel" in fundamental processes.

The fitness of dioxygen as a fuel is shown, from the thermodynamic point of view, by the free energy decrease in the formation of the bonds with hydrogen, which is lower only than that of fluorine. On the other hand, from a kinetic point of view, the reduction of triplet dioxygen can be conveniently controlled when a stepwise reaction is considered. The uptake of the first electron in a chemical reaction with substrates in a singlet states is forbidden by spin restriction, and the "activation" of O_2 molecule is necessary[1].

This situation is reflected by the free energy *increase* accompanying the formation of the first O-H bond in the chemical reduction of O_2 with H_2 to H_2O via $\cdot HO_2$ and $\cdot OH$ radicals with H_2O_2 as intermediate or via HO_2 and HO with H_2O and O as intermediates.

The same situation is found when the standard electrode potentials are considered: the free energy change is positive for both $O_2 + e \rightarrow O_2^-$ and $O_2 + H^+ + e \rightarrow \cdot HO_2$ reactions[2].

Thus, a catalytic step is always necessary to exploit the energy of most oxidation reactions, using O_2 as oxidant.

The oxygen may be activated in different ways. Thermal excitation into the first electronically excited (singlet) state requires considerable energy. Thus, the activation takes place also, and most commonly, by bonding O_2 to an appropriate site. The activation center may be a transition metal atom in a suitable coordination compound or a site at the surface of a solid [3].

In biological systems, the bonding of dioxygen to a metal center occurs, as an essential step, in

A. J. L. Pombeiro and J. A. McCleverty (eds.),
Molecular Electrochemistry of Inorganic, Bioinorganic and Organomettallic Compounds, 391–404.

very different functions such as in dioxygen transport proteins and in oxygenases. In the first case the metal - oxygen interaction is completely reversible, whithout any noticeable increase of the dioxygen reactivity, while, in the second case, some very reactive intermediate, involving coordination of dioxygen and, possibly, of the substrate, is formed.

In the naturally occurring systems as well as in the large number of model compounds which have been proposed for the active center for both types of function (in both cases, most often macrocyclic complexes of iron or cobalt), the factors influencing in the right way the character of the metal - dioxygen interaction are still not completely understood [4].

The most extensively studied chelates proposed as models of the naturally occurring systems utilizing molecular oxygen and most synthetic catalyst for the dioxygen reduction are transition metal phthalocyanines, tetrasubstituted porphyrins and tetraazaannulenes. The N_4 chelates exibit the highest activity.

Dioxygen can be coordinated to transition metal complexes in low oxidations state. Depending on the ability of the metal to donate electrons, the adducts are formulated as superoxo or peroxo complexes, while in the case of bridging molecular oxygen mu-peroxo complexes can be formed.

Transport properties can be expressed by the equilibrium and rate constant of the oxygenation reaction, while the simplest way to probe the activation of coordinated dioxygen is the investigation of the efficiency of the metal complex in the catalytic reduction of molecular oxygen. Investigation of electrochemical reduction of molecular oxygen is a most useful way to obtain evidence for a catalytic reaction and suggestions for a possible mechanism.

Activation of dioxygen seems to be achieved by a still not well understood combination of factors.

A major role in the explanation of the activation, in the electrochemical reduction of O_2 coordinated to a transition metal complex is played by the concept of redox catalysis [5,6].

$$M^{2+} + O_2 \text{----}> M^{3+} - O_2^-$$

$$M^{3+} - O_2^- \text{----}> \text{products} + M^{3+}$$

$$M^{3+} + e \text{----}> M^{2+}$$

Beside the redox catalysis of the first electron uptake, the successive steps are also relevant in determining the pathway of the reaction, the final product of the reduction of O_2 and its function as a "fuel" or as an oxidant.

Depending on catalyst nature and other experimental conditions the pathway and the final product of the electrochemical reduction can be an overall two electron reduction giving H_2O_2 or an overall four electron reduction giving H_2O.

Different concepts have been proposed for a qualitative explanation: MO theory was applied to explain the higher interaction between dioxygen and the metal center, observed with Fe(II) and Co(II) phthalocyanine dioxygen adducts where dioxygen is bonded in a "end on" configuration. In Cobalt and Iron Phthalocyanines the difference in selectivity (formation of H_2O_2 or H_2O) is ascribed to the difference in the negative charge on the O_2 molecule coordinated to the metal atom [3]. A relatively minor difference between electron levels of the electrode and reactant was recognised as a condition for fast electron transfer [7].

It was recently proposed that if the O_2 molecule interacts with only one metal center then H_2O_2 may be produced, while H_2O may be the final product if two metal atoms are available to give an O-O bridge in the adduct [8].

When the four electron reduction is seeked to achieve the maximum energy release by the electrochemical reduction of O_2, as in fuel cells, different strategies have been proposed to get a rapid multiple electron transfer. An efficient four electron electrocatalysis was claimed using dicobalt dimers (face-to face cobalt porphyrins) [7] or complexes containing multiple centers which can serve as electron donors are used [9].

On the other hand it was noted that the geometrical structure alone does not determine the reduction pathway [3].

The opinion was recently expressed [3] that "it is difficult to predict the reduction pathway of dioxygen, based only on the bonding of the molecule to the catalyst".

It is thus necessary to collect further evidence for the major factors influencing the pathway and final product of the catalyzed reduction of molecular oxygen. One strategy consists of the investigation of structurally related models where controlled changes in the electronic and steric effects could be introduced.

2. THE SERIES OF COBALT CHELATES DERIVED FROM COBALOXIME

Cobalt complexes, which were proposed as models of Vitamin B12 group complexes [10], have been investigated in our laboratory for a long time, mainly using electrochemical techniques, with the aim to point out the relevance of the relative stability of the oxidation states Co(III), Co(II), Co(I) and the chemical reactions coupled with the electron transfer, in the mechanism of formation and cleavage of the metal - carbon bond [11]. We further studied the influence of structural changes in the equatorial ligand, and proposed a series of strictly related complexes obtained by substituting the O-H-O bridges of the parent cobaloxime with $-(CH_2)_3-$ and/or $-BF_2-$ [12]: Fig. 1 .

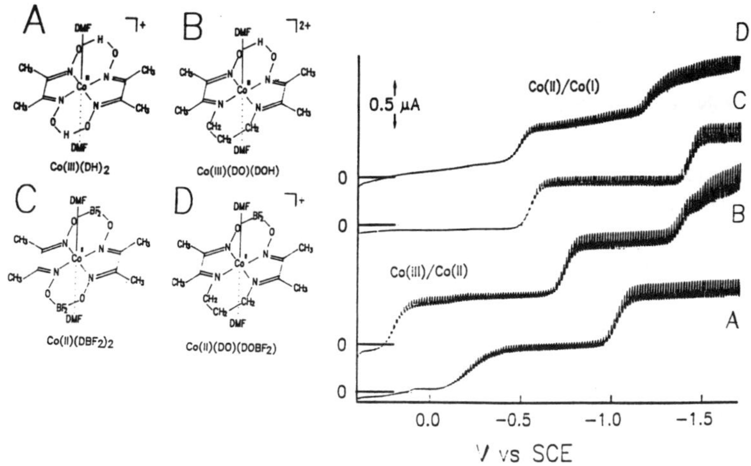

Figure 1. Schematic structures and polarograms in DMF under N_2 of the Co complexes.

More recently we have been interested in oxygen transport and activation properties of these cobalt chelates [12,13].

The changes in the structure of the equatorial ligand turned out to be very useful to probe the influence of equatorial ligand in tuning the reactivity of the central cobalt atom.

This is clearly shown by the electrochemical behaviour in Dimethyl Formamide (DMF) and Tetraethylammonium perchlorate (TEAP) as supporting electrolite. Polarograms of the four complexes show a shift of the Co(II)/Co(I) wave towards more positive values with increasing electron withdrawing power of the equatorial ligand.

The most striking effect is that of the -BF_2- groups, which strongly increase the stability of the Co(II) oxidation state. In fact both complexes C and D are stable in air as Co(II) complexes and their polarograms show the monoelectronic cathodic wave corresponding to a reversible Co(II)/Co(I) electron transfer and (on Pt electrode) the monoelectronic, quasi reversible anodic wave corresponding to the Co(II)/Co(III) electron transfer.

The shift of the $E_{1/2}$ of the Co(III)/Co(II) couple, due to the substitution of the O-H-O bridge with the -BF_2- group, amount to about 475 mV from complex A to complex C and by about 170 mV from complex B to complex D , while the shift due to the substitution of the O-H-O bridge with the -$(CH_2)_3$- group, which increases the charge of the cation, amount to about 300 mV in chelate A relative to chelate B. Finally the substitution of -BF_2- with -$(CH_2)_3$- while increasing the charge of the cation, decreases the withdrawing power of the equatorial ligand. As a result of these conflicting effects the reduction potential is not much changed.

On the other hand, with a less electron withdrawing equatorial ligand, cobaloxime(II) (complex A) is easily oxidized to a Co(III) species and could not be prepared in a pure Co(II) oxidation state by controlled electrolysis at the potential of the Co(III)/Co(II) reduction. This result is consistent with qualitative chemical behaviour: preparation by the Schrauzer method gives a high yield of a solid product but a rapidly decomposing Co(II) solution even when oxygen is carefully excluded, while chelate B, in similar experimental conditions, is much more stable. The formation of the Co(II) species of complex B by controlled potential reduction can be monitored by the increase in absorbance in the visible spectral region at 522 nm. A slow chemical reaction following the electron transfer can be also monitored by spectrophotometry [13a].

3. THE INTERACTION WITH MOLECULAR OXYGEN

The reactivity toward dioxygen of the above series of cobalt chelates shows a strict correlation between the characteristics and extent of metal - O_2 interaction and the oxidation state of the chelate.

Complexes A and B, which are air stable in the Co(III) oxidation state, interacts with dioxygen only when they are reduced to the corresponding formally Co(II) species.

In fact, the kinetics of oxygenation of cobaloxime(II), complex A, could be investigated only in solutions prepared by mixing cobalt perchlorate and potassium dimethylglyoximate in methanol solution in the presence of dioxygen, with stopped flow technique. The formation of a transient μ-peroxo dicobaloxime(III) followed by a disproportionation to a μ-superoxo cobaloxime(III) and solvolytic decomposition to cobaloxime(III) was proposed [14].

In the case of complex B more direct proof could be obtained for the formation of an adduct with molecular oxygen, thanks to the higher stability of the electrogenerated Co(II) oxidation state for this chelate. Several pieces of evidence can be offered: polarography and cyclic voltammetry (CV) of the Co(III)/Co(II) electrode process for the complex in the presence of dissolved oxygen clearly point to a chemical reaction following the electron transfer. The $E_{1/2}$ potential of the Co(III)/Co(II) process is shifted towards less negative values and the cyclic voltammetry confirms, from the ratio of the cathodic to anodic peak intensity (ipc/ipa), that the Co(III)/Co(II) electron transfer is followed by a chemical reaction [12,13].

The influence of dissolved dioxygen on the Co(III)/Co(II) process is confirmed and the regeneration of Co(III) species by the reaction of Co(II) species during the controlled potential reduction, is suggested by the steady increase of charge Q passed, even when the theoretical value for the monoelectronic process is reached.

Moreover, *a new polarographic wave* is observed for both the complexes A and B at potentials more negative than that of the Co(III)/Co(II) process (-0.277 V vs SCE) but less negative than that of the O_2/O_2^- reduction in the same experimental conditions. In the case of complex B the wave is reversible up to p_{O2} = 100 Torr and is associated to the reduction of the dioxygen adduct. The corresponding CV peaks due to the reduction of the dioxygen adducts are shown in Fig.2.

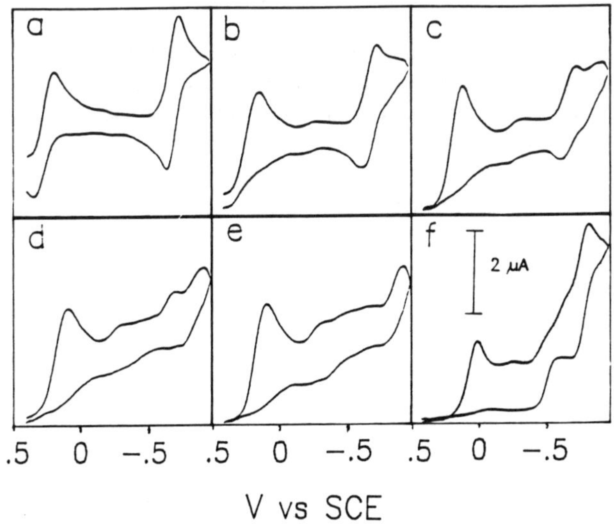

Figure 2. CV of complex B in DMF and TEAP 0.1 M at 100 mV/s on Hg electrode under; a) pure N_2 b) p_{O2} = 2 Torr ; c) p_{O2} = 5 Torr ; d) p_{O2} = 15 Torr ; e) p_{O2} = 25 Torr ; f) p_{O2} = 80 Torr.

CV confirms this finding and suggests, in addition, that the reduction of the dioxygen adduct is chemically irreversible.

The log plot of the ratio of the height of the new wave to that of the Co(II)/Co(I) reduction wave in the absence of dissolved oxygen, is linearly related to the p_{O2} , showing that the dioxygen adduct has a Co/O_2 = 1/1 stoichiometric ratio.

If it is assumed that the coordinated dioxygen is the active site in the reduction of the adducts of chelates A and B, the shift of the potentials of the corresponding polarographic wave (350 and 560 mV, respectively, relative to the reversible potential of the couple O_2/O_2^-) amount to a remarcable activation of the reduction process.

The results obtained with complex B lead to the proposal of the following mechanism of the electrode reaction

$[Co(III)]^{2+}$ + e <---> $[Co(II)]^+$ (1

$[Co(II)]^+$ + O_2 <---> $[Co\text{-}O_2]^+$ (2

$[Co\text{-}O_2]^+$ + e <---> $[Co\text{-}O_2]$ (3

$[Co\text{-}O_2]$ -----> products (4

The formation of the adduct is too fast to be investigated even by stopped flow techniques but the reversible changes in the UV - Visible spectrum can be observed by bubbling dioxygen or inert gas through the solution of the chelate B previously reduced to the Co(II) oxidation state, (Fig. 3) or by spectroelectrochemistry during the formation of the adduct from the electrogenerated Co(II) species by controlled potential reduction in the presence of dissolved oxygen [13] .

In both cases the results refer to the equilibrium concentration of the adduct in the bulk of the solution from electrogenerated B in the Co(II) oxidation state and molecular.

oxygen. The equilibrium constant K_{O_2} = 125 - 130 M^{-1} could be calculated, in complete agreement between the two methods.

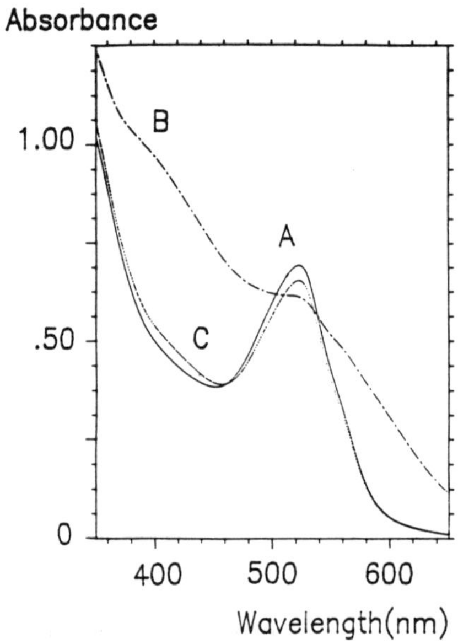

Figure 3. Spectra of complex B in DMF: A) after reduction to Co(II) under N_2; B) after bubbling O_2; C) after bubbling N_2 again.

On the other hand the formation constant was calculated also for the reaction of dissolved oxygen with chelate B in the Co(II) oxidation state, generated *at the electrode surface* at the potential corresponding to the Co(III)/Co(II) reduction. A substantially higher formation constant ($K_{O2} = 7900\ M^{-1}$) was obtained [13].

Contrasting with the high reactivity of complexes A and B in Co(II) oxidation state towards O_2, the complexes C and D, though being formally Co(II) species, are completely inert in air and under pure O_2 both in the solid state, in water or in non aqueous (e.g. DMF solution). The electronic spectra are insensitive to the presence of dissolved O_2 up to $p_{O2} = 760$ torr.

Notwithstanding the lack of evidence for the formation of an adduct with dissolved O_2 in the bulk from spectral data and other conventional analytical tecniques, the electrochemical investigations of complex C reveals a behaviour similar to that of complex B. The reversible formation of new polarographic waves and corresponding CV peaks is observed. By increasing p_{O2}, kinetically controlled waves develops in the polarograms at potentials slightly less negative than that of the reversible Co(II)/Co(I) electrode process.

For both the complexes C and D a new polarographic wave and CV peaks is assigned to the electrochemical reduction of an interaction product with dioxygen. The absence of the anodic counterpart in the CV confirms that these electrode processes are chemically irreversible. Fig.4

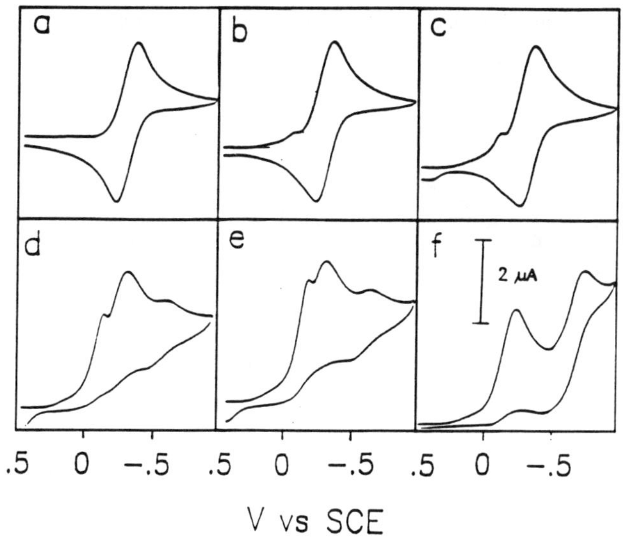

Figure 4. CV of complex C in DMF and TEAP 0.1 M at 100 mV/s on Hg electrode under; a) pure N_2 b) $p_{O2} = 4$ Torr ; c) $p_{O2} = 8$ Torr ; d) $p_{O2} = 20$ Torr ; e) $p_{O2} = 30$ Torr ; f) $p_{O2} = 70$ Torr.

If the peak current in the CV of complex C is used in the Hill's equation for the determination of the oxigenation ratio, the result (K_{O_2} = 7900 M^{-1}) compares well with that obained for the oxygenation ratio of chelate B at the electrode surface.
Two mechanism could be considered for the formation of the adduct:

a) $O_2 + e$ <----> O_2^- (5

$O_2^- + [Co(II)L]^{n+}$ <----> $[Co(II)L\text{-}O_2]^{(n-1)+}$ (6

b) $O_2 + [Co(II)L]^{n+}$ <----> $[Co(II)L\text{-}O2]^{n+}$ (7

$[Co(II)L\text{-}O_2]^{n+} + e$ <----> $[Co(II)L\text{-}O_2]^{(n-1)+}$ (8

If the difference between the the cathodic peak potentials for the O_2 reduction in the presence and absence of the cobalt chelate is considered (see, for a similar problem with a tetraaza-macrocycle, the most reasonable choice is the reaction sequence (7-8.
Thus the new wave observed in the polarograms of chelate C, in the presence of dioxygen is assigned to reaction (8.

4. ELECTROCATALYSIS OF DIOXYGEN REDUCTION

The electrocatalytic reduction of molecular oxygen can be shown by controlled potential reduction of the cobalt chelate in the presence of dissolved oxygen and a suitable acceptor of O_2^- like CF_3COOH.
With chelate B the current corresponding to the Co(III)/Co(II) reduction in aprotic solvents, and in the presence of a protogenic substrate, linearly increases by increasing p_{O_2}, and the concentration of the chelate. (Fig. 5 a and b)

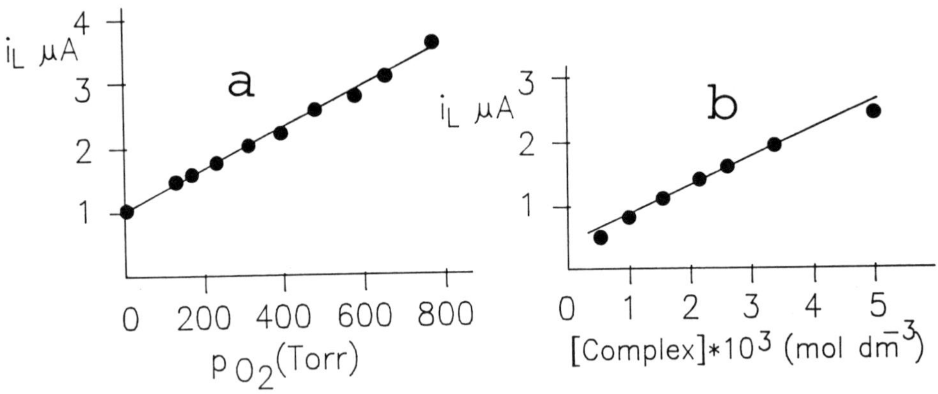

Figure 5. Catalytic current i_L for complex B in DMF + TEAP 0.1 M; a) as function of p_{O_2} with constant concentrations of B and CF_3COOH; b) as function of concentration of B with constant concentrations of O_2 and CF_3COOH.

The regeneration of oxidized species of the chelate is documented by the steady state current, still recorded at the potential of the Co(III)/Co(II) reduction when the stoichiometric amount of Faradays is passed.

Hydrogen peroxide is formed, with 100% yeld in terms of coulombs passed per mol of hydrogen peroxide, in the first turnovers. (Fig.6A ,B)

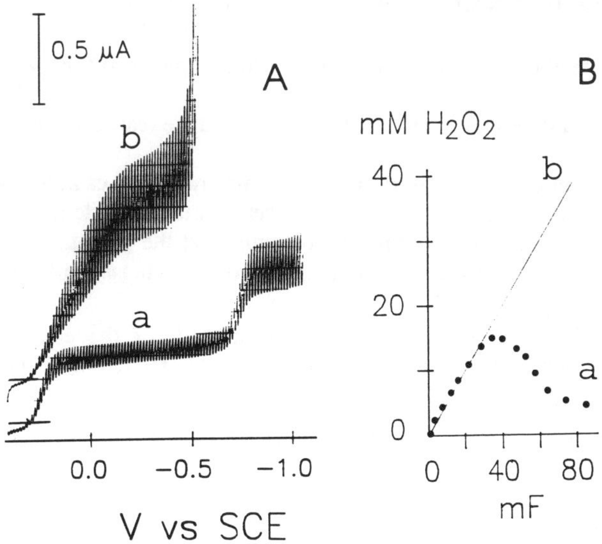

Figure 6. A) Polarography of complex B in DMF: a) under N_2; b) under O_2 and in the presence of CF_3COOH: B) a) Coulombic yield in H_2O_2 electrosynthesis during controlled potential reduction at 0.0 V for complex B; b) theorectical yield.

The formation, in quantitative yield, of H_2O_2 at the potential of the Co(III)/Co(II) redox couple, can be explained with the following overall electrode process:

$[Co(III)]^{n+} + e \leftarrow\!\!\!-\!\!\!-\!\!\!\rightarrow [Co(II)]^{(n-1)+}$ (1

$[Co(II)]^{(n-1)+} + O_2 \leftarrow\!\!\!-\!\!\!-\!\!\!\rightarrow [Co\text{-}O_2]^{(n-1)+}$ (2

$[Co\text{-}O_2]^{(n-1)+} + H^+ \dashrightarrow [Co(III)]^{n+} + \cdot OOH$ (5

$2 \cdot OOH \dashrightarrow O_2 + H_2O_2$

Electrocatalytic reduction of O_2 takes place also at the Co(II) level, i.e. at the potential of the reduction of the $[Co(II)\text{-}O_2]^{(n-1)+}$, intermediate between the Co(III)/Co(II) and Co(II)/Co(I) redox steps.

This species is not stable as is shown by the lack of the anodic counterpart in the CV; in the presence of protons the adduct can be protonated:

$$[Co(II)\text{-}O_2]^{(n-1)+} + H^+ \dashrightarrow [Co\text{-}OOH]^{n+} \qquad (6$$

and reduced at less negative potential relative to the unprotonated form:

$$[Co\text{-}OOH]^{n+} + e \dashrightarrow [Co(II)]^{(n-1)+} + \cdot OOH \qquad (7$$

The shift of the reduction potential of the protonated relative to unprotonated form amounts to about 400-500 mV.

The electrocatalytic process at this potential can be schematized with the reaction sequence (1-2-6-7)

The above proposed mechanism can be invoked also to explain the redox catalysis which is observed in the interaction of chelate C with molecular oxygen at the electrode surface, when controlled potential reduction is carried out in dimethylformamide, at the potential of the CV peak assigned to the reduction of dioxygen in the presence of the chelate and of protons (CF_3COOH).

On the other hand, contrasting with the results above mentioned for the electrocatalytic reduction of O_2 in the same experimental condition with chelate B, the reduction, in the case of chelate C, involves *four electrons per mole of O_2 and produce H_2O* in 100% yield. (Fig.7A,B). The stoichiometry of the process is shown also by the manometric measurement of absorbed O_2 vs Faradays.

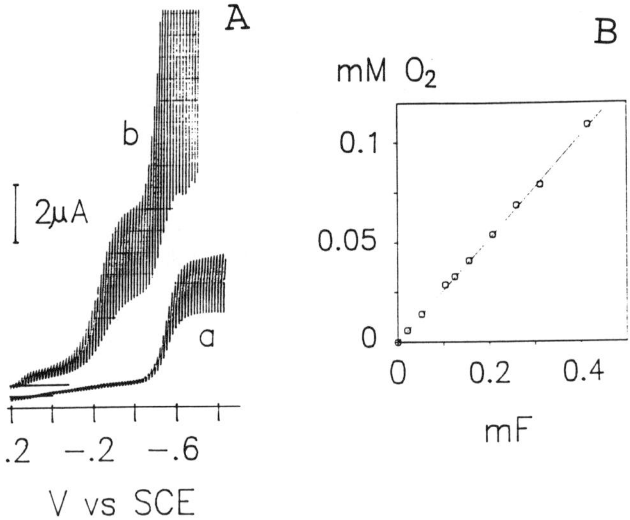

Figure 7. A) Polarography of complex C in DMF: a) under N_2; b) under O_2 and in the presence of CF_3COOH: B) Experimental oxygen adsorption during electrolysis of C at -0.370 V in DMF and in the presence of complex C and CF_3COOH; the straight line corresponds to the theoretical yield of H_2O.

5. THE INITIATION OF A RADICAL MECHANISM

Manometric measurements of the oxygen absorption during the catalytic reduction of molecular oxygen in the presence of complex B, revealed that the above described electrocatalytic process can initiate an independent catalytic oxidation by a radical mechanism.

Figure 8 shows the absorption of oxygen vs time in an electrochemical cell containing a dimethylformamide solution of chelate B, supporting electrolyte, CF_3COOH, in equilibrium, under stirring, with O_2. In the experiment documented in Fig 8, for the first 30-40 minutes the absorption of O_2 is stoichiometrically consistent with the above mechanism for the controlled potential reduction in the correspondence of Co(III)/Co(II) redox process.

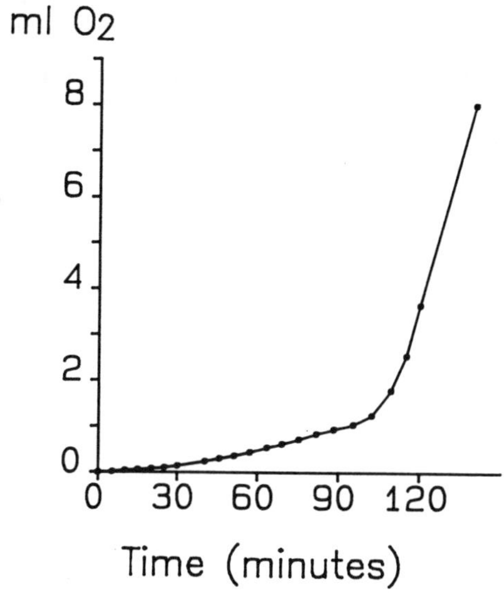

Figure 8. Oxygen absorbtion from a solution of complex B in DMF against time during and after controlled potential reduction in the presence of CF_3COOH

Interruption of electrolysis does not interrupt the oxygen absorption: on the contrary, the rate of absorption increases a bruptly and strongly. Formation of products coming from the autoxidation of dimethylformamide is observed while production of H_2O_2 ceased.

$$HCON(CH_3)_2 + 1/2\ O_2 \text{ ----> } HN(CH_3)_2 + CO_2$$

Evidence for a radicalic mechanism has been obtained by performing the same experiment in the presence of a radical trap (e.g. Ionol). No increase of oxygen absorption was observed after stopping the electrolysis. Fig. 9

Figure 9. Oxygen absorption:
a) as in figure 8
b) in the presence of Ionol

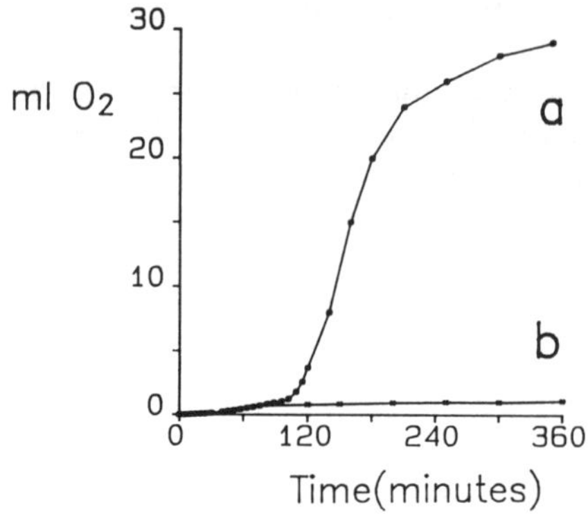

Moreover, the same dioxygen absorption rate was observed when, instead of carrying out the initial electrolysis, a small amount of propionic aldehyde was added.

Finally, the titration of active oxygen containing products shows that beside H_2O_2, hydroperoxidic species are present.

Consistent with all above evidence, a reaction scheme for the complex B, including a number of possibly competitive pathways Fig. 10

1) the starting electrochemical reduction step Co(III)/Co(II)
2) the oxygenation reaction
3) the direct reduction of the dioxygen adduct
4) the protonation of the adduct with formation of hydroperoxidic species
5) the regeneration of the Co(III) complex completing the catalytyc cycle and yielding H_2O_2
6) the hydrogen abstraction from an aldehyde (in the actual case the DMF solvent) either by the hydroperoxy species or by the dioxygen adduct, and the initiation of a radical mechanism (where the Co(III) and Co(II) species could be still intervene.

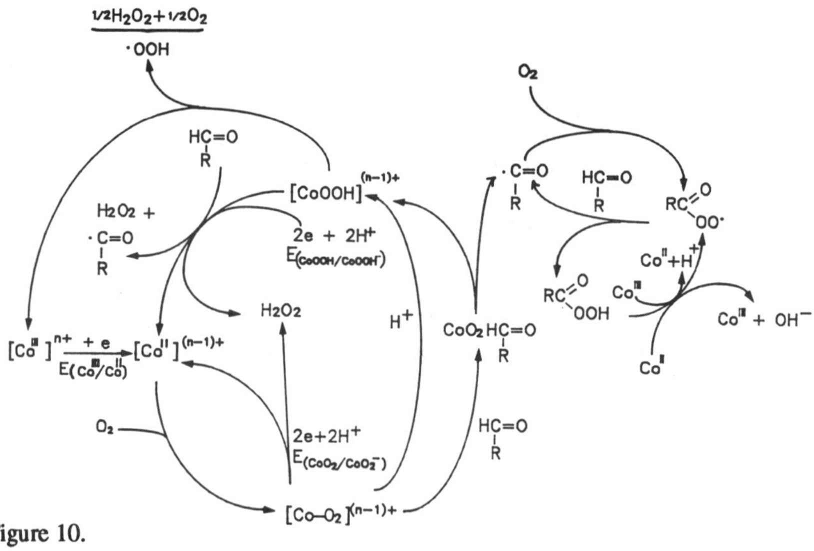

Figure 10.

CONCLUDING REMARKS

The above results show striking differences in the electrochemical behaviour and in the interaction with molecular oxygen, among a series of cobalt complexes whose structure, on the other hand, is strictly related.

The extent of electron withdrawing power of the equatorial ligand appears to dominate, as qualitatively expected, the relative stability of the Co(III)/ Co(II) oxidation states and the strong shifts of Co(III)/Co(II) and Co(II)/Co(I) redox potentials.

The electronic charge on central metal atom also controls the interaction of the chelates with molecular oxygen.

The Co(II) oxidation state of chelate A is unstable in air; that of chelate B as obtained either in the bulk of the solution by controlled potential reduction or at the electrode surface from the Co(III)/Co(II) electrode process, gives evidence for the formation of a dioxygen adduct. In the presence of protons the electrocatalyitic reduction ensues with formation of H_2O_2.

The decreasing charge on cobalt in the formally Co(II) species in the chelates C and D hinder the interaction with O_2 to such a point that they are completely inert in air and under pure oxygen. In spite of this, evidence for the interaction with O_2 was obtained for chelate C *at the surface of the electrode*

An additional distinction should be made on the basis of the oxygenation constants: the values are of the same order of magnitude for the Co(II) chelate B and for the chelate C *at the electrode surface*, but much higher than that found for the adduct formation from the Co(II) chelate B in the bulk of the solution.

An electron surface effect, such as that which was pointed out for the electrocatalytic reduction of O_2 in the presence of adsorbed vs dissolved porphyrins, should be invoked also in the present case.[16]

Reasonable mechanisms could be easily proposed for the redox catalysis in the reduction of molecular oxygen, involving either regeneration of the Co(III) species (i.e. at the potential level of the Co(III)/Co(II) redox step of chelate B) or regeneration of the Co(II) species (i.e. at the potential level of the Co(II)/Co(I) redox step for chelate B and chelate C)

Less easy is the rationalization of the different pathways of dioxygen reduction in the presence of protons (leading to H_2O_2 in the presence of chelate B and H_2O in the presence of chelate C). The four electron reduction could be related with the posible adsorbed state of the catalyst in the case of chelate C.

Finally, the initiation of a radical mechanism was pointed out. This finding supports mechanisms involving hydroperoxo radicals in some intermediate steps of the electrocatalytic oxidation. The autoxidation of DMF as a solvent and more generally the weight of the radical mechanism in the present type of electrochemical oxidation should be considered in future work.

REFERENCES

[1] P.George "The fitness of Oxygen" in "Oxidases and Related Redox Systems" T.E.King, H.S.Mason and M.Morrison Eds. Vol. 1 Pag.3 J.Wiley1965

[2] M.R.Tarasevich, A.Sadkowski and E.Yeager "Oxygen Electrochemistry" in "Comprehensive Treatise of Electrochemistry" B.E.Conway, J, O'M. Bockris, E.Yeager, S.U.M.Khan, and R.E. White Eds. Vol. 7 Pag. 301 Plenum press 1983

[3] A.Elzing, A.Van der Putten, W.Visscher and E.Barendrecht, Rec.Trav.Chim.P.B. 1990 109 31

[4] E.C.Niederhoffer, J.H.Timmons, and A.E.Martell, Chem.Rev. 1984 84 137

[5] A.Van Putten, A.Elzing, W.Visscher, and E.Barendrecht, J.Electroanal.Chem. 1987 221 95

[6] F.Beck, J.Applied Electrochem. 1977 7 239

[7] H.Y.Liu, M.J.Weaver, C.B. Wang, and C.K.Chang, J.Electroanal.Chem. 1983 145 439

[8] M.A. Elmorsi, M.Gabor and E.Barendrecht, J.Applied Electrochem. 1991 21 1005 and references therein.

[9] C.Shi, F.C.Anson, J.Am.Chem.Soc. 1991 113 9564 and references therein

[10] G.N.Schrauzer, Acc.Chem.Res. 1968 1 97

[11] A.Bigotto, G.Costa, G.Mestroni, G.Pellizer, A.Puxeddu, E.Reisenhofer, L.Stefani, and G.Tauzher, Inorg.Chim.Acta 1970 4 41

[12] a) G.Costa, A.Puxeddu, C.Tavagnacco, G.Balducci, and Rakesh Kumar, Gazz.Chim. Ital. 986 116 735

b) G.Costa, A.Puxeddu, C.Tavagnacco, Rakesh Kumar, and G.Balducci, Gazz.Chim. Ital. 1988 118 333

c) G.Costa, C.Tavagnacco, A.Puxeddu, G.Balducci, and Rakesh Kumar , J.Organomet.Chem. 1987 330 185

[13] a)G.Costa, C.Tavagnacco, G.Balducci, J.Electroanal.Chem. 1991 318 183

b) G.Costa, C.Tavagnacco, G.Balducci and Rakesh Kumar in "Dioxygen Activation and Homogeneous Catalytic Oxidation" Proc. of IV Int. Symp. on Dioxygen Actvation and Homogeneous Catalytic Oxidation ,L.I.Simandi Ed., Elsevier 1991 - Pag. 337

[14] L.I.Simandi, C.R.Savage, Z.A.Schelly and S.Nemeth, Inorg.Chem. 1982 21 2765

[15] W.S.Webley, R.R.Durand,Jr and F.C.Anson, J.Electroanal.Chem. 1987 229 273
221 95

[16] Y.O.Su, T.Kuwana, S.M.Chen , J.Electroanal.Chem. 1990 288 177

BEHAVIOUR OF THE IONOMER / METAL INTERFACE TOWARDS THE ELECTROREDUCTION OF OXYGEN

C. PALITEIRO and M. L. PEREIRA
Department of Chemistry
University of Coimbra
P-3049 Coimbra
Portugal

ABSTRACT. Experimental results on the electroreduction of O_2 on Nafion-coated Pt and Au electrodes in 0.5 M H_2SO_4 and 1 M NaOH are presented and analysed. On coated Pt, the amount of O_2 reduced becomes disturbed by a limitation of O_2 transport through the film and the current is decreased at practically all potentials. On coated Au, such transport limitation of O_2 also occurs; however, for small enough film thickness and whenever the Au surface is able to disproportionate H_2O_2 or HO_2^-, the film retards the diffusion of these species to the solution bulk, allows a higher amount of disproportionation to occur and, as a consequence, gives a higher current than that observed on a bare Au electrode.

1. INTRODUCTION

It was recently shown that the interaction of a thin film of Nafion with the metal electrode where it is deposited depends on the nature of the electrode and it was then suggested that this dependence arises because the interaction of Nafion with different metals occurs through different parts of the Nafion microstructure; such interaction apparently occurs with Pt through the ionic cluster region and with Au through the hydrophobic region [*1*].

When in contact with a O_2-containing aqueous solution a Nafion film extracts O_2 into its hydrophobic region [*2*]. Therefore, following the above suggestion, we would expect a film of Nafion to develop a higher availability of O_2 next to a gold surface than to a platinum surface and to influence more the electroreduction of O_2 on gold than on platinum.

In this context, we analysed the electroreduction of O_2 on Nafion-coated platinum and gold electrodes and the results obtained are presented and discussed in this paper.

2. EXPERIMENTAL

Nafion coatings were prepared by droping on the disc electrodes from a microsyringe an adequate volume of a 1.07 % (w/v) solution of Nafion 117 in water and ethanol, allowing the solvent to evaporate and drying the deposited film overnight. Film thickness was calculated assuming that films had cilindrical geometry and taking the density of wet Nafion films (1.58 [*3*]) as the value of film density.

Experiments were performed using standard electrochemical equipment. Potentials were measured against a saturated-calomel or a Hg/HgO/1M HO^- electrode but, hereafter, they are

A. J. L. Pombeiro and J. A. McCleverty (eds.),
Molecular Electrochemistry of Inorganic, Bioinorganic and Organomettallic Compounds, 405–410.

referred to the reversible hydrogen electrode (RHE).

3. RESULTS

3.1. PLATINUM ELECTRODES

From our study of the electroreduction of O_2 in 0.5 M H_2SO_4 on Pt electrodes coated with a Nafion film [*4*], we recall the following conclusions: i) the film does not alter the mechanism of the reduction reaction; ii) the film concentrates O_2 from the solution but, nevertheless, the current measured for the filmed electrodes only increases relatively to the uncoated electrodes near the onset of the reduction where electron transfer controls; once diffusion becomes important, the current decreases because, whatever the thickness, the transport of O_2 inside the film is slowed down; iii) the values found of the O_2 concentration, $C_{f,1}$, and the diffusion coeficient of O_2, $D_{f,1}$, inside the film, showed that films of low thickness (~0.5 μ) behave like Nafion membranes whereas thicker films behave like recast films.

In 1 M NaOH the current is decreased from the onset potential even by a 0.5 μ-thick Nafion film

3.2. GOLD ELECTRODES

Given the higher proportion of the hydrophobic region in the films of low thickness observed for Pt+Naf electrodes, we concentrated in the study of Au electrodes coated with films about 0.5 μ thick. Two surfaces of polycrystalline Au were examined: the normal pc-Au surface and a electrochemically-pretreated pc-Au surface which exhibits a (100)-type behaviour towards O_2 reduction and is therefore referred to hereafter as "100"-Au.

3.2.1. *pc-Au electrodes*

In 0.5 M H_2SO_4, the activity of pc-Au for O_2 reduction is low: the potential of the reduction onset is 400 mV more cathodic than that observed for Pt and only H_2O_2 is produced even at high cathodic potentials. Contrasting with the report of Yeager's group [*1*], we did not observe any significant change of the CV of pc-Au in the presence of a 0.3μ-thick film. Nevertheless, such a film decreases the current in all potential range explored (cf. Fig. 1). This decrease is certainly due to the transport limitation of O_2 within the film as observed for Pt+Naf electrodes.

The polarization curves shown in Fig. 2a) are typical of a pc-Au surface in 1M NaOH cycled up to 1.5 V [*5*]. In such conditions, i) from ~900 mV to ~300 mV, O_2 is reduced to HO_2^-, which then disproportionates (cf. Fig. 3a); the O_2 thus produced recycles; ii) below ~300 mV, HO_2^- is reduced to HO^- and the current raises in a second wave up to the $4e^-$ level; iii) at -200 mV, the current is purely diffusion-controlled (cf. Fig. 4a).

A Nafion film decreases the current at the second wave but increases the current at the first one (cf. Fig. 2b) and c)). The current decrease at the second wave is made evident by the Levich plots at -200mV (see Fig. 4) and is due to the decrease of the rate constant for the transport of O_2 in the film. To confirm that, we applied the method of Gough and Leypoldt [*6*] to estimate $C_{f,1}$, $D_{f,1}$ and the film permeability to O_2, $P_{f,1}$, which expresses the combined effects on the limiting current of the O_2 exchange between the solution and the Nafion film and the rate constant $(K_{D,f})_1$ for the transport of O_2 in the film. The results obtained are summarized in

Table 1. In the calculations we made $D_1 = 1.65 \times 10^{-5}$ $cm^2 s^{-1}$ [8], $C_1 = 8.4 \times 10^{-7}$ mol cm^{-3} [9], and $\nu = 1.0624 \times 10^{-2}$ $cm^2 s^{-1}$ [10]. The values of $P_{f,1}$, $C_{f,1}$ and $D_{f,1}$ obtained are comparable to those found [4] for Nafion films of similar thickness contacting a 0.5 M H_2SO_4 solution; this is what we expect if O_2 diffuses through the hydrophobic region of the Nafion microstructure.

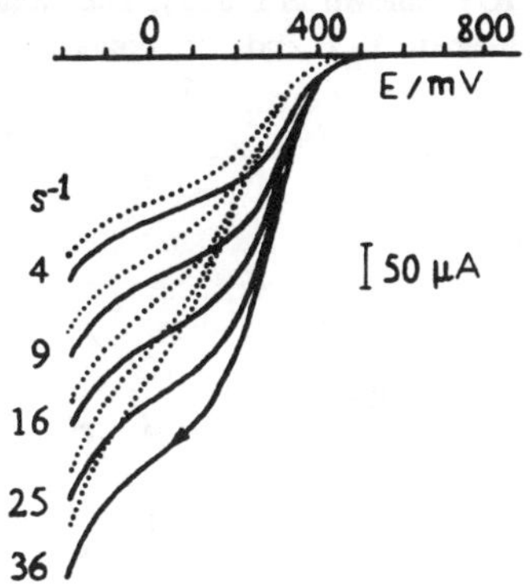

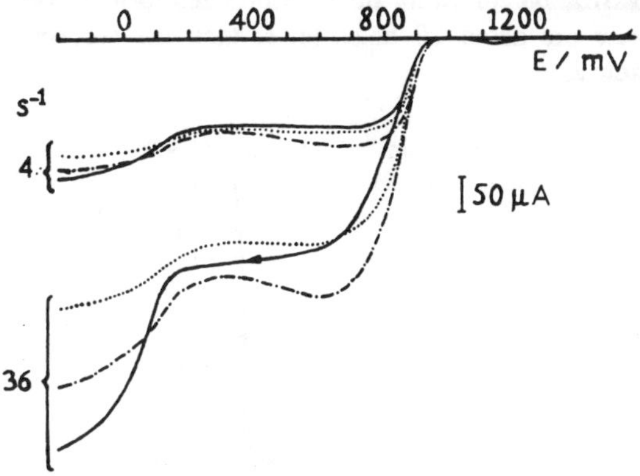

Fig. 1. Polarization curves recorded at 10 mV s^{-1} for O_2 reduction in O_2-saturated 0.5 M H_2SO_4 on a a) (—) pc-Au , b) (···) pc-Au + 0.3 µ Naf electrode.
Fig. 2. Polarization curves recorded at 10 mV s^{-1} for O_2 reduction in O_2-saturated 1 M NaOH on a a) (—) pc-Au , b) (-·-) pc-Au+0.5µ Naf and c) (...) pc-Au+2.5µ Naf electrode.

TABLE 1: Values of $P_{f,1}$, $C_{f,1}$, β_1 and $D_{f,1}$ for Nafion films in 1 M NaOH

d_f (µ)	$P_{f,1}$ (cm s^{-1} x 10^2)	$C_{f,1}$ (mol cm^{-3} x 10^6)	β_1	$D_{f,1}$ (cm^2 s^{-1} x 10^6)
0.5	6	9	11	0.3
2.5	2.4	6	7	0.8

From the results in Table 1 it is easy to verify that, from 6 s^{-1} for the 0.5µ–thick film and from 2 s^{-1} for the 2.5µ–thick film, the rate constant $K_{D,1}$ for the transport of O_2 to the solution becomes greater than $(K_{D,f})_1$; therefore, limitations in the transport of O_2 within the film make the current lower at the second wave.

These limitations also certainly occur at the first wave; nevertheless a current increase was actually observed in this potential range in the presence of Nafion films. We ascribe this increase to the increase of the amount of HO_2^- that disproportionates at the metal / ionomer interface and to the corresponding increase of the amount of O_2 recycled.

In the potential region of the first wave, the increase of the current above the 2e⁻ level observed on the pc-Au surface is the net result of a competition between the rate of HO_2^- disproportionation (measured by the rate constant K_4) and the rate of transport of HO_2^- to the solution bulk (measured by the rate constant $K_{D,2}$). On the uncoated pc-Au, disproportionation loses and, therefore, the current decreases towards the 2e⁻ level as the rotation rate increases (cf. Fig. 5a)). However, on a pc-Au electrode coated with a Nafion film disproportionation wins because, as we demonstrate below, the concentration of HO_2^- within the film, $C_{f,2}$, is much lower than the concentration C_2 of the same solution. As an increased amount of HO_2^- is allowed to disproportionate, the current increases in the 1st wave if the film thickness is low enough to disturb significantly the transport of O_2 to the electrode.

To estimate $P_{f,2}$, the film permeability to HO_2^-, we applied the above-mentioned method of Gough and Leypoldt to the analysis of the influence of Nafion films on the diffusion current of polarization curves for the reduction and for the oxidation of HO_2^- (shown in Fig. 3). The values found are about 20 times lower than the corresponding values of $P_{f,1}$ and are collected in Table 2.

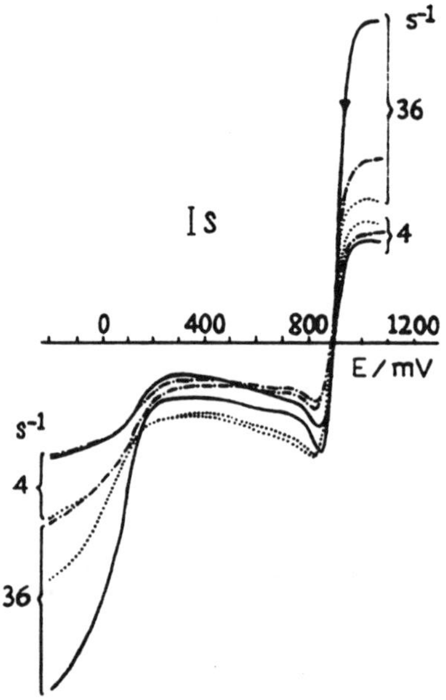

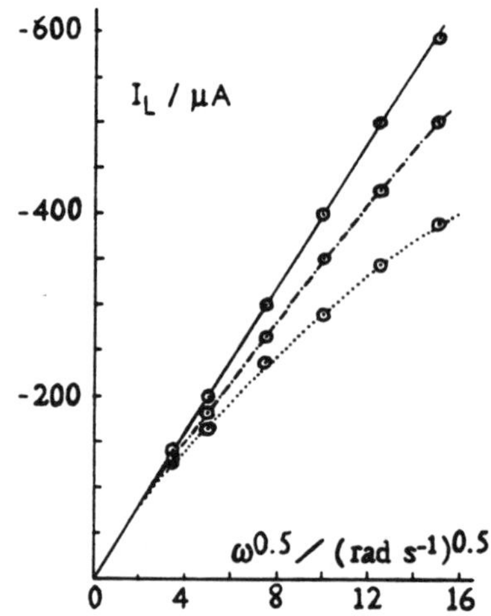

Fig. 3. Polarization curves recorded at 10 mV s^{-1} in argon-saturated 1 M NaOH + 10^{-3} M HO_2^- on a) (—) pc-Au, S = 20 µA, b) (--) pc-Au+0.5µ Naf, S = 10 µA and (···) pc-Au+2.5µ Naf, S = 5 µA.

Fig. 4. Levich plots obtained at -200 mV from the polarization curves of Fig. 2. a) (—) pc-Au, b) (--) pc-Au+0.5µ Naf, c) (···) pc-Au+2.5µ Naf.

TABLE 2: Values of $P_{f,2}$, $C_{f,2}$, β_2 and $D_{f,2}$ for Nafion films in 1 M NaOH

d_f (µ)	$P_{f,2}$ (cm s^{-1} x 10^2)	$C_{f,2}$ (mol cm^{-3} x 10^8)	β_2	$D_{f,2}$ ($cm^2 s^{-1}$ x 10^6)
0.5	0.3	1.8	0.018	8.5
2.5	0.16	-	-	-

3.2.2. *"100"- Au electrodes*

The behaviour of Nafion-coated "100"-Au electrodes was studied only in 1 M NaOH. Fig. 6 a)

shows typical [11] polarization curves for O_2 reduction on the "100"-Au surface in that electrolyte. The observed increase in current in the low overpotential region relatively to the current recorded for a pc-Au surface is related to the presence of partially-discharged adsorbed HO^- in that potential region and results from the heterogeneous disproportionation of HO_2^- catalysed by the adsorbed HO^- followed by the recycling of the O_2 produced. Therefore, the current first raises to the 4e$^-$ level and decreases afterwards as HO^- desorbs.

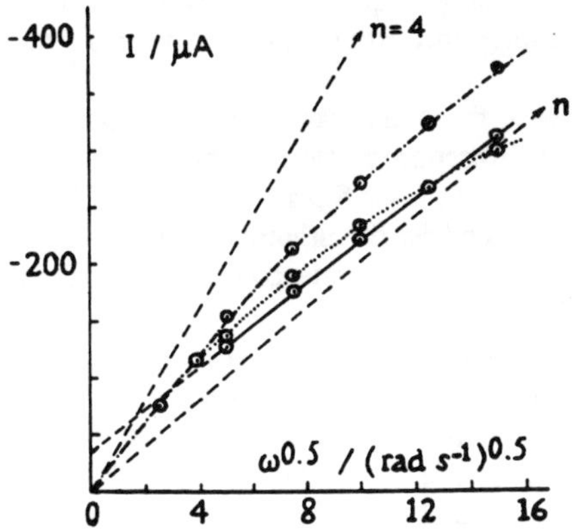

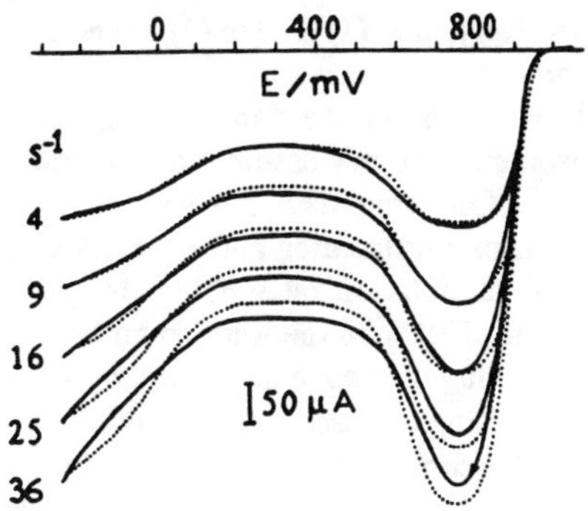

Fig. 5. Levich plots obtained at 600 mV from the polarization curves of Fig. 2. a) (—) pc-Au, b) (-·-) pc-Au+0.5μ Naf, c) (...) pc-Au+2.5μ Naf.
Fig. 6. Polarization curves recorded at 10 mV s^{-1} for O_2 reduction in O_2-saturated 1M NaOH on a a) (—) "100"-Au and b) (···) "100"-Au+0.3μ Naf electrode.

TABLE 3: I_L experimental values vs. calculated I_L values for 1st order and 2nd order disproportionation.

rotation rate (s^{-1})	Au (750 mv)			Au + Naf (750 mv)		
	Exp.	1st order (a)	2nd order (b)	Exp.	1st order (a)	2nd order (b)
1	-	-	-	109	108	106
2	-	-	-	149	150	148
3	-	-	-	180	182	181
4	208	208	207	207	208	207
9	306	302	306	302	302	306
16	388	390	401	393	390	401
25	465	474	495	480	474	495
36	528	555	588	552	555	588

(a) $K_4 = 1.5 \times 10^{-2}$ cm s^{-1}; (b) $K_4 = 7 \times 10^4$ cm^4 mol^{-1} s^{-1}

As the mechanism of HO_2^- disproportionation, we accept that suggested by McIntyre [12] described by the following equations where Au means the Au surface:

$Au + HO^- \longrightarrow AuOH + e^-$
$AuOH + HO_2^- \longrightarrow AuO_2H + HO^-$ (1)
$AuO_2H + HO_2^- \longrightarrow O_2 + AuOH + HO^-$ (2)

The rate constant K_4 for the disproportionation can be determined as the value that fits the equations [*12*]

$$I_L = 2\,F\,A\,Z_1\,C_1\,\varpi^{0.5} + 2\,F\,A\,Z_1 C_1\,\varpi^{0.5}[(K_4/(Z_2\,\varpi^{0.5} + K_4)] \quad (3)$$

$$I_L = 4\,F\,A\,Z_1\,C_1\,\varpi^{0.5} - F\,A\,Z_2^2\,\varpi\;\;[((1+ 4Z_1C_1K_4\,/\,Z_2^2\,\varpi^{0.5})^{0.5})-1] \quad (4)$$

where $Z = 0.62\;\;D^{2/3}\;\nu^{-1/6}$ and which describe the Levich plot at 750 mV,, when the rds is, respectively, step (1) or step (2), therefore, when the disproportionation is, respectively, 1st order or 2nd order.

Table 3 shows the values of K_4 that gave the best fit to the experimental values of I_L measured at 750 mV assumming 1st order or 2nd order disproportionation. In the calculation we used the following data: $F = 96487$ C mol^{-1}, $A = 0.162$ cm^2 ; $D_2 = 8.5 \times 10^{-6}$ cm^2 s^{-1} [*13*]. 1st order disproportionation and $K_4 = 1.5 \times 10^{-2}$ cms^{-1} gave the best fit, although a positive shift is noticed for rotation rates above 16 s^{-1} certainly because then the rate constant $K_{D,2}$ for the transport of HO_2^- to the solution bulk becomes significant relatively to K_4.

Following the above discussion of the data obtained on bare and coated pc-Au electrodes, it is not surprising to observe that, relatively to the bare "100"-Au electrode, a 0.3μ-thick Nafion film raises the current in the low overpotential region (cf. Fig. 6b)). As Table 3 shows, 1st order disproportionation with *the same* K_4 value obtained for the bare "100"-pc electrode (1.5×10^{-2} cm s^{-1}) still gives the best fit to the experimental I_L values and no shift is noticed at the higher rotation rates explored.

It is interesting to note that this best fit was obtained using the value of D_2 as the value of the diffusion coefficient of HO_2^-, $D_{f,2}$, within the film. This suggests that, relatively to the transport in solution, the transport of HO_2^- within the film is not significantly altered. The above-calculated value of $P_{f,2}$ for a 0.5μ-thick film can then be used to estimate the HO_2^- concentration in Nafion films of low thickness in contact with a 10^{-3} M HO_2^- alkaline solution. We found 2×10^{-8} mol cm^{-3}, to which corresponds a partition coefficient $\beta_2 = 0.02$. It is this low β_2 that retards the diffusion of HO_2^- to the solution bulk allowing a greater amount of that ion to disproportionate at higher rotation rates.

4. REFERENCES

1. Chu, D., Tryk, D., Gervasio, D. and Yeager, E. B., *J. Electroanal. Chem.*, (1989) 272, 277.
2. Gottesfeld, S., Raistrick, I. R. and Srinivasan, S., *J. Electrochem. Soc.*, (1987) 134, 1455.
3. Lawson, D. R., Whiteley, L. D. and Martin, C. R., *J. Electrochem. Soc.*, (1988) 135, 2247.
4. Paliteiro, C., Pereira, M. L. and Jorge, A. M., *Port. Electrochim. Acta*, (1991) 9, in press.
5. Paliteiro, C., Hamnett, A. and Goodenough, J. B., *J. Electroanal. Chem.*, (1987) 234, 193.
6. Gough, D. A. and Leypoldt, J. K., *Anal. Chem.*, (1979) 54, 439.
7. Nicholson, R. S. and Shain, I., *Anal. Chem.*, (1964) 36, 706.
8. Gubbins, K. E. and Walker, R. D., *J. Electrochem. Soc.*, (1965) 112, 459.
9. Davies, R. E., Horvath , G. L. and Tobias, C. W. *Electrochim. Acta*, (1967) 12, 287.
10. Washburn, E. W. et al., International Critical Tables 5, McGraw-Hill, New York, 1929.
11. Paliteiro, C., paper in preparation.
12. McIntyre, J. D., *J. Phys. Chem.*, (1967) 71, 1196.

REDOX PROPERTIES OF THE AMAVADINE MODELS $[V(HIDA)_2]^{2-}$ AND $[V(HIDPA)_2]^{2-}$ AND THEIR ELECTROINDUCED REACTIVITY TOWARDS ACTIVATED-THIOLS AND -PHENOLS

JOÃO J. R. FRAÚSTO DA SILVA, M. FÁTIMA C. GUEDES DA SILVA, JOSÉ ARMANDO L. DA SILVA and ARMANDO J. L. POMBEIRO
Centro de Química Estrutural, Complexo I,
Instituto Superior Técnico,
Av. Rovisco Pais 1096 Lisboa codex, Portugal

ABSTRACT. The Amavadine models $[Bu_4N]_2[V(HIDA)_2]$ ($HIDA^{(3-)}$ = N-hydroxy-iminodiacetate) and $[Bu_4N]_2[V(HIDPA)_2]$ ($HIDPA^{(3-)}$ = N-hydroxy-α,α'-iminopropionate), in water or in aprotic medium (NCMe, DMSO, DMF or CH_2Cl_2), undergo, at a Pt or vitreous C electrode, a single-electron reversible oxidation at an oxidation potential which is considerably lower in the organic solvents ($E^{ox}_{1/2}$ in the range 0.1-0.2 V *versus* SCE) than in the aqueous medium ($E^{ox}_{1/2}$ = 0.52 or 0.49 V *versus* SCE, respectively). When the complexes are generated *in situ* in the organic solvent (DMSO or DMF), a chemical oxidation occurs readily to give a mixture of the V(IV) and V(V) species, with conceivable involvement of the ligand. These complexes are shown to act as electron transfer mediators for the electrocatalytic oxidation of activated thiols (*e.g.*, cysteine or mercaptoacetic acid) or phenols (*e.g.*, cathecol or gallic acid), the former substrates undergoing a single-electron oxidation with proton evolution and S-S coupling (formation of cystine from cysteine), whereas a distinct two-electron process is involved in the latter case (oxidation of cathecol).

1. INTRODUCTION

Although the biological role of Amavadine, a natural vanadium complex present in the mushrooms *Amanita muscaria*, has not yet been established beyond any doubt, there is considerable circumstancial evidence that suggests it could be involved in some kind of electron transfer processes [1]. Therefore, a growing interest is developing in the investigation of the redox properties of inorganic models of Amavadine [2,3] which, itself, appears to contain the octacoordinated vanadium(IV) centre $[V(HIDPA)_2]^{2-}$ ($HIDPA^{(3-)}$ = N-hydroxy-α,α'-iminodipropionate) [4] rather than the previously proposed [5] five-coordinated oxovanadium(IV) species $[VO(HIDPA)_2]$ with bidentade

A. J. L. Pombeiro and J. A. McCleverty (eds.),
Molecular Electrochemistry of Inorganic, Bioinorganic and Organometallic Compounds, 411–415.

ligands.

In this work we describe our electrochemical studies, by cyclic voltammetry and controlled potential electrolysis, of the redox behaviour of $[BuN]_2[VL_2]$ (1, L = $HIDPA^{(3-)}$ or $HIDA^{(3-)}$, where $HIDA^{(3-)}$ = N-hydroxy-iminodiacetate) in water and in aprotic media.

2. RESULTS and DISCUSSION

Both complexes $[Bu_4N]_2[VL_2]$ (1, L = HIDPA or HIDA), in water, undergo, by cyclic voltammetry (CV) and controlled potential electrolysis (CPE) at a Pt or vitreous C electrode, a single-electron V(IV)/V(V) reversible oxidation at $E^{ox}_{1/2}$ = 0.49 or 0.52 V *versus* SCE, respectively, the former corresponding to a less anodic value in agreement with the expected stronger electron releasing ability of HIDPA relatively to HIDA. In certain organic solvents (NCMe, DMSO, DMF or CH_2Cl_2) such reversible anodic oxidation occurs at a lower potential [*e.g.*, 0.06 or 0.17 V for (1, L = HIDPA or HIDA), respectively, in DMSO] (Fig. 1a) and, furthermore, when the complexes are prepared *in situ* in those solvents (DMSO or DMF)— by reacting VO^{2+} with the appropriate ligand— a chemical oxidation occurs spontaneously to the corresponding V(V) complexes (Fig. 1b). Moreover, upon anodic CPE of the V(IV) species in DMSO, apart from the expected formation of the corresponding V(V) product, a novel V(IV) species— conceivably with the VO^{2+} group, as suggested by EPR— has been detected.

These and other observations indicate the non-innocent roles played by the ligand/solvent system which, depending on the experimental conditions and on the vanadium source, can act as a deoxygenating (of VO^{2+}), an oxidizing [of VCl_3 or V(IV)] or a reducing [of V(V) species] agent, thus inducing the interconversion of V(IV) and V(V) complexes and conversion into such compounds of lower oxidation state V species.

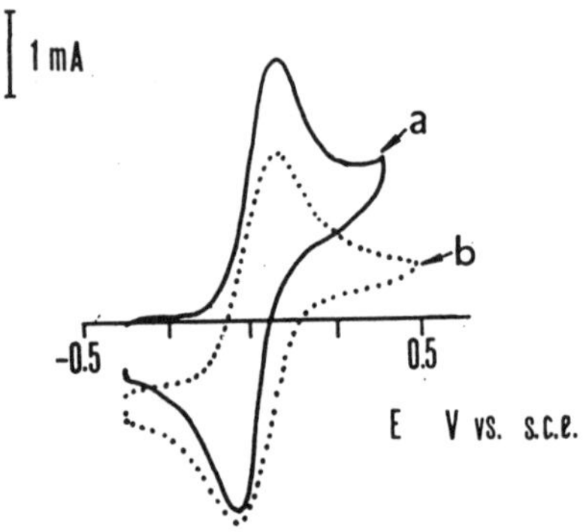

Figure 1. Cyclic voltammograms, at a Pt electrode, in DMSO/0.2M $[Bu_4N][BF_4]$, of $[V(HIDPA)_2]^{2-}$ (previously synthesized) (a), and of the species generated *in situ* from VO^{2+} + HIDPA (1:2) (b). Scan rate 100 mV/s.

The observed ready V(IV)/V(V) interconversion is involved in the electrocatalytic oxidation of thiols mediated by the $[VL_2]^{2-/-}$ systems of our study and may be of biological significance in the natural system.

In fact, both cysteine ($HSCH_2CHNH_2COOH$) and mercaptoacetic acid ($HSCH_2CH_2COOH$) undergo a single-electron electrocatalytic oxidation at the oxidation potential of each of our vanadium(IV) complexes, as indicated by Fig.2 for the oxidation of mercaptoacetic acid mediated by the $[V(HIDA)]^{2-/-}$ system.

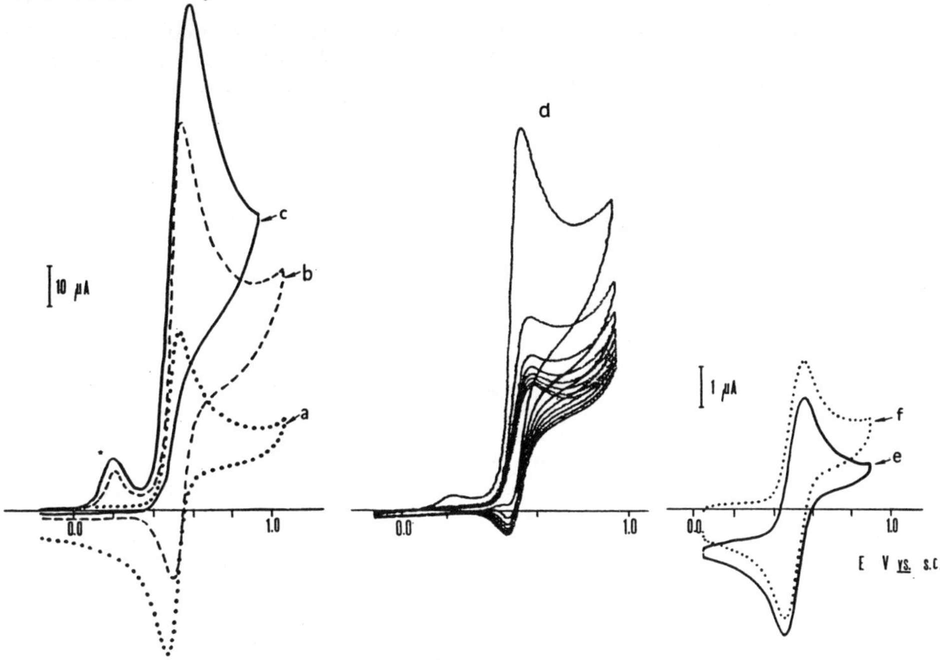

Figure 2. Cyclic voltammograms, at a glassy C electrode, of an aqueous 0.1M KCl solution of $[V(HIDA)_2]^{2-}$ (with mercaptoacetic acid).

Scan rate = 200 mV/s

(a)- Initial complex without thiol

(b),(c)- Solution (a) after addition of thiol in molar ratios 1:1 and 1:2, relative to the V complex (*- unassigned wave)

(d)- Repeated continuous cyclic scans of a 1:8 (V complex:thiol) solution

(e)- 1:8 (V complex:thiol) solution after exhaustive anodic CPE (*ca.* 8 electrons / V involved)

(f)- Regenerated V(IV) complex upon cathodic single-electron CPE of the solution (e)

The catalytic nature of the anodic wave is evident from the enhancement of the anodic peak current upon addition of increasing amounts of the thiol (a, b and c) for which, in the absence of the V system, no direct anodic oxidation has been detected apart from an unidentified irreproducible and weak wave (possibly associated with an adsorption phenomenon). Moreover, repeated continuous cyclic scans (Fig.2 d) regenerate the initial anodic wave for the V system, observed without addition of the thiol, as a result of the consumption of the latter at the vicinity of the electrode by the electrocatalytic V(V)/V(IV) redox pair.

In addition, exhaustive CPE of a solution of the V(IV) complex with the thiol in a molar ratio 1:8 requires the overall transfer of 9 (1+8) electrons per vanadium and the resulting CV exhibits the expected cathodic reversible wave for the final V(V) product (Fig.2 e), which, upon a subsequent cathodic *ca.* one-electron CPE regenerates the parent V(IV) complex (f).

In the case of cysteine, the electrocatalytic preparative oxidation afforded the dimer, cystine (which was isolated and identified by IR spectroscopy), formed upon electroinduced proton evolution followed by S-S coupling (eq. 1)

$$HSR \xrightarrow{-e/-H^+} {}^{\cdot}SR \longrightarrow 1/2\ RS\text{-}SR \qquad (1)$$

Therefore, the electrocatalytic process can be described by the following overall scheme

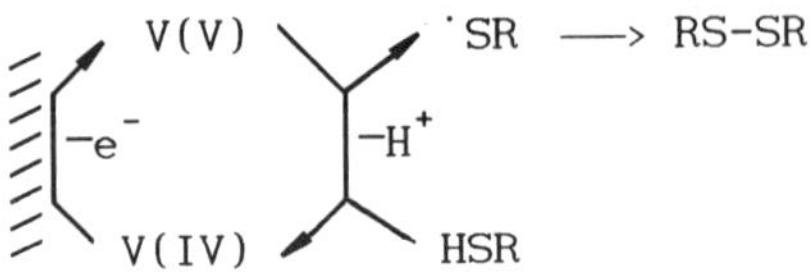

These results are in agreement with the recognized [6] electron-mediating behaviour of Amavadine for the oxidation of glutathione and other biological thiols (cysteine, cysteine methyl ester and penicillamine).

Strutural and electronic effects should play an important role in the electrocatalytic processes described above which appear to require the presence, in the thiol substrate molecule, of an activating -COOH (in cysteine) or of a derived ester group (in cysteine methyl ester[6]). In fact, not only the replacement of -SH at the cysteine molecule (*e.g.*, by -OH or -SMe to form serine or methionine, respectively), but also the elimination of the -COOH group (to give 2-aminoethanethiol) results in complete loss of the electrocatalytic effect which, however, is maintained upon abstraction of the amino group (to form mercaptoacetic acid). Moreover, the replacement of the -COOH moiety in the mercaptoacetic acid formula, *e.g.*, by $-NH_2$, $-CH_3$, $-CH_2SH$ or -OH (to give 2-aminoethanethiol, propanothiol, 1,3-propanodithiol or 2-mercaptoethanol, respectively) also supresses the activity.

A related electron-mediating effect of our vanadium complexes has been observed for the electrocatalytic oxidation of some phenol derivatives (dihydroxy- or trihydroxy- phenol species), such cathecol and gallic acid which, however, undergo a distinct anodic process upon CPE at the vanadium (IV) oxidation wave (involving two electrons in the case of cathecol). In contrast to the case of the thiols, the reversible redox wave of the V(V)/V(IV) pair is not detected upon the exhaustive anodic CPE (a distinct and irreversible wave, at a higher potential, is then generated from cathecol), indicating the occurrence of a subsequent reaction, conceivably involving the products of the oxidation of the substrate and of the vanadium centre.

Phenol itself (as well as the thiophenol), tiron —$(OH)_2C_6H_2(SO_3Na)_2$— and other species with the phenolic OH group, such as ascorbic acid, are not substrates for these electrocatalytic processes; this again shows the selectivity of the vanadium system.

3. FINAL COMMENTS

The results obtained in the present study give some support to the hypothesis presented by one of us [1] on the involvement of Amavadine in some kind of protective/defensive system of the mushrooms. Indeed, the specific oxidation of certain thiols to form disulphide bridges may be a mechanism for the "cross-linking" of protein fibers necessary for the regeneration of damaged tissues. On the other hand, we can not exclude a role of the same compound on the promotion of the polymerisation of certain organic substrates, such as polyphenols and indoles, which are involved in the formation of protective melanine (causing the browning and hardening of plant tissues when damaged) which requires, for instance, the oxidation of dihydroxyphenylalanine (dopa) to dopaquinone. This reaction is catalysed by certain copper metalloenzymes whose range of redox potentials (0.3-0.8 V *versus* NHE) includes those observed in our vanadium model complexes and a possible analogy of functions is worth noticing.

4. REFERENCES

[1] Fraústo da Silva, J. J. R. (1989), Chem. Speciation and Bioavailability, 1, 139.
[2] Nawi, M. A.; and Riechel, T. L.(1987), Inorg. Chim. Acta, 136, 33.
[3] Fraústo da Silva, J. J. R.; Pombeiro, A. J. L.; Silva, J. A. L.; and Silva, M. F. C. G. (1990), Italian-Portuguese-Spanish Meeting in Inorganic Chemistry, Gandia, Spain, B20, p.240.
[4] Carrondo, M. A. A. F. C. T.; Duarte, M. T. L. S.; Pessoa, J. C; Silva, J. A. L.; Fraústo da Silva, J. J. R.; Vaz, M. C. T. A.; and Vilas-Boas, L. F. (1988), J. Chem. Soc., Chem. Commun., 1158.
[5] Bayer, E.; and Kneifel, H. (1972), Z. Naturforsch., 27B, 207. Kneifel, H.; and Bayer, E. (1973), Angew. Chem. Intern. Ed., 12, 508.
[6] Thackrey, R. D; and Riechel, T. L. (1988), J. Electroanal. Chem., 245, 131.

5. ACKNOWLEDGEMENTS

This work has been partially supported by JNICT and INIC (Portugal)

ELECTROCHEMICAL STUDY OF THE OXIDATION CHEMISTRY OF CHROMIUM CARBONYL THIOLATES $RSCr(CO)_5^-$ AND $RSCr_2(CO)_{10}^-$

J. SPRING[a], C. P. JANZEN[a], M. DARENSBOURG[a], J. C. CALABRESE[b], P. J. KRUSIC[b], J. N. VERPEAUX[c], C. AMATORE[c].
[a]*Department of Chemistry, Texas A M University, College Station, Texas 77843, USA.*
[b]*Central Research and DevelopmentDepartment, E.I. Du Pont de Nemours and C°, Experimental Station, Wilmington, Delaware 19880-0328, USA.*
[c]*Laboratoire de Chimie de l'Ecole Normale Supérieure, URA CNRS 1110, 24 rue Lhomond 75231 Paris Cédex 05 France.*

ABSTRACT. The electrochemical oxidation of chromium carbonyl thiolates $RSCr(CO)_5^-$ has been investigated. Preparative scale oxidation yields the disulfide complex $(RSSR)Cr(CO)_5$ as the major product; the mechanism of its formation could be established by cyclic voltammetry experiments: the short lived 17 electron radical initially formed undergoes a fragmentation reaction giving the unsaturated pentacarbonyl chromium and the alkylthiyl radical which readily dimerizes. Recombination of these two species gives the product. A slower reaction was shown to take place in the diffusion layer; it results in the formation of the binuclear thiolate complex $RSCr_2(CO)_{10}^-$ the redox behaviour of which is presented.

Introduction

Chemical reactions such as protonation or electron transfer reactions of transition metal complexes can occur either at the metal center or at one of the ligands. Factors governing this ligand *vs* metal reactivity are not clear and it appears worthwhile to gain information in this area. In this respect, transition metal thiolato complexes are an important family of compounds. For example, the iron derivatives $CpFe(CO)_2SAr$ have been shown to react with electrophiles via the sulfur lone pair[1], suggesting that the HOMO has a strong sulfur 3d character. This was also confirmed by the structure of the dicationic dimer obtained upon one electron oxidation[2]:

$$CpFe(CO)_2SR - e \longrightarrow CpFe(CO)_2SR^+ \longrightarrow 1/2\ [CpFe(CO)_2(\mu RSSR)(CO)_2FeCp]^{2+}$$

However, the isoelectronic iron carbonyl thiolate yields a metal-metal bonded dimer upon oxidation[3]:

$$PhSFe(CO)_4^- - e \longrightarrow PhSFe(CO)_4 \longrightarrow 1/2\ (CO)_3Fe(\mu SPh)_2Fe(CO)_3 + CO$$

and no product resulting from a coupling at the sulfur atom.

In comparison, very little was known about the oxidation chemistry of the related and isoelectronic chromium carbonyl thiolate $RSCr(CO)_5^-$.

A. J. L. Pombeiro and J. A. McCleverty (eds.), Molecular Electrochemistry of Inorganic, Bioinorganic and Organomettallic Compounds, 417–421.

Preparative scale oxidation

Alkylthiolate complexes (R = Et, *i*-Pr, *t*-Bu), prepared by reaction of $ClCr(CO)_5^-$ with sodium mercarptides[4], were oxidized by ferricinium tetrafluoroborate in dichloromethane or THF at -78°C. The oily product could be compared with the product obtained by reaction of the alkyl disulfide with the photochemically generated $Cr(CO)_5THF$ and identified namely by IR spectroscopy[5] as the previously described $(RSSR)Cr(CO)_5$.

Electrochemical studies by means of cyclic voltammetry

ALKYLTHIOCHROMIUM CARBONYL RADICAL

The 17 electron radical $RSCr(CO)_5$ could not be detected by ESR. Attempts to observe it directly by fast cyclic voltammetry were not successful either. Indeed, even at scan rates as high as 500 V/s, no indication of reversibility could be obtained for the oxidation wave of the chromium thiolate anion. It can be deduced that the transient radical has a half life time shorter than 50 microseconds.

FORMATION OF THE DISULFIDE PENTACARBONYL CHROMIUM COMPLEX

Three possible further evolutions could be *a priori* invoked for the highly reactive 17 electron radical:

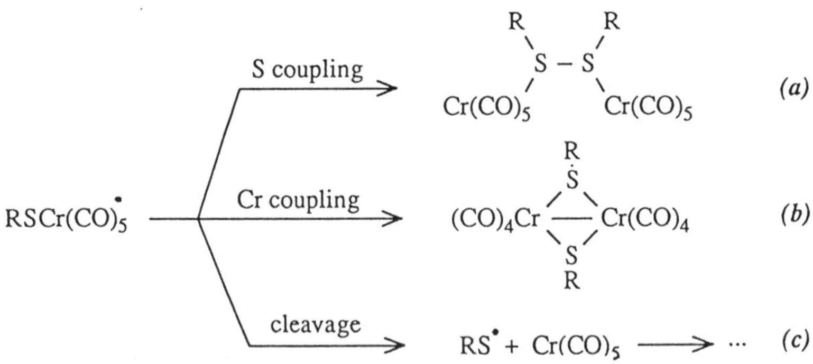

Pathway *(a)* involving a sulfur-sulfur coupling might appear as the most likely since the final product has a disulfide ligand.

However, a kinetic study based on the dependence of the oxidation peak potential on the potential scan rate showed unambiguously that the chemical reaction consuming the radical could not be second order in radical: a 30mV/log v slope was actually obtained and excluded any kind of radical radical coupling since such a reaction has to be first order in radical[6]. The oxidation peak potential was then shown to be independent on the concentration of the starting anion, bringing evidence that the chemical reaction does not involve this starting substrate.

The simplest way to reconcile these data is pathway *(c)* in which the 17 electron radical undergoes a first order cleavage. The so formed thiyl radical RS• dimerizes to yield the organic disulfide which then coordinates to the unsaturated pentacarbonyl chromium fragment:

$$RSCr(CO)_5 \longrightarrow RS^{\bullet} + Cr(CO)_5$$
$$2\ RS^{\bullet} \longrightarrow RSSR$$
$$RSSR + Cr(CO)_5 \longrightarrow (RSSR)Cr(CO)_5$$

FORMATION OF THE BINUCLEAR CHROMIUM THIOLATE

Additional effects confirm this mode of fragmentation. The cyclic voltammogram obtained by oxidation of $RSCr(CO)_5^-$ displays, beside the irreversible oxidation wave O_1, three smaller reversible oxidation waves O_2, O_3 and O_4. The size of waves O_2 and O_3 increases at lower scan rates -see fig. 1A. The apparent greater amplitude for reduction wave R_2 compared to the corresponding oxidation wave O_2 can be explained considering the longer time elapsed since the oxidation process at wave O_1. This is even more clearly demonstrated in fig. 1B, and indicates that the formation of the product oxidized at wave O_2 is slow on the time scale of cyclic voltammetry, and far too slow to explain the irreversibility of wave O_1 anyway.

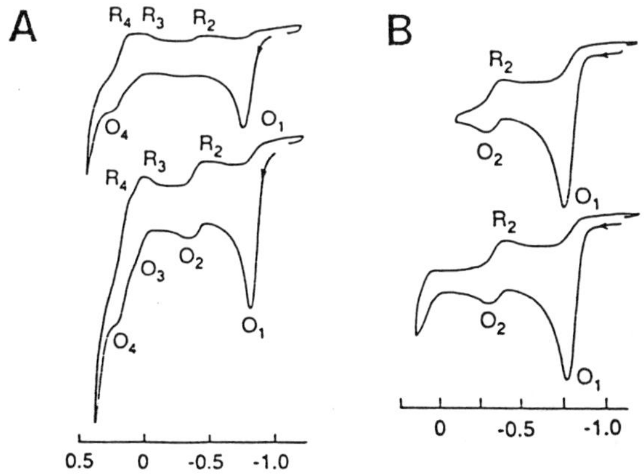

Figure 1. Cyclic voltammograms of $EtSCr(CO)_5^-$ PPN^+ (2 mM solution in THF 0.3 M n-Bu_4N BF_4) recorded at a gold disc electrode. Potentials refer to a $Ag/AgBF_4$ 0.02 M in THF; fig.A: scan rate 1 V/s (upper CV) and 0.1 V/s (lower CV); fig.B scan rate 0.1 V/s for two different turning potentials.

Redox couples O_2/R_2 and O_3/R_3 could be assigned to first and second reversible oxidations of the binuclear complex $RSCr_2(CO)_{10}^-$, see next paragraph. Therefore, the slow reaction taking place in the diffusion layer has to be a coupling between the starting substrate and the coordinatively unsaturated fragment generated at wave O_1:

$$RSCr(CO)_5^- + Cr(CO)_5 \longrightarrow RSCr_2(CO)_{10}^-$$

a reaction very similar to the coupling of $RSCr(CO)_5^-$ with $Cr(CO)_5(THF)$ used to prepare this binuclear complex[4].

REDOX BEHAVIOUR OF $RSCr_2(CO)_{10}^-$

It has been mentionned above that the binuclear thiolate complex undergoes two successive reversible oxidations. The first oxidation wave takes place at a potential more positive than the oxidation of the mononuclear complex, as generally observed in related systems. The corresponding radical is stable on the time scale of cyclic voltammetry. It could also be observed by ESR at temperatures lower than -30°C. The second oxidation leads to a cationic compound $RS[Cr(CO)_5]_2^+$ which can be described as an organometallic sulfonium cation isoelectronic with the previously described[7] phosphinidene species $RP[Cr(CO)_5]_2$.

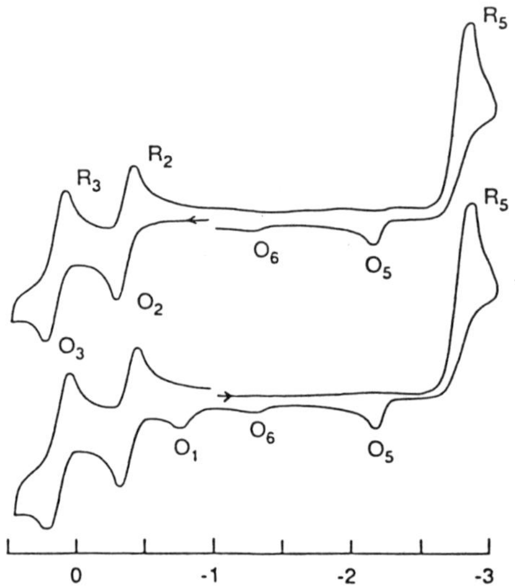

Figure 2. Cyclic voltammograms of $EtSCr_2(CO)_{10}^-$ PPN^+ (2 mM solution in THF 0.3 M $n\text{-}Bu_4N$ BF_4) recorded at a gold disc electrode. Potentials refer to a $Ag/AgBF_4$ 0.02 M in THF; scan rate 0.1 V/s, starting in the anodic (upper CV) or cathodic (lower CV) direction.

Reduction of the PPN^+ salt of the binuclear thiolate was then studied. The cyclic voltammogram exhibits a tall irreversible wave R_5 which could be shown to correspond to the overlap of two waves *(i)* the reduction of the PPN^+ counter cation and *(ii)* the two electron reduction of the thiolate complex producing two anionic mononuclear chromates according to the following equation:

$$RSCr_2(CO)_{10}^- + 2e \longrightarrow RSCr(CO)_5^- + Cr(CO)_5^{2-}$$

The mononuclear chromium thiolate could easily be identified by the oxidation wave O_1 visible in fig. 2, whereas the oxidation of $Cr(CO)_5^{2-}$ gives rise to the oxidation peaks O_5 and O_6 in agreement with independent studies[8] of this particular system:

$$Cr(CO)_5^{2-} - e \longrightarrow Cr(CO)_5^- \qquad \text{wave } O_5$$

$$2\ Cr(CO)_5^- \longrightarrow Cr_2(CO)_{10}^{2-}$$

$$Cr_2(CO)_{10}^{2-} - e \longrightarrow Cr_2(CO)_{10}^- \qquad \text{wave } O_6$$

The experimental details concerning this work as well as ESR, I.R. and crystal structure data can be found in ref. 4

References

1 Ashby, M. T.; Enemark, J. H.; Lichtenberger, D.L. Inorg. Chem. 1988, 27, 191.
2 Treichel, P.M.; Rosenheim, L.D. J. Am. Chem. Soc. 1981, 103, 691. Treichel, P.M.; Rosenheim, L.D.; Schmidt, J.S. Inorg. Chem 1983, 22, 3960.
3 Liaw, W.-F.; Kim, C.; Darensbourg, M.Y.; Rheingold, A.L. J. Am. Chem. Soc. 1989, 111, 3591.
4 Springs, J.; Janzen, C.P.; Darensbourg, M.Y.; Calabrese, J.C.; Krusic, P.J.; Verpeaux, J.N.; Amatore, C. J. Am. Chem. Soc. 1990, 112, 5790.
5 Ehrl, W.; Vahrenkamp, H. Chem. Ber. 1970, 103, 3563. Abel, E.W.; Moss, I.; Orrell, K.G.; Aqreshi, K.B.; Sik, V.; Stephenson, D.J. J. Chem. Soc. Dalton Trans. 1988, 1489.
6 Nadjo, L.; Savéant, J.M.; J. Electroanal. Chem. 1973, 48, 113.
7 Huttner, G.; Evertz, K. Acc. Chem. Res. 1986, 19, 406.
8 Amatore, C.; Krusic, P.J.; Pedersen, S. U.; Verpeaux, J.N. submitted.

NEW DEVELOPMENTS IN VOLTAMMETRIC BEHAVIOUR AND ELECTROCATALYTIC ACTIVITY OF Ni(II) TETRAAZAMACROCYCLIC COMPLEXES

J.TARASZEWSKA and G.ROSLONEK
Institute of Physical Chemistry
Polish Academy of Sciences
Kasprzaka 44/52, 01-224 Warsaw, Poland

ABSTRACT. The electrochemical behaviour of tetraazamacrocyclic Ni(II) complexes containing a pendant amino group has been studied by cyclic voltammetry at glassy carbon electrodes in function of pH and kind of solvent. The internal pH-dependent equilibrium between the open and chelated form of the pendant arm, the influence of solvent on the heterogenous kinetics of Ni(III)/Ni(II) redox couple in these complexes, formation of the modified electrode in strongly alkaline solutions and its application to the electrocatalytic oxidation of simple alcohols have been studied and discussed.

1. Introduction

Azamacrocyclic ligands with pendant arms have recently attracted considerable attention mainly due to their analytical, biological and catalytic applications [1]. Compounds of this class combine the properties of a rigid macrocyclic structural unit with those of a flexible side-chain to which the functional group can be bound. Such systems are especially interesting for studying the influence of axial coordination on the properties of the coordinated metal ion.

In our work we were interested to study: 1) the electrochemical behaviour of selected Ni-tetraazamacrocyclic complexes with pendant amino group in dependence on the pH of solution, 2) the electrocatalytic activity of these complexes and 3) the influence of solvent on the heterogenous kinetics of Ni(II)/Ni(III) redox couple in these complexes.

2. Materials and Methods

The cyclic voltammetric (CV) experiments were caried out in

A. J. L. Pombeiro and J. A. McCleverty (eds.),
Molecular Electrochemistry of Inorganic, Bioinorganic and Organomettallic Compounds, 423–429.

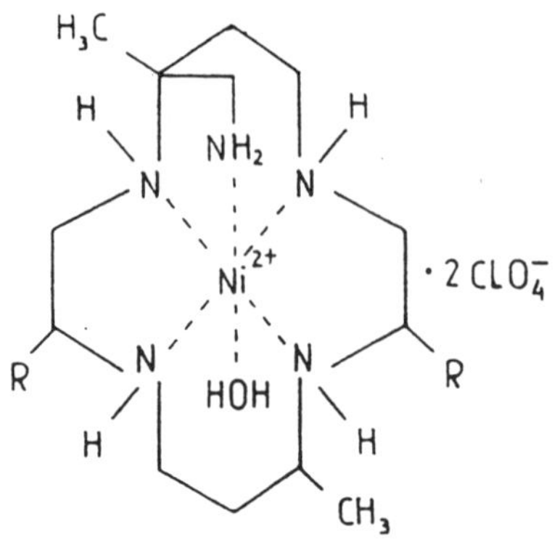

R = H : complex 1
R = CH_3 : complex 2

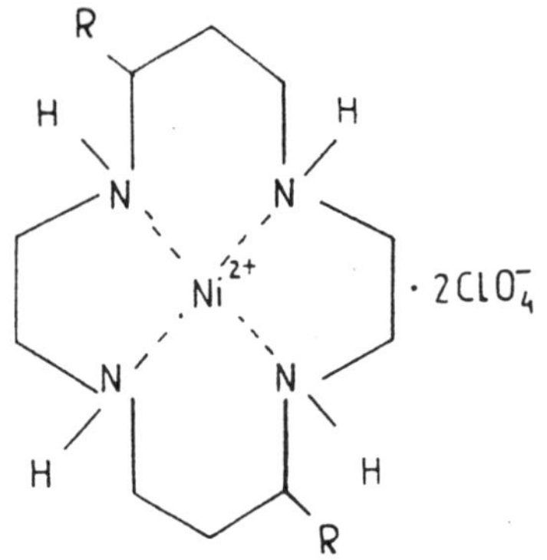

R = CH_3 : Ni-[Me]2-cyclam
R = H : Ni-cyclam

Figure 1. Schematic structures of studied complexes.

a three-electrode system at 25±0.5°C. A glassy carbon GC-30 or GC-20 (Tokai Carbon Co) sealed in a teflon tube was used as the working electrode. The counter electrode was a Pt cylinder or plate (in kinetic studies). A 1M NaCl aqueaus calomel electrode (MCE) was used as the reference electrode. In kinetic studies potentials were referred vs. ferrocene/ferricinium (Fc) internal reference system.

The GCE was polished with 0.3μm alumina slurry to a mirror like surface. After polishing the GCE was immersed in concentrated H_2SO_4 for 1h and througly rinsed with H_2O prior to use. In kinetic studies before each measurement the GCE was cycled in the potential range ±0.3V around the standard potential of the reaction in a given solvent. The electrode modification was accomplished by cycling the potential between 0.0 and +0.6V in freshly prepared deareated solutions of the studied complexes in strongly alkaline solutions.

The CV curves were recorded by using a measuring system consisting of potentiostat, generator and x-y recorder. For rapid scan rates the curves were recorded on the screen of the monitor of an IBM PC-XT computer using a programme written in our laboratory. In kinetic experiments the positive feedback technique was used for cell resistance iR compensation.

All chemicals were p.a. grade reagents. Merck $NaClO_4$ used in kinetic experiments was additionaly dried at 60°C in vacuum, and kept in the dry box. Complexes 1 and 2 were prepared according to [2]. The solvents used were dried and purified for electrochemical use.

3. Results and Discussion

3.1. VOLTAMMETRIC CHARACTERISTICS

Full voltammetric characteristics of studied complexes was performed in ACN solutions of 0.1M TEAP (Fig.2).

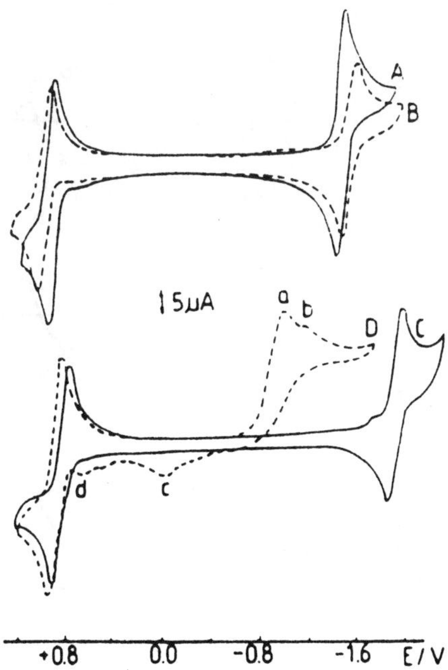

Figure 2. Cyclic voltammograms of Ni(II)-cyclam (A), Ni(II)--[Me]$_2$cyclam (B), complex 1 (C), and complex 2 (D).

Ni(II) in these complexes undergoes one electron oxidation to Ni(III) at potential of about +0.95V and one electron reduction to Ni(I) at potential of about -1.7V.

3.2. BEHAVIOUR IN ACIDIC SOLUTIONS

An interesting aspect of study of these complexes, in aqueous solutions, is the internal pH-dependent equilibrium between the open and chelated form of the pendant arm (Fig.3). With decrease of pH peak a due to the oxidation of Ni(II) in octahedral form decreases and a new peak b corresponding to oxidation of Ni(II) from square-planar form appears. At low scan rates independently of the pH value, only one cathodic peak c, corresponding to the reduction of Ni(III) in complex with the coordinating NH_2 group is observed (Fig.3A). This suggests that Ni(III) from complex in square-planar form must undergo a chemical reaction.

A significant increase in the scan rate has an essential

influence on the cathodic part of the curves (Fig.3B). On curve 1, besides peak c a new peak d appears corresponding to the reduction of Ni(III) from the square-planar form. The kinetic effect is connected with the competition of H^+ and Ni(III) ions for the side-chain NH_2 group [3].

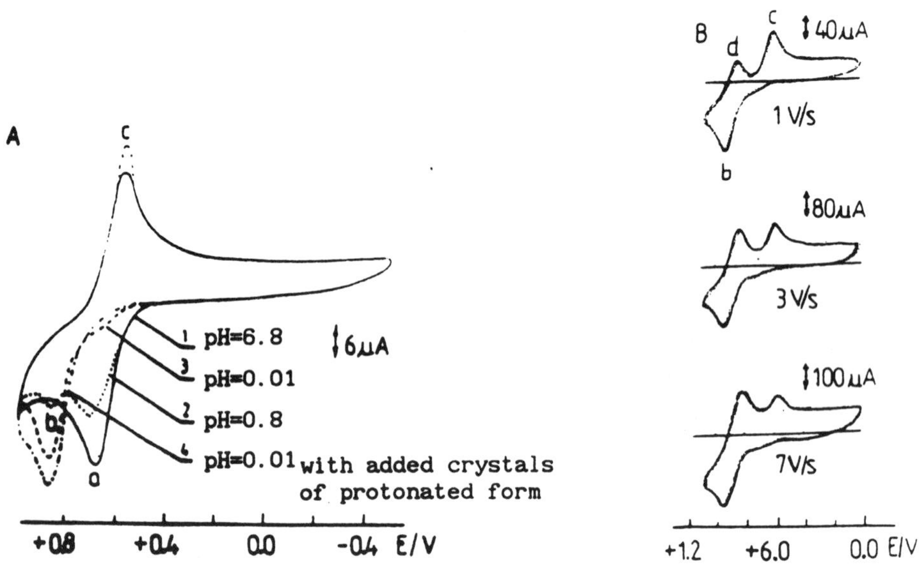

Figure. 3. Electrochemical behaviour of complex 1 as a function of pH (A) and scan rate (B).

3.3 FORMATION OF THE MODIFIED ELECTRODE.

The behaviour of complex 1 was completely different in strongly alkaline solutions. When the potential was continously cycled between 0.0 and +0.6V while the electrode substrate was immersed in a mixture of complex 1 in 0.1M NaOH the electrode surface covered with brown deposit, and it worked as the modified electrode (ME) in solution of pure 0.1M NaOH.

An important step in the formation of the ME is the adsorption of $[Ni(II)(OH)L(H_2O)]^+$ complex formed in alkaline solution by the replacement of the coordinating NH_2 group by a competing ligand such as the OH group. Peak a in Fig.4 is due to this process. The detailed mechanism of ME formation is discussed in [4].

3.4 ELECTROCATALYTIC OXIDATION OF SIMPLE ALCOHOLS

The immobilized redox center acted as a catalyst for the

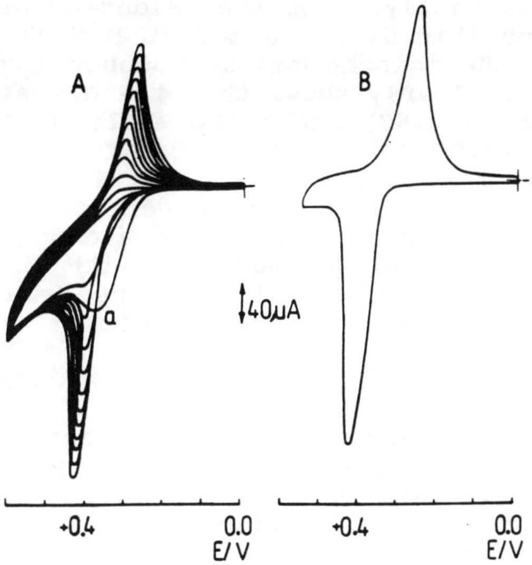

Figure 4. (A) Formation of ME; (B) the voltammetric response of complex 1 ME in pure 0.1M NaOH.

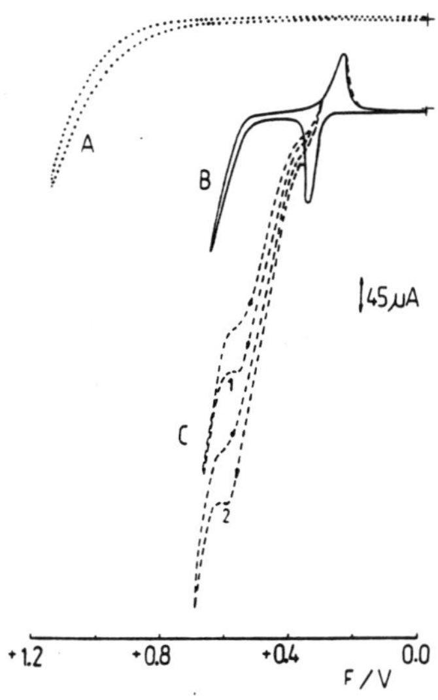

Figure 5. Electrocatalytic oxidation of methanol.

center acted as a catalyst for the oxidation of simple alcohols and OH^- ions (Fig.5). At a bare GCE MeOH is not oxidized prior to the discharge of supporting electrolyte (Fig.5A). Fig.5B, clearly shows that the ME catalyses oxidation of OH^- ions. Fig.5C, shows the electrocatalytic oxidaticn of MeOH at +0.575V.

The behaviour of EtOH and of 1- and 2-PrOH was similar. For all four alcohols the oxidation peak heights were linearly dependent on the concentration of alcohol in solution.

The essential virtue of GCE modified with studied complexes is the lowest potential value of the MeOH oxidation in comparison with other electrode materials.

3.5 SOLVENT EFFECT ON Ni(II)/Ni(III) REDOX COUPLE

The standard, heterogenous rate constants, k_s, of the Ni(II)/Ni(III) redox couple in complexes 1 and 2 were determinated in such solvents as ACN, H_2O, AC, DMF, and DMSO, by Nicholson's method for the quasi-reversible processes. For complex 1 the values of k_s were nearly independent on the $NaClO_4$ concentration (0.1-1M). This independence is extremly important because the measured k_s values can be discussed as "true" values. We suppose that the same is valid for complex 2. For both complexes the linear relation between the $\ln k_s$ and the solvent donor number (DN) has been found (Fig.6) what can be explained by the inner solvation shell effect.

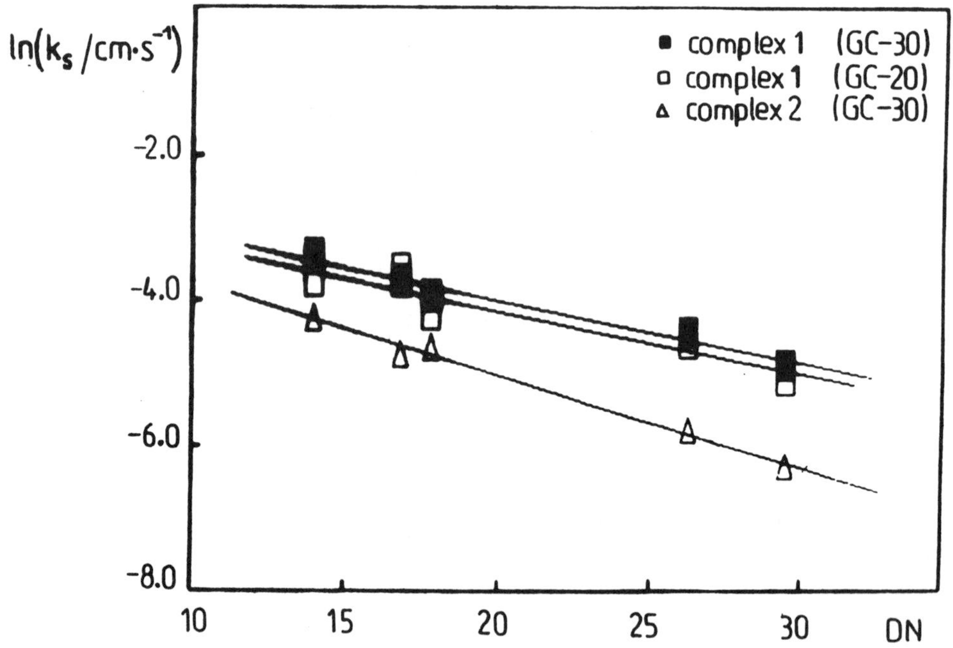

Figure 6. Dependence of $\ln k_s$ on DN for studied complexes.

4. References

1. Kaden,Thomas A. (1984) "Synthesis and metal complexes of azamacrocycles with pendant arms having additional ligatin groups", Topics in Current Chemistry, 121,157-179.

2. Korybut-Daszkiewicz,B. (1988) "Tetraazamacrocyclic nickel(II) complexes containing a pendant amino group", J.Coord.Chem., 19, 101-102.

3. Taraszewska,J., Rosłonek,G., and Korybut-Daszkiewicz,B. (1991) "Voltammetric behaviour of Ni(II) and Co(III) tetraazamacrocyclic complexes containing a pendant amino group", J.Electroanal.Chem., 297, 245-255.

4. Rosłonek,G, and Taraszewska,J. "Electrocatalytic oxidation of alcohols on glassy carbon electrodes electrochemically modified with nickel tetraazamacrocyclic complexes. Mechanism of film formation", J.Electroanal.Chem., (in press).

5. Taraszewska,J.,and Rosłonek,G., "Effect of solvent on the kinetic and thermodynamic redox properties of Ni(III)/ Ni(II) in tetraazamacrocyclic complexes with pendant amino group and cyclam studied on the glassy carbon electrodes", J.Electroanal.Chem.,(in press).

ELECTROOXIDATION AS A ROUTE TO DIHYDROGEN COMPLEXES

B. E. HAUGER[a], M. TILSET[b] and K. G. CAULTON[a]
[a]*Indiana University, Department of Chemistry, Bloomington, Indiana 47405, U.S.A.*
[b]*University of Oslo, Department of Chemistry, P.O.Box 1033 Blindem, N-0315 Oslo 3, Norway*

ABSTRACT. Efforts have been made to measure the oxidation potential of the formally 16 valence electron dihydride $IrH_2(P^tBu_2Ph)_2Cl$ (**1**) by cyclic voltammetry. At a scan rate of 20 V/s the peak potential for the oxidation of **1** (1 mM / 0.1 M $N(C_4H_9)_4PF_6$) was found to be +0.70 V vs. the ferrocene/ferrocenium couple in dichloromethane at a platinum electrode (d = 1.0 mm). The peak for the back reduction is small at 50 V/s. This suggests the presence of a follow up reaction. This conclusion is supported by the observation of reductive peaks at +0.20 V and -0.80 V. Further studies to identify the product(s) of this oxidation are in progress.

Introduction

Since the discovery of the η^2-H_2 ligand, it has been the subject of intense interest due to the numerous catalytic processes which involve dihydrogen activation. We are exploring the technique of electrosynthesis of dihydrogen complexes from the abundantly available metal polyhydrides. It is our hope that this technique will prove to be generally applicable and allow the synthesis and study of complexes previously unobserved.

The oxidation of monohydrides has been observed to result in the production of dihydrogen complexes. One of us has reported[1] that the chemical oxidation of $CpRu(CO)(PPh_3)H$ resulted in the production of the known dihydrogen complex $CpRu(CO)(PPh_3)(\eta^2\text{-}H_2)^+$. The suggested mechanism involved intermolecular proton transfer (eq. 1). We propose to extend this work to

$$CpRu(CO)(PPh_3)H \; + \; CpRu(CO)(PPh_3)H^{+\bullet} \longrightarrow CpRu(CO)(PPh_3)(H_2)^+ \; + \; CpRu(CO)(PPh_3)^\bullet \qquad (1)$$

polyhydrides. The presence of more than one hydride ligand provides the opportunity for intramolecular rearrangement, as well as intermolecular exchange, to produce dihydrogen complexes. This possibility remains relatively unexplored.

A. J. L. Pombeiro and J. A. McCleverty (eds.),
Molecular Electrochemistry of Inorganic, Bioinorganic and Organomettallic Compounds, 431–434.

MATERIALS AND METHODS

All manipulations were carried out under an inert atmosphere using standard Schlenk or glove-box techniques. Methylene chloride was distilled under nitrogen from calcium hydride. Tetrahydrofuran and hexanes were distilled under nitrogen from sodium/benzophenone. Benzene was distilled under nitrogen from sodium/potassium alloy. Toluene was distilled under nitrogen from potassium. Tetra n-butyl ammonium hexafluorophosphate (Fluka) was heated to 100 °C under dynamic vacuum for 24 hours before use. Ferrocene was sublimed before use. ^{1}H and ^{31}P NMR spectra were obtained on a Nicolet NT-360 instrument. Negative ^{31}P chemical shifts are upfield from external 85% H_3PO_4. Cyclic voltammograms were recorded on an EG&G/Parc 273A potentiostat/galvanostat through an external computer workstation using software supplied by EG&G/Parc. The reference electrode was saturated silver/silver chloride in methylene chloride with 0.1 M $TBAPF_6$ as the supporting electrolyte. The auxillary electrode was a platinum wire. The working electrode was of local construction using 1.0 mm diameter platinum wire. The electrochemical micro-cell was purchased from EG&G/Parc. All electrochemical measurements were carried out under a blanket of solvent-saturated argon gas.

Preparation of $IrH_2(P^tBu_2Ph)_2Cl$ (Compound 1). 0.700 grams (0.775 mmol) of $[Ir(cyclooctene)_2Cl]_2$ was combined with 0.725 mL of di-^{t}butyl phenyl phosphine in 100 mL of THF to yield a brown solution. H_2 gas was then bubbled through the solution for 30 minutes. The THF was removed *in vacuo*. The resulting brownish solid was washed with 10 mL of hexanes to yield 0.713 grams (68% yield) of $IrH_2(P^tBu_2Ph)_2Cl$. $^{31}P\{^1H\}$ NMR in C_6D_6: 66.7 (s). ^{1}H NMR in C_6D_6: 8.15-7.10 (m, phenyl protons), 1.43 (vt, P-^{t}Bu = 13 Hz), -32.1 (t, Ir-H, J_{P-H} = 14 Hz).

Preparation of $IrH_2(P^tBu_2Ph)_2(OCH_2CF_3)$. 0.200 grams (0.296 mmol) of $IrH_2(P^tBu_2Ph)_2Cl$ and 0.050 grams (0.410 mmol) of sodium trifluoroalkoxide were weighed into a Schlenk flask. 40 mL of THF was added. The brown solution was stirred for approximately one hour. The THF was removed *in vacuo*. The resulting solid was extracted into hexanes and filtered to yield a clear solution. The solvent was removed *in vacuo* to yield 0.19 grams (0.257 mmol, 87% yield) of $IrH_2(P^tBu_2Ph)_2(OCH_2CF_3)$. $^{31}P\{^1H\}$ NMR in C_6D_6: 65.4 (s). ^{1}H NMR in C_6D_6: 8.15-7.0 (phenyl protons), 4.02 (q, $-OCH_2CF_3$, J_{H-F} = 10 Hz), 1.40 (vt, P-^{t}Bu = 13 Hz), -30.3 (t, Ir-H, J_{P-H} = 14 Hz).

Preparation of $IrH_2(P^tBu_2Ph)_2OPh$. 58 mg of $IrH_2(P^tBu_2Ph)_2Cl$ and 18 mg of KOPh were dissolved in 5 mL of toluene. The solution was stirred for three hours. Solvent was removed *in vacuo* to yield a brown oil. $^{31}P\{^1H\}$ NMR in C_6D_6: 63.4 (s). ^{1}H NMR in C_6D_6: 8.2-6.5 (phenyl protons), 1.37 (vt, P-^{t}Bu = 13 Hz), -32.2 (t, Ir-H, J_{P-H} = 14 Hz).

Preparation of $IrH_2(P^tBu_2Me)_2Cl$. A THF solution containing 0.36 grams of P^tBu_2Me was added to a stirred solution containing 0.507 grams of $[Ir(cyclooctene)_2Cl]_2$. The solution was stirred for one hour. H_2 gas was then bubbled through the solution for one hour. The solvent was removed *in vacuo* to yield an orange powder (0.534 grams, 87% yield). $^{31}P\{^1H\}$ NMR in C_6D_6: 47.3 (s). ^{1}H NMR in C_6D_6: 1.53 (br. s, P-Me), 1.19 (vt, P-^{t}Bu = 12 Hz), -32.05 (t, Ir-H, J_{P-H} = 13 Hz).

Results

At a scan rate of 20 V/s the peak potential for the oxidation of 1 was found to be +0.70 V vs. the ferrocene/ferrocenium couple in dichloromethane at a platinum electrode. At higher scan rates the peak potential shifts to higher voltages indicating the presence of a chemical reaction which follows the electron transfer step. Also characteristic of an EC mechanism is the ratio of the cathodic and anodic peak currents which becomes closer to one as scan speed is increased. Further supporting the conclusion that new compounds are being generated is the observation of new reductive peaks. At slower scan speeds the only new reductive peak is observed at -0.80 V. As the scan speed is increased, a peak at +0.20 gains intensity while the peak at -0.80 loses intensity. Further studies to identify the product(s) of the oxidation are in progress.

Discussion

In addition to the synthetic applications of cyclic voltammetry described above, we propose to utilize the analytical capabilities of this technique to answer some involved questions related to halide and alkoxide π-donation to late transition metal complexes. The model complex $IrH_2(PH_3)_2Cl$ has recently been studied by *ab initio* calculations.[2] This study calculated a ground state based on a trigonal bipyramid to be 40 kcal/mol more stable than a structure based on a square based pyramid. A neutron diffraction study of $IrH_2(P^tBu_2Ph)_2Cl$ is currently underway to confirm the TBP geometry. An X-ray study[3] has determined that the Ir-Cl bond length is 2.401(3)Å in $IrH_2(P^tBu_2Ph)_2Cl$. This value compares well to the 2.412(1)Å found[4] for the related complex $IrH(P^iPr_3)_2(Ph)Cl$ which bears the distorted TBP structure. A previous extended Hückel theory study[5] of $IrH_2(PH_3)_2Cl$ indicated that the distorted TBP structure is favored by "poor σ-donating and good π-donating" ligands. We therefore believe that a systematic study of available π-donating ligands is necessary to qualify the previous statement as well as determine if indeed the distorted trigonal bipyramid structure is rigidly maintained in solution.

Such a study would require the determination of the oxidation peak potential of a series of $IrH_2(PR_3)_2X$ and their 18-electron adducts $IrH_2(PR_3)_2X(L)$ (L = any two electron donor). It is worth mentioning that irreversibility of the oxidation of the species of interest will result in a cathodic shift. However, similar rate constants for the consumption of different electrochemically-generated species will result in fairly small differences in the potential shift at a given bulk concentration and scan speed.[6] We speculate that the reaction which consumes the different electrochemically-generated species will cause only negligible differences in potential shift within the series. Under these limiting assumptions, oxidative peak potentials may be compared within the series. The difference in the oxidation peak potentials of the series of formally 16-electron complexes (delta Ox. 16) should reflect the differences in the sum of the σ- and π-donating capabilities of the X ligands. The differences in the oxidation peak potentials among the analogous 18-electron complexes (delta Ox. 18) should reflect predominantly differences in the σ-donor ability of the X ligands. The differences in these differences (delta Ox. 16 - delta Ox. 18) should then reflect predominantly the π-donor ability of the ligand X. Similarly, crystallographic determination of the differences in metal-X bond lengths for both 16- and 18-electron complexes would reflect the π-donating ability of X. A linear relationship between π-donor ability in solution and π-donor ability in the solid state would confirm that π-donation is maintained in solution.

Acknowledgments. This work was supported by the National Science Foundation and by NATO. We thank Helmut Rothfuss for his extensive technical assistance and expertise.

References

1. Ryan, O.; Tilset, M.; Parker, V. *Organometallics* **1991**, *10*, 298-304.

2. Riehl, J.; Jean, Y.; Eisenstein, O.; Pelissier, M. *Organometallics* **1992**, *11*, 729-737.

3. Albinati, A.; (1992) personal communication.

4. Werner, H.; Hohn, A.; Dzialles, M. *Angew. Chem. Int. Ed. Engl.* **1986**, *25*, 1090.

5. Rachidi, I.; Eisenstein, O.; Jean, Y. *New J. Chem.* **1990**, *14*, 671.

6. Bard, A. J.; Faulkner, L. R. (1980) Electrochemical Methods: Fundamentals and Applications, Wiley, New York.

PART 7. MECHANISTIC STUDIES (INCLUDING NUMERICAL and COMPUTING METHODS)

MECHANISTIC STUDIES IN ORGANO-METALLIC ELECTROCHEMISTRY. THE ROLE OF COMPUTER BASED TECHNIQUES & SIMULATION

N. TAYLOR, T. BODDINGTON, P.HALFORD-MAW, R.LUO and D.MILLS
School of Chemistry, Leeds University, Leeds LS2 9JT England

ABSTRACT. The use of dedicated interactive microcomputer systems for electrochemical measurements & simulations is now feasible. The digitisation of the experimental response allows storage/retrieval etc. & provides the essential facility of post capture treatment of data. The forms of the latter include direct comparisons of data with simulations, countering influences of experimental artifacts including uncompensated electrolyte resistance & charging currents, & the creation of new forms of the data via convolution & other methods. Convolutions allow valid tests for adherence to important common mechanistic schemes, for example involving simple electron transfer E_1 & fast electron transfer E_n reactions, EC_{cat} processes & all or part of a CEC scheme with arbitrary rates of electron transfer & first order chemical kinetics In many cases combinations of whole data sets, for example at different sweep rates, can be employed in a single test thereby paralleling with the transformed data the simplicity of the relationships of polarography. In the case of E_1 processes in chronoamperometry a new & equally exact alternative to convolution methods is described which is simpler to implement & has definite advantages experimentally.

1. Introduction.

The use of computers in both the management & treatment of data of electrochemical experiments & the digital simulation of voltammetric responses for a wide variety of mechanistic schemes has been well established in the last decade & beyond. However early attempts involved the use of mainframe computers & thus excluded widespread use in the interactive sense when experiments were in progress. The advent of cheap computers, in particular the IBM PC & its compatibles, has transformed the situation in that dedicated & interactive systems can be established cheaply & with the speed of the latest products simulations matching previous sophistication also become feasible.

The role of the computer varies with the application but in the widest sense encompasses waveform generation, data capture, treatment & display. Hard copy of graphs etc. is also important & indeed all figures in this paper are produced direct from a data capture program. Computer waveform generation except where special functions of voltage or current are required adds relatively little advantage to experiments & will not be considered further. The role of computers via digitisation of the experimental responses is much more dramatic & in our view, apart from storage & retrieval' provides the essential facility of post-capture treatment of data. The latter takes several forms including direct comparisons of data with simulations, the reduction of the influence of experimental artifacts including uncompensated electrolyte resistance, charging currents etc. & importantly the transformation of data via convolution & other methods. For stationary electrodes these latter transformations can produce relationships in principle independent of the voltammetric technique & experimental conditions used which allow valid tests of whole data sets for adherence to particular mechanistic schemes in some cases paralleling with the transformed data the simplicity of the Hevrovsky-Ilkovitch relationship of polarography. Parameter determination can be made on the basis of reliable assessment of mechanism from the electrochemical data alone or utilising additional information as required.

A. J. L. Pombeiro and J. A. McCleverty (eds.), Molecular Electrochemistry of Inorganic, Bioinorganic and Organomettallic Compounds, 437–452.

Convolution techniques have been established principally by Savéant [1-3] & Oldham [4,5] & for planar & in some cases spherical & hemispherical electrodes provide valid methods of mechanistic & parameter determination for systems involving, for example, consecutive fast electron transfer reactions in an E_n scheme, & all or part of a CEC scheme, (C is a reversible (two way) chemical reaction which exhibits only first order kinetics & E an electrochemical reaction, both reactions of arbitrary rate). In the latter case the computational complexity however increases greatly when reversible chemical reactions with species of differing diffusion coefficients are involved. The simplest EC_{cat} scheme is also fully resolvable for arbitrary parameters & here, somewhat in contrast to the former, the practical implementation of the unequal diffusion coefficient case is relatively straightforward. There are also specific & general advantages for certain other schemes for example, with some restrictions in kinetic regimes, in treatments involving dimerisation etc.. More generally the semi-differentiated cyclic voltammetric response (the deconvolution) is more akin to the symmetrical differentiated polarographic or AC technique peaks than to the rather one sided voltammograms initially obtained. The displacement of the outward & return peaks of the deconvolutions here gives a direct indication of electron transfer rate rather independent of precise details of the overall mechanism involved including the EC_{irrev} case where the return current peak but not the transformed peak may be entirely absent. Likewise, within some limitations, for the four sided square scheme involving two electron transfers & two isomerisations the convolutions of the cyclic voltammetric response provide a precise test for adherence although full parameter determination is not available. Other similar cases exist.

The application of simple mechanistic tests directly to the linear sweep voltammogram is well established [6,7] & remains important but in general is restricted to analysis of peak heights & in particular peak positions as functions of the time-scale of the experiments via the sweep rates. This 'two point' analysis is obviously somewhat dangerous from the outset & contrasts with the convolution methods which have the obvious advantages of potentially utilising all data points of a given experiment & indeed with equal facility can include data from other & not necessarily identical experiments for example as the sweep rate is altered in cyclic voltammetry. The simple current peak analysis of the linear sweep voltammogram also relies more heavily on the experimental system in the sense that the test criteria require that the outgoing peak positions must be exceeded considerably without interference from other reactions before the return sweep is invoked. Furthermore when the sweep form is non-linear with time by design, or if there is appreciable electrolyte resistance in the medium used which cannot be compensated directly at the potentiostat then peak analysis is intractable without the production of specific 'working curves'. In contrast the form of the voltage-time profile is unimportant in a full mechanistic analysis via convolution methods & by means of appropriate rescaling of the voltage axis the techniques in principle work equally well in media of high resistance provided that electromigration remains unimportant in determining the diffusional fields of the species involved.

In chronoamperometry the simplest criteria have relied on adherence or otherwise to the diffusion limited square root of time Cottrell dependence of the current as a test for 'reversible' electron transfer in the important case of the simple E_1 system or, for example, in the EC_{irrev} case have utilised working curves for the response of double step experiments designed to induce an insensitivity to the rate of electron transfer. Convolution methods provide ready access to the above electrochemical & first order chemical kinetics irrespective of rate & in the double step experiments the data are thus treated without the requirement of working curves. Again as with cyclic methods all data gained can be utilised. In chronoamperometric experiments to determine the simple electron transfer kinetics the convolution method does have one practical defect in that the early data which must be emphasised when kinetics are fast leads to errors in the estimation of the convolutions due to the incorporation of charging currents. The 'history dependent' nature of the convolutions makes the effect persist however beyond the charging current timescale thus increasing the minimum useable time window & hence restricting the maximum rate which can be reliably assessed. A new alternative & equally exact method for the E_1 system avoiding this problem will be described.

2. Convolutions in Electrochemistry.

2.1. BRIEF OUTLINE.

In an electrochemical scheme the material to be converted or that produced flows to or from the electrode by virtue of the adjacent diffusional gradients created by the differing concentrations of a given species at the electrode & in the immediate bulk solution. Mass balance requires that the fluxes match & the current remains proportional to these. The concentrations at the electrode are wholly (reversible electron transfer) or partly dependent on the potential of the electrode depending on whether in the necessary charge balance the matching current requirement of the electron transfer process itself can be sustained without significant departure from Nernstian conditions of concentrations at the electrode. Away from the electrode the concentrations & diffusional fields depend both on the bulk values & on the entire history of fluxes at the electrode & furthermore, together with the current, on any processes occurring in the bulk solution. Laplace transform methods allow treatment of the fluxes etc. for a wide variety of electrochemical models including those of single electrochemical processes & such processes accompanied by pre or prior first order chemical reactions to yield electrode concentrations as functions of time. The electrode concentrations are thereby determined from the current history quite independently of the potential but some form of model, normally that due to Butler & Volmer, must be invoked to introduce the latter & thus complete the current-time-potential relationship. In some ways the expressions giving the concentrations of electroactive species at the electrode give a most important insight into what tests might be applied since the idea that at potentials well 'past' the E^0 or 'half wave' potential $E_{1/2}$ the concentration of electroactive reactant at the electrode is forced to be small by the electrode process, with that of the electrochemical product becoming correspondingly large. This will be illustrated in the following cases.

2.2. INTRODUCING THE SEMI-INTEGRAL OF THE CURRENT.

In this section we use the well established theory of simple electron processes to re-emphasise the advantages obtained by implementing convolution techniques as opposed to the more conventional linear sweep & diffusion controlled chronoamperometric criteria based directly on the current. The brief review also serves to stress the feature that such expressions obtained need not depend on any particular experimental voltammetric technique & indeed one can combine data from say cyclic & chronoamperometric experiments as a composite test of adherence to the proposed mechanistic scheme.

For a single step electrochemical reaction of arbitrary rate

$$A \dashrightarrow B + (z_B - z_A)\,e \qquad (1)$$

initially with a bulk phase containing A only the corresponding electrode concentrations C_A^* &

C_B^* are [1-5]

$$C_A^{bulk} - C_A^* = I_1/(z_B-z_A) F A D_A^{1/2} \qquad (2)$$

$$\& \; C_B^* = I_1/(z_B-z_A) F A D_B^{1/2} \qquad (3)$$

respectively where D & z are the diffusion coefficients & charge numbers of the electroactive species & I_1 is the semi-integral or square root of time convolution of the current given by

$$I_1(t) = \pi^{-1/2} \int_0^t i(u)/(t-u)^{1/2} \, du \qquad (4)$$

The convolution at time t is thus akin to an integral of the current but with a factor decreasing the importance of the current contribution at a given time of measurement u by the square root of the time elapsed since that measurement was made. Indeed if the I_1 is similarly convoluted then the result is the charge Q ie. the integral of i(u) & hence the terms semi-integration is used for this particular convolution process. The semi-differential or deconvolution is likewise obtained via semi-integration followed by differentiation.

At potentials lying well beyond that of the position of the polarographic wave the electrode kinetics in the forward direction become fast ultimately & C_A^* approaches zero. Thus I_1 attains the limiting value

$$I_1 = I_{lim} = (z_B-z_A) F A D_A^{1/2} C_A^{bulk} \qquad (5)$$

thus allowing determination of D_A from a cyclic voltammetric or chronoamperometric experiment etc. encompassing this voltage extreme.

In the event of the potential returning to the initial value or 'beyond' such that the C_A^{bulk} is reattained at the electrode & C_B^* tending to zero then I_1 also returns to zero although current flow due to the diminishing but continued diffusion gradients persist. This return of I_1 provides one test for simple E processes via cyclic voltammetry (not necessarily linear sweep) & for example by double step chronoamperometry.

The potential is formally introduced normally via the Butler-Volmer Relationship here conveniently couched in the form

$$i/(z_B-z_A) F A = k_f C_A^* - k_b C_B^* \qquad (6)$$

where k_f & k_b are potential dependent rate constants related to the standard rate constant k_f^0 at potential $E=E^0$ via

$$k_f = k_f^0 \exp(\alpha(E-E^0)F(z_B-z_A)/RT) \qquad (7)$$

$$\& \; k_b = k_f^0 \exp-((1-\alpha)(E-E^0)F(z_B-z_A)/RT) \qquad (8)$$

where α is the symmetry factor. At this stage the chronoamperometric response is easily perceived from [8]

$$i = k_f I_{lim}/D_A^{1/2} - ((k_f/D_A^{1/2})+(k_b/D_B^{1/2}))I_1 \qquad (9)$$

since with k_f & k_b constant the current thus varies linearly with the convolution I_1 quite irrespective of the kinetic factors involved. If the intercept $I_1 = 0$ is measurable (ie. k_f not too large) then k_f & D_A are directly obtained from the above & provided D_B is determined (by for example using B as the reactant species) & the slope of the above is reasonably measurable then all other factors become accessible. For potential regimes where k_f is large the slope of the above is also large & I_1 approaches a constant value given by the relationships

$$(E-E_{1/2})F(z_B-z_A)/RT)=\ln((I_{lim}-I_1)/I_1) \quad (10)$$

& $$E_{1/2}=E^0+(RT/(z_B-z_A))\ln(D_B/D_A)^{1/2}\ . \quad (11)$$

The semi-integral $I_1(t)$ of a current function varying as $i= a.t^{-1/2}$ is readily shown to be $a.\pi^{1/2}$ & thus $i(t)=I_1/(\pi t)^{1/2}$. Constancy of I_1 is thus equivalent to Cottrell behaviour & the slope of i vs $(\pi t)^{-1/2}$ has the shape of the polarographic wave. For the latter kinetic regime the heterogeneous kinetic factors are not available.

For cyclic voltammetry equation (2) is best rewritten as

$$i.(D_A{}^{1/2}/k_f{}^0).\exp(-\alpha(E-E^0)F(z_B-z_A)/RT)=I_{lim}-I_1-I_1.\exp((E-E^{1/2})F(z_B-z_A)/RT) \quad (12)$$

thereby indicating directly additional model dependent tests which can be made. One immediate feature is that if the current crosses zero in a series of multicycling experiments the I_1/E relationship follows that of equation (10) quite irrespective of the electrode kinetics.

At large k_f (& k_b) i.e. reversible electron transfer I_1 adheres likewise to the form of (10) & thus throughout the current variation depends solely on potential & not upon the shape & speed of the variation of potential with time. One consequence is that the overlay of I_1 as potentials are retraced can be used as a test of adherence to this regime of electron transfer & additionally provides for a substance known to exhibit fast electron transfer a simple routine for estimating the uncompensated resistance R_u. This is carried out, for example, by creating new potential axes via $E_{new}=E_{old}$ -i.R with variation of R until overlay is achieved & this value then applies to other experiments in the same medium.

The form of (10) also indicates immediately that the function dI_1/dE also 'overlays' on retracing potentials & its symmetrical sech squared form centred at $E_{1/2}$ has a 'half width' of $(R.T/(F(z_B-z_A)/RT))\ln((2^{1/2})+1)/(2^{1/2}-1))$ i.e. $90.53/(z_B-z_A)/RT)$ mV at 298K. The symmetry of the peaks makes in many ways a more satisfactory 'pictorial 'display of results than the cyclic voltammogram itself & via any displacements indicates any onset of heterogeneous kinetic control. Likewise as will be indicated later any shrinkage of the return peak can indicate the influence of a following irreversible chemical reaction.

In the electrochemically irreversible case k_b is negligible & (9) takes the form

$$i= k_f\ I_{lim}/D_A{}^{1/2}\ -\ (k_f/D_A{}^{1/2})I_1 \quad (13)$$

yielding a logarithmic form now rewritten as

$$E-E^0=-(RT/(\alpha.F.(z_B-z_A)))[\ln((I_{lim}-I_1)/i)-\ln(D_A{}^{1/2}/k_f{}^0)] \quad (14)$$

& thus providing a test on a single curve for a complete data set or combinations of such sets as for example sweep rate is varied. Note that the above allows determination of $k_f{}^0$ only via separate determination of E^0. The quasi-reversible regime can be likewise treated.

2.3. KINETIC CONVOLUTIONS.

If the E_1 reaction (1) is followed by the irreversible first order homogeneous process

$$B \longrightarrow P$$

with rate constant k_c then the convolution treatment is similar to the above except that the concentration of B at the electrode is now given by [9]

$$C_B^* = I_2/ (z_B-z_A)\, F\, A\, D_B^{1/2} \tag{15}$$

where the 'kinetic' convolution I_2 is

$$I_2(t) = \pi^{-1/2}\int_0^t i(u).\exp(-k_c(t-u))/(t-u)^{1/2}\, du \tag{16}$$

C_A^* is given as before but because the diffusional field of B is denuded by reaction it returns after a potential cycle more sluggishly to the bulk value. This is readily seen in the behaviour of I_1.

Since C_B^* tends to zero on return to potentials sufficiently prior to E^0 then one procedure to assess k_c can be based on inserting test values of the rate constant until such a condition is met. Care has to be exercised here in the choice of experimental conditions since the time-scale should not be extended such that much of B is effectively removed otherwise sensitivity of the test procedure is also jeopardised. In cyclic voltammetry this would require some retention of the return peak. Once k_c is established fast electron transfer is indicated if the logarithmic relationship equivalent to (10) i.e.

$$(E-E_{1/2})F(z_B-z_A)/RT)=\ln((I_{1lim}-I_1)/I_2) \tag{17}$$

is found to be held. As in the absence of chemical reaction (17) can also be used via any current crossing points to estimate $E_{1/2}$ independent of the electron transfer kinetics. Simultaneous measurements of both $E_{1/2}$ & k_c can be made if test values of k_c vs the $E_{1/2}$ calculated from (17) are plotted for differing sweep rates & the coincidence found.

In chronoamperometry linearity of the current & the sum of the convolution terms is obtained only with the correct choice of k_c thus providing a route to the latter. Alternatively the return of I_2 to zero in a suitable double pulse experiment provides one procedure.

One additional feature of I_2 should be mentioned at this stage namely that in the limit of large k_c the I_2 value approaches $i(t)/k_c^{1/2}$. This has obvious implications in simplifying all expressions containing I_2 in this kinetic regime.

$C_{rev}E$ & EC_{rev} schemes introduce the equilibrium constant K into expressions for the electrode concentrations, parameter extraction is feasible at the equal (chemical) D limit (see below).

The EC_{cat} reaction scheme [10] in which A is recycled via the first order homogeneous reaction of B with a suitable reagent in excess leads to an identical expression for C_B^* to the EC_{irrev} case. If diffusion coefficients are equal I_2 also replaces I_1 in the expression for C_A^*. This system in which the simple electron transfer expressions have I_2 replacing I_1 is simple to deal with via the ideas discussed. When the diffusion coefficients are unequal C_A^* depends on a more complex convolution described in outline below.

2.4. DEALING WITH UNEQUAL DIFFUSION COEFFICIENTS.

Inequality in diffusion coefficients of electroactive species merely causes matching of fluxes to occur at unequal shifts of concentrations from initial conditions at the electrode. The effect is to shift the axis to be 'centred' on the $E_{1/2}$ potential rather than E^0. The situation for species engaged on both sides of chemical reactions is more complex & in some cases demands rather elaborate forms of convolution to be made. Even the ER_{rev} equal D case replaces the I_2 of equation (17) say with $(I_1+K.I_2)/(1+K)$ thus increasing the complexity of parameter extraction.

In this section we use the EC_{cat} system to illustrate one such solution. Using the Laplace transforms leads to the following expressions for the concentrations of species A & B at the electrode in terms of the convolutions:

$$C_A^* = C_A^{bulk} - (I_1 - I_3)/(z_B - z_A)FAD_A^{1/2} \qquad (18)$$

$$C_B^* = I_2/\,(z_B - z_A)FAD_B^{1/2} \quad \text{i.e. the latter as for } EC_{irrev}. \qquad (19)$$

Here the term I_3 is a new type of convolution defined as follows:

$$I_3(t) = \int_0^t i(u)\,\phi(t-u)\,du \qquad (20)$$

The term $\phi(t)$ has different expressions depending on the relative magnitude of the diffusion coefficients of species A & B:

$$D_A = D_B : \quad \phi(t) = (1-\exp(-k_c.t))/(\pi t)^{1/2} \qquad (21)$$

$$D_A > D_B : \quad \phi(t) = (2a/\pi^{1/2})\,[F(at^{1/2})-(\exp(-k_c t)F(bt^{1/2}))] \qquad (22)$$

$$D_A < D_B : \quad \phi(t) = -(2a/\pi^{1/2})\,[G(at^{1/2})-(\exp(-k_c t)G(bt^{1/2}))] \qquad (23)$$

where a & b are the positive quantities:

$$a = (D_A k_c / |D_B - D_A|)^{1/2} \quad \& \quad b = a.d^{1/2} \quad \text{with} \quad d = (D_B/D_A)^{1/2} \qquad (24)$$

The function $F(x)$ is Dawsons Integral & $G(x)$ is $(1/2)\pi^{1/2}\exp(x^2)\text{erf}(x)$.

In the important limiting case where $|D_A - D_B| \longrightarrow 0$ (i.e. the frequently encountered situation in which the diffusion coefficients of A & B are nearly equal), it becomes very difficult to evaluate the above numerically. As a consequence, the approximate form shown below may be used.

$$D_A \approx D_B : \quad \phi(t) = [(\pi t)^{-1/2}((1-\exp(-k_c t))][1-((d-1)/2k_c t)] \qquad (25)$$

In the special case where $D_A = D_B$ i.e. **d=1**, the above equation reduces exactly to (21). This demonstrates the consistency of the above approximate form with the exact treatment described earlier.

In cyclic voltammetric experiments k_c can be found as described for EC_{irrev} from the return of I_2 towards zero. The requirement of C_A^* approaching zero after the wave i.e. a plateau region of I_1-I_3 allows the diffusion coefficients to be estimated.

2.5. COMPUTER IMPLEMENTATION OF CONVOLUTIONS.

2.5.1 *The I_1 Convolution.* The function $I_1(t)$ must by necessity be evaluated numerically using, for example, the closed Newton-Cotes formulae (the Trapezoidal & Simpson's rules) for equally-spaced abscissae (the time axis is divided into N equal intervals of Δt). This can be done by direct numerical solution of equation (4) or by using special algorithms e.g. the Gamma function ($\Gamma(x)$):

$$I_1(k\Delta t) = \frac{1}{\sqrt{\pi}} \sum_{j=1}^{j=k} \frac{i(j\Delta t - 1/2\Delta t)\sqrt{\Delta t}}{\sqrt{k - j + 1/2}} \tag{26}$$

$$I_1(k\Delta t) = \frac{1}{\sqrt{\pi}} \sum_{j=1}^{j=k} \frac{\Gamma(k - j + 1/2)}{(k - j)!} \sqrt{\Delta t} \cdot i(j\Delta t) \tag{27}$$

where **j** represents the integration index i.e. $\mathbf{j=u/\Delta t}$. The above equations give the approximate value of I_1 at time $\mathbf{t=k\Delta t}$ (k ranges from 0 to N). Generally, as far as computational speed is concerned, the Γ function algorithm is relatively faster than the first & when the current rises from the base line as in a cyclic voltammogram also gives good accuracy (see later). However it was found that the Γ function algorithm converges much slower in situations where the current data at the beginning is very important (e.g. in chronoamperometric data). For comparison the convolution of the Cottrell behaviour $f(t) = at^{-1/2}$ should yield I_1 constant at $a.\pi^{1/2}$. The most recent data can also present a problem since this block is curtailed least & the crudity of the algorithm has a major effect. We endeavoured to limit errors by polynomial fitting early & late data & introducing 100 small intervals to replace these single segments. The results of this are illustrated in figure 1 for the $t^{-1/2}$ variation. Other powers of time from -2 to +2 (all of which yield analytical solutions for I_1) were also used to test accuracy in a variety of situations.

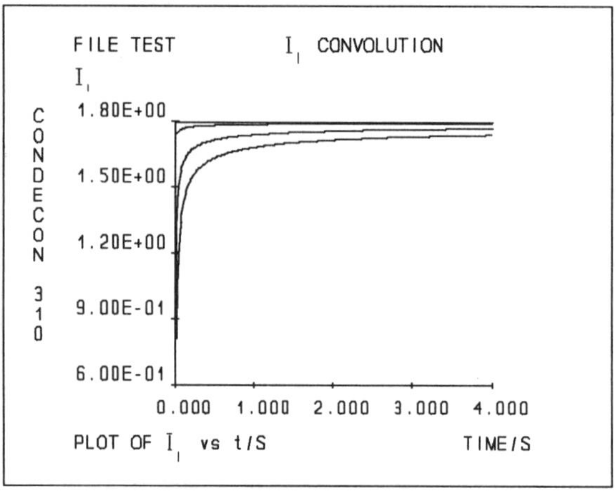

Figure 1. A comparison of I_1 calculated by the various methods. Outwards from the theoretical step function these are positioned in the order (a) modified first algorithm (b) unmodified first algorithm (c) Γ function algorithm.

2.5.2. The I_2 & I_3 Convolutions. The exponential scaling factor additional in the I_2 convolution presents problems when k_c is large since a few 'recent' segments then contribute the majority of the estimate. At the limit of large k_c the $I_2(t)$ approaches $i(t)/k_c^{1/2}$ i.e. the current may be used directly to determine I_2. Sophisticated smoothing & multi-differentiation routines of the Savitsky-Golay type were already available to us & enabled the following highly accurate procedure to be enacted. Taylor expansion within the convolution gives at, large $k_c t$, the form

$$I_2(t) = i(t)/k_c^{1/2} + (1/2k_c^{3/2})(di/dt)_t - (1/8k_c^{5/2})(d^2i/dt^2)_t + \text{higher terms} \qquad (28)$$

& provided, using the three terms shown, a highly accurate routine to extend to the range covered by the straightforward method.

The determination of I_3 requires the re-evaluation of finite integrals at each data point of time. These integrals however, have well established numerical methods associated with them (see for example Numerical Recipes or The Handbook of Mathematical Functions, Abramowitz & Stegun, Dover Press) & presented no difficulties in attaining beyond 10 digit accuracy.

3. Simulation of Data.

From the outset the simulation program developed at Leeds incorporated together linear sweep voltammetry, chronoamperometry & chronopotentiometry techniques using as far as possible common routines. The combination with convolution techniques to assess accuracy reliably over a wide range of parameters provided rapid & essential guides to the ongoing development & yielded ultimately a versatile & accurate tool for our research & teaching activities. At present 14 different mechanisms can be simulated including all the common variants. All parameters can be varied quite independently over the widest possible range & the accuracy checks made give confidence to all areas covered. Direct comparison of simulated & experimental voltammograms or their deconvolutions etc. can be routinely practised since all data is transportable between segments of the electrochemical suite of programs. Production of working curves for any of the voltammetric techniques is available & is particularly important when the mechanism is not amenable to the convolution techniques.

It is beyond the scope of the present article to describe in detail the evolution of our simulation program over a period of some years. The book "Digital Simulation in Electrochemistry" by Britz & references [11-15] provide clear & concise guidelines to implementing the Implicit Crank Nicolson Method used & it was only necessary to develop certain areas to obtain highly accurate procedures. The $\delta x/2$ 'shift' was used together with a 7 point parabolic fit in all implementations but the greatest enhancement in accuracy came in the treatment of chemical reactions where introducing polynomial fitting in iterative routines gave greatly improved accuracy.

4. A New Treatment of Chronoamperometric Data for the E_1 Process.

We have made over a period of some five years use of an alternative general method for treating E_1 chronoamperometric data at planar electrodes which allows simple & accurate 'linear' tests & determination of the various parameters. It generates simple 'linear' tests for model adherence in the same manner as the convolutions but is less demanding in computational requirements. The method has two additional major advantages. Firstly the charging currents do not persist in effect past their own time-scale in the way that they influence the convolutions. This serves to reduce the time at which data treatment becomes accurate & increases the magnitudes of the heterogeneous rate constants which can be reliably assessed. Secondly filtering of regularly varying noise (for example mains 'hum') can be incorporated by suitable choice of the intervals over which integration of the current is carried out. It should be pointed out that the general method will work for a wide variety of other functions & carries the same advantages. A general review of this will be published shortly.

If only A is present in the bulk the chronamperometric current variation follows the well established form

$$i(t)= a.\exp(b^2.t).\mathrm{erfc}(bt) \quad (29)$$

where erfc is the error function complement & a & b are given by

$$a= I_{lim}(k_f/D_A{}^{1/2}) \ \& \ \ b= (k_f/D_A{}^{1/2}) + (k_b/D_B{}^{1/2}). \quad (30,31)$$

The current variation is readily converted to an EXACT & UNIQUE 'step' relationship of the form

$$\Delta Q/ \Delta i=(2/\pi^{1/2})(a/b)(\Delta t^{1/2}/ \Delta i) \ + b^{-2} \quad (32)$$

where the Δ terms are the integration of that variable over a given but arbitrary interval of the data. For example $\Delta i= i_2-i_1$ & $\Delta t^{1/2}= t_2{}^{1/2}-t_1{}^{1/2}$ for the interval between 1 & 2. Likewise the ΔQ term is the charge passed during that same interval & of note is unaffected by charging or other current contributions outside the segment range. It should be stressed that such a solution allows successive or overlapping segments of varying size to be combined in any way to provide a test for the data set as a whole. Choosing fixed time intervals to determine the segments has the obvious application to noise reduction eluded to earlier. The alternative form

$$\Delta Q/ \Delta t^{1/2}=(2/\pi^{1/2})(a/b) \ + b^{-2} \ (\Delta i/\Delta t^{1/2}) \quad (33)$$

where the emphasis of parameters in slope & intercept is reversed also has a changed distribution of a given data set on a linear plot with a potential advantage under given circumstances. The parameters can be extracted in a similar manner to that envisaged for the linear variation of i & I_1 &, as in that case, if both materials are initially present then incorporation of this situation enables both diffusion coefficients & hence E^0 to be determined.

5. Applications.

In this section examples of data treatment are chosen to exemplify the features of the computer based treatments discussed, all of which should have wide relevance in organometallic chemistry. No particular attempt is made to determine the requirements to distinguish between different possible mechanisms from a 'raw' cyclic voltammogram or chronoamperogram but the treatments made rather serve to emphasise the essential features of the individual reaction schemes. Single cyclic voltammograms, & chronoamperograms even more so, are rather devoid of features but 'shape' is portrayed in a more tangible sense by the convolutions etc. & ultimately by any linear relationships if these become available. There is, of course, no substitute for careful technique & this must always ensure that the widest possible range of time-scale is utilised to create regions of sensitivity (& perhaps insensitivy) to the different kinetic factors thought to be involved. Some comments on the individual systems are included but in no way should be considered exhaustive.

5.1. SIMPLE ELECTRON TRANSFER.

5.1.1. *General Features.* The features of voltammetric results for fast electron systems are simple & well understood. Tests for electrochemical 'reversibility' could include ultimately the appropriate logarithmic expression on which the voltammetric & chronoamperometric results, if desired, could be co-displayed. Indications of quasi-reversible electron transfer might arise from the observation of greater displacement of current peaks as sweep rate is increased but the same observation of the displacement of the I_1 convolutions would additionally check via the return to zero that, for example, no chemical reaction was influencing the system. The electrochemically 'irreversible' system in many ways poses most difficulties in testing for adherence & determining the rate parameters individually & the reverse sweep voltage should always be taken as far as possible to attempt experimentally to observe (even) the onset of a reverse wave. Following fast chemical reactions give rather similar looking cyclic voltammetric responses unless the sweep speed here can be increased to reveal a return wave. The accuracy of a simulation is readily checked in an irreversible system by producing the return wave & treating the whole data as a quasi-reversible system. The outward response & its return can be checked via the slow electron transfer logarithmic function. Once mechanism is established kinetic parameters, when accessible, are possibly best determined via chronoamperometry.

Figures 2-5 illustrate the features discussed using simulated data.

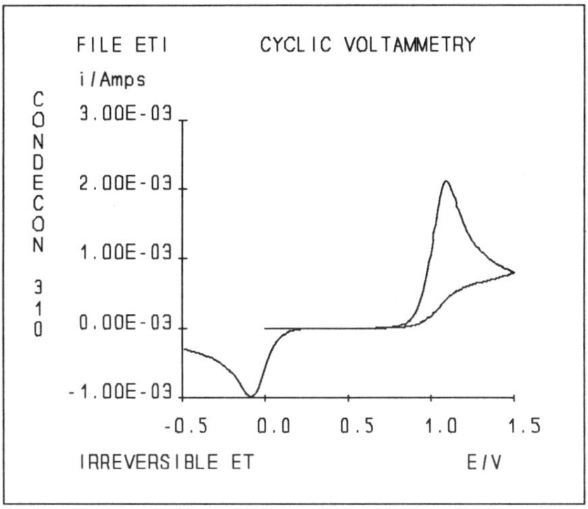

Figure 2. Cyclic Voltammogram for an irreversible ET reaction simulated with α=0.5, k^0=10^{-8}m/s, D=10^{-8}m^2/s & sweep rate v=1V/s. E^0 is 0.5V.

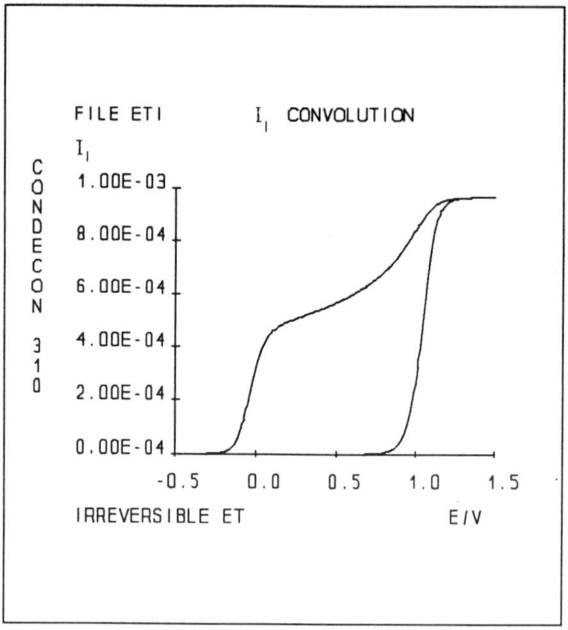

Figure 3. I_1 convolution for the same irreversible ET reaction. Note the return after the second wave.

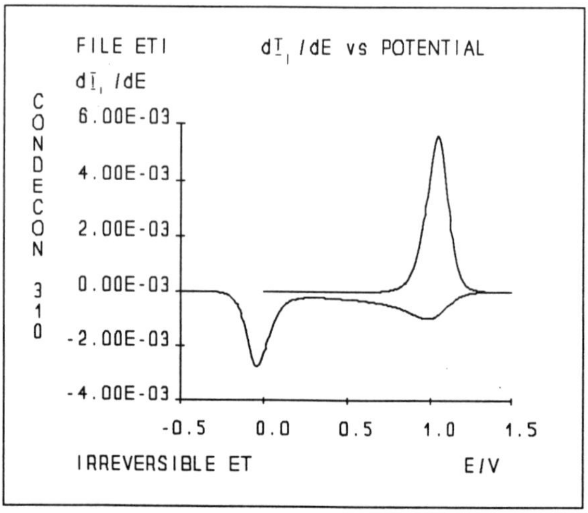

Figure 4. Deconvolution of same irreversible ET reaction. Note the two return peaks.

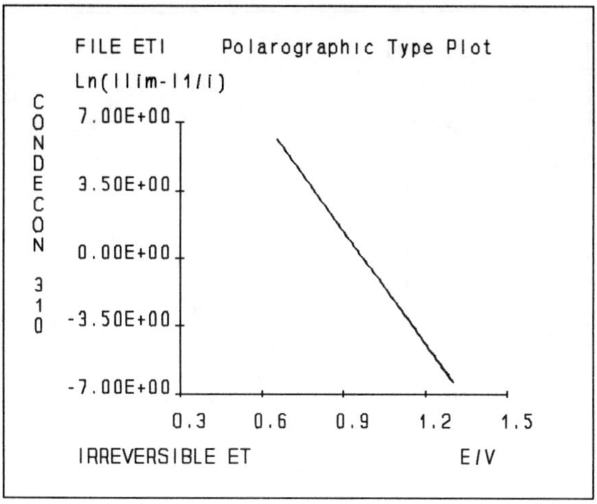

Figure 5. Logarithmic Plot of data of same irreversible ET reaction. Note the data cover both parts of the first wave.

5.1.2. *Determination of heterogeneous kinetics.* This section illustrates in outline the determination of the parameters in the quasi-reversible Fe^{2+}/Fe^{3+} system in aqueous electrolyte using the 'step' relations. In these experiments both species were present & both anodic & cathodic branches were examined in the same set of experiments.

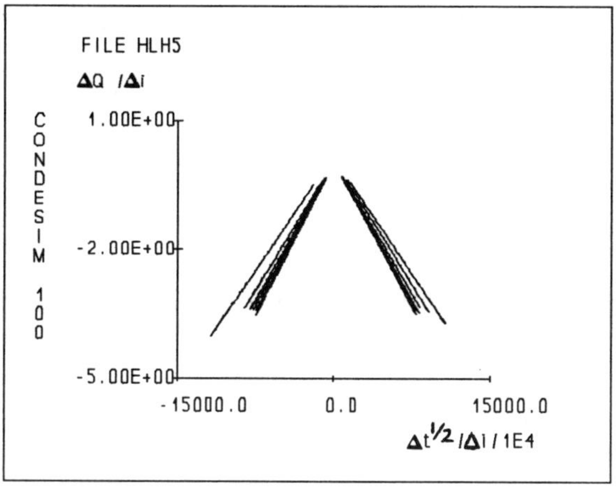

Figure 6. ΔQ/ Δi vs $\Delta t^{1/2}$/ Δi for the quasi-reversible Fe^{2+}/Fe^{3+} system in aqueous electrolyte. Step voltages range over ± 100mV.

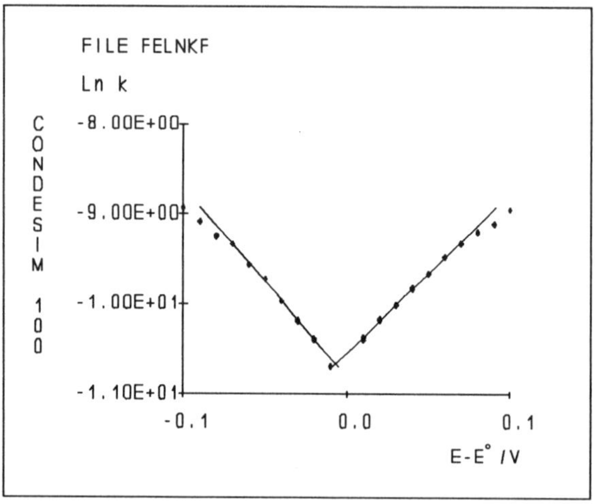

Figure 7. Ln k vs E-E^0 for the quasi-reversible Fe^{2+}/Fe^{3+} system in aqueous electrolyte. Data calculated from step function analysis as described in the text.

5.2. FOLLOWING CHEMICAL REACTIONS.

5.2.1. The EC_{irrev}Case. Since 1983 we have made extensive use [16] of the I_2 convolutions in determining rate constants for following chemical reactions. Studies have involved 21 Cr/Mo/W mono-metallic phosphine carbonyls & some bi-metallic derivatives, a comprehensive group of metallo-borane complexes & a variety of other Cu, Ag, Pd, Rh compounds. In the first group a temperature range was covered & extensive thermodynamic & kinetic information was extracted for the effectively irreversible isomerisation reactions following the first electrochemical process. To illustrate one method we shall again use simulated cyclic voltammetric data since the results derived serve to underline the accuracy of the simulation package used. In this simulation the electron transfer was fast (k^0=10 m/s) on the time-scale of the voltammogram (2s). k_c was 1s^{-1}. Figure 8 illustrates the logarithmic plot which would be incidently common for all $E_{rev}C_{irrev}$ processes except for position on the voltage axis..

5.2.2. *An example of EC_{rev}.* In the equal D case the sum of I_1 & K.I_2 tends to zero at the end of the return wave when the sum of the forward & backward chemical rate constants is inserted for k_c. K is k_{for}/k_{back}. Thus test values of k_c yield K via the above condition & if suitable sweep rates are chosen then the crossing of K vs k_c curves indicates the appropriate values of the individual rate constants. Figure 9 illustrates the recovery of parameters for the case K= k_{for}=k_{back}=1 involving slow electron transfer- k_f^0=1X10^{-6}ms^{-1}.

5.2.3. *The EC_{cat}Case.* This important area has been the subject of much practical & theoretical interest. In the simplest variant (see Section 2.3) the I_2 convolution merely replaces I_1 in the E_1 relationships, for example in the 'polarographic' logarithmic expression. The chronoamperograms & voltammograms after the 'wave' are characterised by plateau regions as the starting material is recycled. No example is given.

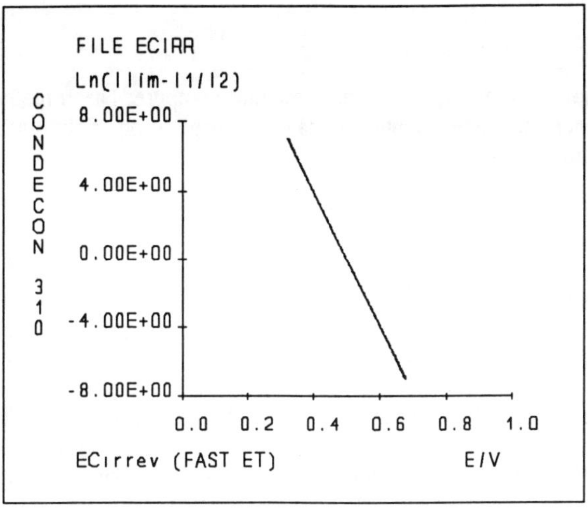

Figure 8. The logarithmic plot E vs $\ln((I_{lim}-I_1)/I_2)$ for the above EC_{irrev} data. The plot is invariant to sweep rate.

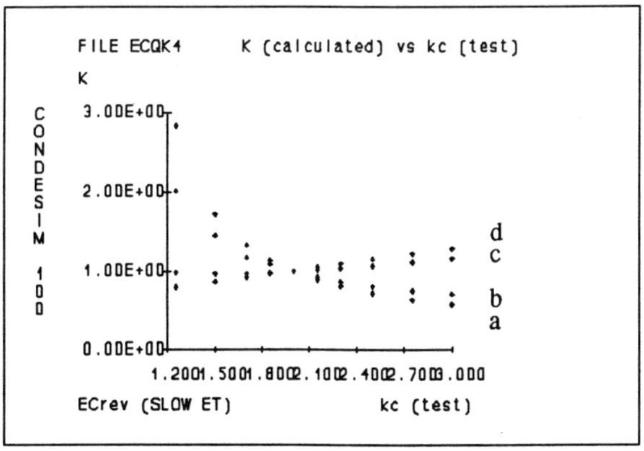

Figure 9 Plots of K vs k_c (test) for simulated data for an EC_{rev} process with slow ET at sweep rates (V/s) of (a) 10 (b) 4 (c) 1 (d) 0.5. Note the well defined intersection.

6. Conclusions.

The paper has attempted to re-emphasise the view that computer based methods can be made to add more to electrochemical experimentation via the incorporation of simulation programs & data treatment by convolution techniques.

7. References.

7.1 GENERAL TEXTS.

The following contain much information on the theoretical aspects of electrochemistry.

Bard, A.J. & Faulkner, L.R. (1980) Electrochemical Methods, Wiley, New York.
Britz, D. (1981) Digital Simulation in Electrochemistry, Springer-Verlag, Berlin.
MacDonald, D.D. (1977) Transient Techniques in Electrochemistry, Plenum Press, New York.
Southampton Group (1985) Instrumental Methods in Electrochemistry, Ellis Horwood, Chichester.
Thirsk, H.R. & Harrison, J. A. (1972) A Guide to the Study of Electrode Kinetics, Academic Press, London.

7.2 PAPERS.

1. Imbeaux, J. C. & Savéant, J-M. (1973) J. Electroanal. Chem., 44, 169.
2. Savéant, J-M. & Tessier, D. (1975) J. Electroanal. Chem., 61, 251.
3. Savéant, J-M. & Tessier, D. (1975); J. Electroanal. Chem., 65, 57.
4. Oldham, K. B.(1972) Anal. Chem., 44, 196.
5. Oldham, K. B.(1986) J. Chem. Soc., Faraday Trans. 1, 82, 1099.
6. Nicholson, R.S. & Shain, I. (1964) Anal. Chem. 36, 706.
7. Matsuda, H. & Ayabe, Y. (1955) Z. Elektrochem. 59, 494.
8. Nadjo, L. & Savéant, J-M. (1973) J. Electroanal. Chem., 48, 113.
9. Oldham, K. B. (1983) J. Electroanal. Chem., 145, 9.
10. Savéant, J-M. & Vianello, E. (1963); Electrochim. Acta, 8, 905.
11. Britz, D. (1987) Anal. Chim. Act. 193, 277.
12. Heinze, J., Storzbach, M. & Mortensen, J. (1984) J. Electroanal. Chem. 165, 61.
13. Britz, D. & Thomsen, K. (1987) Anal. Chim. Act. 194, 317.
14. Britz, D. (1988) J. Electroanal. Chem. 240, 17.
15. Britz, D., Heinze, J., Mortensen, J. & Storzbach, M. (1988) J. Electroanal. Chem. 240, 27.
16. Dobson, I.D., Taylor, N. & Tipping, L.R.H. (1986) Electrochemistry, Sensors & Analysis, Elsevier, Amsterdam, pp. 61-75.

NUMERICAL METHODS IN SYNTHESIS AND ANALYSIS OF ELECTROCHEMICAL RESPONSES

RENATO SEEBER, MARIA I. PILO and GAVINO SANNA
Department of Chemistry, University of Sassari
Via Vienna, 2
07100 Sassari
Italy

ABSTRACT. Though literature reports the trends of the significant quantities of the responses for different potential controlled techniques for many different types of mechanisms, direct comparisons between experimental and theoretical responses are often necessary. For an experimenter who wants to calculate a theoretical response, numerical methods are particularly appealing. In particular, digital simulation methods based on finite difference expressions of the differential equations accounting for diffusion and kinetics of the electrode process are probably the simplest numerical techniques to use. Many different models have been proposed. Nonuniform space or space-time discretizations have been suggested to speed up the simulation procedure. As to the analysis of the response, more or less trivial treatments (filtering of noise, background subtraction, convolution or deconvolution procedures, *etc.*) can lead to signals where the significant quantities can be estimated more accurately. The qualitative and quantitative definition of the operative mechanism can be in part or completely computerized. A key point consists of some general optimization method, such as the Simplex.

Controlled potential techniques [1] are particularly suitable for getting information on an electrode process, *i.e.* for establishing the type of operative mechanism and for determining the values of the relevant thermodynamic and kinetic parameters. In other words, to define the redox behaviour of a chemical compound. Based on an impressive amount of theoretical work, literature reports the expected values of the parameters characterizing the current *vs.* time or current *vs.* potential responses, that are obtained by different electrochemical techniques for the most common types of electrode processes. These values depend on intrinsic (thermodynamic and kinetic) parameters of the chemical system, as well as on experimental variables. In particular, tens of articles have been published on linear sweep and cyclic voltammetry [2], which are often the techniques that furnish the largest quantity of information with this respect. Great attention has also been devoted to single and double potential step chronoamperometry, to controlled potential coulometry, to d.c. and a.c. polarography, to voltammetry with rotating disk or ring-disk electrodes and to different pulse techniques. The theoretical and experimental aspects of coupling the electrochemical measurement with a spectroscopic monitoring of the solution near the working electrode (*e.g.* UV-visible spectroelectrochemistry with OTTLE cells and *in situ* e.s.r. spectroelectrochemistry) have also been extensively analysed.

A first approach to the elucidation of the mechanism of an electrode process usually consists of a comparison between experimental responses and theoretical data reported in the literature. This often allows the identification of the type of mechanism and a proper use of so-called *working curves*, reporting the trend of significant quantities of the responses as a function of a suitable combination of intrinsic and experimental parameters, may also lead to a quantitative definition of the mechanism.

A. J. L. Pombeiro and J. A. McCleverty (eds.),
Molecular Electrochemistry of Inorganic, Bioinorganic and Organometallic Compounds, 453–466.

However, both in a qualitative and in a quantitative study of an electrode process, it is particularly important for an experimenter to be able to compute on his/her own theoretical responses that fit the experimental curves. In principle, one has first to write a suitable set of partial differential equations together with initial and boundary conditions, accounting for: *i)* mass transfer process of electroactive species, intermediates and products; *ii)* nature of chemical reactions possibly coupled to the charge transfer; *iii)* initial concentrations of all species involved in the overall process; *iv)* concentrations at the electrode and in the bulk of the solution as the experiment proceeds. The resulting so-called *boundary value problem* may be referred to either unstirred or stirred solution, depending on the technique used. In any case, an excess of inert salt is supposed to be present in the solution, so that mass transfer by migration can be disregarded. Planar or nonplanar diffusion can be considered; the use of ultramicroelectrodes implies non trivial additional complications. The second stage is to find a solution for the set of equations. To this purpose, three basic approaches are possible: *i)* an analytical solution of the mathematical problem; *ii)* a partial analytical solution, completed by a numerical method such as numerical integration; *iii)* a numerical solution based on the use of a digital computer. Though the first approach would be preferable since it leads to an exact solution, its applicability is however limited to relatively simple systems; in addition, an experimenter is usually unfamiliar with the mathematical instruments required. For the same last reason the second choice is also poorly appealing. Luckily, the third (last) option is practical also for those who are unwilling to learn a lot of mathematics. In this context, two different approaches are possible: either a numerical solution of the set of equations or a *simulation* of the experiment. In the former case - polynomial approximation, orthogonal collocation method - a trial function made up with orthogonal Legendre polynomials can be used to approximate the concentration profile of the species involved in the electrochemical process, which is a general method to solve differential equations. We shall however address our discussion to the latter approach, *i.e.* towards the various *digital simulation* techniques [3,4].

For the sake of simplicity, we shall consider the case of purely diffusion conditions, *i.e.* we shall disregard taking into account hydrodynamic complications due to mass transfer *via* convection or associated with the dropping mercury electrode. More exactly, we shall deal with the case of semiinfinite linear diffusion condition, *i.e.* we shall also disregard considering effects of non-planarity of the electrode surface and of limited diffusion layer thickness. Apart from the cases of a hanging mercury electrode, of ultramicroelectrodes and of thin-layer cells, these assumptions fit well the usual experimental conditions for studies in the field of molecular electrochemistry. Furthermore, as pointed out at the beginning of the present survey, our attention will be mainly devoted to controlled potential techniques such as linear sweep and cyclic voltammetry or chronoamperometry, rather than to constant current techniques, such as chronopotentiometry.

In the case of an uncomplicated charge transfer:

$$A + e^- \underset{k_b}{\overset{k_f}{\rightleftharpoons}} B$$

the calculation of the *i vs. t* or *i vs. E* response implies the evaluation of the flux of electroactive species at the electrode surface at different times from the start of the polarization, or at different potentials in a linear scan. This requires the computation of the concentration profiles for the species involved in the electrode process, $a(x;t)$ and $b(x;t)$ in our simple case. The following system of differential equations (Fick's second law) has to be solved:

$$\left.\begin{aligned}(\partial a/\partial t) &= D_A(\partial^2 a/\partial x^2)\\ (\partial b/\partial t) &= D_B(\partial^2 b/\partial x^2)\end{aligned}\right\}$$

under the appropriate initial :

$$\left.\begin{aligned}a &= a^*\\ b &= b^* \ (=0)\end{aligned}\right\} \qquad t = 0\,;\, x \geqslant 0$$

and boundary conditions :

$$\left.\begin{aligned}a &\dashrightarrow a^*\\ b &\dashrightarrow b^* \ (=0)\end{aligned}\right\} \qquad t > 0\,;\, x \dashrightarrow \infty$$

$$\left.\begin{aligned}D_A(\partial a/\partial x) &= k_f a - k_b b\\ D_B(\partial b/\partial x) &= k_b b - k_f a\end{aligned}\right\} \qquad t > 0\,;\, x = 0$$

where a and b indicate the concentrations of the species A and B, respectively, at time t and distance x from the electrode surface, D_A and D_B are the relevant diffusion coefficients, a^* and b^* are the values of a and b in the bulk of the solution, k_f and k_b are the forward (reduction) and backward (oxidation) charge transfer kinetic constants, respectively; they both depend on the electrode potential, E, on the standard heterogeneous kinetic constant, $k_{s,h}$, and on the charge transfer coefficient, α. The flux of electroactive species at the electrode can be expressed by the following relationships:

$$q_{x=0} = D_A(\partial a/\partial x)_{x=0} = -D_B(\partial b/\partial x)_{x=0} = k_f a_{x=0} - k_b b_{x=0}$$

The digital simulation method which is closest to a physical view of the phenomenon requires looking at the solution as divided in volume elements (*boxes*), equal to one another, extending away

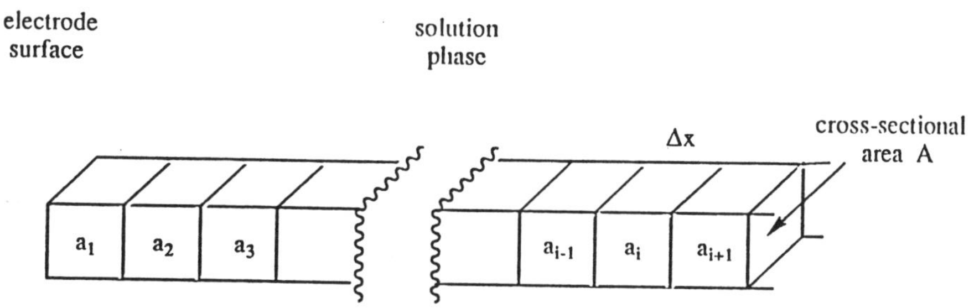

Figure 1. Discrete model of the solution.

from the planar electrode surface, of cross-sectional area (A) parallel to the electrode, of thickness Δx small enough to consider constant the concentration of each species inside [3] - see Figure 1. The time flowing is also thought as discretized in (small) Δt intervals.

If a_{i-1}, a_i, and a_{i+1} indicate concentration values of A diffusing species in the volume elements with index i-1, i, and i+1, respectively, the change in concentration in the i-th element occuring in a time interval Δt is related to the fluxes from or to the right side and to or from the left side, q_r and q_l, respectively, by the following relationship:

$$\Delta a_i/\Delta t = \Delta N_i/(A\Delta x\Delta t) = (q_r - q_l)/\Delta x$$

where ΔN_i is the change in the number of moles occuring in the time interval Δt. Since $q = D(\partial a/\partial x)$, the expression above can be formulated in a finite difference notation, where the flux between two adjacent volume elements is approximated by the difference between the concentration in the elements, divided by the distance of the relevant centres:

$$q_r = q_{i+1,i} \approx D(a_{i+1} - a_i)/\Delta x$$
$$q_l = q_{i,i-1} \approx D(a_i - a_{i-1})/\Delta x$$

Hence:

$$\Delta a_i/\Delta t \approx D[(a_{i+1} - a_i)/\Delta x - (a_i - a_{i-1})/\Delta x]/\Delta x$$

i.e.:

$$\Delta a_i \approx D\Delta t/\Delta x^2(a_{i+1} - 2a_i + a_{i-1})$$

The basic formula which allows the computation of **one new** concentration value from **three old** values, at a Δt time difference is:

$$a'_i = a_i + D\Delta t/\Delta x^2(a_{i+1} - 2a_i + a_{i-1})$$

where the apex indicates a new concentration value; henceforth we will indifferently use the symbolism

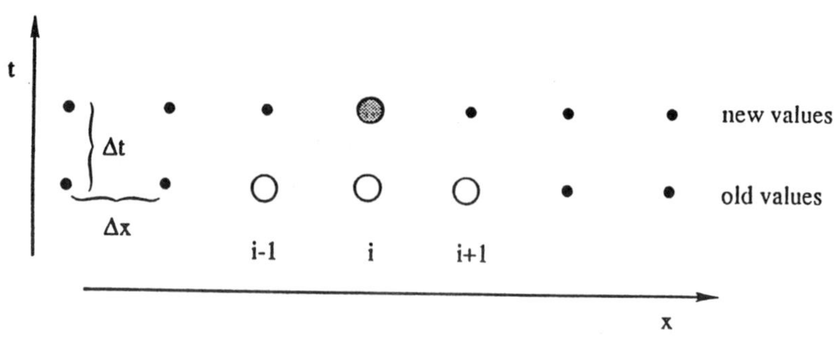

a' or $a_{t+\Delta t}$. The term $D\Delta t/\Delta x^2$ is usually indicated by the symbol β - dimensionless diffusion coefficient . It could be shown that stability in flux and concentration is only achieved when $\beta \leqslant 0.5$.

This is how the *explicit finite difference method* works in solving numerically partial differential equations of parabolic type, as those expressing the Fick's second law.

As a matter of fact, no true knowledge about how to solve partial differential equations was required so far, but we shall see presently how to get a solution by a proper mathematical approach. We have in advance to discuss how to deal with the boundaries, *i.e.* with the electrode surface and with the bulk of the solution. As to the latter boundary, the question is: *how far from the electrode the perturbation arising from the charge transfer has a significant effect?* In the case of a potential step in the limiting current region, *i.e.* in the case of a perturbation that propagates at maximum possible speed, the concentration profile of the species consumed at the electrode is expressed by:

$$a = a^{*}\mathrm{erf}[x/(2D_{\mathrm{A}}^{1/2}t^{1/2})]$$

where erf is the 'error function', *i.e.* the integrated normal error distribution.

It can be calculated that at a distance

$$x = 6D_{\mathrm{A}}^{1/2}t^{1/2}$$

the difference between a and a^* values is less than 0.01%. As a consequence, if m is the number of time intervals since the beginning of the potential step, the diffusion layer consists of a number of boxes given by:

$$n = 6D_{\mathrm{A}}^{1/2}m^{1/2}\Delta t^{1/2}/\Delta x$$

In the $(n+1)$-th box a can always be assumed to be equal to a^*.

To deal with the second boundary the representation of the discretized concentration profile reported in Figure 2 can be useful.

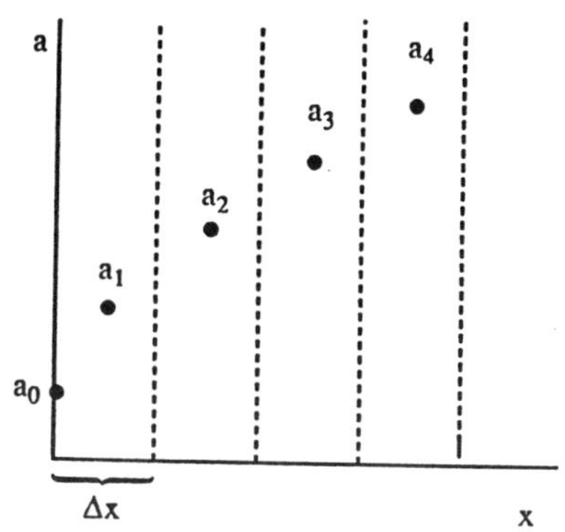

Figure 2. Discrete representation of concentration profile.

The electrode boundary is important with respect to the computation both of the flux at the electrode surface, and hence of the current flowing, and of the concentration in the first box, adjacent to the electrode. The following system of equations has to be solved:

$$q_{A,x=0} = D_A[(a_1 - a_0)/(\Delta x/2)]$$
$$q_{B,x=0} = D_B[(b_1 - b_0)/(\Delta x/2)]$$
$$q_{A,x=0} = k_f a_0 - k_b b_0$$
$$q_{A,x=0} = -q_{B,x=0}$$

which leads to

$$q_{A,x=0} = (\overline{f}a_1 - \overline{b}b_1)/[1/k_{s,h} + \overline{f}\Delta x/(2D_A) + \overline{b}\Delta x/(2D_B)]$$

where

$$\overline{f} = \exp[-\alpha nF(E - E°)/RT] \; ; \; \overline{b} = \exp[(1-\alpha)nF(E - E°)/RT]$$

so that:

$$k_f = k_{s,h}\overline{f} \, ; \; k_b = k_{s,h}\overline{b}$$

and the new concentration values in the first box can be evaluated as:

$$a'_1 = a_1 + \beta_A(a_2 - a_1) - q_{A,x=0}$$

$$b'_1 = b_1 + \beta_B(b_2 - b_1) - q_{B,x=0}$$

Relatively slow homogeneous chemical reactions coupled to the charge transfer can be easily dealt with according to this digital simulation method. As an example, in the case of a first order irreversible chemical reaction involving the electrogenerated species B (EC_{irrev} mechanism), the relevant II Fick's law should be modified as:

$$\partial b/\partial t = D_B \, \partial^2 b/\partial x^2 - k_c b$$

where k_c is the chemical kinetic constant.

The commonly used finite difference expressions accounting for the change in B concentration due to both diffusion and kinetics are the following:

$$b^{\#}_i = b_i + \beta_B(b_{i+1} - 2b_i + b_{i-1})$$
$$b'_i = b^{\#}_i - k_c\Delta t b^{\#}_i$$

If we think at the centres of the boxes as knots where concentration values are computed, there is no reason why the resulting grid should be in some way different near the electrode. It seems then a spontaneous choice to fix between the electrode and the first point in the solution a distance equal to that between two adjacent points inside the solution. In this way, we are working according to the *point method* [4]. The expression of the flux at the electrode has to be changed accounting for a Δx, rather than a $\Delta x/2$ distance between the first point in the solution and the electrode. For the sake of brevity, we shan't discuss here many aspects of the two methods, as some more or less hidden 'fudge factors' of the box method which, in fact, make things 'work' better.

Some considerations about Taylor series become urgent at this point, in order to be able to go a bit further in our survey. *What have Taylor series to do with things like these?* The Taylor series expansion for the function f(x) can be written at the point x_0 as:

forward series

$$f(x_0+h) = f(x_0) + hf'(x_0) + (h^2/2!)f''(x_0) + (h^3/3!)f'''(x_0) + \ldots\ldots\ldots$$

backward series

$$f(x_0-h) = f(x_0) - hf'(x_0) + (h^2/2!)f''(x_0) - (h^3/3!)f'''(x_0) + \ldots\ldots\ldots$$

From these equations corresponding formulas for *forward difference* and *backward difference* expressions of the first derivative can be easily drawn out:

$$f'(x_0) = [f(x_0+h) - f(x_0)]/h + O(h)$$

and

$$f'(x_0) = [f(x_0) - f(x_0-h)]/h + O(h)$$

where $O(h)$ denotes terms containing first and higher order powers of h. Suitable linear combination of the forward and backward series leads to a *central difference* expression:

$$f'(x_0) = [f(x_0+h) - f(x_0-h)]/(2h) + O(h^2)$$

Dropping the $O(h)$ and $O(h^2)$ terms we obtain approximate expressions where the so-called *local truncation error* is of order h for the forward and backward difference formulas and of order h^2 (hence lower) for the central difference expression. Figure 3 (upper abscissa scale) shows the geometric meaning of the different approximations.

By simple addition of forward to backward series expression a *central difference* formula for the second derivative is obtained:

$$f''(x_0) = [f(x_0+h) - 2f(x_0) + f(x_0-h)]/h^2 + O(h^2)$$

A local truncation error of order h^2 is present in the central difference approximation, where the $O(h^2)$ terms are disregarded.

We can now compare the approximated expressions for f ' and f " derived from Taylor series expansions with the finite difference formulation of the II Fick's law derived by following a 'physical' approach. The first derivative with respect to time, $\partial a/\partial t$, is approximated by a forward difference formula:

$$(\partial a/\partial t)_{i,t} \approx (a_{i,t+\Delta t} - a_{i,t})/\Delta t$$

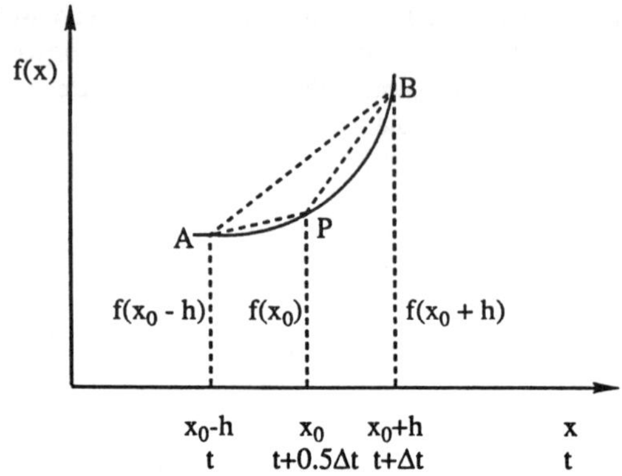

Figure 3. Approximations for the tangent at P (upper abscissa scale).
Different meaning of chord AB in approximating $\partial a/ \partial t$ in classic, Crank-Nicolson or Laasonen method (lower abscissa scale).

with a local truncation error of order Δt. On the other hand, the second derivative with respect to space, $\partial ^2 a/ \partial x^2$, is computed by a central difference approximation:

$$(\partial^2 a/ \partial x^2)_{i,t} \approx (a_{i+1,t} - 2a_{i,t} + a_{i-1,t})/\Delta x^2$$

with a local truncation error of order Δx^2. The result is an unsymmetrical expression with a local truncation error of $O(\Delta t) + O(\Delta x^2)$. It seems hence desiderable to look for an expression where the asymmetry has been removed. The *DuFort-Frankel algorithm* [4,5] is still an explicit method, since **one new** concentration value can be explicitly computed **from three old** values. It works however on three different time levels and is stable for all values of β:

$$(a_{i,t+\Delta t} - a_{i,t-\Delta t})/2 = \beta[a_{i-1,t} - (a_{i,t-\Delta t} + a_{i,t+\Delta t}) + a_{i+1,t}]$$

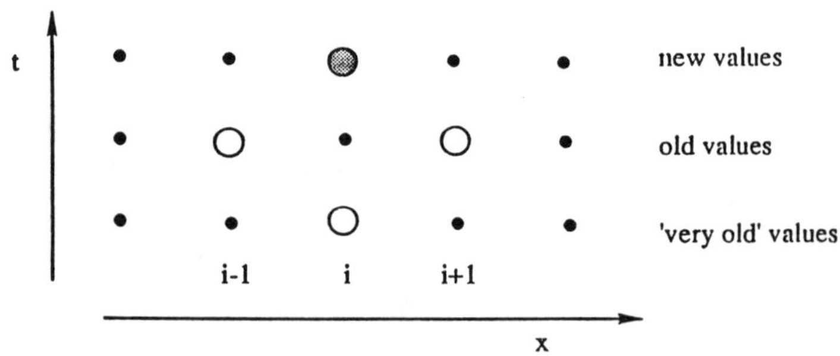

It could be demonstrated that the local truncation error is $O(\Delta t^2) + O(\Delta x^2) + O(\Delta t^2/\Delta x^2)$.

We can get central difference approximations for both time first order and space second order derivatives also without working on three levels, *i.e.* on concentration values at three subsequent iteration steps. The approximation $\partial a/\partial t \approx (a_{i,t+\Delta t} - a_{i,t})/\Delta t$ is, in fact, central difference with respect to the first derivative in the $a_{i,t+0.5\Delta t}$ point. Analogously, the space second derivative can be thought as referred to a_{i-1}, a_i, and a_{i+1} at $t+0.5\Delta t$, fictitiously computed as half-sum of a_{i-1}, a_i, and a_{i+1} at t and $t+\Delta t$, respectively:

$$a_{i,t+\Delta t} - a_{i,t} = (\beta/2)(a_{i+1,t+\Delta t} - 2a_{i,t+\Delta t} + a_{i-1,t+\Delta t} + a_{i+1,t} - 2a_{i,t} + a_{i-1,t})$$

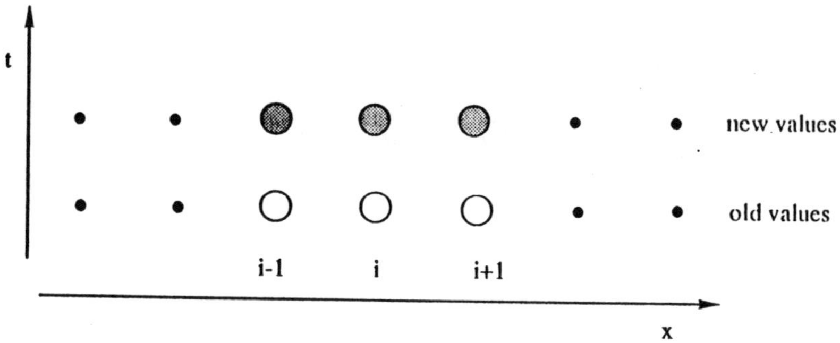

The resulting *Crank-Nicolson formula* [4] is a two-level, six point formula, with a local truncation error of $O(\Delta t^2) + O(\Delta x^2)$; it is stable for all β values. It is an *implicit method*, in the sense that it is.....not explicit: no direct expression allows the computation of a new concentration value from some old ones.

The computational way to solve the problem of diffusion by the Crank-Nicolson method is quite different from that followed in the explicit schemes. Once the old concentration profile is known, three unknowns are still present in the equation written above. If $n+1$ points are required to describe the concentration profile in the diffusion layer, n equations similar to that written above will constitute a system with $n+2$ unknowns (a_0, a_1, a_2, ..., a_{i-1}, a_i, a_{i+1}, ..., a_n, a_{n+1}). The boundary conditions (a_{n+1} = constant value; a_0 involved in Nernst's equation, or some other constraint) allow the solution of the **system of equations**, which exhibits the following general form:

$$a'_0 + da'_1 + a'_2 = e_1$$
$$a'_1 + da'_2 + a'_3 = e_2$$

.

.

$$a'_{i-1} + da'_i + a'_{i+1} = e_i$$

.

.

$$a'_{n-1} + da'_n + a'_{n+1} = e_n$$

where d and e are known quantities, depending on β and on β and old concentration values, respectively.

Looking at Figure 3, considering time on the abscissa (lower scale), $f(x_0 - h) \equiv a_i$ (old value) and $f(x_0 + h) \equiv a'_i$ (new value), the chord AB approximates the slope at A, *i.e.* at a time t, according to a forward difference scheme (classic explicit method); on the other hand, it constitutes a central difference approximation at a time $t + 0.5\Delta t$ (Crank-Nicolson). We can also use it as a backward difference approximation for the slope at B, *i.e.* at a time $t + \Delta t$ (Laasonen, fully implicit method) [4,6]:

$$a_{i,t+\Delta t} - a_{i,t} = \beta(a_{i-1,t+\Delta t} - 2a_{i,t+\Delta t} + a_{i+1,t+\Delta t})$$

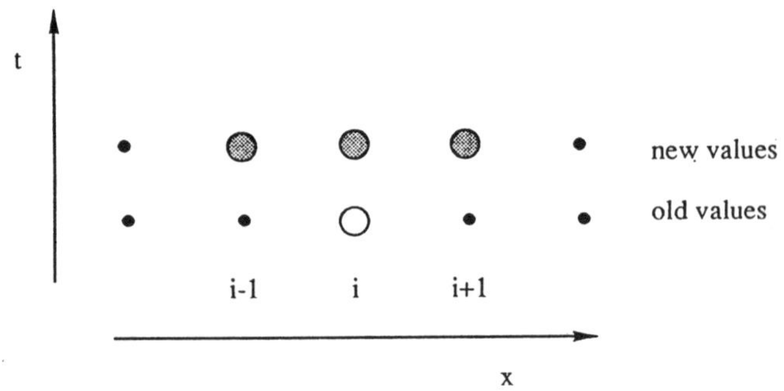

The *backwards implicit approximation*, whose solution requires, in principle, an approach similar to that of the Crank-Nicolson scheme, implies a truncation error of $O(\Delta t) + O(\Delta x^2)$. It is however stable for any value of β.

We shan't illustrate here all modifications of the schemes outlined above, that have been proposed in the abundant literature on this subject. The choice among the different possible methods can be suggested by the preference for accuracy (with respect to current or concentration values) or for short computation times. Many factors have to be taken into account. Although the attention towards implicit methods is growing in the last years more and more, complicated electrode mechanisms are often easier to formulate in the context of explicit methods; these require however deep modifications to the usual schemes when very fast homogeneous chemical reactions have to be simulated.

Remarkable saving in the computer time required by the simulation procedure can be achieved by taking advantage of the fact that the concentration profiles of the species are smoother and smoother at increasing the distance from the electrode surface, where the diffusing perturbation arises. This consideration allows the use of a space grid with wider and wider mesh - *nonuniform space grid* [7]. For the same reason it is also possible to correspondingly update the concentration values at a lower and lower frequency - *nonuniform space-time grid* [7].

In a nonuniform space discretization - box method - a possible formulation of the function defining 'centres' and boundaries of the boxes is as follows:

$$X_{i=q/2} = \Delta x \exp\{[0.5c(q-1)] - 1\}/c \qquad q = 1,2,3,..............$$

where c determines the 'rate' at which the dimension of the space elements changes; even and odd q values define 'mean points' and boundaries of the boxes, respectively. The dimensionless diffusion coefficients become lower and lower at increasing the distance from the electrode, and for every box left and right β should be defined:

$$\beta_{l,i} = D\Delta t/[(X_i - X_{i-1})(X_{i+0.5} - X_{i-0.5})]$$
$$\beta_{r,i} = D\Delta t/[(X_{i+1} - X_i)(X_{i+0.5} - X_{i-0.5})]$$

$X_{i+0.5} - X_{i-0.5}$ represents the width of the i-th box, and X_i is the relevant 'mean point'.

It is clear from the above formulas that β_l and β_r become lower and lower at increasing the distance from the electrode. In the classic explicit method, where their value should not exceed 0.5, this makes possible the use of time increments which also increase at increasing the distance from the electrode. In particular, every $p = \ln 2/(2c)$ space elements the frequency of computation is halved and the β values are correspondingly doubled. A nonuniform space-time grid is shown in Figure 4.

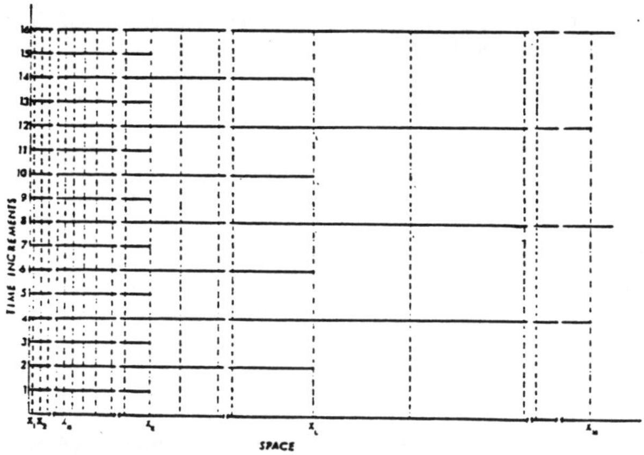

Figure 4. Nonuniform space-time grid.

Though it is clear from the foregoing that the problem of the *synthesis* of an electrochemical response can be solved by following many different approaches (we want to underline once more that the digital simulation methods, briefly discussed above, only represent one possible choice) it is much more difficult to draw general guidelines for the *analysis*. A first preliminary step may consist of some kind of elaboration of the experimental signal. With this respect, we want to cite the abundant literature on the so-called *semiintegral electroanalysis*, essentially applied to linear sweep or cyclic voltammetry; the semiintegral, *i.e.* the integration of one-half order of the current with respect to time is computed, and the analysis of the resulting curve, resembling in shape a d.c. polarographic response, offers some advantages over the direct analysis of the linear sweep or cyclic voltammetric curve:

$$m(t) = d^{-1/2}[i(t)]/dt^{-1/2} = d^{1/2}[q(t)]/dt^{1/2}$$

where $q(t)$ is the charge passed at time t.

In the case of semiinfinite linear diffusion the semiintegration coincides with the simplest case of *convolutive potential sweep voltammetry*:

$$I(t) = \pi^{-1/2} \int_0^t [i(\tau)/(t - \tau)^{1/2}]d\tau$$

Many algorithms have been proposed to perform the semiintegration operation efficiently. *Semidifferentiation* has also been proposed, usually computed as the derivative of the semiintegral curve:

$$e(t) = d^{1/2}[i(t)]/dt^{1/2} = d[m(t)]/dt \quad (\propto 1/\cosh^2(t) \text{ for uncomplicated reversible charge transfer})$$

Convolution with diffusion - $t^{-1/2}$ - or with diffusion and homogeneous kinetics - kinetic convolution - as well as deconvolution of diffusion, can be easily performed by discrete Fourier transform. If $g(t)$ is the convolution (or deconvolution) function (*e.g.* $t^{-1/2}$ if the operation regards semiinfinite linear diffusion), convoluted and deconvoluted responses:

$$c(t) = i(t)*g(t)$$

and

$$d(t) = i(t)*^{-1}g(t)$$

can be computed by the following operations:

$$c(t) = F^{-1}\{F[i(t)]\cdot F[g(t)]\}$$

and

$$d(t) = F^{-1}\{F[i(t)]/F[g(t)]\}$$

where F and F^{-1} indicate direct and inverse Fourier transform, respectively. Obviously, it is practical to deal with sequences rather than with functions, in order to use the *Discrete* (*Fast*) *Fourier Transform* algorithm. The operations imply, independently of the complexity of the sequences involved, simple multiplications - for convolution - and divisions - for deconvolution - of complex numbers representing response and filter in the frequency domain [8]. In particular, it has been shown that the quantities characterizing a cyclic voltammetric response can be more accurately evaluated on the signal after deconvolution of diffusion. Figure 5 reports the set of direct, deconvoluted and convoluted responses, respectively, computed for an $EC_{I\ order,\ irrev}$ mechanism.

As it has been stressed at the beginning of this survey, it is important for an experimenter to be able to calculate theoretical responses. Once the experimental response has been properly manipulated - *e.g.* the noise has been filtered off, the background has been subtracted, convolution or deconvolution operations have been possibly performed - the problem is how to perform the comparison with theoretical responses in order to: *i*) establish the nature of the mechanism; *ii*) determine the relevant thermodynamic and kinetic parameters. The first step, *i.e.* the identification of the type of the operative mechanism, can be simply based on the comparison of the trends of some significant quantities of experimental and computed responses. As an example, in the case of cyclic voltammetric experiments, the forward peak potential, the ratio between backward and forward peak currents, the difference

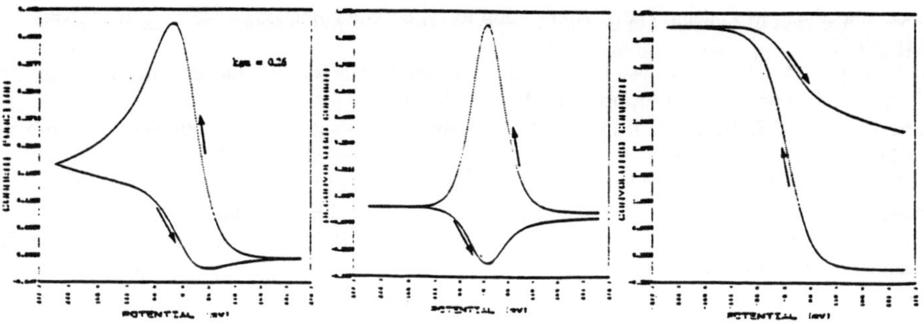

Figure 5. Simulated cyclic voltammetric curve for $EC_{I\ order,irrev}$ mechanism with the relevant deconvoluted and convoluted responses.

between forward and backward peak potentials can be evaluated on experimental and theoretical responses at varying the potential scan rate, for different values of the depolarizer concentration. In a more sophisticated approach, experimental responses are classified by examination of a suitable library of theoretical responses, through some *computerized pattern recognition technique*. Though some very interesting articles have been published on this subject, a lot of work has still to be made, both as to a suitable *feature selection, i.e.* to the choice of the most suitable quantities to characterize the responses and hence defining the hyperspace to work in, and as to the most effective pattern recognition method.

After identifying the type of mechanism, the fitting of the trends of significant parameters of experimental responses with the corresponding trends computed theoretically can both confirm the choice of the mechanism and allow estimates of its thermodynamic and kinetic parameters. A refinement of these can be carried out by fitting some chosen single experimental responses with computed signals. General *numerical optimization methods* such as Simplex can be employed to 'draw' the set of values of the parameters of the mechanism towards that leading to the best fit between theoretical and experimental data.

REFERENCES

1. Bard, Allen J. and Faulkner, Larry R. (1980) Electrochemical Methods. Fundamentals and Applications, Wiley, New York, and references therein.
2. Bontempelli, G., Magno, F., Mazzocchin, G.A. and Seeber, R. (1989) 'Linear sweep and cyclic voltammetry', Ann.Chim. 79, 103-216, and references therein.
3. Feldberg, Stephen W. (1969) 'Digital simulation: a general method for solving electrochemical diffusion-kinetic problems', in A.J.Bard (ed.), Electroanalytical Chemistry, Vol. 3, M.Dekker, New York, pp.199-296.
4. Britz, D. (1980) Digital Simulation in Electrochemistry, Springer-Verlag, Berlin, and references therein ; Britz, D. (1988) Digital Simulation in Electrochemistry, 2nd Edition, Springer-Verlag, Berlin, and references therein.
5. Feldberg , Stephen W. (1990) 'A fast quasi-explicit finite difference method for simulating

electrochemical phenomena. Part I. Application to cyclic voltammetric problems', J.Electroanal.Chem. 290, 49-65.

6. Rudolph, M. (1991) 'A fast implicit finite difference algorithm for the digital simulation of electrochemical processes', J.Electroanal.Chem. 314, 13-22.

7. Seeber, R. and Stefani, S. (1981) 'Explicit finite difference method in simulating electrode processes', Anal.Chem. 53, 1011-1016.

8. Pilo, Maria I., Sanna, G. and Seeber, R. (1992) 'Analysis of cyclic voltammetric responses by Fourier transform-based deconvolution and convolution procedures', J.Electroanal.Chem. 323, 103-115. See references therein for semiintegral and semidifferential electroanalysis, as well as for convolutive potential sweep voltammetry.

MOLECULES WITH IDENTICAL ELECTROACTIVE CENTRES AS A PROBE FOR DIELECTRIC PROPERTIES OF ELECTROLYTES

N. TAYLOR, T. BODDINGTON, I. DOBSON, B. GILL, R. LUO and B. SHAW
School of Chemistry, Leeds University, Leeds LS2 9JT, England

ABSTRACT. Bimetallic complexes of the type *cis, cis*-$M(CO)_4(\mu\text{-dppe})_2M_1(CO)_4$ (where M & M_1 are Cr, Mo or W, dppe is 1,2-bis(diphenylphosphino)ethane) have relatively independent electroactive centres in close enough proximity for coulombic forces to determine in part the interactions occuring. In electrolyte solutions using solvent mixtures the bulk dielectric constants appear to determine the interaction only when large (>16-17). Below this threshold the onset of ion-pairing is suggested to radically alter the dielectric properties involved. This has implications in the use of low dielectric solvents in any comparative study.

1. Introduction.

Molecules containing two or more independent electroactive centres in close proximity, e.g. bimetallic species with substituted phosphine bridging groups, have important electrochemical properties which potentially can be used to elucidate the dielectric properties of the electrolyte medium in use. Such molecules can exhibit successive voltammetric waves the spacing of which is in part determined by coulombic interactions between the centres & thus is influenced by the dielectric properties of the electrolyte.

2. Successive Nernstian Processes.

2.1. THE GENERAL CASE.

The voltammetry of molecules undergoing 'successive' electron transfer processes is well established & in the 'electrochemically reversible' or Nernstian case with uniformity of diffusion coefficient D yields the following expressions.

Consider the reaction scheme

$$X_0 \xrightarrow[E_1^0]{z_0^+} X_1 \xrightarrow[E_2^0]{z_1^+} X_2 \xrightarrow[E_3^0]{z_2^+} \dots$$

where the z_j^+ terms are the successive charge numbers of the X_j species & the E^0 terms the corresponding Standard Potentials of the changes. X_0 is the sole bulk species at time t=0. The above yields directly

$$I_1/F.D^{1/2}.X_0^{bulk} = \sum_{j=1}^{n} \phi_j.\delta z_j \Big/ \sum_{j=1}^{n} \phi_j$$

where I_1 is the semi-integral of the current i(t) with time t (anodic currents are positive)

$$I_1(t) = \pi^{-1/2} \int_0^t i(u)/(t-u)^{1/2}\, du,$$

A. J. L. Pombeiro and J. A. McCleverty (eds.),
Molecular Electrochemistry of Inorganic, Bioinorganic and Organomettallic Compounds, 467–475.

$\phi_j(E)$ is the Nernstian ratio (X_j/X_0) of species j at the electrode given by

$$\phi_j(E) = \exp[-(F/RT)\{\sum_{r=0}^{j} \Delta z_r E^\circ_r - \delta z_j E\}]$$

where for $j = 0$ $\delta z = 0$ otherwise $\delta z_j = z_j - z_0$ i.e. the total increase in charge number up to species j, & also $\Delta z_j = z_j - z_{j-1}$ i.e. the increment in charge number for the given process $X_{j-1} \longrightarrow X_j$. This general relationship between I_1 & E holds for any variation of E with time & thus can be used with a variety of voltammetric techniques.
The above simplifies somewhat for a pair of processes with constant Δz to yield

$$I_1/(\Delta zF.D^{1/2}.X_0^{bulk}) =$$

$$(2+\exp[-(\Delta zF/RT)(E-E^\circ_2)])/(1+\exp[(-\Delta z/RT)(E-E^\circ_2)]+\exp[-(2\Delta z/RT)(E-(E^\circ_1+E^\circ_2))])$$

& further to

$$\mathrm{Ln}((I_{lim}-I_1)/I_1) = (-\Delta zF/RT)(E-((E^\circ_1+E^\circ_2)/2)) + \mathrm{Ln}((1/2\alpha+\beta)/(1+1/2\alpha\beta))$$

where $\alpha = \exp[(-\Delta z/2RT)((E^\circ_1-E^\circ_2)]$, $\beta = \exp[(-\Delta zF/RT)(E-((E^\circ_1+E^\circ_2)/2))]$ & $I_{lim} = 2\Delta zF.D^{1/2}.X_0^{bulk}$.

These forms are discussed in detail by Ammar & Savéant (1). Bard & co-workers (2) have examined polymeric materials where the electroactive centres are potentially identical.

2.2. IDENTICAL ELECTROACTIVE CENTRES.

In the case of $\alpha=2$ (i.e. $E^\circ_1-E^\circ_2=-(RT/\Delta zF).\mathrm{Ln}(4)$) the second logarithmic term disappears & the form of the above becomes identical to that of a single stage process with identical Δz except that I_{lim} & I_1 & the component currents have twice the magnitude. This situation is applicable to symmetrical non-interacting identical centres in a molecule, the formal potential separation arising from the statistical factor involved. Bard (2) has studied such a system extended to multiple centres. Notice the distinct contrast here with the case where the second formal potential well precedes the first in the wave. Here $\alpha \longrightarrow 0$ & the slope of the logarithmic function is doubled to $2.\Delta zF/RT$ i.e. a $2\Delta z$ wave.

When coulombic interactions occur the centres become non-identical & the shape of I_1 as a function of E may still be determined but with further displacement of the formal potentials involved. Furthermore the dielectric properties of the surrounding medium influence the coulombic interactions & changes of the separation of the potentials can be used to monitor these.

With increasing differences of potential between successive formal potentials the function dI_1/dE displays a pair of equal peaks symmetrically dispersed around the mean of the formal potentials. Furthermore peak potentials must correspond also on the return sweep if the system is indeed Nernstian & appropriate compensation for electrolyte resistance has been carried out (see below). This provides the best means of comparing theory with experiment & was widely adopted in this work. Note however that only in the case of completely separated peaks do the peak positions correspond to the formal potentials otherwise the latter must be discerned by, for example, comparison of experimental & simulated results.

3. Experimental.

3.0. APPARATUS.

A conventional 3 electrode system was used for cyclic voltammetry with 1 cm square planar platinum working & counter electrodes & a reference electrode using the same TBAP (tetrabutyl ammonium perchlorate) electrolyte (normally at 0.1 mol dm^{-3}). In dimethyl formamide (DMF) containing solutions the Ag reference electrode was immersed in a 0.01 mol dm^{-3} Ag^+ solution; in chloromethane solutions the electrolyte was saturated with lithium chloride. An EG&G Potentiostat Model 362 was coupled to a computer based data capture & treatment system. The latter included the ability to software compensate for electrolyte resistance between reference & working electrodes & to semi-integrate & semi-differentiate the current responses obtained.

3.1. THE TEST SYSTEM.

The cyclic voltammetry of the series of bimetallic compounds *cis, cis*-$M(CO)_4(\mu\text{-dppe})_2M_1(CO)_4$ (where M & M_1 are Cr, Mo or W, dppe is 1,2-bis(diphenylphosphino)ethane) has been studied in TBAP electrolyte solutions containing mixtures of hexane, methylene chloride, chloroform & dimethyl formamide. The phosphine bridge provides a M M' separation of about 900 pm. The solvents used provide a good range of bulk dielectric constant in the absence of added electrolyte & when incorporated with electrolyte gave sufficient variation of dielectric properties for large changes in the voltammograms to be observed. The corresponding monometallic complexes have been examined by Morse (3) but were re-examined in our experimental system with similar results. All of the bimetallic complexes showed rather similar behaviour in exhibiting successive single electron oxidations, the potential separation of which was least for the homo-bimetallic species. Further oxidation could occur at more extreme potentials. The stability conferred on the first oxidation product by the inclusion of Cr was most noticeable & in these cases allowed room temperature measurements to be made. The three homo-bimetallic species differed greatly in this stability & the successive processes were studied in the cases of Mo & W only at low temperature. In contrast the Cr complex gave 'reversible' voltammograms at room temperatures at moderate scan rates of not greater than 2 Vs^{-1}. The mono complexes $M(CO)_4$(dppePP') had a similar trend in stability. In view of this most attention will focus on the homo-bimetallic Cr complexes.

3.2. COMPUTER TREATMENT OF DATA.

The cyclic voltammetric currents were normally not the main source of attention but rather the semi-integral I_1 of the current was calculated & from this the semi-differential or square root of time deconvolution dI_1/dt. This & dI_1/dE were used for the clearest displays of results. The main difference in application of the latter pair depends on whether a potential ramp linear with time is applied at the working electrode surface i.e. hardware compensated at the potentiostat if necessary for resistance between working & counter electrodes. In this case using the linearly varying potential as an axis dI_1/dt & dI_1/dE are similar in shape & either provides a suitable display. If uncompensated resistance remains then in Nernstian systems I_1 is a function of the appropriate E (E_{comp}) & suitable compensation via $E_{comp} = E_{applied} - i.R_u$ can be applied post capture using the overlay of I_1 on return of the sweep to determine the uncompensated resistance R_u. dI_1/dt does not retain the symmetrical features under these circumstances since the linearity of potential & time is lost.

3.3. COMPUTER SIMULATIONS.

As mentioned earlier the formal potentials in the Nernstian case do not correspond exactly to the peak positions in dI_1/dE when the peaks are not completely separated. One remedy we adopted was to compare the peaks with simulated data until a match was found thus identifying the magnitudes involved. dI_1/dE for this purpose can be obtained directly by computation of the differential of the $I_1(E)$ function using a suitable algorithm. If simulation of the cyclic voltammogram is also required then this need only involve generating I_1 as a function of time via any desired E/time relationship (e.g. a given sweep rate in linear sweep voltammetry) & then convoluting $I_1(t)$ with the inverse square root of time to produce the charge Q. Differentiation of Q produces the cyclic voltammogram. The convolution & differentiation algorithms required are part of a package we have produced for data treatment, the only change required for simulation was to externally construct a suitable file of I_1 as a function of time & import this for the further treatment.

4. Results.

4.1 DICHLOROMETHANE AS SOLVENT.

All six bimetallic & the three corresponding monometallic complexes were examined in dichloromethane solutions containing TBAP at 0.1 mol dm^{-3}. The cyclic voltammetric results mainly at a sweep rate of 2 Vs^{-1} are summarised in table 1. Further waves appear at higher potential but will not be considered here.

Metal Centre		i_p Peaks First Process		i_p Peaks Second Process				
M	M'	E_c/V	E_a/V	E_c/V	E_a/V	E^0_1/V	E^0_2/V	Temperature/K
Cr	Cr	0.699	0.759	0.848	0.906	0.728	0.876	293
Cr	Mo	0.678	0.737	0.935	0.993	0.708	0.964	293
Cr	W	0.720	0.780	1.002	1.060	0.748	1.031	293
Mo	Mo	0.866	0.937	1.014	1.091	0.902	1.052	208
W	Mo	0.888	0.941	1.016	1.083	0.911	1.051	223
W	W	0.874	0.971	1.034	1.109	0.923	1.072	203
Cr	-	0.658	0.740			0.699		293
Mo	-	0.819	0.901			0.860		275
W	-	0.853	0.907			0.880		223

Table 1. Results in dichloromethane for the nine complexes of the type *cis, cis*-$M(CO)_4(\mu$-dppe$)_2M_1(CO)_4$ & $M(CO)_4$(dppePP'). Here the E^0 values were obtained from the mean of the cathodic & anodic peak potentials E_c & E_a.

At sweep rates well below 2 Vs^{-1} even the Cr homo-bimetallic complexes exhibited chemical decomposition after the first peak at room temperature. Figure 1 illustrates this via typical dI_1/dE responses at three sweep rates. The enhancement of the first peak with respect to the second suggests an ECE process is operating with the second E^0 lying 'before' the first in the wave. Note the appearence of the decomposition product on the return sweep. An increase in temperature also emphasises the decomposition causing the detail of the two waves to be lost progressively as temperature is raised until by 45°C a single but extended wave was observed. Here as in other cases the deconvolution still clearly showed the composite nature of the wave. Cooling a system already showing independent waves caused a decrease in the separation of potentials. Some of this may be due to changes in the dielectric properties of the medium as discussed for mixed solvents below.

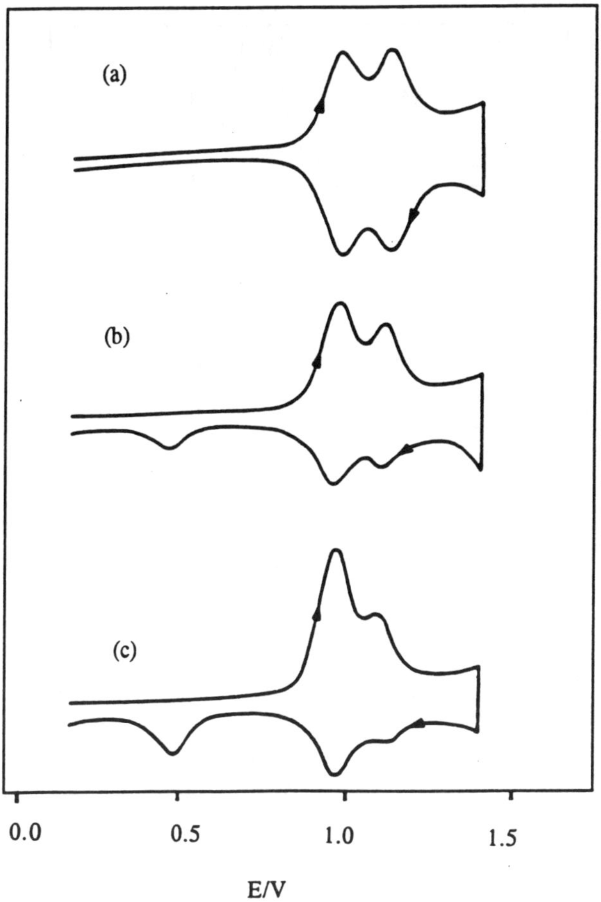

Figure 1. dI_1/dE for the Cr homo-bimetallic complex at 293K.
Sweep rates (a) 1.0 Vs^{-1}, (b) 0.2 Vs^{-1} & (c) 0.1 Vs^{-1}.

4.2. THE Cr HOMO-BIMETALLIC COMPLEX IN MIXED SOLVENTS.

With non-interacting mixed pure solvents the bulk dielectric constant approximates to that of the weight average of the individual dielectric constants. With added electrolyte material the dielectric properties of the solution are not solely determined by the solvent system but nevertheless the averaged solvent value ϵ_{av} provides a tangible guide & will be used as one factor to display results. Solvent systems involved dichloromethane (ϵ_{293K}=9.08), chloroform (ϵ_{293K}=4.81), dimethylformamide (ϵ_{293K}=37) & hexane (ϵ_{293K}=1.89). The initial mixed solvent system contained dichloromethane & dimethyl formamide. Figure 2 illustrates the results for 0.1 mol dm^{-3} TBAP in 5-95% dichloromethane electrolyte at 298K at 2 Vs^{-1} using dI_1/dE for greater clarity of the display.

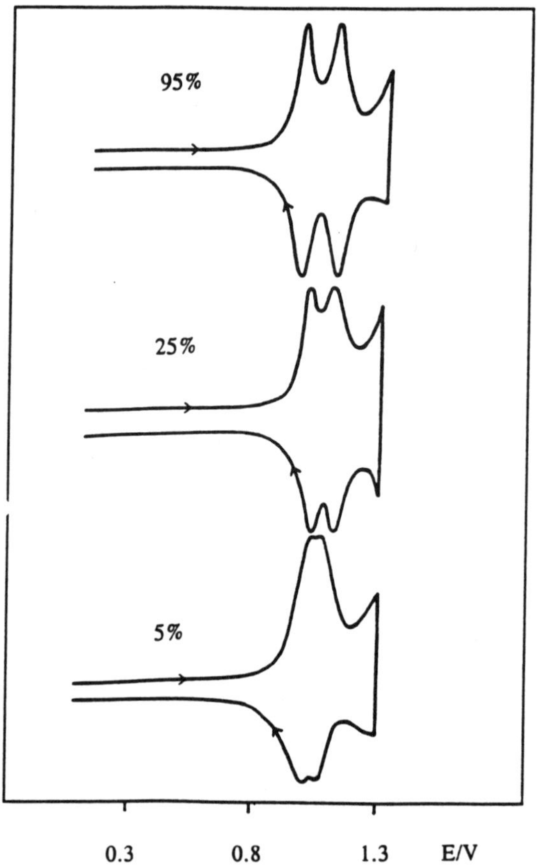

Figure 2. The dI_1/dE response for the Cr homo-bimetallic system in mixed solvent systems of dichloromethane & dimethyl formamide at 293K.
The percentages marked are those of dichloromethane 'weight for weight'.
The ϵ_{av} range over 10.5 (95%), 30.0 (25%) & 35.5 (5%).

Note in the figure how, in the higher dielectric constant media, the separation of potentials is greatly reduced & in all cases required comparison of the experimental response with simulations to assess accurately the E^0 values. The equal heights of the successive peaks indicates little impact of the decomposition under these conditions.

Corresponding studies with hexane dichloromethane & DMF chloroform mixtures were made under similar conditions. Results are presented for all systems in figure 3 as ΔE_{corr} vs ϵ_{av} where ΔE_{corr} is the formal E^0 separation less the statistical (RT/F)Ln4 term (35 mV at 293K).

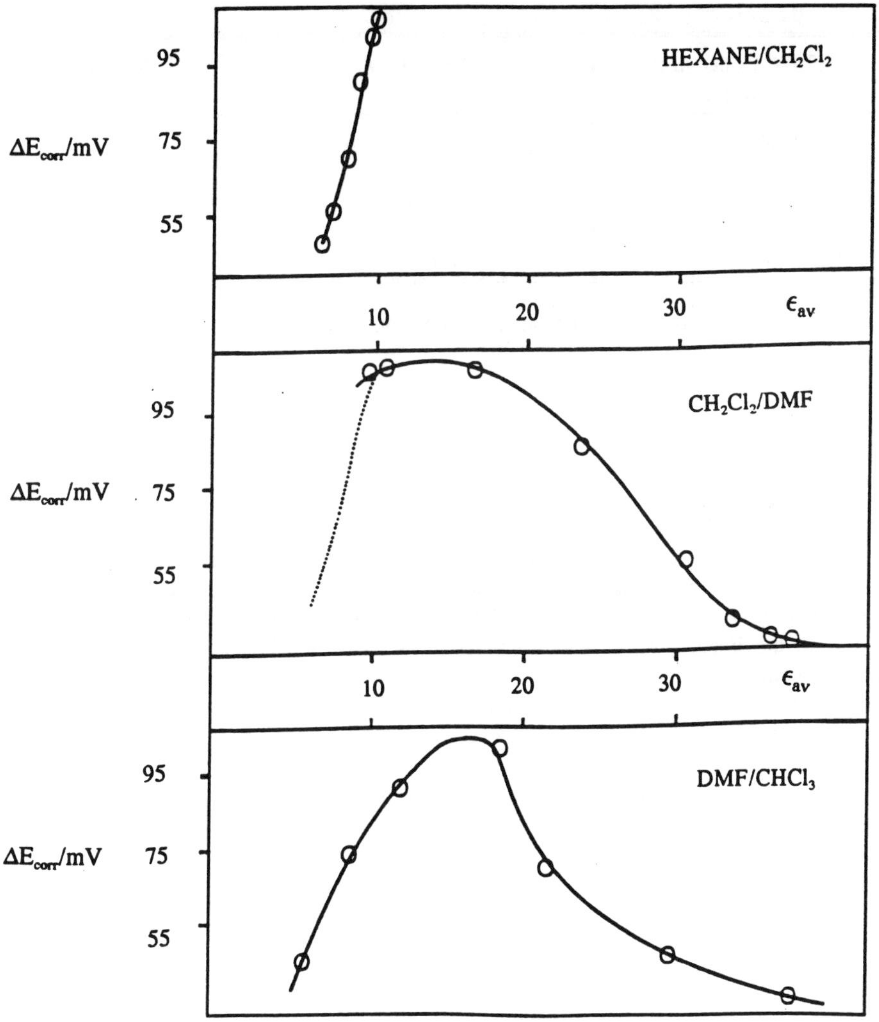

Figure 3. ΔE_{corr} vs ϵ_{av} for the Cr homo-bimetallic system in mixed solvent systems at 293K.

The important features here are that there is a decrease in interaction below about $\epsilon_{av}=15$ & that this does not depend on the solvent system. We tentatively suggest that the apparent increase in dielectric constant at low ϵ_{av} is due to an onset of ion-pairing in the electrolyte. Some evidence for this came from the observation that increases in electrolyte concentration in dichloromethane gave decreases in ΔE_{corr} & such decreases persisted when conductivity data showed much of the extra added electrolyte to be ion-paired. Table 2 contains these results. At higher ϵ_{av} values the ΔE_{corr} & the corresponding interaction energy do follow $1/\epsilon_{av}$ dependence as expected if ϵ_{av} does indeed here largely determine the dielectric properties.

TBPA Concentration /mol dm^{-3}	ΔE_{corr} /mV	Cell Resistance/Ω 1 cm spacing, 1 cm^2 plates
0.05	102	1099
0.10	102	501
0.20	97	294
0.50	85	132
1.00	80	122

Table 2. ΔE_{corr} & Electrolyte Resistance as a function of TBAP concentration.

5. Conclusions.

In many centred molecules the interactions between the electro-active centres in a given molecule modify the spacing of the formal Standard Potentials of the successive processes by an amount that depends in the case of coulombic forces in part on the dielectric properties of the local surrounding electrolyte solution. This feature has been observed in the case of bimetallic complexes in mixed solvent systems of relatively low bulk dielectric constants, & has been used to ascertain the impact of electrolyte concentration & in particular ion-pairing on electrolyte dielectric behaviour.

A necessary pre-requisite to such a study required accurate assessment of the formal potentials involved & this in turn demanded careful appraisal of adherence to a suitable mechanistic scheme. This criterion was satisfied by a combination of convolution techniques applied to cyclic voltammetry & direct comparison with simulation results normally via the symmetrical deconvolution or dI_1/dE peaks produced when the time-scales are such that the system appeared electrochemically reversible.

Results for the Cr homo-bimetallic complexes in the mixed solvents indicate that 'normal' $1/\epsilon$ trends of potential separation with changes in solvent composition occur providing the bulk dielectric constant of such a solvent mix in the absence of electrolyte would be above a threshold value of 16-17. Here then the coulombic interactions decrease as the composition is varied towards the solvent component of higher dielectric constant & vice versa. Below the threshold the situation is progressively reversed until it is apparent that interactions again increase. One feature here is that the transition region is not particularly sensitive to the exact chemical nature of the solvent components & such changes might therefore be induced by the onset of ion-pairing in the low dielectric media. To shed further light on this experiments were performed in which only the amount of added electrolyte was varied. Here considerable variations of potential separation were observed which were entirely in accord with the trend expected from a role of ion-pairing in determining the local dielectric properties of the medium & furthermore correlated with resistance data.

6. References.

1. Ammar, F. & Savéant, J-M. (1973) J. Electroanal. Chem., 47, 215.
2. Smith, W.H. & Bard, A.J. (1977) J. Electroanal. Chem., 76, 19.
3. Cook, R.L. & Morse, J.G. (1984) Inorg Chem 23, 2332.

A COMPARATIVE STUDY OF NUMERICAL METHODS FOR CYCLIC VOLTAMMETRY DIGITAL SIMULATION OF AN ELECTROCHEMICAL PROCESS WITH A COUPLED CHEMICAL REACTION

M. AMÉLIA N.D.A. LEMOS, ARMANDO J.L. POMBEIRO
Centro de Química Estrutural
Complexo I
Instituto Superior Técnico
1096 Lisboa Codex
Portugal

ABSTRACT. In order to select the most efficient integration method for the digital simulation of a cyclic voltammetric process with a coupled chemical reaction, for example in an ECEE mechanism, five different known numerical methods were tested, the modified Euler, the third and fourth order Runge-Kutta, the Adams-Moulton and the Milne methods. The calculation time, stability and error for each method were evaluated at different scan rates and for a wide range of the chemical homogeneous rate constant as a function of the integration step values. The results of these calculations are discussed and indicate, e.g., that the higher order methods are, usually, as expected, much more accurate than the lower order ones. Significant exceptions are the third order Runge-Kutta method, which always behaves worse than the corresponding second order scheme, and the Milne method that is unstable for all the conditions tested. Although for some sets of parameters all the methods presented an acceptable accuracy, for more demanding conditions only the fourth-order Runge-Kutta was accurate and stable.

1. Introduction

In the investigation of electrochemical mechanisms, the use of digital simulation techniques opens new routes to the determination of kinetic parameters that would be difficult to obtain otherwise [1, 2]. Within our electrochemical studies of hydride complexes, we have recognized, by cyclic voltammetry, the possible involvement of several redox couples with one or more

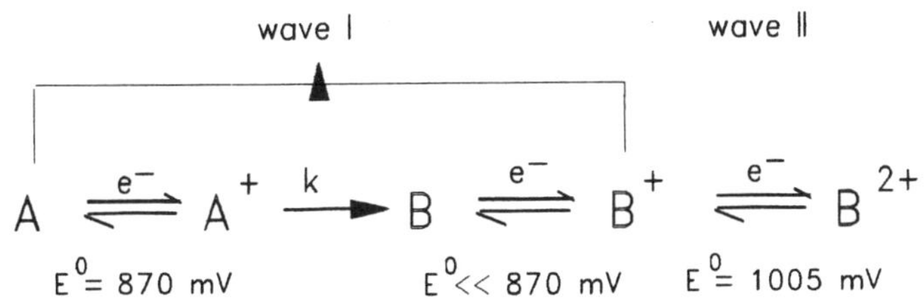

Figure 1. ECEE mechanism considered in this work, with the standard redox potentials (E^0) used in the simulated voltammograms.

A. J. L. Pombeiro and J. A. McCleverty (eds.),
Molecular Electrochemistry of Inorganic, Bioinorganic and Organometallic Compounds, 477–482.

chemical steps, namely of the ECEE or ECECE type mechanism, as observed [3] for trans-$[FeH(CNR)(dppe)_2]^+$ (R = alkyl or aryl; dppe = 1,2 bis-diphenylphosphynoethane).
Typical cyclic voltammograms of the ECEE system indicated in Fig. 1, exhibit two waves whose relative intensities vary with the scan rate as depicted in Fig. 2 (for two different homogeneous rate constants k).

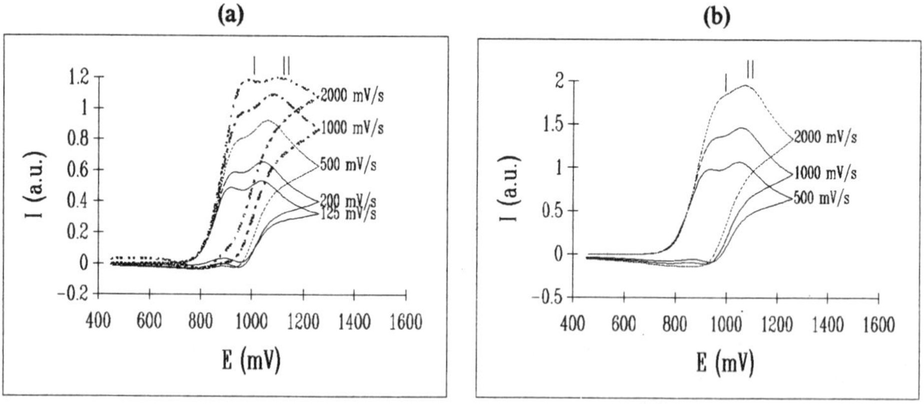

Figure 2. Typical simulated voltammograms for the ECEE mechanism shown in Figure 1, for two values of the homogeneous rate constant [(a) k = 10 s^{-1}, (b) k = 1000 s^{-1}]. I (current intensity) in an arbitrary unit.

2. Digital Simulation

Several different methods have been used to solve the partial differential equations describing the diffusion processes that take place in a cyclic voltammetric experiment [1, 4 - 6]. The use of methods that discretize the space coordinates separately from the time coordinate and then solve the corresponding ordinary differential equations, like the box method, have the distinct advantage, over the finite difference methods, of requiring much less computer memory, and thus allowing the convenient use of a personal computer.

We have used the box method and focused our attention in the choice of a suitable numerical integration scheme taking into account its *stability*, *computation time* and *error*. The number of boxes used was twenty with a fixed length, except for the first one that had half the length of the other ones to increase accuracy in the computation of the current.

To integrate the ordinary differential equations resulting from space discretization we tried the modified Euler method (which is equivalent to a second-order Runge-Kutta scheme), the third and fourth order Runge-Kutta as well as the Adams-Moulton and Milne predictor-corrector schemes [7, 8]. The Milne method was eliminated from the start, since it was impossible to obtain stability (*i.e.*, convergence to the desired solution) for the step values that were tried.

The Adams-Moulton method, being a predictor-corrector scheme, involved the use of the corrector formula until the difference in two successive iterations was below a predetermined value, but in most cases it only used the corrector formula once.

Since no accurate data was found for this system, an 'exact' solution was obtained by successively halving the step size using the fourth-order Runge-Kutta method, until no significant differences were observed in the computed voltammogram. Relative errors in the currents were computed for points with ten millivolt spacing and averaged over the entire voltammogram.

The program to compute all the voltammograms was written in Turbo Pascal Version 6 (from Borland Int.) and run on a 386SX/25MHz based IBM compatible personal computer, equipped with a Fasmath Cyrix math co-processor. The time taken by each method to complete the computation of each voltammogram was measured by the program itself and includes overheads due to file writing and to on-line graphical display of the computed voltammogram.

3. Results and Discussion

For a fixed step size, the average error varied in the expected way, decreasing with the increase of the method's order, except for the case of the third-order Runge-Kutta scheme, which always gave an unexpectedly high error and the Milne integration formula that proved to be unstable for all the conditions tested. The observed order for the errors is as follows: third-order Runge-Kutta > modified Euler > fourth-order Runge-Kutta > Adams-Moulton; moreover, the computation time increases in the following order: modified Euler < third-order Runge-Kutta < Adams-Moulton < fourth-order Runge-Kutta, as expected.

Computation times also follow the expected evolution, increasing with the number of function evaluations per step required by the method. The Adams-Moulton scheme, however, which requires only two evaluations per step, like the modified Euler scheme, was only slightly faster than the fourth-order Runge-Kutta due to the amount of computation involved in the predictor and corrector formulas. The former also provided extremely low errors, when applicable, but showed a tendency to become unstable at higher step sizes.

The relation between the average error (computed as indicated above) and the computation time is shown in figure 3.

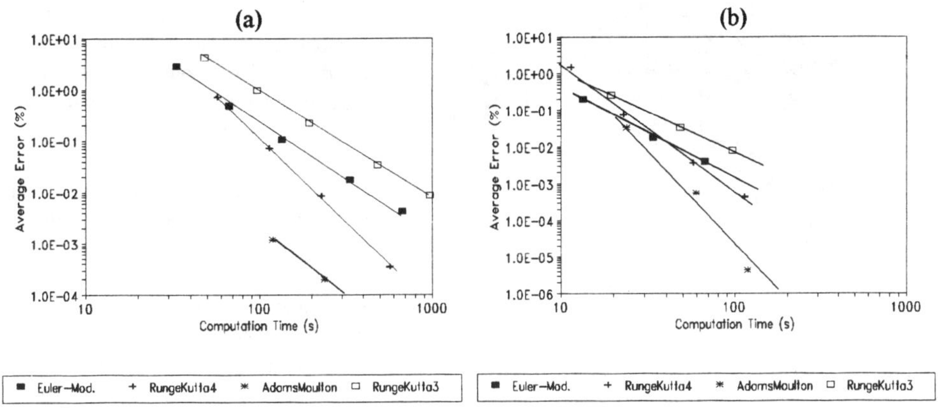

Figure 3- Comparison of the average error versus computation time, for the modified Euler, 3rd and 4th order Runge-Kutta, and Adams-Moulton integration methods, applied to an ECEE scheme [(a) $k = 10\ s^{-1}$, scan rate = 200 mV/s; (b) $k = 100\ s^{-1}$, scan rate = 1000 mV/s].

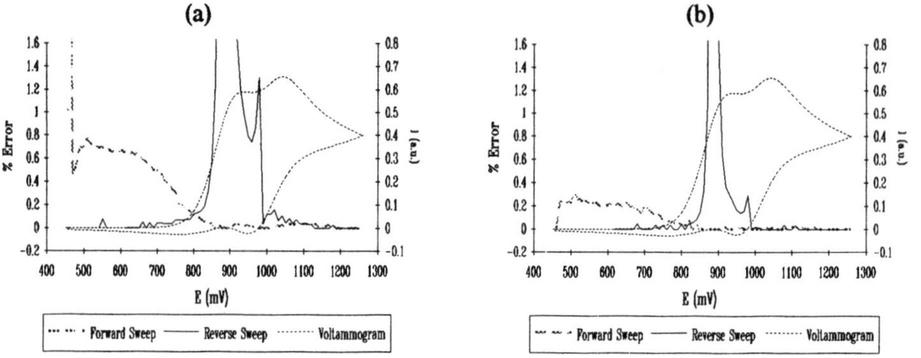

Figure 4. Plot of the relative error (in percentage) as a function of the potential for (a) the modified Euler method and (b) the fourth-order Runge-Kutta method. (Voltammograms simulated for k=10 s^{-1}, scan rate = 200 mV/s and step size = 2 mV).

However, it is noteworthy to mention that the relative errors are not evenly distributed along the voltammogram, and affect mainly the portions were the current is nearly null. This is depicted in figure 4, were the percent error is plotted along the voltammogram. It can be clearly seen that the error rises sharply each time the voltammogram approaches zero current. It is also apparent that the modified Euler method produces larger errors than the fourth-order Runge-Kutta. However these correspond to insignificant absolute errors when comparing the theoretical voltammograms with the experimental data.

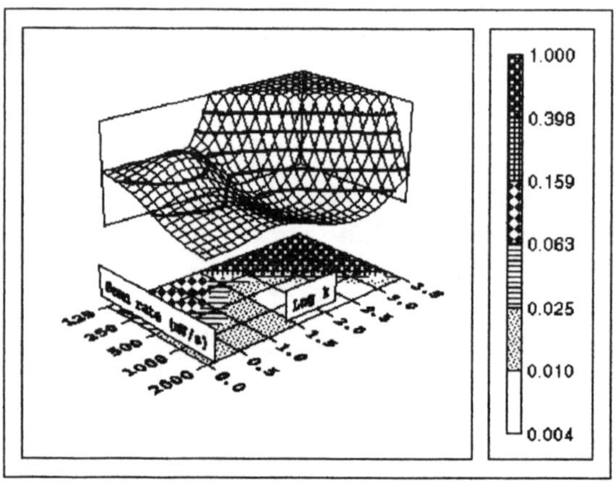

Figure 5. Surface representing the average error (%) as a function of the scan rate (log scale) and the logarithm of the homogeneous rate constant k obtained for the Euler method with a 1 mV step size.

A more important difference is related to the stability of the methods. This was evaluated by computing voltammograms with a wide range of values for the scan rate and the homogeneous rate constant, using a fixed step size of one millivolt, which was considered to be a reasonable step size for most cases. The surface representing the error *versus* the scan rate and the homogeneous rate constant was obtained for all the methods (see figure 5 for an example). The error regions for the four methods are shown in Figure 6, revealing that the third and fourth order Runge-Kutta schemes have a wider stability region, in relation to the two above-mentioned variables, than either the modified Euler or the Adams-Moulton method.

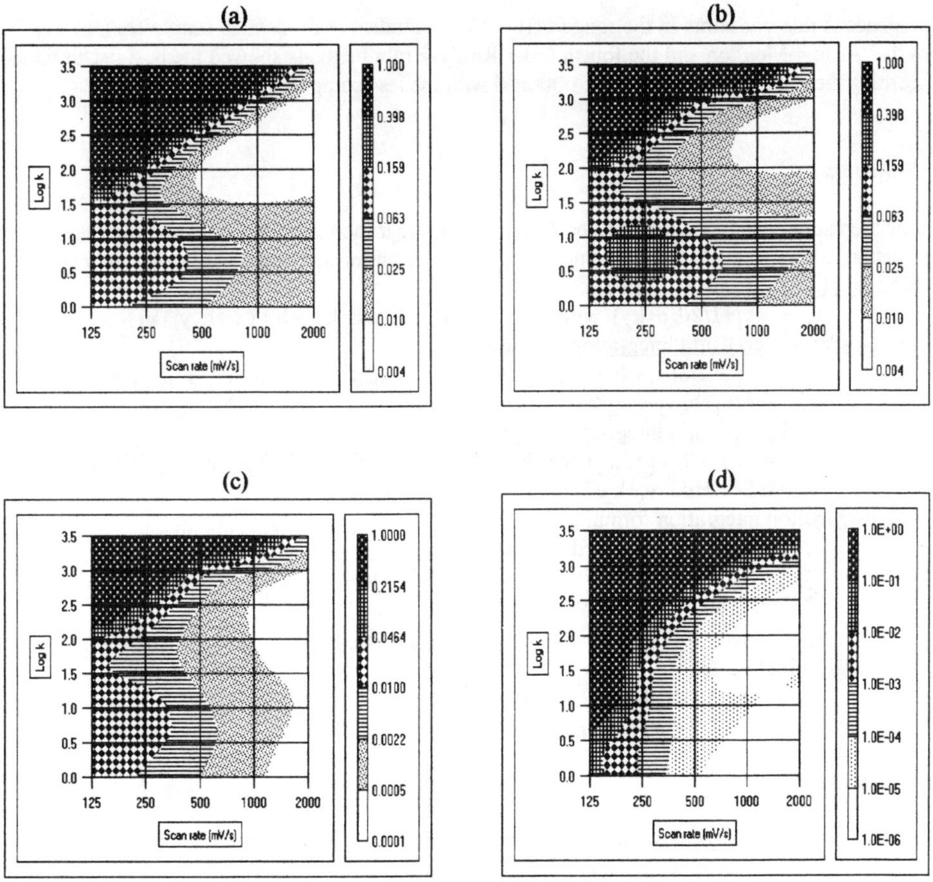

Figure 6. Average error regions (%) for the modified Euler (a), third order Runge-Kutta (b), fourth order Runge-Kutta (c) and the Adams-Moulton (d) methods. Areas indicated as having 1% error correspond to unstable regions.

Stability is very important when one considers a general application program and ensures that larger step sizes can be used, even in difficult conditions. It should be noted that the step size was kept at 1 mV, implying that the actual integration time interval decreased with increasing scan rates, leading to the observed decrease in error.

As a conclusion we may say that, in general, all four methods tested appear to be suitable for the integration of the equations involved in the studied process. However, for more demanding conditions, as the stiffness of the problem increases (*e.g.*, cases with high homogeneous rate constants at moderate scan rates) third and fourth order Runge-Kutta methods are more capable of remaining stable than the other two methods. Runge-Kutta methods also provided very low error levels which would make them particularly suited for wide range application programs. For example, in the system analysed in reference [1], $[FeH(CNR)(dppe)_2]^+$, which involved homogeneous rate constants in the range 1/10 - 10 s^{-1}, studied at moderated scan rates (see Fig.3), both the Adams-Moulton and the fourth order Runge-Kutta methods showed the best performance concerning the minimum average error obtained with the less computational time.

4. Appendix

The mathematical equations used for each numerical integration method [7, 8] are summarized in this section, where y stands for a nonlinear ordinary differential equation and h for the step size.

Modified Euler integration formula:

$$y_{i+1} = y_i + 1/2(k_1+k_2) \text{ with } k_1 = h\, f(x_i, y_i) \text{ and } k_2 = h\, f(x_i+h, y_i+k_1)$$

Third-order Runge-Kutta integration formula:

$$y_{i+1} = y_i + 1/6(k_1+4k_2+k_3) \text{ with } k_1 = h\, f(x_i, y_i),\ k_2 = h\, f(x_i+h/2, y_i+k_1/2) \text{ and } k_3 = h\, f(x_i+h, y_i+2k_2-k_1)$$

Fourth-order Runge-Kutta integration formula:

$$y_{i+1} = y_i + 1/6(k_1+2k_2+2k_3+k_4) \text{ with } k_1 = h\, f(x_i, y_i),\ k_2 = h\, f(x_i+h/2, y_i+k_1/2)\ k_3 = h\, f(x_i+h/2, y_i+k_2/2) \text{ and } k_4 = h\, f(x_i+h, y_i+k_3)$$

Adams-Moulton integration formula:

a) Adams-Bashforth predictor formula:

$$y_{n+1} = y_n + h\,(55f_n - 59f_{n-1} + 37f_{n-2} - 9f_{n-3})/24$$

b) Fourth order Adams-Moulton corrector formula:

$$y_{n+1} = y_n + h\,(9f_{n+1} + 19f_n - 5f_{n-1} + f_{n-2})/24$$

Milne integration formula:

$$y_{n+1} = y_{n-3} + 4h\,(2f_n - f_{n-1} + 2f_{n-2})/3$$
$$y_{n+1} = y_n + h\,(f_{n+1} - 4f_n + f_{n-1})/3$$

References

[1] - Britz, D. (1988) Digital Simulation in Electrochemistry, 2nd Ed., Springer Verlag, Berlin.

[2] - Arévalo, M.C., Farnia, G., Severin, M.G. and Vianello, E. (1987) 'Kinetic analysis of a slow electron transfer coupled with a father-son reaction', J. Electroanal. Chem. 220, 201-211.

[3] - Lemos, M.A.N.D.A. and Pombeiro, A.J.L. (1992) 'Electrochemical behaviour of trans-$[FeH(CNR)(dppe)_2]^+$. Kinetic parameters determined by digital simulation of cyclic voltammetry', J. Organometal. Chem., submitted to publication.

[4] - Feldberg, S.W. (1990) 'A fast quasi-explicit finite difference method for simulating electrochemical phenomena. Part 1. Application to cyclic voltammetric problems', J. Electroanal. Chem. 290, 49-65.

[5] - Kavanaugh, T.C., Friedrichs, M.S., Friesner, R.A. and Bard, A.J. (1990) 'A new approach to electrochemical simulations based on eigenvalue-eigenvector solutions of the diffusion equation. Part II. Cyclic voltammetry and heterogeneous kinetics', J. Electroanal. Chem. 283, 1-14.

[6] - Gosser, D.K., and Rieger, P.H. (1988) 'Treatment of homogeneous kinetics in electrochemical digital simulation programs', Anal. Chem. 60, 1159-1167.

[7] - Constantinides, A. (1987) Applied numerical methods with personal computers, McGraw-Hill, New York.

[8] - Conte, S.D. and de Boor, C. (1972) Elementary numerical analysis, 2nd Ed., McGraw-Hill Kogakusha, Tokyo.

Acknowledgement - This work has been partially supported by JNICT and INIC.

MECHANISM OF THE ELECTROACTIVATION OF THE METAL-HYDRIDE BOND IN $[ReClH(NCR)(Ph_2PCH_2CH_2PPh_2)_2][BF_4]$

M. Fátima C. GUEDES da SILVA, João J.R. FRAÚSTO da SILVA and Armando J.L. POMBEIRO
Centro de Química Estrutural, Complexo I, Instituto Superior Técnico
Av. Rovisco Pais, 1096 Lisboa codex, Portugal

Christian AMATORE and Jean-Nöel VERPEAUX
Laboratoire de Chimie, URA CNRS 110, Ecole Normale Superieure,
24 rue Lhomond, F-75231 Paris cedex 05, France

ABSTRACT. The rhenium-hydride complexes $[ReClH(NCR)(dppe)_2][BF_4]$ (1, R = aryl, dppe = $Ph_2PCH_2CH_2PPh_2$), obtained by protonation of the parent neutral nitrile compounds, undergo, in an aprotic medium and at a Pt electrode, metal-hydrogen bond cleavage upon electrochemical oxidation or reduction, and the mechanisms of these distinct electroactivation processes have been investigated (for 1, R = C_6H_4Cl-4) by preparative scale electrolysis, cyclic voltammetry and simulation techniques.

1. INTRODUCTION

Transition metal hydrides constitute an important class of compounds with a relevant role in coordination and organometallic chemistry, namely as intermediates in catalysis [1]. However, their electrochemical behaviour has not yet been sufficiently investigated, in spite of their expected ready activation by electrochemical methods [2-6], e.g., via the electrogeneration of reactive (paramagnetic) species.

Within the course of our studies on the activation (towards protonation) of small unsaturated species by electron-rich metal sites [7], we have observed that the complexes cis-$[ReCl(NCR)(dppe)_2]$ are susceptible to proton attack at either the nitrile ligand or the rhenium atom (eq. 1) to give the methyleneamido [8] or the hydrido [9] complexes $[ReCl(NCHR)(dppe)_2]^+$ or $[ReClH(NCR)(dppe)_2]^+$ (1), respectively, and now we report the preliminary investigation of the electrochemical behaviour of the latter compounds, undertaken by controlled potential electrolysis and cyclic voltammetry.

A. J. L. Pombeiro and J. A. McCleverty (eds.),
Molecular Electrochemistry of Inorganic, Bioinorganic and Organomettallic Compounds, 483–487.

$$\underline{cis}\text{-}[ReCl(NCR)(dppe)_2] + HBF_4 \longrightarrow$$

$$\longrightarrow [ReClH(NCR)(dppe)_2][BF_4] \qquad (1)$$

2. RESULTS AND DISCUSSION

2.1. Preparative Electrolysis

Preparative electrolytic reduction at the cathodic wave (see below) of (1, $R = C_6H_4Cl\text{-}4$), onward denoted by $\underline{ReH}^+$ where $\underline{Re}$ stands for $\{ReCl(NCR)(dppe)_2\}$, in 0.3 M $[Bu_4N][BF_4]$/THF, at a Pt electrode, was shown to consume one Faraday per mole yielding quantitatively trans-$[ReCl(NCR)(dppe)_2]$ [10], in a simplified way denoted by $\underline{Re}$ (eq.2). Therefore, cathodically induced dehydrogenation has occurred, but the fate of hydrogen has not been elucidated.

$$\underline{ReH}^+ + e^- \longrightarrow \underline{Re} + (H^{\bullet}) \qquad (2)$$

It was further established that the same complex, $\underline{Re}$, was obtained by deprotonation of $\underline{ReH}^+$ with a base such as $[Bu_4N]OH$.

Moreover, preparative electrolytic oxidation, at the anodic wave of $\underline{ReH}^+$ (see below), involved the consumption of two Faradays per mole; in addition, acid-base titration of the electrolyzed solution indicated that one proton per mole of $\underline{ReH}^+$ was released during this anodic process which was shown to give the dioxidized nitrile complex $\underline{Re}^{2+}$ (eq. 3).

$$\underline{ReH}^+ - 2e \longrightarrow \underline{Re}^{2+} + H^+ \qquad (3)$$

2.2. Cyclic Voltammetry

The cyclic voltammetric studies of $\underline{ReH}^+$ were performed in THF or CH_2Cl_2, at a platinum disc electrode (e.g., with 125 μm diameter), at various scan rates (in the range 0.05 to 300 V/s).

It undergoes anodic and cathodic processes, at $E^o_{\underline{ReH}^{2+}/\underline{ReH}^+} = 0.89$V and $E^o_{\underline{ReH}^+/\underline{ReH}} = -0.99$ V (in CH_2Cl_2), respectively, which are both chemically reversible and monoelectronic at sufficiently high scan rates.

However, a pronounced decrease of the reversibility of both the anodic and the cathodic waves was observed upon lowering the scan rate.

Therefore, e.g., at 100 mV/s, although the reduction wave remained monoelectronic, the deprotonated complex $\underline{Re}$ was already formed to an appreciate extent on this time scale as shown by the detection of its characteristic first anodic wave at $E^o_{\underline{Re}^+/\underline{Re}} = -0.31$ V (Fig.).

Moreover, at low scan rates, oxidation of $\underline{ReH}^+$ brough about a more complicated pattern: it tended, to some extent, to a two-electron process, chemically irreversible, to give $\underline{Re}^{2+}$ as indicated by the appearance of the corresponding reduction waves for the $\underline{Re}^{2+/+}$ and $\underline{Re}^{+/o}$ couples on the return scan (Fig.)

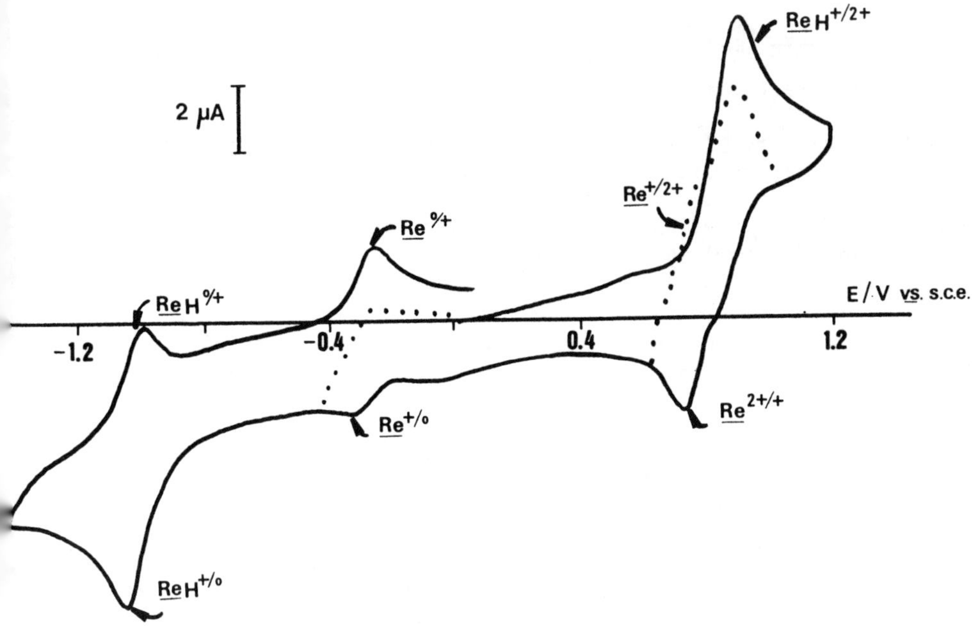

Figure — Cyclic voltammograms for 0.8 mM $[ReHCl(NCC_6H_4Cl\text{-}4)(dppe)_2]^+$ ($\underline{ReH}^+$) in 0.3 M $[Bu_4N][BF_4]/CH_2Cl_2$, at a Pt electrode (ν = 100 mV/s).

Cyclic voltammetry experiments, performed at different concentrations of $\underline{ReH}^+$, showed that an increase of the concentration of this species resulted in an increase of the scan rate required to restore the reversibility of its oxidation wave. Such an observation allowed us to rule out an ordinary ECE process (eqs. 4) which would not display such a concentration effect.

$$\underline{ReH}^+ \underset{}{\overset{-e}{\rightleftharpoons}} \underline{ReH}^{2+} \xrightarrow{-H^+} \underline{Re}^+ \overset{-e}{\rightleftharpoons} \underline{Re}^{2+} \qquad (4)$$

A DISP2-type mechanism [11] was then considered and investigated by simulation [12] of the cyclic voltammograms at various concentrations and different scan rates. A close agreement with the experimental set of voltammograms was observed, indicating that the oxidation of $\underline{Re}^+$, formed in the anodic process, is an homogeneous reaction performed by $\underline{Re}H^{2+}$ (eqs. 5-7, where B denotes a basic species).

$$\underline{Re}H^+ - e \rightleftharpoons \underline{Re}H^{2+} \quad (5)$$

$$\underline{Re}H^{2+} + B \overset{K}{\rightleftharpoons} \underline{Re}^+ + HB^+ \quad (6)$$

$$\underline{Re}^+ + \underline{Re}H^{2+} \xrightarrow{k} \underline{Re}^{2+} + \underline{Re}H^+ \quad (7)$$

The simulation techniques allowed the determination of the overall rate constant for the formation of $\underline{Re}^{2+}$, i.e., $k_{ap} = kK[B]/[HB^+]$, although an independent determination of both k and K was not possible in view of the steady state behaviour of transient $\underline{Re}^+$.

As expected, k_{ap} was found to increase the basicity of the medium: 100 or 250 $M^{-1} s^{-1}$ for CH_2Cl_2 or THF solvent, respectively, whereas a further increase was observed upon addition of pyridine.

The simulation techniques were also applied to the reduction of $\underline{Re}H^+$, for which neither the abovementioned concentration dependence nor the medium basicity effect were observed. However, the simplest EC scheme did not provide an acceptable agreement with the experimental cyclic voltammograms, and a bimolecular chemical reaction (involving an hydrogen atom abstractor) should be involved, although the detailed analysis was precluded by difficulties encountered in getting a reproduceable reversibility degree for the cathodic wave.

Nevertheless, our studies indicate that electroinduced metal--hydrogen bond cleavage, either upon oxidation or reduction, can occur through bimolecular processes, following pathways which are considerably more complex than a simple intramolecular rupture of such a bond.

ACKNOWLEDGEMENTS

This work was partially supported by J.N.I.C.T. (The National Board for Scientific and Technological Research) and I.N.I.C. (The National Institute for Scientific Research), Portugal, as well as by CNRS (Centre National de la Recherche Scientific) and ENS (Ecole Normale Superieure), France.

REFERENCES

[1] See, e.g., Masters C. (1981) Homogeneous Catalysis, John Wiley and Sons, New York; Crabtree R.H. (1988) The Organometallic Chemistry of the Transition Metals, John Wiley and Sons, New York.

[2] Lemos M.A.N.D.A., and Pombeiro A.J.L. (1987), J. Organometal. Chem. 332, C17.

[3] Chen L., and Davies J.A. (1990), Inorg. Chim. Acta 175, 41

[4] Detty M.R., and Jones W.D. (1987), J. Am. Chem. Soc. 109, 5666.

[5] Klinger R.J., Huffman J.C., and Kochi J.K. (1980), J. Am. Chem. Soc. 102, 208.

[6] Bruce D.W., Cole-Hamilton D.J., and Pogorzelec P. (1987), J. Chem. Soc., Dalton Trans., 2941; Koelle U., and Ohst S. (1986), Inorg. Chem. 25, 2689.

[7] Pombeiro A.J.L. (1989), in Advances in Metal Carbene Chemistry, Schubert U. (ed.), Kluwer Academic Publ., Dordrecht, p.79-99; Pombeiro A.J.L. (1988) J. Organometal. Chem. 358, 273; Pombeiro A.J.L., and Richards R.L. (1990), Coord. Chem. Rev. 104, 13.

[8] Pombeiro A.J.L., Hughes D.L., and Richards R.L. (1988), J. Chem. Soc., Chem. Commun, 1052.

[9] Silva M.F.C.G., Pombeiro A.J.L., and Richards R.L. (1991), Internat. Conf. Inorg. Chem., The Royal Society of Chemistry/Gesellschaft Deutscher Chemiker, Brighton, P.21.

[10] Pombeiro A.J.L., Silva M.F.C.G., Hughes D.L., and Richards R.L. (1989), Polyhedron 8, 1872

[11] Amatore C., and Savéant J.-M. (1979), J. Electroanal. Chem. 102, 21; Amatore C. (1991), in Organic Electrochemistry, 3rd ed., Baizer M.M. and Lund H. (eds.), M. Dekker, New York, Ch. II, p. 11-119.

[12] Gosser D.K. (Jr.), and Zhang F. (1991), Talanta 38, 715.

ELECTROCHEMISTRY AND *CIS/TRANS*-ISOMERIZATION OF $Cr^{III/II}$-AMINOCARBOXYLATE COMPLEXES

R. MEIER, M. BODDIN, M. HECHT
TU of Leipzig
Department of Natural Sciences
P.O. Box 66
D-O-7030 Leipzig
Germany

F. A. SCHULTZ
Purdue University
School of Science at Indianapolis
Indianapolis
IN 46205
U.S.A.

ABSTRACT. Cr^{III}-aminocarboxylate complexes in which the nitrogen atoms of the Cr^{III}-N_2O_4 chromophore coordinate in a *cis* configuration undergo kinetically facile one electron reduction at -1.2 to -1.4 V vs SCE. The corresponding *trans* isomers are reduced at more negative potentials with a much slower rate of electron transfer. The different isomeric preferences of Cr^{III} and Cr^{II} give rise to the occurence of square schemes when N-alkyl substituted iminodiacetates are ligands. The difference in electron transfer rate for the *cis* and *trans* reductions is believed to result from a difference in character and spatial orientation of the $Cr^{III/II}$ redox orbitals between *cis* and *trans* isomers.

1. Introduction

It is well known from the literature that $Cr^{III/II}$ complexes with EDTA-type ligands have rate constants of the heterogeneous eletron transfer (k_{ob}) which are more than 10^8 times faster than that of the $Cr(H_2O)_6^{3+/2+}$-couple [1,2]. Furthermore, it is known from Tanaka's work [1] that k_{ob} of $Cr^{III/II}$-CyDTA is one order of magnitude smaller than that of $Cr^{III/II}$-EDTA. Ogino et al. [3] have shown that the formal redox potentials ($E^{o\prime}$) of $Cr^{III/II}$-CyDTA, $Cr^{III/II}$-EDTA, $Cr^{III/II}$-1.3-PDTA and $Cr^{III/II}$-EDDaDp span a range of about 200 mV and are shifted negatively in the given order. On the other hand, Davies and Earley [4] established a LFE-relationship for certain $Cr^{III/II}$-aminocarboxyalte complexes (including $Cr^{III/II}$-EDTA and $Cr^{III/II}$-EDDaDp) while studying the rate constants of the homogeneous oxidation Cr^{II}-Cr^{III}. According to their findings, the oxidation rate increases as the $E^{o\prime}$-values of the $Cr^{III/II}$-couples become more negative.

Therefore, one problem we intended to clarify was the question of whether the k_{ob}-values for $Cr^{III/II}$-CyDTA, $Cr^{III/II}$-EDTA, $Cr^{III/II}$-1.3-PDTA and $Cr^{III/II}$-EDDaDp change in a manner with their $E^{o\prime}$-values.

In the case of electrochemical studies of the $Co^{III/II}(IDA)_2^{-/2-}$-couple it was found [5,6] that the *cis* isomers are reduced much more rapidly

A. J. L. Pombeiro and J. A. McCleverty (eds.),
Molecular Electrochemistry of Inorganic, Bioinorganic and Organomettallic Compounds, 489–494.

than the *trans* isomers. The isomeric preferences of Co^{III} and Co^{II} give rise to the occurence of a square scheme.

Because the electrochemistry of $Cr^{III/II}$-complexes with tridentate aminocarboxyalte ligands has never been studied before, we have started a detailed investigation of various complexes of this type with different alkyl substituents at the amino-nitrogen atom. We report here some preliminary results of this study.

2. Materials and Methods

All complexes and those ligands not available commercially (1.3-PDTA, EIDA, n-PIDA and n-BIDA) were prepared according to methods given in the literature [3,8]. Their purity was proved by microanalysis and the values gained show satisfactory agreement with literature values.

The following abbreviations for the ligands are used throughout the paper: CyDTA = trans-cyclohexanediaminetetraacetate, EDTA = ethylenediaminetetraacetate, 1.3-PDTA = 1.3-propanediaminetetraacetate, EDDaDp = ethylenediamine-N,N'-dipropionate-N,N'-diacetate, IDA = iminodiacetate, MIDA = N-methyl-iminodiacetate, EIDA = N-ethyl-iminodiacetate, n-PIDA = N-n-propyl-iminodiacetate, n-BIDA = N-n-butyl-iminodiacetate.

All electrochemical measurements were done at a hanging mercury drop electrode. The equipment used was the same as described previously [9].

3. Results and Discussion

3.1. Electrochemistry of $Cr^{III/II}$-couples with sexadentate ligands

The electrode reaction of $Cr^{III/II}$-CyDTA has been studied in detail by Tanaka et al. [1b]. They report a value of k_{ob} = 0.029 cm s^{-1}. We have tried to estimate this k_{ob} in the pH-range 4 to 8 where $E^{o'}$ is independent of pH. Using cyclic voltammetry (CV) and faradaic impedance measurements we confirm that k_{ob} is smaller than 0.1 cm s^{-1} and α (charge transfer coefficient) $\leq$ 0.4. Up to now we are unable to get exact values for these constants because a chemical pre-equilibrium of unidentified origin interferes with the charge transfer process. Further work is underway to clarify this point.

In the case of $Cr^{III/II}$-EDTA, Tanaka et al. [1] gave k_{ob} = 0.21 cm s^{-1} and Bard et al. [2] have estimated k_{ob} = 0.45 cm s^{-1}. Using CV with sweep rates up to 125 V s^{-1} and impedance measurements with a frequency of 5 kHz the electrochemical reaction exhibits close to Nernstian behavior. Thus, we conclude under our conditions (complex concentration 2 mM, pH 5.0, 0.333 M Na_2SO_4) that k_{ob} $\geq$ 1 cms^{-1}. Measurements will be done using appropriate methodology to get the exact value for k_{ob}.

Similar behavior is shown by $Cr^{III/II}$-1.3-PDTA and $Cr^{III/II}$-EDDaDp, but here k_{ob} seems to be even higher than that for $Cr^{III/II}$-EDTA.

3.2. Electrochemistry of $Cr^{III/II}$-couples with tridentate ligands of the iminodiacetate type

Two isomeric forms of 2:1 complexes of Cr^{III} with IDA and MIDA have been characterized [8]. In *cis*-$[Cr^{III}(IDA)_2]^-$, IDA behaves as a tridentate ligand, adopting a facial cis geometry with respect to the nitrogen atoms [7]. In the Cr^{III} bis complexes containing MIDA, EIDA, n-PIDA and n-BIDA formation of the facial trans isomers is favoured [8].

The structures of the Cr^{III} complexes with IDA and MIDA are shown in Figure 1.

Figure 1.

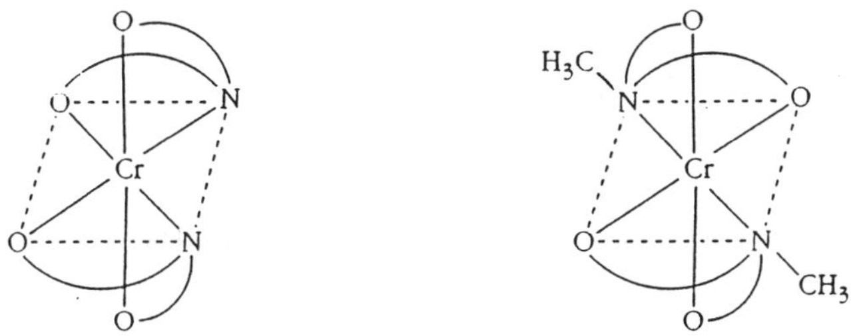

cis - $[Cr(III)(IDA)_2]^-$ (I)

trans - $[Cr(III)(MIDA)_2]^-$ (II)

CV measurements of the $Cr^{III/II}(IDA)_2^{-/2-}$ couple show that the electrode raction is reversible both chemically and electrochemically (k_{ob} > 0.1 cm s^{-1}).

CV's recorded for reduction of the *trans*-$[Cr^{III}(RIDA)_2]^-$ complexes show that the different coordination preferences of Cr^{III} and Cr^{II} give rise to a square scheme (Scheme 1). This behavior becomes visible paricularly when two successive CV-scans are recorded. As an example, a CV for the reduction of *trans*-$[Cr^{III}(MIDA)_2]^-$ is given in Figure 2.

The reduction of the original Cr^{III} *trans* complex is followed by immediate rearrangement of *trans*-$[Cr^{II}(RIDA)_2]^{2-}$ into *cis*-$[Cr^{II}(RIDA)_2]^{2-}$. The isomerization in the Cr^{II}-state is so fast that no anodic wave for the oxidation of the Cr^{II} *trans* species is observed for all RIDA-complexes at sweep rates up to 200 V s^{-1} .

The second scan shows that now a mixture of Cr^{III} *cis* and *trans* isomers is reduced. Their relative peak currents depend on sweep rate. This feature can be used to estimate the kinetics of the *cis* to *trans* isomerization in the Cr^{III} state. Evaluation of these parameters using a fit of experimental CV's to simulated curves is in progress and the results will be given elsewhere.

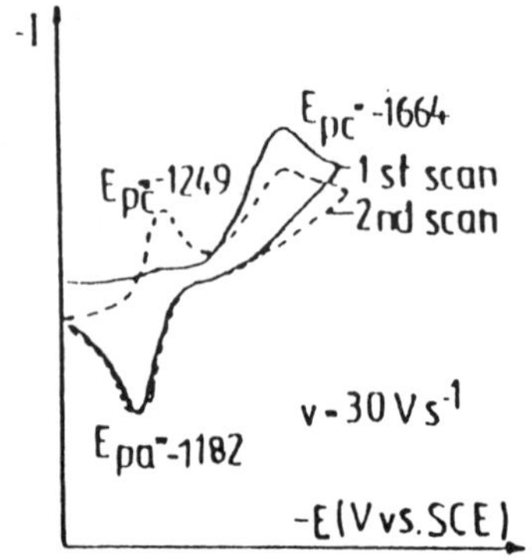

Figure 2.

CV reduction of *trans* - $[Cr(III)(MIDA)_2]^-$ at 30 V s^{-1} at HMDE. pH 9, 0.1 M MIDA, 1 M ionic strength Na_2SO_4.

Scheme I

$$trans\text{-}[Cr(III)(MIDA)_2]^- + e^- \xrightarrow{-1.4\ V} trans\text{-}[Cr(II)(MIDA)_2]^{2-}$$

$$K^{III}_{tc} \qquad\qquad\qquad\qquad K^{II}_{tc}$$

$$cis\text{-}[Cr(III)(MIDA)_2]^- + e^- \underset{-1.2\ V}{\rightleftharpoons} cis\text{-}[Cr(II)(MIDA)_2]^{2-}$$

Nevertheless, the follwing preliminary conclusions can be drawn from the sweep rate dependece of the wave heights:

- all RIDA-complexes exist completely in the trans form as Cr^{III} ($K_{tc} >> 1$)
- the magnitude of the rate constant for *cis* to *trans* conversion in the Cr^{III} state decreases in the order of complexes with n-BIDA > n-PIDA > EIDA > MIDA
- $K_{tc} << 1$ for all Cr^{II}–RIDA complexes.

Another interesting observation is the fact that the heterogeneous charge transfer of *trans*-$[Cr^{III}(RIDA)_2]^-$ reduction is sluggish compared to that of *cis*-$[Cr^{III}(IDA)_2]^-$ and *cis*-$[Cr^{III}(RIDA)_2]^-$. This diffenrence is believed to result from a difference in character and spatial orientation of the $Cr^{III/II}$ redox orbitals between the *cis* and *trans* isomers (Figure 3) [8b].

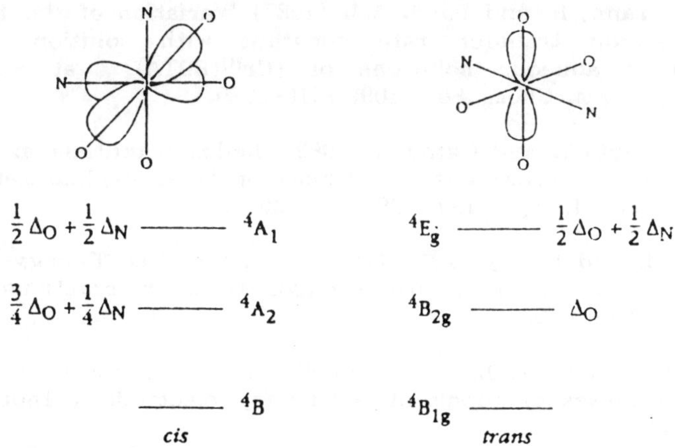

Figure 3.

Electronic structure and spatial orientation of LUMO of *cis* and *trans* isomers of Cr(III)-N_2O_4 chromophore. (Ligand field parameters from AOM analysis of Hoggard and Schmidtke, BBPC *76*, 1013 (1972)).

Both LUMO's are antibonding with respect to metal-ligand bonds, but nuclear displacements must be larger in the case of trans-CrN_2O_4 to account for slow electron transfer. The present state of evaluation of the kinetics of the electrochemical *trans*-reduction allows the conclusion that for all RIDA-couple $k_{ob} < 0.001\ cms^{-1}$ and $\alpha \cong 0.3$.

4. REFERENCES

[1a] Yamada,A. and Tanaka,N. (1969) 'Determination of kinetic parameters of the electrode reaction of ethylenediamintetraacetatochromate(III) ions', Bull. Chem. Soc. Jpn. 42, 1600-1603.

[1b] Tanaka,N., Tomita,T. and Yamada, A. (1970) 'Polarographic behaviors and potentiostatic determination for the electrode reaction of trans-1,2-cyclohexanediamineteraacetatochromate(III)-complexes', Bull. Chem. Soc. Jpn. 43, 2042-2045.

[1c] Tanaka, N. (1975) 'Mechanisms and characteristics of electrode reactions of analytical interest', Pure & Appl. Chem. 44, 627-644.

[2] Zhang, X., Yang, H. and Bard, A.J. (1987) 'Variation of the heterogeneous electron transfer rate constant with solution viscosity: reduction of aqueous solutions of $[Cr^{III}(EDTA)]^-$ at a mercury electrode', J. Am. Chem. Soc. 109, 1916-1920.

[3] Ogino,H., Nagata,T. and Ogino,K. (1989) 'Redox potentials and related thermodynamic parameters of (diaminopolycarboxylato)metal(III/II) redox couples', Inorg. Chem. 28, 3656-3659.

[4] Davies, K.M. and Earley, J.E. (1976) 'Linear Free Energy Relationships for some inner-sphere electron-transfer reactions', Inorg. Chem. 15, 1074-1079.

[5] Rader, R.R. and Mc Millin, D.R. (1979) 'Electrochemistry of iminodiacetate complexes of cobalt at stationary electrodes', Inorg. Chem. 18, 545-549.

[6] Ohsaka,T., Oyama,N., Yamaguchi, S. and Matsuda, H. (1981) 'Electrode kinetics of the redox couples of Co(III/II) complexes with ethylenediamine-N-acetate, iminodiacetate and diethylenetriamine', Bull. Chem. Soc. Jpn. 54, 2475-2481.

[7] Mootz,D. and Wunderlich,H. (1980) 'Potassium cis-bis-(iminodiacetato) chromate (III) trihydrate', Acta Cryst. B36, 445-447.

[8a] Weyh,J.A. and Hamm, R.E. (1968) 'Iminodiacetato, methyliminodiacetato and 1,3-propandiaminetetraacetato complexes of Chromium(III)', Inorg. Chem. 7, 2431-2435.

[8b] Hoggard, P.E. and Schmidtke, H.-H. (1972) 'A spectroscopic investigation of chromium(III)complexes with iminodiacetic acid and methyliminodiacetic acid', Ber. Bunsenges. Phys. Chem. 76, 1013-1021.

[8c] Wernicke,A., Schmidtke,H.-H. and. Hoggard,P.E. (1977) 'Synthesis and characterisation of trans-bis(N-alkyl-iminodiacetato) chromium(III) complexes', Inorg. Chim. Acta 24, 145-148.

[9] Meier, R., Werner, G. and Otto, M. (1990) 'Voltammetric study of the reduction process of ethylendiaminetetraacetato vanadium(IV)', Anal. Chim. Acta 236, 315-324.

THREE DIMENSIONAL ELECTROCHEMISTRY: AN ALTERNATIVE EXPERIMENTAL WAY FOR THE STUDY OF ELECTROCHEMICAL PROCESSES

N. PAPADOPOULOS, K. HASIOTIS and S. SOTIROPOULOS
Laboratory of Physical Chemistry, Department of Chemistry,
Aristotle University of Thessaloniki,
Thessaloniki 54006, Greece

Abstract. We have demonstrated recently that electrochemical reactions can be studied with the use of 3D Voltammetry , a computerized electrochemical method developed in our Laboratory. The method relies on the principle that the complete electrochemical behavior of a system can be obtained through a series of steps to different potentials with recording of the current time curves to yield a three dimensional surface. Intersections of this three dimensional surface with planes vertically oriented to the t axis produce polarographic type i-E curves for different times. It is also possible to use three dimensional voltammetry based on double potential step chronoamperometric data collection to obtain information on the reaction products of the electrochemical reaction. Three dimensional Voltammetry can also be used for the study of the slowly proceeding adsorption of an organic molecule on an electrode surface. In the latter case we monitor how the capacitance of the double layer changes because of the adsorption of an organic molecule in the electrified interface.

The electrochemical oxidation and reduction reactions of organic molecules often comprise complex sequences of electrochemical and chemical steps. In addition to the electron transfer step, coupled homogeneous chemical reactions are frequently involved.[1,2] The study of chemical reactions or intermediates which are not a major species (in bulk solution) but are only formed by a chemical process, after an electrochemical reaction, has become a major application of several techniques e.g. cyclic

A. J. L. Pombeiro and J. A. McCleverty (eds.),
Molecular Electrochemistry of Inorganic, Bioinorganic and Organomettallic Compounds, 495–500.

voltammetry, rotating ring-disk electrodes, spectroelectrochemistry and double potential step chronoamperometry. These electrochemical techniques for the elucidation of reaction mechanisms and the determination of kinetic parameters rely on making current measurements as a function of parameters such as sweep rate ,time or electrode rotation rate. In this kind of studies several of the techniques will be equally applicable but there are situations where one technique has a unique advantage. Obviously a technique is more powerful for the elucidation of the mechanism of an electrochemical reaction, when its time window gives us the opportunity to investigate the reaction at very short and at very long times as well. In particular it would be useful to be able to identify reaction intermediates and monitor their time dependence concentration. Furthermore complexity of mathematics needed to draw a conclusion from experimental data and obscurity of the shape of experimental curves are two vital restrictions on a method's applicability.

With these thoughts in mind, we have proposed very recently that one can clearly investigate the electrochemical behavior of a process employing three dimensional voltammetry.[3] In this technique, a current potential time surface is constructed from chronoamperometric measurements with very closely spaced potential steps. The validity of this methodology does not lie of course on the display of the i-E-t surface in the 3D space, but on the information that the reconstructed i-E curves can provide. The latter have the advantages of a chronoamperometric technique i.e. potentiostatic conditions and an extended time-window together with those of voltammetric techniques, i.e. a clear picture of the dependence of the electrochemical reaction on potential. That is, Three Dimensional Voltammetry can be considered as a kind of a polarography with a fine-resolved time window.

Three dimensional voltammetry unites three conventional electrochemical techniques, classical polarography, pulse polarography and chronoamperometry into one very powerful experimental approach. 3D Voltammetry possess all the advantages of these methods without inheriting their limitations. Classical polarography and pulse polarography can be used to get a feel of the electrochemical behavior of a new system but both techniques suffer from a restricted time window. Both techniques discard much of the information contained in the i-t curve from each drop. Chronoamperometry possess an extended time window. However this technique can be applied once the reaction mechanism has been definitely established by some other technique, since a chronoamperometric curve has the form of a falling transient and information is only obtained by a detailed mathematical analysis of its shape. 3D Voltammetry combines an extended time window and convenience in the analysis and interpretation of the experimental results. With 3D Voltammetry any electrode that has an easily renewable and constant surface can be used as working electrode with the advantage that a polarographic response can be obtained even on a solid electrode. Polarographic response is very convenient for quantitative measurements because current plateaus are obtained in the mass-transfer limited region of each wave. The theory of classic polarography for reactions complicated by chemical steps or catalytic effects is in a highly developed state. The mathematical description of polarography has been developed sufficiently to enable kinetic parameters to be determined for a wide variety of mechanisms.

In fig 1 is displayed the three dimensional surface obtained for the oxidation of tetrahydroxy-1,4-benzoquinone on a Bi (upd) Pt electrode. In fig 1 are displayed intersections of this three dimensional surface with planes vertically oriented to the t axis and the first derivative of these curves. From these figures it appears that the mechanism of oxidation of tetrahydroxy-1,4- benzoquinone follows the follow path depicted in the scheme in fig 1

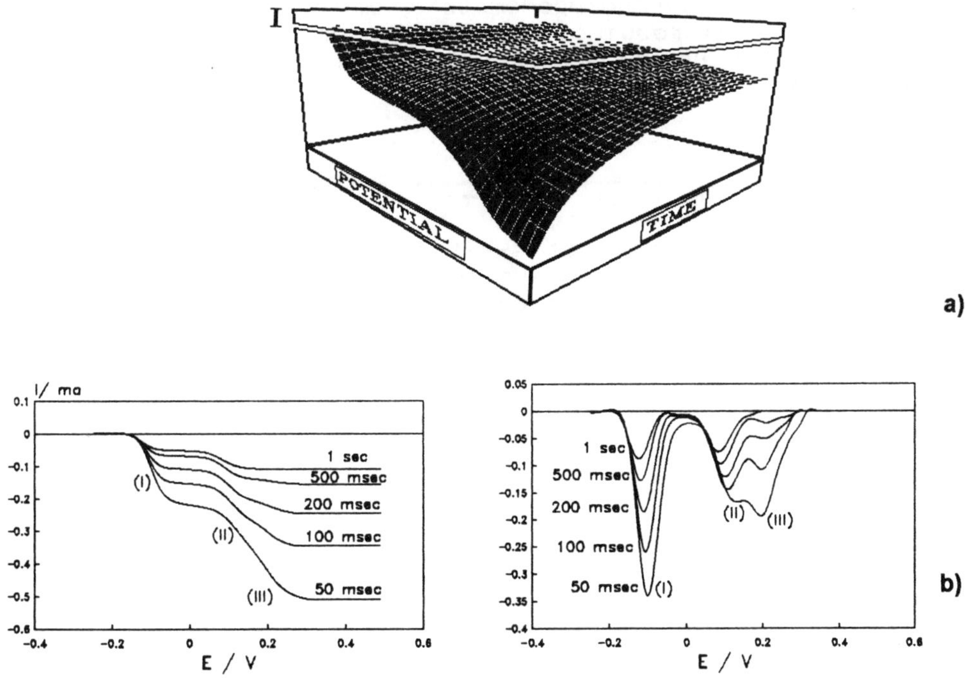

Figure 1. (a) Constrution of the I-E-t surface for the oxidation of tetrahydroxy-1,4-benzoquinone from chronoamperometric measurements on a Pt (Tl upd) electrode. (b) Intersections of the three dimensional surface with planes vertically oriented to the t axis.

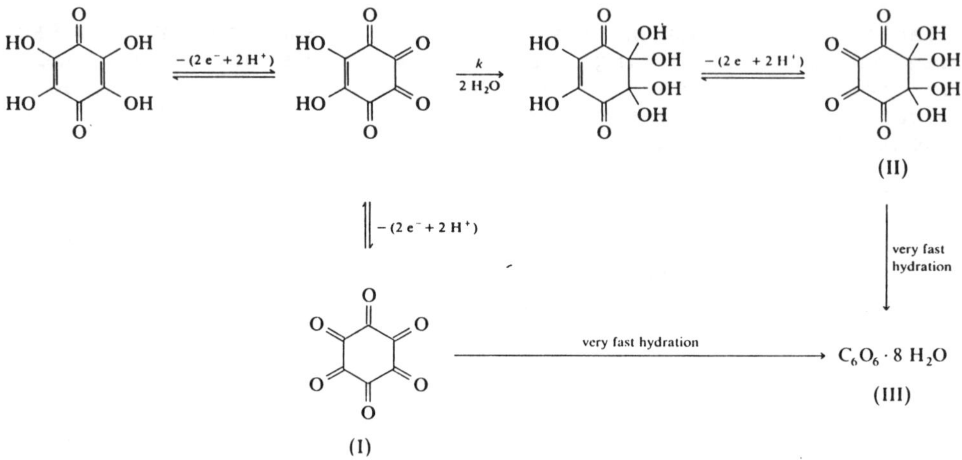

Figure 1. (c) Reaction scheme.

In fig 2 is displayed the first derivative of the sampled voltammograms derived from single and double potential step chronoamperometric measurements for the reduction of 4-nitroquinoline-N- oxide on a Tl upd platinum electrode. The reduction of heterocyclic nitro compounds follows an E.C.E. mechanism.

$$R\text{-}NO_2 \underset{}{\overset{2e^- + 2H^+}{\rightleftharpoons}} R\text{-}N(OH)_2 \xrightarrow[-H_2O]{X} R\text{-}NO \overset{2e^- + 2H^+}{\rightleftharpoons} R\text{-}NHOH$$

However in the case of 4-nitroquinoline -N- oxide the chemical step is fast relative to the rate of mass transport in the time scales of our experiment and the electrochemical reaction is controlled by diffusion. The product of the first reduction step of p-nitroquinoline -N- oxide undergoes very rapidly a chemical change to form R-NO which is further reduced to form R-NHOH. Single step chronovoltammetry gives no direct evidence for the existence of this reaction.

In the case of double potential step measurements a potential step at -.5 V vs SME was applied for three seconds to the working electrode in order to form a layer of the product of the electrochemical reaction in the region near the electrode surface. After this pulse a second potential step was applied , and the voltage was increased each time to cover all the available voltage range. Using this procedure two peaks appeared in the resulting derivative chronovoltammograms one for the reduction of R-NO_2, that still reached the electrode surface, to R-NHOH and one for the oxidation of R-NHOH to R-NO

It is also possible to use 3 D Voltammetry for the elucidation of adsorption processes of organic molecules on

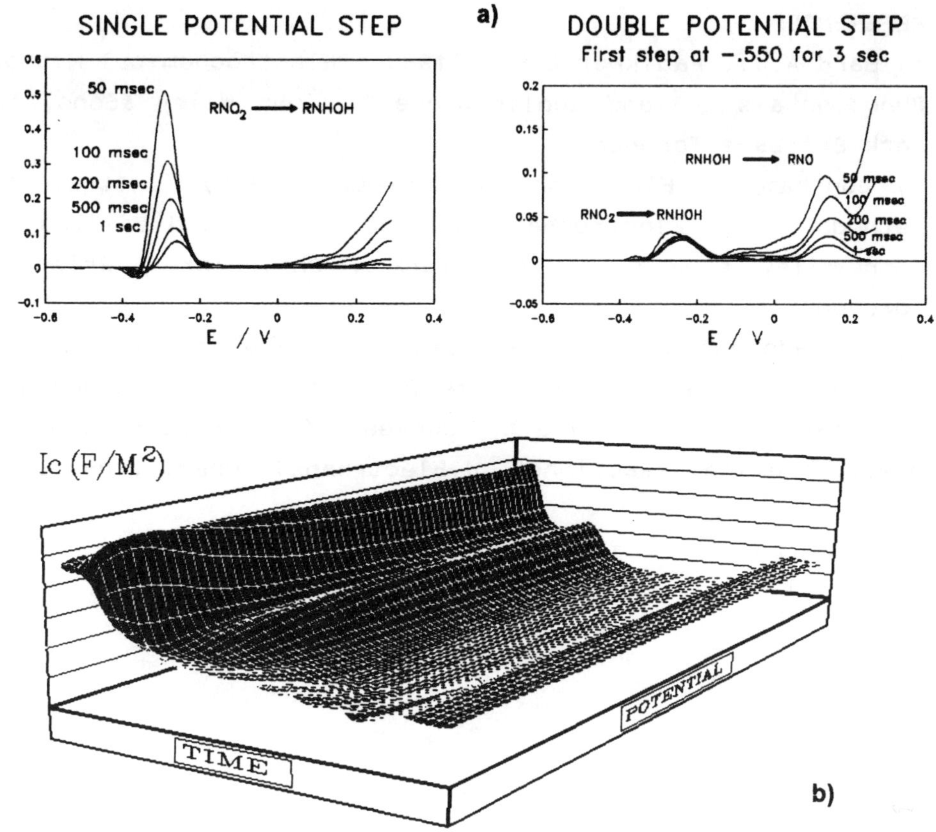

Figure 2 (a) Reduction of 4-nitoquinoline-n-oxide. (b) C-E-t surface for the interface between Hg and an aqueous solution of SDS (sodium dodecyl sulphate)

an electrode's surface especially when the development of the adsorbed film is characterized by strong kinetic effects. Kinetic effects in the case of adsorption processes can be attributed to diffusion controlled processes and to phase changes leading to a compact layer formations, surface aggregation phenomena or polylayer formation. In this case the double layer capacitance changes are monitored from ac phase sensitive chronoamperometric measurements. In fig 2 is given the C-E-t surface for the interface formed between a hanging mercury drop electrode and a aqueous solution of sodium dodecyl sulfate.

REFERENCES

1) Bard A.J., Faulkner L.R. (1980) "Electrochemical Methods Fundamentals and applications ", John Wiley &Sons, New York Brisbane,Torondo

2) Southampton Electrochemical Group, (1985) "Instrumental Methods in Electrochemistry", Ellis Horword Limited and John Wiley & Sons New York , Chilester , Brisbane, Torondo

3) Papadopoulos N. ,Hasiotis C., Kokkinidis G. and Papanastasiou G. (1991) "Three Dimensional Electrochemistry Utilisation of i-E-t curves for elucidation of electrochemical reactions" J.Electroanal. Chem. ,308 ,83-96

PART 8. SPECTROELECTROCHEMISTRY

SPECTROELECTROCHEMICAL (UV/VIS/NEAR IR, EPR) STUDIES OF COORDINATION COMPLEXES

S.A. MACGREGOR, E. McINNES, R.J. SORBIE and L.J. YELLOWLEES
Department of Chemistry, University of Edinburgh,
Edinburgh EH9 3JJ, Scotland, U.K.

ABSTRACT. The design of variable temperature spectroelectrochemical cells for use in uv/vis/nir and epr spectrometers is detailed. The systems investigated using the spectroelectrochemical cells are binuclear mixed-valence ruthenium and osmium complexes and mononuclear Pt species. Intervalence charge transfer (IVCT) bands of the mixed-valence compounds are studied *in situ* with the M(II),M(III) state electrochemically generated from the parent M(II),M(II) complex. The symmetric species $[M_2(\mu\text{-Cl})_3L_6]^{2+}$ where M = Ru,Os and L is a tertiary phosphine ligand are class III delocalised systems whereas the asymmetric complex $[Cl(PPh_3)_2Ru(\mu\text{-Cl})_3Ru(PPh_3)_2CS]^+$ is a class II, valence trapped compound. Removal of electrons from the symmetric complex results in metal-metal interaction as confirmed by single crystal X-ray determinations of $[Os_2(\mu\text{-Cl})_3(PEt_3)_6]^+$ and $[Os_2(\mu\text{-Cl})_3(PEt_3)_6]^{2+}$. The thermal electron transfer rate constant is determined for the mixed-valence asymmetric complex from the IVCT band; k_{th} = 2.9 x 10^4 s^{-1}. The complex $[Pt(bpy)Cl_2]$ has a reversible one-elctron reduction. The electronic character of the semi-occupied molecular orbital (SOMO) of $[Pt(bpy)Cl_2]^-$ is studied by epr and uv/vis/nir spectroelectrochemistry and is determined to be primarily bpy-based. Epr data are also given for $[Pt(dpk)Cl_2]^-$ and $[Pt(B_{10}H_{12})_2]^{3-}$, the SOMO of the former complex is ligand-based but has considerably greater Pt character in the second.

1. Introduction

The development of variable temperature *in situ* spectroelectrochemical techniques has enabled us to probe the electronic characteristics of the frontier orbitals of redox-active materials. The important feature of these methods is that the electrosynthesised species is generated inside the cavity of the spectrometer. Thus the electron transfer product, which is usually air and/or moisture sensitive, can be studied directly by the chosen spectroscopic method **without** the necessity of transporting the unstable solution from the electrosynthesis cell to the spectrometer. Most spectroscopic methods can be coupled with electrochemical techniques, for example, infrared, raman, resonance raman, uv/vis, epr have all been reported[1].

The cell designs we have successfully utilised with uv/vis/nir and epr spectrometers are detailed briefly. The variable temperature control of our cells is an important feature since many redox products are stable only at low temperatures. The range of systems which can be studied spectroelectrochemically are extensive. This paper considers only redox processes which are

A. J. L. Pombeiro and J. A. McCleverty (eds.),
Molecular Electrochemistry of Inorganic, Bioinorganic and Organometallic Compounds, 503–517.

chemically reversible. That is, we are concerned with examining the changes to compounds which accompany changes in electronic populations rather than changes in stoichiometry. The uv/vis/nir and epr spectroelectrochemical techniques are discussed with particular reference to
(i) mixed-valence binulear compounds of ruthenium and osmium, and
(ii) mononuclear platinum complexes.

The uv/vis/nir spectra of mixed-valence species exhibit unique low-energy intervalence charge transfer (IVCT) transitions which are absent in the fully reduced and oxidised complexes. Class II[2] (valence localised) mixed-valence systems have broad, weak, structureless IVCT bands which are sensitive to molecular environment (eg solvent[3]). Detailed study of these bands in trapped valence species allows os to calculate intramolecular electron transfer rates. Class III (valence delocalised) systems have more intense low-energy transitions which are often structured and are largely insensitive to solvent[4]. We will consider the degree of metal-metal interactions within class III species by analysis of low-energy absorption transitions.

The ruthenium and osmium binulear complexes discussed in this paper have a face-sharing bioctahedral (FSBO) structure as shown below:

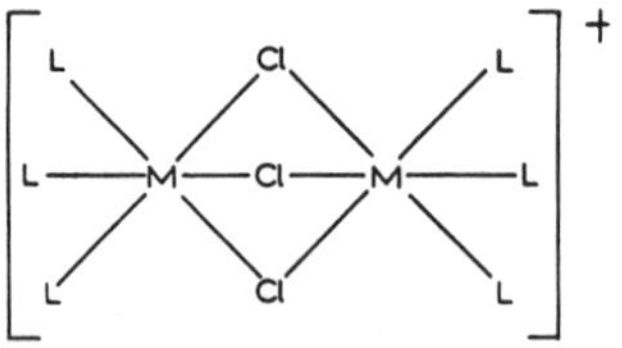

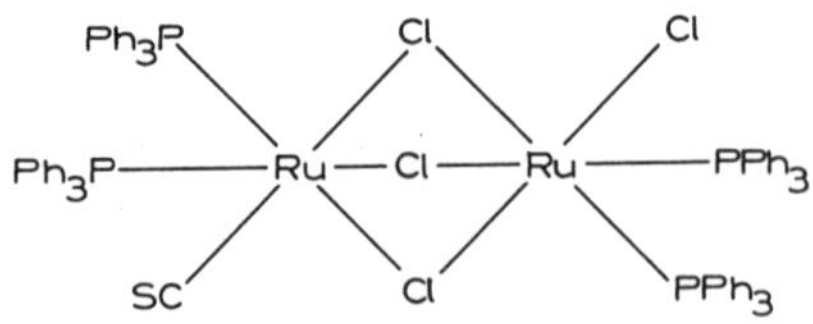

The symmetric species $[L_3M(\mu\text{-Cl})_3ML_3]^+$ have tertiary phosphine terminal ligands and bridging chloride ligands. They are prepared with the metal centres both in the +2 oxidation state, that is, d^6,d^6 complexes. The mixed-valence state is generated by a one-electron exhaustive electrolysis. The asymmetric complex $[Cl(PPh_3)_2Ru(\mu\text{-Cl})_3Ru(PPh_3)_2CS]$ is also of the II,II type but the II,III state is formed electrochemically.

Paramagnetic platinum-containing complexes are rare[5]. Most isolated Pt(I) complexes are binuclear diamagnetic species. We report here the electrochemical and spectroelectrochemical (uv/vis/nir and epr) study of one-electron reduction products of some simple Pt(II) containing complexes.

2. Experimental

All reactions were carried out under an atmosphere of dry, oxygen-free nitrogen using standard Schlenk techniques. The binulear species $[M_2(\mu\text{-Cl})_3L_6]^+$, M = Ru, L = PMe_2Ph; M = Os, L = PEt_3 were prepared according to the method of Chatt[6], M = Os, L = PMe_2Ph using the method of Coombe[7]. The asymmetric complex $[Cl(PPh_3)_2Ru(\mu\text{-Cl})_3Ru(PPh_3)_2CS]$ was prepared according to the method of Stephenson[8]. The monomeric platinum complexes $Pt(bpy)Cl_2$, $Pt(dpk)Cl_2$, $[Pt(bpy)py_2](BF_4)_2$ and $[Pt(bpy)_2](BF_4)_2$ were all prepared following the method of Morgan and Burstall[9]. The literature preparation for $[Pt(B_{10}H_{12})_2]^{2-}$ was used[10] with metathesis of the $(Me_4N)^+$ counter ion by $(BTMA)^+$ using BTMACl. All compounds gave elemental analysis (C,H,N) in excellent agreement with their stated formulations.

Dichloromethane was pretreated with KOH pellets and distilled over P_2O_5 immediately prior to use. All other solvents were used as obtained without further purification. $[^nBu_4N]BF_4$ was recrystallised from water/methanol.

All electrochemical studies were carried out using a PAR model 170 potentiostat and programmer linked to a Hewlett-Packard 7045A X-Y recorder. Positive feedback ohmic compensation was applied for all cyclic voltammograms recorded. Electrochemical experiments employed a Pt micro-working electrode, a Pt counter electrode and Ag/AgCl reference electrode against which the ferrocene/ferricinium couple was measured at +0.56V. Cells were thermostated using a Haake model Q thermostated bath circulator. Coulometric experiments utilised an H-type cell with the Pt basket working electrode fritted from the Pt counter electrode. Inert electrolyte, $[^{n}Bu_4N]BF_4$, was used in 0.5M concentration in CH_2Cl_2 and $CHCl_3$ and 0.1M concentration in all other solvents. All solutions were purged with Ar for 20 minutes prior to study. The scan rate used for all reported results was $100 mVs^{-1}$. D_o was determined using a rotating glassy carbon disc electrode (A = $0.3848 cm^2$). Electrogeneration potentials were 150 mV more positive than the $E_{1/2}$ for the couple.

The optically transparent electrode cell (O.T.E.) for use in uv/vis/nir spectrometers is shown in Figure 1[11]. The Pt/Rh gauze working electrode (transparency ~40%) is housed in a quartz uv/vis/nir cell of 0.05cm pathlength such that the spectrometer beam passes through the emtal grid. At the top of the cell is a quartz extension which functions as a resevoir for the solution into which is placed the Pt counter electrode and the Ag/AgCl reference electrode both of which are fritted from the test solution. The temperature of the cell may be altered by passing prechilled nitrogen gas over the quartz surfaces.

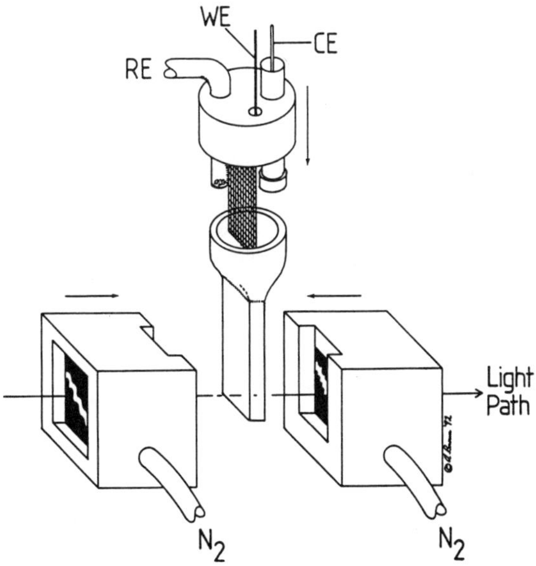

Figure 1. Schematic representation of the O.T.E. cell.

During the course of an electrogeneration experiment spectra are recorded every five minutes. Upon completion, the regeneration of the statring material was done routinely to confirm the chemical integrity of the system. Uv/vis/nir spectra were recorded on a Perkin-Elmer λ9 spectrophotometer.

Electrogenerations were carried out in the electron paramagnetic resonance *in situ* electrochemical, or E.P.RES.I.S.E, cell shown in Figure 2[12]. The cylindrical Pt working electrode and insulated Ag/AgCl reference electrode are contained in a wide bore quartz epr tube fitted at the bottom with a quartz frit to separate the Pt counter electrode. The cell is thermostated using prechilled N_2 gas.

Epr spectra were recorded on an X-band Bruker ER200D-SCR spectrometer fitted with a wide bore 4107WZ cylindrical cavity.

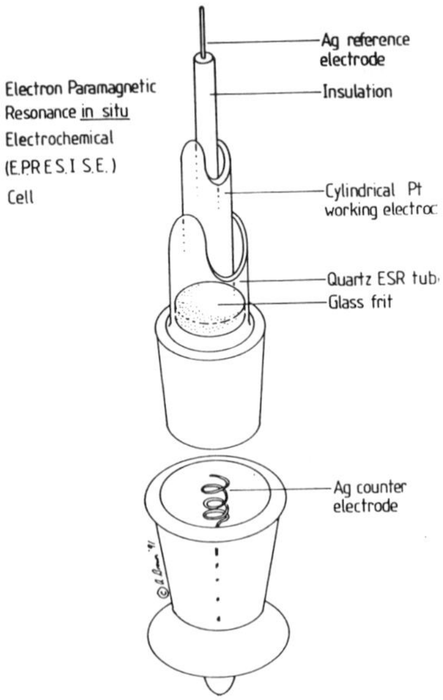

Figure 2. Schematic Representation of the E.P.RES.I.S.E. cell

Extended Huckel Molecular Orbital (E.H.M.O.) calculations employed the ICON 8 program[13] with the standard basis set of the valence orbitals and electrons of the elements present. All H_{ii}'s were charge iterated. The experimental models, with full molecular symmetry were used in all calculations.

3. Results and Discussion

3.1. BINUCLEAR COMPLEXES

3.1.1. *Symmetric* $[L_3M(\mu\text{-}Cl)_3ML_3]^+$ *species.* The symmetric binuclear complexes $[L_3Ru(\mu\text{-}Cl)_3ML_3]^+$ where L = tertiary phosphine ligand have been well characterised in the literature[14], the most recent publication detailing the single crystal X-ray structural determination of $[Ru_2(\mu\text{-}Cl)_3(PBu_3)_6]^+$ [15]. The Ru centres in the parent species are in oxidation states Ru(II),Ru(II) with a d^6,d^6 electronic configuration and hence are expected to have no Ru-Ru bond.The complexes undergo two well-separated reversible one-electron oxidations and the mixed-valence Ru(II),Ru(III) sysytems can be studied spectroelectrochemically. Preliminary studies on $[Ru_2(\mu\text{-}Cl)_3(PEt_2Ph)_6]^{2+}$ assigned the complex as a delocalised, class III, system[16] with a weak Ru-Ru interaction (formal bond order of 1/2).

We have undertaken the detailed study of the binuclear symmetric osmium complexes with the view of addressing the following questions:

(i) Have the diosmium species similar structures and electrochemical behaviours to the diruthenium

compounds?

(ii) To what extent does changing from Ru to Os in these compounds influence the metal-metal interaction?

The techniques used to answer these questions include spectroelectrochemistry, X-ray crystallography and kinetic studies.

The single crystal X-ray structure of $[Os_2(\mu\text{-}Cl)_3(PMe_2Ph)_6]PF_6$, $I^+PF_6^-$, confirmed the FSBO structure[17]. The metal-metal distances, average MClM and PMP angles are compared in Table 1 for the two complexes $[M_2(\mu\text{-}Cl)_3(PMe_2Ph)_6]^+$, M = Ru, Os.

TABLE 1. Selected structural data for $[M_2(\mu\text{-}Cl)_3(PMe_2Ph)_6]^+$

M	M-M(Å)	MClM(°)	PMP(°)
Os	3.440(1)	87.0	95.6
Ru[18]	3.39	86.0	95.3

Changing the identity of the metal has therefore had no impact on the overall structure of these symmetric bimetallic compounds and we might therefore predict that the diosmium complexes will have a similar redox behaviour to the Ru_2 species. Figure 3 shows the cyclic voltammogram obtained for I^+ and Table 2 compares the half-wave potentials for the analogous bimetallics.

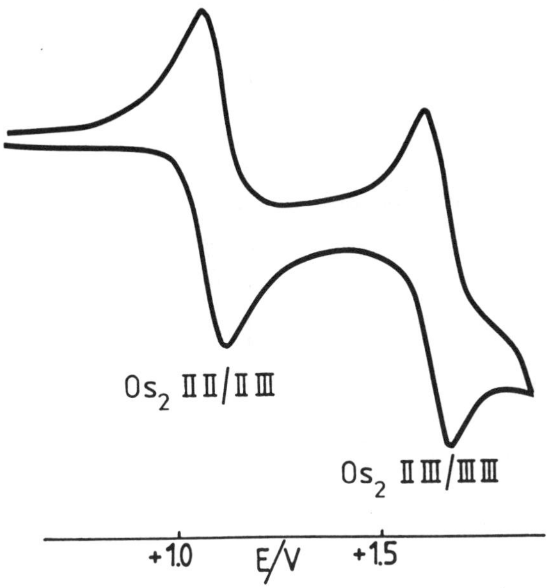

Figure 3. Cyclic voltammogram of $[Os_2(\mu\text{-}Cl)_3(PMe_2Ph)_6]^+$ in CH_2Cl_2/0.5M$[^nBu_4N]BF_4$ at 290K

The oxidation processes were each confirmed to involve one electron by coulometric experiments. No reductive processes were observed in the potential range studied (-2.5V). The diosmium species is approximately 200mV easier to oxidise than the corresponding Ru_2 complex which makes it an easier sysytem to work with experimentally.

TABLE 2. Electrochemical and spectroelectrochemical data for $[M_2(\mu\text{-}Cl)_3L_6]^+$ in $CH_2Cl_2/0.5M[^nBu_4N]BF_4$

		$E_{1/2}/V^b$		$\nu_{max}/cm^{-1}(\epsilon/mol^{-1}cm^{-1}dm^3)$	
M	L	II,II/II,III	II,III/III,III	II,III	III,III
Ru	PMe_2Ph	1.23(0.070)[a]	1.75(0.100)	5000(2610)[c] 7770(2860)	
Os	PMe_2Ph	1.05(0.060)	1.57(0.080)	5000(4900) 9200(650)	11480(7220)[d] 15000(2120)
Os	PEt_3	0.90(0.060)	1.46(0.070)	4900(4800) 7800(800)	11900(10500) 14900(770)

(a) $E_p^f - E_p^r$ (V), (b) T = 290K, (c) T = 266K, (d) T = 233K

The mixed-valence state $[Os_2(\mu\text{-}Cl)_3(PMe_2Ph)_6]^{2+}$, is air and moisture sensitive but is indefinitely stable under an atmosphere of nitrogen at 267K. Figure 4a shows the absorption spectral changes measured in an O.T.E. cell associated with the conversion of $\mathbf{I}^+ \rightarrow \mathbf{I}^{2+}$ and Figure 4b shows the further one-electron oxidation $\mathbf{I}^{2+} \rightarrow \mathbf{I}^{3+}$.

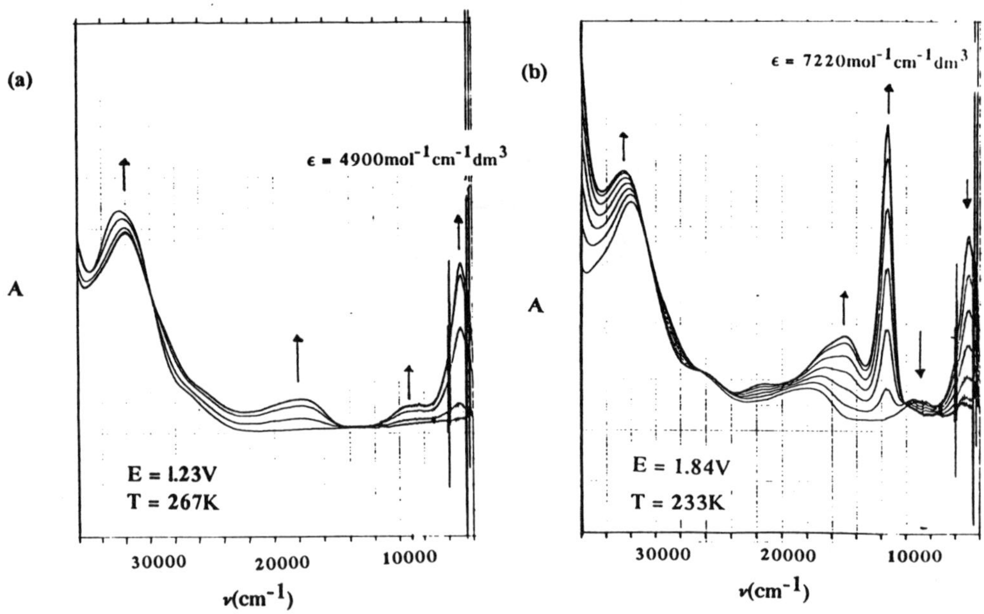

Figure 4. Absorption spectral changes accompanying the sequential one-elctron oxidations of $[Os_2(\mu\text{-}Cl)_3(PMe_2Ph)_6]^+$ in $CH_2Cl_2/0.5M[^nBu_4N]BF_4$, (a) $\mathbf{I}^+\rightarrow\mathbf{I}^{2+}$ at 267K, (b) $\mathbf{I}^{2+}\rightarrow\mathbf{I}^{3+}$ at 233K. Spikes at low energy are solvent overtone absorptions.

Note that the doubly oxidised species is stable at 233K whereas $[Ru_2(\mu\text{-}Cl)_3(PMe_2Ph)_6]^{3+}$ cannot be characterised as it is extremely unstable. The absorption spectra were recorded over the range 3125-50000cm^{-1} but only the nir region will be discussed here. The higher energy part of the spectra are dominated by charge transfer transitions involving the metal centre and the Cl^- and phosphine ligands.

As I^{2+} is generated from I^+, we observe intense bands growing in from the essentially featureless background of the parent $[Os_2(\mu\text{-}Cl)_3(PMe_2Ph)_6]^+$spectrum. The nir region of the spectrum of I^{2+} is dominated by two bands which shift to higher energy on further oxidation. Band positions and extinction coefficients for the diosmium and diruthenium species are given in Table 2. Concentrating on the low-energy bands of the mixed-valence Os(II),Os(III) state, we have determined that the position of the band maxima depend on the nature of the metal, bridging ligand and terminal ligand[17] but not on the properties of the solvent. Hence a plot of the energy of the band maximum against $1/D_{op}$ - $1/D_s$ shows no dependence on solvent, see Figure 5. On the same graph a plot of ν_{max} versus $1/D_{op}$ - $1/D_s$ for the well established class II species $[(bpy)_2ClRu(4,4'\text{-}bpy)RuCl(bpy)_2]^{3+}$ [19] is also shown.

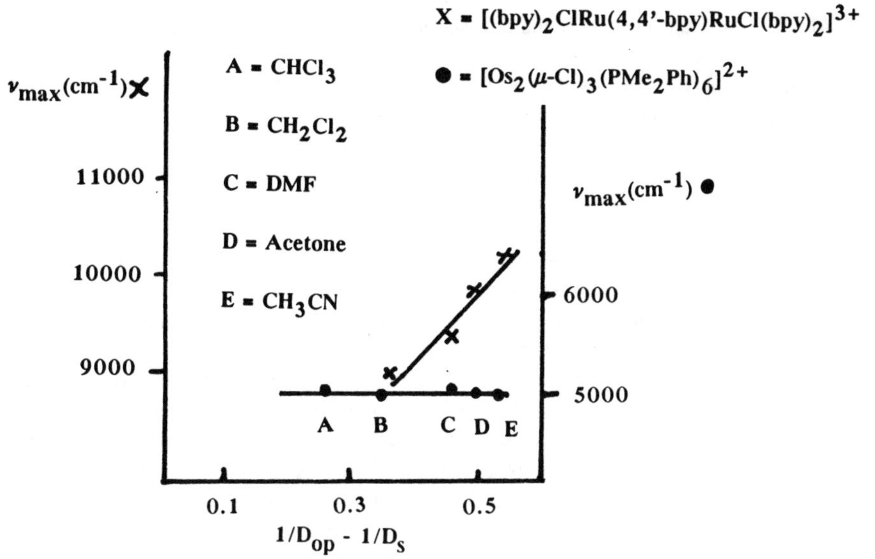

Figure 5. ν_{max} versus $1/D_{op}$ - $1/D_s$ for $[Os_2(\mu\text{-}Cl)_3(PMe_2Ph)_6]^{2+}$ and $[(bpy)_2ClRu(4,4'\text{-}bpy)RuCl(bpy)_2]^{3+}$

In summary then, the behaviour of the low-energy transitions of the mixed-valence $[Os_2(\mu\text{-}Cl)_3(PMe_2Ph)_6]^+$ show i) no solvent dependence and ii) intense, structured bands. Thus we suggest that the diosmium species belongs to class III mixed-valence materials with the definite possibility of metal-metal interactions occurring in the oxidised states.

The origin of the two nir bands is explained by consideration of the molecular orbital scheme for the interaction of metal-based d-orbitals within a FSBO D_{3h} structure[20]. A qualitative molecular orbital diagram is shown in Figure 6.

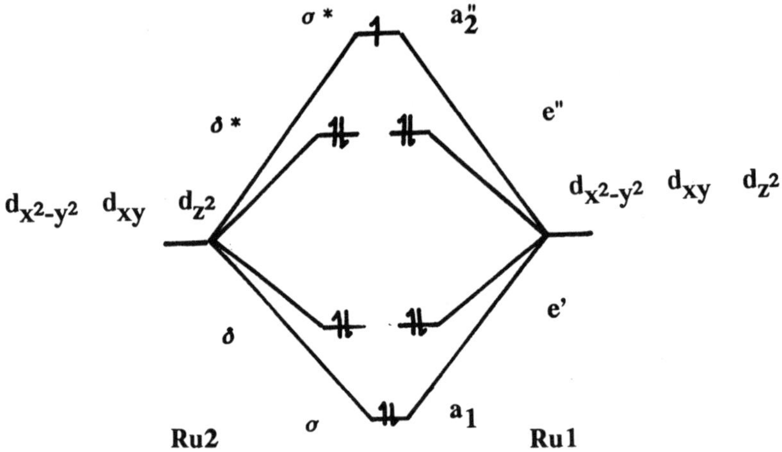

Figure 6. Molecular orbital scheme for the interaction of d orbitals in D_{3h} symmetry.

For the Os(II),Os(II) parent complex the six molecular orbitals will be full and hence the low-energy region of the spectrum should be blank which is as observed (see Figure 4a). The mixed-valence Os(II),Os(III) is generated by removing an electron from the antibonding a_2'' orbital leaving it singly occupied. Thus, electronic transitions from the metal-based molecular orbitals to the partially occupied a_2'' orbital can be envisaged. Application of electronic selection rules determine that only two transitions are allowed and will therefore carry measureable intensities: $e'' \rightarrow a_2''$ (x,y polarised) and $a_1' \rightarrow a_2''$ (z polarised). We therefore assign the two nir bands in the spectrum of $\mathbf{I}^{2+}$ to the $e'' \rightarrow a_2''$ ($\delta^* \rightarrow \sigma^*$) and $a_1' \rightarrow a_2''$ ($\sigma \rightarrow \sigma^*$) transitions.

Continuation of the argument for the Os(III),Os(III) oxidation state means removal of the second a_2'' electron. The selection rules permit the same two transitions which we then assign to the lowest energy transitions in $\mathbf{I}^{3+}$. Note that on oxidation the bands move to higher energy and become more intense. We suggest that these observations indicate an increasing metal-metal interaction on removal of antibonding electrons, that is, the metal-metal separation is decreasing on oxidation.

The species $\mathbf{I}^{2+}$ was too unstable to allow full single crystal X-ray studies. We therefore prepared $[Os_2(\mu\text{-}Cl)_3(PEt_3)_6]^+$ which has a more accessible oxidation potential than its PMe_2Ph analogue, see Table 2. The spectroelectrochemical behaviour of the PEt_3 species was similar to $\mathbf{I}^{2+}$ and $\mathbf{I}^{3+}$. $[Os_2(\mu\text{-}Cl)_3(PEt_3)_6]^+$ could be chemically oxidised using perchloric acid to the dication which was relatively stable. X-ray crystal structures of $[Os_2(\mu\text{-}Cl)_3(PEt_3)_6]^+$ and $[Os_2(\mu\text{-}Cl)_3(PEt_3)_6]^{2+}$ were determined, full structural details are available[17]. A view of the mono- and dication are shown in Figure 7 with relavent bond distances and angles.

Both mono- and dication have FSBO structures. In the dicationic species the two metal centres are indistinguishable crystallographically which supports delocalisation of the odd electron between the two metal centres. Comparison of the two structures shows that on oxidation the metal-metal distance decreases as does the angle at the OsClOs bridge. Thus we suggest that removing electrons from d^6,d^6 bimetallic complexes causes the metal centres to become interacting, 1/2 a formal metal-metal bond in the d^6,d^5 species and a full bond in the d^5,d^5 compounds.

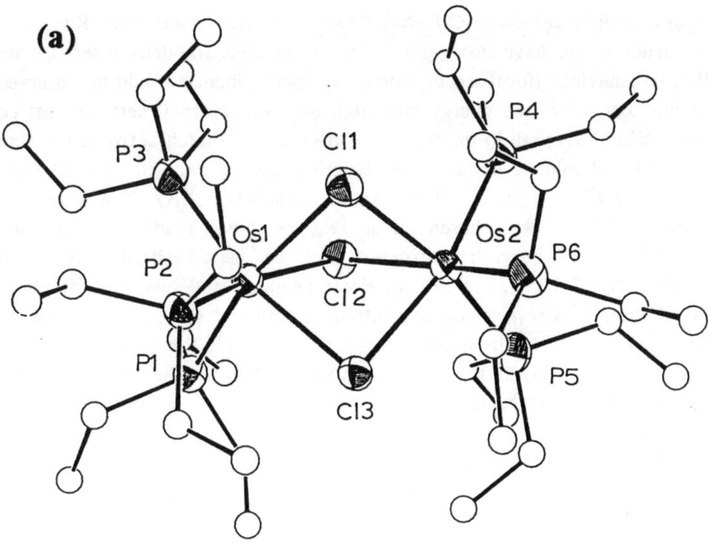

Ox. St.	Os--Os (pm)	Os-Cl	Os-P	OsClOs
1+	347.3(1)	249.2(4)	230.4(4)	88.25(12)
2+	340.6(1)	247.2(5)	235.0(6)	87.01(17)

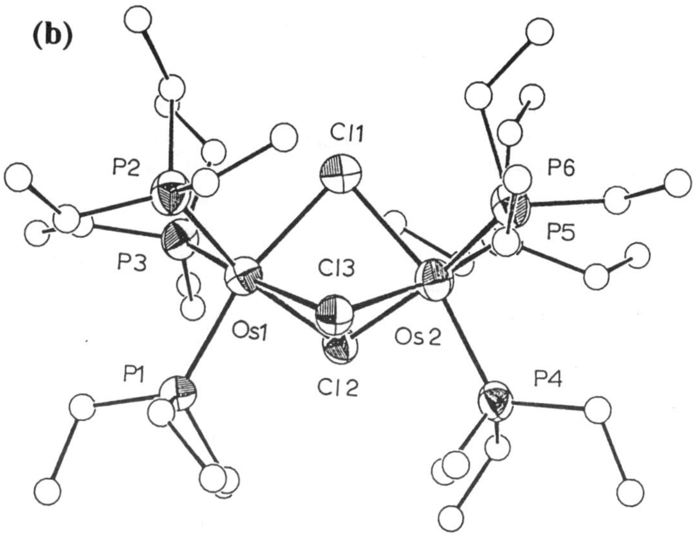

Figure 7. A view of (a) I^+ and (b) I^{2+} with relevant structural parameters.

3.1.2. *Asymmetric Complex $[Cl(PPh_3)_2Ru(\mu\text{-}Cl)_3Ru(PPh_3)_2CS]$.* All symmetric Ru and Os complexes with a FSBO structure we have investigated have been class III, delocalised species[17]. We argued that class II-type behaviour (localised or valence trapped systems) would be observed, if at all, for asymmetric complexes where the energy mismatch between the two metal centres could be sufficient to minimise d-orbital overlap. In order to maximise the mismatch between the centres we thought that one Ru centre should be bound to π-accepting ligands stabilising Ru(II) and the other Ru centre bound to π-donor ligands which would stabilise Ru(III). The title species $[Cl(PPh_3)_2Ru(\mu\text{-}Cl)_3Ru(PPh_3)_2CS]$, **II**, was chosen for study since it was readily available in our laboratory and it fulfilled the above criteria. The structure of **II** has been verified as triply-chloro bridged[21] with a Ru-Ru distance of 3.35Å. Preliminary electrochemical and spectroelectrochemical investigations[16] of **II** in thf revealed a reversible one-electron oxidation at +0.70V and a weak band at 14800cm^{-1} (ϵ = 100mol^{-1}cm^{-1}dm^{3}) in the absorption spectrum of **II**$^+$. We decided to make an in depth study of this complex to confirm the class II nature of **II**$^+$ and hence to measure an intramolecular rate constant for self-electron transfer.

The complex **II**$^+$ was studied spectroelectrochemically in an O.T.E. cell in a variety of solvents. The characteristics of the IVCT band in the different solvents are given in Table 4 together with the solvent parameters.

TABLE 4. IVCT band characteristics of **II**$^+$ as a function of solvent at T = 258K

Solvent	ν_{max}(cm^{-1})	ϵ(mol^{-1}cm^{-1}dm^{3})	$\Delta_{1/2}$(cm^{-1})^{a}	D_s	D_{op}
CH_2Cl_2	14500	470	5200	8.9	2.028
$CHCl_3$	12600	450	4900	4.7	2.085
DMF^{b}	16000	470	5850	36.7	2.0458
P.C.^{b}	15600	484	5700	65.1	2.0133

(a) Band width at half-height, (b) DMF = dimethylformamide, P.C. = propylene carbonate

The position of the IVCT band maximum, ν_{max}, is found to depend on the solvent and an approximate linear correlation is achieved for ν_{max} versus $1/D_{op} - 1/D_s$. Solubility problems prevented further study in other solvent systems. The dependence of ν_{max} on the solvent system suggested that **II**$^+$ is a localised mixed-valence system with a weak IVCT transition in the visible region of the spectrum. Application of Hush's theory[22] for class II-type species to the absorption spectrum of **II**$^+$ confirms this assignment with α, the degree of delocalisation approximately equal to 0.07. (α for a class II species should be < 0.25 and cannot be absolutely determined for this system since the metal-metal distance in **II**$^+$ is unknown.)

Further application of Hush's[22] treatment for an asymmetric species permits us to determine several important parameters pertaining to the thermal electron transfer from one Ru centre to the other, see equation 1.

$$[Cl(PPh_3)_2Ru(III)(\mu\text{-}Cl)_3Ru(II)(PPh_3)_2CS]^+ \xrightarrow[k_{th}]{} [Cl(PPh_3)_2Ru(II)(\mu\text{-}Cl)_3Ru(III)(PPh_3)_2CS]^+ \quad \underline{1}$$

Thus analysis of the IVCT band profile allows us to calculate ν_{et}, the frequency factor for electron transfer, ΔG^*_{th}, the free energy barrier for thermal electron transfer and k_{th}, the rate constant for the thermal electron transfer. These parameters are tabulated in Table 5 for **II**$^+$ and for some single-bridged diruthenium complexes also exhibiting class II-type behaviour[23].

TABLE 5. Electron transfer parameters for class II mixed-valence compounds.

Complex	$\nu_{max}(cm^{-1})$	$\nu_{et}(s^{-1})$	$\Delta G^*_{th}(V)$	$k_{th}(s^{-1})$
II^{+} [a]	14500	2.08×10^{15}	0.64	2.9×10^{4}
$[L_2ClRu(pyz)RuClL_2]^{3+}$ [b,c]	7340	2.5×10^{14}	0.23	3.9×10^{9}
$[(NH_3)_5Ru(pyz)RuClL_2]^{4+}$ [b,c]	10406	5.2×10^{14}	0.41	5.7×10^{7}
$[L_2ClRu(4,4'B)RuClL_2]^{3+}$ [b,c]	10163	4.3×10^{13}	0.32	1.9×10^{8}

(a) CH_2Cl_2, (b) CH_3CN, (c) L = 2,2'-bipyridine, pyz = pyrazine, 4,4'B = 4,4'-bipyridine

In our triple chloro-bridged species both ν_{max} and ΔG^*_{th} are greater than in the single bridged complexes. We suggest that this is a consequence of the greater asymmetry of the two metal centres in **II**$^{+}$ than in the single bridged compounds. The single bridged species are known to undergo electron transfer *via* an inner sphere mechanism through π^* orbitals on the bridge. In **II**$^{+}$ the thermal electron transfer rate constant is smaller than those for the singly-bridged species. We would argue that in **II**$^{+}$ the electron transfer is also inner sphere, not through the bridging Cl^- ligands, but *via* minimal metal d-orbital overlap. Further analysis of the results determines that inner sphere reorganisation energies are similar for the single and triple bridged mixed-valence species.

The spectroelectrochemical study of the IVCT transition in the asymmetric complex **II**$^{+}$ has confirmed that this species is a class II mixed-valence compound and several physical parameters pertaining to thermal electron transfer rates have been evaluated. Further studies on other asymmetric FSBO complexes are now in progress to enhance the work presented here.

3.2 MONOMERIC PLATINUM COMPLEXES

A cyclic voltammogram of $[Pt(bpy)Cl_2]$, III, in DMF/0.1M$[^{n}Bu_4N]BF_4$ reveals a reversible one-electron reduction step at -1.17V and a further quasi-reversible reduction process at more negative potentials. Full electrochemical data are given in Table 6. Coulometric studies confirm the first redox process to involve one electron.

TABLE 6. Electrochemical data of Pt(II) complexes in DMF/0.1M$[^{n}Bu_4N]BF_4$ at 290K

Complex	$E_1{}^{a}/V$	E_2
$[Pt(bpy)Cl_2]$	-1.17(0.070)[b]	-1.82(0.110)
$[Pt(dpk)Cl_2]$	-0.72(0.060)	-1.37(0.120)
$[Pt(B_{10}H_{12})_2]^{2-}$	-1.65(0.080)	——

(a) $E = (E^f_p + E^r_p)/2$, (b) $(E^f_p - E^r_p)$

The uv/vis spectrum of III is dominated by the intense $\pi \rightarrow \pi^*$ intraligand transisitions of coordinated bpy centred at 31600cm^{-1}. The other two features of the absorption spectrum have been assigned as a spin allowed metal-to-ligand charge transfer band near 25600cm^{-1} which is solvent dependent and a d-d transition at 21000cm^{-1} [24]. Thus electronic spectroscopy suggests that the lowest unoccupied molecular orbital (LUMO) on III is metal-based.

The electrochemical reduction of III at -1.32V affords a highly coloured olive green solution. The uv/vis/nir spectrum of III$^-$ can be studied spectroelectrochemically in an O.T.E. cell at 240K. The spectrum of III^- bears a striking resemblence to that of $[Ir(bpy)_3]^{2+}$ [25] which has been formulated as $[Ir(III)(bpy^o)_2(bpy^-)]^{2+}$. The nir and vis bands in the spectrum of $[Ir(bpy)_3]^{2+}$ have been assigned to intraligand transitions of bpy^- and are thus assigned in III^-. We are therefore suggesting that the semi-occupied molecular orbital (SOMO) of III^- is bpy-based.

The spin-1/2 ^{195}Pt nucleus is in a natural abundance of 34%, so that platinum nuclear hyperfine structure might be expected in the epr spectrum of III^- if the SOMO contains metal character. Measurement of the metal nuclear hyperfine coupling tensor can give information about the location of the spin density. The isotropic epr spectrum of III^- consists of a single line, due to the Pt isotopes with zero spin, and two satellites, the ±1/2 hyperfine lines of ^{195}Pt. Our spectrum is in good agreement with that reported by Kochi[26]. The g value and Pt hyperfine coupling constant, A_{iso}, are given in Table 7.

TABLE 7. Epr parameters of Pt(I) species generated by one-electron reduction of Pt(II) complexes.

Complex		g_{iso} [a]	g_1 [b]	g_2	g_3	A_{iso}	A_1	A_2	A_3
$[Pt(bpy)Cl_2]^-$ [c]	exp [d]	1.998	2.038	2.009	1.935	58	59	101	25
	sim	——	2.038	2.011	1.938	—	60.3	85.8	23.0
$[Pt(dpk)Cl_2]^-$	exp	2.011	2.052	2.010	1.970	49	——	33	67
	sim	——	2.049	2.009	1.970	—	40.0	29.6	73.0
$[Pt(B_{10}H_{12})_2]^{3-}$	exp	——	1.895	2.021		—	214	268	

(a) temperature of isotropic measurements 280K, (b) temperature of rhombic measurements 77K, (c) DMF/0.1M[$^nBu_4N$]BF_4, (d) exp = experimental values, sim = computer simulation values.

The spectrum of III^- generated in the E.P.RES.I.S.E. cell at 240K and then frozen to 100K is shown in Figure 8a. Figure 8B is the 77K spectrum of III^- chemically generated by reduction of III by $NaBH_4$ in DMF. Care must be excercised with the chemical reduction as III has an accessible chemically irreversible second reduction step. The two spectra are clearly identical. The three principal values of the g tensor are observed but the six ^{195}Pt satellites are not all resolved. However sufficient may be measured to permit unambiguous experimental determination of the hyperfine coupling constants, as given in Table 7. Note that the averaged $\langle g \rangle$ and $\langle A \rangle$ values correspond well with the observed g_{iso} and A_{iso} values and hence we can assign the same sign to all coupling constants. A computer-simulated spectrum, based on the paprmeters of Table 7, is shown in Figure 8c. The refined g and A values from the computer simulation, which give an almost identical spectrum to the observed one, are also quoted in Table 7.

E.H.M.O. calculations were performed using the coordinates of the atoms of $[Pt(bpy)Cl_2]$ as given by Osborn and Rogers[27], with the z axis perpendicular to the molecular plane and the y axis bisecting the two Pt-Cl bonds. The results of the M.O. considerations suggest that the SOMO of III^- is of b_2 symmetry with a $5d_{yz}$ contribution from the Pt. Following Maki's treatment[28] of this case but using the nomenclature of Rieger *et al*[29] we use equation 2, in which we have neglected any Pt 6p contributions, to evaluate a^2, the metal spin density.

$$A_x - \langle A \rangle = P_d[-4/7a^2 + 2/3\Delta g_x - 5/42(\Delta g_y + \Delta g_z)] \qquad 2$$

Substituting experimental values of A_x, $\langle A \rangle$, the Δg_i values ($\Delta g_i = g_i - g_e$) and P_d = 526.4G for

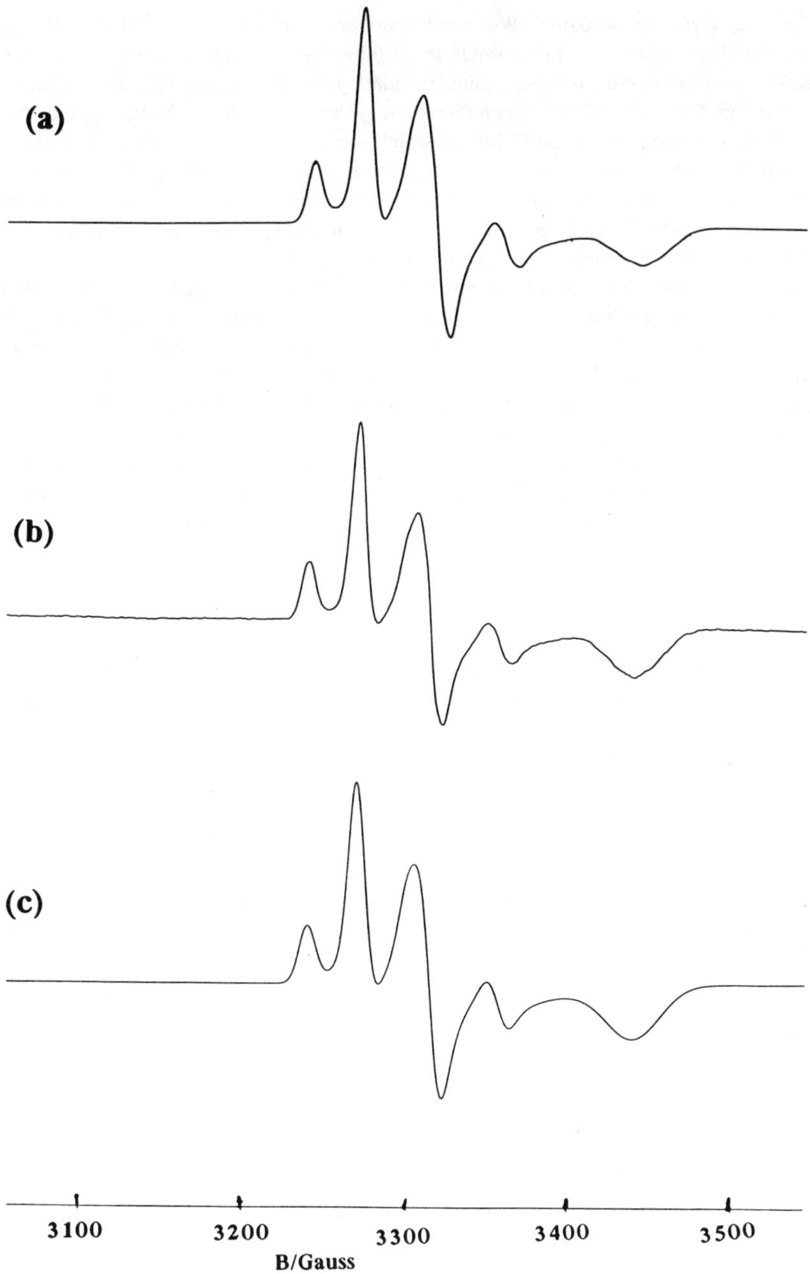

Figure 8. Epr spectra of $[Pt(bpy)Cl_2]^-$ (a) T = 100K, electrochemically generated in DMF, (b) T = 77K, chemically generated by $NaBH_4$, (c) computer-simulated spectrum using the parameters in Table 7. The computer simulation employed Gaussian line shapes with a half-line width of 10G, 12G and 30G for g_1, g_2 and g_3 respectively.

^{195}Pt [30] into $\underline{2}$ gives a^2 = 0.042. We must therefore conclude from analysis of the epr spectroelectrochemical data that the SOMO of $[Pt(bpy)Cl_2]^-$ is not primarily metal d-based. E.H.M.O. considerations back up this conclusion calculating the total percentage metal character in the SOMO of III^- as 5.9%. The calculations indicate that a π*-antibonding orbital of bpy contributes more than 90% to the character of the SOMO of which approximately 34% is located on the two N atoms. This implies that the hyperfine coupling to nitrogen in III^- should be of the order of 2G which is smaller than our experimental line widths and therefore will not be resolved. This may, in part, be reponsible for the fairly broad signals observed in the epr spectrum. Experiments are in progress with mixed-solvent systems to try and improve the resolution.

Thus the spectroelectrochemical results suggest that the SOMO of $[Pt(bpy)Cl_2]^-$ is primarily based on the π* antibonding orbital of bpy and III^- should be formulated as $[Pt(II)(bpy^-)Cl_2]^-$. Spectroelectrochemical experiments on the related complex $[Pt(dpk)Cl_2]$, dpk = dipyridyl ketone, suggest that the SOMO of the monoreduced species is also ligand based. Interestingly, the sandwich complex $[Pt(B_{10}H_{12})_2]^{2-}$ which may be electroreduced at -1.82V (see Table 6) to $[Pt(B_{10}H_{12})_2]^{3-}$ has a very different epr spectrum[31]. The axial spectrum has much larger Pt hyperfine coupling constants, see Table 8. E.H.M.O. calculations on $[Pt(B_{10}H_{12})_2]^{2-}$ suggest that the LUMO has 45% metal character, that is, $[Pt(B_{10}H_{12})_2]^{3-}$ has considerably more Pt(I) metal character in it's SOMO than $[Pt(bpy)Cl_2]^-$ or $[Pt(dpk)Cl_2]^-$. Further work on this system is in progress.

4. Acknowledgements

We thank the SERC (RJS and SAM) and the University of Edinburgh for financial support. We gratefully acknowledge Dr A.J. Welch for E.H.M.O. calculations, Drs A.J. Blake and R.O. Gould for X-ray crystallographic determinations and Drs D. Collison and K.J. Taylor (Manchester University) and R.E.P. Winpenny for helpful discussion and computer simulation of the epr results.

5. References

1. For recent general reviews, see (a) C.A. Widrig, M.D. Porter, M.D. Ryan, T.G. Strein, A.G. Ewing, Anal. Chem., (1990), 62, 1R, (b) K.M. Kadish, X. Mu, Pure Appl. Chem., (1990), 62, 1051.
2. M.B. Robin, P. Day, Adv. Inorg. Chem. and Radiochem., (1967), 10, 247.
3. (a) C. Creutz, Prog. Inorg. Chem., (1983), 30, 1, (b) B.S. Brunschwig, S. Ehrenson, N. Sutin, J. Phys. Chem., (1986), 90, 3657.
4. C. Creutz, M.H. Chou, Inorg. Chem., (1987), 26, 2995.
5. P.S. Braterman, J.-I. Song, C. Vogler, W. Kaim, Inorg. Chem., (1992), 31, 222.
6. J. Chatt, R.G. Hayter, J. Chem. Soc., (1961), 896.
7. V.T. Coombe, PhD Thesis, University of Edinburgh, 1985.
8. P.W. Armit, W.J. Sime, T.A. Stephenson, J. Chem. Soc., Dalton Trans., (1976), 2121.
9. G.T. Morgan, F.H. Burstall, J.Chem. Soc., (1934), 965.
10. F. Klanberg, P.A. Wegner, G.W. Parshall, E.L. Muetterties, Inorg. Chem., (1968), 7, 2073.
11. G.A. Heath, L.J. Yellowlees, P.S. Braterman, J. Chem. Soc., Chem. Commun. (1981), 287.
12. M.R. Low PhD Thesis, University of Edinburgh, 1987.
13. J. Howell, A. Rossi, D. Wallace, K. Haraki, R. Hoffmann, ICON, Quantum Chemistry Program Exchange, Univ. Indiana, (1977), no. 344.
14. E.A. Seddon and K.R. Seddon, The Chemistry of Ruthenium, Elsevier, Amsterdam, (1984)

15. F.A. Cotton, R.C. Torralba, Inorg. Chem., (1991), 30, 2196.
16. G.A. Heath, A.J. Lindsay, T.A. Stephenson, D.K. Vattis, J. Organomet. Che., (1982), 233, 353.
17. R.J. Sorbie, PhD Thesis, University of Edinburgh, 1989.
18. M. Laing, L. Pope, Acta Crytallogr., (1976), 332,1547.
19. M.J. Powers, T.J. Meyer, J. Am. Chem. Soc., (1980), 102, 1239.
20. N.S. Hush, J.K. Beattie, V.M. Ellis, Inorg. Chem., (1984), 23, 3339.
21. A.J. Fraser, R.O. Gould, J. Chem. Soc., Dalton Trans., (1974), 1139.
22. N.S. Hush, Prog. Inorg. Chem., (1967), 8, 391.
23. T.J. Meyer, Chem. Phys. Lett., (1979), 64, 417.
24. V.M. Miskowski, V.H. Houlding, Inorg. Chem., (1989), 28, 1529.
25. V.T. Coombe, G.A. Heath, A.J. MacKenzie, L.J. Yellowlees, Inorg. Chem., (1984), 23, 3423.
26. R.J. Klinger, J.C. Huffman, J.K. Kochi, J. Am. Chem. Soc., (1982), 104, 2147.
27. R.S. Osborn, D. Rogers, J. Chem. Soc., Dalton Trans., (1974), 1002.
28. A,H, Maki, N. Edelstein, A, Davidson, R.H. Holm, J. Am. Chem. Soc., (1964), 86, 4580.
29. (a) N.G. Connelly, W.E. Geiger, G,A. Lane, S.J. Raven, P.H. Rieger, J. Am. Chem. Soc., (1986), 108, 6219, (b) J.A. De Gray, W.E. Geiger, G.A. Lane, P.H. Rieger, Inorg. Chem., (1991), 30, 4100.
30. J.S. Morton, K.J. Preston, J. Magn. Reson., (1978), 30, 577.
31. S.A. Macgregor, PhD Thesis, University of Edinburgh, 1992.

ONE- AND TWO-ELECTRON OXIDATIONS OF BIMETALLIC FULVALENE COMPLEXES STUDIED BY VOLTAMMETRY AND IR SPECTROELECTROCHEMISTRY

WILLIAM E. GEIGER, CHRISTOPHER G. ATWOOD, TEEN T. CHIN
Department of Chemistry
University of Vermont
Burlington, VT 05405 USA

ABSTRACT. The oxidation of a series of fulvalene (Fv) bimetallic complexes has been studied by several electrochemical and spectroscopic methods, including cyclic voltammetry, coulometry, and infrared spectroelectrochemistry. $Fv[Mn_2(CO)_4(PPh_3)_2]$ and $CH_2[(C_5H_4)_2Mn(CO)_2(PPh_3)]_2$ each oxidize in two $1e^-$ steps, the monocations being Class II mixed-valent. Oxidation of the doubly-linked complex $Fv[Mn_2(CO)_4(\mu$-dppm] yields a monocation with stronger site-site interactions. A Mn-Mn distance of 4.07 A has been characterized by X-ray crystallography of the monocation. Analogous complexes of rhodium have also been studied. $Fv[Rh_2(CO)_4]$ and $Fv[Rh_2(CO)_2(PPh_3)_2]$ each oxidize in a single $2e^-$ wave to corresponding dications, which have a Rh-Rh bond. The tetracarbonyl dication undergoes facile decomposition, but it is characterized by IR and optical spectroscopies. The doubly-bridged complex $Fv[Rh_2(CO)_2(\mu$-dppm] displays nearly Nernstian two-electron voltammetry and gives a dication in which the Rh-Rh distance has decreased from 4.51 A to 2.83 A. The X-Ray crystallographic results were obtained by A.L. Rheingold.

1. Introduction

Fulvalene dimetal complexes have been profitably employed in studies of metal-metal interactions across a spacer ligand with two nominally aromatic components. Work on the oxidation of bi-ferrocene, **1**, bis(fulvalene)diiron, **2**, and their analogues occupy a central focus in these investigations.[1] The ferrocenyl moiety

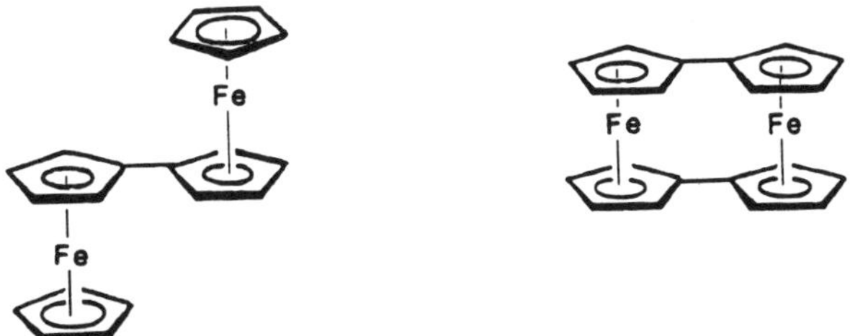

1 **2**

is valuable in these studies owing to its inertness to substitu-

A. J. L. Pombeiro and J. A. McCleverty (eds.),
Molecular Electrochemistry of Inorganic, Bioinorganic and Organomettallic Compounds, 519–532.

tion in either the Fe(III) (d^5) or Fe(II) (d^6) oxidation state. Formally mixed-valent ions are most often studied by optical and Mossbauer spectroscopies and by measurement of their magnetic properties. A detailed picture continues to emerge on how the site delocalization is affected by the nature of the organic linkage and the local environmental effects such as ion pairing, solvation, and lattice structures.[2]

One expects that the conditions under which metal-metal delocalization is observed will depend on the nature of the metal and the ancillary ligands, but this area is only beginning to be explored. In designing complexes with such studies in mind, one places a high priority on attaching spectroscopic labels with which the mixed valency of the cations may be easily probed. We have previously used this strategy with biphenyl complexes of chromium.[3-5] In **3** and **4**, charge localization of the radical cation is easily probed through the IR stretching frequency of the CO ligand and the hyperfine coupling constant of the P atom (^{31}P, 100%, I = 1/2) on the phosphine ligand.

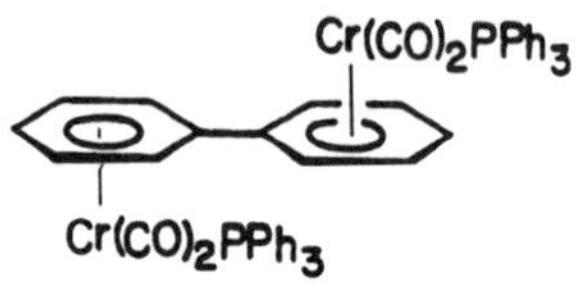

3

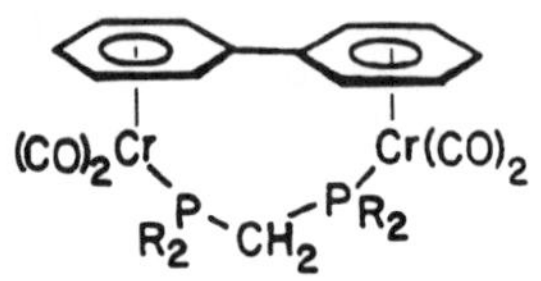

4

1.1 SIGNIFICANCE OF SHIFTS IN CARBONYL STRETCHING FREQUENCIES

In brief, a one-electron oxidation centered on a single Cr site results in an increase of <u>ca</u> 110 cm^{-1} in ν_{CO}(sym) and a somewhat lesser increase for ν_{CO}(asym).[4,6] The change arises from decreased back-bonding into the CO antibonding orbital. A trapped-valent monocation for a complex like $\mathbf{3}^+$ therefore has one pair of CO bands for the Cr(I) site, $\Delta\nu_{CO}$(sym) ≈ 100-110 cm^{-1} and one pair for the Cr(0) site, $\Delta\nu_{CO}$(sym) ≈ 0-20 cm^{-1}. The small but usually measurable shift for the Cr(0) site is thought to arise from σ-bond charge transmission between the two halves of the molecule. The monocation $\mathbf{4}^+$, which has an apparently delocalized ground state, has only two CO bands, $\Delta\nu_{CO}$(sym) = 79 cm^{-1}. [3]

1.2 SIGNIFICANCE OF ΔE^{o} VALUES OF SUCCESSIVE OXIDATIONS

If the two redox sites in the fulvalene complex are chemically identical but strictly non-interacting, the electrode potential necessary to extract an electron from the monocation will be 0.036 V positive of that necessary to extract an electron from the neutral complex, i.e., ΔE^{o} (= E^{o}_{2}-E^{o}_{1}) = 36 mV.[7] The monocation would be called Class I mixed-valent in the Robin-Day classification scheme.[8]. If the monocation has a totally delocalized ground state (Class III), ΔE^{o} is much larger: e.g., 540 mV for **4** and 590 mV for **2**, systems known to have a delocalized mixed-valent cation.[3,9]. Class II systems, which are delocalized under certain conditions (such as in photoexcitation) are expected to have intermediate values of ΔE^{o}.

Given the fact that the ΔE^{o} values depend not only on delocalization energies but also on a variety of structural and electronic factors [10], it is unlikely that any particular value of ΔE^{o} above 36 mV will be diagnostic of a particular type of mixed-valency. Rather, ΔE^{o} values contribute to the overall picture when piecing together the mixed-valency studies. Biferrocene, which has a Class II cation with moderately strong M-M interactions, has ΔE^{o} = 350 mV in CH_2Cl_2. [11]

1.3 ONE- vs. TWO-ELECTRON ANODIC PROCESSES

The idea that homonuclear dimetallic fulvalene d^6, d^6 complexes will oxidize in separate one-electron steps assumes that the structure of the complex does not undergo a profound change during the redox processes. Large or irreversible structural changes might effect the separation of the oxidation waves and, in the extreme, might lead to a preferential stabilization of the dication such that $E^{o}_{2} < E^{o}_{1}$, in which case a single $2e^-$ wave might be observed.

The present contribution discusses the oxidation and mixed-valency of dimanganese (and, to a lesser extent, dicobalt) fulvalene complexes which oxidize in discrete $1e^-$ steps. Different behavior is observed for dirhodium fulvalene complexes, in which $2e^-$ waves are observed in oxidations that involve reversible making and breaking of M-M bonds.

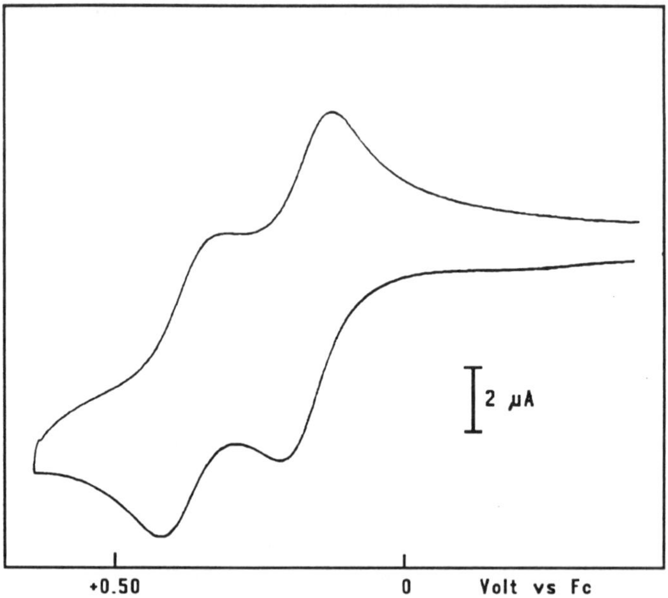

Figure 1. CV of cmpd 5 (0.50mM) in CH_2Cl_2, v=0.20 V/s

2. Results

2.1 Oxidation of Dimanganese Fulvalene Complexes

Cyclic voltammetry (CV) scans of **5** and **6** (Fig 1) reveal two $1e^-$, nearly Nernstian, oxidations. The first wave of **5** has about the same potential as does the mononuclear complex $CpMn(CO)_2(PPh_3)$, **7**, E^o = +0.23 V. (All potentials in this paper are referenced to Ferrocene and are collected in Table 1). The ΔE^o value for the oxidations of **5** is 210 mV. Complex **6**, in which a methylene

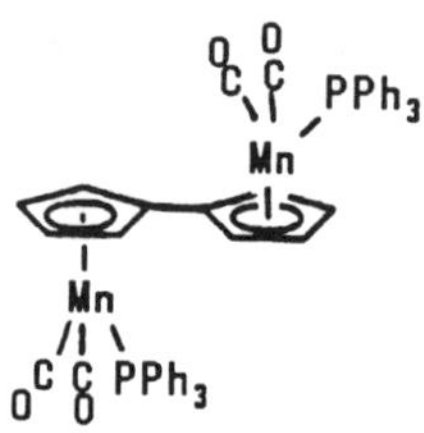

5

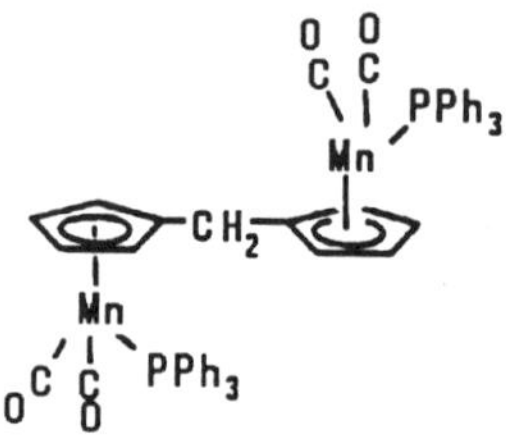

6

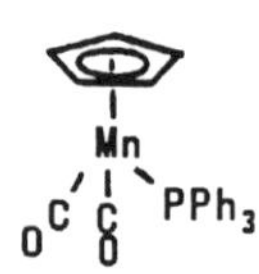

7

group is interposed between the two cyclopentadienyl rings, has a much smaller value, ΔE^o = 78 mV. This situation is reminiscent of bis(ferrocenyl)methane compared to biferrocene.[11,12] The two monocations **5**$^+$ and **6**$^+$ display trapped valence on the vibrational time scale, with one Mn(II) site and one Mn(I) site. The oxidations are conveniently studied by transmission-mode IR spectroelectrochemistry. Figure 2, which monitors IR spectra during the oxidation of 2 mM **7** at 273 K, is typical. The pair of CO stretches for **7** is cleanly replaced by another pair for **7**$^+$ at spectral shifts of +117 cm^{-1} for $\nu(CO)_{sym}$ and +102 cm^{-1} for $\nu(CO)_{asym}$ (Table 2). These shifts may be taken as diagnostic of a $1e^-$ metal-based oxidation, as originally noted by Connelly.[13]

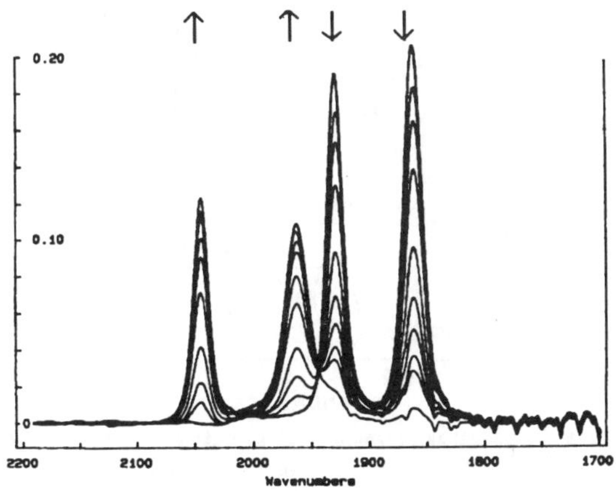

Figure 2 Spectroelectrochemical Oxidation of 2 mM $CpMn(CO)_2PPh_3$ in CH_2Cl_2 ; T=273 K

TABLE 1. Formal Potentials (vs Fc) of Fulvalene Dimetallic Complexes in CH_2Cl_2/0.1 M Bu_4NPF_6

Complex		$E^o{}_1$(0/+) V	$E^o{}_2$(+/2+) V	ΔE^o mV
$CpMn(CO)_2(PPh_3)$	**7**	+0.23		
$Fv[Mn_2(CO)_4(PPh_3)_2]$	**5**	+0.19	+0.40	210
$CH_2[(C_5H_4)Mn(CO)_2(PPh_3)]_2$	**6**	+0.19	+0.26	70
$Fv[Mn_2(CO)_4(\mu$-dppm]	**8**	-0.07	+0.38	450
$Cp_2Mn_2(CO)_4(\mu$-dppm)	**9**	+0.19	+0.31	120
$Fv[Co(C_8H_{12})]_2$	**10**	-0.30	-0.08	220
$Fv[Co(C_2B_4H_6)]_2$	**11**	+1.03	+1.30	270
$Fv[Rh_2(CO)_4]$	**12**	-0.18 (n = $2e^-$)		
$Fv[Rh_2(CO)_2(PPh_3)_2]$	**13**	-0.46 (n = $2e^-$)		
$Fv[Rh_2(CO)_2(\mu$-dppm]	**14**	-0.75 (n = $2e^-$)		

The monocation **5**$^+$ has four CO absorptions (Fig 3). The highest frequency band has shifted +102 cm^{-1}; the lowest has shifted +16 cm^{-1}. This is clearly compatible with a SOMO (singly-occupied MO) localized on one side of the complex (predominantly on Mn). The dication of **5** returns again to having only two CO bands, since both sites are again equivalent, $\Delta\nu_{CO}$(sym) = +121 cm^{-1},$\Delta\nu_{CO}$(asym) = +110 cm^{-1}, diagnostic of full 1e$^-$ oxidation of both metals.

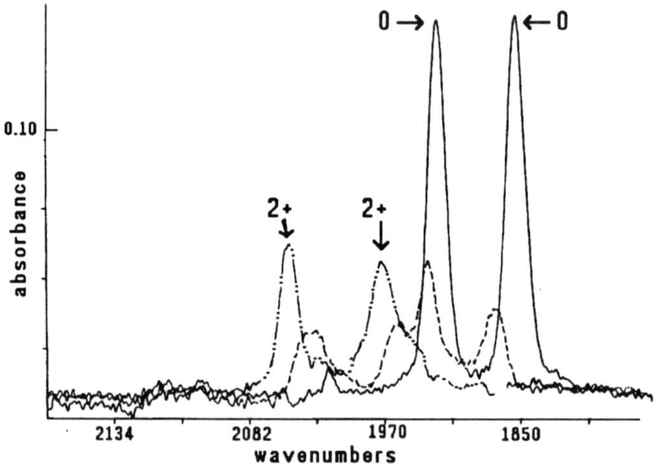

Figure 3. Spectroelectrochemical oxidation of cmpd 5 (1mM) in CH_2Cl_2. —— = neutral complex; - - - = cation; —·—·— = dication. T= 243K

Similar results are obtained for **6**, in which the redox sites are separated by an additional saturated carbon. For **6**$^+$ the largest CO shift is +113 cm^{-1}, the smallest +6 cm^{-1}. We conclude that even in Mn sites separated by bis(cyclopentadienyl)methane, there is a small amount of site interaction. Oxidation of **6** to a dication again produces a pair of CO bands with the value of ν_{CO}(sym) shifting by an additional +9 cm^{-1} from the cation.

Table 2. Carbonyl IR Absorptions for Compounds and Ions Studied in CH_2Cl_2

Complex	Charge	$\nu_{CO}(cm^{-1})$	$\Delta\nu_{CO}(sym)(cm^{-1})$
7	0	1864,1931	
	+1	1966,2048	117
5	0	1862,1927	
	+1	1878,1934,1959,2029	102
	+2	1972,2050	123
6	0	1860,1928	
	+1	1866,1932,1959,2041	113
	+2	1967,2050	122
8	0	1865,1928	
	+1	1885,1929,1952,2003	75
	+2	2022,2051	123
12[a]	0	1973,2033	
	+2	2114,2146	113
13[b]	0	1937,1959	
	+2	2049,2069	110
14	0	1945	
	+2	2052(sh),2083	138

[a] Data on compound **12** in acetone [b] From ref 15(c)

Voltammetry of the doubly-bridged complex **8**, while still showing two reversible 1e⁻ oxidations, reveals two important

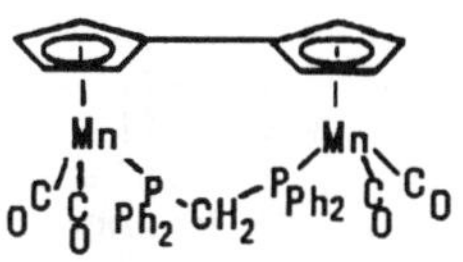

8

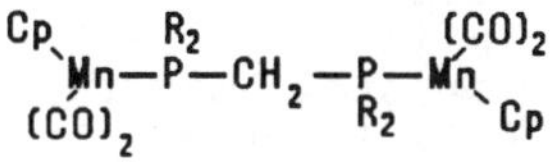

9

changes compared to **5** and **6**. The first oxidation is more facile than that of mononuclear **7** (by <u>ca</u> 250 mV), and the ΔE^o value increased to 450 mV (Table 1). Since the latter is more typical of highly interacting redox centers (<u>vide</u> <u>ante</u>) it is not surprising that IR spectra of $\mathbf{8}^+$ are qualitatively different than those of $\mathbf{5}^+$ or $\mathbf{6}^+$.

The highest frequency band in $\mathbf{8}^+$ is only 75 cm^{-1} above that of the neutral complex. A further shift of 48 cm^{-1} occurs in formation of the <u>dication</u>. The total shift in $\nu_{CO}(sym)$ of

123 cm^{-1} in going from **8** to **8**$^{2+}$ is internally consistent with the other Mn_2 complexes studied, but now the $\Delta\nu_{CO}$(sym) changes are more evenly spread out between the first and second oxidation steps. Detailed interpretation of these shifts awaits another forum, but it may be concluded that there is a greatly increased interaction between redox centers in **8**$^+$ as compared to **5**$^+$ or **6**$^+$.

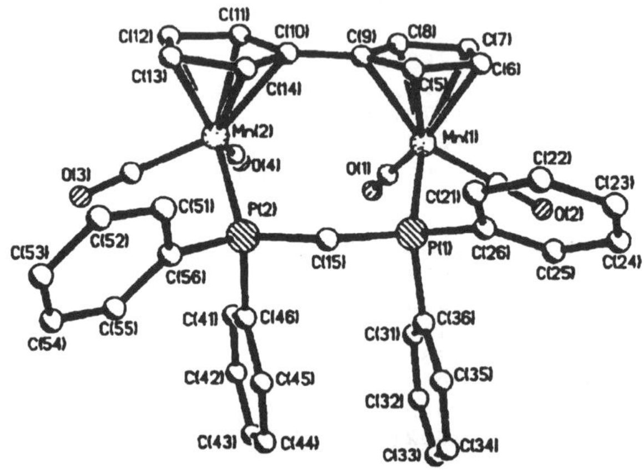

Figure 4. ORTEP diagram of $Fv[Mn_2(CO)_4(\mu\text{-dppm})]^+$

Since the M-M interactions are weak in both **5**$^+$ and **9**$^+$ (unpublished data), each of which contains one of the bridging ligands found in **8**$^+$, it would appear that the increased redox site interaction in **8**$^+$ must arise from the syn arrangement of the metals, rather than from delocalization through a bridging ligand. It is therefore most likely that a through-space, direct, metal-metal interaction is responsible for the increased site interaction. The X-ray crystal structure of **8**$^+$ (Fig 4) reveals a Mn-Mn separation of 4.07 A. Other physical methods are being pursued to better quantify the delocalization and mixed-valency of **8**$^+$.

2.2 Oxidation and Reduction of Fulvalene Dicobalt Complexes

The fulvalene complex disubstituted with $Co(\eta^4\text{-}C_8H_{12})$, **10**, has anodic behavior similar to that of biferrocene, with two 1e$^-$ oxidations, ΔE^o = 228 mV. The fulvalene bis[Co(carboranyl)] complex **11** [14] shows similar behavior, with ΔE^o = 270 mV for **11**/**11**$^+$/**11**$^{2+}$. These complexes lack the convenient CO infrared label present in **5-9** and present a greater challenge for the diagnosis of mixed-valent interactions. The ΔE^{o} values suggest that $\mathbf{10}^{+}$ and $\mathbf{11}^{+}$ will be Class II systems.

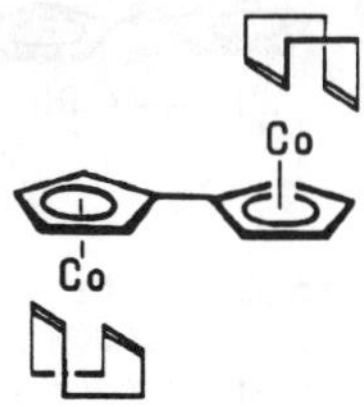

10

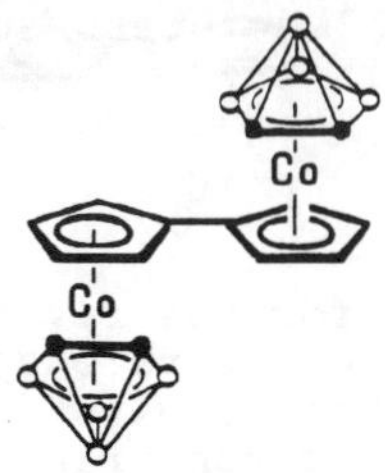

11

Interestingly, the second one-electron <u>reduction</u> of **11** is considerably more negative than the first reduction, with ΔE^{o} = 460 mV, possibly indicative of a delocalized monoanion.

2.3 Oxidation of Fulvalene Dirhodium Complexes

Complexes **12-14** present a more complex redox picture than do the foregoing Mn_2 and Co_2 complexes, in that they oxidize in a single $2e^-$ wave. Connelly and co-workers showed earlier that the

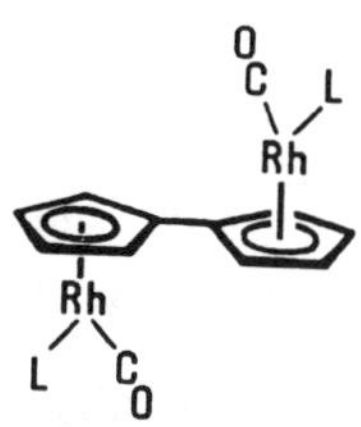

12, L = CO ; **13**, L = PPh_3

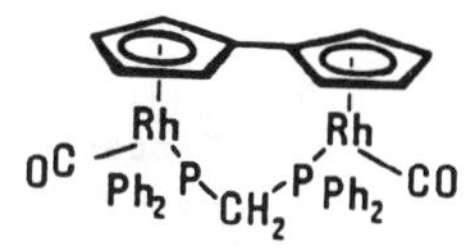

14

mixed carbonyl-phosphine complex **13** oxidized through Eq 1, which involves making and breaking of a metal-metal bond coupled to an overall two-electron process. [15] An important aspect of this process is the rotation of the metals from an anti to syn conformation. Our recent high-speed voltammetry experiments on **13** reveal no evidence of intermediates below scan rates of 200 V/s, implying that the electron-transfer processes and bond making and breaking processes in $\mathbf{13}/\mathbf{13}^{2+}$ are closely linked. This gives

Eq 1

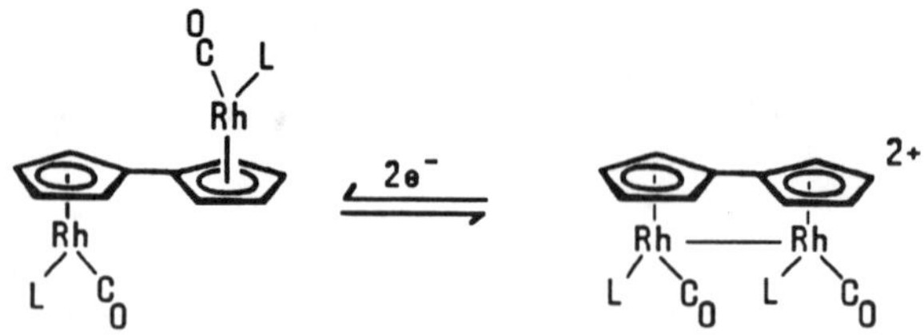

a rationale for the reversible anodic voltammetry of **12** and **13**, the mononuclear versions of which [e.g., $CpRh(CO)_2$] give irreversible oxidations, with rapid coupling or decomposition reactions following electron transfer. [15,16]

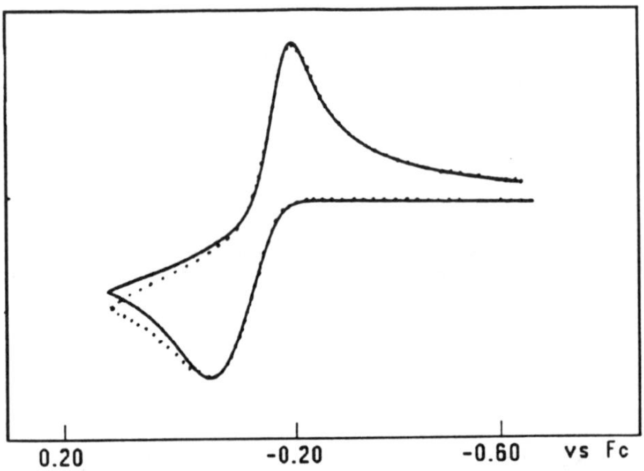

Figure 5. Experimental (dots) and theoretical (solid) CV scan (v=1 V/s) of cmpd 12 in acetone; E^o_1 =-0.07V, E^o_2 =-0.24V; k_{s1} =0.03 cm/s;k_{s2} = 0.1 cm/s. Conc = 0.44 mM.

Digital simulations reproduce the observed CV curves of **12** with the EE mechanism of Eqs 2 and 3. $E^o{}_2$ is slightly negative of $E^o{}_1$, but the first oxidation (Eq 2) has the slower heterogeneous charge-transfer rate (Fig 5). It is possible, indeed likely,

Eq 2 $$Fv[Rh_2(CO)_4] \rightleftharpoons Fv[Rh_2(CO)_4]^+ + e^- \qquad E^o{}_1$$

Eq 3 $$Fv[Rh_2(CO)_4]^+ \rightleftharpoons Fv[Rh_2(CO)_4]^{2+} + e^- \qquad E^o{}_2$$

that the first oxidation is followed by a rapid isomerization of the metals into a syn conformation, with formation of at least a partial metal-metal bond (Eq 4).

Eq 4

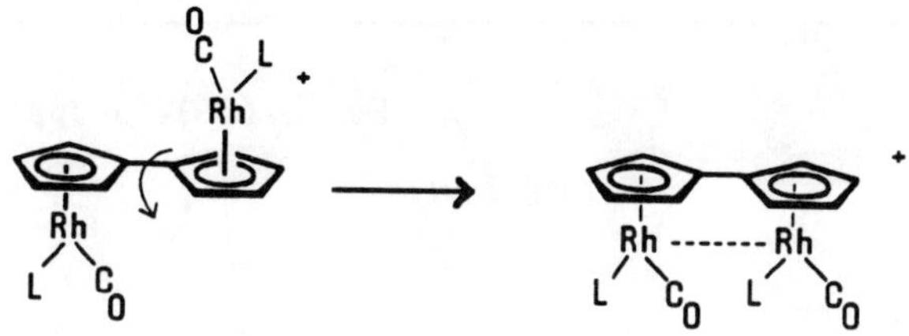

A sequence of Eqs 2, 4, 3 constitutes an ECE process. The present data cannot distinguish between the EE and ECE models; however, they do show that the anti/syn rotation is rapid and reversible on the voltammetric time scale.

The dication $\{Fv[Rh_2(CO)_4]\}^{2+}$, $\mathbf{12}^{2+}$, is quite sensitive to nucleophilic attack. Care must be exercised in order to obtain its physical properties. Spectroelectrochemical experiments at 243 K in CH_2Cl_2 gave the vibrational spectrum, $\Delta\nu_{CO}$(sym) = 113 cm^{-1}, $\Delta\nu_{CO}$(asym) = 141 cm^{-1} followed by decomposition of the dication.[17]

The occurrence of a metal-metal bond in $\mathbf{12}^{2+}$ was confirmed by optical spectroscopy. A general characteristic of transition metal complexes having a M-M bond is the presence of a σ to σ^* transition in the 300 to 500 nm range with molar aborbances of (6-20) x 10^3 $M^{-1}cm^{-1}$.[18] Such a band is present in $\mathbf{12}^{2+}$ but absent in **12**. For these spectra the dication was obtained as its bis-PF_6 salt through oxidation by two moles of $[Cp_2Fe][PF_6]$ and rapid precipitation.

Results on the doubly-bridged complex **14** are relevant in order to assess whether or not the anti-to-syn rotation (Eq 4) is necessary to obtain a "direct" $2e^-$ transfer in these systems. In mechanistic language, is an ECE process (with C being the rotation) a necessary condition for net $2e^-$ waves in fulvalene redox processes?[19]

Voltammetry of **14** answers this question in the negative. CV scans in CH_2Cl_2 reveal a nearly Nernstian $2e^-$ wave (Fig 6). Crystals of the stable dication $\mathbf{14}^{2+}$ were grown from acetone/ether and analyzed by A.L. Rheingold, who provided the structure shown in Fig 7. Comparison with the starting material showed that the Rh-Rh distance had decreased from 4.51 Å to 2.83 Å upon two-electron oxidation. The latter is a typical single bond length for Rh.

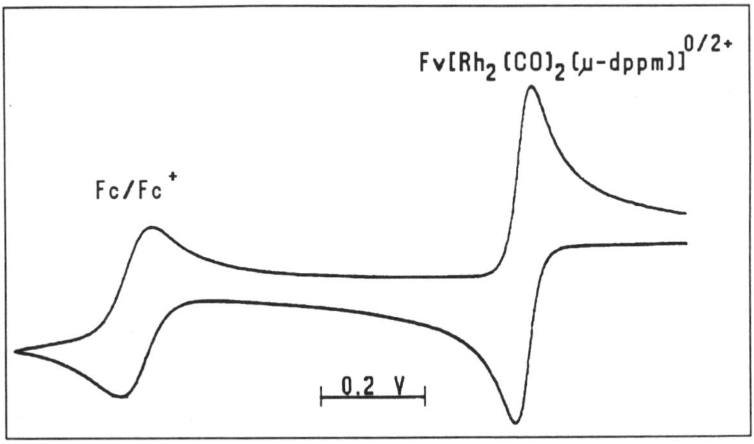

Figure 6. CV scan (v=0.1 V/s) of mixture of cmpd 14 and Fc in CH_2Cl_2, each _ca_ 1mM

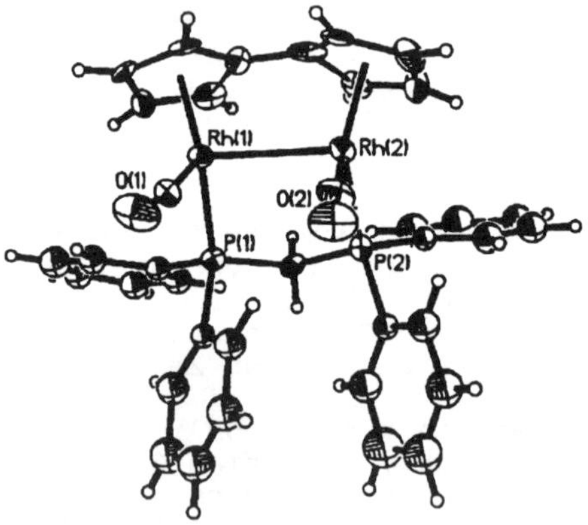

Figure 7. ORTEP diagram of $Fv[Rh_2(CO)_2(\mu\text{-dppm})]^{2+}$

Acknowledgements

The authors gratefully acknowledge the financial support of the National Science Foundation (CHE 86-03732 and 91-16332). The Co-carboranyl work was done in part by Dr. Ian Quarmby on samples provided by Prof. Russell Grimes. Dr. Peter Sheridan obtained much of the voltammetric data on the Mn complexes. X-Ray analyses were done at the University of Delaware by Dr. A.L. Rheingold, whom we thank for permission to include these results.

References

1. (a) Hendrickson, D.N.; Oh, S.M.; Dong, T-Y.; Moore, M.F. (1985), Comments Inorg. Chem., 4, 329 (b) Cowan, D.O.; LeVanda, C.; Park, C.; Kaufman, F. (1973), Accounts Chem. Research, 6, 1

2. (a) Webb, R.J.; Geib, S.J.; Staley, D.L.; Rheingold, A.L.; Hendrickson, D.N. (1990), J. Am. Chem. Soc., 112, 5031 (b) Dong, T-Y.; Chou, C-Y. (1990), J. Chem. Soc., Chem. Commun., 1332 (c) Blackbourn, R.L.; Hupp, J.T. (1988), J. Phys. Chem., 92, 2817

3. Van Order, Jr., N.; Bitterwolf, T.E.; Rheingold, A.L.; Geiger, W.E. (1987), J. Am. Chem. Soc., 109, 5680

4. Geiger, W.E.; Van Order, Jr., N.; Pierce, D.T.; Bitterwolf, T.E.; Rheingold, A.L.; Chasteen, N.D. (1991), Organometallics, 10, 2403

5. Pierce, D.T. (1991), Ph.D. Dissertation, Univ. of Vermont

6. Connelly, N.G.; Demidowicz, Z.; Kelly, R.L. (1975), J. Chem. Soc., Dalton Trans., 2335

7. Ammar, F.; Saveant, J.M. (1973), J. Electroanal. Chem., 47, 215

8. Robin, M.B.; Day, P. (1967), Adv. Inorg. Chem. Radiochem., 10, 247

9. (a) Kirchner, R.F.; Loew, G.H.; Mueller-Westerhoff, U.T. (1976), Inorg. Chem., 15, 2665 (b) Hillman, M.; Kvick, A. (1983), Organometallics, 2, 1780

10. Richardson, D.E.; Taube, H. (1984), Coord. Chem. Rev., 60, 107

11. LeVanda, C.; Bechgaard, K.; Cowan, D.O. (1976), J. Org. Chem., 41, 2700

12. (a) Morrison, Jr., W.H.; Krogsrud, S.; Hendrickson, D.N. (1973), Inorg. Chem., 12, 1998 (b) Shu, P.; Bechgaard, K.; Cowan, D.O. (1976), J. Org. Chem., 41, 1849

13. (a) Connelly, N.G.; Kitchen, M.D. (1977), J. Chem. Soc., Dalton Trans., 931 (b) Connelly, N.G.; Freeman, M.J.; Orpen, A.G.; Sheehan, A.; Sheridan, J.B.; Sweigart, D.A. (1985), J. Chem. Soc., Dalton Trans., 1019

14. Davis, J.H.; Sinn, E.; Grimes, R.N. (1989), J. Am. Chem. Soc., 111, 4784

15. (a) McKinney, R.J. (1980), J. Chem. Soc., Chem. Commun., 603 (b) Connelly, N.G.; Lucy, A.R.; Payne, J.D.; Galas, A.M.R. (1983), J. Chem. Soc., Dalton Trans., 121 (c) Connelly, N.G.; Lucy, A.R.; Payne, J.D.; Galas, A.M.R.; Geiger, W.E. (1983), J. Chem. Soc., Dalton Trans., 1879.

16. Gennett, T.; Grzeszczyk, E.; Jefferson, A.; Sidur, K.M. (1987), Inorg. Chem., 26, 1856

17. The dication of $FvRh_2(CO)_4$ as recently reported appears to have been contaminated with decomposition products: see Bitterwolf, T.E.; Spink, W.C.; Rausch, M.D. (1989), J. Organometal. Chem., 363, 189

18. (a) Miskowski, V.M.; Gray, H.B. (1987), "Understanding Molecular Properties", Avery, J.(ed), D. Reidel Publishing Co. (b) Wrighton, M.S.; Ginley, D.S. (1975), J. Am. Chem. Soc., 97, 2065 (c) Abrahamson, H.B.; Palazzotto, M.C.; Reichel, C.L.; Wrighton, M.S. (1979), J. Am. Chem. Soc., 101, 4123

19. (a) Moulton, R.; Weidman, T.W.; Vollhardt, K.P.C.; Bard, A. J. (1986), Inorg. Chem., 25, 1846 (b) Vollhardt, K.P.C.; Weidman, T.W. (1984), Organometallics, 3, 82

SPECTRO-ELECTROCHEMISTRY AND ELECTROCHEMICAL SPECTROSCOPY

Graham A. HEATH
Research School of Chemistry, Australian National University
G.P.O. Box 4, Camberra ACT 2601, Australia

ABSTRACT. The electron-transfer reactions available to transition metal complexes are intimately linked to their molecular and electronic structures. Electrochemistry provides a direct way of defining these redox processes, thereby mapping the metal and ligand frontier orbitals in a way which complements electronic spectroscopy. For co-ordination complexes, the general possibility of reconciling electrochemical (i.e. equilibrated) and optical charge-transfer data is receiving renewed attention. In the case of hexahalo complexes $[MX_6]^{z-}$ of d^n ions, the halide to metal charge-transfer envelopes are complicated but generally well understood. The charge-transfer energies for the accessible $[M^VCl_6]^{1-}$ and $[M^{IV}Cl_6]^{2-}$ complexes have now been recorded and mapped against the corresponding $d^n \rightarrow d^{n+1}$ (V/IV or IV/III) couple. A striking linear correlation emerges between $h\upsilon_{max}$ (1mct) and E^0 in each series, with a gradient of 1.35 eV/V in both cases. The implications of this orderly linear relationship are discussed and an argument presented that such $E^0/h\upsilon$ correlations have a respectable physical basis after all. The present analysis can be extended fruitfully to trends in progressively substituted complexes $MX_{6-n}L_n$ ($L = \pi$-acceptor ligand) where reconciliation of the electrochemical and spectroscopic data yields a conceptual representation of the relative frontier orbital energies of M, X, and L as a function of stoichiometry.

1. Introduction

"Optical spectroscopists like to think they *shed light* on electronic structure...
Electrochemists have ***the potential*** to do the same....."

(Thoughts on first setting eyes on Sintra..)

If we are to attach bonding significance to the electrode potentials we measure, they must reflect thermodynamic values (be reversible) and we must know the electronic nature/molecular structure of the the electro-active species as a function of oxidation state.

This is where *in situ* spectro-electrochemistry comes in. Optical spectro-electrochemistry is a formidable technique in this regard because it can both guarantee the integrity of the one-electron transfer as a chemically and electrochemically reversible step, and it can fingerprint the product. Furthermore, the resultant spectrum will generally tell us quite a lot about the constitution of the product.

The detailed technology is important to success, but Dr Yellowlees will speak of this. Lesley was once my PhD student, but truly she taught me the technique not the other way around, so you will be hearing from the proper source

A. J. L. Pombeiro and J. A. McCleverty (eds.),
Molecular Electrochemistry of Inorganic, Bioinorganic and Organomettallic Compounds, 533–547.

I think together we learned the alchemy of orbitall mapping.

Electrochemical spectroscopy is the larger ambition; to me it means is the elucidation of electronic structure by marshalling all the electrochemical and spectro-electrochemical evidence. I have borrowed this graphic term from Heinze who focussed on the power of voltammetry and invested with my own meaning. **He is free to make a disclaimer.**

If one takes the optimisic view (as we do) that both metal and ligand frontier orbitals can be mapped from systematic electrode potential data then testable predictions of the optical charge-transfer band energies for a given complex must follow. The present paper is devoted to exploring this challenge.

In our laboratory, we have for several years being exploring and exploiting the naive scheme.shown below. In principle the energy of an electronic charge transfer transition can be related to the notional electrode potentials of the appropriate donor and acceptor orbitals as shown.

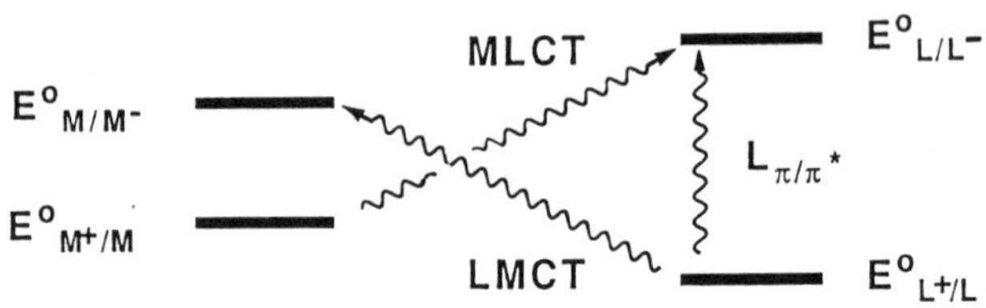

For example: $h\upsilon_{MLCT}$ (electron volts) $\alpha \left[E^o_{M^+/M} - E^o_{L/L^-}\right]$

In practice linear relationships between with α not far from unity very frequently emerge in diverse systems. Below we illustrate some particularly instructive ones, particularly concerning halide to metal charge-transfer (lmct) which has received less attention, and discuss why this fortunate situation prevails.

2. Metal Bipyridyls - Probing Redox-Active Ligands

The first task is to illustrate the proposition that both ligand and metal orbital energies are faithfully mapped by voltammetry. The metal bipyridyl complexes provide the exampole par excellence of our ability to probe redox-active ligands and the essentially localized nature of the each reduction is well known. There is a remarkable linear relationship between the first localised reduction in tris bipyridyl complexes and the formal charge on the metal centre (ranging from (-II to +III), which effectively tunes the ligand electrode potential over a wide range. Presumably this is an important and general phenomenon, and we will encounter it again in discussing the optical spectra of hexahalide complexes.

The equally triumphant mapping of metal valence orbitals is described in the next section.

3. Hexa-halometallates - Probing 4d^n and 5d^n Valence Shells

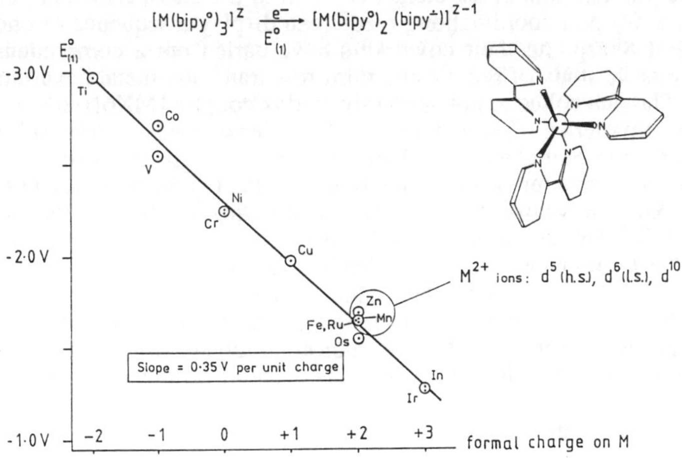

Figure. Effect of co-ordinated metal ion charge on ligand frontier orbitals. This is a quite moderate sensitivity; e.g. porphyrins shift by about 0.8V and acac⁻ by about 0.75V per unit metal charge (see later).

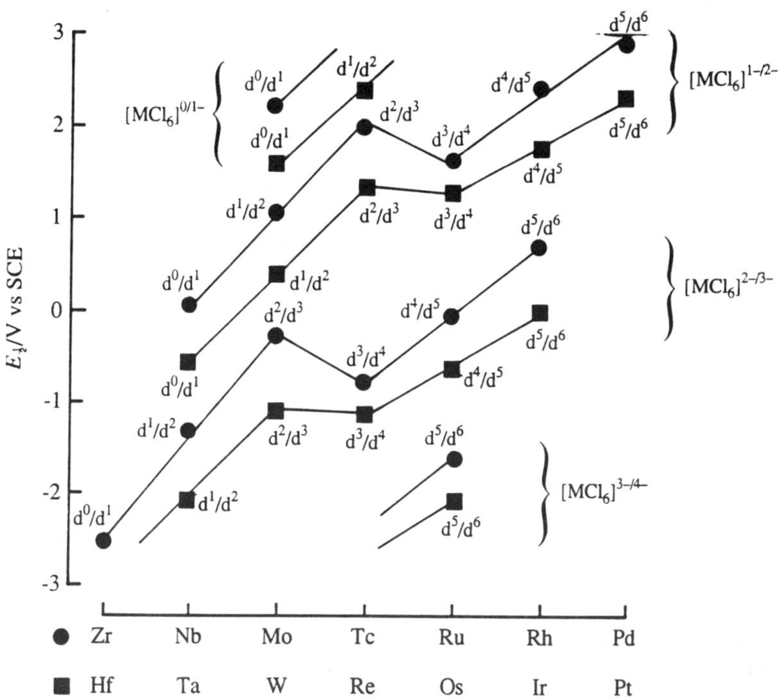

Figure. Trends in half-wave potentials, $E_{\frac{1}{2}}$ in V, of $[MCl_6]^{z/z-1}$ couples in CH_2Cl_2;
●: 4d-hexachlorometallates; ■: 5d-hexachlorometallates.

$[MX_6]$ complexes provide the simplest vehicle for studying the periodicity of redox behaviour in the 4dn and 5dn block elements. Within the class, each compound when dissolved in ultra-dry non-coordinationg media can display a sequence of one-electron steps. Heath and Sharp and their coworkers have caried out a comprehensive non-aqueous voltammetric study of second and third row transition metal hexafluorides and hexachlorides. The rich collection of successive redox couples [MX6]z/z-1 (z = 0,-1,-2,-3) follow periodic patterns. (See Figure) The orderly trends across each series are related to the increasing central ion core charge, and to the progression in electron/electron correlations (exchange and repulsion)between dn and dn+1 configurations within the t2g subshell. (Thinking ahead, we might encounter something similar in optical *halide to metal* charge transfer energies of the dn complexes.)

In fact the Heath Sharp series is identical in form and origin to the classical progression in ionisation energies for the p-block elements (B to Ne and Al to Ar). This arises through the strict isomorphism of pn and t2n configurations. It is a remarkable relationship; it spans light element atoms vs molecular metal complexes and it spans gas phase spectroscopic data vs equilibrium (electrochemical) measurements in solution.

No-one can doubt the faithfulness of this orbital mapping.

The numerical data from fig can be manipuilated in many instructive ways to reveal influence of core-charge on electron-affinity, influence of mounting electron-electron correlation with increasing n on a given element, nephelauxetic effects.of ligands

The summation of these effects determines the Heath Sharp progression along any isovalent series.

Table 1

Equilibrium type	Cell reaction (example)	E_{cell}
(a) Congeneric iso-electronic couples[a]	$[TaCl_6]^{2-} + [NbCl_6]^{1-} \rightarrow [TaCl_6]^{1-} + [NbCl_6]^{2-}$ $d^1 \quad d^0 \quad d^0 \quad d^1$	+ 0.65
(b) Adjacent iso-electronic couples[b]	$[TaCl_6]^{2-} + [WCl_6] \rightarrow [TaCl_6]^{-} + [NbCl_6]^{-}$ $d^1 \quad d^1 \quad d^0 \quad d^1$	+ 2.15
(c) Successive couples (disproportionation)[a]	$[WCl_6]^{-} + [WCl_6]^{-} \rightarrow [WCl_6] + [WCl_6]^{2-}$ $d^1 \quad d^1 \quad d^0 \quad d^2$	- 1.20
(d) Adjacent isovalent	$[TaCl_6]^{2-} + [WCl_6]^{-} \rightarrow [TaCl_6]^{-} + [WCl_6]^{2-}$	+ 0.95

[a] Charge redistribution and solvation terms for reactions a and d are self-cancelling. [b] For reactions b and c calculations using the Born equation yield values of ca. 0.1 e.V,[1] substantially less than the effects considered here.

The numerical data are best regarded in terms of thermodynamic cell reactions, by combining the electrode potentials in various significant ways as exemplified in reactions a to d (Table 1).

Reaction a defines the driving force for electron-transfer from $[TaCl_6]^{2-}$ to $[NbCl_6]^{1-}$, and typifies the relative stability of the higher oxidation states in the 5d series. Combination of neighbouring couples within one series, such as $[TaCl_6]^{1-/2-}$ and $[WCl_6]^{0/1-}$ in reaction b, illustrates the preference of the transferred electron for the species of higher atomic number and shows the profound influence of core charge on iso-electronic energy levels. Equally, comparison of successive couples for one element, such as $[WCl_6]^{0/1-}$ and $[WCl_6]^{1-/2-}$ in reaction c, reflects the unfavourable energy of disproportionation (*i.e.*, electronic reorganisation at fixed core charge). Reactions such as d exemplify the anticipated increase in oxidising strength of isovalent ions across each series, which is seen to arise from the superimposed influence of core charge and electronic reorganisation.

The contribution of interelectronic correlation terms to the separation between E^0 (d^{n-1}/d^n) and E^0 (d^n/d^{n+1}) is uniformly $[r + k]$ for $n = 1, 2, 4$, and 5 but $[R - 2k]$ for $n = 3$,† where r is the repulsion energy and k the exchange energy between any two parallel 4d (or 5d) electrons and R is the repulsion energy for electrons paired within one orbital.[3] The resulting prediction of approximately linear progression in E^0 values, interrupted at the d^3/d^4 couple, is not altered by smooth variation of these parameters with changing atomic number or valency.

Historically, we perceived it was important to move on from voltammetry to spectro-electrochemistry in this context to underwrite the integrity of our couples. Especially since many of them imply hitherto uncharacterised species. For example,we have recently described the fully authenticated spectra of four very reactive pentavalent complexes $[ReCl_6]^{1-}$, $[RuCl_6]^{1-}$, $[IrCl_6]^{1-}$ and $[IrBr_6]^{1-}$ in a cryostatted spectro-electrochemical cell. From a purely operational viewpoint, it is invariably the intense charge-transfer spectrum that one is dealing with, as considered in the next section.

4. Halide-to-metal charge-Transfer Spectra, $[MX_6]^{2-}$

Halide to metal charge transfer (hmct) transitions tend to dominate the uv/visible spectrum in these complexes.. The manifolds are complex at first sight but generally well understood, and historically thay have contributed greatly to our understsnding of bonding in tsuch complexes. One of the themes of this paper is the beauty and informative power of hmct transitions in varied systems.

Their value lies particularly in the success of a straightforward one-electron promotion model for the excited state. For a d^n ion, the observed manifold reflects the accessible electronic states of the d^{n+1} metal ion, and the energies of the electron-donating halide Symmetry Orbitals.

$$[M^{IV}\{X_6\}^{6-}]^{=} \xrightarrow{h\nu} *[M^{III}\{X_6\}^{5-}]^{=}$$

---

† The same conclusion may be rigorously derived in terms of Racah parameters $\{[r + k] = A - 5B, [R - 2k] = A + 10B + 5C\}$. Note that, although each electron added to the t_{2g} subshell contributes an increment 4 Dq in ligand field stabilisation energy this term is contained in every d^n/d^{n+1} couple and does not contribute to the *separation* in successive E° values, even if 10 Dq varies smoothly across the series.

The spectrum of $[IrCl_6]^{2-}$ (d5) is shown. Low-spin d^5 ions are an ideal case, since electron-promotion yields the unique closed-shell configuration (dp^6, S = 0, L = 0). In this case , the observed spectral manifold directly maps both the energy and the complexity of the halide symmetry orbitals. For this reason, despite its structural simplicity the spectrum shows *three* characteric Cl(π) to t2g absorption bands. These arise by promotion from the three uppermost electronic levels of the six-halide array, t1g (Laporte forbidden), t1u, and t2u. Notice how only four chlorides are represented at a time. The underlying principle is that the transitions achieve intensity mainly through coplanar interaction between the donor orbitals and the unfilled metal acceptor orbital. This may be arbitrary here but becomes decisive in MX5L and MX4L2 complexes.

In accord with the notion of hmct as a one-electron internal reduction, iso-electronic systems (eg IrIV, OsIII) tend to have spectra of very similar appearance displaced /*in energy* along the frequency axis according to the optical electronegativity (= reducibility) of the central ion. Even with differing dn configurations, comparable transitions can be identified readily. The chloride to metal charge-transfer band positions for all 14 accessible $[MCl_6]^{1-}$ and $[MCl_6]^{2-}$ complexes have now been mapped against their respective V/IV and IV/III couples. .Here we have plotted the strongest band in the leading manifold (πt2u) in every case.

Beautiful linear progressions emerge in each case. Both lines have have a gradient 1.35 eV/V. The transition energy is effectively raised by 12,500 cm-1 (1,5 eV) between the M(V) and M(IV) plots. Clearly our cavalier equation is vindicated in this system and takes the form hn = $a(E_{1/2}M - E^*_{1/2}Cl)l$, with $\alpha = 1.35$. $E^*_{1/2}Cl$ is not directly measureable but we expect it to be essentially constant within either series, that is fixed for a given central ion charge.

Notice the remarkable observation that α is greater than unity, i.e. that the shifts in charge-transfer energy *amplify* the underlying shifts in E1/2M. Intuitively we felt that this represents the systematically greater energisation of the photo-excited (non-equilbrated) states. This connects with suggestions elsewhere as discussed below.

5. Reconciliation of Electrochemical and Spectroscopic Data (hexahalides)

As noted above, in linear plots of hν versus electrode potential, we observe a slope of 1.35 for LMCT in $[MCl_6]^{2-}$ systems, taking $E^*_{1/2}(Cl^{0/1-})$ to be invariant (Figure), and Meyer found 1.16 for MLCT excitation in numerous osmium(II) di-imine systems. This factor is typically somewhat greater than unity in accord with the intrinsic additional vibronic energy of the non-equilibrated optically excited states.

$$h\nu = \Delta G + \lambda$$
$$= (1 + \rho)\Delta G \text{ (see below)}$$

But whence the linearity? Meyer has already noted with other authorities that intramolecular and solvent dipole contributions to this reorganisation energy (ρ_i and ρ_o) are themselves proportional to the energy gap between ground state and and excited state. It is of course intuitively reasonable that as we advance along the distortion coordinate (q) and the excitation energy increases, λ will increase as $h\nu_0$ (=ΔG) does.

Moreover, to describe charge-transfer and intervalence transitions in which the electron is largely localized in initial and final states, Hush proposes a vibronic coupling constant ρ

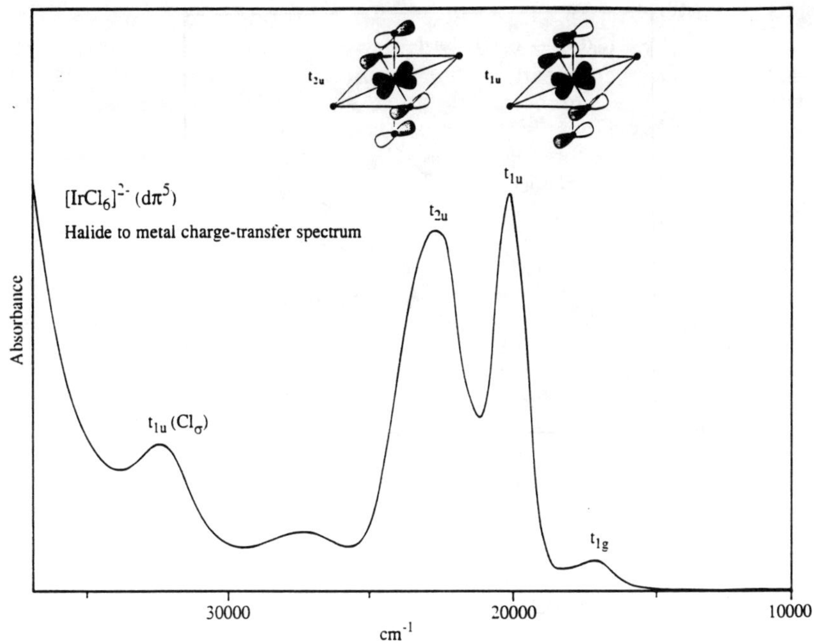

In-Plane Halide to Metal Charge Transfer Transitions

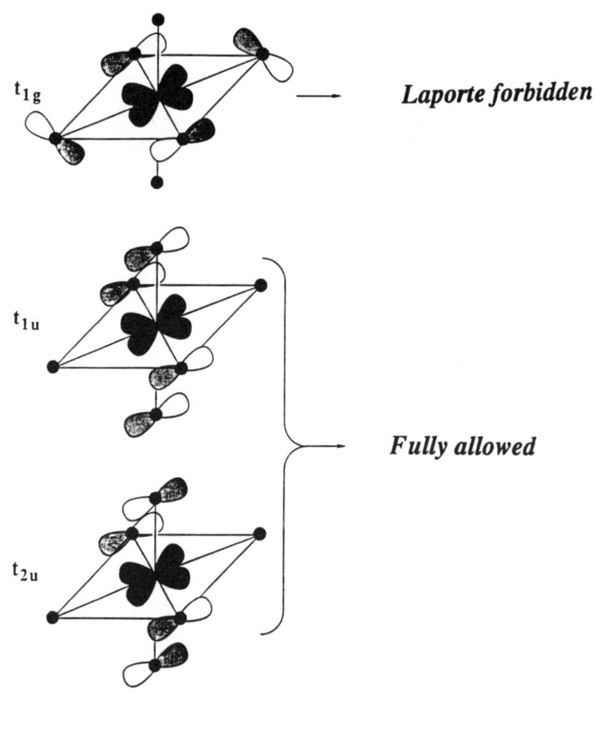

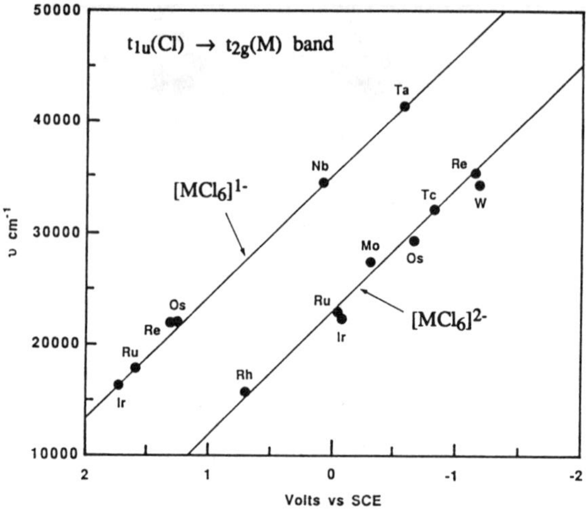

Figure. Reduction potential/optical charge-transfer data correlations for $4d^n$ ∉ $5d^n$ $[MCl_6]^-$ e $[MCl_6]^{2-}$ complexes

$$h\nu = \Delta G + \lambda$$

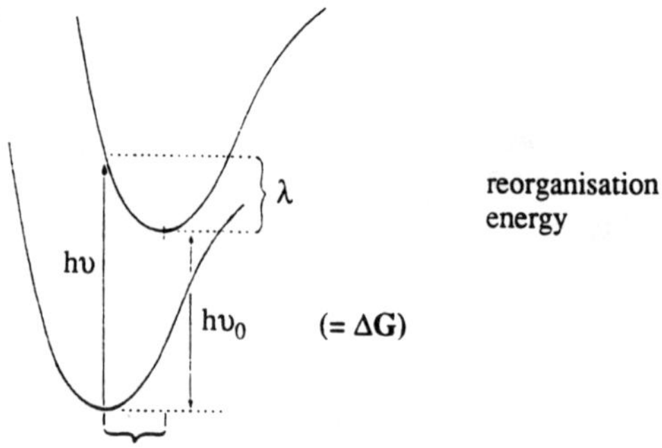

Figure. Reconciliation of optical and electrochemical data. ΔG refers to the energy change in reaching the *thermally equilibrated* excited state in solution. λ is the nett vibrational reorganisation term. As ΔG increases, the added energisation, λ, tends to increase as well.

such that $\lambda = \rho\Delta G$ and $h\nu = (1 + \rho)\Delta G = \alpha\Delta G$, where $\alpha = (1 + \rho)$.
Then, equivalently,:

$h\nu$ (in eV) = $\alpha\Delta E_{1/2}$(excitn)

Here we have set $nF\Delta E_{1/2}$ equal to ΔG for the electronic promotion This is precise if we define $E^*_{1/2}Cl$ for the oxidation of the chloride anion in the presence of the reduced metal as the appropriate half-cell reaction. We will return to this point after discussing mixed nitrile halide complexes, where the electrode potentials are again experimentally inaccessible for both ligand types.

To recap, the linear dependence of charge transfer energies for the numerous hexahalides (Fig) is a very striking result because, at a stroke, a wealth of scattered data fall into place. Moreover it is a purely empirical finding, a correlation determined by Nature rather than a human construct. What does it tell us? It strongly suggests that the halide ligand has fixed oxidation potential, It also appears that within the structural/electronic domain prevailing. the reorganisation energy is itself linearly proportinal to the equilibrium promotion energy.

That is, $hn = (1 + \rho)h\nu_0$ where $\rho = 0.35$

This is beginning to look very exciting because it suggests that far from worrying over the impropriety of merging electrochemical and optical data, or trying to guesstimate the reorganisation terms (or select systems free of significant reorganisation) *in fact we have a direct means to assess them in many systems of interest.*

6. Substituted Halide Complexes $Mx_{6-n}Ln$

It is important to generalise the observation of the rational periodic patterns in electronic behaviour and to extend the applicability of voltammetric orbital mapping beyond the classic case of the hexahalides. We have a perceptiion of not just one Heath-Sharp series (for $[MX_6]^{z-}$) but a corresponding progression in many systems where the ligand set is fixed, and only M varies. Indeed that this is the real importance of those pioneering observations. To this end we have studied *substituted* lower symmetry complexes where successive halide ligands are replaced by a specified ligand L.

We examined an exhaustive series of substituted ruthenium complexes [RuX6]2- through $[RuX_{6-n}(RCN)_n]$ to $[Ru(RCN)_6]^{2+}$. The objective was to define the systematic variation in electrode potentials (Ru V/IV,,Ru V/III, and Ru III/II) and the accompanying trends in both halide-to-metal and metal-to-nitrile optical charge-transfer spectra.

Electrochemical characterization of the complete series $[RuX_{6-n}(RCN)_n]^z$. (n = 0 - 6, X = Cl or Br) has provided confirmation of the systematic correlation between stoichiometry and central ion reduction potential, previously noted for the $[OsCl_{6-n}(py)_n[^z$ (n = 0 - 4) system . A graph of successive E^o values ($Ru^{IV/III}$ and $Ru^{III/II}$) vs. the extent of halide ligation (6 - n) shows that, upon each substitution of halide by nitrile, both couples shift by approximately +0.6 volts.

Using spectro-electrochemical techniques, it was possible to record the optical spectra of each complex in multiple oxidation levels, defining the states RuIV for n = 0 to 2, RuIII for n = 0 to 5, and RuII for n = 2 to 6. and We find that within each $[MX_{6-}$

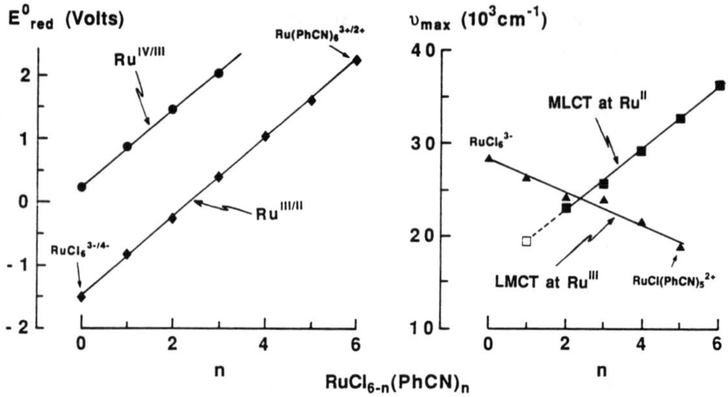
E⁰red (Volts)
Ru IV/III
Ru(PhCN)6 3+/2+
Ru III/II
RuCl6 3-/4-
n
υmax (10³cm⁻¹)
MLCT at Ru II
RuCl6 3-
LMCT at Ru III
RuCl(PhCN)5 2+
n
RuCl6-n(PhCN)n

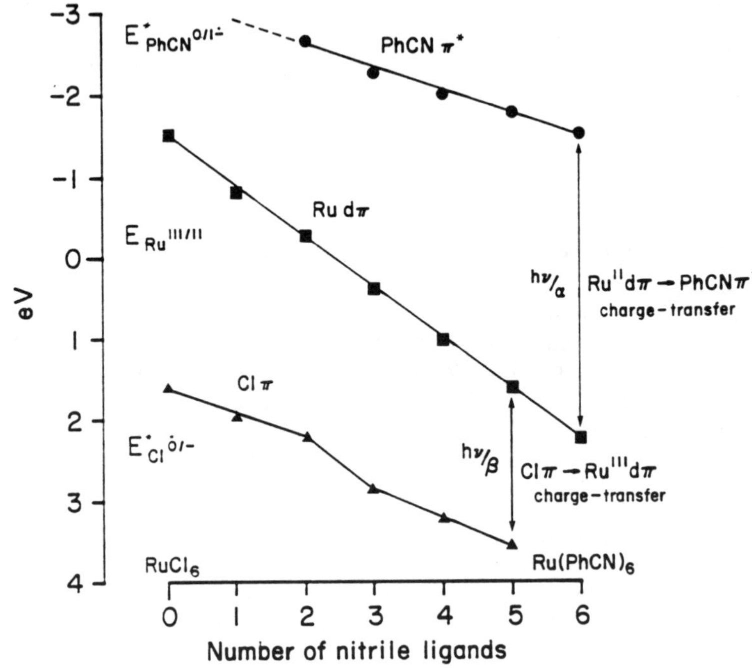
E PhCN 0/1-
PhCN π*
E Ru III/II
Ru dπ
eV
Cl π
E Cl 0/-
hν/α
Ru II dπ → PhCN π*
charge-transfer
hν/β
Cl π → Ru III dπ
charge-transfer
RuCl6
Ru(PhCN)6
Number of nitrile ligands

$_n(RCN)_n]^z$ series the pattern of charge transfer spectra is unmistakably complementary to the progression in $E_{1/2}$ data. Absorption energies for both ligand-to-metal charge transfer ($Cl \rightarrow Ru^{III}$) and metal-to-ligand charge transfer ($Ru^{II} \rightarrow RCN$) vary in an orderly way with halide ligation, as shown in the Figure. These measurements provided an unusual opportunuty to document the characteristic trends in appearance and location of both halide to metal (RuIII) and metal to ligand (RuII) as a function of stoichiometry. Thus with progressively substituted mixed halide/nitrile compounds at our disposal, we are able to study empirical hν - $E_{1/2}$ relationships for both LMCT and MLCT spectra that prevail simultaneously, i.e. in the one body of compounds.

Frontier orbital mapping (mixed complexes) The systematic relationships presented above, for example plotting electrode potentials or optical transition energies against stoichiometry are deliberately empirical. However, it is instructive to merge these electrochemical and spectroscopic observations into one conceptual framework. We ascribe notional redox potentials ($E^*_{1/2}$) to the nitrile acceptor and halide donor, as before. It is then possible to compile the optical and electrochemical data so as to map the relative metal and ligand frontier-orbital domain in each $[RuX_{6-n}(RCN)_n]^z$ complex. Defined in this way, the relative frontier-orbital energies (Ru dπ, Cl^- or Br^- π, and RCN π^*) are shown to change dramatically as n increases from zero to six. . See figure.

Frontier orbital mapping Thus far we have presented systematic relationships which are purely empirical. However, it is instructive to merge these electrochemical and spectroscopic observations into one conceptual framework. We ascribe notional redox potentials ($E^*_{1/2}$) to the nitrile acceptor and halide donor, as before, and so express the charge-transfer transition energies as follows.

MLCT: $$hn(Ru^{II}ÆRCN) = \alpha[E_{1/2}(Ru^{III/II}) - E^*_{1/2}(RCN^{0/1-})]$$ Equation 1

LMCT: $$hn(X^-ÆRu^{III}) = \alpha[E^*_{1/2}(Cl^{0/1-}) - E_{1/2}(Ru^{III/II})]$$ Equation 2

These equations are in accord with the wide-spread empirical observation of linear relationships between charge-transfer band energies and electrochemical potentials, and they have a rational theoretical foundation as explained above.

The spectro-electrochemical proportionality factor (α) has been found to be constant in a given chemical context. For example, we observe a value of 1.3 for LMCT in $[MCl_6]^{2-}$ systems, taking $E^*_{1/2}(Cl^{0/1-})$ to be invariant (Figure), and Meyer found 1.16 for MLCT excitation in numerous osmium(II) di-imine systems. The factor is typically somewhat greater than unity in accord with the intrinsic additional energization of the non-equilibrated optically excited states.

The figure shows a frontier orbital map of metal and ligands constructed for the $[RuCl_{6-n}(PhCN)_n]^z$ family, by use of equations 1 and 2. Note that the vertical axis is related to electrode potential; i.e. these data are electron-transfer couples, not orbital energies, and as such they are particularly appropriate for describing both redox and optical charge-transfer phenomena. For example, the metal-centered electron/electron correlation terms, which require explicit estimation in other treatments, are automatically encompassed in the experimental $E_{1/2}$ values.

If there were no change in the overall energy of the halide or nitrile orbitals upon progressive substitution, the rate of change of the charge-transfer transition energies would match that of the metal-based electrode potentials. By contrast, in circumstances where replacement of halide by a neutral ligand stabilizes the metal and ligand orbitals to the same extent, the charge-transfer bands will not shift with varying stoichiometry. The Jorgenson optical electronegativity model is equally restrictive since, in a specified oxidation state, neither metal nor ligand X_{opt} is considered to vary.

For the binary halide/nitrile complexes of ruthenium studied here, it is clear that none of these models is sufficient; replacement of each chloride by benzonitrile stabilizes the ruthenium orbitals to the greatest degree (0.6 eV), affects the π-donor levels of the remaining coordinated chlorides to a smaller extent (0.4 eV overall mean), and lowers the nitrile π-acceptor orbitals still less (0.3 eV). This may be compared with theoretical estimates of 0.25 to 0.5eV for halide/halide interactions in *cis* -$[RuX_2(NH_3)_4]^+$ (X = Cl^- or Br^-).

It is interesting to note that in the $OsX_{6-n}(NH_3)_n$ family (X = Cl^-, Br^-, I^-; n = 2,3,4), where the measured change in the $Os^{IV/III}$ couple is of the same order (0.33 V/X^-), the halide to metal charge-transfer bands do not shift appreciably as the stoichiometry alters. Clearly, this is well accomodated by our model. Thus, in addition to the familiar effects of the ligands on the metal center, a "ligand field theory for ligands" is required. The present analysis is an early step in the quantification of these terms.

Conclusions In complexes of general form $MX_{6-n}(RCN)_n$, systematic relationships have been demonstrated between the extent of halide ligation and the electrode potentials characterizing the coordinated metal ion. The same dependence of $E_{1/2}$ on n is found for both chloride and bromide, is common to congeneric $4d^n$ and $5d^n$ metal ions, and is shared by iso-valent metal ions as diverse as niobium and ruthenium.

As noted in the introduction, one corollary of such wide-ranging ligand-additivity relationships is that any family of complexes where the stoichiometry is fixed and the metal varies widely should show extended periodic patterns in electrode potentials. For the hexahalometallates themselves, the optical halide to metal charge-transfer energies are now shown to systematically reflect the corresponding central ion reduction potentials. The spectra then provide an equivalent periodic expression of the underlying progressions in core charge, electronic correlation terms, and relativistic effects.

The study of the $RuX_{6-n}(RCN)_n$ systems has included investigation of their accessible Cl→Ru^{IV}, Cl→Ru^{III}, and Ru^{II}→RCN charge-transfer bands. These spectra also follow trends fully complementary to the orderly trends in the appropriate central ion electrode potential, once the accompanying variations in halide and nitrile frontier orbitals are taken into account. This approach offers an empirical and integrated analysis of charge-transfer spectra, quantified by electrochemical measurements on the complexes concerned rather than by appeal to generalized optical electronegativity parameters.

7 The MX4L2 Chromophore

In the realm of substituted halide complexes one structure is coming to the fore in our laboratory and that is the trans-MX_4L_2 (X = Cl, Br) chromophore. It merits further comment.

The first point is that for a given L, be it MeCN, CO., PR_3, etc, we have every hope of being able to assemble a wide cast of complexes with extensive different M, so that the periodic dependence of electrode potentials can be explored..

Furthermore as exemplified by handsome spectrum of $[RuCl_4(MeCN)_2]^{1-}$ (dp^5) there is only a unique allowed Cl-to-metal CT band, rather than hexahalide manifold,. Remarkably, the equatorial MCl_4 chromphore contains all the chemical information of the progenitor MCl_6 complexes in a exquisitely simplified form, because the eu symmetry donor level is the direct physical counterpart of t1u and t2u.
This does not seem to have been properly recognized before, perhaps symptomatic of the general tendency to dismiss CT envelopes as uninteresting or unhelp[ful. This is a ubiuitous type, seen for example in $IrX_4(PR_3)_2$, $OsX_4(CO)_2$. It it even turns up in the uv domain of face to face $[Re_2Cl_8]^{2-}$Returning to monomeric $MX4L_2$ chromophores

Here dxy (b_{2g}) is the singly occupied orbital. We believe we can identify not only Xp to M dp CT but also Xp to ds which allows us to estimate 10dq and also Xs to dp to give further insight into the energetics of the halide/metal marriage. Moreover, in the near infrared region the weaker direct d-d bands dxz, dyz to dxy provide an unambiguous measure of the tetragonal element of the crystal field.

Of what relevance is this to electrochemistry? At a pragmatic level many of the complexes we wish to study have to be **driven** to the apppropriate oxidation state for example to the d_5 ion or even eletrogenerated and characterised *in situ* via redox-induced halide substitutions.

More importantly, in the context of orbital mapping or "electrochemical spectroscopy" the same redox-active orbital (dxy) figures in the primary CT process and in interconfigurational d-d excitations and in the d5/d6 reduction.. Do we find a faithful and illuminating correlation? We have tested this on an archetypal system OsX_4L_2. This system starkly illustrates how the incremental electronic response to progressive substitution can be linear or non-linear depending on the nature of L (and X).

8 Ligand additivity in one form or another is one of the cornerstones of intuitive inorganic chemistry.

In the series $[OsX_6]^{3-}$, $[OsX_5(CO)]^{2-}$, *trans*- $[OsX_4CO)_2]^{1-}$ (X = Cl and Br) and in the series *trans* -$[OsX_4(MeCN)_2]^{1-}$, *trans* -$[OsX_4(MeCN)(CO)]^{1-}$, *trans* -$[OsX_4(CO)_2]^{1-}$, osmium IV/III and III/II redox couples shift two to three times further for the first introduction of CO than for the second, contrary to recent predictions; the corresponding halide to Os^{III} charge transfer spectra confirm the strongly non-linear accumulation of ligand electronic effects. These results contrast with the regular progression found upon stepwise replacement of X by MeCN

Comment on the use of notional $E^*_{1/2}$ ligand electrode potentials

It is helpful to define the half-cell reactions relevant to metal-to-ligand charge-transfer processes in solution. Representing the ground state complex as [M-L] and visualizing the stable oxidized and reduced forms as $[M^+\text{-}L]^+$ and $[M\text{-}L^-]^-$, and the equilibrated excited state as $[M^+\text{-}L^-]^*$, we have:

a $[M^+\text{-}L]^+ + e^-$ Æ $[M\text{-}L]$ $E_{1/2}(M^+/M)$

b $[M\text{-}L] + e^-$ Æ $[M\text{-}L^-]^-$ $E_{1/2}(L/L^-)$

c $[M^+\text{-}L]^+ + e^-$ Æ $[M^+\text{-}L^-]^*$ $E^*_{1/2}(L/L^-)$

The distinction between **b** and **c** is that in the latter, voltammetrically unmeasurable process, the ligand is reduced in the presence of the oxidized metal ion. We can now combine half-cells and specify $\Delta G°$ for the following thermally equilibrated solutions processes, setting $\Delta E_{1/2}$ equal to minus the standard cell potential to avoid intrusion of negative terms.

(**c**-**a**) excitation:

$[M\text{-}L]^+ + [M\text{-}L]$ Æ $[M\text{-}L]^+ + [M\text{-}L]^*$ $DE_{1/2}$(excitn)

(**b**-**a**) disproportionation:

$[M\text{-}L] + [M\text{-}L]$ Æ $[M\text{-}L]^+ + [M\text{-}L]^-$ $DE_{1/2}$(dispro)

(**c**-**b**) comproportionation/promotion:

$[M\text{-}L]^+ + [M\text{-}L^-]^-$ Æ $[M\text{-}L] + [M\text{-}L]^*$ $DE_{1/2}$(compro)

For the excitation reaction, the term $\Delta E_{1/2}$(excitn) is equivalent to the term $[E_{1/2}(Ru^{III/II}) - E^*_{1/2}(RCN^{0/-})]$ in equation above. Naturally, a corresponding set of cell reactions exists to deal with ligand-to-metal charge-transfer. Prompted by the availability of both $E_{1/2}(M^+/M)$ and $E_{1/2}(L/L^-)$ data, especially for the bipyridyl systems, there has been a tendency elsewhere to focus on $\Delta E_{1/2}$(dispro) in correlating redox and optical charge-transfer behavior, . This introduces an explicit discrepancy in electron correlation terms. Reaction (**c**-**b**) measures the difference between disproportionation and the true excitation reaction. Thus it seems that, rather than being a retrograde step, the use of E* values may be preferred, particularly since we have defined $E^*_{1/2}(L/L^-)$ such that $nf\Delta E_{1/2}$ is an exact expression for the free energy of electronic promotion in solution.DG, referred to earlier.

9. α-Dioxime Complexes - Blue Cu(I) Species

The poor solubility of dimethyl-glyoxime complexes has always restricted study of their interesting physical properties (crystalline $Ni(dmgH)_2$ is a familiar red gravimetric precipitate as well as a one-dimensional electrical conductor). We have reported the synthesis of a series of freely soluble carbocyclic bis-dioxime complexes of structural types **1** to **4**.

The *bis*-α-dioximes **1** to **4** above are evidently structurally related to the macrocyclic complexes **5**, **6**, and **7**, and it is important to discover whether this analogy extends to their electronic properties as well. The Cu^{2+} and Ni^{2+} complexes readily undergo reversible one-electron reduction in every case, and spectroscopic characterisation of the products (by uv/vis./near-ir, NMR and ESR) establishes that the former yield true Cu^{I} complexes but that the nickel complexes are reduced at the ligand (L), contrary to suggestions elsewhere. Accordingly, whereas the reversible electrode potential $E^o\{[Cu^{II}(L)]/[Cu^{I}(L)]\}$ has its usual significance, we take the view that the corresponding couple for the nickel compounds, $E^o\{[Ni^{II}(L)]/[Ni^{II}(L-)]\}$, systematically maps the trend in the energy of the ligand-based frontier orbital (LUMO) across the series **1** to **7**.

All the Cu^{I} complexes are characterised by a striking absorption band responsible for their intense blue colour and ranging from υ_{max} of 16,600 cm^{-1} (**1**) to 13,300 (**7**). We assign this band to a favoured metal to ligand charge-transfer (MLCT) transition. The above analysis implies that the optical charge-transfer energy, $h\nu_{MLCT}$ $\{CU^{I}(L) \rightarrow Cu^{II}(L-)\}$ should be linearly related to ΔE^o, where $\Delta E^o = E^o\{[Cu^{II}(L)]/[Cu^{I}(L)]\} - E^o\{[Ni^{II}(L)]/[Ni^{II}(L-)]$, since this difference in electrode potentials reflects the frontier orbital gap for the individual complexes.

APPLICATION OF "IN SITU" REFLECTANCE SPECTROSCOPY TO MOLECULAR ELECTROCATALYSIS

Claude LAMY
Laboratoire de Chimie I, Electrochimie et Interactions,
URA au CNRS n° 350, Université de Poitiers,
40 avenue du Recteur Pineau, 86022 Poitiers cedex, France

ABSTRACT

The basic principles of reflectance spectroscopy is first recalled, emphasizing what is particular to each method (vibrational fingerprints for Electro-Modulated Infrared Reflectance Spectroscopy -EMIRS-, and electronic spectra for UV-visible Reflectance Spectroscopy -UVERS-). Experimental set-ups are also presented for each technique.

Several examples illustrate the application of reflectance spectroscopies to the elucidation of reaction mechanisms in Molecular Electrocatalysis :

- adsorption of dyes (viologen derivatives,...) at a platinum electrode, as observed by UVERS ;
- UVERS study of the formation of electron conducting polymer films at a gold electrode, and of their modification by intercalation of metal aggregates or of iron phthalocyanine ;
- EMIRS investigation of the reactive and poisoning species formed during the chemisorption of methanol at platinum metal electrodes.

I - INTRODUCTION

Molecular electrocatalysis deals with the heterogeneous catalysis, by the electrode material, of electrochemical reactions involving organic molecules [1]. Thus the role of the solid electrode is of primary importance, particularly through its chemical nature, and its geometric and electronic structures [2]. Such electrochemical reactions are generally complex, because they proceed through many reaction pathways with several reaction intermediates and products, so that the elucidation of the reaction mechanism can be relatively difficult. In molecular electrocatalysis, as in heterogeneous chemical catalysis, the catalytic surface affects greatly the reaction kinetics mainly through the adsorption of reactants, intermediates and products. The adsorbed species, which result from strong interactions of the reactive organic molecule with the electrode surface, play a key role in the overall reaction, as they determine the reaction selectivity (reactive intermediates) and the electrocatalytic activity (poisoning intermediates) [3]. The primary goal for a detailed kinetic study of electrocatalytic reactions is therefore the determination of the nature, of the structure and of the surface concentration of adsorbed intermediates under working conditions. This can only be achieved by using convenient "in situ" spectroscopic techniques able to provide detailed information on adsorbed species at the molecular level. Among them, reflectance

A. J. L. Pombeiro and J. A. McCleverty (eds.),
Molecular Electrochemistry of Inorganic, Bioinorganic and Organometallic Compounds, 549–571.

spectroscopies, either in the UV-visible range [4] or in the Infrared range [5] [6], are particularly well appropriate to investigate the structure of the electrode/electrolyte interface, and to observe the different adsorbed species, even in the presence of strong absorbing media, such as aqueous solutions.

In this paper, the basic principles of reflectance spectroscopy will be first discussed, emphasizing what is particular to each method : electronic spectra for UV-visible Reflectance Spectroscopy (UVERS), vibrational fingerprints for Electrochemically Modulated Infrared Reflectance Spectroscopy (EMIRS). After a short presentation of the experimental set-up for each technique, various examples, taken mainly from our laboratory, will be given.

II - BASIC PRINCIPLES OF REFLECTANCE SPECTROSCOPY

In reflectance spectroscopy, the incident light of intensity I° is reflected upon the electrode surface (external reflection mode), which is either free of absorbing species - reflected intensity $I(o) = R(o)\, I^\circ$ - or covered by an absorbing layer of thickness d - reflected intensity $I(d) = R(d)\, I^\circ$ -, where R(o) and R(d) are the reflectivity coefficients of the electrode surface, without and with the absorbing layer, respectively. The relative reflectivity change, $\delta R/R$, of the surface, is thus defined as :

$$\frac{\delta R}{R} = \frac{R(d) - R(o)}{R(o)} = \frac{I(d) - I(o)}{I(o)} \tag{1}$$

On the other hand, the variation of absorbance of the surface due to the presence of the absorbing layer is :

$$\delta A = \ln \frac{I^\circ}{I(d)} - \ln \frac{I^\circ}{I(o)} = - \ln \frac{I(d)}{I(o)} \tag{2}$$

where the absorbance A is derived, as usual, from the Beer-Lambert law in transmission spectroscopy [$I = I^\circ e^{-\alpha(\nu) C d}$, with $\alpha(\nu)$ the linear absorption coefficient at frequency ν, and C the molar concentration of absorbing species], i.e. :

$$A = \log_{10} (I^\circ/I) = \epsilon\, C\, d \tag{3}$$

with $\epsilon = \alpha(\nu)/\ln(10)$ the molar absorption coefficient. For thin layers or adsorbed films $\delta R/R << 1$, so that eqn.(2) can be easily transformed, using eqn.(1) :

$$\delta A = - \ln \frac{I(d)}{I(o)} = - \ln \left(1 + \frac{\delta R}{R}\right) \approx - \frac{\delta R}{R} \tag{4}$$

This relation shows that the relative reflectivity change, $\delta R/R$, is equal to the opposite of the absorbance variation, δA, due to the presence of a thin film at the electrode surface, so that the absorbance spectrum of the species can be obtained by reflectivity measurements. This is the so-called Reflection-Absorption Spectroscopy (RAS).

The relative change of reflectivity, $\delta R/R$, of the electrode surface, can be evaluated in the three-phase model for a film of small thickness d ($d << \lambda$ where λ is the wavelength of the incident light) [7]. As a consequence of light reflection at a highly conductive surface, the s-polarized component (i.e. the electric field vector perpendicular to the plane of incidence, defined by the propagation direction and the vector normal to the surface) of the electromagnetic wave is inactive, whereas the p-polarized component (parallel to the plane of incidence) interacts strongly with vibrational modes associated with electrical dipole oscillating perpendicular to the surface (this is the so-called "surface selection rule"). Therefore $\delta R/R$ is only significant for p-polarized light, and, for a not too high absorbing film, it reduces to :

$$\left(-\frac{\delta R}{R}\right)_p \approx \frac{4 \sin \phi_1 \tan \phi_1}{n_2^3} \alpha_2 d \quad \text{and} \quad \left(-\frac{\delta R}{R}\right)_s \approx 0 \qquad (5)$$

where ϕ_1 is the angle of incidence, n_2 the refractive index of the film, and $\alpha_2 = 4 \pi k_2 / \lambda$ its absorption coefficient (see the Beer-Lambert law), with k_2 the imaginary part of a complex refractive index $\tilde{n}_2 = n_2 - i k_2$, i.e. the extinction coefficient. The quantity, $A_2 = \alpha_2 d$, is the absorbance of the film, so that the change δA in the absorbance of the electrode surface due to the presence of the film is :

$$\delta A = \left(-\frac{\delta R}{R}\right)_p \approx \frac{4 \sin \phi_1 \tan \phi_1}{n_2^3} A_2 \qquad (6)$$

This last relation shows that, for high angles of incidence ($\phi_1 > 80°$), the factor " $4 \sin \phi_1 \tan \phi_1 / n_2^3$ " is much greater than unity, so that it behaves as an enhancement factor for the absorbance in a Reflectance-Absorbance Spectroscopy experiment, as compared to the measurement in Transmission Spectroscopy.

III - UV-VISIBLE REFLECTANCE SPECTROSCOPY (UVERS)

III-1 - Experimental

UV-visible Reflectance Spectroscopy (UVERS) can be easily performed with a dispersive UV-visible spectrophotometer and the weak relative reflectivity changes ($\delta R/R \approx 10^{-2}$ to 10^{-5}), due to the presence of a very thin film or that of an adsorbed monolayer, can be monitored by potential modulation and detection of the useful signal with a Lock-In Amplifier. However, for time evolving films, or for rapid electrochemical processes, it is better to use a rapid scan spectrometer, or a spectrometer with a diode array detector, where the improvement of the signal-to-noise ratio can be made with a signal averager after accumulation of hundreds to thousands of reflectance spectra [8] [9]. In this way, Differential Reflectance Spectra can be easily recorded, the averaged spectrum taken at a given reference potential being subtracted from the averaged spectra at various electrode potentials.

An other way to obtain reflectance spectra is to record simultaneously, as a function of the electrode potential E, the current intensity (cyclic voltammogram), and the relative reflectivity change of the electrode surface $(\delta R/R)_\lambda$, at a given wavelength, λ, (the so-called reflectogram, which can be averaged several hundred times, with a rapid scan spectrometer, in order to improve the signal-to-noise ratio). Then the reflectograms $(\delta R/R)_\lambda = f(E)$, collected at various wavelengths λ, are plotted in a three-dimension diagram $\delta A = (-\delta R/R)_\lambda = f(E,\lambda)$, which, if cut at constant E, gives the usual absorbance spectra $(\delta A)_E = f(\lambda)$ at different potentials

E [10]. In this way the evolution of the absorbance of the surface film with the electrode potential can be easily monitored.

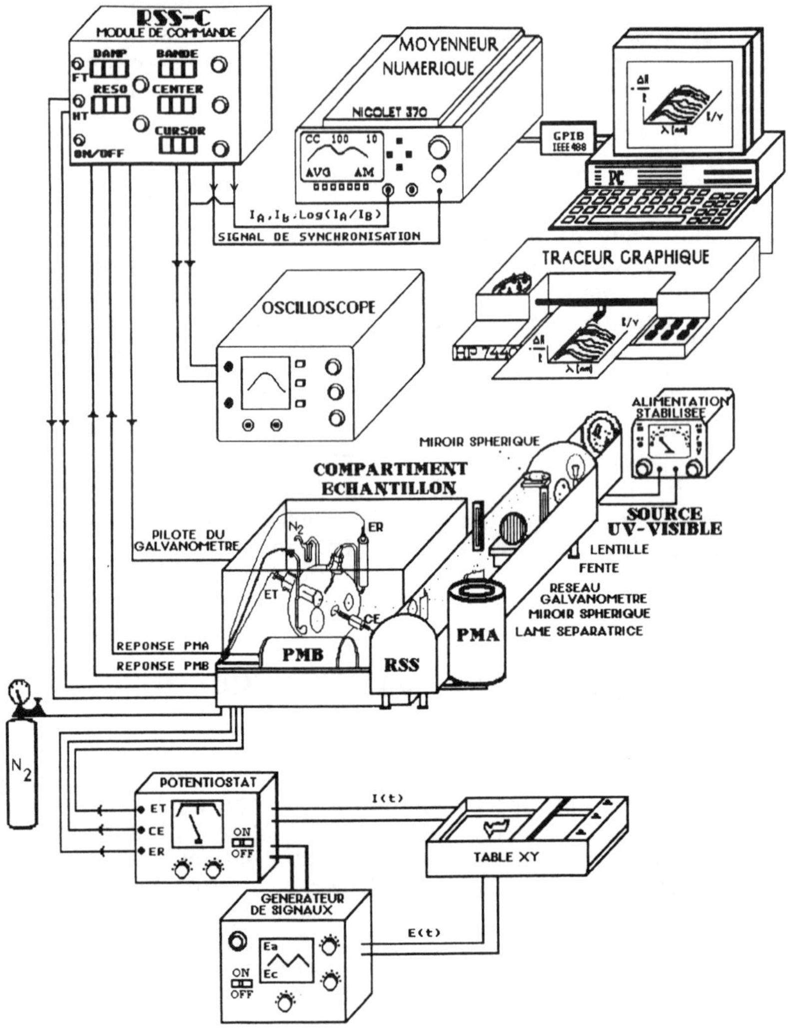

Figure 1 : Schematic diagram of the UVERS experiment.

The UVER Spectra were recorded "in situ" with a Rapid Scan Spectrometer (Harrick RSS-C). The complete diagram of the spectrometer coupled with the electrochemical set-up is shown in Figure 1, which also displays the data acquisition and processing system based on a signal averager (Nicolet 370) and a microcomputer (PC AT 286). The three-electrode spectro-electrochemical cell includes two quartz windows allowing the incident beam to be single-reflected at the electrode surface with an angle of incidence of about 45°. The working electrode consists of a metal disc (area A $\approx$ 0.5 cm^2), carefully polished to mirror finish with fine alumina powders down to 0.05 μm. A reference electrode and a counter-electrode complete the electrochemical arrangement.

III-2 - Adsorption of dyes at a platinum electrode

As a first example to illustrate the way with which the absorbance spectra can be obtained from three-dimension reflectograms, the adsorption of viologen derivatives at a platinum electrode was investigated [11] [12].

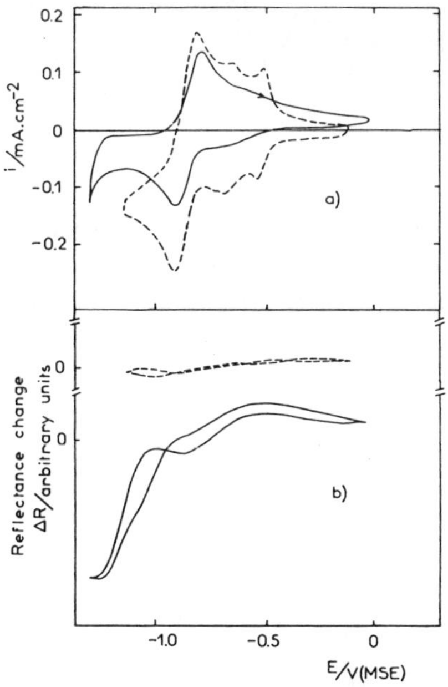

Figure 2 : (a) Voltammograms, and (b) reflectograms recorded at λ= 400 nm, of a platinum electrode in 0.1 M K_2SO_4 (25 °C ; 300 m V s^{-1}) : without (- - -) and with (-----) 4.6x10^{-5} M methylviologen.

In the case of methylviologen (MV^{2+}) dissolved in a neutral solution (0.1 M K_2SO_4 ; pH $\approx$ 3.9), the cyclic voltammograms and reflectograms at a fixed wavelength λ (λ= 400 nm) were simultaneously recorded at a sweep rate of 300 mV s^{-1} (Fig.2) [11]. In the presence of MV^{2+}, the voltammogram shows clearly the adsorption of both the di-cation MV^{2+}, and the radical cation $MV^{+\cdot}$, since the hydrogen adsorption peaks are greatly reduced in the presence of methylviologen (Fig. 2a). The change in the reflectogram is much more important for potentials lower than the half wave potential ($E_{MV^{2+}/MV^{+}}$ = - 1.09 V/MSE), where the adsorption of the blue-coloured cation radical $MV^{+\cdot}$ causes a strong decrease of the electrode reflectivity as a result from strong light absorption at λ= 400 nm (Fig. 2b). By recording 36 reflectograms $(\delta R/R)_{\lambda}$ = f(E), at various close wavelengths λ in the entire spectral range (300-750 nm), it is possible to draw a three-dimension diagram $\delta A = (-\delta R/R)_{\lambda} = f(E,\lambda)$ (Fig.3).

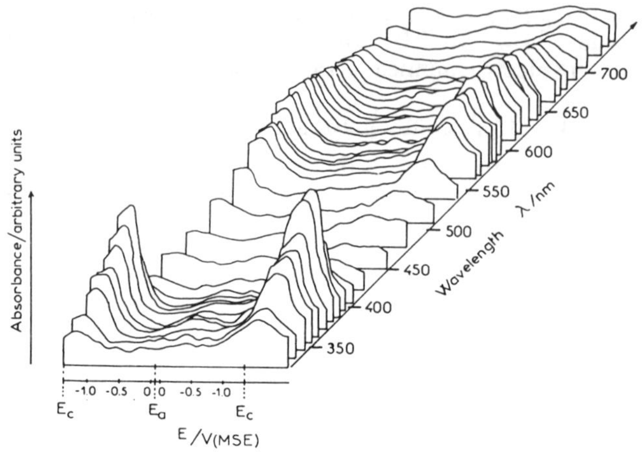

__Figure 3__ : 3-D diagram giving the absorbance change δA of a Pt electrode covered with adsorbed methylviologen, vs. potential E, recorded at various fixed wavelengths λ(reflectograms). 1.25x10^{-4} M MV^{2+}; 0.1 M K_2SO_4; 25 °C; 300 mV/s.

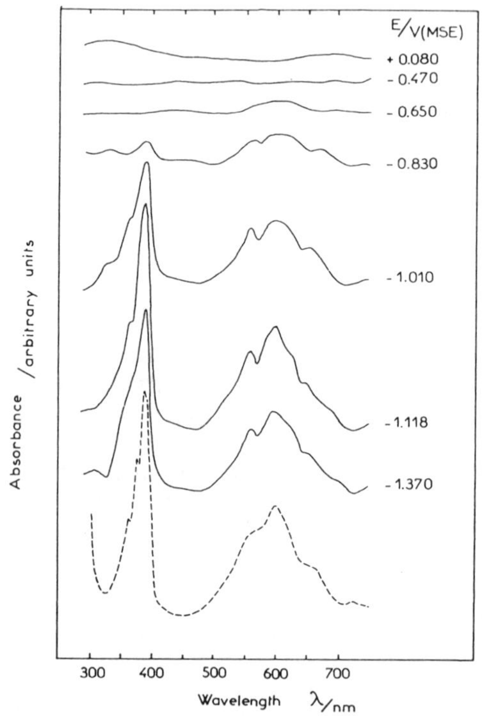

__Figure 4__ : Absorbance spectra δA(λ) of $MV^{+\cdot}$ radicals adsorbed at a Pt electrode, as obtained from Fig. 3, during the positive sweep, by cutting the 3-D diagram at various constant electrode potentials. Same conditions as in Fig. 3. The dashed line corresponds to bulk $MV^{+\cdot}$ produced in solution by photolytic reduction of MV^{2+}.

This diagram contains all the information needed to study the adsorbed species. At potentials close to the lower limit E_c, the absorbance $\delta A(\lambda)$ greatly varies with λ, displaying a rather narrow peak at 400 nm and a wider one at around 600 nm, associated with the absorbance spectrum of the cation radical $MV^{+\cdot}$, whereas at potentials close to the upper limit E_a, where the cation radical is oxidized to MV^{2+}, there is nearly no absorbance variation. This is more easily seen by cutting the 3-dimension diagram $\delta A(E,\lambda)$ at constant potentials E, and drawing, in a 2-dimension plot, the absorbance spectra $\delta A(\lambda)_E$ at the various chosen potentials E (Fig.4). The attribution of these spectra to light absorption by $MV^{+\cdot}$ is easily done by comparison with the transmission spectrum of $MV^{+\cdot}$ produced in solution by photoreduction of MV^{2+} (see the bottom spectrum in dashed line of Fig.4). One can notice a small positive wavelength shift (i.e. at lower energy) of the spectra of the adsorbed species, which may result from specific interactions of the adsorbed radicals with the electrode surface.

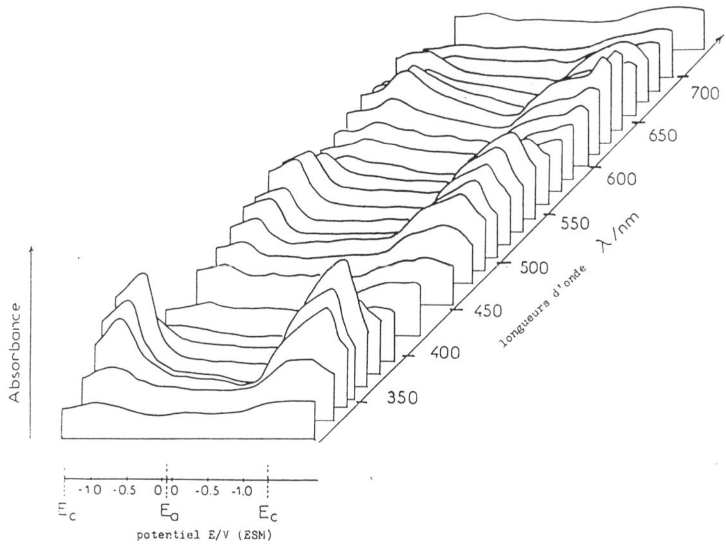

__Figure 5__ : 3-D plot of the absorbance δA due to an adsorbed layer of o,o'-DV^{4+} at a platinum electrode, as a function of electrode potential E, recorded at different fixed wavelengths λ. $6.2x10^{-5}$ M o,o'-DV^{4+} ; 0.1 M K_2SO_4 ; 25 °C ; 300 mV s^{-1}.

The same behaviour at a platinum electrode was more or less observed with other viologen derivatives, such as diheptylviologen C_7V^{2+}, tetradecyl-methyl viologen $C_{14}MV^{2+}$, dibenzylviologen BV^{2+}, and viologen compounds with two bipyridinium groups bonded through one (o-DV^{4+}, m-DV^{4+}) or two (o,o'-DV^{4+}, m,m'-DV^{4+}) xylene radicals in ortho- or meta- position [12]. The case of the diviologen derivatives (DV^{4+}) is particularly interesting, since these compounds can undergo successive reductions, depending on the electrode potential E/V(MSE)

$$DV^{4+} + e^- \xrightarrow{-0.65} DV^{3+} + e^- \xrightarrow{-0.72} DV^{2+} + e^- \xrightarrow{-1.45} DV^{+}$$

(the given potentials correspond to the reduction of o,o'-DV^{4+}). The 3-dimension diagram $\delta A = f(E,\lambda)$, obtained after recording several reflectograms $\delta R/R$ at a sweep rate of 300 mV/s and at various constant wavelengths λ (in the range from 300 to 750 nm), displays several absorption peaks (Fig.5), which are better seen in the 2-dimension absorbance spectra $\delta A = f(\lambda)$ as a function of potential (Fig. 6). The absorption spectra are relatively complex, since both DV^{3+} and DV^{2+} adsorbed species contribute to the absorbance, as seen by comparison with the transmission spectra of DV^{3+} and DV^{2+} species produced by photoreduction of DV^{4+} into the bulk of the solution (dashed curves in Fig.6). Both species result from electroreduction, each one being characterized by an absorption band at 650 nm and 500-550 nm for DV^{3+} and DV^{2+} respectively. Here again, a small red-shift (20 to 30 nm) is observed for the adsorbed species, as a result from strong π-d interactions between the diviologen derivatives and the platinum surface. The time evolution of the adsorbed species can also be deduced from these experiments, e.g. the life time of DV^{3+} can be evaluated to be 2-3 s at potentials greater than $E°_{DV^{4+}/DV^{3+}}$.

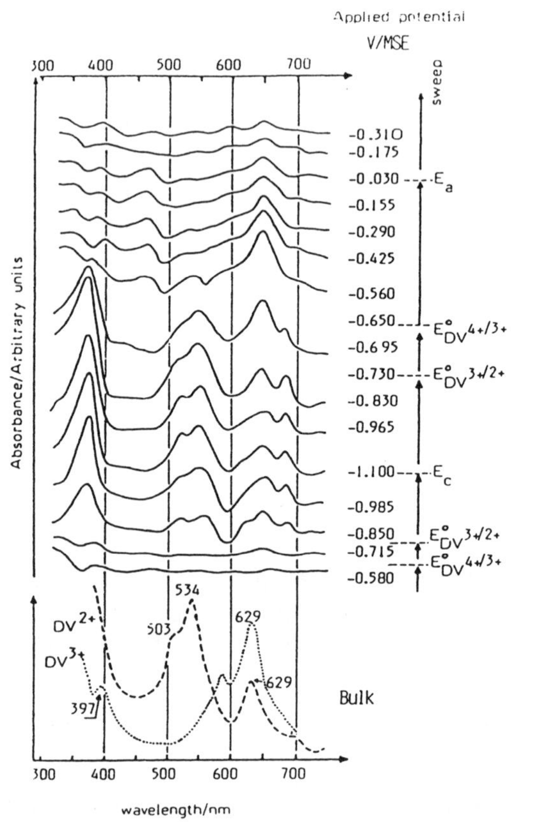

Figure 6 : Absorbance spectra $\delta A(\lambda)$ of o,o'-DV^{4+} derived radicals adsorbed at a Pt electrode, as obtained from the 3-D diagram given in figure 5, at various electrode potentials. Same conditions as in figure 5. The absorbance spectra of DV^{3+} and DV^{2+} radicals are given in dashed lines at the bottom of the figure.

III-3 Dispersion of electrocatalysts into a conducting polymer modified electrode

Electron conducting polymers, such as polyaniline (PAni), polypyrrole (PPy), or polythiophene, can be electrodeposited at inert (glassy carbon,...) or active (platinum, gold,...) electrodes to form relative thick films (from a few hundred Å to a few μm thickness). These films by themselves do not display much catalytic properties, either for the oxidation of small organic molecules [13], or for oxygen reduction [14]. However the incorporation and dispersion of catalysts, such as active metals (Pt,...), or macrocycle compounds (metal phthalocyanine) into the conducting polymer matrix, lead to very electroactive materials. For example the dispersion of a very small amount of platinum (between 0.1 and 1 mg cm^{-2}) allows to reach very high oxidation currents, e.g. for methanol electrooxidation, hundred times greater than those obtained with smooth platinum electrodes (Fig.7) [15].

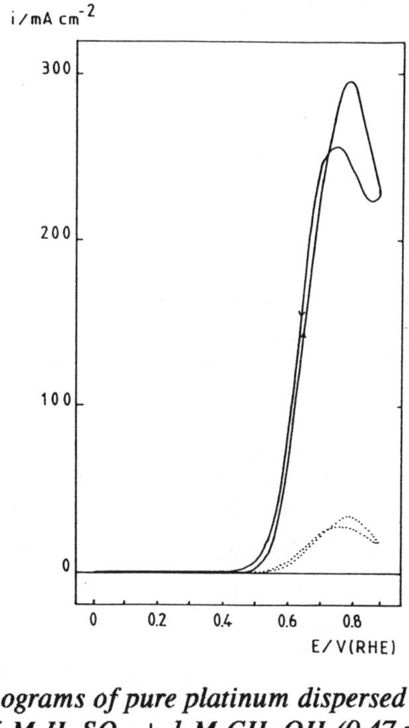

Figure 7 : Voltammograms of pure platinum dispersed in a polyaniline coated gold electrode in 0.5 M H_2SO_4 + 1 M CH_3OH (0.47 mg Pt/cm^2, 5 mV s^{-1}) : (...) 25 °C ; (---) 65 °C.

Similarly, the incorporation of iron phthalocyanine (FePc) into a conducting polymer (PAni, or PPy) gives very active electrocatalysts for oxygen reduction in acid medium (Fig.8) [16].

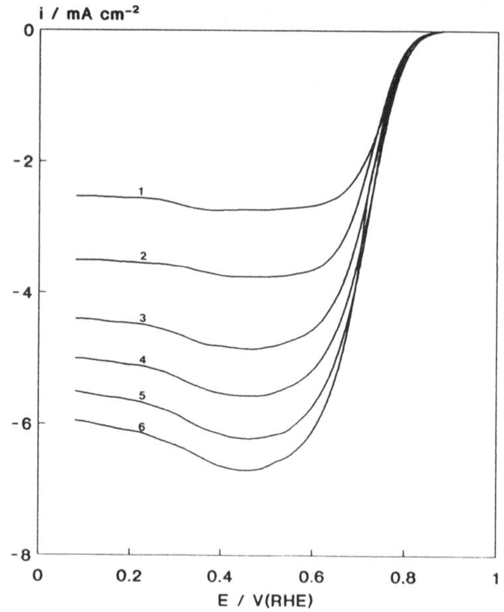

Figure 8 : Quasi steady state polarization curves I(E) of oxygen reduction at a Au/PAni/FePc rotating disc electrode in 0.5 M H_2SO_4 (25 °C ; 2 mV s^{-1}). Rotation speed (rpm) : (1) 500, (2) 1000, (3) 2000, (4) 3000, (5) 4000, (6) 5000.

These examples demonstrate the interest of conducting polymers as matrices to disperse active electrocatalysts, so that a detailed investigation of their electroformation and of the catalyst incorporation process was undertaken at the molecular level, using "in situ" UV-visible Reflectance Spectroscopy (UVERS).

III-3-a Electroformation of a conducting polymer film

The electropolymerisation of aniline or pyrrole was usually realized on a gold disc (0.5 cm diameter), which is an excellent reflecting substrate, and which is usually inactive towards the electrocatalytic reaction under study.

The early steps of polymerisation, carried out under cyclic voltammetry conditions (5 mV s^{-1}), can be easily followed by UVERS for both PAni (Fig.9) and PPy. In both cases the first spectrum recorded, as soon as the polymerisation process starts in a very sharp potential range (0.95 to 1 V/RHE), is relatively different from the successive spectra. The first spectrum can be attributed to the monomer radical cation, whereas the successive spectra, which stabilize after a few tens cycles, correspond to the different oligomers, before the formation of the polymeric film itself. Then, during the film growth the spectra do not change greatly, except for their intensity which continuously increases. The effect of potential is clearly seen on a 3-D diagram with the electrode potential on the x-axis and the wavelength on the y-axis, i.e. in a reflectogram representation (Fig. 10). After a few cycles, the two redox processes, displayed on the voltammogram of the polyaniline film, are better characterized by their absorption bands on spectra represented in a 2-D plot for various electrode potentials (Fig. 11). The absorption bands observed at 440 nm and 560 nm correspond to the first and second redox systems respectively starting from the reduced form [17] [18].

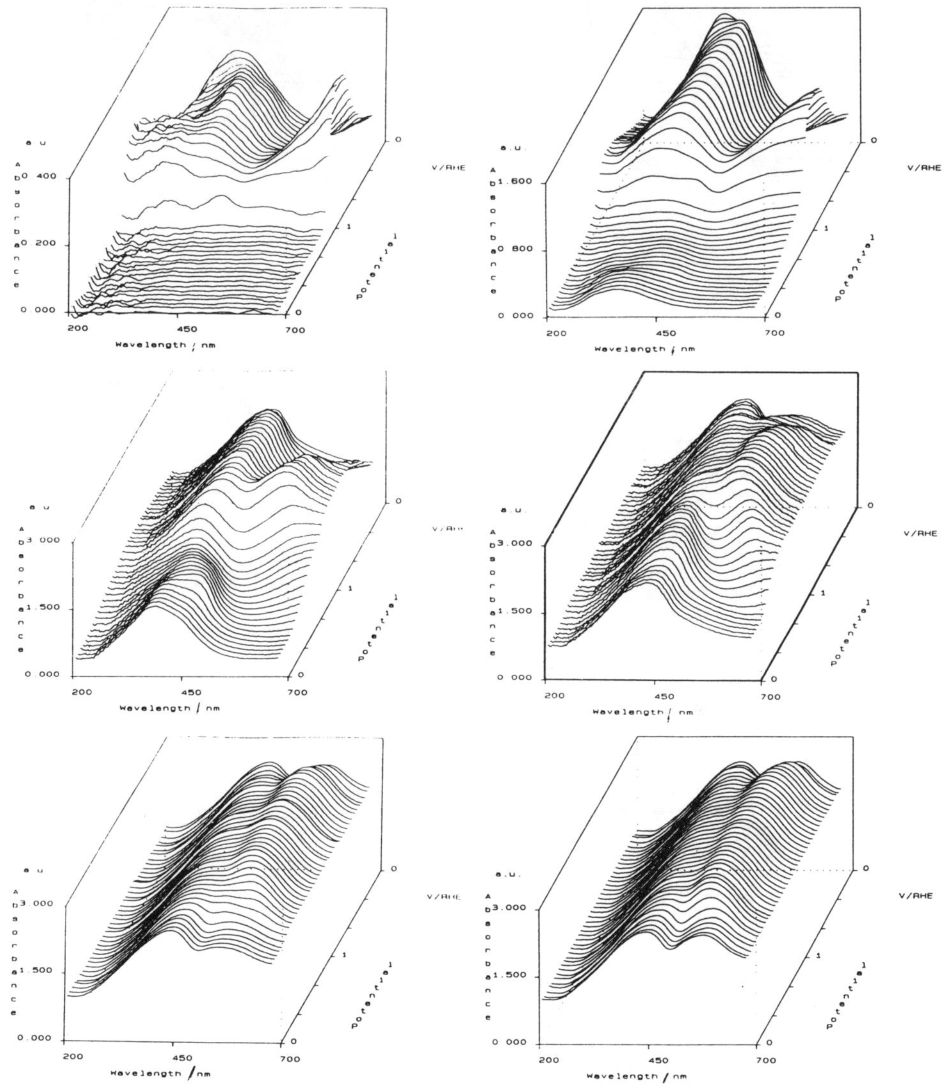

<u>Figure 9</u> : 3-D diagram showing the variation of the UVERS absorbance spectra $\delta A(\lambda,E)$ as a function of the electrode potential E during the electroformation of a polyaniline film at a gold electrode during continuous potential cycling in 0.5 M H_2SO_4 + 0.1 M aniline (25 °C ; 5 mV s^{-1}) : 1st to 6th cycle.

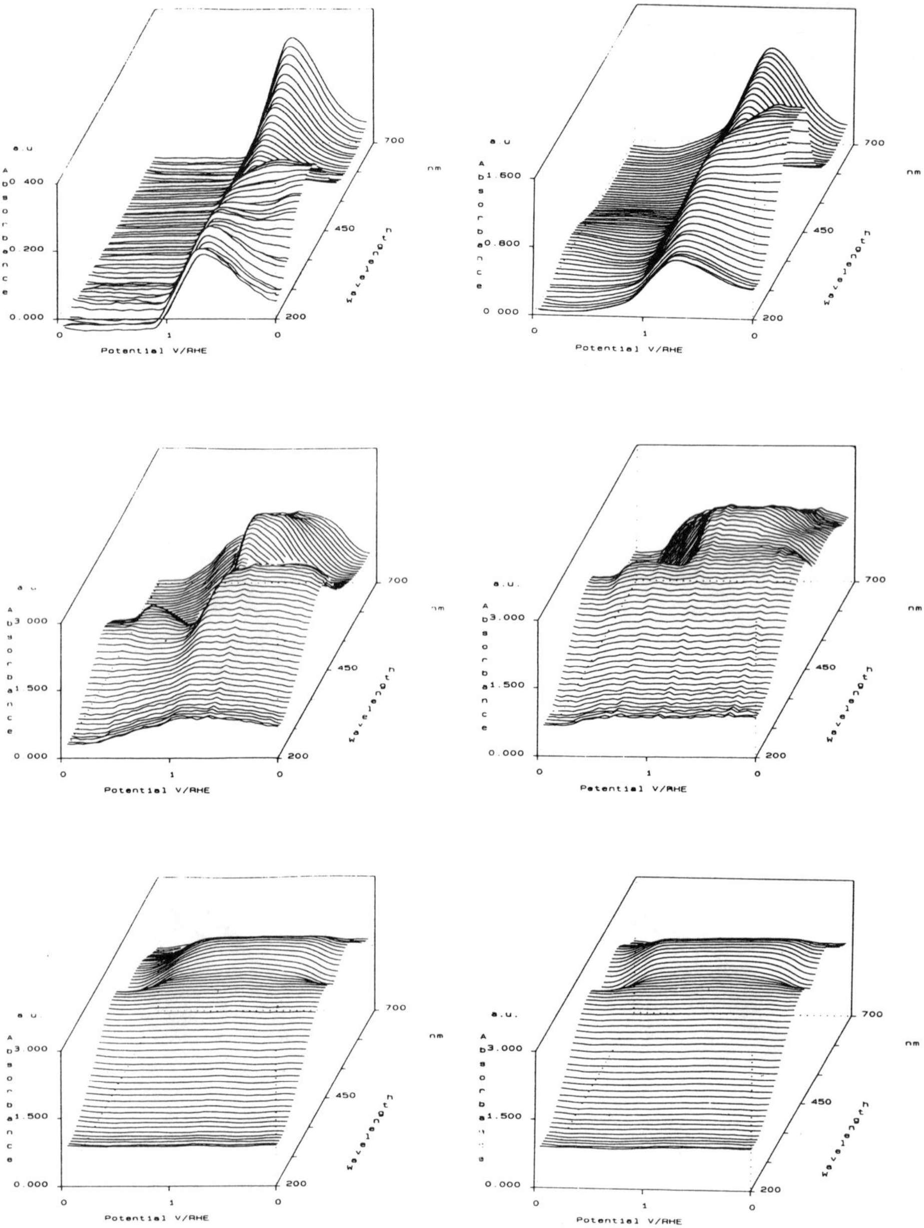

<u>Figure 10</u> : 90° rotation of the 3-D diagram of figure 9 allowing to observe the reflectograms at various wavelengths λ. Same conditions as in figure 9.

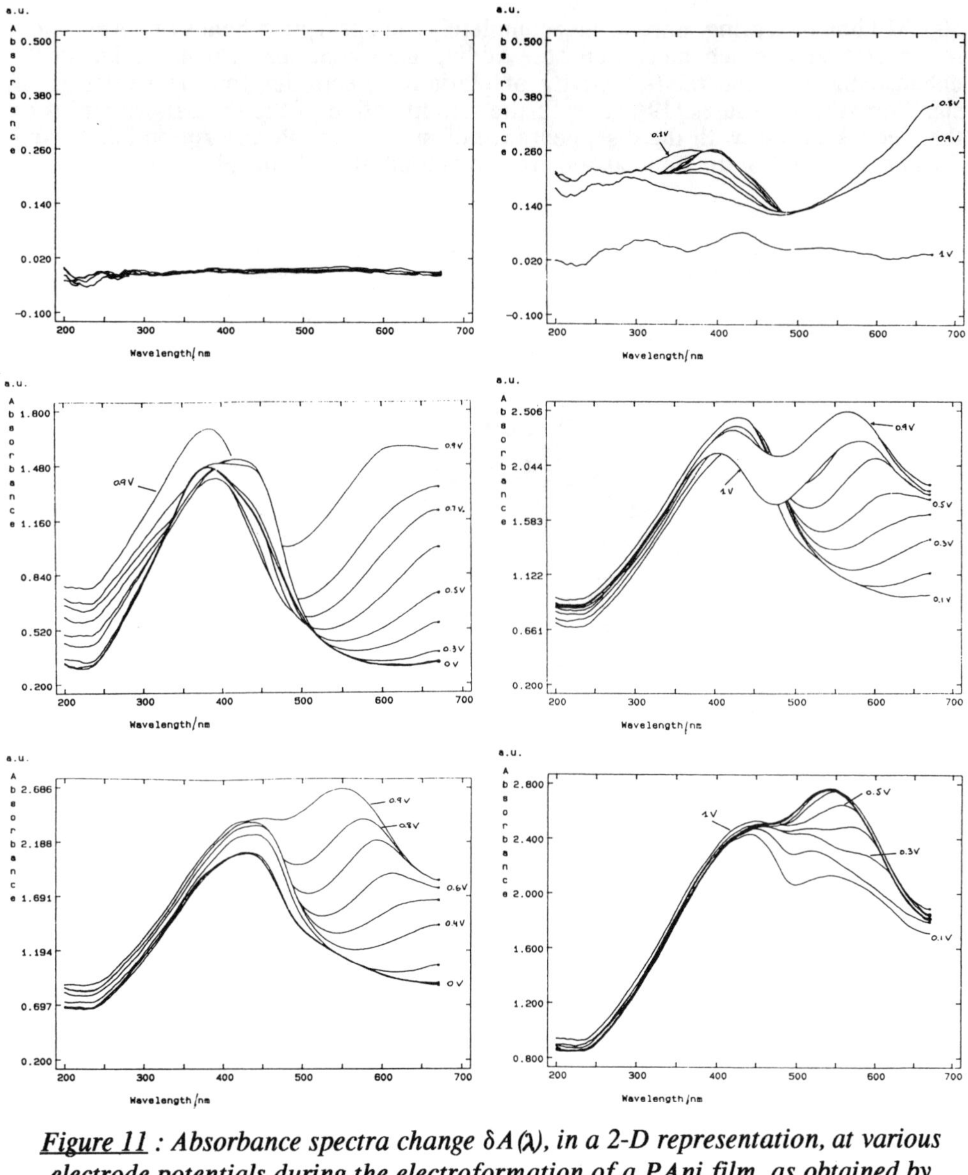

Figure 11 : Absorbance spectra change δA(λ), in a 2-D representation, at various electrode potentials during the electroformation of a PAni film, as obtained by cutting the 3-D diagram of figure 10 at constant potentials. Only the 1st, 3rd and the 4th cycles are given during both the positive and the negative sweeps.

The incorporation of platinum particles inside polyaniline by electrochemical reduction of a platinum salt changes greatly the absorbance spectra, as seen in a 3-D representation (Fig. 12a). New bands appear at around 300, 400,

500, 620 nm, obscuring more or less completely the absorption bands of PAni. Their interpretation is much more delicate. Adding lead adatoms, which are known to enhance the electrocatalytic activity of platinum electrodes for the oxidation of small organic molecules [19], gives further modification of the optical properties of the electrocatalyst, with the disappearance of most of the Pt absorption bands, and the enhancement of the band at 560 nm corresponding to PAni (Fig. 12b).

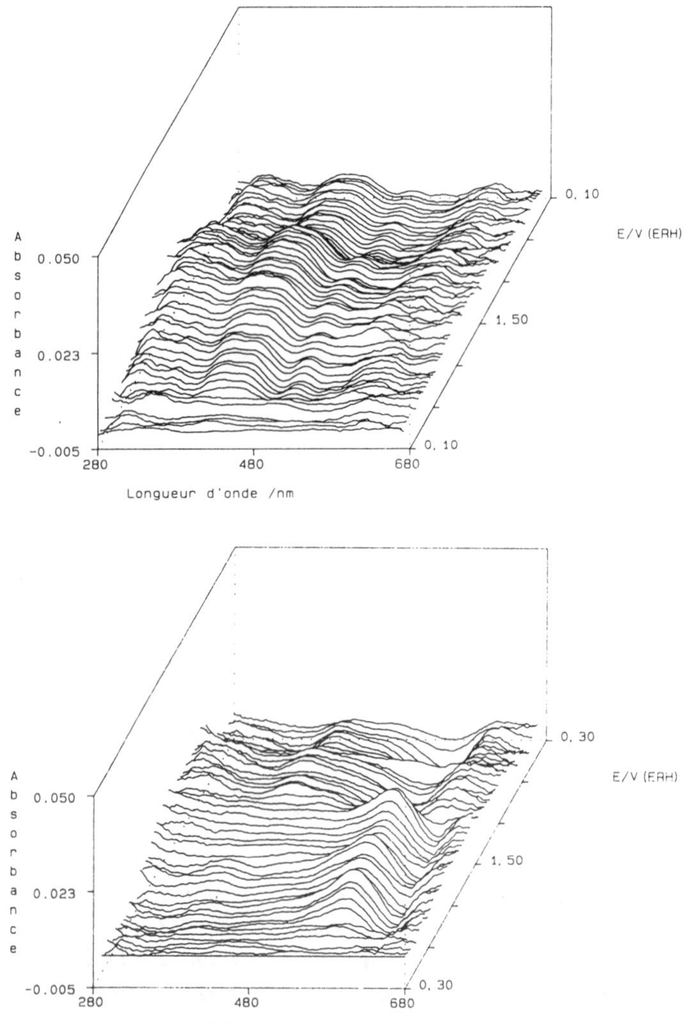

<u>Figure 12</u> : 3-D absorbance spectra showing the effect of electrodeposition, inside a PAni film grown at a gold electrode, of Pt particles (a) and, of Pt+Pb particles (b) (0.5 M H_2SO_4 + 0.1 M aniline ; 25 °C ; 50 mV s^{-1}).

III-3-b Incorporation of macrocycles in a conducting polymer modified-gold electrode

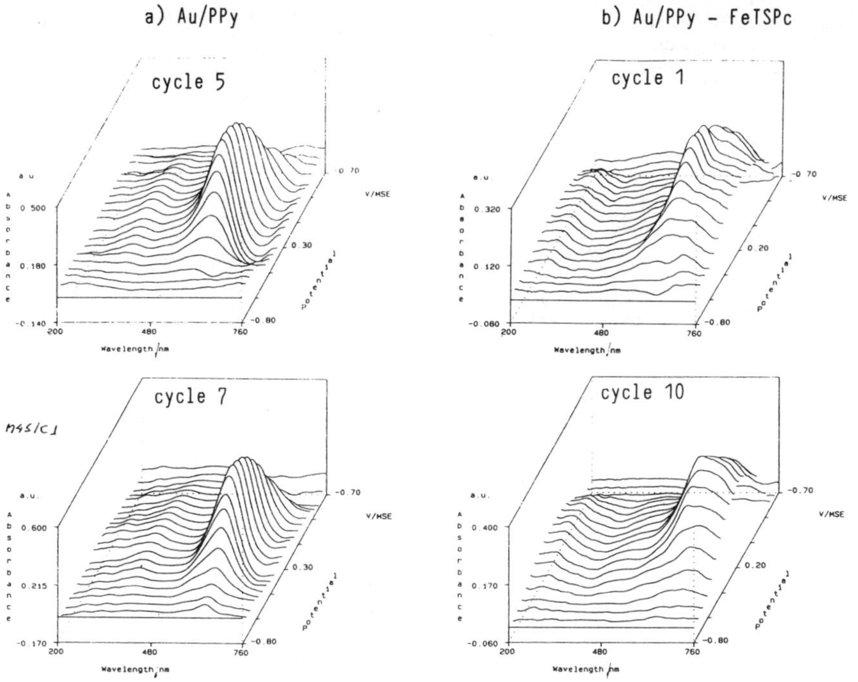

Figure 13 : 3-D absorbance spectra of a polypyrrole film electroformed at a gold electrode at a constant potential (0.3 V/MSE) in 0.5 M H_2SO_4 + 0.1 M pyrrole : (a) without and (b) with incorporation of tetrasulfonated iron phthalocyanine.

Metallophthalocyanine, such as iron phthalocyanine (FePc) can be easily incorporated into a conducting polymer film (either PAni or PPy), as tetrasulfonated form (FeTSPc), during polymer electroformation at a gold electrode [16]. This leads to very good electrocatalysts for oxygen reduction in acid medium. The incorporation process and the electrocatalytic properties of such modified electrodes can be easily investigated "in situ" by UVERS. For example, Fig.13 shows the 3-D absorbance spectra of a PPy film without (Fig.13a) and with (Fig.13b) the incorporation of FeTSPc, recorded during potential cycling at 5 mV s^{-1}. The incorporation process is better seen on a 2-D representation of the absorbance spectra (Fig.14), where the incorporation of the FeTSPc anion is evidenced by absorption bands at 440, 500, 610 640 and 670 nm, whereas the absorption bands of the PPy film are attenuated (bands at 370, 440, 520-540 nm).

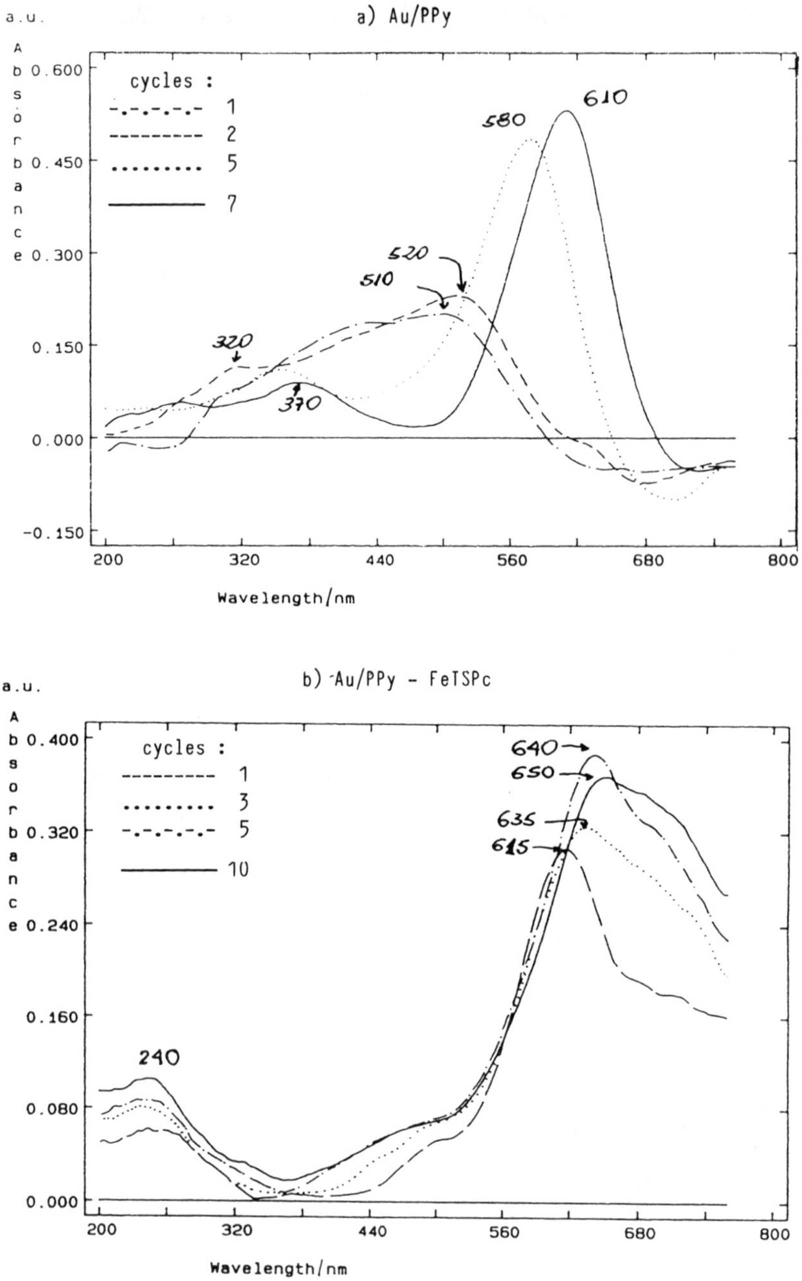

<u>Figure 14</u> : 2-D representation of the absorbance spectra of a polypyrrole film coated gold electrode :
(a) without FeTSPc, (b) with the incorporation of FeTSPc in the film.

IV - ELECTROCHEMICALLY MODULATED INFRARED REFLECTANCE SPECTROSCOPY (EMIRS)

Since this spectroscopic technique and its application to electrocatalysis has been widely described elsewhere [5] [6] [20] [21] [22] , it will be briefly discussed here, and only a typical example will be presented.

IV-1 - Experimental

EMIRS spectra were recorded with an EMIRS III grating spectrometer (Hi-Tek, GB), coupled with an electrochemical equipment (potentiostat and waveform generator) allowing a rigourous control of the electrode potential during the spectroscopic experiments. Using a 3-electrode spectro-electrochemical cell, it is thus possible to record voltammograms before and after acquisition of the spectroscopic data.

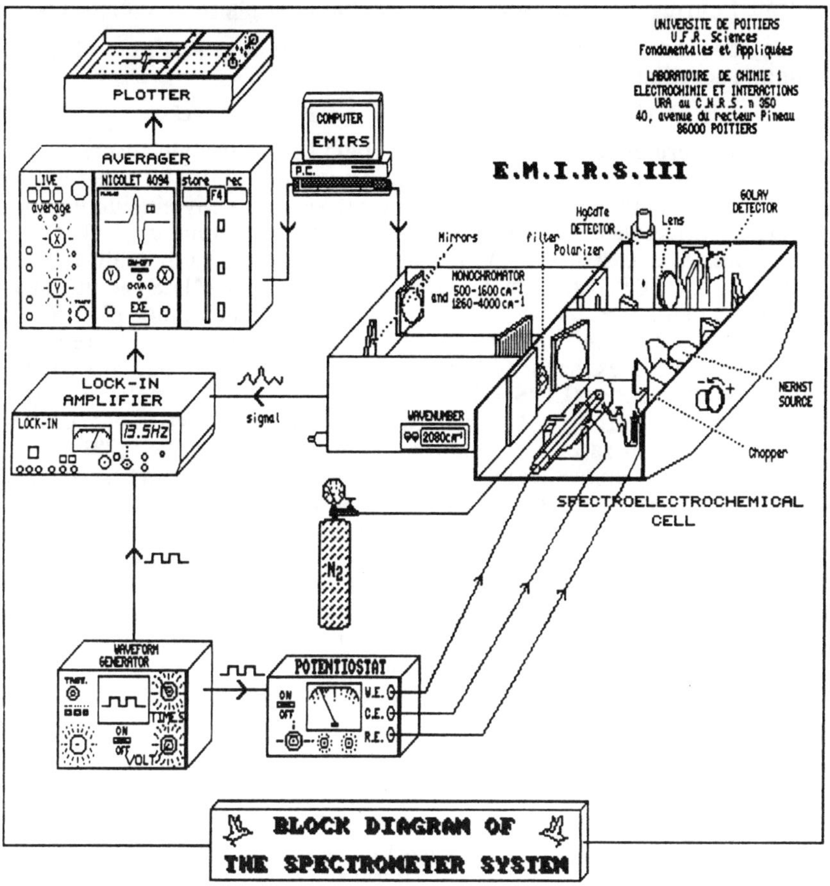

Figure 15 : Block diagram of the EMIRS experiment.

The block diagram of the overall system is shown in Fig.15. The electrode potential is modulated at a low frequency (typically 13 Hz), so that the intensity of the reflected infrared beam is modulated at the same frequency. This allows, by lock-in amplification of the signal, and numerical averaging, followed by integration of the spectra, in order to improve further the signal-to-noise ratio, to monitor very small reflectivity changes ($\delta R/R \approx 10^{-3}$ to 10^{-5}) of the electrode surface. This is sufficient to observe monolayers or even submonolayers (coverage degree below 10 %) of intermediate species adsorbed at a small area electrode ($A \approx 0.5$ cm^2). The EMIRS spectra are displayed as the relative reflectivity change $\delta R/R$ of the electrode surface vs. the wavenumber $\bar{\nu}$ of the incident light (500 cm^{-1} $< \bar{\nu} <$ 4000 cm^{-1}).

The working electrode is a metallic disc (0.7 to 0.8 cm diameter) carefully polished with fine alumina (down to 0.05 μm). A reference electrode and a counter electrode complete the 3-electrode arrangement.

IV-2 - Investigation of the reactive intermediates and of the poisoning species formed during the chemisorption of methanol at platinum electrodes

The reaction mechanisms of methanol oxidation at platinum electrodes in acid medium, according to the overall reaction,

$$CH_3OH + H_2O \dashrightarrow CO_2 + 6\,H_{aq}^{+} + 6\,e^-$$

involve several intermediate species, some of them being strongly adsorbed and thus behaving as catalytic poisons of the electrode active sites [23]. The exact nature and structure of the poisoning species were only relatively recently elucidated, using "in situ" infrared reflectance spectroscopy. Both linearly-bonded -CO (main IR absorption bands at c.a. 2080 cm^{-1}) and bridge-bonded >CO or multi-bonded CO (small bands at c.a. 1850-1950 cm^{-1}), were detected, by EMIRS, at the electrode surface as a result of the dissociative chemisorption of methanol [24]. Since that time, these results were confirmed by several Laboratories, using different infrared reflectance spectroscopies, such as EMIRS [25], Linear Potential Sweep Infrared Reflectance Spectroscopy (LIPSIRS) [26], Subtractively Normalized Interfacial Fourier Transform Infrared Reflectance Spectroscopy (SNIFTIRS) [27], or Sweep Potential Absorbance Infrared Reflectance Spectroscopy (SPAIRS) [28].

With relatively concentrated methanol solutions ($\geq$ 0.1 M), and during the large times (from a few tens of minutes to a few hours) generally necessary for spectral acquisition, the CO adsorbates block progressively the electrode active sites, preventing the adsorption of reactive intermediates which are weakly bonded to the electrode surface. Therefore by using lower methanol concentrations (10^{-3} to 10^{-2} M), and short acquisition times (a few minutes allowing to acquire only one spectral scan, without averaging), it was thus possible to detect many adsorbed species, as shown in Figure 16 [29] [30]. Apart from very small bands at 2060 cm^{-1} and 1940 cm^{-1}, corresponding to linearly- and bridge-bonded CO respectively, there are several bands in the spectral range 500 to 3000 cm^{-1}. Among them, are clearly seen the absorption band of $(CO_2)_{ads}$ (at 2350 cm^{-1}), the CH-stretching mode ν(C-H) of either $(CH_3OH)_{ads}$ or CH_3O_{ads} (at 2900 cm^{-1}), or of $HCOO_{ads}$ (at 2820 cm^{-1}) or CHO_{ads} (at 2600 cm^{-1}), the stretching mode ν(C=O) of the carbonyl group >C=O, either in CHO_{ads} (at 1720-1740 cm^{-1}), or in $HCOO_{ads}$ (at 1640 cm^{-1}), the bending mode δ(O-H) of adsorbed H_2O (at 1620 cm^{-1}), the bending mode δ(C-H) of CH_3OH_{ads}, CH_3O_{ads}, $HCOO_{ads}$ or CHO_{ads} (in the range 1360 to 1480 cm^{-1}), the stretching mode ν(C-O) of either adsorbed formate $HCOO_{ads}$

(1320 to 1330 cm^{-1}), or CH_3O_{ads} (at 1030 cm^{-1}), or CH_3OH_{ads} (at 1000 cm^{-1}), the ClO_4^- anion absorption band (at 1100 cm^{-1}), the bending mode of δ(OCO) of $HCOO_{ads}$ (at 770 cm^{-1}), and the bending mode δ(OH) of CH_3OH_{ads} at 600 cm^{-1}.

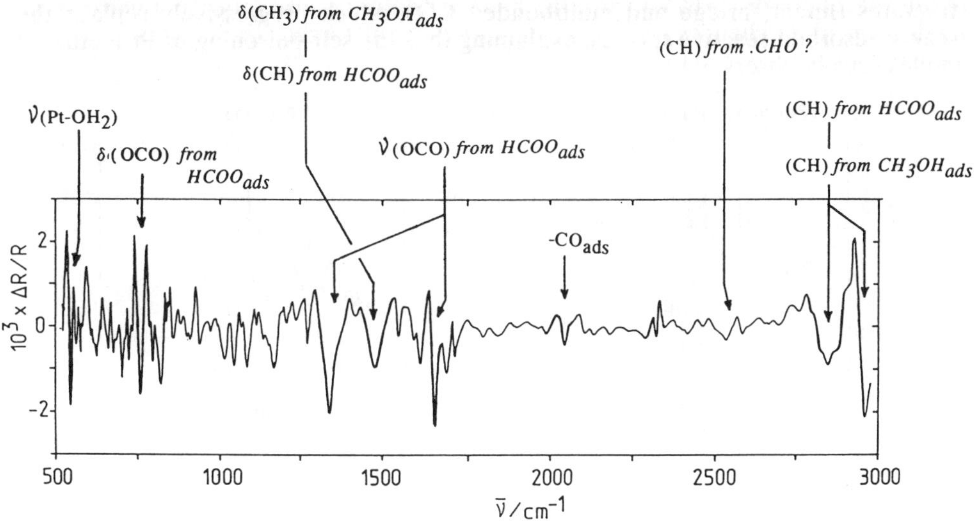

Figure 16 : EMIRS spectra of the adsorbed species resulting from methanol dissociation and oxidation at a polycrystalline electrode ; 10^{-3} M CH_3OH in 0.5 M $HClO_4$; $\delta E = 400$ mV ; $f = 13.5$ Hz ; $\bar{E} = 0.2$ V/RHE ; 10 averaged scans.

The position of the band centres can be better determined by integration of the EMIRS spectra, which greatly reduces the noise, as shown in Fig.17, for the adsorption of 10^{-3} M CH_3OH at a polycrystalline electrode and at a Pt(111) single crystal, respectively. The time evolution of the spectra are also shown after 5 averaged scans (Fig.17c). During the first scan, particularly at a Pt(111) electrode, the strong absorption band at around 1675-1687 cm^{-1} corresponds to the (C=O) stretching mode of methyl formate [31]. The band at 1471 cm^{-1} (polycrystalline Pt), or at 1465 cm^{-1} [Pt(111)], corresponds to the bending mode δ(C-H) of CH_3O_{ads}. The bands at ca. 1740-1750 cm^{-1} can be attributed to adsorbed CHO_{ads}. These bands disappear as the number of scans, i.e. the adsorption time increases, as a result of the formation of reaction products, such as $HCOO_{ads}$ (bands at 1650 cm^{-1}), and of the poisoning species, CO, multibonded to the surface (band at ca. 1880 cm^{-1}), particularly for Pt(111).

To summarize these results, it becomes now clear that EMIR Spectroscopy is particularly well suited to follow the fate of the different small adsorbed organic residues, resulting from the chemisorption of small organic molecules, such as CH_3OH. The nature and the quantity of adsorbed species depend strongly on the structure of the catalytic surface, on the concentration of methanol in solution, on the adsorption time, on the applied electrode potential,... All these various experimental conditions lead to a great variety of adsorbed species, and control their surface distribution. According to these spectroscopic data, the reaction oxidation mechanisms of methanol adsorption and oxidation at platinum electrodes are better understood : after the initial adsorption of methanol, different reactive primary species are formed, such as methoxy, formyl, formate radicals. Surface reactions between these intermediates lead either to product such as formaldehyde, formic acid, methyl formate, or to poisoning species, adsorbed CO, with different structures (linear, bridge and multibonded CO), which progressively replace the weakly adsorbed reactive species, explaining thus the self-poisoning of Pt methanol anodes usually observed.

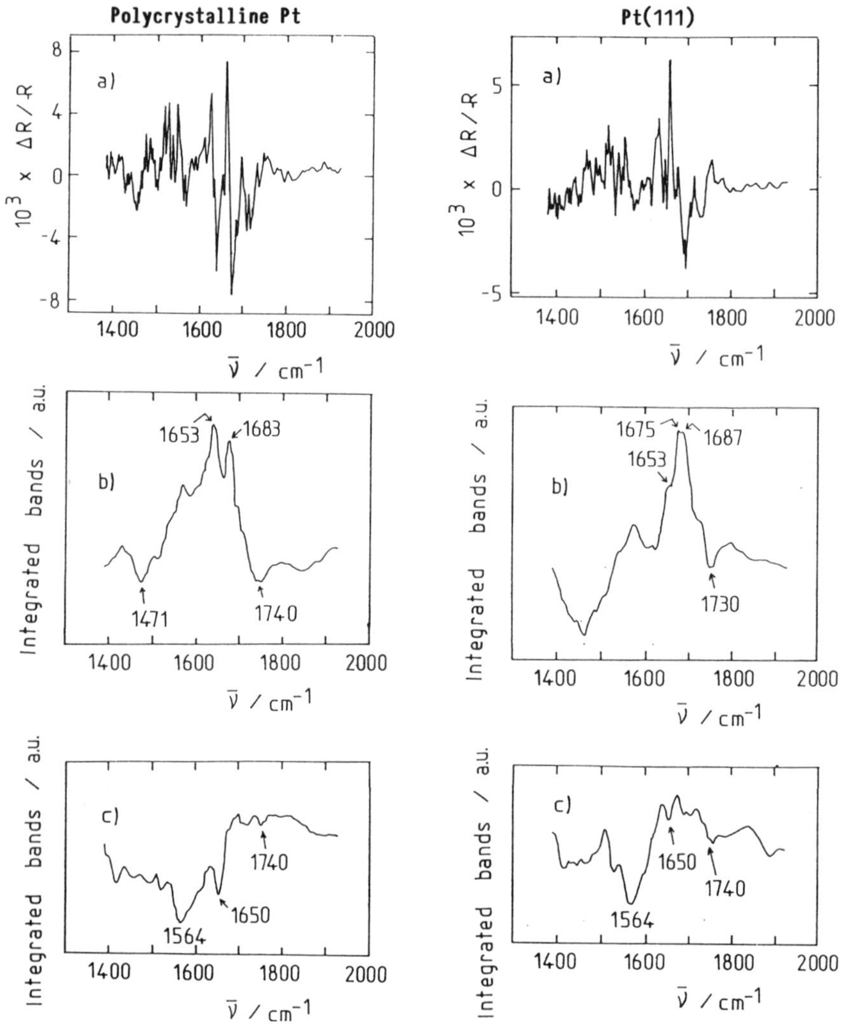

Figure 17 : Original a), and integrated b) and c), EMIRS spectra of the adsorbed species resulting from methanol dissociation at a polycristalline platinum and at a Pt(111) electrode (10^{-3} M CH_3OH in 0.5 M $HClO_4$, $\bar{E}$ = 0.3 V/RHE , δE = 300 mV , f = 13.5 Hz) : a) and b) 1 scan ; c) 5 scans.

V - CONCLUSIONS

The different examples discussed in this paper clearly show the great interest, for investigations in the field of "Molecular Electrocatalysis", of "in situ" Reflectance Spectroscopic Techniques, both in the UV-visible range (UVERS) and in the Infrared range (EMIRS).

UVERS is particularly useful to study the adsorption of monolayers or submonolayers of strongly absorbing dyes, such as viologen derivatives. In this case, the electronic spectra are relatively easy to interpret by comparison with similar bulk species produced "in situ" by photolysis, although the absorption bands can be slightly shifted in wavelength due to specific interactions with the catalytic surface. UVERS also allows to study the formation of coloured films, such as conducting polymer films (polyaniline, polypyrrole,...) at reflecting smooth electrode surfaces, and to follow the incorporation of either catalytic metal particles (platinum particles,...), or macrocycle compounds (iron phthalocyanine, for example), in such polymeric substrates. However the spectra are relatively more difficult to interpret due to a lack of reference spectra for conducting polymers.

On the other hand, EMIRS is an unique tool to identify "in situ", as a consequence of fingerprinting through their vibrational spectra, the adsorbed intermediates produced, even at very low electrode coverages (a few percent), by the chemisorption of small organic molecules, either the reactive species, or the poisoning species. In most cases the latter were identified as adsorbed CO, either linearly- or bridge- and multi-bonded to the electrode surface.

In the case of methanol adsorption and oxidation, which has been thoroughly studied since several years, and which was taken as a typical example, apart from the poisoning CO species, which are always present at the electrode surface, particularly for high methanol concentration ($\geq$ 0.1 M), there are many other adsorbed species at the electrode surface, which behave as reactive intermediates, namely methyl, formyl, formate radicals, and even methyl formate. The surface distribution of the different adsorbed species was shown to depend strongly on the electrode structure, electrode potential, concentration of the electroreactive species, temperature and pH of the electrolytic solution.

ACKNOWLEDGMENTS

Most of the results presented here were obtained in our Laboratory by Prof. Bernard Beden, Drs. Pierre Crouigneau, Françoise Hahn, Jean-Michel Léger, and Ph.D. students (Abderrahim El Hourch, Hervé Laborde, Paulo Olivi, André Rakotondrainibe). Some postdoctoral fellows were also involved in these experiments : Prof. Auro Tanaka (IFQSC, USP, Sao Paulo, Brazil), Prof. Inès Fonseca and Prof. Maria Irene Lopes (CECUL, Universidade de Lisboa, Portugal).

REFERENCES

[1] J. O'M. Bockris and A.K.N. Reddy, Modern Electrochemistry, vol.2, Plenum Press, New York (1972), p.1141.

[2] A.J. Appleby, Electrocatalysis, in "Modern Aspects of Electrochemistry", B.E. Conway and J. O'M. Bockris (Eds.), vol.9, Plenum Press, New York (1974), p.369.

[3] G.P. Sakellaropoulos, Surface Reactions and Selectivity in Electrocatalysis, in "Advances in Catalysis", D.D. Eley, H. Pines and P.B. Weisz (Eds.), vol.30, Academic Press, New York (1981), p.218.

[4] D.M. Kolb, UV-visible Reflectance Spectroscopy, in "Spectro-electrochemistry- Theory and practice", R.J. Gale (Ed.), Plenum press, New York and London (1988), chap.4, p.87.

[5] A. Bewick and B.S. Pons, Infrared spectroscopy of the electrode-electrolyte solution interface, in " Advances in Infrared and Raman Spectroscopy", R.J.H. Clark and R.E. Hester (Eds.), vol.12, Wiley Heyden, London (1985), chap.1, p.1.

[6] B. Beden and C. Lamy, Infrared Reflectance Spectroscopy, in "Spectroelectrochemistry - Theory and Practice", R.J. Gale (Ed.), Plenum Press, New York (1988), chap.5, p.189.

[7] J.D.E. McIntyre and D.E. Aspnes, Surf. Sci., 24 (1971) 417.

[8] D.M. Kolb and R. Kötz, Surf. Sci., 64 (1977) 698.

[9] J.D.E. McIntyre, in "Optical Properties of Solids - New Developments", B.O. Seraphin (Ed.), North-Holland, Amsterdam (1976), p. 555.

[10] F. Hahn, B. Beden, M.-J. Croissant and C. Lamy, Electrochim. Acta., 31 (1986) 335.

[11] B. Beden, O. Enea, F. Hahn and C. Lamy, J. Electroanal. Chem., 170 (1984) 357.

[12] P. Crouigneau, O. Enea and B. Beden, J. Electroanal. Chem., 218 (1987) 307.

[13] H. Laborde, J.-M. Léger, C. Lamy, F. Garnier and A. Yassar, J. Appl. Electrochem., 20 (1990) 524.

[14] A. Elzing, A. Van der Putten, W. Visscher and E. Barendrecht, J. Electroanal. Chem., 200 (1986) 313 ; idem, 233 (1987) 113.

[15] P. Ocon-Esteban, J.-M. Léger, C. Lamy and E. Geniès, J. Appl. Electrochem., 19 (1989) 462.

[16] A. El Hourch, S. Belcadi, P. Moisy, P. Crouigneau, J.-M. Léger and C. Lamy, J. Electroanal. Chem., 1992, in press.

[17] E.M. Geniès, A. Boyle, M. Lapkowski and Tsintavis, Synth. Met., 36 (1990) 139

[18] Y.-B. Shim, M.-S. Won and S.-M. Park, J. Electrochem. Soc., 137 (1990) 538

[19] R.R. Adzic in "Advances in Electrochemistry and Electro-chemical Engineering", vol.13, H. Gerischer (Ed.), Wiley & Sons, New York (1984), p.159.

[20] C. Lamy, F. Hahn and B. Beden, Portugaliae Electrochim. Acta, 7 (1989) 435.

[21] B. Beden, Application of EMIRS and Related Techniques to Electrocatalysis, in "Spectroscopic and Diffraction Techniques in Interfacial Electrochemistry", C. Gutiérrez and C. Melendres (Eds.), Nato ASI Series, vol.320, Kluwer Academic Pub., Dordrecht (1990), p.103.

[22] B. Beden, J.-M. Léger and C. Lamy, Electrocatalytic Oxidation of Oxygenated Aliphatic Organic Compounds at Noble Metal Electrodes, in "Modern Aspects of Electrochemistry", Vol.22, J. O'M. Bockris, B.E. Conway and R.E. White (Eds.), Plenum Press, New York (1992), chap.2, p.97, in press.

[23] B.D. McNicol, Direct Methanol/Air System, in "Power Sources for Electric Vehicles", Vol. 11, B.D. McNicol and D.A.J. Rand (Eds.), Elsevier, Amsterdam (1984), p.807.

[24] B. Beden, C. Lamy, A. Bewick and K. Kunimatsu, J. Electroanal. Chem., 121 (1981) 343.

[25] B. Beden, F. Hahn, S. Juanto, C. Lamy and J-M. Léger, J. Electroanal. Chem., 225 (1987) 215.

[26] K. Kunimatsu, J. Electroanal Chem., 140 (1982) 205.

[27] S. Pons, J. Electroanal. Chem., 160 (1984) 369 ; K. Ashley and S. Pons, Chem. Rev., 88 (1988) 673.

[28] D.S. Corrigan and M.J. Weaver, J. Electroanal. Chem., 241 (1988) 143.

[29] B. Beden, F. Hahn, J.-M. Léger, C. Lamy and M.-I. Lopes, J. Electroanal. Chem., 258 (1989) 463.

[30] M.-I. Santos-Lopes, B. Beden, F. Hahn, J.-M. Léger and C. Lamy, J. Electroanal. Chem., 313 (1991) 323.

[31] M.-I. Lopes, I. Fonseca, P. Olivi, B. Beden, F. Hahn, J.-M. Léger and C. Lamy, J. Electroanal. Chem, submitted.

LOW TEMPERATURE ELECTROCHEMISTRY AND SPECTROELECTROCHEMISTRY OF CATALYTICALLY IMPORTANT TUNGSTEN(0) COMPLEXES

MICHAŁ WILGOCKI, TERESA SZYMAŃSKA-BUZAR,
MAREK JAROSZEWSKI and JÓZEF J. ZIÓŁKOWSKI

Institute of Chemistry, University of Wroclaw,
F. Joliot-Curie 14, 50-383 Wroclaw, Poland.

ABSTRACT. The construction of the variable temperature electrochemical cryostat equipped with non-aqueous reference half-cell with liquid junction and its utilization for the characterization of organometallic complexes is reported. New tetracarbonylbis(olefine) tungsten(0) complexes of the formula *trans*-$W(CO)_4L_2$ have been synthetized (where L is terminal or cyclic olefine). The electrochemical oxidation of $W^{0/+1}$ complexes have been studied by the low temperature: cyclic voltammetry, micro bulk-electrolysis and *in situ* ESR spectroelectrochemistry.

1. Introduction

The construction of the variable temperature electrochemical cryostat and its application for the characterization of catalytically important tungsten complexes is presented. It has been established [1-3] that also under voltammetric conditions the most precise and very stable potential of the working electrode can be maintained using the Kawai's type reference half-cell with liquid junction [4]. Thus we have used this reference electrode to examine the substitution reactions with contribution of the active complex forms, electrochemically generated in acetonitrile [5]. In the present paper a new cell for low temperature electrochemistry (cyclic voltammetry and micro-bulk electrolysis) and low temperature ESR spectroelectrochemistry is described, in which the high quality data could be obtained in the high resistance medium .

Olefin complexes of metal carbonyl fragments are intermediates in catalytic reactions of alkenes [6]. The reactions catalyzed by Group 6 the metal carbonyl compounds produce the $M(CO)_5L$ and $M(CO)_4L_2$ species (L denotes alkene) [7]. The *cis* and *trans* stereoisomers are possible for the $M(CO)_4L_2$ type complexes. In most cases the disubstituted d^6 metal carbonyls, synthesized in the photosubstitution reaction, have the *cis* geometry. There are, however, the systematic exceptions: the *trans* iso-

A. J. L. Pombeiro and J. A. McCleverty (eds.),
Molecular Electrochemistry of Inorganic, Bioinorganic and Organomettallic Compounds, 573–582.

mer is obtained, as the more stable than the *cis* one, when **L** are olefinic or carbene ligands [8]. The *trans* oriented olefin or carbene ligands could not participate in catalytic rearrangement of olefin. This requires the *cis* geometry of the $W(CO))_4L_2$ intermediate. We have decided to find out whether the electrochemical oxidation would induce the isoisomerization of *trans*-tetracarbonyl-bis(alkene) tungsten(0) complexes to the cis species in the CH_2Cl_2 medium.

2. Experimental

Reagent grade dichloromethane, hexane and olefins were dried over calcium hydride and distilled under nitrogen.

Trans-$W(CO)_4L_2$ complexes were synthetized in photochemical reaction of $W(CO)_6$ with excess of the respective olefine (terminal or cyclic) in hexane. The crude products were purified chromatographically using silica gel, and finely, were recrystallized from n-hexane. All manipulations of air-sensitive materials were conducted under argon using a Schlenk apparatus. IR and NMR spectroscopy were used for their identification. The existence of only one strong CO streching band in the IR spectrum confirms the *trans*-geometry in the $W(CO)_4L_2$ complex obtained (D_{4h}) symmetry; ν(CO): a_1, b_1(IR inactive), e(IR active) [8]. In all IR measurements no evidence for the formation of the *cis*-$W(CO)_4l_2$ species, characterized by four IR active ν(CO) vibrations, were found.

A voltammetric studies were carried out on a Laboratorni Pristroje Polagraphic Analyzer (PA3 and PA4) for slow scan rates (up to 0.2 V/s) and on an EP-22 potentiostat and an EG-20 function generator (Elpan, Poland) for faster scan rates (up to 100 V/s). The final curves were collected using AC module type LC-020 (Ambex, Poland) and IBM PC XT computer. All variable temperature measurements (up to *ca.* -80°C) were performed in the presented here (Figure 1) electrochemical cryostat, under argon atmosphere, using Pt (planar electrode), Pt auxiliary electrode (wire) and the reference half-cell with liquid junction: Ag/AgCl, satd. LiCl + 0.1 M TBAP in CH_2Cl_2 || 0.1 M TBAP in CH_2Cl_2. The same type of electrodes were used in the micro-bulk electrolysis and spectroelectrochemical studies. The micro-bulk electrolysis were carried out with OH 404A potentiostat associated with an OH 404C digital integrator unit (Radelkis, Hungary). All these measurements were performed under argon in 1.5-1.0 ml, and finaly in 0.2 ml home made cryostats with continous temperature control. The *in situ* ESR spectroelectrochemical measurements were carried out also in home made quartz cryostats for 1 ml, and finaly for 0.2 ml of sample solution. ESR spectra of the solutions, and of the frozen solution, were recorded on ESR-Radiopan (Poland) Spectrometer.

3. Results and discussion

3.1. THE VARIABLE TEMPERATURE ELECTROCHEMICAL CRYOSTAT EQUIPPED WITH REFERENCE HALF-CELL WITH LIQUID JUNCTION

Most often to the high purity and dry solvent the "aqueous" reference electrode with one or two fritted glass (at the end of the reference electrode and/or of the intermediate salt bridge) is introduced. Such a

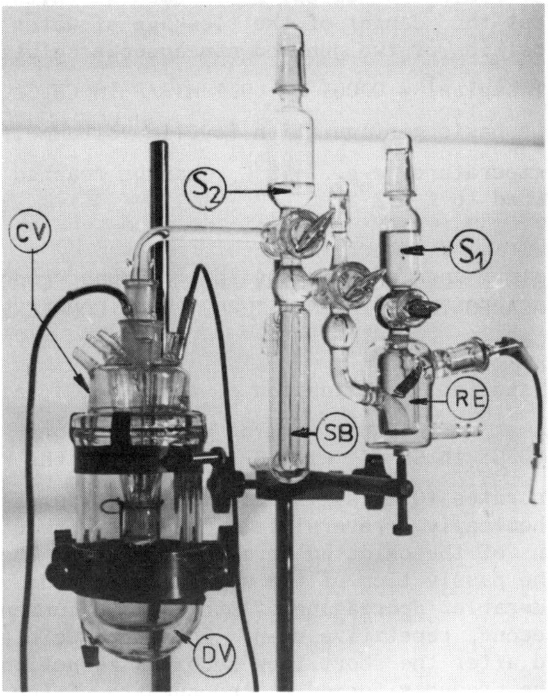

Figure 1. The variable temperature electrochemical cryostat equipped with non-aqueous reference half-cell with liquid junction. The configuration for low temperature cyclic voltammetry. CV - voltammetric cell; DV - Dewar vessel; SB - thermostated salt bridge compartment; RE - thermostated reference electrode compartment (Ag/AgCl); S_1- reference electrode solution (LiClsatd. + 0.1M TBAP in CH_2Cl_2); S_2- salt bridge solution (0.1M TBAP in CH_2Cl_2).

procedure may cause the inevitable leakage of H_2O, Cl^- or Ag^+ to a sample solution. Every frit increases the ohmic drop of the electrochemical cell. In consequence, it could result in the apparent irreversibility of the electrode process, or in the extreme case, in the complete change of the electrode reaction mechanism (*e.g.*, radical cation scavenging by nucleophilic solvent impurities).

In the here reported electrochemical cryostat, Figure 1, we have applied the potentiometric, somewhat modified Kawai's reference electrode with J-shaped liquid junction end. This construction minimizes the ohmic drop (no fritted disks are used). Application of the reference electrode with liquid junction allows to carry out the micro-bulk electrolysis and spectroelectrochemical measurements (with sample volume *ca.* 0.2 mL) without the danger of the leakage of water, Cl^- or Ag^+, because the composition of two juncted non-aqueous solutions is practically the same ([complex] ≃ 0.001 M + 0.1 MTBAP in CH_2Cl_2 || 0.1 M TBAP in CH_2Cl_2). In its basic configuration (cyclic voltammetry, Figure 1) a a steady state temperature, *e.g.*, -75°C, can be reached within few minutes and maintained to ± 0.2°C.

3.2. LOW TEMPERATURE ELECTROCHEMISTRY AND SPECTROELECTROCHEMISTRY OF TRANS-TETRACARBONYLBIS(ALKENE) TUNGSTEN(0) COMPLEXES

Cyclic voltammograms for the oxidation of *trans*-$W(CO)_4(alkene)_2$ complexes were measured at a Pt-planar working electrode, over the temperature range 20 to -80°C, in CH_2Cl_2 with 0.1 M TBAP. At the room temperatures and slow scan rates (0.2V/s) the electrooxidation of the complexes in question is chemically irreversible (no reverse-reduction current); The decomposition of the oxidated form is observed. The decomposition products cause the passivation of the working electrode. As a consequence, the considerable decreasing of the anodic current is observed, already in the second, repetitive scan. The reproducible voltammograms could be obtained after the short (few seconds) reduction (at negative potentials) of the products formed at the surface of the platinum electrode (Figure 2 and 4).

It is important to note that the scan rate increase and/or temperature decrease leads to the formation of the additional (chemically irreversible) peak (Figures 2, 4; Table 1). Both - the scan rate increase (at the constant temperature) and the temperature decrease (at the constant scan rate) is indicative of the chemically irreversible character of the first electrode process and chemically reversible nature of the second electrode process (at low temperatures); However, it would not result in the complete separation of the overlapping peaks (under scan rate studied and temperature range accessible). It should be also nonoted that at the scan rate increase and/or temperature decrease the the first peak would grow but never over the height of the second one.

TABLE 1. Summary of IR, ESR and CV data for trans-$W(CO)_4(alkene)_2$ complexes.

trans-$W(CO)_4L_2$ parent complex	IR $\nu(CO)$ [cm^{-1}]	ESR (after electrooxidation of parent complex) g_1	g_2	g_3
L = 1-pentene	1943	2.254	2.137	1.889
L = 1-hexene	1947	2.258	2.143	1.896
L = cyclopentene	1920	g = 2.070 (in solution)		
L = cyclohexene	1930	2.288	2.164	1.874
L = cycloheptene	1940	2.255	2.139	1.874
L = cyclooctene	1940	2.256	2.154	1.894

CYCLIC VOLTAMMETRY E_{p_a} vs Ag/AgCl, [mV]	
trans-$W(CO)_4(1\text{-pentene})_2$	$E_{p_a}^{trans\ 0/+1} \simeq 1056^{a,c}$
	$E_{p_a}^{cis\ 0/1+} = 1173^{a,c}$
trans-$W(CO)_4(1\text{-hexene})_2$	$E_{p_a}^{trans\ 0/+1} = 955^{b,c}$
	$E_{p_a}^{cis\ 0/+1} = 1085^{b,c}$

a: 10 V s^{-1}; b: 0.2 V s^{-1}; c: -25°C.

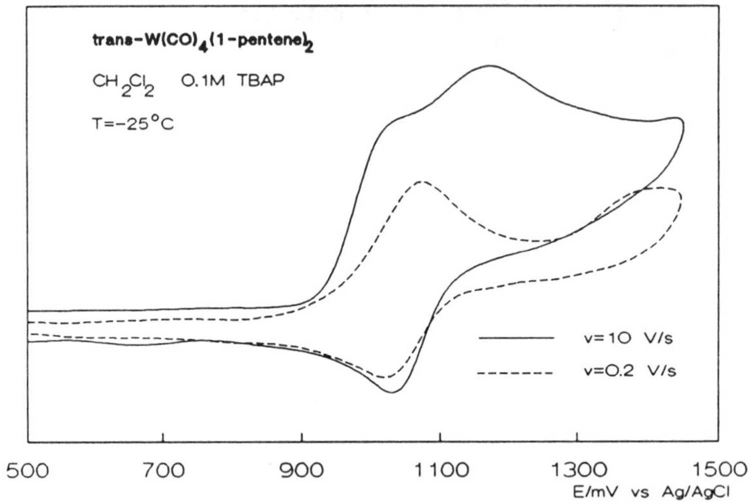

Figure 2. Cyclic voltammograms of *trans*-$W(CO)_4$(1-pentene)$_2$ in CH_2Cl_2 containing 0.1 M TBAP at platinum electrode and -25°C. Scan rate: .... 0.05, - - - 0.2, —— 0.4 V/s.

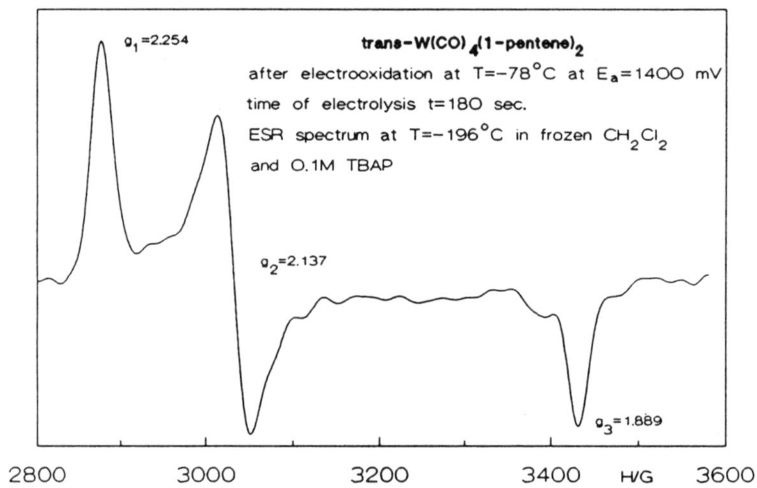

Figure 3. X-band ESR spectrum of frozen solution obtained after *in situ* micro-bulk electrolysis (at E_a = 1.4 V and -78°C) of the parent *trans*-$W(CO)_4$(1-pentene)$_2$ complex in CH_2Cl_2 containing 0.1 M TBAP. Electrolysis time: 180 s.

The micro-bulk (V_{sample} = 0.2 mL) low temperature (up to -80°C) electrolyses revealed the pink colour of the oxidation products of the parent complexes with the cyclic alkenes (except of cyclohexene complex) and the yellow colour of those with the chain alkenes. These measurements allowed the proper choise of temperature and electrooxidation time for the ESR spectroelectrotrochemical studies of the particular complexes. The products of electrochemical oxidation of the parent complexes (Table 1) were investigated mainly as a frozen CH_2Cl_2 solution. The main results are illustrated in Figures 3, 5 and 6. The characteristic feature of these spectra are three g values close to 2, without the hyperfine coupling, because of the spin-lattice relaxation (Table 1). Roughly, these spectra could be regarded as the results of the rhombic distortion of the octahedral tungsten(+1) compounds, which can be expected for *cis* configuration of the 17-electron $[W(CO)_4(alkene)_2]^+$species [10]. The deviation from the octahedral symmetry is much higher for *cis*-$M(CO)_4L_2$, (C_{2v}), than for *trans*-$M(CO)_4L_2$ one; For the tetragonally distorted D_{4h} symmetry of the *trans*-$[M(CO)_4L_2]^+$ cation, two g values and $g_\perp > g_\parallel$ should be expected [11].

A consideration of the ESR and cyclic voltammetric results suggests that: (i) on the longer time scale of spectroelectrochemistry the *cis*-$[W(CO)_4(alkene)_2]^+$ cation is detected; (ii) on the short time scale of cyclic voltammetry the formation of unstable, *trans*-$[W(CO)_4(alkene)_2]^+$, species is also observed, (iii) the first peak corresponds to $\mathbf{trans}^{0/+1}$ oxidation and the second one (at more positive potential) to $\mathbf{cis}^{0/+1}$, (iv) the isomerization reaction of *trans*$^+$- to *cis*$^+$-species proceeds rapidly.

The electrochemical behaviour of the here reported complexes is analogous to that of the earlier investigated dicarbene tungsten (0) one [12]. The basic feature of the alkenes complexes is the smaller difference in values of the standard potentials, $E^o_{cis} - E^o_{trans}$, and high instablity of the 17-electron organometallic radicals. The identification of the latter is important for catalysis and thus is widely investigated [13-16].

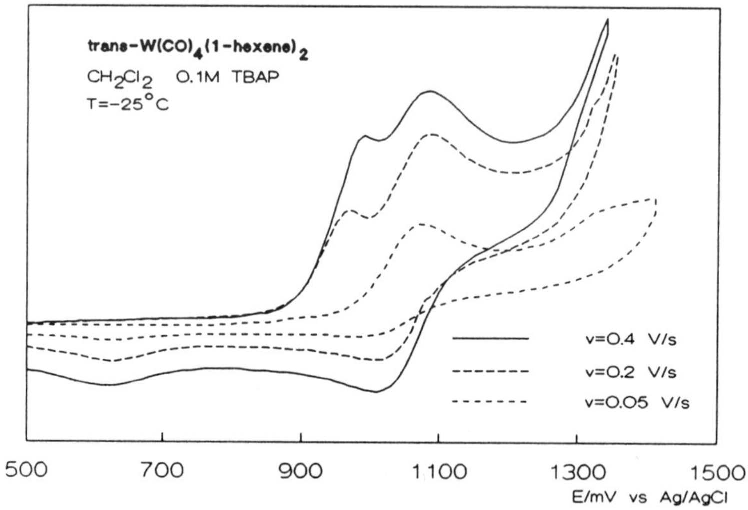

Figure 4. Cyclic voltammograms of *trans*-$W(CO)_4(1\text{-hexene})_2$ in CH_2Cl_2 containing 0.1 M TBAP at platinum electrode at -25°C. Scan rate: ... 0.05, - - - 0.2, —— 0.4 V/s.

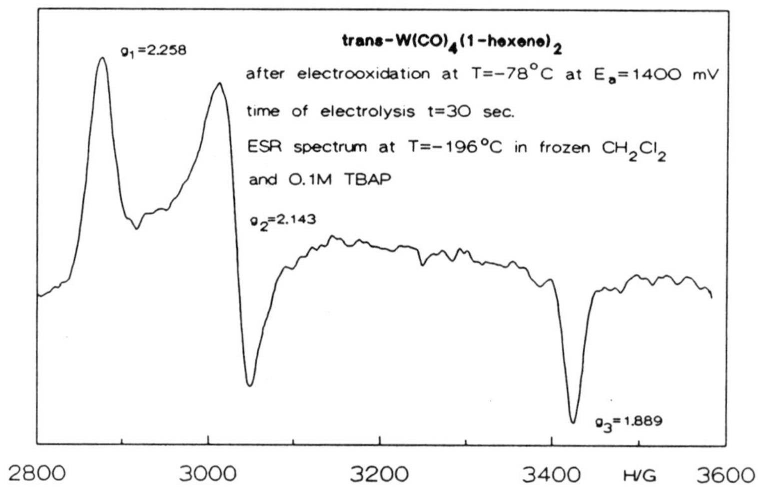

Figure 5. X-band ESR spectrum of frozen solution obtained after *in situ* micro-bulk electrolysis (at E_a = 1.4 V and -78°C) of the parent *trans*-$W(CO)_4(1\text{-hexene})_2$ complex in CH_2Cl_2 containing 0.1 M TBAP. Electrolysis time: 30s.

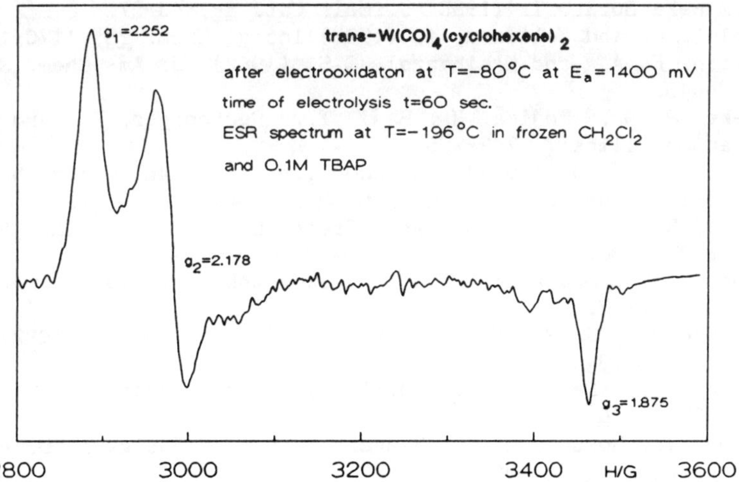

Figure 6. X-band ESR spectrum of frozen solution obtained after *in situ* micro-bulk electrolysis (at E_a = 1.4 V and -80°C) of the parent trans--W(CO)$_4$(cyclohexene)$_2$ starting complex in CH_2Cl_2 containing 0.1 M TBAP. Electrolysis time: 60 s.

Acknowledgement

This work was supported by the Committe for Scientific Research (Poland) under grant no KBN 207359101.

References

1. Wilgocki, M. (1985) J. Coord. Chem. 14, 39-47.
2. Wilgocki, M. (1985) J. Coord. Chem. 14, 151-162.
3. Wilgocki, M. (1988) J. Coord. Chem. 16, 357-374.
4. Tsukuda, H., Kawai, T., Maeda, M. and Ohtaki, H. (1975) Bull. Chem. Soc. Japan 48, 691-695.
5. Wilgocki, M., Szymańska-Buzar, T., Jaroszewski, M. and J. J. Ziółkowski (1987) Proc. 2nd Beijing Conf. and Exhib. on Instrum. Analysis 1229-1230.
6. Salomon, R. G. (1988) Tetrahedron 39, 485-575.

7. Szymańska-Buzar, T. (1988) J. Mol. Cat. 48, 43-57.
8. Daniel, C. and Veillard, A. (1988) Inorg. Chem. 28, 1170-1173.
9. Cotton, F. A. and Kraihanzel, C.S. (1962) J. Am. Chem. Soc. 84, 4432-4438.
10. Rieke, R. D., Kojima, H., Saji, T., Rechberger, P. and Öfele K. Organometallics 7, 749-755.
11. Ballhausen, C.J. (1962) Introduction to Ligand Field Theory, Mc Graw-Hill Book Company, Inc., p. 99.
12. Rieke, R. D., Kojima, H. and Öfele, K. (1976) J. Am. Chem. Soc. 98, 6735-6736.
13. Bond, A. M., Grabarić, B. S. and Jackowski, J. (1978) Inorg. Chem. 11, 2153-2157.
14. Hershberger, J. W., Klinger, R. J. and Kochi, J. K. (1982) J. Am. Chem. Soc. 104, 3034-3043.
15. Vallat, A., Person, M., Roullier, L. and Laviron, E. (1987) Inorg. Chem. 26, 332-335.
16. Zhang, Y., Gosser, D. K., Rieger, P. H. and Sweigart, D. A. (1991) Inorg. Chem. 113, 4062-4068.

ELECTROCHEMISTRY AND SPECTROELECTROCHEMISTRY OF DINUCLEAR RUTHENIUM AND OSMIUM COMPLEXES.

R. HAGE[a], J.G. HAASNOOT[b], J. REEDIJK[b] and J. G. VOS[c],
[a]*Unilever Research Laboratory, Olivier van Noortlaan 120, 3133 AT, Vlaardingen, The Netherlands*
[b]*Department of Chemistry, Gorlaeus Laboratories, Leiden University, P.O. Box 9502, 2300 RA, The Netherlands*
[c]*School of Chemical Sciences, Dublin City University, Dublin 9, Ireland*

ABSTRACT. In this contribution the electrochemical properties of a series of dinuclear ruthenium compounds based on the bridging ligands 3,5-bis(pyridin-2-yl)-1,2,4-triazole (Hbpt) and 3,5-bis(pyrazin-2-yl)-1,2,4-triazole (Hbpzt) are discussed. The redox potentials observed for the metal-based oxidation in these compounds suggest that relatively strong interaction exists between the two metal centres. Using spectroelectrochemical techniques, the intervalence transitions of the mixed-valence dinuclear compounds have been studied. The extent of electron delocalisation, expressed as the parameter α^2, is roughly the same for both ligands, with a value of 0.016 for the bpt^- and of 0.019 for the $bpzt^-$ mixed-valence compounds. This suggests strongly that for these compounds the HOMO of the bridging ligand and not the LUMO dictates the electronic interaction between the metal centres.

1. Introduction.

In the last few years there have been many studies of the electrochemical properties of ruthenium and osmium polypyridyl complexes.[1] The compounds have attracted much attention because of their potential application as photo- and redox- catalysts and the reversibility of the electrochemical processes greatly facilitates these investigations.[2] Because of the availability of a well-documented synthetic chemistry many compounds containing different ligands have been reported. Two types of redox processes are normally observed in these compounds: a metal-based oxidation and ligand-based reductions. Both processes are greatly affected by the nature of the ligands present.[1] In this contribution the electrochemical properties of a series of mononuclear and dinuclear ruthenium and osmium bis(2,2'-bipyridyl) complexes containing substituted triazoles as ligands are discussed. The ligands investigated are 3,5-bis(pyridin-2-yl)-1,2,4-triazole (Hbpt) and 3,5-bis(pyrazin-2-yl)-1,2,4-triazole (Hbpzt), see Figure 1. For the dinuclear

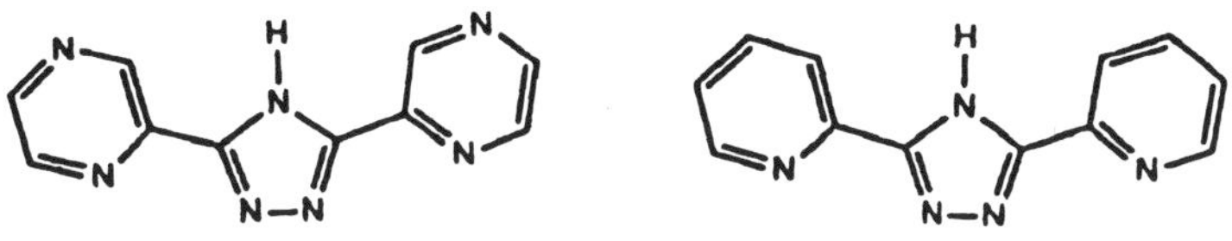

Figure 1. Structure of the ligands Hbpzt (left) and Hbpt(right)

A. J. L. Pombeiro and J. A. McCleverty (eds.),
Molecular Electrochemistry of Inorganic, Bioinorganic and Organomettallic Compounds, 583–588.

compounds the metal-based reduction processes are evaluated in order to estimate the amount of interaction between the metal centres. Spectroelectrochemical techniques have been used to investigate the properties of the mixed-valence compounds.

2. Discussion

2.1 GENERAL

An important feature of these ligands is that as bridging ligands they are asymmetric. Detailed investigations on the related 3-(pyridin-2-yl)-1,2,4-triazole compounds have shown that the N2 atom of the triazole ring has substantially different electronic properties from the N2 site. This is particularly clear from the electrochemical and acid-base properties of these compounds.[3,4] As a result the two metal centres in a dinuclear compound are not in the same environment. The preparation of the complexes and the electrochemical instrumentation used have been reported in previous publications.[5-7] The compounds discussed are of the types $[M(bpy)_2(BL)]^+$ and $[(M(bpy)_2)_2(BL)]^{3+}$, where bpy = 2,2'-bipyridyl, BL = Hbpt ot Hbpzt and M = Ru or Os. The compounds are abbreviated as Ru(bpt), Os(bpzt) etc, and Ru(bpt)Ru and Ru(bpt)Os etc. Because of the asymmetry of the bridging ligands the complex Ru(bpt)Os is different from Os(bpt)Ru. The mononuclear complexes containing the protonated triazole ligands are abbreviated as Ru(Hbpt) etc. The abbreviations given above indicate that both triazole ligands deprotonate upon coordination. The structure of the compound Ru(bpzt)Ru is given in Figure 2.

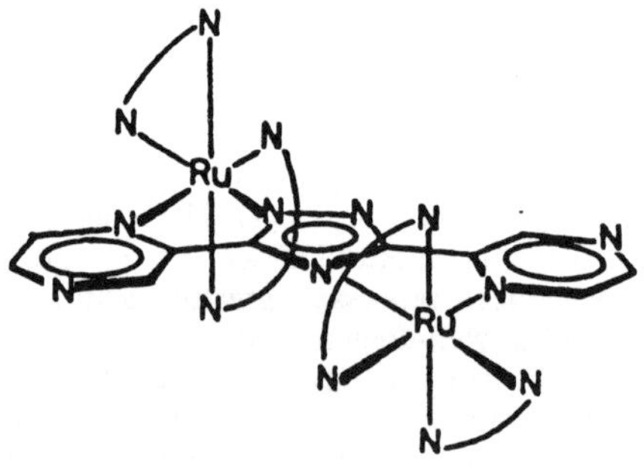

Figure 2. Structure of Ru(bpzt)Ru

2.2 ELECTROCHEMICAL PROPERTIES

A list of the metal-based M(II/III) oxidation potentials and of the ligand-based reduction potentials is given in Table 1. Electrochemical data for the complexes containing the protonated triazole ligands were obtained by the addition of acid to the measuring solution. For this reason the ligand based reduction potentials could not be measured. A typical set of data including a cyclic voltammogram for the metal-based oxidation and a differential pulse polarogram for the ligand-based reductions are shown in Figure 3.

As expected the oxidation potentials of the osmium compounds are 0.4 V lower than those observed for the corresponding ruthenium compounds. The data in Table 1

TABLE 1. Electrochemical data (V vs SCE) between +2.0 and -2.0 V for the Hbpt and Hbpzt complexes as obtained in acetonitrile containing 0.1 M tetrabutylammoniumperchlorate.

Compound	M(II/III)		ligand based reduction potentials				
Ru(bpt)	0.85			-1.47		-1.72	
Ru(Hbpt)	1.06						
Os(bpt)	0.49			-1.41		-1.69	
Os(Hbpt)	0.80						
Ru(bpzt)	0.99			-1.42		-1.62	-1.96
Ru(Hbpzt)	1.24						
Ru(bpt)Ru	1.04	1.34		-1.40		-1.62	-1.67
Ru(bpt)Os	0.73	1.20		-1.33	-1.41	-1.59	-1.69
Os(bpt)Ru	0.65	1.30		-1.36		-1.63	
Os(bpt)Os	0.64	0.85		-1.34		-1.57	
Ru(bpzt)Ru	1.16	1.46	-1.26	-1.39	-1.55	-1.63	-1.85

further show that protonation has the same effect on the first metal-based oxidation as does the formation of a dinuclear species. This suggests that the observed increase in the first oxidation potential upon the formation of the dinuclear species can be attributed to the decrease in electron density on the bridging ligand. Another important observation is that the difference between the two metal-based oxidations in the dinuclear compounds is rather large.

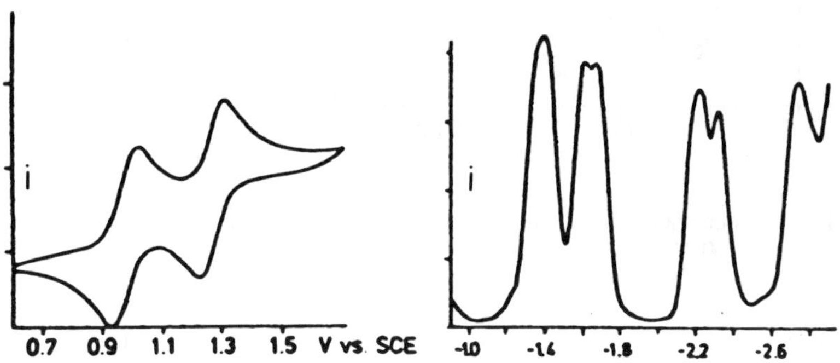

Figure 3. Cyclic voltammogram of the metal-based oxidation (left) and differential pulse polarogram of the ligand-based reductions for Ru(bpt)Ru

This large difference can only partly be explained by the asymmetry in the ligand. This asymmetry only results in a maximal difference of about 100 mV (see redox data for the mixed metal compounds). The values obtained therefore suggest a substantial interaction between the two metal centres in the dinuclear compounds.

The values obtained for the ligand-based reductions suggest strongly that these processes are bpy-based for most compounds. For the bpzt-based dinuclear compounds the first reduction appears to be based on the bridging ligand. This is indeed confirmed by resonance Raman studies.[7] This difference in location is most likely caused by the presence of a pyrazine ring in the latter compounds.

2.3 MIXED-VALENCE COMPOUNDS

The electron delocalisation present in the dinuclear compounds can be investigated by monitoring the intervalence band in the mixed-valence compounds. Since there is a rather large difference between the two metal-based oxidation potentials, the mixed-valence compounds can be prepared readily *in situ*. The data given in Table 2 were therefore obtained using standard spectroelectrochemical techniques. Figure 4 gives some typical intervalence spectra. The bands observed at about 1900 cm^{-1} in this figure have been assigned to Os(III)-based dπ-dπ transitions. In the discussion given below a correction has been applied to the characteristics of the intervalence band to allow for the presence of these bands.

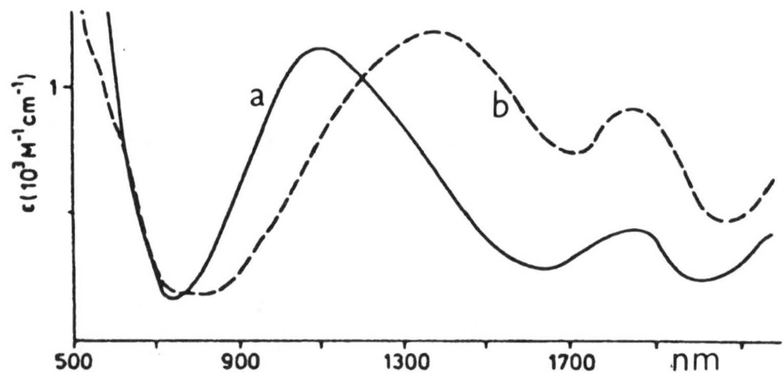

Figure 4. Intervalance bands obtained for Os(bpt)Ru (a) and Ru(bpt)Os (b) in acetonitrile after oxidation at 1.00 V.

The electron delocalisation parameter α^2 can be obtained from equation 1, where E_{op} is the energy of the intervalence band in cm^{-1}, ε its extinction coefficient, and $\Delta\nu_{1/2}$ the width at half height of the same transition.

$$\alpha^2 = \frac{(4.2 \times 10^{-4})\varepsilon_{max}\,\Delta\nu_{1/2}}{d^2 E_{op}} \qquad 1$$

where d is the distance between the metal centres which for the compound Ru(bpt)Ru has been determined as 6.184 Å.[6] The rate constant for thermally-induced electron transfer, k_{th}, can be evaluated from the characteristics of the intervalence band using equations 2 and 3 [8-10]

$$E_{th} = \frac{(E_{op})^2}{4(E_{op} - \delta E)} - H_{AB} \qquad 2$$

$$k_{th} = \upsilon_n \exp(-E_{th}/ kT) \qquad 3$$

where E_{th} is the activation barrier for electron transfer, δE a measure for the asymmetry in the system (estimated from the electrochemical data as 100 mV for the RuRu compounds 170 mV for the Ru(bpt)Os and 350 mV for the Os(bpt)Ru), H_{AB} the degree of electronic coupling (cm^{-1}) and υ_n the frequency factor taken as 5 x 10^{12} s^{-1}. As the errors in the values obtained for k_{th} are rather large (50%) they have to be considered as rough estimates only.

TABLE 2. Characteristics of the intervalence transition of the mixed-valence compounds. Solvent acetonitrile.

Compounds	E_{op}	ε_{max}	$\Delta\nu_{1/2}$	α^2	k_{th}
	cm^{-1}	$M^{-1}cm^{-1}$	cm^{-1}		s^{-1}
$Ru^{III}(bpt)Ru^{II}$	5556	2400	3300	0.016	$6x10^{10}$
$Ru^{III}(bpt)Os^{II}$	7143	1200	3800	0.007	$2x10^9$
$Os^{III}(bpt)Ru^{II}$	8772	1140	4300	0.006	$2x10^7$
$Os^{III}(bpt)Os^{II}$	8333	460	4500	0.003	$3x10^8$
	5000	440	3500	0.003	$3x10^{10}$
$Ru^{III}(bpzt)Ru^{II}$	5400	2200	4200	0.019	$27x10^{10}$

In all compounds the extent of electron delocalisation is rather large. For the osmium containing compounds electron delocalisation is smaller, most likely because of the reduced overlap for the Os-bpt orbitals. The data in Table 2 and Figure 4 clearly show that the intervalence bands of two mixed metal compounds possess quite different characteristics. This is a clear indication of the asymmetry in the bridging ligand. This asymmetry is further reflected in the different oxidation potentials observed for the two compounds (See Table 1) and in the values obtained for k_{th} (See Table 2).

The values obtained for k^{th} are quite different for the five complexes. As expected, complexes with a lower activation barrier yield higher rate constants. The results obtained for the mixed metal compounds suggest that small differences in the oxidation potentials of the compounds can have a significant effect on the rate of electron transfer. Compared with values obtained for other similar systems [10,12] the rates of electron transfer observed are rather high.

Another important point is the similarity of the α^2 values of the bpt and bpzt compounds. This similarity, and a comparison with the electrochemical and spectroelectrochemical and spectroscopic data of these compounds with those of other dinuclear ruthenium poly(pyridyl) compounds [2,11], suggest that in dinuclear species based on bpt and bpzt bridging ligands interaction is not through the π^*-orbital of the bridging ligand, but through the ground-state .

REFERENCES

1 Juris, A., Balzani, V., Barigelletti, F., Campagna, S., Belser, P. and Zelewsky, A. von (1988) Coord. Chem. Rev. 84, 85-277

2. Balzani, V. and Scandola, F. (1991) Supramolecular Photochemistry, Ellis Horwood, Chichester

3. Buchanan, B.E., Wang, R., Vos, J.G., Hage, R., Haasnoot, J.G. and Reedijk, J. (1990) Inorg. Chem. 29, 3263-3265

4. Buchanan, B.E., Vos, J.G., Kaneko, M., Putten, W.J.M. van der, Kelly, J.M., Hage, R., Graaff, R.A.G. de, Prins, R., Haasnoot, J.G. and Reedijk, J. (1990) J. Chem. Soc.,Dalton Trans. 2425-2431.

5. Hage, R., Dijkhuis, A.H.J., Haasnoot, J.G., Prins, R., Reedijk, J., Buchanan, B.E. and Vos, J.G. (1988) Inorg. Chem. 27, 2185-2189

6. Hage, R., Haasnoot, J.G., Nieuwenhuis, H.A., Reedijk, J., Ridder, D.J.A. de, and Vos, J.G. (1990) J. Am. Chem. Soc. 112, 9245-9251

7. Hage, R., Haasnoot, J.G., Reedijk, J., Wang, R. and Vos, J.G. (1991) Inorg. Chem. 30, 3269-3269

8. Hush, N.S. (1967) Prog. Inorg. Chem. 8, 391-444

9 Hush, N.S. (1968) Electrochim. Acta 13, 1005-1023

10. Creutz, C. (1983) Prog. Inorg. Chem. 30, 1-75

11. Ernst, E., Kasack, V., and Kaim, W. (1988) Inorg. Chem. 27, 1146-1148.

12. Curtis, J.C., Bernstein, J.S., and Meyer, T.J. (1985) Inorg. Chem. 24, 385-397.

THE OPTICAL SPECTRA AND VOLTAMMETRY OF MIXED HALIDE / NITRILE / CARBONYL OSMIUM(III) COMPLEXES. ADDITIVE AND NON-ADDITIVE EFFECTS

Graham A. HEATH and David G. HUMPHREY
Research School of Chemistry
Australian National University
G.P.O. Box 4
Canberra ACT 2601
Australia

Abstract

In the series $[OsX_6]^{3-}$,$[OsX_5(CO)]^{2-}$, *trans*-$[OsX_4(CO)_2]^{1-}$ (X=Br) and the series *trans*-$[OsX_4(MeCN)_2]^{1-}$, *trans*-$[OsX_4(MeCN)(CO)]^{1-}$, *trans*-$[OsX_4(CO)_2]^{1-}$, osmium IV/III and III/II redox couples shift two to three times further for the first introduction of CO than for the second, contrary to recent predictions; the corresponding halide to Os^{III} charge transfer spectra (X=Cl and Br) confirm the strongly non-linear accumulation of ligand electronic effects.

In recent times there has been an increased interest in the additive effect ligands have on electrochemical potentials. Linear correlations have been found to exist between redox potentials and stoichiometry, in a diverse range of organometallic and coordination compounds[1,2]. Fenske had earlier suggested that the oxidation potentials in the $[Mn(CO)_{6-n}(CNMe)_n]^{1+}$ series of complexes could be linearly correlated with the energy of the HOMO, as calculated by Fenske-Hall methods[3]. Bursten built upon these ideas, in a model for trends in the E^o for binary complexes $[ML_{6-n}L'_n]$ which focussed on the change in d_p orbital energy of the redox-active orbital through interaction with ligands L and L'[1]. Pickett *et al.* established the concept of ligand parameters, in order to quantify the overall donor capabilities of a range of ligands[4]. Lever recently used similar principles to define parameters for a large number of ligands, and using these parameters predicted the redox potentials for a wide and varied range of organometallic and coordination compounds[5,6]. Common to each of the above models is the assumption that the effects of the ligands are arithmetically additive, so that successive substitutions lead to equal shifts in electrode potentials.

With these models in mind, the attention of the reader is drawn to a seemingly simple series of complexes (see Table 1) recently studied in our laboratory. The relationship between $E_{1/2}$ and composition has been observed to vary in either a linear or non-linear fashion, depending upon the nature of the ligands occupying the axial sites of an octahedron.

A. J. L. Pombeiro and J. A. McCleverty (eds.),
Molecular Electrochemistry of Inorganic, Bioinorganic and Organomettallic Compounds, 589–593.

Table 1 *Summary of Electrode Potentials and Charge-Transfer Maxima for Monomeric Osmium Complexes.*

Complex[a]		$E_{1/2}$ (V)[b]		υ_{max}, (cm^{-1})[c]
		IV/III	III/II	$X(\pi)\rightarrow Os^{III}$ CT
$[OsBr_6]^{3-}$	(Cl)	-0.37 (-0.49)	-2.28 (*d*)	28 700 (35 020)
$[OsBr_5(MeCN)]^{2-}$	(Cl)	+0.25 (+0.21)	-1.45 (-1.47)	26 530 (32 720)
$[OsBr_4(MeCN)_2]^{1-}$	(Cl)	+0.82 (+0.79)	-0.77 (-1.00)	24 190 (30 060)
$[OsBr_5(CO)]^{2-}$	(Cl)	+1.16 (+1.20)	-0.36 (-0.60)	20 360 (29 640)
$[OsBr_4(CO)_2]^{1-}$	(Cl)	+1.84 (+1.64)	+0.27 (-0.10)	19 320 (26 660)
$[OsBr_4(CO)(MeCN)]^{1-}$	(Cl)	+1.53 (+1.59)	-0.08 (-0.15)	20 390 (26 480)
$[OsBr_5(CNBu^t)]^{2-}$	(Cl)	+0.50 (+0.37)	-0.95 (-1.00)	23 000 (29 520)
$[OsBr_4(CNBu^t)_2]^{1-}$	(Cl)	+0.94 (+1.01)	-0.66 (-0.69)	22 420 (28 570)
$[OsBr_4(CO)(CNBu^t)]^{1-}$	(Cl)	+1.47 (+1.55)	-0.07 (-0.19)	20 440 (26 480)

[a] *All complexes are listed as Os^{III} species; data in parentheses is for the corresponding chloro complex,* [b] *In 0.5 mol. dm^{-3} $^nBu_4NPF_6/CH_2Cl_2$ at 213 K, versus Ag/AgCl, against which $[Fe(C_5H_5)_2]^{+/0}$ occurs at +0.55 V,* [c] *For those complexes not isolated as Os^{III}, this species was electrogenerated in 0.5 mol. dm^{-3} $^nBu_4NPF_6/CH_2Cl_2$ in an OTTLE cell,* [d] *Not observed.*

The variation of $E_{1/2}$ (both $Os^{III/II}$ and $Os^{IV/III}$) with composition for the $[OsBr_{6-n}(MeCN)_n]^{z-}$ (n=0, 1, 2(*t*)) complexes is plotted in Figure 1 (a). This series of nitrile complexes shows essentially linear variation of $E_{1/2}$ with n. The $Os^{III/II}$ couple shifts by ~0.8 V at the first substitution of Br by MeCN, and then by ~0.7 V at the second substitution. The corresponding shifts in the $Os^{IV/III}$ couple are 0.62 and 0.57 V.

Correlations between electrochemical potentials and charge-transfer spectroscopic data have been previously noted by a number of authors[2] Figure 1 (b) maps the movement of the $Br(\pi)\rightarrow Os^{III}(d_\pi)$ charge-transfer maxima (υ_{max}) for the complexes $[OsBr_6]^{3-}$, $[OsBr_5(MeCN)]^{2-}$ and *trans*-$[OsBr_4(MeCN)_2]^{1-}$. The charge-transfer manifold clearly moves linearly with composition.

Treating the electrochemical and spectroscopic data for the halide/carbonyl and halide/isonitrile complexes in the same way as that of the halide/nitrile complexes, reveals distinctly different behaviour of $E_{1/2}$ and υ_{max} with composition.

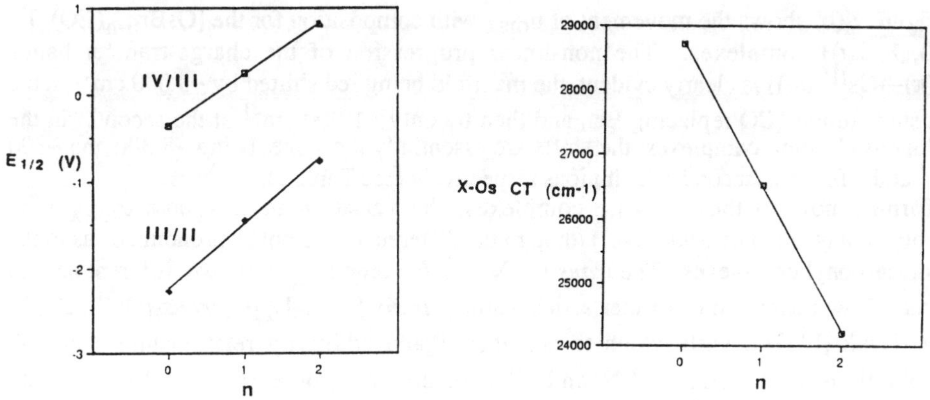

(a) (b)

Figure 1

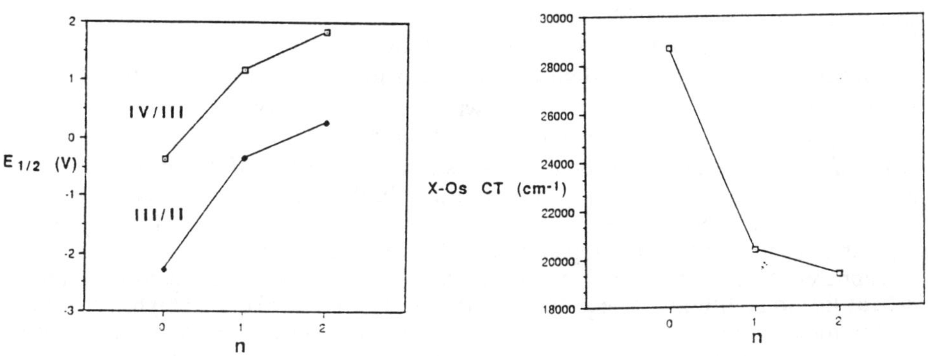

(a) (b)

Figure 2

A plot of $E_{1/2}$($Os^{III/II}$ and $Os^{IV/III}$) versus stoichiometry for the halide/carbonyl compounds $[OsBr_6]^{3-}$, $[OsBr_5(CO)]^{2-}$ and *trans*-$[OsBr_4(CO)_2]^{1-}$. is shown in Figure 2 (a). The $Os^{III/II}$ couple shifts by ~1.9 V upon the first substitution of bromide by carbonyl, and then by only ~0.6 V at the second substitution. This distinctly non-additive effect is also observed in the $Os^{IV/III}$ couple. Similar shifts in $E_{1/2}$ occur in the analogous chloride complexes. Such behaviour is in direct contrast to what is observed in the halide/nitrile complexes, and is *not predicted by current ligand additivity models.*

Figure 2 (b) shows the movement of u_{max} with composition for the $[OsBr_{6-n}(CO)_n]^{z-}$ (n=0, 1, 2(*t*)) complexes. The non-linear progression of the charge-transfer bands ($Br(\pi) \rightarrow Os^{III}(d_\pi)$) is clearly evident, the manifold being red shifted by ~8 000 cm^{-1} at the first substitution (CO replacing Br), and then by only ~1 000 cm^{-1} at the second. In the analogous chloride complexes, the shifts are essentially the same, being ~8 300 and ~300 cm^{-1} at the first and second substitutions respectively (see Table 1).

Turning now to the isonitrile complexes, the variation of $E_{1/2}$ and υ_{max} with composition is again non-additive, although the differences are not as pronounced as in the halide/carbonyl complexes. The *trans*-$[OsX_4LL']^{z-}$ complexes with two different neutral ligands allow formation of further series, namely *trans*-$[OsX_4L_2]^{z-}$, *trans*-$[OsX_4LL']^{z-}$, *trans*-$[OsX_4L'_2]^{z-}$, with which to consider ligand additivity relationships. For all combinations of L=CO, MeCN and $CNBu^t$, the variation of $E_{1/2}$ and υ_{max} with composition is again non-linear.

Electrochemical ($E_{1/2}$) and charge-transfer spectral data (υ_{max}) has been used to establish two distinctly different types of behaviour associated with the $\{OsX_4\}$ site, depending on the identity of the ligands L and L'. For the $[OsX_{6-n}(MeCN)_n]^z$ (X=Cl, Br; n=0, 1, 2) series of complexes, $E_{1/2}$ and υ_{max} vary in a *linear* manner with composition. However, for those complexes involving the replacement of halide by strong π-acid ligands, $[OsX_{6-n}(L)_n]^z$ (X=Cl, Br; n=0, 1, 2; L=CO, $CNBu^t$) $E_{1/2}$ and υ_{max} move in a distinctly *non-linear* fashion with composition.

This contrasting behaviour can be accounted for in terms of the differing π-donor/acceptor strengths of the ligands L and L', and the polarizability of the $\{OsX_4\}^{z-}$ site. Much of the previous work concerned with ligand additivity has involved complexes with strong π-acid ligands, often in conjunction with first row transition metals. The presence of strong π-acid ligands on the metal site attenuates the effects of other incoming ligands, through decreasing the site polarizability. In the case under discussion here, the π-donor halides contribute to the electron density on the metal centre. This, coupled with having a third row transition metal, produces a relatively electron rich site.

It is apparent that certain combinations of ligands and metals do not comply with recent models of ligand additivity. The purpose of this discussion has been to highlight some of the limitations of these models, which in turn can only lead to the improvement of our understanding of a concept as fundamental to inorganic chemists, as ligand additivity.

References

1. Bursten, B.E. and Green, M.R. (1988), 'Ligand Additivity in the Vibrational Spectroscopy, Electrochemistry, and Photoelectron Spectroscopy of Metal Carbonyl Derivatives', *Prog. Inorg. Chem.*, **36**, 393-485.
2. Duff, C. and Heath, G.A. (1991), 'Stepwise Ligand-Additivity Effects on Electrode Potentials and Charge-Transfer Spectra in Hexahalide, Mixed Halide/Nitrile, and Hexakis(nitrile) Complexes of Ruthenium (IV), -(III), and-(II)', *Inorg. Chem.*, **30**, 2528-2535, and references therein.
3. Fenske, R.F. and Sarapu, A.C. (1975), 'The Transition Metal-Isocyanide Bond. An Approximate Molecular Orbital Study', *Inorg. Chem.*, **14**, 247-253.

4. Chatt, J., Kan, C.T., Leigh, G.J., Pickett, C.J., and Stanley, D.R. (1980), 'Transition-metal Binding Sites and Ligand Parameters', *J. Chem. Soc., Dalton Trans.*, 2032-2038.
5. Lever, A.P.B. (1990), 'Electrochemical Parametrization of Metal Complex Redox Potentials, Using the Ruthenium(III)/Ruthenium(II) Couple To Generate a Ligand Electrochemical Series', *Inorg. Chem.*, **29**, 1271-1285.
6. Lever, A.P.B. (1990), 'Electrochemical Parametrization of Rhenium Redox Couples', *Inorg. Chem.*, **30**, 1980-1985.

BUCKMINSTERFULLERENE AND ITS ELECTROGENERATED ANIONS C_{60}^{1-}, C_{60}^{2-} AND C_{60}^{3-}

G. A. HEATH[a], J. E. McGRADY[a] and R. L. MARTIN[b]
[a]*Research School of Chemistry, The Australian National University*
[b]*Chemistry Department, Monash University, Australia*

Despite the fundamental difference between equilibrated and spectroscopic data, there is renewed interest in the reconciliation of electrode potentials and optical transition energies to map the valence levels of molecules[1].
It has been shown that the transition energy, hν for first optical transition can be approximated by the relationship

$$h\nu = (1 + \rho)\Delta E$$

where ρ is a vibronic coupling constant, typically 0.3 or less, which reflects the internal and external reorganisation processes associated with electron transfer[2], and ΔE is the difference between oxidation and reduction potentials. Buckminsterfullerene, C_{60}, provides an interesting test of this relationship, as both internal and external reorganisation is likely to be minimal in such a rigid, non polar molecule.

We combine extensive spectroscopic data for the series C_{60}^{n-}, (n=0,1,2,3) in conjunction with electrochemical data and Molecular Orbital calculations performed within the Multiple Scattering X-Alpha (MS-Xα) approximation[3], to provide a self consistent map of the valence energy levels of C_{60} and its derivatives.
The species C_{60}^{n-} (n=0,1,2,3) have been electrogenerated in 0.5M $Bu^n{}_4NBF_4$ in CH_2Cl_2 solution at -60°C The UV and near-IR Spectra were recorded (in the range 5000cm^{-1} to 50000cm^{-1}) in an Optically Transparent Thin Layer Electrochemical (OTTLE) cell.
The spectra are presented in Figure 1

The valence level M.O. diagram for C_{60}, as calculated by the MS X-α method is presented in figure 2. Bond lengths used to generate the structure were those of Lüthi and Almlöf[4].
C_{60} is idiosyncratic in that both the HOMO ($4h_u$) and LUMO ($5t_{1u}$), and so the HOMO-LUMO transition is forbidden. However, in concentrated solution, the onset of absortion can be detected at 16300 cm^{-1} (see figure 1).
Scheme 1 illustrates the relationship between electrode potentials and spectroscopic data for C_{60}.

A. J. L. Pombeiro and J. A. McCleverty (eds.),
Molecular Electrochemistry of Inorganic, Bioinorganic and Organomettallic Compounds, 595–600.

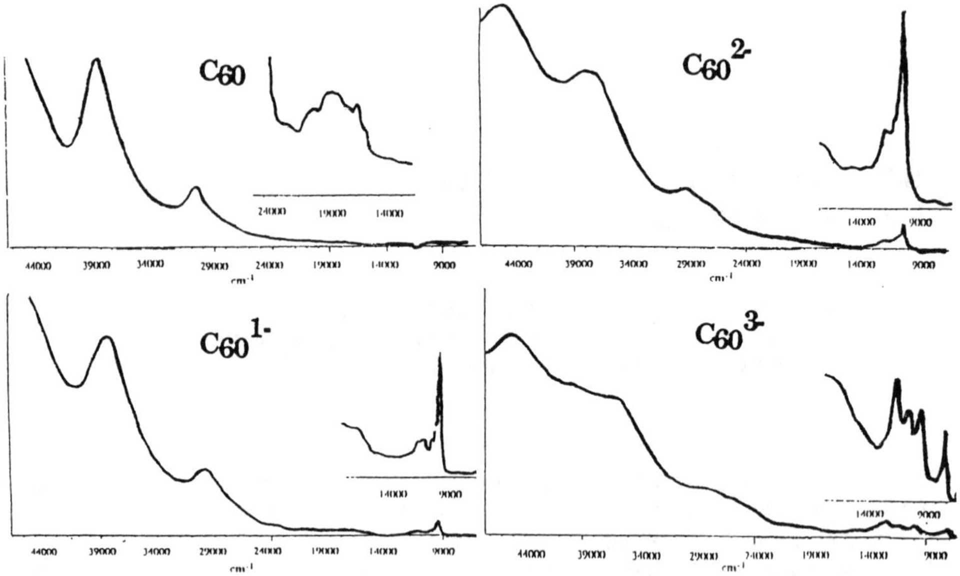

Figure 1. UV and Near-IR spectra of C_{60} and its derivatives C_{60}^{1-}, C_{60}^{2-} and C_{60}^{3-}.

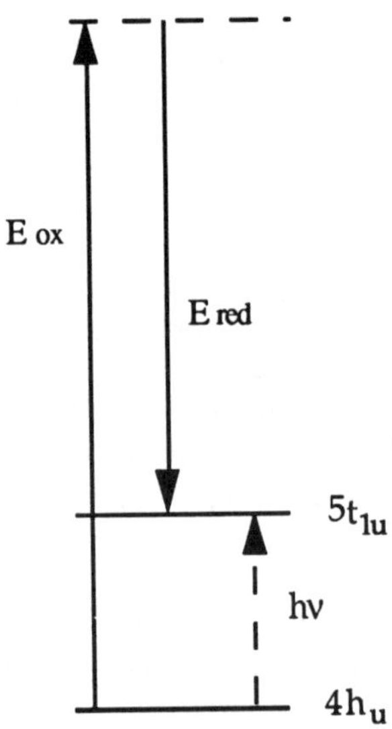

Eox = 1.76V v SCE[5]
Ered = -0.45V v SCE
Therefore ΔE = 2.2 V
hν = 16300cm^{-1} = 2.0. eV

As suggested earlier, ρ is small because internal and external reorganisation energies are small.

Scheme 1. Relathionship between electrochemical and spectroscopic data for C_{60}.

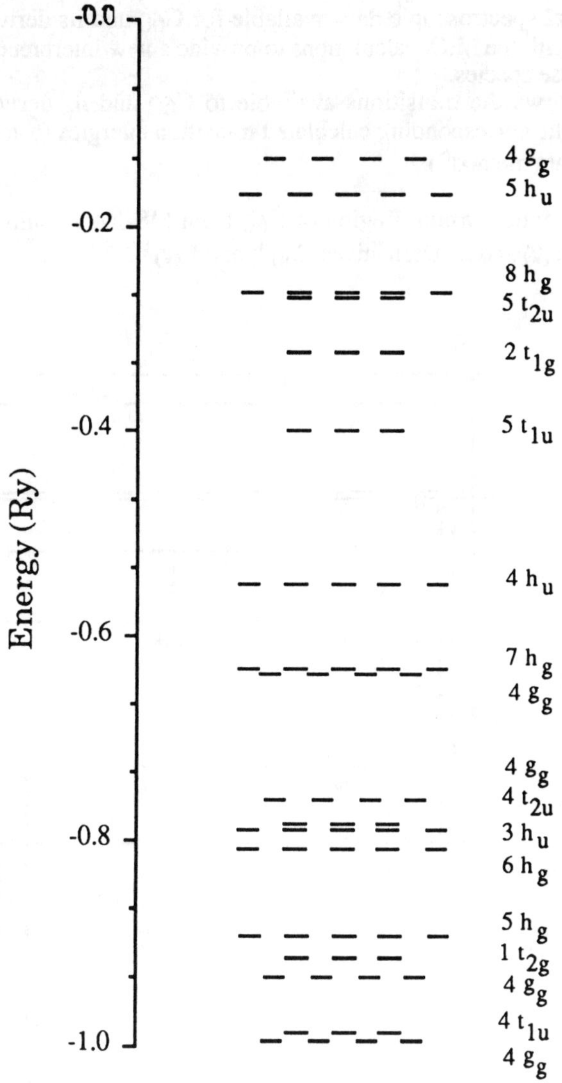

Figure 2. Valence level M.O. diagram for C_{60}.

The additional spectroscopic data available for C_{60} and its derivatives can be taken in conjunction with the M.O. calculations to provide a new interpretation of the broad range spectra of these species.
Scheme 2 shows the transitions available to C_{60} and its derivatives, while Table 1 summarises the corresponding calculated transition energies (calculated using the Slater Transition State method[6]).

Scheme 2 Frontier Orbital Region of C_{60} from MS-Xα Results, Showing Transitions Available to C_{60} and its Derivatives C_{60}^{n-} and C_{60}^{1+}.

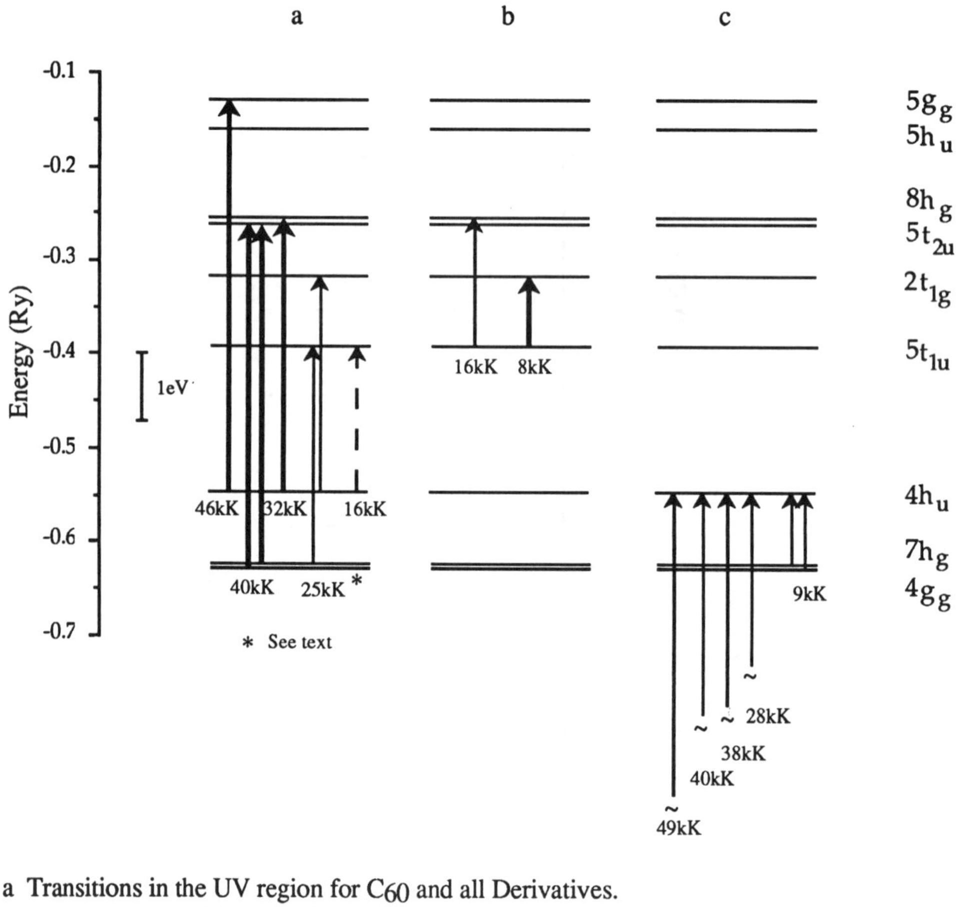

a Transitions in the UV region for C_{60} and all Derivatives.
b Additional Transitions for C_{60}^{n-} from the Partially Occupied $5t_{1u}$ Orbital.
c Additional Transitions for C_{60}^{1+} into the Partially Occupied $4h_u$ Orbital.
Bold lines indicate intense transitions.
Dashed line indicates forbidden transition.

Table 1 Calculated Transition Energies of C_{60} and Derivatives

	C_{60}^{+}	C_{60}	C_{60}^{1-}	C_{60}^{2-}	C_{60}^{3-}
$4h_u \rightarrow 5t_{1u}$ *	16500	16500	16600	16600	16700
$4h_u \rightarrow 2t_{1g}$ #	24900	24900	24900	24800	24800
$7h_g \rightarrow 5t_{1u}$ #	25500	25500	25600	25600	25700
$4h_u \rightarrow 8h_g$	31600	31700	31700	31800	31900
$7h_g \rightarrow 5t_{2u}$	39800	40000	40100	40200	40300
$4g_g \rightarrow 5t_{2u}$	40300	40400	40500	40700	40900
$4h_u \rightarrow 5g_g$	45300	45500	45800	46000	46200
$5t_{1u} \rightarrow 2t_{1g}$	-	-	8300	8200	8100
$5t_{1u} \rightarrow 8h_g$	-	-	15200	15200	15200

* Forbidden Transition.

\# See text.

On the basis of the calculated transition energies, we propose the following assignments of the bands in the spectrum of C_{60} and its anions.

$46000 cm^{-1}$	$4h_u \rightarrow 5g_g$
$40000 cm^{-1}$	$4g_g \rightarrow 5t_{2u}$
$32000 cm^{-1}$	$4h_u \rightarrow 8h_g$
$16300 cm^{-1}$	$4h_u \rightarrow 5t_{1u}$
$9300 cm^{-1}$	$5t_{1u} \rightarrow 2t_{1g}$ (anions only)

We suggest that the absence of any feature at $25000 cm^{-1}$, corresponding to the two predicted transitions ($4h_u \rightarrow 2t_{1g}$ and $7h_g \rightarrow 5t_{1u}$) is due to mutual configuration interaction between the two states[7]. The result of this configuration interaction is to symmetrically displace the two bands about their energetic barycentre, with concomitant intensity stealing, such that the band at higher energy gains intensity at the expense of that at lower energy[8]. We propose that the weak band lies at around $18000 cm^{-1}$, while the strong band is displaced symmetrically about $25000 cm^{-1}$, and lies coincident with the band at $32000 cm^{-1}$.

Reference to Scheme 2c indicates that near-IR features are expected in the spectrum of C_{60}^{1+}, due to $7h_g \rightarrow 4h_u$ and $4g_g \rightarrow 4h_u$ transitions. These transitions are calculated to occur at $9010 cm^{-1}$ and $9050 cm^{-1}$ respectively. The near-IR spectrum of C_{60}^{1+} has been reported in matrix isolation[9], and a band at $10200 cm^{-1}$ has been assigned to the $7h_g \rightarrow 4h_u$ transition, in reasonable accord with our calculated transition energy.

The observation of this $7h_g \rightarrow 4h_u$ band in C_{60}^{1+} means that now the $7h_g \rightarrow 4h_u$, $4h_u \rightarrow 5t_{1u}$ and $5t_{1u} \rightarrow 2t_{1g}$ transitions have all been observed experimentally, in C_{60}^{1+}, C_{60}^{0} and C_{60}^{1-} respectively. This enables us to map the frontier orbital domain of C_{60} using entirely experimental results, as illustrated in Scheme 3.

Scheme 3 Frontier Orbital Region of C_{60} Mapped by Experimental Results.

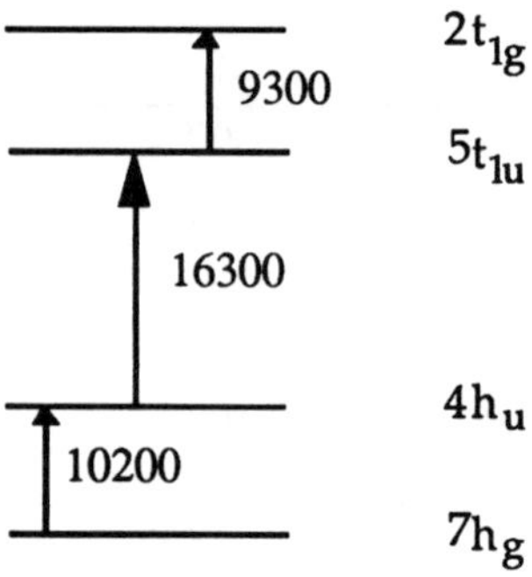

This scheme is in excellent quantitative agreement with the theoretical picture illustrated by Scheme 2(a, b and c), and gives strong experimental evidence that the $4h_u \rightarrow 2t_{1g}$ and $7h_g \rightarrow 5t_{1u}$ separations lie in the region of 25000cm^{-1}, and therefore the absence of a feature in this region must be due to configuration interaction.
We have illustrated the utility of combining electrochemical, spectroscopic and theoretical data in mapping the frontier orbitals of C_{60} , and in doing so provided a new and self consistent assignment of the broad range spectrum of C_{60}.

Acknowledgements.

We would like to thank Dr. A.M. Vassallo for supply of a sample[10] of C_{60}, and Dr. R.G. Raptis for technical assistance with the measurement of spectra. JEM would like to thank the SERC for an overseas studentship.

References.

1 a)Lever, A.B.P., *Inorganic electronic spectroscopy*, 2nd ed.; Elsevier: New York 1986: p776.
b)Duff, C.M. and Heath, G.A., *Inorg. Chem.*, 1991, 30, 2528.
2 Hush, N.S., *Electrochim. Acta* 1968, 13, 1005.
3 a)Case, D.A., *Ann. Rev. Phys. Chem.*, 1982, 33, 151.
b)Cook, M. and Case, D.A., *QCPE* 1982, No. 465.
4 Lüthi, H.P., and Almlöf, J., *Chem. Phys. Lett.*, 1987, 137, 501.
5 Dubois, D., Kadish, K.M., Flanagan, S. and Wilson, L.J., *J. Am. Chem. Soc.*, 1991, 113, 7773
6 Slater, J.C., *Adv. Quantum Chem.*, 1972, 6,1.
7 Heath, G.A., McGrady, J.E. and Martin, R.L., submitted to *J. Chem. Soc. Chem. Commun.*
8 Shelnutt, J.A. and Ortiz, V., *J. Phys. Chem.*, 1985, 89, 4733.
9 Dubois, D., Kadish, K.M., Flanagan, S., Haufler, R.E., Chibante, L.B.F. and Wilson, M.A., *J. Am. Chem. Soc.*, 1991, 113, 4364.
10 Vassallo, A.M., Palmisano A.J., Pang, L.S.K. and Wilson, M.A., *J. Chem. Soc. Chem. Commun.*, 1992, 60.

PART 9. EXPERIMENTAL ASPECTS

PRACTICAL ASPECTS OF ELECTROCHEMISTRY AND APPLICATIONS TOWARDS THE SOLVING OF CHEMICAL PROBLEMS

Karl M. KADISH* and Dominique DUBOIS
Department of Chemistry,
University of Houston
Houston, Texas 77204-5641
USA

Abstract. Electrochemical techniques such as cyclic voltammetry are now routinely used to probe the chemical reactivity of neutral, oxidized and reduced inorganic, organic and organometallic compounds under a variety of different solution conditions. A number of practical considerations associated with these voltammetric measurements, such as the selection of a given solvent/supporting electrolyte system, are discussed in this paper. Examples are illustrated for the case of Buckminsterfullerene, C_{60}, a compound which undergoes up to five one-electron reductions in nonaqueous media. This compound has only recently become available in relatively large quantities and is an ideal molecule to investigate the effects of media and temperature on multiple redox reactions involving a highly studied and most important species whose electrochemistry is still relatively unknown.

Introduction.

The application of electroanalytical techniques towards the solving of chemical problems has long been successfully demonstrated in the fields of inorganic, organic and organometallic chemistry. These techniques have been applied to the study of numerous chemical systems and have provided a variety of data for workers in a number of different experimental areas.

Voltammetric techniques are the most widely used among the available electroanalytical methods and, of these, cyclic voltammetry is the most widespread as might be suggested from papers appearing in this present volume. The popular use of cyclic voltammetry stems in part from the fact that the methodology is simple yet provides a wealth of information about a given investigated system using low cost and readily available instrumentation. Many electrochemical setups are now computerized for routine operation and, in these cases, the main concern of the researcher, other than picking the type of voltammetric technique (cyclic, differential pulse, thin-layer, rotating disk electrode, etc.), is to select the proper experimental conditions.

A number of parameters can be varied in a given voltammetric experiment and several examples are as follows:

A. J. L. Pombeiro and J. A. McCleverty (eds.),
Molecular Electrochemistry of Inorganic, Bioinorganic and Organometallic Compounds, 603–613.

(1) Solvent (type).

(2) Supporting electrolyte (type and concentration).

(3) Temperature.

(4) Time-scale of the measurement (or scan rate).

(5) Type and nature of working electrode (geometry, size and material).

(6) Reference electrode (real or internal reference).

The selection of a specific combination of experimental parameters may be straightforward for experts in the field of electrochemistry but this is not always the case and, if the wrong selection is made, one may end up carrying out essentially useless experiments. The chemistry of an investigated compound should in large part determine selection of given solution/electrode conditions but even with a good knowledge of the compound's chemistry, it is not always possible to select the "ideal" conditions to run a given voltammetric experiment.

In addition to selecting the "best" or "right" solution and electrode conditions one must first determine the most relevant experiments which will get the most out of a given measurement, i.e. one must first ascertain which measurements are needed to provide the most essential information and also to solve a specific chemical problem. Along these lines, one must decide what type of instrumentation to use, how much electrochemical or chemical expertise is needed and also if there are alternative (e.g. cheaper, easier, faster) experiments which might be carried out in order to solve a given chemical or electrochemical problem or to obtain a given piece of chemical/electrochemical information.

This present paper addresses some of the above points. Examples are provided using electrochemical data for a newly available and until recently totally unknown compound, Buckminsterfullerene, C_{60} [1-4]. Buckminsterfullerene was selected as the "Molecule of the year" by *Science* in 1991 [5] and provides an excellent example to illustrate how a researcher might design voltammetric experiments in order to optimize the data and information output.

The examples provided along these lines are not meant to be exhaustive nor do they address major scientific issues which are covered in the current literature. They are only intended as practical hints or suggestions as to how one might carry out meaningful voltammetric experiments when electrochemically characterizing a totally new compound. Samples of C_{60} are now available in relatively large quantities but this was not the case as recently as mid-1991 when only very limited amounts were available to the scientific community and even today its high cost is still a factor limiting its widespread use.

C_{60}: Five reductions and still counting... How many electrons is enough?

Theory predicts that C_{60} may be reduced by six electrons [6] and the experimental verification of this prediction is an interesting example of how science often advances in a stepwise manner. C_{60} was first shown to be reducible by two electrons [7] after which three [8] then four [9] and finally five electrons [10] were shown to be added as is qualitatively shown in Figure 1.

November 1990 (a)

January 1991 (b)

May 1991 (c)

September 1991 (d)

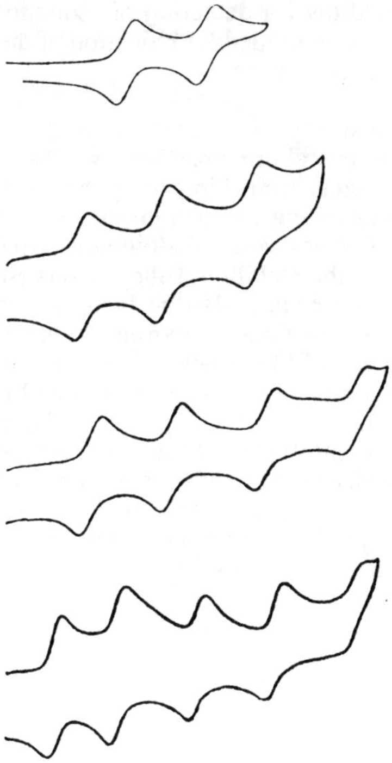

1992 **6 reductions ?**

Figure 1. Cyclic voltammograms of C_{60} as a function of their publication date.

Can C_{60} be reduced further? The answer is yes [11]. Do we need such information and should we look for new solvents or mixed solvent systems where all six reductions are easily observed? Perhaps. However, if we only want to show that C_{60}, as predicted [6], has a high electron affinity and is converted to C_{60}^- at a quite positive potential, then the use of media which have quite limited electrochemical windows will suffice. Under these conditions a number of different non-aqueous solvents may be utilized, with the choice depending upon what is available, what solubilizes C_{60} and what can be most easily purified.

On the other hand, if one wants to show that not only can C_{60} be easily converted to C_{60}^-, but also that its LUMO can accept six electrons, then conditions are needed which will allow for the obtaining of $E_{1/2}$ values at very negative potentials. When this can be accomplished, the resulting voltammograms provide experimental verification that a fifth electron can be added to C_{60} at relatively low energy and this will prove that the C_{60} LUMO is triply degenerate (or nearly degenerate) as suggested.

Going even further, if it must be demonstrated that a specific alkali metal can reduce C_{60} to C_{60}^{6-} in a given medium, then experimental conditions must be found which will allow for detection of reduction processes at potentials which may be as negative as those needed for reduction of the corresponding alkali cation under the same experimental conditions.

The ability to observe electrode reactions of a given electroactive molecule depends on several principal factors. The most important is the potential "observation" window which is limited in part by the electrochemical inactivity limit of the media (i.e. solvent, supporting electrolyte and nature or amount of impurities present in the mixture). The observed window will also depend upon the type of working electrode as well as on the stability of the various electrogenerated species. Potential scan rate and temperature may also be factors to consider with respect to observing cyclic voltammetric reactions at "extreme" negative or positive potentials. For example, the third reduction of C_{60} is not fully reversible in dichloromethane at low scan rates but a well-defined reversible wave is obtained in this solvent as the scan rate is increased above 3 V/s (see Figure 2). Similarly, the fourth reduction of C_{60} can be easily detected in dichloromethane at scan rates of 10-20 V/s [9] (Figure 1c) but this reaction is *totally* undetectable at scan rates of ca. 0.1 V/s (voltammogram not shown). It is also worth noting that a well-defined fourth reduction of C_{60} can be observed by cyclic voltammetry at low scan rates if one simply lowers the temperature, and this was nicely demonstrated in a paper given in the present workshop by Heath *et al.* [12].

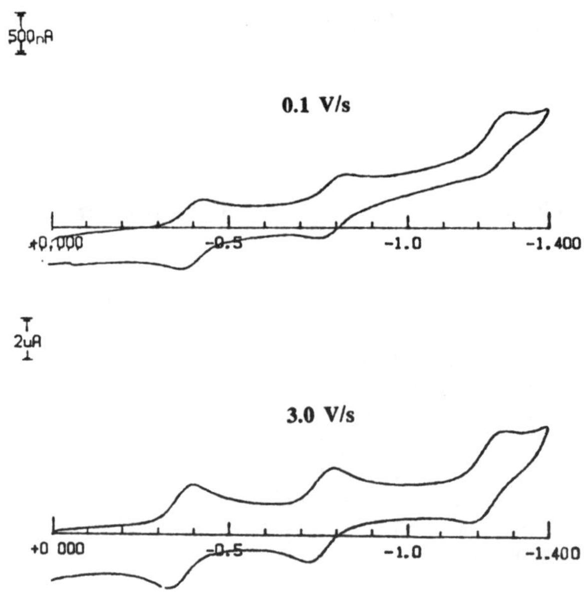

Figure 2. Cyclic voltammograms of C_{60} in dichloromethane, 0.1 $TBAClO_4$ at 0.1 V/s (top) and 3.0 V/s (bottom).

Half-Wave Potentials.

Reversible half-wave potentials determined from cyclic voltammograms generally relate closely to thermodynamic E^o values for the redox system under study. However, many types of interactions may exist between the surrounding medium and the neutral or electrogenerated species, and because of this, the determined values of $E_{1/2}$ reflect thermodynamics only in the particular medium used for observation. Moreover, the half-wave potentials are meaningless in themselves and must be related (or compared) to some arbitrary standard (such as a reference electrode potential) in order to acquire significance.

In attempting to measure values of $E_{1/2}$ one must first ascertain the best conditions for making the measurements. One must also select a specific reference potential for reporting the data (popular examples include SCE, NHE and the Fc^+/Fc couple). Since $E_{1/2}$ values reflect the thermodynamics of a system, one must also select the best or "correct" temperature (e.g. 25 °C, "room temperature" or low temperature) for obtaining and reporting results of the measurement. As will be seen on the following pages, a selection of the best conditions is not always straightforward and may require the carrying out of a large number of experiments under different solution conditions in order to obtain data which is deemed to be the most useful for the solving of a given chemical problem.

Temperature of Measurement.

$E_{1/2}$ values are now most generally reported at room temperature (whatever that is) but earlier measurements were often made using the thermodynamically standard temperature of 25 °C. Low temperature values of $E_{1/2}$ have also been reported for a variety of inorganic and organometallic species. This is generally the case when a coupled chemical reaction is involved and when reversible $E_{1/2}$ values are needed to characterize a given redox couple in the absence of associated chemistry which might render the overall process irreversible.

As a first approximation, it is tempting to assume that the measured $E_{1/2}$ values are not significantly affected by variations of temperature. This is of course not true, even if one assumes that the entropy of electron transfer is small. Although $E_{1/2}$ values may be virtually temperature independent for many redox systems, it is not altogether evident that this would be the case when dealing with a totally unknown system.

An example is given in Figure 3 which shows how $E_{1/2}$ for the first three reductions of C_{60} varies with temperature in benzonitrile (T = 0 to 80 °C) and dichloromethane (T = -30 to 10 °C) [13]. All three reductions of C_{60} are affected by temperature to a different extent but the shift is largest for the third reduction where a fifty degree increase in temperature results in a decrease of $E_{1/2}$ by 80 mV. This shift in potential is substantial and is associated with a large negative entropy of electron transfer which simply cannot be ignored.

As seen in Figure 3 the first reduction of C_{60} does not significantly depend upon temperature and a single value of $E_{1/2}$ can generally be used to describe this process over a wide range of temperature. In contrast, $E_{1/2}$ values for the following two processes vary significantly with temperature and as the temperature is increased a larger potential difference is observed between the first and second or first and third reductions. For example, the absolute potential difference between the first and third reductions amounts to about 1.0 V at 80 °C in benzonitrile and about 0.8 V at -30 °C in dichloromethane. This difference might lead to different chemistry at the two temperature extremes and this fact should not be overlooked. In summary, it is quite clear from the data that when an $E_{1/2}$ is quoted and/or used to solve a chemical problem, the appropriate temperature at which the data was obtained must be given.

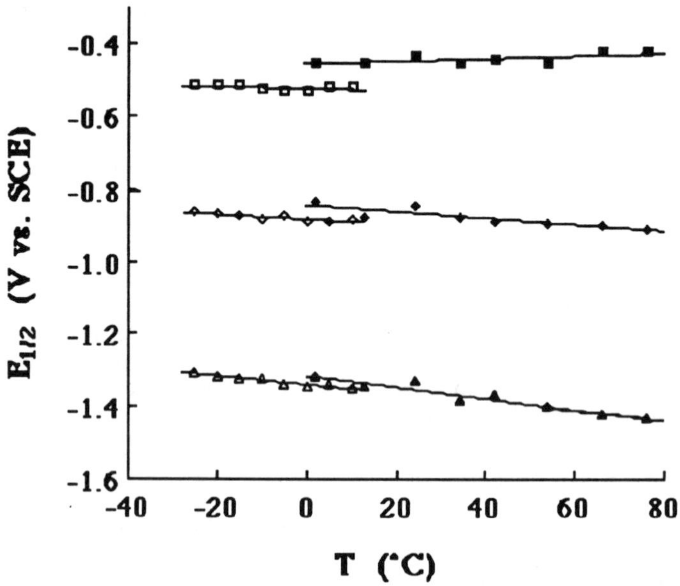

Figure 3. Plot of half-wave potentials for reduction of C_{60} as a function of temperature in benzonitrile, 0.1 M $TBAClO_4$ (solid points) and dichloromethane, 0.1 M $TBAClO_4$ (open points).

Solvent Dependence.

It is generally well understood that values of $E_{1/2}$ can vary markedly as a function of solvent. Numerous papers have been published in this area and they often include correlations between $E_{1/2}$ and one or more solvent parameters such as the Gutmann donor number, the Gutmann acceptor number, or the Dimroth-Reichardt parameter [14]. In general, values of $E_{1/2}$ have been shown to correlate with the Gutmann donor number for electrode reactions of the type M^n/M^{n-1} where n is a positive number and with the Gutmann acceptor number or Dimroth-Reichardt parameter when n is 0 or a negative number [15]. However, this is not always the case as shown in Figure 4 which illustrates how $E_{1/2}$ for the first two reductions of C_{60} vary as a function of the solvent Gutmann donor number in six different solvents [13]. The $E_{1/2}$ for the first reduction varies linearly with the donor number (Figure 4) but the slope is positive rather than negative. This is totally unexpected since donor solvents are expected to preferentially stabilize the oxidized and not reduced form of the C_{60}/C_{60}^- couple which would lead to a negative slope. On the other hand, the deviation is suggestive of either unexpected chemistry or unexpected interactions involving the electroactive species and the solvent [13]. This might not be observed without plots of the type given in Figure 4 and the value of these plots is self-evident.

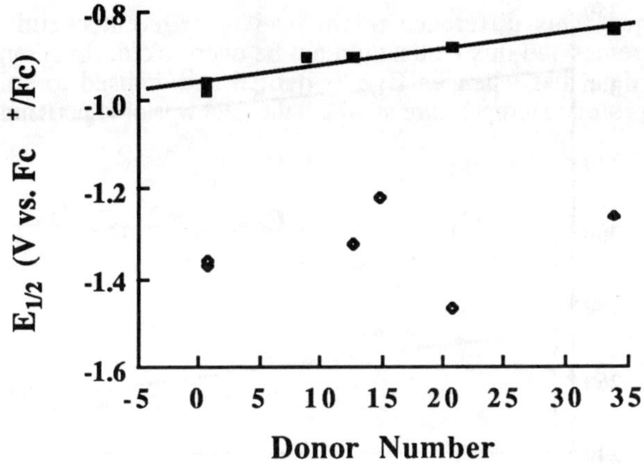

Figure 4. Plot of $E_{1/2}$ Gutmann donor number for the first (squares) and second (diamonds) reduction of C_{60} in solvents containing 0.1 M $TBAClO_4$.

Correlations Between $E_{1/2}$ and Spectroscopic Data.

It is well known that half-wave potentials can be correlated with other physicochemical characteristics of a compound such as the UV-visible data. Figure 5 shows the spectra of C_{60}, C_{60}^- and C_{60}^{2-} which were obtained by thin-layer spectroelectrochemistry in dichloromethane, 0.05 M $TBABF_4$ [9].

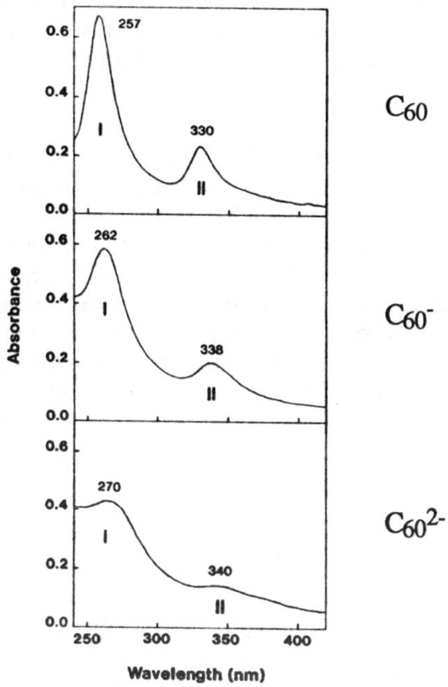

Figure 5. UV-visible spectra of C_{60}, C_{60}^- and C_{60}^{2-} in dichloromethane containing 0.05 M $TBABF_4$ [16].

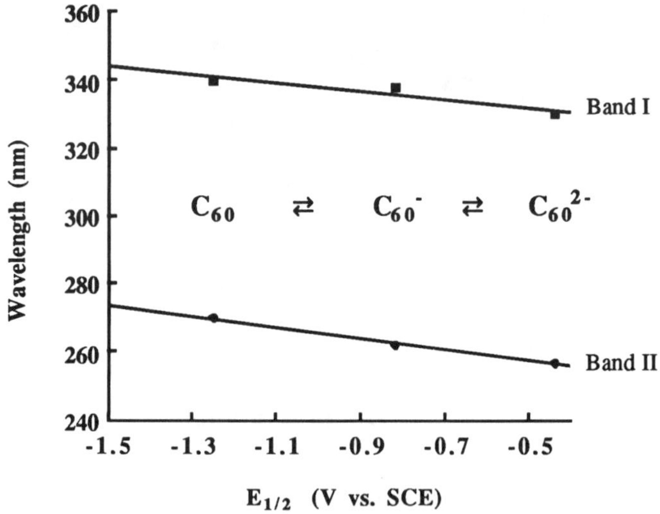

Figure 6. Relationship between maximum wavelengths of C_{60}, C_{60}^- and C_{60}^{2-} and half-wave potentials for formation of C_{60}^-, C_{60}^{2-} and C_{60}^{3-}.

The maximum wavelengths of the two monitored bands correlate well with $E_{1/2}$ for formation of C_{60}^-, C_{60}^{2-} and C_{60}^{3-} under the same experimental conditions and this is shown in Figure 6 where bands I and II are identified in Figure 5. Plots of the type given in Figure 6 can be used to predict $E_{1/2}$ values for a given compound whose spectrum is already known or alternatively to predict spectral maxima for a compound whose $E_{1/2}$ values have been measured but whose UV-visible spectra are lacking. The data can also be used to determine the actual orbitals involved in the specific electron transfer reaction.

A good correlation between $E_{1/2}$ and λ_{max} is obtained for the first three C_{60} reductions, but if it not altogether clear whether this correlation between the two sets of data is theoretically justified. On the other hand, both sets of data were obtained under the very same conditions (temperature, solvent and supporting electrolyte) and, because of this, the plots are certainly valid for predictive purposes. It should be noted, however, that the two sets of data might not be consistent if different temperatures and/or different solvents were used for the two types of measurements. This can often be the case when spectral data are obtained in a solvent which does not contain supporting electrolyte. Along the same lines, the electrochemical and spectroscopic data should preferentially be taken in solutions having the same ionic strength. If this is not done, extreme care must then be taken in using correlations involving this data.

Supporting Electrolyte.

One of the most often overlooked variables in electrochemistry is the supporting electrolyte which is often selected on the basis of availability, cost, solubility, or perhaps just precedent (my electrochemical colleagues use a given salt, why shouldn't I?). The most commonly utilized salts in nonaqueous media have generally been tetraalkylammonium perchlorates, tetrafluoroborates or hexafluorophosphates and these are usually utilized in 0.05 to 0.2 M concentrations. Chemistry is sometimes a consideration in selection of the supporting electrolyte but in most cases the electrochemist has assumed that a salt is a salt and has spent little or no time investigating how $E_{1/2}$ of a given redox process might change upon changing this parameter in the electrochemical experiment.

The variation of supporting electrolyte type may lead to very significant changes in half-wave potentials for oxidation or reduction of a given compound and this is illustrated in Table I and Figure 7 which presents data for the first two reductions of C_{60} in solvents containing different supporting electrolytes.

Table I. Half-wave potentials for the first two reductions of C_{60} in THF as a function of the supporting electrolyte [13]

Supporting Electrolyte 0.1 M	First Reduction (V vs. SCE)	Second Reduction (V vs. SCE)
$NaClO_4$	-0.61	-1.32
$LiClO_4$	-0.51	-1.16
$TBAClO_4$	-0.33	-0.92
$THAClO_4$	-0.29	-0.89
$TOAClO_4$	-0.24	-0.84
TBPBr	-0.16	-0.79

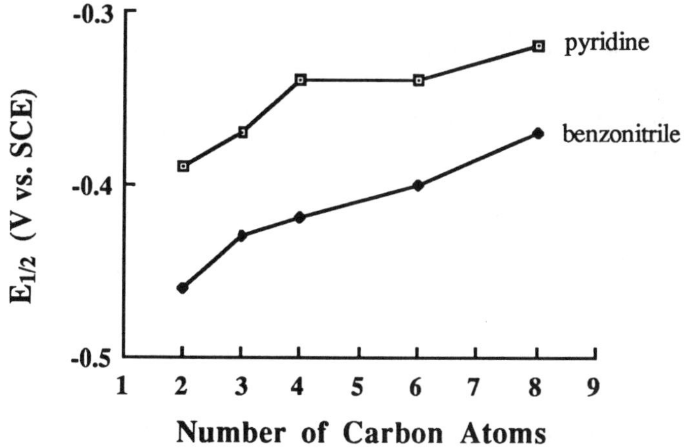

Figure 7. $E_{1/2}$ for the first reduction of C_{60} vs. the number of carbon atoms of the alkyl chain in solutions of 0.1 M tetraalkylammonium perchlorate supporting electrolytes [13].

The shift of $E_{1/2}$ is large upon changing from Na^+ to a tetraalkylammonium or tetraalkylphosphonium cation in the supporting electrolyte and large shifts are also obtained when one changes the size of the tetraalkylammonium cation as shown in Figure 7 for solutions containing 0.1 M $TRA^+ClO_4^-$ where $R = (CH_2)_nCH_3$ and n = 1, 2, 3, 5, 7). Thus, as seen by this data, the value of $E_{1/2}$ depends on the type of supporting electrolyte, even at a constant ionic strength.

Summary.

We have attempted to briefly show some of the variables associated with measurements of potentials for the reduction of C_{60} in nonaqueous media. Five one-electron reductions are observed and the corresponding half-wave potentials can be modulated by up to 600 to 700 mV with changes in solvent, supporting electrolyte or temperature of the measurement. The observation of a given electrode reaction, as well as its reversibility, will depend upon the specific solution conditions, and in a given solvent/supporting electrolyte mixture, can also be modulated to a certain extent by the use of variable temperature measurements or variable scan rates with cyclic voltammetry.

Acknowledgements.

The support of the National Science Foundation (Grant # CHE-8822881) and the Robert A. Welch Foundation (Grant # E-680) is gratefully acknowledged.

References

[1]. Kroto, H. W.; Heath, J. R.; O'Brien, S. C.; Curl, R. F. and Smalley, R. E. (1985) *Nature* ,*318*, 162-163.

[2]. Krätschmer, W.; Lamb, L. D.; Fostiropoulos, K.; Huffman, D. R. (1990) *Nature* , *347*, 354-358

[3] Liu, S.; Lu, Y. -J.; Kappes, M. M.; Ibers, J. A. (1991) *Science* , *254*, 408-410.

[4]. Kroto, H. W.; Allaf, A. W.; Balm, S. P. (1991) *Chem. Rev.*, *91*, 1213-1235 and references therein.

[5]. (1991)*Science* , *254*, 1705-1707.

[6]. Haddon, R. C.; Brus, L. E.; Raghavarachi, K. (1986) *Chem. Phys. Lett.*, *125*, 459-464.

[7]. Haufler, R. E.; Conceicao, J.; Chibante, L. P. F.; Chai, Y.; Byrne, N. E.; Flanagan, S.; Haley, M. M.; O'Brien, S. C.; Pan, C.; Xiao, Z.; Billups, W. E.; Ciufolini, M. A.; Hauge, R. H.; Margrave, J. L.; Wilson, L. J.; Curl, R. F.; Smalley, R. E. (1990) *J. Phys. Chem.*, *94*, 8634-8636.

[8]. Allemand, P. -M.; Koch, A.; Wudl, F.; Rubin, Y.; Diederich, F.; Alvarez, M. M.; Anz, S. J.; Whetten, R. L. (1991) *J. Am. Chem. Soc.*, *113*, 1050-1051.

[9]. Dubois, D.; Kadish. K. M.; Flanagan, S.; Haufler, R. E.; Chibante, L. P. F.; Wilson, L. J. (1991) *J. Am. Chem. Soc.*, *113*, 4364-4366.

[10]. Dubois, D.; Kadish, K. M.; Flanagan, S.; Wilson, L. J. (1991) *J. Am. Chem. Soc.*, *113*, 7773-7774.

[11]. Jehoulet, C.; Bard, A. J. Submitted for publication.

[12]. Heath, G. A.; McGrady, J. E.; Martin, R. L. Presented at "NATO Advanced Research Workshop: Molecular Electrochemistry of Inorganic, Bioinorganic and Organometallic Compounds", March 25-29, 1992, Sintra, Portugal.

[13]. Dubois, D.; Moninot, G.; Kutner, W.; Jones, M. T.; Kadish, K. M. Submitted for publication.

[14]. Reichardt, C. (1988) *Solvents and Solvent Effects in Organic Chemistry* VCH Publishers, New York, and references therein.

[15]. Kadish, K. M.; Cornillon, J. -L.; Yao, C. -L.; Malinski, T.; Gritzner, G. (1987) *J. Electroanal. Chem.*, *235*, 189-207.

[16]. Indicated wavelengths are not the actual maxima but, for consistency, have been taken at the intersection of tangents taken at inflexion points nearest the maxima.

CHARGE TRANSPORT PROPERTIES OF ELECTRODES MODIFIED WITH OSMIUM CONTAINING POLYMER FILMS.

R.J. FORSTER and J. G. VOS
School of Chemical Sciences, Dublin City University, Dublin 9, Ireland

ABSTRACT. The charge transport behaviour of glassy-carbon electrodes modified with the osmium containing polymer $[Os(bipy)_2(PVI)_nCl]Cl$, where bipy = 2,2'-bipyridyl, PVI is poly-N-vinylimidazole and n = 5, 10, 15, 20, 25 is reported. The polymer coatings behave like ideal surface bound redox couples and standard theoretical models have been used to evaluate the charge transport process. The charge transport parameters are found to be strongly dependent on the nature and concentration of the electrolyte. The effect of the loading of the redox centre (n) is also considered. The diffusion coefficients obtained depend on the time domain used for the measurement. Activation parameters have been obtained to probe the nature of the rate-limiting step in the charge-transport process. The results are compared with those obtained for the analogous polymers $[Os(bipy)_2(PVP)_nCl]Cl$, where PVP = poly-4-vinylpyridine.

1. Introduction.

The extensive investigation of modified electrodes is to a large extent motivated by their potential application in areas such as sensor design and redox catalysis. The recent developments in this area of research have been extensively reviewed.[1-3] One of the most important aspects of these electrodes is the nature of the charge transport process, the manner in which electrons and ions involved in the redox processes are transported within and into the polymer layer. The study of the charge transport process in modifying layers and of the factors which affect it are therefore not only of great importance from the fundamental point of view but are also important for the development of modified electrodes as novel sensors and in other electrochemical devices. In this contribution the charge transport characteristics of the osmium containing polymers $[Os(bipy)_2(PVI)_nCl]Cl$ in HCl and NaCl are reported, (for a structure see Figure 1) where bipy = 2,2'-bipyridyl, PVI is poly-N-vinylimidazole and n = 5, 10, 15, 20, 25. The homogeneous charge transport rates are investigated using both cyclic voltammetry and short-timescale potential step methods. The charge transport process is investigated as a function of the nature and concentration of the electrolyte, the temperature and the loading of the redox centre in the polymer (n). The influence of the polymer back-bone is also discussed.

2. Experimental Part.

The metallopolymers were obtained by refluxing $[Os(bipy)_2Cl_2]$ with the appropriate amount of PVI (mol weight 110,000 g/mol). By varying the amount of polymer, osmium containing polymers of the type $[Os(bipy)_2(PVI)_nCl]Cl$ with different values of n could be obtained. A detailed description of the synthesis and characterisation of the polymers is reported elsewhere.[4] Adherent polymer layers with surface coverages of typically 2 - 4×10^{-8} mol/cm^2 were obtained by evaporation of the required volume of a 1% solution of

A. J. L. Pombeiro and J. A. McCleverty (eds.),
Molecular Electrochemistry of Inorganic, Bioinorganic and Organomettallic Compounds, 615–621.

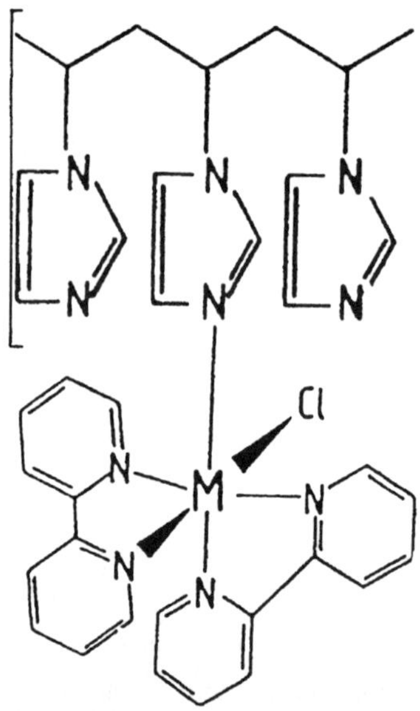

Figure 1. Molecular structure of [Os(bipy)$_2$(PVI)$_n$ Cl]Cl

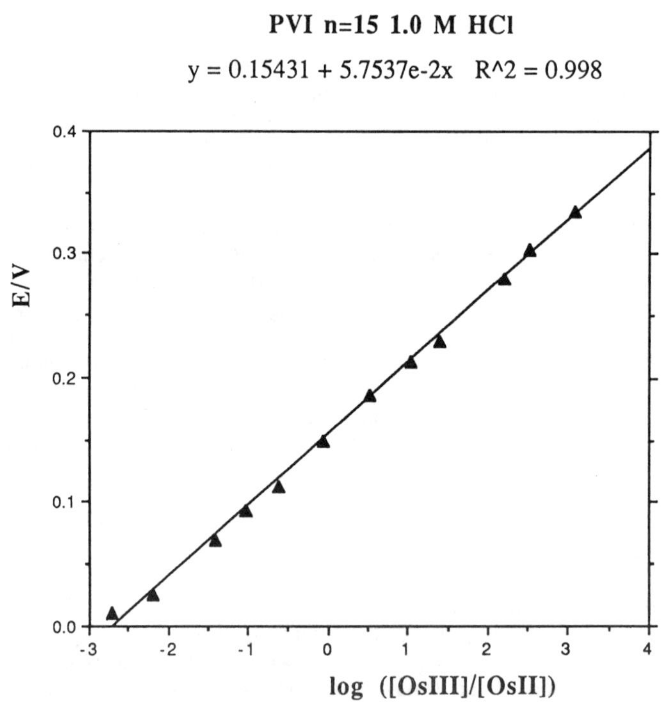

Figure 2. Nerst plot for [Os(bipy)$_2$(PVI)$_{15}$ Cl]Cl obtained in 1.0M HCl

the metallopolymer in methanol on the electrode surface. Cyclic voltammetry and chronoamperometry were carried out using procedures and instrumentation as described in earlier work.[5,6] IR drop was compensated via positive feedback circuitry. The concentration of the osmium centres in the film was approximated from the density of the dry polymers as measured by flotation in non-swelling solvents.[5,6] The values obtained in this way are: n=25 0.35 M; n=20, 0.37 M; n=15, 0.45 M; n=10, 0.65 M; and n=5 1.05 M. The diffusion coefficients and activation parameters are reproducible to within 2% on a single coating and to within 10% for different coatings of the same polymer.

3. Results.

The ideal behaviour of these polymer coatings is illustrated by a typical Nernst plot in Figure 2. The $Os^{II/III}$ redox couple is found at about 150 mV vs SCE, depending both on the nature and the concentration of the electrolyte. In Table 1 the apparent diffusion coefficients for the metallopolymers in HCl and NaCl have been listed. Diffusion coefficients were obtained using the Cottrell equation [7] for the potential step experiments, as D_{ct}(PS), and from the Randles-Sevcik equation [8] for the cyclic voltammetry experiments, as D_{ct}(CV). The activation parameters are given in Table 2.

4. Discussion.

The cyclic voltammograms observed are typical for reversible, single electron redox reactions of surface bound species. Ideal Nernst plots (See Figure 2) and small (< 60 mV at 500 mV/s) peak-to-peak separation support this. Under the conditions employed for both CV and PS measurements no evidence of a contribution from migration to the overall current response was observed. Therefore if migration is present it is unlikely to adversely affect the determination of D_{ct}

Significant differences are observed in the charge transport rates as measured by potential step and cyclic voltammetry. This suggest that the diffusion coefficients obtained depend on the time domain used in the experiments. Alternatively, concentration changes within the polymer could also be responsible for such variations. However, this is thought to be less likely as substantial differences in swelling in the polymer layer (over 300 %) would have to occur. To investigate this behaviour in more detail, potential step measurements were carried out over a longer time scale. Two typical plots are given in Figure 3. Figure 3a shows that for the case of $[Os(bipy)_2(PVI)_5Cl]Cl$ in 0.1 M NaCl two linear regions are observed in the Cottrell plot, possibly indicating the presence of two homogeneous charge transport rates. It is noted however that only the short time data exhibit the zero intercept expected for a diffusion controlled process. On the other hand a similar experiment carried out for the n = 15 polymer in 1.0 M HCl only one linear section is observed. Interestingly, in the latter case the diffusion coefficients observed for CV and PS measurements are very similar, while for the case illustrated in Figure 3a substantial differences are observed between CV and PS values. For data as obtained in Figure 3a the traditional Cottrell analysis can only be applied to the short time data, as only these data show a zero intercept. However at longer timescales where the Cottrell line has a non-zero intercept a diffusion coefficient can also be extracted from the slope. The value calculated from the second linear section of the Cottrell line is equal within experimental error to D_{ct} as obtained from the Randles-Sevcik equation. Data of the type presented in Figure 3a clearly show that current flow is time dependent, while this most likely reflects a time dependency in the homogeneous charge transport rate, the long time scale data do not exhibit a zero current intercept and cannot therefore be analysed using

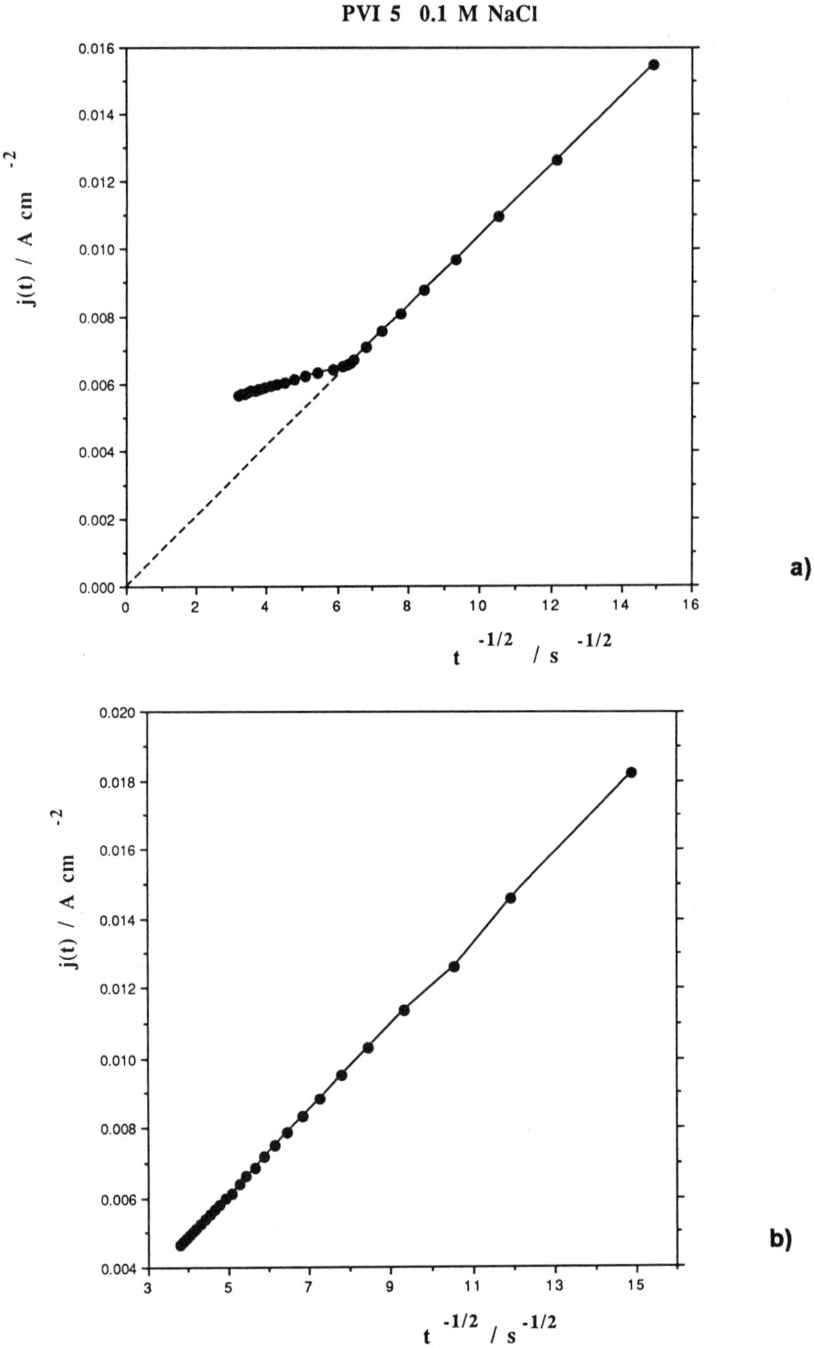

Figure 3. Cottrell plots obtained for a) $[Os(bipy)_2(PVI)_5\ Cl]Cl$ in 0.1M NaCl and b) $[Os(bipy)_2(PVI)_{15}\ Cl]Cl$ in 0.1M HCl

Table 1. The effect of redox site loading and electrolyte on the homogeneous diffusion coefficients of glassy-carbon electrodes coated with $[Os(bipy)_2(PVI)_nCl]Cl$ films.

		HCl		NaCl	
loading n	el.conc. M	$D_{ct}(PS)/10^{-9}$ cm^2s^{-1}	$D_{ct}(CV)/10^{-10}$ cm^2s^{-1}	$D_{ct}(PS)/10^{-9}$ cm^2s^{-1}	$D_{ct}(CV)/10^{-10}$ cm^2s^{-1}
5	0.1	0.46	2.3	1.2	3.9
	1.0	1.15	4.1	2.3	4.5
10	0.1	0.75	2.3	3.0	1.8
	1.0	1.23	9.7	20.1	2.7
15	0.1	1.03	2.6	2.0	1.7
	1.0	1.6	11.0	11.0	10.0
20	0.1	1.3	2.9	1.2	2.5
	1.0	2.0	9.6	3.1	10.4
25	0.1	1.5	3.2	0.5	2.7
	1.0	2.3	7.8	1.2	10.8

Table 2. Activation parameters for charge transport through $[Os(bipy)_2(PVI)_nCl]Cl$ films as obtained by potentialstep measurements and cyclic voltammetry, (CV values in brackets).

Loading	Elec.Conc. M	E_a kJ/mol	$\Delta H^\#$ kJ/mol	$\Delta S^\#$ $Jmol^{-1}K^{-1}$	$\Delta G^\#$ kJ/mol
$[Os(bipy)_2(PVI)_nCl]Cl$ in HCl					
5	0.1	45(15)	43(18)	-16(-97)	47(47)
	1.0	43(24)	41(22)	-15(-84)	45(47)
10	0.1	43(31)	40(29)	-31(-80)	50(53)
	1.0	38(41)	35(38)	-45(-36)	49(49)
15	0.1	39(99)	36(97)	-49(140)	51(55)
	1.0	34(125)	31(122)	-62(227)	50(55)
20	0.1	31(110)	28(108)	-80(176)	55(55)
	1.0	27(131)	24(129)	-88(256)	51(53)
25	0.1	22(114)	20(112)	-111(182)	53(58)
	1.0	19(135)	17(132)	-117(262)	52(54)
$[Os(bipy)_2(PVI)_nCl]Cl$ in NaCl					
5	0.1	12(36)	10(31)	-130(-63)	48(50)
	1.0	14(61)	11(59)	-108(29)	43(49)
10	0.1	14(42)	11(40)	-126(-39)	49(52)
	1.0	15(61)	12(59)	-112(29)	46(50)
15	0.1	13(114)	11(112)	-133(191)	51(55)
	1.0	15(61)	13(59)	-118(21)	48(53)
20	0.1	12(116)	9(113)	-143(193)	52(56)
	1.0	20(128)	18(126)	-106(242)	50(54)
25	0.1	12(116)	9(114)	-144(188)	53(58)
	1.0	24(134)	21(131)	-98(257)	51(55)

the conventional Cottrell equation.

The D_{ct} values increase with electrolyte concentration. Surprisingly the largest diffusion coefficients are not observed for highest loadings, but in general for the n=15 polymer. This clearly shows that a high concentration of redox centres does not necessarily lead to optimum charge transport.

The nature of the rate detemining step in the charge transport process can be probed by an analysis of the activation parameters. Three rate determining processes can occur, electron hopping, ion movement and polymer movement. For the first two processes activation energies of less than 60 kJ/mol are expected, coupled with negative entropy terms. For polymer movement larger activation energies are expected together with positive entropies.[9] The values observed for D_{ct} are considered too low to represent electron hopping processes. This process is therefore not further considered in the following analysis.

Based on these assumptions the activation energies obtained by potential step experiments, E_a(PS), in Table 2 suggest that under these experimental conditions ion movement remains the rate determining step for all loadings, both at low and at high electrolyte concentrations. The activation parameters decrease with decreasing loading suggesting improved ion transport. No major differences are observed as a function of electrolyte concentration. Somewhat surprisingly E_a(PS) values in NaCl are lower than in HCl. They suggest ion transport as the rate determining process and do not show a significant dependence on the electrolyte concentration. The activation energies as obtained by cyclic voltammetry, E_a(CV), suggest that at longer time scales, ion movement is rate determining for the higher loadings (n= 5,10) and polymer movement for the lower loadings. This is supported by the positive entropy values observed at lower loadings. There is relatively little variation as a result of the electrolyte concentration.

5. Concluding Remarks.

The results obtained in this study clearly show that the charge-transport process is strongly dependent on both the nature and the concentration of the electrolyte. The relatively small effect of a change of pH on the charge transport parameters is somewhat surprising. In HCl protonation of the free imidazole groups is expected to result in an increased swelling. This is possibly explained by the high buffering capacity of PVI.

Recently similar studies have been reported for the analogous polymer [Os-(bipy)$_2$(PVP)$_n$Cl]Cl, where PVP = poly-4-vinylpyridine.[5,6,10,] A comparison with these data clearly shows that the charge transport process is also dependent on the nature of the polymer back bone. The charge transport parameters observed for both systems are of the same order of magnitude. There are however significant variations in their variation with both redox centre loading and electrolyte. In NaCl D_{ct}(CV) increases with decreasing loading for the PVI polymers but for the PVP systems this parameter decreases after a maximum at n=15. For the PVP polymers charge transport is faster in HCl than in NaCl for high electrolyte concentrations while for low electrolyte concentrations the values obtained in both electrolytes are similar. For the PVI polymers however D_{ct} is larger in NaCl than in HCl for the higher loadings (n<20). The charge transport parameters obtained by CV are similar for both polymer systems with no clear trends. Differences are also observed in the activation energies. For the PVI polymers in both 0.1 and 1.0 M HCl E_a(PS) decreases with decreasing metal loading, but for PVP an increase is observed at 1.0 M while at 0.1 M the activation energy is constant. The activation parameters as obtained from CV are similar for both systems, polymer movement taking over from ion movement as the rate detemining step at loadings below n=15.

Finally, one of the most interesting aspects is the difference observed between the diffusion coefficients obtained by CV and PS methods. The results obtained suggest that different processes may be rate determining at different timescales. This might suggest that at the PS time scale mass transport is via highly mobile counterions already present in the film, while total equilibrium involving counter ion transport into the film, solvent movement and polymer movement is only reached at longer timescales. The conditions of this equilibrium are most likely controlled by the structure imposed on the polymer by the electrolyte.

6. Acknowledgements.

Financial support from EOLAS, the Irish Science and Technology Agency, is gratefully acknowledged. Johnson Matthey UK is thanked for a generous loan of potassium hexachloroosmate.

References.

1 Hillman, A.R.(1987) in Linford, R.G. (ed) Electrochemical Science and Technology of Polymers, Vol 1, Chapter 5, Elsevier, Amsterdam, pp 103-284.
2 Murray, R.W. (1984) Electroanal. Chem. 13, 191-368.
3 Forster, R.J. and Vos, J.G. (1992) in Smyth, M.R. and Vos, J.G. (eds) Analytical Voltammetry, Comprehensive Analytical Chemistry Vol XXVII, series Editor, Svehhla, Elsevier, Amsterdam, pp465-531.
4 Forster, R.J. and Vos, J.G. (1990) Macromolecules, 23, 4372-4377.
5 Forster, R.J., Kelly, A.J., Vos, J.G., and Lyons, M.E.G. (1990) J. Electroanal. Chem., 270, 365-379.
6 Forster, R.J., Vos, J.G., and Lyons, M.E.G. (1991) J. Chem. Soc., Faraday Trans. 87, 3761-3767.
7 Cottrell, F.G. (1902) Z. Phys. Chem. 42, 385.
8 Sevcik, A., (1948) Collect. Czech. Chem. Commun. 44, 327.
9 Inzelt, G. (1989) Electrochim. Acta 34, 83-91.
10 Forster, R.J. and Vos. J.G. (1992) Electrochim. Acta 37, 159-167.

PART 10. ELECTROCHEMISTRY in NON - CONVENTIONAL CONDITIONS. USE of ULTRAMICROELECTRODES

UNCONVENTIONAL ELECTROCHEMISTRY AT ULTRAMICROELECTRODES: NEW APPROACHES FOR THE INVESTIGATION OF CHEMICAL REACTIVITY

Christian AMATORE
Ecole Normale Supérieure
Département de Chimie. URA CNRS 1110
24 rue Lhomond.
75231 Paris Cedex 05
France.

ABSTRACT. Several aspects of electrochemistry at ultramicroelectrodes are presented and discussed in relevance to their application to the analysis of chemical reactivity. The limits of fast scan cyclic voltammetry are examined, and the method shown to allow kinetic investigations in the nanosecond time scale. On the other hand, the dual nature of steady state at ultramicroelectrodes is explained, and it is shown how steady state currents may be used, in combination with transient chronoamperometry, for the determination of absolute electron stoichiometries in voltammetric methods. Finally the interest of electrochemistry in highly resistive conditions for discussion and investigation of chemical reactivity is presented.

1. Introduction

Electrochemical experiments are classically conducted under specific conditions that are particular to the discipline. These stem from the triple necessity of (i) carrying a current through a liquid solution, (ii) controlling the electrode potential, and (iii) being able to derive quantitative kinetic or thermodynamic informations from current-potential curves. Therefore conditions that minimize the solution resistance (solvents with large dielectric constants, presence of a large concentration of an inert dissociated ionic salt) are used to satisfy the two first requirements. Similarly, most kinetic or thermodynamic data are extracted from experimental current-potential curves through the relay of theories that generally do not consider migration as a significant mass transport mode. Therefore, one uses generally a large excess of an inert salt *vis a vis* the substrate concentration, even when a smaller concentration of the salt may have been sufficient to satisfy the two above exigencies.

Obviously there is nothing to be said against such experimental conditions *per se*, when the purpose is purely analytical. However, whenever general kinetic knowledge (*viz.* non specifically related to electrochemical conditions) is wanted, one must remember that the reactivity of any chemical species may be extremely sensitive on the medium where it is studied. This is particularly true for organometallic reactivity that generally highly depends on the nucleophilicity or ligating ability of the medium, as well as on ionic strength and salt effects.

In this respect, one may easily understand why kinetic data obtained under classical electrochemical conditions are sometimes looked at "suspiciously" by "homogeneous" chemists. Indeed, the requirement of a solvent with a large dielectric constant, able to solvate ions, is generally tantamount to saying that the solvent possess a significant dipole, and is accessible to a metal center. Therefore, it is prone to participate nucleophilically or electrophilically in the reaction course, or at least may stabilize some poorly ligated transient species

A. J. L. Pombeiro and J. A. McCleverty (eds.),
Molecular Electrochemistry of Inorganic, Bioinorganic and Organometallic Compounds, 625–644.

by coordinating them (compare *e.g.* DMF, ACN, *etc*, which are strong ligands in organometallic chemistry). Furthermore, most classical electrochemical solvents also may alter a reaction course owing to their acid-base properties, as well as to their capability of transferring atoms to unsaturated metal centers. Similarly, usual supporting electrolytes, besides imposing a large ionic strength and thus being prone to modify a reaction by salt effects, also may be involved specifically in a reaction mechanism through proton transfers (*e.g.* Hoffman reaction from tetraalkyl ammonium cations) or hydrogen atom transfers,[1] but also by transfer of anionic nucleophilic moieties such as fluoride ions (*e.g.* from BF_4^-) or σ-phenyl anions (*e.g.* from BPh_4^-).[2]

From what precedes, it is understood that whenever the goal is to obtain general kinetic informations, in relevance to "homogeneous" chemistry, it is highly desirable that such informations could be derived under conditions that reproduce as closely as possible those encountered for the "homogeneous" system of interest. Unfortunately, up to the introduction of ultramicroelectrodes, this seemed to have to remain a wish. However, the considerable decrease of ohmic drop at these electrodes[3] has offered some experimental reality to this wish. Indeed, over the past decade, several groups have established experimentally as well as theoretically, that at these electrodes, meaningful electrochemical data could be obtained under conditions that were almost unthinkable before the past decade.[3] For example, and within the context of the above discussion, one may recall that electrochemistry without added electrolyte[4] or in solvents with low dielectric constants such as arenes[5,6] or hexane[7] is no more a fantasy.

Such progresses have obviously opened new frontiers to electrochemistry. Yet, several other areas have also become accessible to the discipline because of the unique properties of ultramicroelectrodes. Among these, and besides the two above examples, we wish to restrict our presentation to electrochemistry in the nanosecond time scale, in the one hand, and to diffusional steady state voltammetry in the other hand. These topics will be discussed based on examples taken from our research in organic or organometallic reactivity.

Finally, to conclude this introduction, and to avoid any possible confusion in the terminology, we wish to define briefly what is an ultramicroelectrode (at least in our sense!). When their interfacial properties are to be considered identical with those of any other electrode of a larger dimension, ultramicroelectrodes must remain much larger than the double layer thickness. This sets a lower dimension of a few tens of Å for ultramicroelectrodes.[8,9] On the other hand, if diffusional steady state voltammetry has to be observed without significant interference of convection, they must be smaller than convective layers, which sets an upper limit of a few tens of μm.[10] Between these limits, all ultramicroelectrodes possess identical ***intrinsic*** physico-chemical properties. However, their behavior (*viz.* ohmic drop, steady state or transient currents, *etc*) obviously depends on the medium and the time-scale considered.[3]

2. Nanosecond Time Resolved Cyclic Voltammetry

2.1. Factors that distort a cyclic voltammogram

For a disk working electrode of considerably smaller dimensions than those of other electrodes and cell, the cell resistance is that of a cone of solution limited by the working electrode surface. It has been established[11] that the resistance is then given by $R = \rho/4r_0$, where ρ is the specific resistivity and r_0 the radius of the working electrode. On the other hand the current, i, that flows through the cell is the sum of two contributions:[12] $i = i_F + i_C$, where i_F and i_C are respectively the Faradaic and capacitive components. Although these may interfere for voltammetric signals of a frequency close to or larger than $1/RC_d$,[13] where C_d is the double layer capacitance, it is sufficient for our purpose here to consider these currents as additive (*vide infra*). For cyclic voltammetry, at a scan rate v, each current is given by:[12]

$$i_C = C_d v = \mathcal{C}(\pi r_0^2)v \quad (1)$$

$$i_F = 0.446 \times F(\pi r_0^2)C^o(FvD/RT)^{1/2} \quad (2)$$

where $\mathcal{C}$ is the specific capacitance of the electrode/solution interface, C^o the concentration of the electroactive species (n = 1) and D its diffusion coefficient. From these expressions, it is seen that the ohmic drop, *iR*, is the sum of two components. The first one, proportional to $\rho r_0(v)^{1/2}$, is due to the Faradaic current, while the other, due to the capacitive current, is proportional to $\rho r_0 v$. Both expressions show that ohmic drop increases with the scan rate, but decreases upon diminishing the electrode radius.

However, independently of any related technological problems, the radius of an electrode cannot be decreased indefinitely if one wants that the contribution of edge currents remains minimal.[14,15] Indeed, observation of an undistorted cyclic voltammogram requires that diffusion remains planar, which happens only when the diffusion layer is extremely small as compared to the electrode radius. The diffusion layer thickness being about $\delta = (\pi DRT/Fv)^{1/2}$, for a scan rate v, this imposes that the electrode radius must be larger than this limit: $r_0 \gg (\pi DRT/Fv)^{1/2}$.

Based on these requirements, one may define a range where the electrode radius must be chosen at any given scan rate, so that the voltammogram is not severely affected by ohmic drop or edge effects. This is represented on Figure 1,[16] next page, for a maximal ohmic drop of 10 mV,[17] and a maximal error of 5% on the peak current due to edge diffusion,[15] for conditions featuring anthracene, 5 mM, in acetonitrile, 0.6 M NBu_4BF_4, at a gold disk electrode. At a scan rate of 10,000 Vs^{-1}, this plot shows that the electrode radius must be selected between 1.75 and 12.5 μm, whereas it must be chosen between 0.15 and 0.50 μm, at v = 1,000,000 Vs^{-1}, so that the above requirements on ohmic drop and edge effects are fulfilled. Furthermore, the same plot shows that no electrode exists that may satisfy these requirements for v ≥ 10,000,000 Vs^{-1}. However, if one considers in addition the feasibility of on line electronic compensation of ohmic drop,[18-20] the upper limits of the above domain are pushed forward, allowing the use of larger electrodes still with a maximal ohmic drop of 10 mV. Figure 1 gives these limits for 90% compensation of ohmic drop.

Obviously, this diagram needs to be completed by considering any instrumental distortions of a voltammogram. These have been examined in great details;[18-22] it suffices here to mention that they result primarily from transit times in the potentiostat electronics and from the fact that the potentiostat/current amplifier device behaves as a filter. It follows that a limit corresponding to a maximal frequency, *viz.* to a maximal scan rate above which the voltammogram will be instrumentally distorted, has to be added on the diagram in Figure 1. Obviously, this frontier, being of a technological origin, can be expected to be pushed forward owing to the design of better potentiostats and the use of better operational amplifiers. As such, it is important to decide to which limits this frontier needs to be pushed.

Indeed, there is a last phenomenon, which has not been considered by previous authors,[21,22] but imposes also an upper limit, yet, now of intrinsic electrochemical origin. It is because increasing the scan rate amounts to decrease proportionally the time allowed for diffusion, and therefore, to decrease the thickness of the diffusion layer. For example, under the conditions of Figure 1, at v = 1,000,000 Vs^{-1} the diffusion layer width is about 50 Å, *i.e.* becomes comparable in size to the double layer.[12,23] Then, the usual approximation, based on Frumkin's seminal work,[24] that amounts to disconnect the transport equations of an electroactive species in the two adjacent layers (*viz.* thermodynamic equilibrium for migration in the double layer, and out of equilibrium diffusion in the diffusion layer), no more holds. The theoretical treatment of the coupling between the diffuse and diffusion layers[9] shows that,

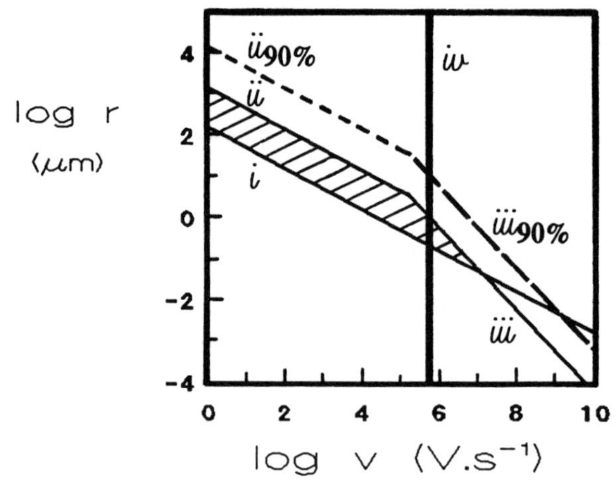

Figure 1. Range of ultramicroelectrodes radii (r_0 in μm) to be used to obtain an undistorted voltammogram at a given scan rate (v in $V.s^{-1}$) as adapted from ref.16. (*i*): limit for least edge diffusion interference (5% error on peak current, from ref.15). (*ii*): limit for least ohmic drop due to Faradaic current (10 mV).[17] (*iii*): limit for least ohmic drop due to capacitive current (10 mV) or cell constant. For a 90% on-line ohmic drop compensation boundaries *ii* and *iii* are pushed upward to $ii_{90\%}$ and $iii_{90\%}$, respectively. All limits are established for a one electron transfer at 20°C, based on errors given above and: D = 10^{-5} $cm^2.s^{-1}$, ρ = 20 Ω.cm, C = 10 $\mu F.cm^{-2}$ and C^0 = 5 mM. Above limit (*iv*) coupling between diffusion layer and double layer is predicted to occur. The thick horizontal line represents the location of the voltammograms shown in Figure 2.

for the "experimental" conditions considered in Figure 1, the predictions of classical[12] theories of voltammetry will hold only if the scan rate remains less than *ca.* 300 to 500 kVs^{-1}. Above this limit severe distortions will affect the voltammograms, which are more disastrous than those due to ohmic drop.[9] Therefore, the quest for undistorted cyclic voltammograms has an experimental significance only for scan rates less than a few hundreds of a kVs^{-1}.

Based on this conclusion, we have limited our current research for better instrumental performances to this range.[20] Several potentiostats with on line electronic compensation of ohmic drop have thus been developed in our group,[18,20] the latest one being able to afford virtually undistorted voltammograms up to scan rates of 150 kVs^{-1}, a range that is easily extended to *ca.* 500 kVs^{-1}, upon using effecting a simple on line arithmetical correction for correction of the slight potential shift due to partial first-order filtering by the instrument.[20,21] Figure 2 represents the voltammetry of anthracene in acetonitrile, 0.6 M NEt_4BF_4, at a 5 μm radius gold disk electrode, as determined with this potentiostat up to 550 kVs^{-1}.[20] From these data a rate constant of 4.8±0.4 $cm.s^{-1}$ could be determined for electron transfer to anthracene.[20]

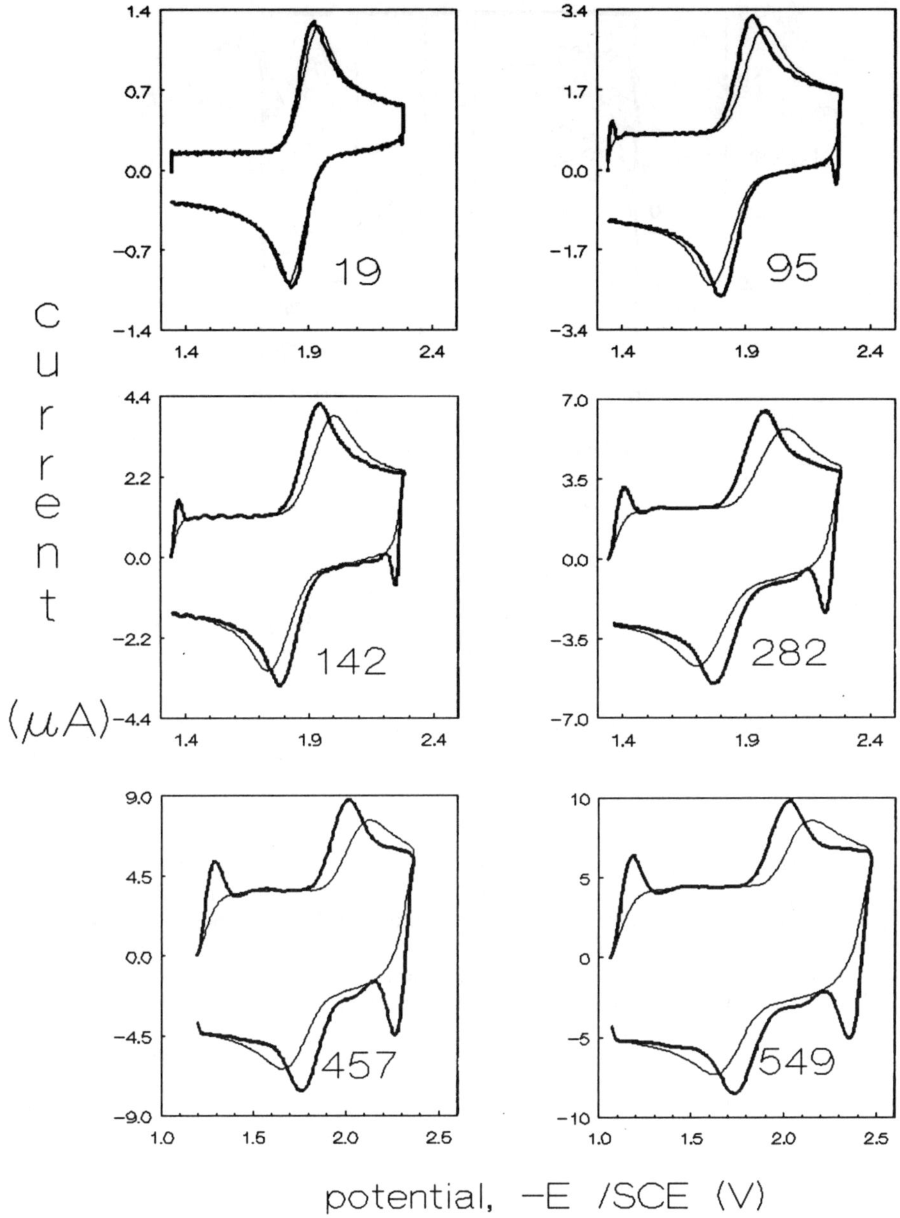

Figure 2. Cyclic voltammetry of anthracene, 10 mM, in acetonitrile, 0.6 M NEt_4BF_4, at a gold disk (r_0 = 5 μm) ultramicroelectrode, with and without ohmic drop compensation, at different scan rates as indicated in $kV.s^{-1}$ on each set of curves.[20]

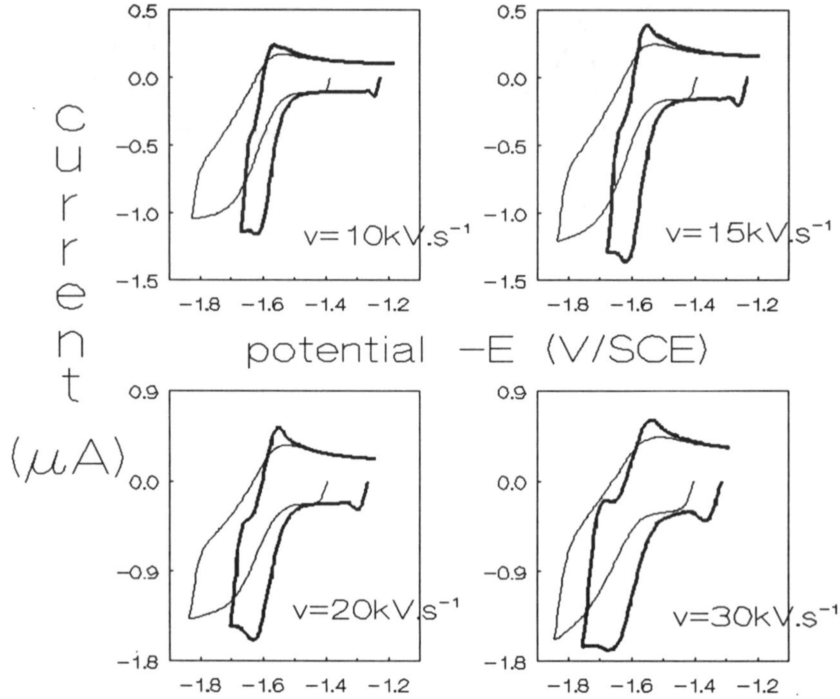

Figure 3. Cyclic voltammetry of hexamethylbenzene, 10 mM, in acetonitrile, 0.6 M NBu_4BF_4, at a 5 μm radius gold disk ultramicroelectrode, at the scan rates given in $kV.s^{-1}$ on each set of on-line ohmic drop compensated or uncompensated voltammograms. 22°C.[27]

2.2. Fast scan cyclic voltammetry for the determination of submicrosecond kinetics

2.2.1. *Kinetic Acidities Of Methylbenzenes.* One electron oxidation of methylbenzenes affords very acidic cation radicals that undergo fast deprotonations by a series of pyridine bases (Py).[25,26] One electron oxidation of the benzylic radical thus formed yields the benzylic cation that adds to the nucleophilic pyridine to give the pyridinium adduct:[25,26]

$$(Me)_nC_6H_{6-n} - e \Longleftrightarrow (Me)_nC_6H_{6-n}^{\bullet +} \qquad (E^0) \qquad (3)$$

$$(Me)_nC_6H_{6-n}^{\bullet +} + Py \longrightarrow (Me)_{n-1}C_6H_{6-n}\text{-}CH_2^{\bullet} + PyH^+ \qquad (k) \qquad (4)$$

$$(Me)_{n-1}C_6H_{6-n}\text{-}CH_2^{\bullet} - e \longrightarrow (Me)_{n-1}C_6H_{6-n}\text{-}CH_2^+ \qquad (E^0_2 \ll E^0) \qquad (5)$$

$$(Me)_{n-1}C_6H_{6-n}\text{-}CH_2^+ + Py \longrightarrow (Me)_{n-1}C_6H_{6-n}\text{-}CH_2Py^+ \qquad (6)$$

The very exergonic proton transfers in eq.4 are extremely fast. However their rate constants could be determined experimentally if the standard oxidation potential of the arenes (eq.3) could be determined independently under the same experimental conditions (acetonitrile, 0.1 M ionic strength).[25,26] To achieve such measurements we resorted to fast scan cyclic voltammetry. Figure 3 illustrates the merit of on line ohmic drop compensation for the evaluation of such important thermodynamic parameters; these were determined as the half sum of the oxidation and reduction peak potentials of compensated voltammograms. Thus E^0 = 1.58, 1.69, 1.75 and 1.77 V *vs* SCE (with E^0 = 0.371 V, for ferrocene on the same scale) were measured for hexamethylbenzene (HMB), pentamethylbenzene (PMB), durene (DUR) and 1,2,3,4-tetramethylbenzene (TMB), respectively.[27] From these values and previously[25] reported rate constants for the homogeneous oxidation of the same methylarenes by a series of phenantrolineiron(III) complexes in the presence of pyridine bases, we were able to determine the rate constants in eq.4, whose values are reported in Table 1,[27] and compared to those determined by nanosecond laser spectroscopy for some of these systems.[28]

Table 1. Kinetic acidity of methylbenzenes with different substituted pyridine bases (HMB: hexamethylbenzene, PMB: pentamethylbenzene, DUR: durene, TMB: 1,2,3,4-tetramethylbenzene) in acetonitrile, 0.1 M NBu_4ClO_4, at 22°C.[a,b]

Z-pyridine		Deprotonation Rate Constant (log k, k in $M^{-1}s^{-1}$)[a,b]					
Z	pK_a	$HMB^{\bullet +}$	$HMB^{\bullet +}$ (*d*-18)	$PMB^{\bullet +}$	$DUR^{\bullet +}$	$DUR^{\bullet +}$ (*d*-14)	$TMB^{\bullet +}$
2-fluoro	4.2	3.95					
2-chloro	6.3	4.80					
3-cyano	7.0	5.00					
4-cyano	8.0	5.28	4.63	5.90	6.05		5.51
3-chloro	9.0	5.72	5.05	6.19	6.21	5.58	5.93
3-fluoro	9.4	5.75					
hydrogen	12.3	6.71	6.20	7.15	7.43	6.94	6.67
2-methyl	14.0	7.14					
4-methyl	14.3	7.01					
4-methoxy	15.0	6.97					
2,6-dimethyl	15.4	7.03 (7.32)	6.46 (6.87)	7.56 (7.60)	8.05 (7.76)		7.46
2,4,6-trimethyl	16.8	7.39 (7.54)					
2,6-di-*t*-butyl	11.8	4.94					

[a]: As obtained using the standard potentials determined in ref.27 and the experimental apparent rate constants measured in ref.25. [b]: Values in parentheses are previous direct determinations by nanosecond spectroscopy reported in from ref.28.

2.2.2. *Dimerization Of Pyrylium Radicals. Determination Of Rate Constants Close To The Diffusion Limit.* One electron reduction of 4-unsubstituted pyrylium cations affords radicals which are known to undergo dimerization[29,30] with rate constants close to the diffusion limit. Thus, on the basis of flash-photolysis experiments, the dimerization of the 2,5-diphenyl-pyrylium radical, **(1)**, has been estimated to proceed with a rate constant of 1.2×10^9 $M^{-1}s^{-1}$:[30]

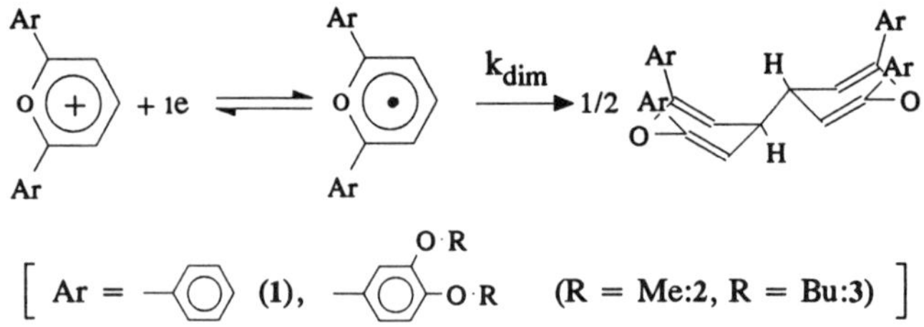

(7)

Such rate constants are formally accessible to fast scan cyclic voltammetry at scan rates in the hundred of kilovolts per second range. To establish this point, we undertook a systematic study of the mechanism in eq.7 for a series of three pyrylium derivatives that differ only by their substituents at the 3 and 4 positions on the phenyl moieties [Z = H (**1**), OMe (**2**) or OBu (**3**); see eq.7]. Figure 4b shows that ohmic drop compensated cyclic voltammetry at *ca.* 150 kV.V^{-1} allows to identify, upon scan reversal, a significant anodic current due to the reoxidation of the pyrylium radical formed upon reduction of the parent cation. This is better seen on a background subtracted voltammogram[21] as shown in Figure 4c, since the Faradaic current is now displayed without being buried within the large capacitive one. Note that because these voltammograms are ohmic drop compensated, the procedure of background subtraction affords a true Faradaic current, since any coupling between capacitive and Faradaic currents is suppressed.[20]

Simulation of the background subtracted voltammograms allows to determine k_{dim} = 4.0×10^9 (**1**),[29d] 2.7×10^8 (**2**) and 4.2×10^8 $M^{-1}s^{-1}$ (**3**), for the respective dimerization rate constants of each radical in eq.7. The rate constant for **1** being very close from that estimated for the diffusion limit in acetonitrile,[31] it was of interest to examine whether the decrease by nearly a factor ten of the dimerization rate when going from **1** to **2** or **3** originated from a diffusion (*viz.* increase of the molecule size) or an activation control. Presuming that all rates are diffusion limited, they should parallel the variations in the diffusion coefficients of the radicals. These were then determined from steady state experiments and, in agreement with usual expectations, shown to decrease upon increasing the size of the substituents of phenyl moieties: D = 1.2×10^{-5} (**1**, Z = H), 1.0×10^{-5} (**2**, Z = OMe) and 0.85×10^{-5} $cm^2.s^{-1}$ (**3**, Z = OBu). However, such a variation is a factor of ten too small to account for the variation of the rate constants. We are thus led to conclude that, except may be for **1**, the dimerization rate constants reflect mainly an activation control of the reaction, thus showing a stabilization of the pyrylium radical by the alkoxy groups borne by the phenyl substituent at their *para* positions. Such a conclusion is further supported by the fact that an identical effect is observed on the E^0 of the three pyrylium cations reduction, but now in the inverse order [E^0 = -0.38 (**1**), -0.50 (**2**) and -0.51 V *vs* SCE (**3**)], in agreement with the fact that an alkoxy group stabilizes more effectively a positive charge than a radical:

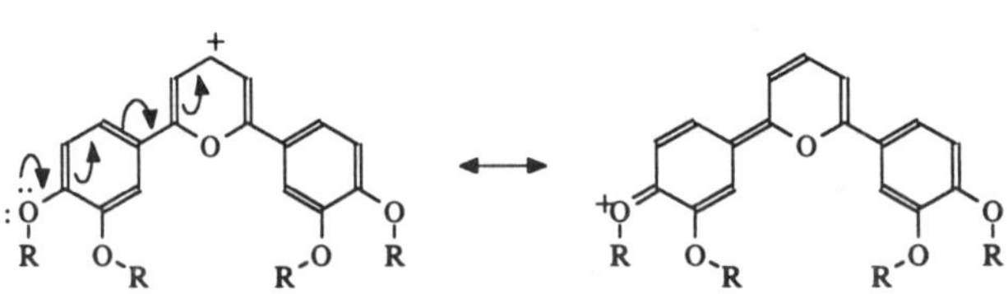

Besides its direct implication for the reactivity of pyrylium radicals, this result is of importance for the corresponding bipyranylidenes, which are formed by an overall dihydrogen

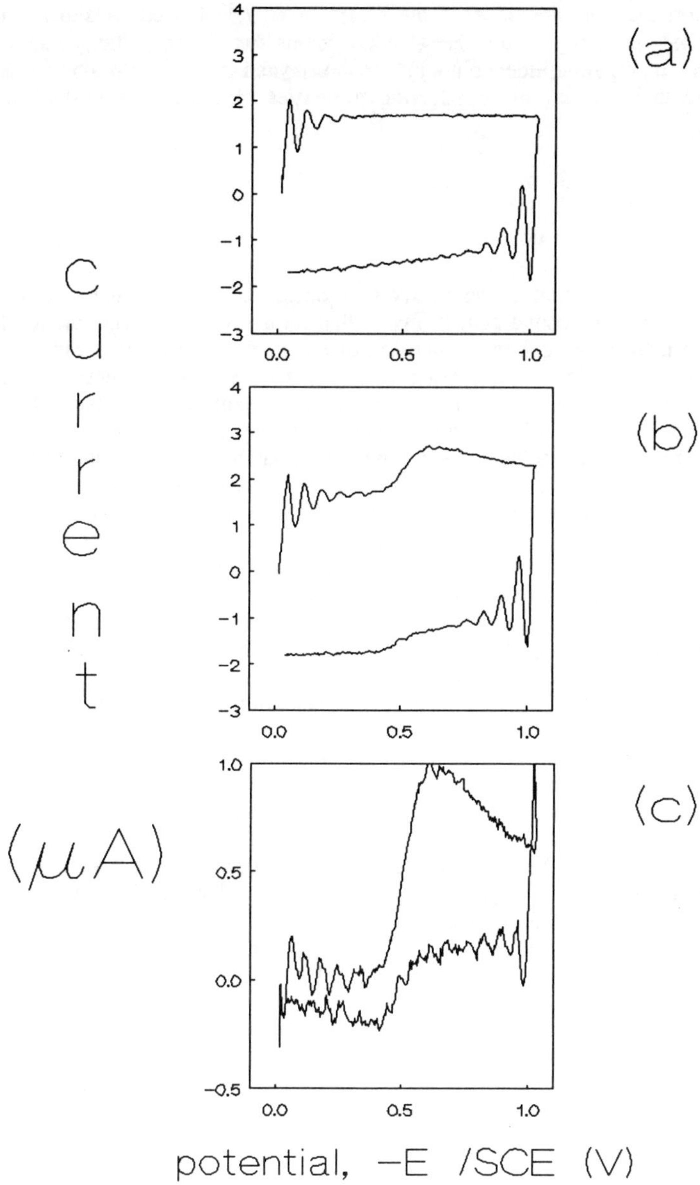

Figure 4. On line ohmic drop compensated cyclic voltammetry of 2,5-di-(*p*-anisole)pyrylium perchlorate (**3**), 5 mM, in acetonitrile, 0.6 M NBu_4BF_4, at a 5 μm radius gold disk ultramicroelectrode and a scan rate of 153 $kV.s^{-1}$. 20°C. (a): in the absence or (b) in the presence of **1**. (c): Background subtracted voltammogram (b - a).

elimination from the 4,4' positions of the dimer in eq.7.[32] Indeed, it shows that, besides providing enough flexibility to the lateral alkyl chains for allowing the formation of stable columnar stacks of bipyrannilidenes units,[32,33] the oxygen moieties are also prone to play a significant rôle in the electronic conducting properties of the aromatic structures in their oxidized states.[33]

3. Steady State At Ultramicroelectrodes

3.1 WHAT IS IN STEADY STATE AND WHAT ISN'T.

It is generally considered that a true steady state occurs at ultramicroelectrodes of spherical geometries because of the convergent diffusion that occurs at times large enough for the size of the diffusion layer to exceed the dimension of the electrode. In fact it can be shown[3] that such a convergent diffusion results is a real steady state for the current at electrodes with diffusion layers of spherical symmetries (*e.g.* disks or hemispherical electrodes) whereas a quasi steady state is observed for the current when the diffusion layer has a cylindrical geometry (*e.g.* bands or cylindrical electrodes). However these steady state currents are often misinterpreted as resulting from steady state diffusion, which is actually not true. Indeed, the steady state at an ultramicroelectrode results from the steadiness of concentration profiles that exists only over distances from the electrode surface that are comparable to the smallest dimension of the electrode. Such a behavior imposes obviously constant gradients at the electrode surface, and therefore steady state currents. However, at distances that exceed the electrode smallest dimension, diffusion remains time dependent, so that the diffusion layer expands toward the solution, thus meeting eventually the convective layer limit at large enough times. This is unequivocally evidenced in Figure 5a which shows the theoretical concentration profile of ferrocene upon its chronoamperometric oxidation at a hemispherical ultramicroelectrode. We want to show in the following that this phenomenon also may be visualized experimentally, on the basis of electrochemiluminescence (ECL) experiments at double band ultramicroelectrodes.[34]

When a system of two parallel band electrodes, of width ω, separated by an insulating gap of thickness δ, is used for ECL generation, one band (the anode) electrogenerates a strong oxidizing species, whereas the other electrode (the cathode) produces a strong reducing one. Recombination of both species within the inter-electrodes plane of symmetry, occurs *via* an exergonic electron transfer that affords an excited state, whose de-excitation occurs instantly, within the time scale of diffusion, with emission of a visible photon. For example, considering ECL from 9,10-diphenylanthracene (DPA), one has:[34]

anode:	$DPA - e \longrightarrow DPA^{\bullet +}$	(8)
cathode:	$DPA + e \longrightarrow DPA^{\bullet -}$	(9)
recombination:	$DPA^{\bullet +} + DPA^{\bullet -} \longrightarrow DPA^{*} + DPA$	(10)
photon emission:	$DPA^{*} \longrightarrow DPA + h\nu$	(11)

Under these circumstances, the constant regeneration of DPA, that occurs in the inter-electrodes space because of eqs.10 and 11, results in a true steady state current at each electrode. This can be observed experimentally[34] and established theoretically as shown in Figure 5b. However, the same Figure shows that the ECL emission do not obey steady state, but instead presents a sharp initial rise (with a time constant comparable to that of the current decrease) followed by a slow steady increase with time. This dual behavior (*viz.* steady state

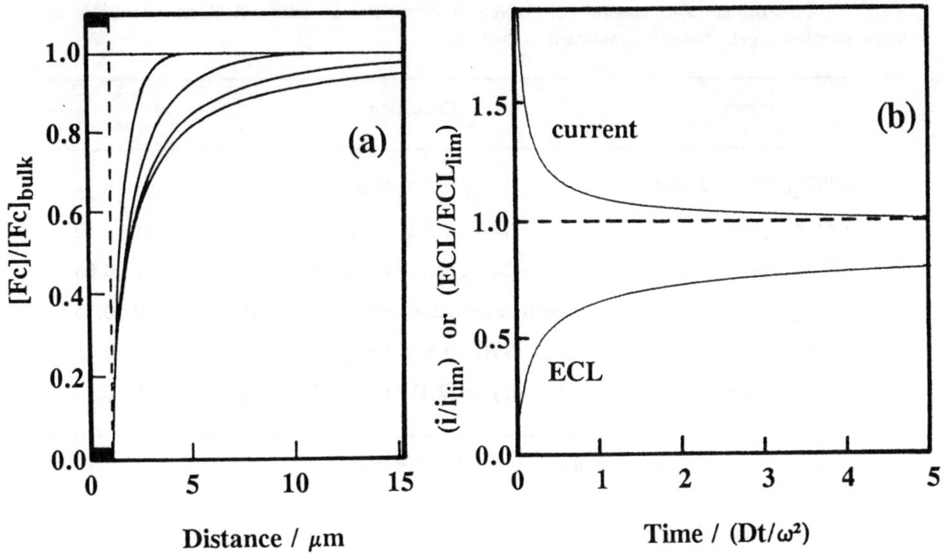

Figure 5. Steady state at ultramicroelectrodes. (a) Normalized theoretical concentration profiles for ferrocene chronoamperometric oxidation at a hemispherical electrode (r_0 = 1 μm) under conditions where a steady state current is observed. From top to bottom, t = 0, 10^{-3}, 10^{-2}, 0.1 and 1 s (b) Theoretical simulation of ECL at a double band assembly, showing the current and ECL intensities (i_{lim} and ECL_{lim} are the limits at infinite time).

current *vs* time dependent ECL) occurs because, the concentration profiles of DPA, $DPA^{\bullet-}$ and $DPA^{\bullet+}$ obey rapidly (*i.e.* for[35] $t \geq t_{1/2} \approx \delta^2/D$) diffusional steady state over distances from the double band that are comparable to the electrode inter-space, but vary slowly with time at distances that are much important (compare Figure 5a). The current that reflects the gradients of concentration profiles at electrode surfaces is then time independent as soon as $t \geq t_{1/2}$. Since most of ECL emission arises from this space domain, one observes a fast rise of ECL intensity, that parallels the rapid initial current drop. However, those $DPA^{\bullet-}$ and $DPA^{\bullet+}$ that diffuse toward the solution along time dependent concentration profiles, undergo delayed recombination. This is reflected in the slow increase of ECL emission with time, that follows the initial sharp rise. Also, because of this dual process, the maximal ECL emission cannot be achieved before infinite times. Note that with the representation used in Figure 5b, this limit at infinite times coincides with the steady state current limit.

3.2. Use of steady state voltammetry for absolute determination of electron stoichiometries in voltammetric experiments.

In voltammetric or chronoamperometric experiments, the determination of an absolute electron stoichiometry is difficult to achieve when the exact mechanism and diffusion coefficient of the electroactive species are unknown, *i.e.* whenever the matter of determining

Table 2. Determination of absolute electron stoichiometries in transient electrochemistry for some selected organometallic systems in reduction.

System	Conditions	n
$Pd^{II}Cl_2(PPh_3)_2$, 2 mM	THF/ 0.3 M NBu_4BF_4	2.04 ± 0.15
$Pd^{II}Cl_2$(dppe), 2 mM	THF/ 0.3 M NBu_4BF_4	1.11 ± 0.12
$Ni^{II}Cl_2$(dppe), 1-3 mM	THF-HMPA (2:1) / 0.3 M NBu_4BF_4	0.95 ± 0.05
$Ni^{II}(BF_4)_2(bipy)_3$, 2 mM	*N*-methylpyrrolidone / 0.1 M NBu_4BF_4	2.10 ± 0.15
C_6H_5-$Pd^{II}I(PPh_3)_2$, 2 mM	THF/ 0.3 M NBu_4BF_4	1.90 ± 0.15
C_6H_5-$Ni^{II}Br$(dppe), 2 mM	THF-HMPA (2:1) / 0.3 M NBu_4BF_4	1.05 ± 0.10

dppe: 1,2-bis(diphenylphosphino)ethane. bipy: 2,2'-bipyridine

an electron stoichiometry is crucial for the kinetic analysis of an unknown mechanism. Indeed, under such conditions, one cannot rely on the result of coulometries that occur over a much larger time scale than most of voltammetric experiments. For example, in voltammetry pyrylium cations reduction is a one electron process (eq.7);[36] yet, in electrolysis a 2/3 of an electron consumption is observed,[32] because of the slow reaction between a pyrylium cation with the dimer formed in eq.7, to afford ultimately a bipyranylidene and a 4H-pyrane. Similarly, reduction of iron pentacarbonyl in THF involves two electrons in voltammetry, because of the fast ECE sequence (*vide infra*):[37]

$$Fe(CO)_5 + e \longrightarrow Fe(CO)_5^{\bullet -} \qquad (12)$$

$$Fe(CO)_5^{\bullet -} \longrightarrow Fe(CO)_4^{\bullet -} + CO \qquad (13)$$

$$Fe(CO)_4^{\bullet -} + e \longrightarrow Fe(CO)_4^{2-} \qquad (14)$$

but in bulk electrolysis, one electron[37] is consumed because of the process in eq.15 which occurs on a time scale larger than that of cyclic voltammetry ($1\ M^{-1}s^{-1} \leq k \ll 50\ M^{-1}s^{-1}$):

$$Fe(CO)_5 + Fe(CO)_4^{2-} \longrightarrow Fe_2(CO)_8^{2-} + CO \qquad (15)$$

Under such circumstances, one is tempted to rely on "calibration" techniques, *i.e.* to compare the current observed to that of a known "reference" compound, *e.g.* ferrocene in organometallic electrochemistry. This is obviously an invalid procedure, as is best shown by the fact that in the past,[37] on the basis of such calibrations, we reported an erroneous (n = 1) electron consumption for $Fe(CO)_5$ reduction in cyclic voltammetry at small scan rates; moreover, such an error made us overestimate the rate constant in eq.15 by a factor 10^6!

The difficulty in determining an electron consumption stems from the fact that only $n \times D$ (steady state current plateau at a disk ultramicroelectrode of radius r_0) or $n \times D^{1/2}$ (transient chronoamperometry) values can be determined from voltammetric experiments. However, whenever one is able to measure both parameters ***within the same time scale*** (*viz.* for $\Theta_{chron.} \approx \Theta_{stead.st.} \approx r_0^2/D$), both n and D values ensue.[36] The above results for pyrylium cation ($n = 0.98 \pm 0.12$)[36] or ironpentacarbonyl ($n = 1.95 \pm 0.06$) were determined through this procedure. Table 2 reports some other examples investigated using this method.

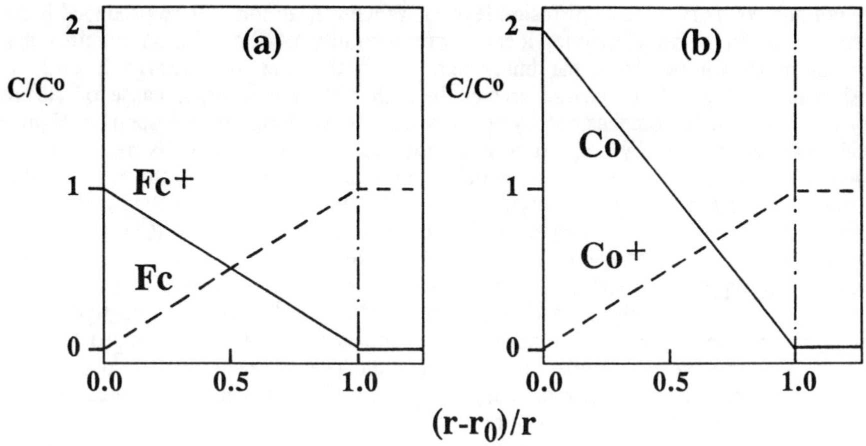

Figure 6. Normalized concentration profiles (C/C°) in the diffusion layer for[4a] (a) oxidation of a neutral (Fc: ferrocene) or (b) reduction of a cationic (Co^+: cobalticinium) species, for a negligible concentration of supporting electrolyte ($C^s \ll C^o$). The excess negative charge concentration profiles required to fulfill local electroneutrality are identical with those of Fc^+ or Co^+.

4. Voltammetry Under Highly Resistive Conditions

The considerable decrease of ohmic drop at ultramicroelectrodes has allowed significant electrochemical data to be obtained from voltammetry in highly resistive media.[3] Thus voltammetry can be performed in usual electrochemical solvents, but without purposefully added supporting electrolyte.[4] Also, steady state voltammograms can be obtained in solvents with very low constants,[5-7] such as alcanes or arenes, if a small concentration of an inert electrolyte is added. In the following we want to present two examples of application of voltammetry under these conditions, to the unravelling of organometallic reactivity.

4.1. VOLTAMMETRY AT LOW IONIC STRENGTH

4.1.1. *Is there Any Meaning In Decreasing The Supporting Electrolyte Concentration Much Below That Of The Electroactive Material?* We want to begin this section by discussing a fundamental, but often misunderstood, aspect of electrochemistry in the absence of added electrolyte. Indeed, in a kind of race for the display of electrochemical data under more and more extreme conditions, several groups have investigated electrochemistry at lower and lower ionic strength. Yet, provided there is enough electrolyte to ensure the building up of double layers at each electrode, one does not observe any significant change in electrochemical behaviors upon decreasing the supporting electrolyte concentration under *ca.* one tenth of the electroactive substrate concentration (C°); most phenomena observed besides this value are easily rationalized in terms of a progressive distortion of voltammograms by the increased ohmic drop.[4,38]

Voltammetric changes of purely electrochemical origin are only observed within a supporting electrolyte concentration range comprised between a few times C° and *ca.* $0.1\times C^o$, because of the electroneutrality law.[4] Indeed, let us assume, for the sake of simplification, that the supporting electrolyte concentration is almost null in the bulk solution (considered of infinite volume *vis a vis* those of diffusion layers). Whenever an ion will be produced because of electron transfer at an electrode, local electroneutrality of the solution requires that a counter ion is drawn out from the bulk solution, so that the local charge is null. This establishes that the excess of charge created in each diffusion layer, because of electrode reactions, is necessarily compensated by an identical excess charge of the opposite sign, due to ionic material educed by migration from the bulk solution. This is represented for oxidation of ferrocene or reduction of cobalticinium, in Figure 6. Therefore, it should be understood that the supporting electrolyte concentration within each diffusion layer is almost independent of that of the supporting electrolyte in the bulk of the solution (C^s), as soon as C^s is smaller than C^o. On the other hand, when C^s is a few times larger than C^o, the relative variations of the supporting electrolyte concentration in the diffusion layers is small, which is another way of saying that migrational components to the overall current are negligible.

This simple reasoning, which summarizes the conclusions of a more sounded theoretical analysis of the problem,[4] establishes that, for almost any reactivity purpose, there is positively no interest in decreasing the supporting electrolyte concentration much below *ca.* one tenth of C^o, since its actual concentration in the diffusion layer, *i.e.* in the very place where the reactivity is examined, never decreases below a value comparable to the substrate concentration in the bulk. Therefore, in terms of reactivity, no clear-cut advantage (except may be glory!) may be earned for $C^s \leq 0.1 \times C^o$. At the opposite, this may result in rather severe disadvantages due in particular to ohmic drop and double layer phenomena.[38] The only two exceptions we see to this rule are (i) when one wants to examine double layer effects on the electrochemical and/or chemical properties of a system, or (ii) when the electroactive species itself is unstable in the presence of supporting electrolyte. In the following, we want to describe an example of the second situation.

4.1.2. *Application to the mechanism of formation of Collman's reagent.* Collman's reagent, $Fe(CO)_4^{2-}$,2 Na^+, is usually prepared upon reduction of ironpentacarbonyl by two molar equivalents of sodium[39] or *e.g.* sodium naphthalenide.[40] Noteworthy, the use of only one equivalent of reducing agent yields quantitatively the dimer $Fe_2(CO)_8^{2-}$.[40] This had been tentatively accounted for by considering that one electron transfer to $Fe(CO)_5$ affords the 17e radical $Fe(CO)_4^{\bullet-}$, whose dimerization gives the binuclear iron carbonyl dianion:[41]

$$Fe(CO)_5 + e \longrightarrow Fe(CO)_4^{\bullet-} + CO \tag{16}$$

$$2\ Fe(CO)_4^{\bullet-} \longrightarrow Fe_2(CO)_8^{2-} \tag{17}$$

The main conclusion to be drawn from this mechanism, is that electrochemical generation of Collman's reagent should be impossible; indeed, one should obtain the dimer, as in eq.17; since the latter is not electrochemically reducible to $Fe(CO)_4^{2-}$, no electrochemical access to Collman's reagent should exist from $Fe(CO)_5$. However, we questioned this conclusion because several factors suggested that the above mechanism could be erroneous.[37]

From a previous work[42] we knew that one electron reduction of iron pentacarbonyl affords the 19e radical $Fe(CO)_5^{\bullet-}$, rather than directly the 17e derivative as featured in eq.16 (dissociative electron transfer). Indeed, $Fe(CO)_5^{\bullet-}$ is efficiently trapped in the presence of $(n\text{-Bu})_3SnH$, to give the corresponding formyl anion:[42]

$$Fe(CO)_5^{\bullet-} + (n\text{-Bu})_3SnH \longrightarrow Fe(CHO)(CO)_4^- + (n\text{-Bu})_3Sn^\bullet \tag{18}$$

$$2\ (n\text{-Bu})_3Sn^\bullet \longrightarrow (n\text{-Bu})_3Sn\text{-}Sn(n\text{-Bu})_3 \tag{19}$$

However, the chemical irreversibility of the voltammetry of $Fe(CO)_5$ up to scan rates of 300,000 $V.s^{-1}$, illustrates the vanishing small life time of this 19e intermediate

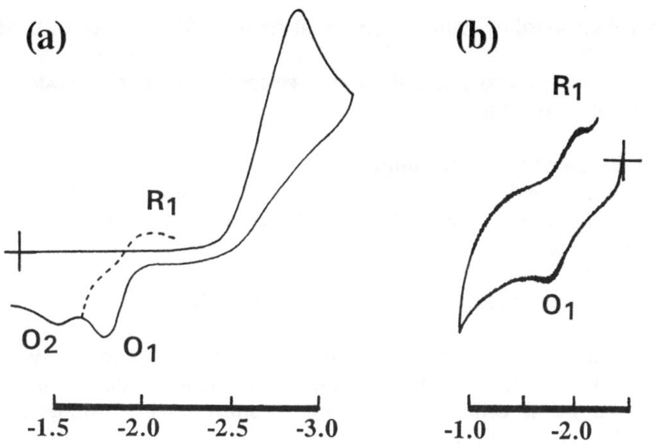

Figure 7. (a) Voltammetry[37] (20 $V.s^{-1}$) of iron pentacarbonyl, 1 mM, reduction in THF, 0.3 M NBu_4BF_4, at a gold disk electrode (r_0 = 0.5 mm). (b) Voltammetry (20 $V.s^{-1}$) of Collman's reagent, 2 mM, in THF/HMPA (95:5, v:v) without added supporting electrolyte, at a gold disk electrode (r_0 = 62.5 μm). 20°C. Note that current scales are different in (a) and (b).

($t_{1/2} \ll 0.3$ μs). Of interest is the fact that at moderate scan rates, two oxidation waves (*viz.* O_1 and O_2 in Figure 7a) are observed upon scan reversal. The more positive one, O_2, is readily identified by its comparison to an authentic sample, and features the one electron chemically reversible oxidation of $Fe_2(CO)_8^{2-}$,[37] that seems to support the mechanism in eqs.15-16. However, double step chronoamperometry, performed with a first step at a potential located on the plateau of $Fe(CO)_5$, and a second step on the plateau of wave O_1 (*viz.* between waves O_1 and O_2) indicates that wave O_1 accounts quantitatively for all reduction products of $Fe(CO)_5$. Such a result establishes that $Fe_2(CO)_8^{2-}$, oxidized at O_2, is not a primary reduction product, but rather originates from the oxidation undergoing at wave O_1. This suggests that the product oxidized at O_1 is the dianion $Fe(CO)_4^{2-}$, formed upon reduction of $Fe(CO)_5$ by an ECE sequence (eqs.12-14). Its reoxidation at wave O_1, would then afford $Fe(CO)_4^{\bullet -}$, whose dimerization in eq.17 ($k_{dim} = 5\times10^5$ $M^{-1}s^{-1}$) provides $Fe_2(CO)_8^{2-}$.

The only way to validate such hypothesis is to compare wave O_1 to the oxidation wave of an authentic sample of Collman's reagent. Unfortunately, this strong base is not sufficiently stable in the presence of tetraalkylammonium salts (Hoffman reaction), to allow its voltammetry to be performed. We thus resorted to voltammetry without added electrolyte. Yet, to obtain a peak-shaped voltammogram and not a sigmoïd shaped one requires the use of scan rates that are moderately large to suppress edge effects (see above and compare *e.g.* Figure 1). The best compromise between reduced ohmic drop and minimization of edge effects under these conditions, is achieved with a 125 μm diameter disk electrode. Figure 7b presents the oxidation voltammogram of $Fe(CO)_4^{2-}$,2 Na^+ under these conditions.[37] Note that, besides providing an easy identification of wave O_1, and therefore validating the mechanism in eqs.12-14, this voltammogram allowed also, and for the first time, a measure of Collman's reagent standard oxidation potential at E^0 = -1.95 V *vs* Ag/0.02 M $AgBF_4$.

4.2. Voltammetry in highly resistive solvents. Application to the mechanism of oxidative addition to zerovalent palladium

Oxidative addition is one of the most ubiquitous reactions in organometallic chemistry and in catalysis of organic reactions by organometallic reagents.[43] Indeed, it allows the activation of a selected bound in a molecule, through the overall insertion of a zerovalent transition metal center, $\mathcal{M}^0$, between two atoms:

$$\text{Substrate-X} + \mathcal{M}^0 \longrightarrow \text{Substrate-}\mathcal{M}^{II}\text{-X} \qquad (20)$$

The organometallic species thus formed is, for example, easily substituted by nucleophiles (at the X moiety), or, after reduction, is prone to undergo a second oxidative addition. Reductive elimination from the metal center then affords a substituted substrate or a homo- or heterocoupling product, accordingly.

Despite the importance of these processes in everyday organic and organometallic chemistry, the exact nature of the oxidative addition in eq.20 remains unknown, *i.e.* except those seldom cases where radical pathways have been identified.[44] Two main class of mechanisms have been proposed for this reaction,[43,45] that can be summarized as follows:

– ***non-ionic mechanisms***, represented for example by a concerted insertion:

– ***ionic mechanisms***, most typically represented by electron transfer ones:

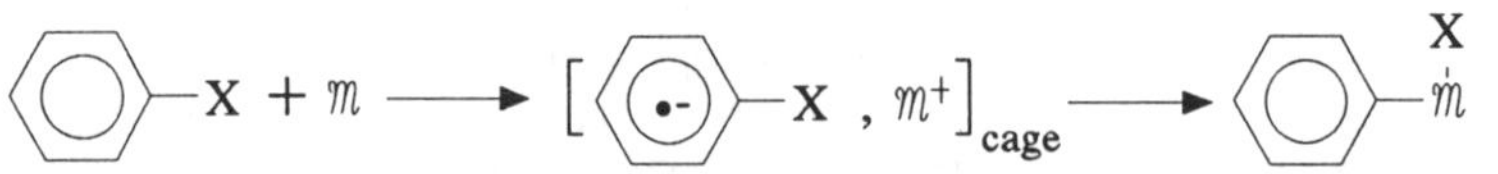

Several attempts have been performed to decide between these two main classes. Yet, the main difficulties of most chemical strategies devised insofar stem from the fact that they are all based on the identification of products originating from cage escape along a possible ionic mechanism. As such, even when products such as ArH, arising from the aromatic halide anion radical, are detected, there are observed at trace levels, owing to the important selectivity of the overall oxidative addition.[43,45] Therefore, in our sense, any conclusion is venturous. Indeed, if for example traces of ArH are detected, one cannot conclude whether they arise from a very few cage escapes taking place along a main reaction path passing through the above ionic mechanism, or if they represent the result of 100% cage escape for all system that seldom react through an ionic mechanism.

In fact, the present available data on aromatic halide anion radicals reactivity seem to support the second hypothesis, since their life time is generally much greater than that corresponding to that of a solvent cage.[46] On the other hand, it must be realized that the known reactivity on aromatic halide anion radicals rests only on cases where the anion radical is a free species, whereas, in the above ionic mechanism, the anion radical is considered to be coordinated at least partially by the metallic moiety. This may induce several bond distortions *vis a vis* a free anion radical, which may help to decrease considerably the life time of the

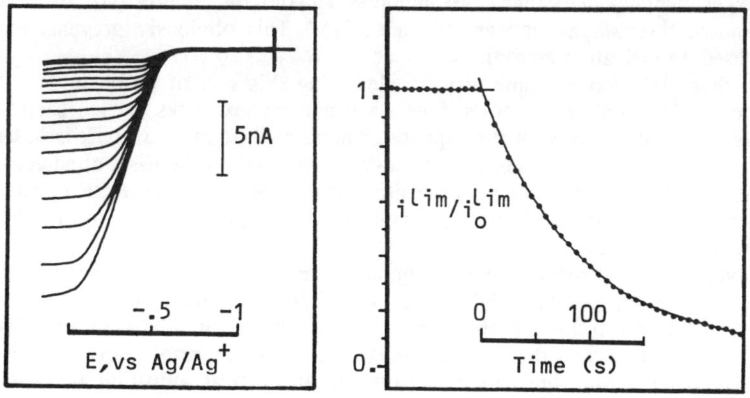

Figure 8. Oxidative addition of iodobenzene (3 mM) to $Pd^0(PPh_3)_4$ (1.5 mM) in the presence of PPh_3 (50 mM) at 25°C in toluene, 0.06 M $NHex_4BF_4$, as followed by the decay of the zerovalent palladium steady state oxidation wave at an ultramicroelectrode (r_0 = 5 μm, 0.2 $V.s^{-1}$).[6]

species. For example, coordination of the metal center to the negative charge in the π^* orbital and to the halide moiety, may well "bend" the carbon-halogen bond, thus allowing a strong coupling between the π^* and the carbon-halogen σ^* orbitals. This may decrease considerably the activation barrier for the carbon-halogen bond cleavage *vis a vis* that determined for free anion radicals.[46] It is thus understood that any discussions based on side-product traces, or on classical knowledge of anion radicals chemistry, are highly hazardous.

Other strategies have been proposed for addressing similar situations (*e.g.* SET *vs* SN_2, or SET *vs* SE_2), but they result non-applicable here. For example, one may use arguments based on linear free energy correlations (*e.g.* Hammett correlations), yet, and in agreement with experimental observations,[43] both mechanisms predict an increase of the rate constants upon increasing either electron accepting capabilities of the organic moiety or electron donating ones for the transition metal center. However, since the theoretical correlation slope (*viz.* ρ in Hammett correlations) for each mechanism is obviously unknown, no diagnostic can be inferred from such experimental correlations. Similarly, one may conceive that, since the geometric requirements in the ionic mechanisms are less than in the non-ionic ones, this should be reflected by a smaller activation entropy.[47] Unfortunately, experimental access to such data is precluded here by the fact that the transition metal moiety that participates in oxidative addition is not the main bulk species, but a low coordinated transient one formed upon a very endergonic equilibrium of unknown equilibrium constant. For example if one considers oxidative addition of iodobenzene to the zerovalent $Pd^0(PPh_3)_4$, the reaction (eq.23) involves the low ligated $Pd^0(PPh_3)_2$ transient species, formed along eqs.21-22:[6,48]

$$Pd^0(PPh_3)_4 \longrightarrow Pd^0(PPh_3)_3 + PPh_3 \qquad (21)$$

$$Pd^0(PPh_3)_3 \Longleftrightarrow Pd^0(PPh_3)_2 + PPh_3 \qquad (K \ll 1) \qquad (22)$$

$$Pd^0(PPh_3)_2 + C_6H_5\text{-}I \longrightarrow C_6H_5\text{-}Pd^{II}I(PPh_3)_2 \qquad (23)$$

Then, the experimentally determined $\Delta S^{\#}$ involves kinetic contribution (*viz.* from eq.23) as well as unknown thermodynamic ones (from eq.22).[6] This obviously prevents any viable discussion based on activation entropies.

However, there is a more simple way of addressing this sort of problems (*viz.* ionic *vs* non-ionic mechanism) that all chemists have learned from textbooks, which seems to have unfortunately been lost recently, at the expense of more sophisticated ones. It is based on the classical SN_1 (ionic) *vs* SN_2 (non-ionic) distinction, performed on the basis of solvent polarity effects. Thus an SN_2 reaction is almost insensitive to changes of the dielectric constant of the solvent, whereas an SN_1 process is highly dependent on this factor, owing to the stabilization of an intermediate carbocationic species.

Taking advantage of the reduced ohmic drop at ultramicroelectrodes, the oxidative addition of a series of *para*-substituted iodobenzenes to $Pd^0(PPh_3)_4$ could then be monitored in toluene,[6] ($\varepsilon_s = 2.4$) by following the decay of the zerovalent oxidation wave (compare Figure 8, previous page), and the results compared to those reported previously for THF[48] ($\varepsilon_s = 7.6$). The reaction was thus shown to have almost identical activation parameters (*e.g.*, $\Delta H^{\#} = 18\pm1$ kcal.mol^{-1}, $\Delta S^{\#} = 1.7\pm1$ cal.mol^{-1}.K^{-1}, in toluene) and lead to identical Hammett correlations (*e.g.*, $\rho = 2.3\pm0.2$, in toluene) in both solvents, which supports a non-ionic mechanism.

5. Conclusion

In this short presentation we have tried to demonstrate that the introduction of ultramicroelectrodes in electrochemistry has opened new frontiers to the method, particularly for what regards applications to organometallic reactivity in relation with "homogeneous" chemistry. We hope that some of the strategies proposed among this limited series of examples may provide useful guidelines for developing other researches in the area.

6. Acknowledgements

Most of this work was supported in parts by CNRS (URA 1110, Activation Moléculaire) and Ecole Normale Supérieure. It is also my great pleasure to acknowledge the essential contributions of my co-workers at ENS, Drs. A. Jutand, J.N. Verpeaux and C. Lefrou, as well as those from Drs. R.M. Wightman (Chapel Hill, North Carolina) and P.J. Krusic (du Pont de Nemours, Wilmington, Delaware) that were made possible because of two independent NATO grants for international research. Finally, I want also to express my gratitude to those students and post-docs whose constant efforts only gave a reality to this work: Drs. M. Azzabi, P. Calas, B. Fosset, S. Pedersen, F. Pflüger, Y. Rollin, at ENS, and Drs. J. Bartelt, M.R. Deakin and S.M. Drew, from R.M. Wightman's research group.

7. References.

1. (a) B.A. Narayanan, C. Amatore, C.P. Casey, J.K. Kochi, *J. Am. Chem. Soc.*, *105*, **1983**, 6351. (b) F. M'Halla, J. Pinson, J.M. Savéant, *J. Am. Chem. Soc.*, *102*, **1980**, 4120.
2. See *e.g.*: M. Catellani, G.P. Chiusoli, V. Fornasari, *Gazz. Chim. Ital.*, *120*, **1990**, 779.
3. R.M. Wightman, D.O. Wipf, *Voltammetry at Ultramicroelectrodes*, in *Electroanalytical Chemistry*, A.J. Bard, Ed., Dekker, New York **1989**. Vol.15, pp.267-353.

4. (a) C. Amatore, M.R. Deakin, R.M. Wightman, *J. Electroanal. Chem.*, *220*, **1987**, 49. (b) C. Amatore, B. Fosset, J. Bartlelt, M.R. Deakin, R.M. Wightman, *J. Electroanal. Chem.*, *256*, **1988**, 255.
5. R. Lines, V.D. Parker, *Acta Chem. Scand.*, *Ser.B*, *31*, **1977**, 369.
6. C. Amatore, F. Pflüger, *Organometallics*, *9*, **1990**, 2276.
7. (a) J.O. Howell, R.M. Wightman, *J. Phys. Chem.*, *88*, **1984**, 3915. (b) L. Geng, A.G. Ewing, J.C. Jernigan, R.W. Murray, *Anal. Chem.*, *58*, **1986**, 852. (c) L. Geng, R.W. Murray, *Inorg. Chem.*, *25*, **1986**, 3115.
8. J.D. Norton, H.S. White, S.W. Feldberg, *J Phys Chem.*, *94*, **1991**, 6772.
9. C. Amatore, C. Lefrou, *J. Electroanal. Chem.*, *296*, **1990**, 335.
10. See *e.g.*: P. Pastore, F. Magno, I. Lavagnini, C. Amatore, *J. Electroanal. Chem.*, *301*, **1991**, 1.
11. J. Newman, *J. Electrochem. Soc.*, *117*, **1970**, 198.
12. A.J. Bard, L.R. Faulkner, *Electrochemical Methods*, J. Wiley & Sons, New York, **1980**.
13. J.C. Imbeaux, J.M. Savéant, *J. Electroanal. Chem.*, *28*, **1970**, 325.
14. D.O. Wipf, A.C. Michael, R.M. Wightman, *J. Electroanal. Chem.*, *269*, **1989**, 15.
15. J. Heinze, *Ber. Busenges. Phys. Chem.*, *85*, **1981**, 1096.
16. Figure 1 is adapted from previous similar considerations in: W.J. Bowyer , E.E. Engelman, D.H.Evans, *J. Electroanal. Chem.*, *262*, **1989**, 67.
17. This value is selected, since at the scan rates considered a voltammogram half-width exceeds generally 60 mV, the theoretical value for a Nernstian one electron voltammogram,[12] owing to limited rate constants of electron transfer (compare Figures 2-4). Therefore, a 10 mV error represents generally an error comparable to or even smaller than the actual experimental uncertainty on defining the peak potential position.
18. C. Amatore, C. Lefrou, F. Pflüger, *J. Electroanal. Chem.*, *270*, **1989**, 43.
19. D. Garreau, P. Hapiot, J.M. Savéant, *J. Electroanal. Chem.*, *281*, **1990**, 73.
20. C. Amatore, C. Lefrou, *J. Electroanal. Chem.*, *in press*, **1992**.
21. R.M. Wightman, D.O. Wipf, *Acc. Chem. Res.*, *23*, **1990**, 64.
22. C.P. Andrieux, P. Hapiot, J.M. Savéant, *Chem. Rev.*, *90*, **1990**, 723.
23. P. Delahay, *Double Layer and Electrode Kinetics*, Interscience, New York, **1965**, 153.
24. A.N. Frumkin, *Z. Phys. Chem.*, *Ser. A*, *164*, **1933**, 121.
25. C.J. Schlesener, C. Amatore, J.K. Kochi, *J. Phys. Chem.*, *90*, **1986**, 3747.
26. C. Amatore, J.K. Kochi, in *Advances in Electron Transfer Chemistry*, JAI Press, **1991**. Vol.1, pp.55-148.
27. C. Amatore, C. Lefrou, *J. Electroanal. Chem.*, *in press*, **1992**.
28. J.M. Masnovi, S. Sankararaman, J.K. Kochi, *J. Am. Chem. Soc.*, *111*, **1989**, 2263.
29. (a) C. Fabre, R. Fugnito, G. Strzelecka, *C. R. Acad. Sci. Fr.*, *Ser. C*, *282*, **1976**, 175. (b) F. Pragst, U. Seydewitz, *J. Prakt. Chem.*, *319*, **1977**, 952. (c) F. Pragst, R. Ziebig, U. Seydewitz, G. Driesel, *Electrochim. Acta*, *25*, **1980**, 341. (d) C. Amatore, A. Jutand, F. Pflüger, *J. Electroanal. Chem.*, *218*, **1987**, 361.
30. H. Kawata, Y. Suzuki, S. Niizuma, *Tetrahedron Lett.*, *27*, **1986**, 4489. Compare to $k_{dim} = 2.5\times10^{9}\ M^{-1}s^{-1}$, determined in ref.29d.
31. $k_{diff} = 2\times10^{10}\ M^{-1}s^{-1}$, see: C.P. Andrieux, C. Blockman, J.M. Dumas-Bouchiat, F. M'Halla, J.M. Savéant, *J. Am. Chem. Soc.*, *102*, **1980**, 3806.
32. C.A. Amatore, A. Jutand, F. Pflüger, C. Jallabert, H. Strzelecka, M. Weber,

Tetrahedron Lett., *30*, **1989**, 1383.
33. V. Gionis, R. Fugnitto, G. Meyer, H. Strzelecka, J.C. Dubois, *Mol. Cryst. Liq. Cryst.*, *90*, **1982**, 153.
34. J.E. Bartelt, R. M. Wightman, *J. Electrochem. Soc.*, *in press*, **1992**.
35. B. Fosset, C. Amatore, J.E. Bartelt, A.C. Michael, R.M. Wightman, *Anal. Chem.*, *63*, **1991**, 306.
36. C. Amatore, M. Azzabi, P. Calas, A. Jutand, C. Lefrou, Y. Rollin, *J. Electroanal. Chem.*, *288*, **1990**, 45.
37. C. Amatore, J.N. Verpeaux, P.J. Krusic, *Organometallics*, *7*, **1988**, 2426.
38. S.M. Drew, R.M. Wightman, C. Amatore, *J. Electroanal. Chem.*, *317*, **1991**, 117.
39. J.P. Collman, *Acc. Chem. Res.*, *8*, **1975**, 342.
40. H. Strong, P.J. Krusic, J. San Filippo, *Inorg. Synth.*, *24*, **1984**, 157.
41. (a) C.J. Pickett, D. Pletcher, *J. Chem. Soc., Dalton Trans.*, **1975**, 879. (b) N. El Murr, A. Chaloyard, *Inorg. Chem.*, *21*, **1982**, 2206.
42. (a) B.A. Narayanan, C. Amatore, J.K. Kochi, *Organometallics*, *5*, **1986**, 926. (b) B.A. Narayanan, J.K. Kochi, *J. Organomet. Chem.*, *272*, **1984**, C49.
43. J.K. Kochi, in *Organometallic Mechanisms and Catalysis*, Academic Press, New York, **1978**. Part I, Chap. 7, pp.156-183.
44. A. Kramer, J.A. Labinger, J.S. Bradley, J.A. Osborn *J. Am. Chem. Soc.*, *95*, **1973**, 7145.
45. (a) T.T. Tsou, J.K. Kochi, *J. Am. Chem. Soc.*, *101*, **1979**, 6319. (b) I.E. Elson, D.G. Morrell, J.K. Kochi, *J. Organomet. Chem.*, *84*, **1975**, C7.
46. C.P. Andrieux, J.M. Savéant, D. Zann, *Nouv. J. Chim.*, *8*, **1984**, 107.
47. see: D. Lexa, J.M. Savéant,K.B. Su, D.L. Wang, *J. Am. Chem. Soc.*, *110*, **1988**, 7617, for mechanistic diagnostic based on this argument in a similar context (*viz.* SET *vs* SN_2).
48. J.F. Fauvarque, F. Pflüger, M. Troupel, *J. Organomet. Chem.*, *208*, **1981**, 419.

A MICROELECTRODE STUDY OF METAL HEXACARBONYLS IN APROTIC SOLVENTS

M.I. MONTENEGRO and M.A. QUEIRÓS
Centro de Química Pura e Aplicada da Universidade do Minho, Largo do Paço, 4719 Braga Codex, Portugal

ABSTRACT. The effect of the electrolyte concentration and solvent on the electrochemical oxidation of $Cr(CO)_6$ at microelectrodes has been investigated. It is shown that reproducible data are obtained in media without the deliberate addition of electrolyte or in very resistive solvents such as toluene. The voltammetric responses are interpreted in the light of ohmic drop, migration, kinetics and double layer effects.

1. Introduction

The electrochemical behaviour of metal carbonyls in aprotic solvents was investigated by Pickett *et al* (1) and Seurat *et al* (2). The electrochemical oxidation of several metal carbonyls was also described in a mixture of the high Lewis acid, room temperature molten salt, composed of aluminium chloride and ethylpyridinium bromide (2:1 molar ratio) and benzene (50% v/v) (3). In these studies $Cr(CO)_6$ was found to be reversibly oxidised, at a potential of 1.53 V vs SCE, to the seventeen electron cation $Cr(CO)_6^+$ which is stable on the time scale of many seconds. The cyclic voltammogram of $Cr(CO)_6$ shows a second oxidation peak at a potential of 2.06 V for further oxidation of the cation $Cr(CO)_6^+$ and the proposed mechanism is (1):

$$Cr(CO)_6 - e \longrightarrow [Cr(CO)_6]^+$$

$$2[Cr(CO)_6]^+ \xrightarrow{\text{slow}} Cr(CO)_6 + [Cr(CO)_6]^{2+}$$

$$[Cr(CO)_6]^{2+} \xrightarrow{\text{-CO}} Cr^{2+}$$

The electrochemical oxidation of $Mo(CO)_6$ and $W(CO)_6$ does not show the same pattern and no evidence was found for the formation of stable carbonyl cations, most probably due to the greater tendency of these bulkier carbonyls to 7 coordination; hence, the cations formed are prone to nucleophilic attack by the solvent.

A. J. L. Pombeiro and J. A. McCleverty (eds.),
Molecular Electrochemistry of Inorganic, Bioinorganic and Organomettallic Compounds, 645–648.

It is well recognized that the addition of a large excess of an inert electrolyte has an effect on the kinetics and thermodynamics of electrochemical systems and in some cases it would be highly desirable to determine the influence of the electrolyte nature and concentration, and of the solvent, on the behaviour of such systems. The unique properties of microelectrodes allow electrochemical studies to be carried out in non-conventional media, such as very resistive solvents or in the absence of electrolyte (4).

Hence, we set out to investigate the influence of the electrolyte concentration on the oxidation of $Cr(CO)_6$, in acetonitrile, at platinum microelectrodes. The electrolyte used was Bu_4NBF_4 (TBAB) and the effect of decreasing its concentration down to zero was investigated. Furthermore, other solvents with lower donor numbers were used such as toluene and xylene.

2. Results and Discussion

The I-E curve for the oxidation of $Cr(CO)_6$ at a platinum microelectrode, in acetonitrile, presents two oxidation waves with half wave potentials at 1.45 and 2.04 V vs SCE, as shown in figure 1a).

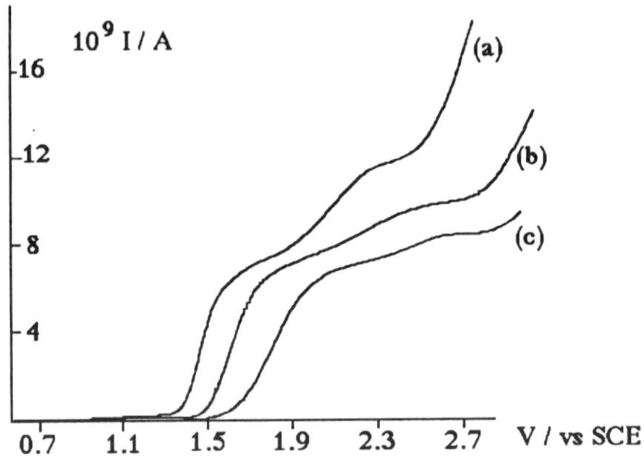

Figure 1. I-E curves for the oxidation of $Cr(CO)_6$, $2x10^{-3}$ M at a Pt microdisc (r = $5x10^{-4}$ cm) in acetonitrile containing (a) 0.1 M TBAB; (b) 10^{-3} M TBAB ; (c) no electrolyte.

The first wave corresponds to a one-electron reversible process as suggested by the Tomes criterion, i.e., $|E_{3/4}-E_{1/4}|$ = 60 mV. The diffusion coefficient, determined by the limiting current and for n=1, was $1.4x10^{-5}$ cm^2 s^{-1}, in agreement with the value obtained in cyclic voltammetry for scan rates above 300 mV/s, when the process is reversible. Furthermore, the $E_{1/2}$ values were the same ($\pm$ 5 mV) when different electrode radii were used (5, 12.5 and 25 μm) confirming the reversibility of the process (5).

As the concentration of the electrolyte decreases, the limiting current is unaltered but, in contrast, the half wave potential shifts to more negative potentials, as shown in figure 1b). When the electrolyte concentration is zero this shift is about 200 mV, a much higher value than that predicted by Oldham (6). This is probably an indication of slow kinetics when the electrolyte concentration decreases as a result of double layer effects.

The second wave is not so well defined but it is clear that it almost completely vanishes when the electrolyte concentration approaches zero. This process corresponds to the oxidation of a positively charged species which will be repelled from the electrode surface as the electrolyte concentration diminishes. On the other hand, it is likely that ion pair formation with the electrolyte anion will favour the second oxidation process and, therefore, absence of the electrolyte will hinder this process. These effects may explain the observed decrease in the limiting current. Table 1 summarizes the voltammetry data for solutions of $Cr(CO)_6$ $2x10^{-3}$ M in acetonitrile, containing different concentrations of TBAB. The corresponding voltammetric data is shown in table 1.

TABLE 1. Voltammetry data for the oxidation of $Cr(CO)_6$ $2x10^{-3}$ M in acetonitrile at a Pt microdisc (d = $5x10^{-4}$ cm), containing different concentrations of electrolyte.

[TBAB]/ M	$E_{1/2}$/ V vs SCE	$\lvert E_{3/4}-E_{1/4}\rvert$ / mV	10^{-9} I_d/ A
0	1.67	130	6.4
$2x10^{-4}$	1.59	120	6.8
$2x10^{-3}$	1.51	100	6.4
$2x10^{-2}$	1.45	69	6.5
$2x10^{-1}$	1.45	60	6.0

Replacing acetonitrile by toluene or xylene, containing TBAB, the $E_{1/2}$ value of the first wave shifs to more positive values, as shown in figure 2, but reversibility is still observed.

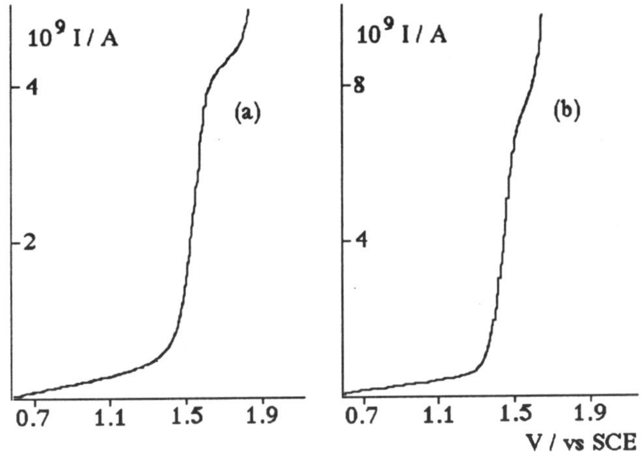

Figure 2. I-E curves for the oxidation of $Cr(CO)_6$ $2x10^{-3}$ M in (a) a mixture of toluene / acetonitrile 9:1; (b) a mixture of xylene / acetonitrile 6:1, containing TBAB 0.1M.

Table 2 summarizes the voltammetry data for the oxidation of $Cr(CO)_6$ in mixtures of toluene / xylene

TABLE 2. Voltammetry data for the oxidation of $Cr(CO)_6$ $2x10^{-3}$ M at a Pt microdisc (d = $5x10^{-4}$ cm), in mixtures of toluene / acetonitrile, containing TBAB 0.1M.

[Toluene]/ %	$E_{1/2}$/ V vs SCE	$\lvert E_{3/4}-E_{1/4} \rvert$ / mV	10^{-9} I_d/ A
0	1.45	60	6.4
50	1.48	70	5.0
70	1.51	70	4.7
90	1.60	70	3.0

The decrease in the limiting current as the toluene percentage increases, results from the lower diffusion coefficients observed in the more viscous media.The shift on the half wave potential is probably a consequence of the lower donor numbers of these solvents which render the oxidation process more difficult (7).

(1) C.J. Pickett and D.Pletcher (1975), J. Chem. Soc., Dalton Trans. 879.
(2) A. Seurat, P. Lemoine, and M. Gross (1980), Electrochimica Acta, 25, 675.
(3) H.L. Chum, D. Koran, and R.A. Osteryoung, (1977), J. Organomet. Chem. 140, 349.
(4) M.I. Montenegro, M.A. Queirós, and J. Daschbach (eds.) (1991), "Microelectrodes: Theory and Applications", NATO ASI Series, Kluwer, Dordretch.
(5) A.M. Bond, K.B. Oldham, and C.G. Zoski (1988), J. Electroanal. Chem. 245, 71.
(6) K.B. Oldham (1988), J. Electroanal. Chem. 250, 1.
(7) I. Chadwick, C. Díaz, G. González, M.A. Santa Ana, and N. Yutronic (1986), J. Chem. Soc., Dalton Trans. 1867.

Acknowledgement

This work was supported by Junta Nacional de Investigação Cienítifica e Tecnológica (project CEN-QUIM 65-90).

THE USE OF MICROELECRODES IN KINETIC STUDIES OF HOMOGENEOUS CHEMICAL REACTIONS OF MOLYBDENUM COMPLEXES

A. M. FONSECA and M. A. QUEIRÓS
Centro de Química Pura e Aplicada, Universidade do Minho
4719 Braga Codex, Portugal

A. R. DIAS
Centro de Química Estrutural, Complexo I, IST, Av. Rovisco Pais
1096 Lisboa Codex, Portugal

ABSTRACT. The rate constants of chemical reactions of the cations obtained after electrochemical oxidation of biscyclopentadienylmolybdenum complexes have been determined, using microelectrodes. Further information from low scan experiments and from the products of chemical oxidation have allowed the proposal of the reaction mechanisms.

1. Introduction

The use of microelectrodes in cyclic voltammetry was first introduced by Wyghtman et al. [1] in 1984 and since then quite a few examples appeared in the literature of their use in fast homogeneous chemical reactions [2,3].

Microelectrodes, due to their small size in the micrometer range, allow the study of fast chemical reactions coupled with heterogeneous electron tranfer reactions. In fact, the very small RC time constant and low solution iR drop associated with microelectrodes, had allowed the achievement of high diffusion rates in fast scan techniques. This means that accuarate voltammetric data can be obtained in high speed voltammetry.

We now report results obtained for the fast chemical reactions that follow the electrochemical oxidation of the complexes $[MoCp_2(ArN_3Ar')]$ $[PF_6]$ and $[MoCp_2(NHCRNH_2)]$ $[PF_6]$.These studies were carried out in CH_3CN solutions with $[Bu_4N][BF_4]$. The standard oxidation potentials were determined from the reversible cyclic

A. J. L. Pombeiro and J. A. McCleverty (eds.),
Molecular Electrochemistry of Inorganic, Bioinorganic and Organomettallic Compounds, 649–652.

voltammograms, the lifetimes being calculated by the Nicholson and Shain's method [4] for an EC process.

2. Results and Discussion

The complexes [MoCp2(ArN3Ar´)] [PF6] undergo a disproportionation reaction after loss of one electron, according to the previously proposed mechanism [5]:

$$[MoCp_2(ArN_3Ar')]^{+} \longrightarrow [MoCp_2(ArN_3Ar')]^{2+} + 1e$$

$$[MoCp_2(ArN_3Ar')]^{2+} \longrightarrow [MoCp_2(ArN_3Ar')]^{+} + [MoCp_2(ArN_3Ar')]^{3+}$$

$$[MoCp_2(ArN_3Ar')]^{3+} \longrightarrow X$$

When Ar = Ar´= C_6H_5 the chemical reaction is very fast and irreversible cyclic voltammograms where obtained at conventional electrodes. Figure 1 shows cyclic voltammograms for this complex 2.0 mmol dm^{-3} in CH_3CN / TBAB 0.2 mol dm^{-3} at a Pt microelectrode (ø = 10 μm).

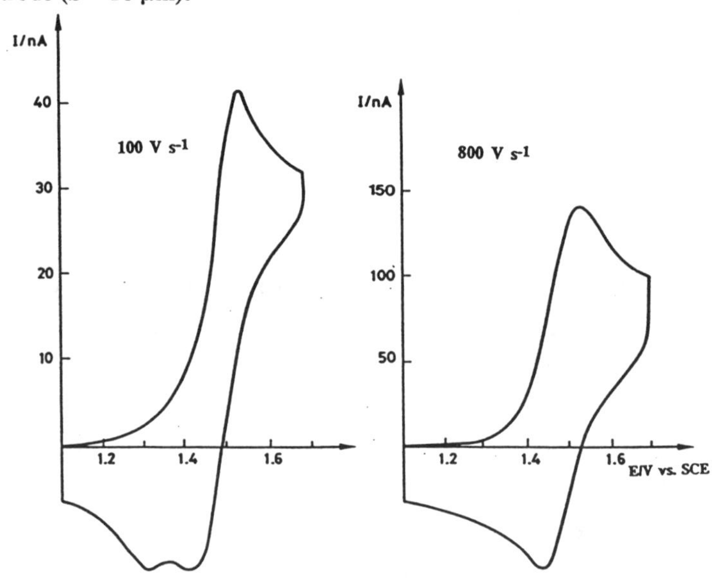

Figure 1. Cyclic voltammograms of [MoCp2(ArN3Ar´)] [PF6] 2.0 mmol dm^{-3} in CH_3CN / TBAB 0.2 mol dm^{-3} at a Pt microelectrode (ø =10 μm).

Table . Kinetic data for complexes [Mo Cp_2 (Ar N_3 Ar´)] [PF_6], A, and [Mo Cp_2 Cl (NHC (C_6H_5) NH_2)] [PF_6], B, 2.0 mmol dm^{-3} in CH_3CN / TBAB 0.2 mol dm^{-3} at a Pt microelectrode (ø=10 μm).

Complex	Ep^a [1] (V)	v (Vs^{-1})	Ip^c/Ip^a	lifetime (s)
A	1.35	100	0. 78	0. 007
Ar,Ar´=C_6H_5		300	0. 89	0. 006
		500	0. 94	0. 006
B	0.87	100	0. 70	0. 019
		200	0. 80	0. 017
		350	0. 93	0. 025

1. Potentials vs. SCE

The kinetic data obtained is summarised in the table.

The complexes [$MoCp_2Cl(NHCRNH_2)$][PF_6], undergo a chemical reaction after electrochemical oxidation. When R = CH_3 this reaction is slow on the time scale of cyclic voltammetry. Evidence for the electroactive product is only found in cyclic voltammograms run after exhaustive electrolysis, as a second oxidation wave is shown, with a peak potential of 1. 15 V . However, when R = C_6H_5 the chemical reaction refered to above is much faster and the cyclic voltammograms show a second oxidation wave with a peak potential of 1.18 V, whose current is dependent on the potential scan rate. The lifetime of the 17-electron species was assessed by cyclic voltammetry at a Pt disc microelectrode. Figure 2 shows cyclic voltammograms obtained from solutions 2.0 mmol dm^{-3} in CH_3CN / TBAB at 100 and 300 V s^{-1}. The lifetimes, calculated according to the Nicholson and Shain's method [4], are shown in the table.

Chemical oxidation of B with [NO] [BF_4] led, after work-up, to the isolation of a product whose oxidation potential was exactly the same as the one observed for the product obtained after electrolysis of B. This compound was characterised by its infrared spectrum and elemental analysis as the benzonitrile derivative of B.

According to these results the following mechanism is proposed:

$[MoCp_2Cl(NHCRNH_2)]^+ \longrightarrow [MoCp_2Cl(NHCRNH_2)]^{2+} + 1e$

$[MoCp_2Cl(NHCRNH_2)]^{2+} \longrightarrow [MoCp_2Cl(N{\equiv}CR)]^{2+} + NH_3$

$[MoCp_2Cl(N{\equiv}CR)]^+ \longrightarrow [MoCp_2Cl(N{\equiv}CR)]^{2+} + 1e$

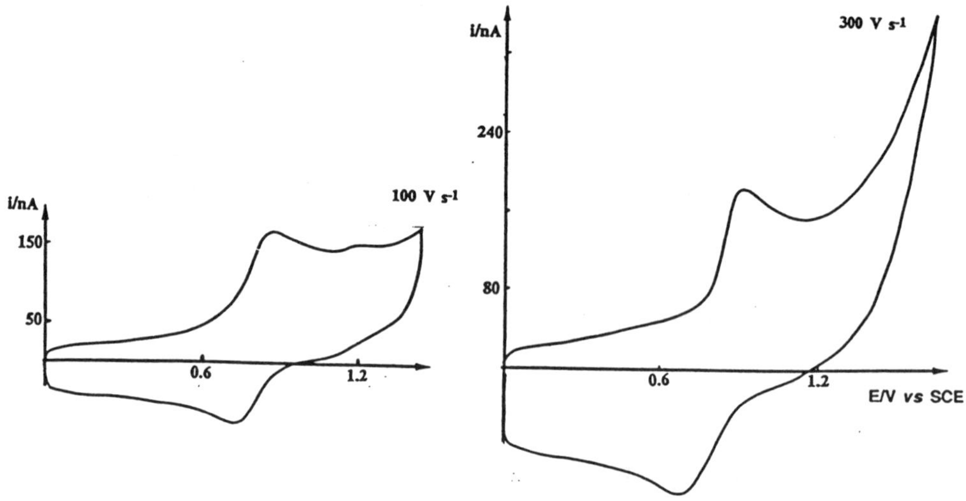

Figure 2. Cyclic voltammograms of [Mo Cp_2 (NHC (C_6H_5)NH_2)] [PF_6] 2.0 mmol dm^{-3} in CH_3CN / TBAB 0.2 mol dm^{-3} at a Pt microelectrode (ø=10 μm).

Acknowledgements

We are grateful to the INIC and Universidade do Minho for financial support

References

1. J. O. Howell and R. M. Wightman (1984) "Anal. Chem." 56, 5.
2. J. M. Savéant (1991), in M.I. Montenegro et. al.(eds.), Microelectrodes: Theory and Applications, Kluwer Academic Publishers, Dordrecht, pp. 307-340.
3. R. M. Wightman and D. O. Wipf (1990) " Ac. Chem. Res." 23, 64.
4. R. S. Nicholson and I. Shain (1964) " Anal. Chem." 36, 706.
5. M. A. Queirós, J. E. Simão and A. R. Dias (1987) "J. Organomet. Chem." 85, 329.

PART 11. GENERAL DISCUSSION, CONCLUSIONS and PROSPECTS

This final section combines the reports on general discussions and main conclusions of the NATO ARW, as well as personal comments (by scientists with complementary expertises) on its contents and significance, and on the prospects of the field of Molecular Electrochemistry of Inorganic, Bioinorganic and Organometallic Compounds.

REPORT 1

ROUND TABLE DISCUSSION

DENNIS TUCK
Department of Chemistry and Biochemistry
University of Windsor, 401 Sunset Ontario
Canada N9B 3P4

On the final afternoon of the Workshop, the participants convened to discuss the posters which were on exhibition, and the discussion eventually became a more general debate, concentrating on essentially three main topics.

The first of these concerned the relationship between electrochemical results, especially redox potentials and cyclic voltammetric data, and molecular structure. An important point, much emphasized, is that a potential relates two states, differing by one electron in total, and it is therefore perhaps somewhat surprising that we should find the many relationships which have been established between redox potentials and such measures of structural property as carbonyl infrared stretching frequencies, bond lengths, NMR frequencies, etc. It is clear that much more understanding is required if these relationships are to be generally accepted. It was suggested that in looking at the development of this field, one should remember that it is in a relatively early stage, and that the early days of NMR spectroscopy generated the same doubts and debates as are presently going on in this field of inorganic chemistry. In summary, it seemed generally agreed that such an approach to structural investigation is useful if the results help us to understand structural relationships and changes in them. Equally, if one can predict the range of conditions under which studies of as yet unprepared species can be carried out, as for example in selecting a solvent, then indeed this is useful practical information. Furthermore, if the various anomalies which appear in the present treatments can be identified and explained, then the whole question of a redox potential-structural relationship should be based upon firmer ground than is presently the case.

A second topic of discussion was whether or not it was important for electrochemists to impress the importance of the field upon those working in other branches of chemistry. It seemed generally agreed there is some reluctance amongst non-electrochemical scientists to attempt electrochemical measurements or electrochemical syntheses, and that this reflects a certain distress which may be traceable to the ways in which electrochemistry is sometimes taught, which is to emphasize the thermodynamic aspects rather than the applications. While everyone

A. J. L. Pombeiro and J. A. McCleverty (eds.),
Molecular Electrochemistry of Inorganic, Bioinorganic and Organomettallic Compounds, 655–656.

accepts that the thermodynamic approach to chemistry and electrochemistry is vitally important, it is also true that an undue emphasis of these matters, particularly in first year courses, can lead to an ignorance of the very dynamic aspects of electrochemistry, whose importance has been reflected in the papers at this workshop.

Finally, there was a general wide ranging discussion on what is meant by the subject of electrochemistry itself. One definition which could have been deduced from the discussion, and from the papers presented, is that electrochemistry is that area of chemical investigation in which the controlled use of electrical energy, and the measurement of electrical parameters, give rise to molecular structural information, and to synthetic methods. This sounds very profound, but can be seen properly in context if one goes back to a much earlier definition, offered by Samuel Glasstone in his classical textbook *The Electrochemistry of Solutions*, first published in 1930. In the opening paragraph of this book, which eventually ran to three editions, Glasstone defined electrochemistry as the study of the interconversion of electrical and chemical energy. This simple statement contains everything that is set out more pompously in the definition proposed above. It may be that if we could urge the use of this simple definition, more chemists would realize that electrochemistry offers a chance of studying synthetic and structural chemistry, in much the way as the many branches of spectroscopy help us to understand what we are trying to do as chemists, which is essentially to understand the relationship between molecular structure and chemical reactions.

REPORT 2

GENERAL DISCUSSIONS AND PROSPECTS

ANTONIN A. VLCEK
J. Heyrovsky Institute of Physical Chemistry & Electrochemistry
Dolejskova 3
18223 Prague 8
Czechoslovakia

The object of molecular electrochemistry, especially of coordination compounds, has been defined as the elucidation of the complete mechanism of the overall process in electrochemical terms, translation of these terms into chemical species and explanation of the observed mechanism on the basis of molecular and submolecular structure.

The discussion emphasized that electrochemistry is a science with excellent theoretical and experimental background making it possible to solve most of the present day problems and to meet the future challenges. There is no need to defend electrochemistry as science as its position is unquestionable, based on the results and contributions to the welfare of mankind. However, warning has been raised to improper use of electrochemistry which might spoil its image in broader scientific community. It has to be welcome that electrochemistry is being more and more widely used as one of the physicochemical methods of investigation of new compounds and processes. However, any such use has to be based on the understanding that electrochemistry investigates dynamics of the system, *i.e.* it deals with reactions and rates rather than static quantities. The latter ones can be extracted from the results of electrochemical measurements, however, only after the kinetic problems have been properly taken care of. It is possible to see many papers in the literature which do not realize this specificity of electrochemistry and the conclusions drawn from experiments (mostly not quite adequate) are either doubtful or wrong. It has been stressed that more has to be done to bring this point to general knowledge in chemical community.

The meeting concentrated also on the role of chemical reactions coupled with electrode reactions, *i.e.* more on chemistry studied electrochemically. Electrochemistry is an excellent method for studying chemistry of short living intermediates and the newly developed — or developing — procedures make it possible to study these processes in more detail. The use of ultramicroelectrodes combined with very high sweep rates and the possibility of using these electrodes in

A. J. L. Pombeiro and J. A. McCleverty (eds.),
Molecular Electrochemistry of Inorganic, Bioinorganic and Organomettallic Compounds, 657–658.

inert media open new prespectives for the study of chemical reactions of products, and/or to cut off any follow-up reactions and thus to extract the electrode reaction, and its chemistry, in its simplest form.

The very high sweep rates make it possible to measure fast electron transfer (ET) at the electrode surfaces. By proper development of the theory of ET, nowadays developed predominantly for homogeneous or photochemically induced ET, to the electrochemically induced ET it might be possible to study structural and solvent dependencies of the electrode reaction rates. The application of the ET theory is of importance also in the newly begun studies of vectorial ET in orientated assemblies, such as oriented films at the electrode surfaces, the films beeing composed of bigger and longer molecules carrying several distinct redox centres. Such studies may be expected to grow significantly in importance in the future. The theory of electrochemically induced ET in these systems would be of great value.

The discussion dealt also with the problem of tailoring of redox potentials, and most recent approaches were presented, mainly based on empirical parameters. Taking into account the simplifying assumptions made in these correlations, the fit of experimental and calculated values is remarkably good. It has, of course, to be emphasized that the parameterisations are mainly for predictions of the region where to look for the redox potential of the given species and, unless understood in more detail, should not be used for any deeper analysis of redox potentials. The discussion mentioned that tuning of electrode potentials should be extended and the more reasonable approach seems to be the tuning of the electrode process, *i.e.*, not only of the redox potential but of the specific reactivity of the products also.

Spectroelectrochemical studies revealed some new results on the nature and stability of products and intermediates. Progress has been made in these predominantly thin layer studies, by extending their use to low temperatures and a wide variety of nonaqueous solvents. It seems thus to be possible in the future to study spectroelectrochemically rather unstable products. The spectroelectrochemistry has, of course, the obligation not only to measure spectra of new compounds produced at electrodes but also to interpret them. Spectroelectrochemical studies have proven to be of value in the detailed investigation of sequential multielectron processes, as emphasized in several lectures and posters and also in the discussion.

The future trends, as they emerged from the proceedings of the Workshop, emphasize the need for the continuation and development along those lines which have shown their usefulness in producing results of general interest for chemistry (and other branches of science). Among those, homogeneous electrochemical catalysis must be mentioned, especially when carried out on electrodes with anchored catalyst. This latter field promises further development, especially when the current tendency to tailor properly not only the catalyst but the whole catalytic cycle is developed to its full potential. The synthetic use of electrochemistry, mainly in the area of organometallics, represents a field with good expectations. The knowledge of processes taking place is improving and provides a well developing theoretical background for future studies. Last, but not least, the possibility of using electrochemically active compounds with multiple redox centers as components of molecular electronic devices has been outlined as a challenge for future development of molecular electrochemistry of coordination compounds.

REPORT 3

GENERAL COMMENTS

CHRISTIAN AMATORE
École Normale Supérieure, Laboratoire de Chimie,
URA CNRS 1110
24 rue Lhomond, F-75231 Paris cedex 05
France

There are at least two ways chemistry can be viewed. A first one, that may be defined as ***static***, deals with structures, *viz.* molecular architectures and their relationships to particular properties. The second one, that we may term ***dynamic***, focusses on the reactivity of molecules on the unravelling of the intricacies of their transformations. These two aspects are usually "worshiped" in different "churches", according to different "ceremonies". It is the merit of the NATO ARW held in Sintra to have allowed chemists belonging to these two different "churches" to meet together, discuss (sometimes very lively, as often it is on matters of "religions" and "semantics", but always with mutual profit), and realize that they possibly were "worshiping" the same "god".

Indeed, the Conference was a subtle and equilibrated blend of lectures from those chemists deploying sophisticated experimental, as well as theoretical, strategies for unravelling the mysteries of fine metal-ligand interactions, with particular emphases on electron transfer or electrochemical properties, and from those chemists trying to understand how the same effects influence the way these molecules react and how (and why?) they possess the talent of affecting and controlling the reactivity of other molecules in catalytic reactions. This delicate combination was reflected also in the audience or in the poster session, as it is now in this book, with the obvious result of stimulating important and constructive discussions. However, I will restric my considerations to the ***dynamical*** aspects.

Odd electron organometallic or inorganic complexes were once viewed as isolated, and often transient violations of the 18-electron rule. Obviously, the electrochemical behavior of several transition metal complexes has been extensively investigated since the leading works of Vlcek and Dessy, showing that 17-electron or 19-electron complexes could be prepared and their reactivity studied by electrochemistry. Yet, with a very few exceptions, these species have not been recognized as the important key intermediates they actually are, up to the last decade when the influential book "*Organometallic Mechanisms and Catalysis*" by J.K. Kochi

A. J. L. Pombeiro and J. A. McCleverty (eds.),
Molecular Electrochemistry of Inorganic, Bioinorganic and Organometallic Compounds, 659–660.

appeared and stressed the crucial role of radical processes in organometallic reactivity and catalysis of organic reactions. It is thus not a surprise that the field of organometallic and inorganic electrochemistry was initially mainly restricted to the study of reversible electron transfers, with a main accent on thermodynamics *viz.* on ***static*** properties.

Unfortunately, restricting electrochemical techniques only to such applications discards their unique potentialities for solving mechanistic problems of molecular reactivity, *viz.* ***dynamic*** properties. These capabilities stem from the two following properties of electrodes. On one hand, they can be viewed as extremely large reagents, whose reducing or oxidizing properties can be tuned continuously and with precision to adjust at will the driving force of a reaction. On the other hand, the ensuing current flow through the electrode is a direct measurement of the kinetics stimulated by the electron exchange at the electrode. It is noteworthy and certainly unique that the same device (the electrode) can be used to generate intermediates (potential) and to analyse at the same time (current) their chemical reactivity. This, associated with the fact that the accessible time scales extend now from several seconds to a few tens of a nanosecond with direct electrochemical techiques (electrodes), or even to a few picoseconds with indirect methods (mediated electrochemistry), evidences the unique suitability of electrochemical techniques for the investigation of chemical reactivity of inorganic, organometallic and bioinorganic compounds.

Initially, electrochemical methods were restricted to rather conducting media (solvents with large dielectric constants, large ionic strength), because of problems related to a precise control of an electrode-solution potential difference. Since the rapid development of ultramicroelectrodes in the past decade, this is no more the case, and electrochemistry in solvents of low dielectric constants (arenes, alkanes,*etc*) or in the absence of deliberately added electrolytes is feasible routinely. Both the theoretical and experimental aspects of such unconventional electrochemistries are now well mastered and the techniques ready for use and application. We have no doubt that the breaking of these historical frontiers of electrochemistry will allow a rich harvest of mechanistic studies performed under conditions that may now approach those used in homogeneous reactions and in catalysis.

Obviously these developments will not be possible by relying on electrochemistry only. Indeed, as we have explained above, electrochemistry is a splendid and powerful tool for the unravelling of mechanistic intricacies that are hardly accessible to other physicochemical methods. However, its Achilles' heel is the fact that electrochemistry is rather blind to chemical structures. Transient electrochemical methods will certainly be able to establish with a considerable subtlety how an intermediate reacts, yet for the most cases they will not be able to tell the structure of this intermediate. This requires then its use in conjunction with other chemical techniques and strategies. In this respect, the knowledge based on structural (*viz.* ***static***) studies will certainly be valuable. However, the coupling of ***dynamic*** electrochemical techniques with "classical" spectroscopic methods (NMR, IR, ESR, *etc*) seems also highly worthwhile and desirable. It is then one of the interests of this NATO ARW Conference to have provided several circumstances where these aspects have been examined and discussed.

INDEX

Acetylide complexes (see Alkynyl complexes)
Adiabatic pathways 195
Adsorption of dyes 549, 553-555
Alcohol 351,
electrocatalytic reduction 423, 426
Aldehyde 351
Alkene complexes 573-581
Alkoxide complexes 78
Alkyl complexes 351, 354
Alkyne based radicals 320
Alkyne complexes 317-326
Alkyne oligomerization,
redox induced 317, 325
Alkyne-derived complexes 331, 332
Alkynes 352
Alkynyl complexes 322, 335
Allene complexes 331, 332, 335
Aluminium compounds 255
Amavadine models 411
Ambidentate nitrogen ligands 258
Amide complexes 247
Amine-substituted benzo-crown ethers 79
Aminoacids 411, 413
Aminocarboxylate complexes 489-493
Aminocarbyne complexes 331, 332, 335, 338, 339, 341,357, 368, 369
Ammonia, electrosynthesis 369
Anilido species 79
Antenna-sensitizer complexes 227
Anthracene and derivatives 628, 629, 634
Aryl complexes 351, 354
Aryl halides (see Organohalides)
Autoxidation 391, 401, 403
Azametallic electrochemistry 357
Azurin 135
π-Back donation 258, 262
Benzidine diamide ligands 250
Bifunctional ligands 89-92
Bioelectrochemistry 133
Biphenylene quinone diimine 247, 249
2,2'-Bipyridine complexes 42, 83, 218, 258, 534, 583, 615, 616
Bis(pyrazole)bis(thioether) ligand 171-176
Blue copper proteins
synthetic models of the type I site 171-176
Bond cleavage or formation (see ET induced reactions)
Boron compounds 255
Buckminsterfullerene,
electroreduction 595, 603, 608
spectroelectrochemistry 595, 609
multiple cattering X-alpha calculations 595
UV and near-IR spectra 595, 596, 598
frontier orbitals 597, 598
redox potential-electron spectroscopic data correlations 610
Bulk electrolysis 5, 57, 63, 114, 125, 299, 301, 311, 331, 338, 340, 345, 347, 353, 354, 398, 399, 400, 401, 413, 414, 483, 484, 519, 573, 579
Carbene complexes (see also Vinylidene complexes) 57, 59, 345-348
Carbon dioxide
activation 307, 309, 310
electrocatalytic reduction 295, 311, 314
Carbon monoxide
oxidoreductase 141
Carbonyl complexes 33-38, 41, 42, 46, 51, 53, 57-66, 77, 95, 96, 193, 194, 202, 203, 205, 233-243, 256-266, 269-274, 277, 284-291, 301-304, 315-318, 332, 335-338, 358, 417-420, 431, 467, 469, 470 , 519, 520, 522, 523, 525-530, 545, 573-581, 590-592 , 636, 638, 639, 645-648
Carbonyl IR stretching frequencies 203, 204, 520, 523, 524, 527, 529
Carbyne complexes (see also Aminocarbyne complexes) 322, 331, 332, 342
Catalysis (see also Electrocatalysis)
redox 295, 391, 392, 403
chemical 295
molecular 295
Catalytic efficiency 295, 299, 301
Chain reaction 264, 265
Charge change localization 4,5,8
Charge separation 83, 85
Chronoamperometry 437
Clusters (see High nuclearity metal clusters)
Co-substrate
magnesium ion 295, 301

A. J. L. Pombeiro and J. A. McCleverty (eds.),
Molecular Electrochemistry of Inorganic, Bioinorganic and Organomettallic Compounds, 661–667.

Cobaloxime-derived complexes 391, 393
Collman's reagent 638, 640
Computer based methods 437,
  convolution methods 437, 438, 439
  convolution-deconvolution 444, 464, 469, 470, 474
  digital simulation 269, 271, 340, 341, 437, 445, 454, 470, 477-481, 483, 528
  EC process 450
  implicit method 461, 462
  interactive treatment 437
  local truncation method 459
  numerical methods 453, 477-481,
    Adams-Moulton 477, 479, 481
    Euler modified method 477, 479, 480, 481
    Milne 477
    Runge-Kutta 477, 479, 480, 481
  simple ET 447
Comproportionation 256
Conducting polymer 549, 557, 558, 563
Convolution-deconvolution techniques (see Computer based methods)
Copper proteins
  electrochemistry 152
  properties and functions 152
  synthetic models 153
  types 151
  voltammetry of type I models 154
  voltammetry of type II models 158
Coulombic interaction 467, 474
Coulometry (see Bulk electrolysis)
Counter-ion effect 285
Crank-Nicholson formula 461
Creatine kinase 143
CV simulation studies (see Computer based methods)
Cyanamide complexes 331, 338, 339, 341
Cyanide complexes 33, 34, 36, 357, 368
Cyclic polyether 79
Cyclic voltammetry 57, 58, 63, 64,114, 125, 138, 218, 255, 280, 296, 300, 311, 320, 325,331, 333, 338, 345-348, 352, 414, 417-420, 423, 431-433, 470, 474, 477, 478, 483, 484, 522, 527, 573, 576-578, 603, 625,
  background subtraction 632, 633
  derivative 269-271, 274
  fast scan 630
  high resistive conditions 632, 633
  nanosecond time resolved 626
  ohmic drop compensation 633
  steady state 635
  submicrosecond kinetics 630
Cyclic voltammogram 58, 70, 103, 126, 141, 187, 174, 175, 222, 224, 226, 248, 249, 250, 312, 315, 348, 382, 385, 387, 395, 397, 412, 413, 425-427, 485, 492, 507, 522, 528, 585, 553, 557, 629-630, 633
Cyclooctadiene complexes 526
Cysteine pyrrole co-polymers 373
Cytochrome b_5 139
Cytochrome c 135
  peroxidase 140
  sulphide: cytochrome c oxidoreductase 140
Deprotonation [see ET (electron transfer) induced reactions]
Deshydrogenation (see ET induced reactions)
Diazenide complexes 357, 363
Diazoalkane complexes 357, 363
Differential pulse polarogram 583
Diffusion coefficients 443
Digital simulation (see Computer based methods)
Dihydrogen complexes 431
α-Diimines 264
Dinitrogen complexes 63-66
  17- and 16- electron complexes 359
  17-electron intermediates 361
  electroreduction 357
  electrosynthesis 357
  ET reactions 359
Dinuclear complexes 257, 321-326, 347, 519-530
  cobalt 526
  manganese 522, 523, 525
  molybdenum 83, 89
  osmium 503, 507-511, 583-588
  rhodium 528, 529
  ruthenium 503, 506-509, 512, 513, 583-588
Diols or diolates 15, 23
α-Dioxime complexes 547
Dioxygen
  adducts 391, 394-396, 398
  electrocatalytic reduction 107, 116-118, 381, 386-389, 391, 392, 398-400, 403, 406-410, 558,
  radical mechanism 401, 403
  reduction 202
2,2'-Dipyridyl complexes (see 2,2'-Bypiridine complexes)
Dithiocarbamate ligand 284
Donor-acceptor electronic coupling 196
Double-square scheme 260, 262

DuFort-Frankel algorithm 460
EC mechanism 69-71
ECE mechanism 529
ECEE mechanism 477-479
Edge currents 627
Edge-plane pyrolytic graphite 133
EECEEEEC mechanism 250
Efficiency (see also Electrochemical and Catalytic efficiency) 279, 280, 284, 288, 290
Electrocatalysis (see also Catalysis) 295, 658
  carboxylation 307-310
  ligand substitution 34, 269, 323
  molecular 549, 557
  reduction,
    carbon dioxide 295, 311, 314
    dioxygen (see Dioxygen)
    organohalides (see Organohalides)
    phenols 411
    thiols 411
Electrochemical cryostat 575
Electrochemical efficiency 19
Electrochemical interfaces 193, 194
Electrochemical irreversibility 5, 7
Electrochemical and homogeneous phase kinetics 194, 201
Electrochemical parameters (see Parameters)
Electrochemical spectroscopy 533
Electrochemiluminescence 7, 634
Electroinduced reactions [see ET (electron transfer) induced reactions]
Electrolytes, dielectric properties 474
Electron "deficiency" 78
Electron delocalization parameter (see Parameters)
Electron self-exchange 209, 211, 213
Electron stoichiometries,
  absolute determination 635, 636
Electron transfer (see ET)
Electron tunneling 195
Electronic coupling 193
  matrix element 195
Electronic spectroscopy 114, 122, 123, 125, 218, 286, 538, 541
Electronic transmission coefficient 195
Electrosynthesis 33, 71, 658
  bidentate ligand complexes 20, 21
  cyanide and isocyanide complexes 33
  halide complexes 18
  in liquid ammonia 26
  low oxidation state species 22-25
  metal halides 17
  phosphido complexes 19, 21
  selenolates 19, 21
Emission spectra 53
ENDOR studies 367, 369
Energy barrier 8, 9
Energy convertor 10
Enzymes, electrochemistry 133, 139, 141
E.P.RES.I.S.E. cell 505, 514
ESR spectroscopy 122, 123, 126, 127,249, 282, 287, 319, 320, 367, 412,503, 514-516
ET (electron transfer) 256, 258, 658
  distance dependence of 211, 213
  dynamical factors 193, 194, 196
  dynamical solvent effects 196
  electronic factors 193, 194
  intramolecular 9
  kinetics 179, 185, 489, 492, 493, 504
  mediator 411
  parameters 513
  structural change from 179
  vectorial 8, 9, 658
ET induced reactions
  alkyne oligomerization 317, 325
  bond cleavage
    C-C 317, 326
    C-H 317, 325, 331, 339, 345, 347, 351, 352
    C-halide 351, 352
    metal-C 351, 353
    metal-H 331, 340, 351, 353, 483, 486
    metal-halide, -O, -S or -N 34, 189
    metal-metal 317, 323, 324, 521, 527-529
    N-H 331,338, 339, 345, 347
  bond formation
    C-C 247, 317, 322, 325, 326
    metal-C 351
    metal-H 351
    metal-halide 351
    metal-metal 521, 527-529
  catalysis 264, 265
  chain catalysis (ETC)
    coupling with organometallic catalysis 277
    initiation 278, 287, 288
    propagation 278, 284
    ligand exchange 279
    ligand chelation 284
    polymerization of terminal alkynes 289
  dinucleation 260
  isomerization 255, 261, 331, 342, 489, 492, 579

nucleophilic attack 331, 341
S-S coupling 411, 414, 418
substitution 71, 72, 255, 263
Explicit finite difference method 457
Extended Hückel M.O. calculations 338
Ferredoxins 135, 373
Ferredoxin electrodes 373, 374
Ferrocenes as mediators 133
Ferrocenyl electron donor substituent 82
Fick's second law 454
Flavodoxin 135
Fluorescence-quenching 85
Free-energy relationships 212
Frontier orbitals 533-535, 543
Fulvalene complexes 519, 522, 525, 526, 527
Fungal laccase 140
Genetic series 6
Gutman donor number 608
Halide-to-metal charge transfer 537
Halide complexes
chromium 15, 18, 23
gallium 15, 18, 23
second and third row transition metals 533-545
thorium 15, 18
Hammett constants 50, 51, 79, 104, 115, 334, 341
High nuclearity metal clusters 229
antimony/nickel carbonyls 241
bismuth/nickel carbonyls 240
cobalt carbonyls 233, 239
cobalt-sulfur 231
cobalt/nickel carbonyls 237
cyclic voltammetry 230-243
iridium carbonyls 233
iridium/iron carbonyls 235
iron-sulfur 229, 231, 373
iron/platinum carbonyls 236
nickel-sulfur 232
palladium/ruthenium carbonyls 238
platinum carbonyls 193, 194, 202, 205 243
structural changes upon ET 229, 237, 243
two-electron processes 235, 239
X-ray structures 230, 243
Hoffmann elimination 353
Homogeneous catalysis,
electrochemical methods (see also Electrocatalysis) 8
Hydrazine complexes 357, 365,
cleavage of the N-N bond 365
Hydride complexes 53, 63-66, 331, 340, 341, 351
iron 481
rhenium 483, 486
Hydrogen complexes 53
Hydrogen peroxide, disproportionation 407
Hydrogenases 373
Hyperconjugation 265
Imide complexes 357, 365, 366
Indium diolates and thiolates 15, 23-24
Indole complexes 57, 59
Inner-shell activation barrier 181
Inner-sphere *vs.* outer-sphere mechanisms 295, 296
Intersecting-state model 207
Intervalence bands 586
Intervalence charge transfer 503, 504, 512, 513
Ion-pairing 467, 474
effects 282, 287
Ion-exchange polymers 373
Isocyanide complexes, 33, 34, 36, 37 , 63-66, 331-337
mechanism of formation 333
phosphonium-functionalized 57-59
Lead diolates and thiolates 15, 18, 23
Ligand π-donation 433
Ligand additivity 74, 338, 544, 545
Ligand parameterisation (see Parameters)
Ligand parameters (see Parameters)
Longitudinal relaxation time 196
Lutetium sandwich complexes 101, 105
Lysine oxidase 140
Main group organometallic compounds 255
Marcus theory 198, 207-211
Marcus-Hush theory 179
Metal centre parameters (see Parameters)
Metal-metal bond (see also ET induced reactions)
as an electron sink 322
Metal-metal interaction 89, 503, 504, 506, 507, 509, 510
steric factors 90
Metallo-sulphur compounds 371
Metallocenes 197, 637
Metalloporphyrins
cobalt 296-298, 381, 389
extra-ring 83
germanium 85
iron 180-185, 211, 213, 295, 296, 298, 302-304
manganese 180, 182, 184
zinc 85
Methanol electrooxidation 549, 566, 567
Methylamine dehydrogenase 140

Methylbenzenes 630, 631
Methyleneamide complexes 331, 332, 335, 338, 339, 341, 366, 268
Microelectrode (see Ultramicroelectrode)
Mitochondria 133
Mixed solvents 472, 474
Mixed-valence compounds 247, 257, 503, 504, 506, 508, 509, 512, 521, 523, 525, 583, 587, 586
M.O. calculations 317, 322
Modified electrodes 426, 557, 563, 615
charge transport 617, 620
cyclic voltammetry 617, 621
osmium containing polymer 615, 616
potential step experiments 620, 621
Molecular electrocatalysis (see also Electrocatalysis) 549, 557
Molecular electrochemistry 3, 4, 657
dynamical aspects 656, 657, 659, 660
Molecular electronic devices 8, 9, 10, 658
Molecular electronics 77
Molecular hysteresis 261
Molecular redox reagents 198
Molybdenum-enzymes 179
Multielectron transfer 5, 188, 658
Multiple metal carbon bonds, complexes (see Aminocarbyne, Carbene, Carbyne, η^2-Vinyl and Vinyledene complexes)
Multiproton liberation 346, 347
Nafion-coated electrodes 406-410
on graphite electrodes 381-384, 386, 387
Naphthalocyanine ligands 101, 102
Nitric oxide ligand 357
electroactivation 372
Nitride complexes 357, 365
Nitrile ligands 335
Nitrogen fixation 357
Nitrogenase substrates 33, 373
electrogeneration of metal binding sites 33, 34, 38
Nitrosyl complexes 43, 51, 52, 77, 78, 89-92, 95
NLO materials 81
NLO properties
molybdenum and tungsten complexes 82
NMR spectroscopy 333, 353
Non-adiabatic pathways 195
Non-conventional electrochemistry (see Ultramicroelectrodes)
Non-cyclic tetraaza ligands 107-110
Non-uniform space grid 462
Non-uniform space-time grid 462, 463
O.T.E. cell 125, 505, 512, 513
O.T.T.L.E. cell 5, 595
Ohmic drop, electronic compensation 627
Olefin hydroformylation 352
Optical electronegativity 544
Optical spectra 589
Optically transparent electrode (see O.T.E. cell)
Orbital mapping 536, 543
Organohalides 15, 23
electrocatalytic carboxylation 307-310
electrocatalytic reduction 295, 297
oxidative addition 640
Overdamped solvent motion 196
Oxidative coupling 249
Oxygen (see Dioxygen)
p-cresolmethylhydroxylase 140
Parameterisation (see Parameters)
Parameters
electrochemical 41-53, 57, 59-67, 60, 63, 65, 71, 74, 331, 334-338, 658
electron delocalization 583, 586
ligand 41-57, 59-67, 71, 74, 331, 334-338
dependence on the metal centre 63-67, 335
metal centre 60, 63, 65, 334-338
oxidation state 48
slopes and intercepts 45, 47
Phenols
electrocatalytic reduction 411
Phenoxy species 79
Photoelectron energies 50
Phthalocyanine ligands 101, 102
Polarography 395-397, 400, 437
Polyvinylimidazole complex 615, 616
Porphyrin complexes (see Metalloporphyrins)
Practical aspects 603
Precursor equilibrium constant 195
Preparative scale electrolysis (see Bulk electrolysis)
Promoters 133
Protein-enzyme complexes 133
Proteins-protein complexes133
Pyridines 264
Pyridyltriazole ligands 583, 584
Pyrylium
radical dimerization 631
cation reduction 636
Quinone complexes 43
Redox catalysis (see Catalysis and Electrocatalysis)
Redox orbital 7, 258
Redox potential
correlations

with electronic IR spectroscopic data 318, 334, 540, 590, 592, 655
with Hammett's constants 104, 334
effect of substituents 101, 104
ligand/metal and structural effects 4, 63 121, 331, 333, 591, 592, 655
prediction 41-53, 331, 336, 337
solvent effects 603, 608
supporting electrolyte effects 610
temperature effects 603, 607
trends 535, 542
Redox properties (see also Redox potential)
protic effects 107, 112, 115
topographic effects 107, 112, 179
Redox proteins, electrochemistry 133, 139, 141
Redox series 218, 221
Reference half-cell, non-aqueous 575
Reflectance spectroscopy
UV-visible 549, 550, 551
IR 549, 550, 565
Rubredoxin 135
Ruthenium dioxide 137
Sacrificial electrodes 15
Schiff-base complexes 311
Secondary harmonic generators 81
Selectivity 299, 300, 304
Self-exchange kinetics 199
Semi-integral electroanalysis 463
Sensor
for alkali metal ions 78
for cholesterol 133, 143
for glucose 133, 143
Series
spectrochemical 41
electrochemical 41
redox 218, 221
Silicon compounds 255
Single-crystal metal electrodes 193, 194
Solvent effects (see also Redox potential) 428
Solvent friction effects 193
Spectroelectrochemistry 102, 125, 503, 507-516, 523, 529, 533, 583, 589
ESR 577-581
infrared 193, 202, 503
UV/VIS 503, 513
variable temperature 503
Square scheme 113, 321, 492
Stationary voltammetry 114
Stereo- and regio-selective synthesis 322
Stereochemistry 46
Steric protection 255
Supramolecular electrochemistry 217
Supramolecular systems 8
Taylor series 459
Temperature effect 295, 301
Tetraazamacrocyclic complexes 84, 86, 102, 108, 296, 384, 393, 423-428
Tetracyanoethylene 261, 265, 266
Theoretical study 207
Thermal electron transfer 512
Thioether macrocyclic complexes 121-128
gold 121, 123, 124, 125, 127
molybdenum 372
silver 121, 122, 123
Thiolate complexes 15,18, 19, 21, 23, 34, 38, 417-420
Thiols 15, 23
electrocatalytic reduction 411
Thomas criterion 647
Three dimensional,
electrochemistry 495-499
voltammetry 495-499
Tin diolates and thiolates 15, 18, 19, 23
Tolman-Bartik factors 50, 51
Transition metal complexes
binuclear (see Dinuclear complexes)
chromium, 41, 45, 51, 260, 261, 265, 317, 321, 417-420, 489-493, 645-648
cobalt 107, 112, 118, 311, 314, 315, 317, 321-323, 326, 381, 389, 526, 637
copper 15, 171-176, 547
group VI 57-62
group VI dinuclear 467
iridium 431, 432, 514
iron 45, 107, 112, 114, 279, 323, 331, 341, 637, 638, 640
iron phthalocyanine 563
manganese 49, 265, 266, 269, 522, 523, 525
molybdenum 34, 38, 51, 77, 97, 186, 317, 325, 261, 264, 331, 333, 334, 335, 337, 372, 649-652
nickel 107, 112, 118, 307, 311-313, 423-428
niobium complexes 45
osmium 43, 45, 69-74,503, 504, 506-511, 583-592, 615, 616
palladium 52, 307, 317, 323, 324, 345-348, 351, 355
platinum 52, 345-348, 351, 355, 503, 504, 513-516
rhenium 45, 48, 52, 264, 331-341
rhodium 259, 528, 529

ruthenium 42, 45, 49, 200, 218, 325, 381-388, 503, 504, 506, 508-510, 512, 513, 583-588
second and third row 535, 536, 539-545
tantalum 45
technetium 52, 63-66
tungsten 289, 257, 258, 261, 264, 573-581
vanadium 411-414
zinc 255
Triazene-1-oxides 50
Trinuclear or tetranuclear complexes (see also High nuclearity metal clusters) 91, 217, 218, 247-250, 323
Tris(pyrazolyl)borate complexes 77, 95, 96
as electron withdrawing groups 82
Two-electron reservoir 258
Two-electron processes 326, 521, 527, 529, 109, 112, 113, 118, 259, 307, 338, 339, 340, 345, 347
Ultramicroelectrodes 625-642, 645-652 , 657, 660
double band 634
double layer effects 645
electrolyte effects 645, 646
kinetics 645, 647, 649, 651, 652
ohmic drop 645
solvent effects 645, 646, 648
steady state at 634
voltammetry 645-648, 650, 652
highly resistive conditions 637-642
Unconventional electrochemistry (see Ultramicroelectrodes)
η^2-vinyl complexes 331, 332, 335, 338, 340, 342
Vinylidene complexes 320-322, 331, 332, 335
anodically induced alkyne rearrangement 317, 320
Viologen adsortion 553, 554, 555
Wacker process 324
X-ray structure 72, 83, 91, 97, 98, 122, 124, 125, 322, 323, 367, 369, 511, 526, 529